W9-AZW-699

STUDENT COMPANION

to accompany
Biochemistry, Seventh Edition

Frank H. Deis
Rutgers University

Nancy Counts Gerber
San Francisco State University

Richard I. Gumport
*College of Medicine at Urbana-Champaign,
University of Illinois*

Expanded Solutions to Text Problems
contributed by
Roger E. Koeppe, II
University of Arkansas at Fayetteville

W. H. Freeman and Company
New York

ISBN-10: 1-4292-3115-7
ISBN-13: 978-1-4292-3115-2

© 2012 by W. H. Freeman and Company

All rights reserved.

Printed in the United States of America

W. H. Freeman and Company
41 Madison Avenue
New York, NY 10010
Houndmills, Basingstoke RG21 6XS, England
www.whfreeman.com

CONTENTS

PART IV: RESPONDING TO ENVIRONMENTAL CHANGES

Opening a comprehensive biochemistry text for the first time can be a daunting experience for a neophyte. So much detailed material is presented that it is natural to wonder if you can possibly master it in one or two semesters of study. Of course, you can't learn everything, but experience indicates that you can, indeed, learn the fundamental concepts in an introductory biochemistry course. We have written this Student Companion for *Biochemistry* to ease your entry into the exciting world of biochemistry.

Your goal is to "know" and "understand" biochemistry. Unfortunately, awareness of these grand goals offers no practical help in reaching them, because they are such high-level and complex intellectual processes. In addition, it is difficult for you to know to what extent you have attained them. We have found that, by subdividing these goals into simpler ones and expressing them in terms of demonstrable behaviors, you can begin to approach them and, in addition, can readily assay your progress toward reaching them. Thus, a part of each chapter consists of Learning Objectives that ask you to do things that will help you to begin to understand biochemistry. When you can master the objectives, you are well on your way to learning the material in the chapter. It is important to add a cautionary note here. Being able to respond to all the objectives adequately does not mean that you know biochemistry, for they are a limited sampling of all the possible objectives; more to the point, they do not explicitly require such higher-level activities as creation, analysis, integration, synthesis, problem-solving, evaluation, application, and appreciation. These more advanced skills will develop to varying levels as you continue your studies of biochemistry beyond the introductory stage.

Each chapter in the Companion consists of an introduction, Learning Objectives, a Self-Test, Answers to Self-Test, Problems, and Answers to Problems. The introduction sets the scene, places the chapter material in the context of what you have already learned, and reminds you of material you may need to review in order to understand what follows. The Learning Objectives are presented in the order that the information they encompass appears in *Biochemistry*. Key Words—important concepts or vocabulary—are italicized in the objectives. Self-Test questions, requiring primarily information recall, are followed by the answers to the questions. A Problems section, in which more complex skills are tested, is followed by answers to the problems. Finally, Expanded Solutions to end-of-chapter problems in the text are presented in a separate section in the Companion.

The Companion may be used in many ways, and as you begin your studies you will develop the "system" that is best for you. Over 30 years of experience teaching introductory biochemistry has suggested one pathway that you should consider. Start by reviewing the prerequisite chapters mentioned in the introduction and skim the Learning Objectives to obtain an overview of what you are to learn. Some students like also to skim the Self-Test questions at this time to form an impression of the levels of difficulty and the kinds of questions that will be asked. Next, read the chapter in *Biochemistry*, using the Learning Objectives to help direct you to the essential concepts. Note the Key Words and look up those you don't know. Then attempt to meet the objectives. When you cannot satisfy an objective, reread the relevant section of the text. You should now take the Self-Test to check your ability to recall and apply what you have learned. Finally, solve the Problems, which have been designed to further test your ability to apply the knowledge you have gained. It is not sufficient simply to read the problems and look at the answers to see if you would have done them the same way. You must struggle through the solutions yourself to benefit from the problems. As you are using the Companion, you will, of course, be integrating what you have learned from your studies and your lectures or laboratory exercises.

Besides helping you to learn biochemistry, you will find the Companion useful in studying for examinations. Go over each of the Learning Objectives in the chapters covered by an exam to ensure that you can respond to it knowledgeably. Similarly, review the Key Words. Decide which chapter topics you feel uncertain about and reread them. This protocol, coupled with a review of your lecture and reading notes, will prepare you well for examinations.

It is important to talk about biochemistry with others in order to learn the pronunciation of scientific terms and names and to help crystallize your thinking. Also, realize that although biochemistry has a sound foundation and we understand much about the chemistry of life, many of our concepts are hypotheses that will require modification or refinement as more experimental evidence accrues. Alternative and sometimes contradictory explanations exist for many biochemical observations. You should not regard the material in *Biochemistry* or the Companion as dogma, and you should, wherever possible, attempt to read about any given topic in at least two sources. Try to follow up topics that particularly interest you by reading about them in the scientific literature. References are given in *Biochemistry*, and your instructor can help you locate research and review articles. In this way, you can begin to appreciate the diversity of opinion and emphasis that exists in the field of biochemistry.

The authors welcome readers' comments, especially those drawing our attention to errors in the text. Comments should be sent to:

Dr. Frank H. Deis
Department of Molecular Biology and Biochemistry
Rutgers University
Nelson Laboratories
604 Allison Rd.
Piscataway, NJ 08854-8082
e-mail deis@rci.rutgers.edu

ACKNOWLEDGMENTS

This book, the *Student Companion,* owes its structure to the vision of the late Dr. Richard I. Gumport. His decades teaching first-year medical students in the College of Medicine at the University of Illinois at Urbana-Champaign led him to analyze how students learn a complex subject, and then to create this book as a tool for learning. Through various editions and revisions, it is good to remember that much of the writing in this book is his.

He was always gracious enough to thank his former co-authors, Ana Jonas, Richard Mintel, and Carl Rhodes, as well as many others who had given him useful ideas, particularly including his students at Illinois and elsewhere who took the time to criticize the Student Companion.

In producing the present edition, thanks are due to Wendy Pogozelski at SUNY Geneseo and Laura Schick Zapanta at the University of Pittsburgh. FHD would like to thank his students at Rutgers, and Peter Lobel in the Department of Pharmacology at UMDNJ. REK, II thanks Denise Greathouse, Olaf Andersen, Frank Millett, and his students at the University of Arkansas in Fayetteville. Finally, NCG thanks the faculty and students in the San Francisco State University Department of Chemistry and Biochemistry for support and feedback.

Frank H. Deis
Nancy C. Gerber
Richard I. Gumport
Roger E. Koeppe, II

January 2011

Biochemistry: An Evolving Science

Chapter 1

The introductory chapter of *Biochemistry* begins by describing the unity of biochemistry. This is an important concept. It means that we can learn about human biochemistry by studying mice, yeast, bacteria, or any living organism. One important feature shared by all cellular organisms is the use of DNA as the genetic material. Specific structural components of DNA are discussed. Many biochemical interactions depend on weak noncovalent interactions. Because the great majority of biochemical processes occur in water, the properties of water and their effects on biomolecules are also described. Then follows a discussion of entropy, energy, and the laws of thermodynamics. This provides a basis for understanding hydrophobic interactions and formation of the DNA double helix. The authors then discuss pH and buffers. This material is important for understanding macromolecules in solution, both DNA in the present chapter and proteins in later chapters. Finally there is a discussion of the human genome, which has been completely sequenced. The implications of understanding the genome are enormous, both for medicine and for biology.

LEARNING OBJECTIVES

When you have mastered this chapter, you should be able to accomplish the following objectives.

Biochemical Unity Underlies Biological Diversity (Text Section 1.1)

1. Differentiate between *Archaea*, *Eukarya*, and *Bacteria*.

DNA Illustrates the Interplay Between Form and Function (Text Section 1.2)

2. Recognize, name, and draw the four bases used in DNA, and explain the structure of the sugar phosphate backbone.

3. Describe how the DNA bases pair with each other. Notice that "larger" (purine-based) always pairs with "smaller" (pyrimidine-based).

4. Explain how *base pairing* provides an accurate means for reproducing DNA sequences.

Concepts from Chemistry Explain the Properties of Biological Molecules (Text Section 1.3)

5. Define the term *covalent bond*, and discuss the strength of typical covalent bonds.

6. List the three kinds of *noncovalent bonds* that mediate interactions of biomolecules and describe their characteristics.

7. Describe how the properties of *water* affect the interactions among biomolecules.

8. Explain the origin of *hydrophobic attractions* between *nonpolar molecules* and give examples of their importance in biochemical interactions.

9. State the *first* and *second laws of thermodynamics*. Define the *entropy (S)* and *enthalpy (H)* of a *system*, and give their mathematical relationship.

10. Describe how the various types of chemical bonds affect the formation of the DNA double helix.

11. Discuss how the laws of thermodynamics affect the formation of the DNA double helix.

12. Define pH and pKa, and explain how the DNA double helix is affected by the pH of the solution surrounding it.

13. Know and use the *Henderson-Hasselbalch Equation*.

14. Understand how buffers respond to changes in $[H^+]$ or $[OH^-]$ concentration.

The Genomic Revolution Is Transforming Biochemistry and Medicine (Text Section 1.4)

15. Discuss the importance of *genome sequencing* of humans and other organisms for science and medicine.

16. Describe the size of the human genome in terms of the number of base pairs and the number of genes.

17. Discuss how the variation in individual genomes leads to individual differences, sometimes including hereditary diseases.

18. Define epigenetic factors and genetic imprinting and explain how they can affect gene expression.

Appendix: Depicting Molecular Structures

19. Explain the uses of different *molecular models*.

20. Relate the planar *Fischer projection* to the tetrahedrally arrayed substituents around a carbon atom.

SELF-TEST

Biochemical Unity Underlies Biological Diversity

1. Which of the following molecular patterns or processes are common to both bacteria and humans?

   (a) development of tissues
   (b) information flow from proteins to DNA
   (c) the "energy currency"
   (d) genetic information flow
   (e) similar biomolecular composition

2. The distinguishing feature of the Eukarya is that

   (a) they are all multicellular.
   (b) they have tough cell walls around each cell.
   (c) they have a well-defined nucleus within each cell.
   (d) they are more primitive than the Archaea or the Bacteria.

DNA Illustrates the Interplay Between Form and Function

3. Which base is NOT found among DNA's building blocks?

   (a) uracil
   (b) thymine
   (c) cytosine
   (d) guanine
   (e) adenine

4. The DNA sequence AAA would pair with the sequence
   (a) AAA.
   (b) GGG.
   (c) CCC.
   (d) TTT.

5. Watson-Crick base pairs are held together by

   (a) van der Waals forces
   (b) hydrogen bonds
   (c) covalent bonds
   (d) hydrophobic attraction

Concepts from Chemistry Explain the Properties of Biological Molecules

6. The properties of water include

   (a) the ability to form hydrophobic bonds with itself.
   (b) a disordered structure in the liquid state.
   (c) a low dielectric constant.
   (d) being a strong dipole, with the negative end at the O atom.
   (e) a diameter of 5 Å.

7. For the bonds or interactions in the left column, indicate all the characteristics in the right column that are appropriate.

   (a) electrostatic interaction
   (b) hydrogen bond
   (c) van der Waals bond
   (d) hydrophobic interaction

   (1) requires nonpolar species
   (2) involves charged species only
   (3) requires polar or charged species
   (4) involves either O and H or N and H atoms
   (5) involves polarizable atoms
   (6) exists only in water
   (7) optimal at the van der Waals contact distance
   (8) has an energy between 4 and 20 kJ/mol in H_2O
   (9) has an energy between 2 and 4 kJ/mol
   (10) is weakened in water

8. Biological membranes are made up of phospholipids, detergent-like molecules with long nonpolar chains attached to a polar head group. When isolated phospholipids are placed in water, they associate spontaneously to form membrane-like structures. Explain this phenomenon.

9. If two molecules had a tendency to associate with each other because groups on their surfaces could form hydrogen bonds, what would be the effect of putting these molecules in water? Explain.

10. Which of the following statements is correct? The entropy of a reaction refers to

    (a) the heat given off by the reaction.
    (b) the tendency of the system to move toward maximal randomness.
    (c) the energy of the transition state.
    (d) the effect of temperature on the rate of the reaction.

11. What are major noncovalent interactions that contribute to the formation of the DNA double helix?

12. The double helix is stablest

    (a) above pH 9.5
    (b) around pH 7
    (c) below pH 5

The Genomic Revolution Is Transforming Biochemistry and Medicine

13. The human genome contains roughly how many protein coding genes?

    (a) 3 billion
    (b) 100,000
    (c) 25,000
    (d) 2,000

14. Humans inherit two copies of nearly every gene (except for those on the X and Y chromosomes). Does it make a difference whether a gene comes from the mother or the father?

    (a) No, it never makes any difference.
    (b) "Genetic imprinting," based on which parent the gene came from, can make a difference.
    (c) Expressing both copies of both genes means the epigenetic factors will cancel out.

Appendix: Depicting Molecular Structures

15. Match the types of molecular models in the left column with the appropriate application in the right column.

(a) space-filling model
(b) ball-and-stick model
(c) skeletal model

(1) shows the bond framework in macro molecules
(2) indicates the volume occupied by a biomolecule
(3) shows the bonding arrangement in small biomolecules

16. Hydrogen atoms are frequently omitted from ball-and-stick models and skeletal models of biomolecules. Explain why.

ANSWERS TO SELF-TEST

1. c, d, e

2. c

3. a. Uracil is not mentioned in this chapter but it is used in RNA. See later chapters.

4. d

5. b. AT base pairs have two hydrogen bonds, and GC base pairs have three.

6. d

7. (a) 2, 8 (b) 3, 4, 8, 10 (c) 5, 7, 9 (d) 1, 6

8. When the nonpolar chains of the individual phospholipid molecules are exposed to water, they form a cavity in the water network and order the water molecules around themselves. The ordering of the water molecules requires energy. By associating with one another through hydrophobic interactions, the nonpolar chains of phospholipids release the ordered water by decreasing the total surface area and hence reduce the energy required to order the water. Such coalescence stabilizes the entire system, and membrane-like structures form.

9. Because of the high dielectric constant of water and its ability to form competing hydrogen bonds, the interaction between the molecules would be weakened.

10. b

11. All four of the non-covalent forces described in Chapter 1 are important in formation of the double helix—electrostatic interactions, hydrogen bonds, van der Waals forces, and the hydrophobic effect.

12. b. At higher or lower pH, guanine and adenine (the "purine" bases of DNA) ionize, and the two strands can't properly form Watson-Crick base pairs.

13. c. 25,000. This low number surprised many people when the human genome was first sequenced.

14. b

15. (a) 2 (b) 3 (c) 1

16. The ball-and-stick model and skeletal model best show the bonding arrangements and the backbone configurations of biomolecules; the inclusion of the numerous hydrogen atoms would obscure the very features revealed by these models.

PROBLEMS

1. As will be seen in succeeding chapters, enzymes provide a specific binding site for substrates where one or more chemical steps can be carried out. Often these sites are

designed to exclude water. Suppose that at a binding site, a negatively charged substrate interacts with a positively charged atom of an enzyme.

(a) Using Coulomb's equation, show how the presence of water might affect the interaction. What sort of environment might be preferable for an ionic interaction? Note that a numerical answer is not required here.

(b) How would an ionic interaction be affected by the distance between the oppositely charged atoms?

2. Water molecules have an unparalleled ability to form hydrogen bonds with one another. Water also has an unusually high heat capacity, as measured by the amount of energy required to increase the temperature of a gram of water by 1°C. How does hydrogen bonding contribute to water's high heat capacity?

3. The Second Law of Thermodynamics states that the entropy (disorder) of a system and its surroundings always increases for a spontaneous process. So why does DNA fold up spontaneously? It is evident that DNA folding moves from a disorderly state (randomly unfolded) to an orderly state (double helix). Explain.

4. What is the molarity of pure water? Show that a change in the concentration of water by ionization does not appreciably affect the molarity of the solution.

5. When sufficient H^+ is added to lower the pH by one unit, what is the corresponding increase in hydrogen ion concentration?

6. You have a solution of HCl that has a pH of 2.1. What is the concentration of HCl needed to make this solution?

7. Human blood maintains a very tight pH range (about 7.35–7.45). It accomplishes this homeostasis partly through what is known as the bicarbonate buffering system. Blood has large amounts of bicarbonate, HCO_3^- , a weak acid with a pKa near physiological pH. As the equilibrium below shows, bicarbonate can pick up H^+ and form carbonic acid. Carbonic acid is degraded by enzymes to form CO_2, which can exit via the lungs.

$$CO_2 + H_2O \longleftrightarrow H_2CO_2 \longleftrightarrow H^+ + HCO_3^-$$

Uncontrolled diabetes can result in ketoacidosis, a condition in which metabolic acids such as β-hydroxybutyric acid (a substance known as a ketone body) build up. β-hydroxybutyric acid has a pKa below that of physiological pH. A characteristic of ketoacidosis is deep and rapid breathing. Why would ketoacidosis lead to this symptom? Explain.

8. A woman is brought into the emergency room experiencing nausea and vomiting. She is also hyperventilating. She says that she ingested a bottle of aspirin about two hours earlier. Aspirin (acetylsalicylic acid) is converted to salicylic acid in the stomach. Assuming 100% conversion to salicylic acid, calculate the percentage of protonated and unprotonated forms of salicylic acid in her stomach, given a pH of 2.0 for the stomach.

9. One way in which scientists look for evidence of extraterrestrial life is to examine meteorites and to see if enantiomeric distributions mimic those found on earth. One amino

acid found on the Murchison meteorite in Australia was α-methylnorvaline, shown below. Draw the L-configuration of this structure.

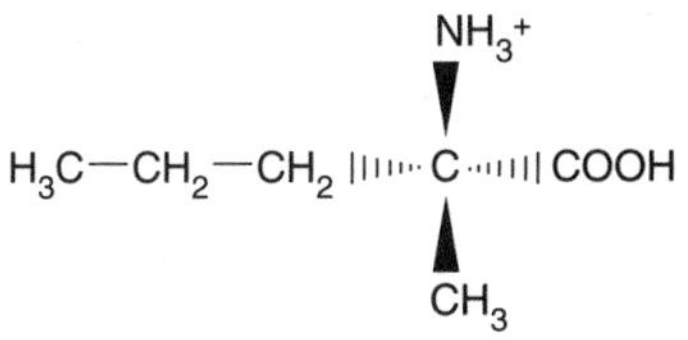

$$H_3C—CH_2—CH_2 \text{ |||}\cdots C \cdots\text{|||| COOH}$$

with NH_3^+ above C and CH_3 below C.

10. Spontaneous processes are those that increase in entropy. Explain how a freezer then can make ice, a process that reduces entropy in the system.

11. It is surprising for some people to think that "clean things" like soaps and detergents can be made from components such as beef fat and ashes. Beef fat contains stearic acid, an 18-carbon carboxylic acid, whereas wood ashes contain potassium hydroxide. When these two substances are mixed and heated, sodium stearate is produced. Sodium stearate is a major component of soap. Explain why soap works. Why is it able to disperse greasy substances in water?

ANSWERS TO PROBLEMS

1. (a) The magnitude of the electrostatic attraction would be diminished by the presence of water because D, the dielectric constant, is relatively high for water. Inspection of Coulomb's equation shows that higher values of D will reduce the force of the attraction. Lower values, such as those for hydrophobic molecules like hexane, allow a higher value for F. We shall see that many enzyme active sites are lined with hydrophobic residues, creating an environment that enhances ionic interaction.

   (b) Inspection of Coulomb's equation also reveals that the force between two oppositely charged atoms will vary inversely with the square of the distance between them.

2. When water is heated, considerable energy is required to break the hydrogen bonds. Only after a large percentage of bonds are broken are the molecules more mobile and the temperature raised. This buffering capacity of water is very important to cells, which can resist changes caused by increases in temperature because of water's high heat capacity.

3. As described in the chapter, all of the "weak forces" commonly involved in biochemistry help in formation of the double helix. Hydrogen bonds allow formation of Watson-Crick base pairs. Van der Waals forces allow "pancake" stacking of base pairs. Ionic forces keep the phosphates distant from each other, and the hydrophobic effect favors putting the flat hydrophobic surfaces of base pairs inside where they don't interact with the surrounding water. But this question is phrased in terms of thermodynamics so we have to remember the equation $\Delta G = \Delta H - T\Delta S$. The entropy term (S) must be unfavorable because obviously the change from two chains with randomly coiled conformation to one precisely formed double helix is a loss of disorder, a gain in orderliness. Thus, because we know that the double helix forms spontaneously in water, there must be a compensating change in heat or enthalpy (H). And indeed a substantial amount of heat is released when the double helix forms. This could be broken down into components, one of which would be the formation of hydrogen bonds mentioned above. The "weak forces" allow bonding which leads to favorable loss of heat energy.

4. The molarity of water equals the number of moles of water per liter. A liter of water weighs 1000 grams, and its molecular weight is 18, so the molarity of water is

$$M = \frac{1000}{18} = 55.6$$

At 25°C, K_w is 1.0×10^{-14}; at neutrality, the concentration of both hydrogen and hydroxyl ions is each equal to 10^{-7} M. Thus, the actual concentration of H_2O is (55.6 − 0.0000001) M; the difference is so small that it can be disregarded.

5. Because pH values are based on a logarithmic scale, every unit change in pH means a tenfold change in hydrogen ion concentration. When pH = 2.0, $[H^+] = 10^{-2}$ M; when pH = 3.0, $[H^+] = 10^{-3}$ M.

6. Assume that HCl in solution is completely ionized to H^+ and Cl^-. Then find the concentration of H^+, which equals the concentration of Cl^-.

$$pH = -\log\left[H^+\right] = 2.1$$

$$\left[H^+\right] = 10^{-2.1}$$

$$= 10^{0.9} \times 10^{-3}$$

$$= 7.94 \times 10^{-3} \, M$$

$$\text{Thus, } \left[H^+\right] = \left[Cl^-\right] = \left[HCl\right] = 7.94 \times 10^{-3} \, M$$

7. The breathing is the body's attempt to return to normal pH. If β-hydroxybutyric acid builds up, it loses its H^+ and increases the concentration of H^+ in the blood. This drives the equilibrium to the left, as the H^+ combines with bicarbonate, generating carbonic acid, which degrades to CO_2 and H_2O. The CO_2 is expelled via the lungs.

8.
$$pH = pKa + \log \frac{[A-]}{[HA]}$$

$$2 = 3 + \log \frac{[A-]}{[HA]}$$

$$-1 = \log \frac{[A-]}{[HA]}$$

$$\frac{[A-]}{[HA]} = 0.1$$

Let x = % A^-

1 − x = % HA

$$\frac{x}{1-x} = 0.1$$

$$x = 0.09$$

$$\%A^- = 9\%$$

$$\% \text{ HA (protonated)} = 91\%$$

9.

$$H_3C-H_2C-H_2C-\underset{\underset{CH_3}{|}}{\overset{\overset{NH_3^+}{|}}{C}}-COOH$$

10. A freezer increases the entropy around it. Work is required to lower the entropy of the water in making ice, and this process occurs at the expense of a freezer moving electricity around and generating heat. Entropy can be made to decrease locally, but only at the expense of useful work. The work causes the entropy of the universe to increase.

11. Soap contains an ionic end that interacts with water and a long nonpolar chain that interacts with nonpolar items such as grease. The soap surrounds the grease particle, creating a micelle. The charged portions interact with the solvent, enabling the micelle to disperse in the solvent and be washed away.

Protein Structure and Composition

Proteins are macromolecules that play central roles in all the processes of life. Chapter 2 begins with a discussion of key properties of proteins and continues with a description of the chemical properties of amino acids—the building blocks of proteins. It is essential that you learn the names, symbols, and properties of the 20 common amino acids at this point, as they will recur throughout the text in connection with protein structures, enzymatic mechanisms, metabolism, protein synthesis, and the regulation of gene expression. Amino acids are ionizable, and it is important to understand the pKa values provided. Following the discussion of amino acids, the chapter turns to peptides and to the linear sequences of amino acid residues in proteins. Next, it describes the folding of these linear polymers into the specific three-dimensional structures of proteins. The primary structure (or sequence of amino acids) dictates the higher orders of structure including secondary (α, β, etc.), tertiary (often globular), and quaternary (with multiple chains). You should note that the majority of functional proteins exist in water and that their structures are stabilized by the weak forces and interactions you learned about in Chapter 1. Hydrophobic groups are folded into the interior of most globular proteins. Some diseases are caused by prions, including mad cow disease and Creutzfeldt-Jakob disease. These are diseases of protein folding—wrongly folded proteins precipitate and form damaging plaque in the brain. The chapter concludes with a discussion of the theory of how proteins fold, including attempts to predict protein folding from amino acid sequences.

LEARNING OBJECTIVES

When you have mastered this chapter, you should be able to accomplish the following objectives.

Introduction

1. List the key properties of proteins.

2. Explain how proteins relate one-dimensional gene structure to three-dimensional structure in the cell, and their complex interactions with each other and various substrates.

Proteins Are Built from a Repertoire of 20 Amino Acids (Text Section 2.1)

3. Draw the structure of an *amino acid* and indicate the following features, which are common to all amino acids: *functional groups, side chains,* and *ionic forms.*

4. Classify each of the 20 amino acids according to the side chain on the α *carbon* as *aliphatic, aromatic, sulfur-containing, aliphatic hydroxyl, basic, acidic,* or *amide derivative.*

5. Give the name and one-letter and three-letter *symbol* of each amino acid. Describe each amino acid in terms of *size, charge, hydrogen-bonding capacity, chemical reactivity,* and *hydrophilic* or *hydrophobic* nature.

6. Describe the way that interactions between *pH* and *pKa* affect the *ionization state* of any given amino acid or its side chain in a protein.

Primary Structure: Amino Acids Are Linked by Peptide Bonds to Form Polypeptide Chains (Text Section 2.2)

7. Draw a *peptide bond* and describe its *conformation* and its role in *polypeptide* sequences. Indicate the *N-* and *C-terminal residues* in *peptides.*

8. Define *main chain, side chains,* and *disulfide bonds* in polypeptides. Give the range of *molecular weights* of proteins.

9. Explain the origin and significance of the unique *amino acid sequences* of proteins.

10. State four reasons why knowing the amino acid sequence of a protein is important.

11. Understand why nearly all peptide bonds are *trans.*

12. Define the ϕ and ψ angles used to describe a peptide bond, and be able to read a *Ramachandran plot.*

Secondary Structure: Polypeptide Chains Can Fold into Regular Structures Such as the Alpha Helix, the Beta Sheet, and Turns and Loops (Text Section 2.3)

13. Differentiate between two major *periodic structures* of proteins: the α *helix* and the β-*pleated sheet.* Describe the patterns of hydrogen bonding, the shapes, and the dimensions of these structures.

14. List the types of interactions among amino acid side chains that stabilize the *three-dimensional structures* of proteins. Give examples of *hydrogen bond donors* and *acceptors.*

15. Describe the role and structure of β *turns* or *hairpin turns* and *omega loops* in the structure of common proteins.

16. Understand the structures of fibrous proteins including α-*keratin,* made of α *helical coiled coils,* and *collagen,* which has a tight triple helix.

Tertiary Structure: Water-Soluble Proteins Fold into Compact Structures with Nonpolar Cores (Text Section 2.4)

17. Using *myoglobin* and *porin* as examples, describe the main characteristics of native folded protein structures.

18. Rationalize the conformational preferences of different amino acids in proteins and polypeptides.

19. Explain what is understood about the role of protein folding in *mad cow disease* and *Alzheimer's disease*.

20. Describe the *nucleation-condensation* model of protein folding.

21. Give evidence that protein folding appears to be a cooperative transition, and explain why that means it is an "all-or-none" process.

22. Explain how *protein folding* proceeds through stabilization of *intermediate states* rather than through a sampling of all possible conformations.

23. Distinguish between *motifs* and *domains* in protein structure.

Quaternary Structure: Polypeptide Chains Can Assemble into Multisubunit Structures (Text Section 2.5)

24. Describe the *primary, secondary, tertiary,* and *quaternary structures* of proteins.

The Amino Acid Sequence of a Protein Determines Its Three-Dimensional Structure (Text Section 2.6)

25. Using *ribonuclease* as an example, describe the evidence for the hypothesis that all of the information needed to specify the three-dimensional structure of a protein is contained in its amino acid sequence.

26. Discuss the methods and advances in the prediction of three-dimensional structures of proteins.

27. List examples of the *modification* and *cleavage* of proteins that expand their functional roles.

SELF-TEST

Introduction

Proteins Are Built from a Repertoire of 20 Amino Acids

1. (a) Examine the four amino acids given below:

A   B   C   D

Indicate which of these amino acids are associated with the following properties:

(a) aliphatic side chain
(b) basic side chain
(c) three ionizable groups
(d) charge of +1 at pH 7.0
(e) pK ~10 in proteins
(f) secondary amino group
(g) designated by the symbol K
(h) in the same class as phenylalanine
(i) most hydrophobic of the four
(j) side chain capable of forming hydrogen bonds

(b) Name the four amino acids.
(c) Name the other amino acids of the same class as D.

2. Draw the structure of cysteine at pH 1.

3. Match the amino acids in the left column with the appropriate side chain types in the right column.

(a) Lys
(b) Glu
(c) Leu
(d) Cys
(e) Trp
(f) Ser

(1) nonpolar aliphatic
(2) nonpolar aromatic
(3) basic
(4) acidic
(5) sulfur-containing
(6) hydroxyl-containing

4. Which of the following amino acids have side chains that are negatively charged under physiologic conditions (i.e., near pH 7)?

(a) Asp
(b) His
(c) Trp
(d) Glu
(e) Cys

5. Why does histidine act as a buffer at pH 6.0? What can you say about the buffering capacity of histidine at pH 7.6?

Primary Structure: Amino Acids Are Linked by Peptide Bonds to Form Polypeptide Chains

6. How many different dipeptides can be made from the 20 L amino acids? What are the minimum and the maximum number of pK_a values for any dipeptide?

7. For the pentapeptide Glu-Met-Arg-Thr-Gly,

(a) name the carboxyl-terminal residue.
(b) give the number of charged groups at pH 7.
(c) give the net charge at pH 1.
(d) write the sequence using one-letter symbols.
(e) draw the peptide bond between the Thr and Gly residues, including both side chains.

8. If a polypeptide has 400 amino acid residues, what is its approximate mass?

(a) 11,000 daltons
(b) 22,000 daltons
(c) 44,000 daltons
(d) 88,000 daltons

9. Which amino acid can stabilize protein structures by forming covalent cross-links between polypeptide chains?

   (a) Met
   (b) Ser
   (c) Gln
   (d) Gly
   (e) Cys

Secondary Structure: Polypeptide Chains Can Fold into Regular Structures Such as the Alpha Helix, the Beta Sheet, and Turns and Loops

10. Discuss the significance of *Ramachandran diagram*. Contrast the conformational states of Gly and Pro in proteins compared with other amino acid residues.

11. Which of the following statements about the peptide bond are true?

    (a) The peptide bond is planar because of the partial double-bond character of the bond between the carbonyl carbon and the nitrogen.
    (b) There is relative freedom of rotation of the bond between the carbonyl carbon and the nitrogen.
    (c) The hydrogen that is bonded to the nitrogen atom is *trans* to the oxygen of the carbonyl group.
    (d) There is no freedom of rotation around the bond between the α carbon and the carbonyl carbon.

12. Which of the following statements about the α helix structure of proteins is correct?

    (a) It is maintained by hydrogen bonding between amino acid side chains.
    (b) It makes up about the same percentage of all proteins.
    (c) It can serve a mechanical role by forming stiff bundles of fibers in some proteins.
    (d) It is stabilized by hydrogen bonds between amide hydrogens and amide oxygens in polypeptide chains.
    (e) It includes all 20 amino acids at equal frequencies.

13. Which of the following properties are common to α-helical and β pleated sheet structures in proteins?

    (a) rod shape
    (b) hydrogen bonds between main-chain CO and NH groups
    (c) axial distance between adjacent amino acids of 3.5 Å
    (d) variable numbers of participating amino acid residues

14. Explain why α helix and β pleated sheet structures are often found in the interior of water-soluble proteins.

Tertiary Structure: Water-Soluble Proteins Fold into Compact Structures with Nonpolar Cores

15. Which of the following amino acid residues are likely to be found on the inside of a water-soluble protein?

    (a) Val
    (b) His
    (c) Ile
    (d) Arg
    (e) Asp

16. Which of the following statements about the structures of water-soluble proteins, exemplified by myoglobin, are NOT true?

    (a) They contain tightly packed amino acids in their interior.

    (b) Most of their nonpolar residues face the aqueous solvent.

    (c) The main-chain NH and CO groups are often involved in H-bonded secondary structures in the interior of these proteins.

    (d) Polar residues such as His may be found in the interior of these proteins if the residues have specific functional roles.

    (e) All of these proteins contain β sheet structural motifs.

17. Which one of the following amino acids interrupts α helices, and also disrupts β sheets?

    (a) Phe   (d) His

    (b) Cys   (e) Pro

    (c) Trp

18. Mad cow disease, Creutzfeldt-Jakob disease, and Alzheimer's disease are caused by

    (a) Viruses

    (b) Bacteria

    (c) Misfolded proteins

19. If we know that a solution of protein is half-folded, what will we find in solution?

    (a) 100% half-folded protein

    (b) 50% fully folded, 50% unfolded

    (c) 33% fully folded, 34% half-folded, and 33% unfolded

Quaternary Structure: Polypeptide Chains Can Assemble into Multisubunit Structures

20. Match the levels of protein structures in the left column with the appropriate descriptions in the right column.

    (a) primary

    (b) secondary

    (c) tertiary

    (d) quaternary

    (1) association of protein subunits

    (2) overall folding of a single chain, can include α-helical and β sheet structures

    (3) linear amino acid sequence

    (4) repetitive arrangement of amino acids that are near each other in the linear sequence

21. Hemoglobin has the structure $\alpha_2\beta_2$. Hemoglobin is a protein. What are α and β called?

The Amino Acid Sequence of a Protein Determines Its Three-Dimensional Structure

22. Which of the following statements are true?

    (a) Ribonuclease (RNase) can be treated with urea and reducing agents to produce a random coil.

    (b) If one oxidizes random-coil RNase in urea, it quickly regains its enzymatic activity.

    (c) If one removes the urea and oxidizes RNase slowly, it will renature and regain its enzymatic activity.

    (d) Although renatured RNase has enzymatic activity, it can be readily distinguished from native RNase.

23. When most proteins are exposed to acidic pH (e.g., pH 2), they lose biological activity. Explain why.

24. Several amino acids can be modified after the synthesis of a polypeptide chain to enhance the functional capabilities of the protein. Match the type of modifying group in the left column with the appropriate amino acid residues in the right column.

(a) phosphate   (1) Glu
(b) hydroxyl   (2) Thr
(c) γ-carboxyl   (3) Pro
(d) acetyl   (4) Ser
                                            (5) N-terminal
                                            (6) Tyr

25. How can a protein be modified to make it more hydrophobic?

ANSWERS TO SELF-TEST

1. (a) (a) C (b) D (c) B, D (d)D (e) B, D (f) A (g) D (h) B (i) C (j) B, D
   (b) A is proline, B is tyrosine, C is leucine, and D is lysine.
   (c) histidine and arginine (basic amino acids)

2. See the structure of cysteine. At pH 1, all the ionizable groups are protonated. Thus the carboxyl is uncharged (COOH), and the amino is positive (NH_3^+). The mercaptan side chain remains –SH.

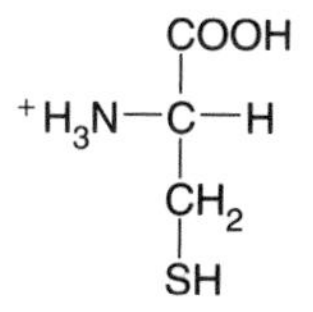

$$
\begin{array}{c}
COOH \\
| \\
^+H_3N-C-H \\
| \\
CH_2 \\
| \\
SH
\end{array}
$$

Cysteine

3. (a) 3   (b) 4   (c) 1   (d) 5   (e) 2   (f) 6

4. a, d, the "acidic" amino acids.

5. Histidine acts as a buffer at pH 6.0 because this is the pK of the imidazole group (see Table 2.1). At pH 7.6, histidine is a poor buffer because no one ionizing group is partially protonated and therefore capable of donating or accepting protons without markedly changing the pH.

6. The 20 L amino acids can form 20 × 400 dipeptides. The minimum number of pK_a values for any dipeptide is two; the maximum is four.

7. (a) glycine
   (b) 4, namely the 2 carboxyl groups of glutamate, the R group of arginine, and the alpha amino group of glycine
   (c) +2, contributed by the N-terminal amino group and the arginine residue
   (d) E-M-R-T-G
   (e) See the structure of the peptide bond below.

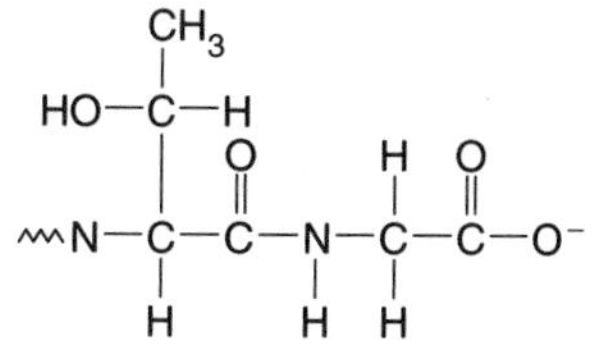

$$
\begin{array}{c}
CH_3 \\
| \\
HO-C-H \\
| \quad\quad O \quad\quad H \quad O \\
\text{\textasciitilde}N-C-C-N-C-C-O^- \\
| \quad\quad\quad\quad | \quad | \\
H \quad\quad\quad H \quad H
\end{array}
$$

Peptide bond

8. c

9. e

10. A Ramachandran plot gives the possible ϕ and ψ angles for the main polypeptide chain containing different amino acid residues. The fact that glycine lacks an R group means that it is much less constrained than other residues. In Figure 2.23, the left-handed helix region, which occurs rarely, generally includes several Gly residues. In contrast to glycine, proline is more highly constrained than most residues because the R group is tied to the amino group. This fixes ϕ at about $-65°$. In Figure 2.20, the rare *cis* form of the peptide bond is shown as occurring about half of the time in X-Pro peptide bonds.

11. a, c

12. c, d

13. b, d

14. In both α-helical and β sheet structures, the polar peptide bonds of the main chain are involved in internal hydrogen bonding, thereby eliminating potential hydrogen bond formation with water. Overall the secondary structures are less polar than the corresponding linear amino acid sequences.

15. a, c. Specific charged and polar amino acid residues may be found inside some proteins, in active sites, but most polar and charged residues are located on the surface of proteins.

16. b, e. Statement (b) is incorrect because globular, water-soluble proteins have most of their nonpolar residues buried in the interior of the protein. Statement (e) is incorrect because not all water-soluble proteins contain β sheet secondary structures. For example, myoglobin is mostly α-helical and lacks β sheet structures.

17. e

18. c

19. b

20. (a) 3   (b) 4   (c) 2   (d) 1

21. Hemoglobin is a protein, α is a protein, β is a protein. A protein is a macromolecule composed of amino acids whether it is one chain or many. α and β are also called subunits.

22. a, c

23. A low pH (pH 2) will cause the protonation of all ionizable side chains and will change the charge distribution on the protein; furthermore, it will impart a large net positive charge to the protein. The resulting repulsion of adjacent positive charges and the disruption of salt bridges often cause unfolding of the protein and loss of biological activity.

24. (a) 2, 4, 6   (b) 3   (c) 1   (d) 5

25. The attachment of a fatty acid chain to a protein can increase its hydrophobicity and promote binding to lipid membranes. Acetylation of surface groups would also work.

PROBLEMS

1. The net charge of a polypeptide at a particular pH can be determined by considering the pK_a value for each ionizable group in the protein. For a linear polypeptide composed of 10 amino acids, how many α-carboxyl and α-amino groups must be considered?

2. For the formation of a polypeptide composed of 20 amino acids, how many water molecules must be removed when the peptide bonds are formed? Although the hydrolysis of a peptide bond is energetically favored, the bond is very stable in solution. Why?

3. Where stereoisomers of biomolecules are possible, only one is usually found in most organisms; for example, only the L amino acids occur in proteins. What problems would occur if, for example, the amino acids in the body proteins of herbivores were in the L isomer form, whereas the amino acids in a large number of the plants they fed upon were in the D isomer form?

4. Each amino acid in a run of several amino acid residues in a polypeptide chain has φ values of approximately −140° and ψ values of approximately +147. What kind of structure is it likely to be?

5. A survey of the location of reverse turns in soluble proteins shows that most reverse turns are located at the surface of the protein, rather than within the hydrophobic core of the folded protein. Can you suggest a reason for this observation?

6. Wool and hair are elastic; both are α-keratins, which contain long polypeptide chains composed of α helices twisted about each other to form cablelike assemblies with cross-links involving Cys residues. Silk, on the other hand, is rigid and resists stretching; it is composed primarily of antiparallel β pleated sheets, which are often stacked and interlocked. Briefly explain these observations in terms of the characteristics of the secondary structures of these proteins.

7. In a particular enzyme, an alanine residue is located in a cleft where the substrate binds. A mutation that changes this residue to a glycine has no effect on activity; however, another mutation, which changes the alanine to a glutamate residue, leads to a complete loss of activity. Provide a brief explanation for these observations.

8. Glycophorin A is a glycoprotein that extends across the red blood cell membrane. The portion of the polypeptide that extends across the membrane bilayer contains 19 amino acid residues and is folded into a helix. What is the width of the bilayer that could be spanned by this helix? The interior of the bilayer includes long acyl chains that are nonpolar. Which of the 20 L amino acids would you expect to find among those in the portion of the polypeptide that traverses the bilayer?

9. (a) Early experiments on the problem of protein folding (notably Anfinsen's experiments on ribonuclease, described in Section 2.6 of the text) suggested that the native three-dimensional structure of a protein was an automatic consequence of its primary structure. Cite experimental evidence that shows that this is the case.

   (b) Later, the discovery that proteins are synthesized directionally on ribosomes, from the amino to the carboxy terminus, complicated the earlier view of protein folding. Explain what the complicating circumstance might be.

   (c) The discovery of chaperone proteins allows both earlier views to be reconciled. Explain how that might be the case.

10. Suppose you are studying the conformation of a monomeric protein that has an unusually high proportion of aromatic amino acid residues throughout the length of the polypeptide chain. Compared with a monomeric protein containing many aliphatic residues, what might you observe for the relative α-helical content for each of the two types of proteins? Would you expect to find aromatic residues on the outside or the inside of a globular protein? What about aliphatic residues?

11. As more and more protein sequences and three-dimensional structures become known, there is a proliferation of computer algorithms for the prediction of folding based on

sequence. How might it be possible to winnow through the possibilities and find the best computer programs? Bear in mind that if the sequence and the structure are available, it is too easy to "reverse engineer" a routine that will produce the correct answer.

12. In its discussion of protein modification and cleavage, the text refers to the synthesis and cleavage of a large polyprotein precursor of virus proteins, as well as to the synthesis of multiple polypeptide hormones from a single polypeptide chain. Is there an advantage to synthesizing a large precursor chain and then cleaving it to create a number of products?

13. The charged form of the imidazole ring of histidine is believed to participate in a reaction catalyzed by an enzyme. At pH 7.0, what is the probability that the imidazole ring will be charged?

14. Calculate the pH at which a solution of cysteine would have no net charge.

15. When diseases such as mad cow disease and scrapie (a similar disease in sheep) were first discovered, many people thought that they were caused by "slow viruses." On the one hand it seemed clear that the diseases were transmissible, but on the other it could take ten years after "infection" for the disease to manifest itself clinically. Carleton Gajdusek visited Papua New Guinea in the 1950's and described a similar human brain disease called kuru. The Fore tribe had customs which brought the living into contact with recently deceased relatives, and somehow kuru was being passed from the dead to the living. How would you show that the disease is caused by infection and not by some sort of genetic transmission? How would you prove that the agent of infection is not a tiny bacterium or a "slow virus"?

16. Lantus[1] is a long-lasting insulin that many diabetics inject just once a day to mimic the basal (background) release of insulin by the pancreas. Lantus is also known as glargine. Lantus/glargine is identical to human insulin except that Arginine at position A21 is replaced by Glycine; in addition, two Arginines are added to the C-terminus of the B-chain. (Normal human insulin's B chain usually ends in Threonine.) Draw the new tripeptide sequence (Thr-Arg-Arg) that is formed at the C-terminus of the B-chain of Lantus.

17. Jello (or "jelly" if you're from Britain or a British tradition) is composed of collagen derived from animal bones, hides, and inedible connective tissue. Why is jello not a particularly useful source of dietary protein?

18. Humalog is an analog of fast-acting human insulin that is also known as Lispro. The difference between Humalog/Lispro and normal human insulin is that two amino acids are switched on the B-chain at positions 28 and 29. The result is an insulin that is absorbed faster under the skin and has a shorter duration of action. Which of the following would you expect to be true about Lispro? (The name is a big clue.)

    (a) L and P on the A chain are reversed.
    (b) K and P on the B chain are reversed.
    (c) L and P on the B chain are reversed.
    (d) L and R on the B chain are reversed.
    (e) The B chain becomes kinked.

19. Monosodium glutamate (MSG) is well-known as common ingredient in Chinese food, but it is also used as a flavor in many other food products. "Yeast protein" or

---

[1]It is an analog of human insulin that is specifically designed to have low solubility at physiological pH. At pH 4, the solution in which it is injected, Lantus/glargine, is soluble, but upon injection into subcutaneous tissue, it is neutralized and it precipitates. Small amounts of the glargine then diffuse slowly from the precipitate, resulting in a relatively slow, constant release. This slow release enables the insulin to be injected just once a day.

"hydrolyzed vegetable protein" in fact means that MSG has been added. Which of the following represents glutamate at physiological pH?

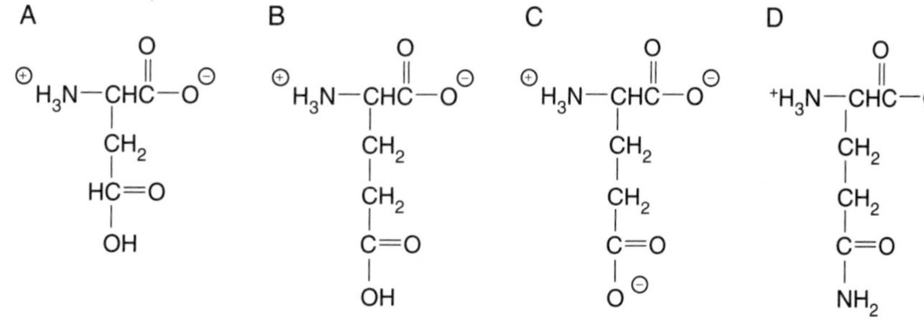

ANSWERS TO PROBLEMS

1. Only the N-terminal α-amino group and the C-terminal α-carboxyl group will undergo ionization. The internal groups will be joined by peptide bonds and are not ionizable. Naturally ionizable R-groups will also contribute but this question focuses on the beginning and the end of the chain.

2. For a peptide of n residues, $(n - 1)$ water molecules must be removed (19 in this case). A significant activation energy barrier makes peptide bonds kinetically stable.

3. All metabolic reactions in an organism are catalyzed by enzymes that are generally specific for either the D or the L isomeric form of a substrate. If an animal (an herbivore in this case) is to be able to digest the protein from a plant and build its own protein from the resulting amino acids, both the animal and the plant must make their proteins from amino acids having the same configuration. In this hypothetical case the L-amino acid animal would be poisoned by its D-amino acid food.

4. From the Ramachandran plot in Figure 2.23 of the text, we see that β conformation is accommodated by φ values of approximately −140° and ψ values of approximately +147°. The structure is most likely a β sheet. In fact, the "low" numbers here imply that it is an antiparallel beta sheet. The parallel β sheet would have higher numbers, more like φ = −160° and ψ = +160°.

5. Figure 2.36 in the text shows that in a reverse turn the CO group of residue 1 is hydrogen-bonded to the NH group of residue 4. However, there are no adjacent amino acid residues available to form intrachain hydrogen bonds with the CO and NH groups of residues 2 and 3. These groups cannot form hydrogen bonds in the hydrophobic environment found in the interior portion of a folded protein. They are more likely to hydrogen bond with water on the surface of the protein.

6. When the α helices in wool are stretched, intrahelix hydrogen bonds are broken as are some of the interhelix disulfide bridges; maximum stretching yields an extended β sheet structure. The Cys cross-links provide some resistance to stretch and help pull the α helices back to their original positions. In silk, the β sheets are already maximally stretched to form hydrogen bonds. Each β pleated sheet resists stretching, but since the contacts between the sheets primarily involve van der Waals forces, the sheets are somewhat flexible.

7. Both alanine and glycine are neutral nonpolar residues with small side chains, whereas the side chain of glutamate is acidic and bulkier than that of alanine. Either feature of the glutamate R group could lead to the loss of activity by altering the protein conformation or by interfering with the binding of the substrate.

8. Since each residue in the α helix is 1.5 Å from its neighbor, the length of the chain that spans the membrane bilayer is 19 × 1.5 Å = 28.5 Å, which is also the width of the membrane. One would expect to find nonpolar amino acid residues in the polypep-

tide portion associated with the membrane bilayer. These would include Phe, Ile, Leu, Val, Met, and Ala (FILMV + A). The actual sequence of the buried chain is

I-T-L-I-I-F-G-V-M-A-G-V-I-G-T-I-L-L-I

9. (a) In Anfinsen's experiment, native ribonuclease is treated with mercaptoethanol to disrupt disulfide bonds and with urea as a denaturant, it unfolds, as indicated by the fact that it becomes enzymically inactive. When urea is removed by dialysis and disulfide bonds reform by oxidation, it regains enzymic activity, suggesting that its native structure has been restored. Note that the refolding occurs in the absence of any cellular proteins or other biosynthetic machinery.

   (b) The discovery that proteins are synthesized directionally on ribosomes beginning at the amino terminus complicates matters somewhat because folding of the amino end of the polypeptide chain could begin before the carboxyl end had been synthesized. Such folding could represent the most stable conformation over a short range, but there would be no guarantee that it would be part of the energy minimum for the entire molecule.

   (c) Chaperone proteins could bind to an initially synthesized polypeptide and prevent it from undergoing final folding until the entire molecule was synthesized. Note that the final fold of the protein is still specified by the primary structure.

10. The higher the proportion of aromatic side chains (such as those of phenylalanine) in the protein, the more likely that steric hindrance among closely located residues could interfere with the establishment of the regular repeating structure of the α helix. Smaller aliphatic side chains like those of leucine, isoleucine, and valine would be less likely to interfere. Structural studies on many proteins reveal that the number of aromatic residues in α-helical segments is relatively low, while the content of aliphatic side chains in such segments is unremarkable, compared to that of other nonhelical regions of a folded protein. Both aliphatic and aromatic side chains (especially that of phenylalanine) are hydrophobic, so that many of them are buried inside a globular protein, away from water molecules.

11. Protein scientists have devised a competition called CASP, or Critical Assessment of Techniques for Protein Structure Prediction, which is held every other year. Laboratories that are working on determination of three-dimensional structure by x-ray crystallography (or nmr) announce that they expect to release the structure in a few months. They give a description of the sequence of the protein and its use in the cell, and withhold the actual structural coordinates until a certain date. In the meantime, laboratories with predictive algorithms publicly post the structure they think the protein will have. The success or failure of the prediction takes place in a public arena, and the better predictors have bragging rights. CASP-4 in 2000 showed that there are several effective programs available, notably ROSETTA, used by David Baker of the University of Washington (sometimes on a server known as ROBETTA). Results of the competition are published in the journal *Protein* and online (in technical language) at the website http://predictioncenter.org/. CASP-6 in 2004 was won by Krzysztof Ginalski of the University of Texas, using ROSETTA along with 3D-Jury and Meta-BASIC.

12. The primary advantage of precursor chain synthesis is that the production of related proteins can be coordinated. This could be important in viral infection, and it may also be important for coordinated synthesis of hormones with related activities. It is worth noting that there are other reasons for the synthesis of polyprotein precursors. For example, the genome of the poliovirus consists of a single RNA molecule that acts as a messenger on entering the cytoplasm of the host. In eukaryotic cells a messenger RNA molecule can be translated into only one polypeptide chain. Therefore the poliovirus can reproduce only by synthesizing its proteins by sequential cleavages.

13. Use the Henderson-Hasselbalch equation to calculate the concentration of histidine, whose imidazole ring is ionized at neutral pH. The value of pK for the ring is 6.0 for a histidine residue in a protein (see Table 2.1).

$$pH = pK + \log \frac{[His]}{[His^+]}$$

$$7.0 = 6.0 + \log \frac{[His]}{[His^+]}$$

$$\log \frac{[His]}{[His^+]} = 1.0$$

$$\frac{[His]}{[His^+]} = 10$$

At pH 7.0, the ratio of uncharged histidine to charged histidine is 10:1, making the probability that the side chain is charged only 9%.

14. To see which form of cysteine has no net charge, examine all the possible forms, beginning with the one that is most protonated:

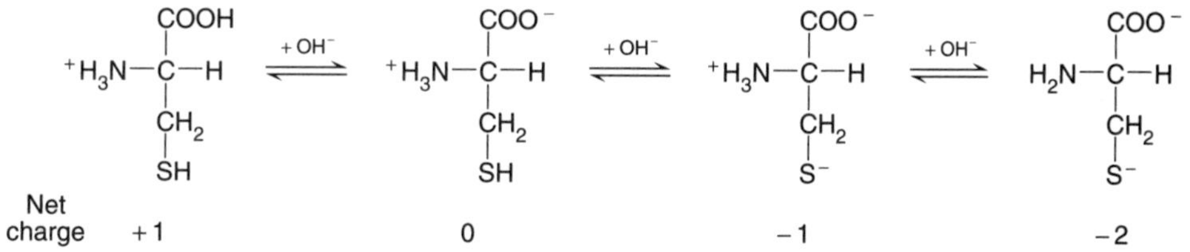

The pH of the cysteine solution at which the amino acid has no net charge will be that point at which there are equal amounts of the compound with a single positive charge and a single negative charge. This is, in effect, the average of the two corresponding pK values, one for the α-carboxyl group and the other for the side chain sulfhydryl group. Thus, (1.8 + 8.3)/2 = 5.05. This value is also known as the *isoelectric point*.

15. The genetic hypothesis was ruled out by injecting material from Kuru-infected brains into chimpanzees. After a period of time, they came down with the disease. It was known that the incubation time was quite long. In fact it took decades to rule out the nanobacteria and "slow virus" hypotheses. Probably there are still a few scientists clinging to those ideas. Stanley Prusiner devoted his career to proving that diseases could be caused by infectious proteins (which he called prions). As he was able to purify the infectious agent (not an easy task—how do you assay during purification?) he reached a point where he could demonstrate that there were no nucleotides present in the solution. As you saw in Chapter 1 of the text, all cells and viruses contain DNA or RNA, which is made up of nucleotides. Adenine and guanine have to be present, and both are built on the bicyclic purine ring. Purines absorb ultraviolet light at certain characteristic frequencies but there is nothing in proteins that absorbs light at those frequencies.

When the scrapie prion was sequenced, it was a surprise to learn that the protein was a normal body protein which had somehow become an agent of disease. Here is a Web page that contains more of the interesting history of this research:

http://www.stanford.edu/group/virus/prion/prion2.html

16.

```
        O        H     O     H      O
        ‖        |     ‖     |      ‖
 —HN—CH—C—O—N—CH—C—N—CH—C—O⁻
       |             |            |
      CH—OH         CH₂          CH₂
       |             |            |
      CH₃           CH₂          CH₂
                     |            |
                    CH₂          CH₂
                     |            |
                    NH           NH
                     |            |
                    C=NH         C=NH
                     |            |
                    NH₂          NH₂
```

17. Collagen is highly enriched in Glycine (1/3 of all residues) and Proline (or Hydroxyproline; 1/6 of all residues). Jello thus does not contain a sufficient variety of amino acids to be a rich nutritional source of amino acids.

18. (b) ("Lis" indicates Lysine, K; "Pro" indicates Proline, P)

19. (c) At physiological pH, the pKas of both carboxylic acid groups have been exceeded, so these are predominantly ionized/deprotonated. The amino group has a pKa below that of physiological pH, so it remains protonated.

Exploring Proteins and Proteomes

Chapter 3 extends Chapter 2 by introducing the most important methods used to investigate proteins. Many of these were essential in discovering the principles of protein structure and function presented in the preceding chapter. These methods also constitute the essentials of the armamentarium of modern biochemical research and underlie current developments in biotechnology. First, the authors define the concept of the proteome, the sum of functioning proteins in the cell and their interactions. Then they outline methodological principles for the analysis and purification of proteins. Next they describe methods of sequencing the amino acids in proteins, and explain why the knowledge revealed by these techniques is so important. They continue with a discussion of antibodies as highly specific analytical reagents, followed by a discussion of the uses of peptides of defined sequence and how they are chemically synthesized. Next, the authors describe powerful new ways to utilize mass spectrometry to analyze proteins and peptide fragments. To close the chapter, there is a discussion of the use of x-ray crystallography and nuclear magnetic resonance spectroscopy to determine the three-dimensional structures of proteins.

LEARNING OBJECTIVES

When you have mastered this chapter, you should be able to accomplish the following objectives.

Introduction

1. Distinguish between the genome and the proteome, and define both terms.

The Purification of Proteins Is an Essential First Step in Understanding Their Function (Text Section 3.1)

2. Describe how a quantitative enzyme assay can be used to calculate the specific activity during protein purification.

3. Define differential centrifugation, and describe how it would be used to produce a protein mixture from a cell homogenate.

4. List the properties of proteins that can be used to accomplish their separation and purification, and correlate them with the appropriate methods: gel-filtration chromatography, dialysis, salting out, ion-exchange chromatography, and affinity chromatography. Describe the basic principles of each of these methods.

5. Describe the principle of electrophoresis and its application in the separation of proteins.

6. Explain the determination of protein mass by SDS-PAGE: sodium dodecyl sulfate-polyacrylamide gel electrophoresis.

7. Define the isoelectric point (pI) of a protein and describe isoelectric focusing as a separation method.

8. Explain the quantitative evaluation of a protein purification scheme.

9. Define the sedimentation coefficient S, and give its common name. Note the range of S values for biomolecules and cells.

10. Describe zonal centrifugation and sedimentation equilibrium and explain their applications to the study of proteins.

11. Know three ways that modern recombinant DNA technology can help with protein purification: producing large amounts of the desired protein, adding tags to allow easier separation of the protein, and altering the structure of the protein.

Amino Acid Sequences Can Be Determined Experimentally (Text Section 3.2)

12. Give examples of the important information that amino acid sequences reveal. For example, comparison of sequences can reveal relationships in function, or evolutionary relationships.

13. Outline the steps in the determination of the amino acid composition and the amino-terminal residue of a peptide.

14. Describe the sequential Edman degradation method and the automated determination of the amino acid sequences of peptides.

15. List the most common reagents used for the specific cleavage of proteins. Explain the application of overlap peptides to protein sequencing.

16. Describe the additional steps that must be used for sequencing disulfide-linked polypeptides and oligomeric proteins.

17. Explain, in general terms, how recombinant DNA technology is used to determine the amino acid sequences of nascent proteins. Note the differences between a nascent protein and one that has undergone posttranslational modifications.

Immunology Provides Important Techniques with Which to Investigate Proteins (Text Section 3.3)

18. Define the terms antibody, antigen, antigenic determinant (epitope), and immunoglobulin G.

19. Contrast polyclonal antibodies and monoclonal antibodies and describe their preparation.

20. Outline methods that use specific antibodies in the analysis or localization of proteins.

21. Describe how the use of fluorescent markers allows direct observation of changes within living cells.

Mass Spectrometry Provides Powerful Tools for Protein Characterization and Identification (Text Section 3.4)

22. Outline the application of mass spectrometry to the analysis of proteins and compare the merits of the various methods of determining the molecular weights of proteins.

23. Define MALDI and TOF and explain how the techniques work together to allow application of mass spectrometry to proteins.

24. Explain how peptides can be sequenced using mass spectrometry.

25. Describe tandem mass spectrometry.

26. Explain why it is necessary to have the genome sequence in order to identify proteins using cleavage, chromatography, and MALDI-TOF.

Proteins Can Be Synthesized by Automated Solid-Phase Methods (Text Section 3.5)

27. List the most important uses of synthetic peptides.

28. Outline the steps of the solid-phase method for the synthesis of peptides.

Three-Dimensional Protein Structure Can Be Determined by X-Ray Crystallography and NMR Spectroscopy (Text Section 3.6)

29. Describe the fundamentals of the method and basic physical principles underlying nuclear magnetic resonance spectrometry as applied to protein structure determination.

30. Provide a similar description of the x-ray crystallographic analysis of a protein, and give the basic physical principles underlying this technique.

31. Compare the relative advantages and disadvantages of x-ray crystallography and NMR spectroscopy for protein structure determination.

SELF-TEST

Introduction

1. The genome sequence tells us all of the proteins an organism can make. Are all of these proteins expressed?

The Purification of Proteins Is an Essential First Step in Understanding Their Function

2. The following five proteins, which are listed with their molecular weights and iso-electric points, were separated by SDS-polyacrylamide gel electrophoresis. Give the order of their migration from the top (the point of sample application) to the bottom of the gel.

		Molecular weight (daltons)	pI
(a)	α-antitrypsin	45,000	5.4
(b)	cytochrome c	13,400	10.6
(c)	myoglobin	17,000	7.0
(d)	serum albumin	69,000	4.8
(e)	transferrin	90,000	5.9

Top ————————— Bottom

3. If the five proteins in Question 2 were separated in an isoelectric-focusing experiment, what would be their distribution between the positive (+) and negative (−) ends of the gel? Indicate the high and low pH ends.

    Cathode (−) ———————— (+) Anode

4. Which of the following statements are NOT true?
    (a) The pI is the pH value at which a protein has no charges.
    (b) At a pH value equal to its pI, a protein will not move in the electric field of an electrophoresis experiment.
    (c) An acidic protein will have a pI greater than 7.
    (d) A basic protein will have a pI greater than 7.

5. SDS-polyacrylamide gel electrophoresis and the isoelectric-focusing method for the separation of proteins have which of the following characteristics in common? Both
    (a) separate native proteins.
    (b) make use of an electrical field.
    (c) separate proteins according to their mass.
    (d) require a pH gradient.
    (e) are carried out on supporting gel matrices.

6. Before high-performance liquid chromatography (HPLC) methods were devised for the separation and analysis of small peptides, electrophoresis on a paper support was frequently used. Separation was effected on the basis of the charge on a peptide at different pH values. Predict the direction of migration for the following peptides at the given pH values. Use C for migration toward the cathode, the negative pole; A for migration toward the anode, the positive pole; and O if the peptide remains stationary.

	pH			
	2.0	4.0	6.0	11.0
(a) Lys-Gly-Ala-Gly				
(b) Lys-Gly-Ala-Glu				
(c) His-Gly-Ala-Glu				
(d) Glu-Gly-Ala-Glu				
(e) Gln-Gly-Ala-Lys				

7. How would the time required for the separation described in question 6 be changed if all the solutions that were used during the electrophoresis contained 100 g/L of table sugar (sucrose)?

8. Examine Table 3.1 in the text, evaluating a protein purification scheme. Does "total activity" go up or down as the protein is purified? Would it have been a good idea to try affinity chromatography at an earlier stage of purification?

9. The molecular weight of a protein can be determined by SDS-polyacrylamide gel electrophoresis or by sedimentation equilibrium. Which method would you use to determine the molecular weight of a protein containing four subunits, each consisting of two polypeptide chains cross-linked by two disulfide bridges? Explain your answer.

Amino Acid Sequences Can Be Determined Experimentally

10. Treating a protein with 6 M HCl at 110° for 24 hours will have what effect?
    (a) The protein will unfold (denature)
    (b) All peptide bonds will be destroyed, releasing the amino acids
    (c) All amino acids will be destroyed

11. Which of the following statements concerning the Edman degradation method are true?
    (a) Phenyl isothiocyanate is coupled to the amino-terminal residue.
    (b) Under mildly acidic conditions, the modified peptide is cleaved into a cyclic derivative of the terminal amino acids and a shortened peptide (minus the first amino acid).
    (c) Once the PTH amino acid is separated from the original peptide, a new cycle of sequential degradation can begin.
    (d) If a protein has a blocked amino-terminal residue (as does N-formyl methionine, for example), it cannot react with phenyl isothiocyanate.

12. When sequencing proteins, one tries to generate overlapping peptides by using cleavages at specific sites. Which of the following statements about the cleavages caused by particular chemicals or enzymes are true?
    (a) Cyanogen bromide cleaves at the carboxyl side of threonine.
    (b) Trypsin cleaves at the carboxyl side of Lys and Arg.
    (c) Chymotrypsin cleaves at the carboxyl side of aromatic and bulky amino acids.
    (d) 2-Nitro-5-thiocyanobenzoate cleaves on the amino side of cysteine residues.
    (e) Chymotrypsin cleaves at the carboxyl side of aspartate and glutamate.

13. What treatments could you apply to the following hemoglobin fragment to determine the amino-terminal residue and to obtain two sets of peptides with overlaps so that the complete amino acid sequence can be established? Give the sequences of the peptides obtained.

    Val-Leu-Ser-Pro-Ala-Lys-Thr-Asn-Val-Lys-Ala-Ala-Trp-Gly-Lys-Val-Gly-Ala-His-Ala-Gly-Glu-Tyr-Gly-Ala-Glu-Ala-Thr-Glu

14. Which of the following techniques are used to locate disulfide bonds in a protein?
    (a) The protein is first reduced and carboxymethylated.
    (b) The protein is cleaved by acid hydrolysis.
    (c) The protein is specifically cleaved under conditions that keep the disulfide bonds intact.
    (d) The peptides are separated by SDS-polyacrylamide gel electrophoresis.
    (e) The peptides are separated by two-dimensional electrophoresis with an intervening performic acid treatment.

15. Which of the following are important reasons for determining the amino acid sequences of proteins?
    (a) Knowledge of amino acid sequences helps elucidate the molecular basis of biological activity.
    (b) Alteration of an amino acid sequence may cause abnormal functioning and disease.
    (c) Amino acid sequences provide insights into evolutionary pathways and protein structures.
    (d) The three-dimensional structure of a protein can be predicted from its amino acid sequence.
    (e) Amino acid sequences provide information about the destination and processing of some proteins.
    (f) Amino acid sequences allow prediction of the DNA sequences encoding them and thereby facilitate the preparation of DNA probes specific for the regions of their genes.

16. In spite of the convenience of using recombinant DNA techniques for determining the amino acid sequences of proteins, chemical analyses of amino acid sequences are frequently required. Explain why.

Immunology Provides Important Techniques with Which to Investigate Proteins

17. Match the terms in the left column with the appropriate item or items from the right column.

    (a) antigens
    (b) antigenic determinants
    (c) polyclonal antibodies
    (d) monoclonal antibodies

    (1) immunoglobulins
    (2) foreign proteins, polysaccharides, or nucleic acids
    (3) antibodies produced by hybridoma cells
    (4) groups recognized by antibodies
    (5) heterogeneous antibodies
    (6) homogeneous antibodies
    (7) antibodies produced by injecting an animal with a foreign substance
    (8) epitopes

18. The methods used to localize a specific protein in an intact cell are
    (a) Western blotting.
    (b) solid-phase immunoassay.
    (c) enzyme-linked immunosorbent assay.
    (d) immunoelectron microscopy.
    (e) fluorescence microscopy.

19. Explain why immunoassays are especially useful for detecting and quantifying small amounts of a substance in a complex mixture.

Mass Spectrometry is a Powerful Technique for Protein Characterization and Identification

20. After isolating and purifying to homogeneity a small enzyme (110 amino acids long) from a culture of bacteria, you are confused as to whether you grew wild-type bacteria or a mutant strain that produced the enzyme with a valine residue at position 66 instead of the glycine found in the wild-type strain. How could you quickly determine which protein you had?

21. If you run a protein of MW = 30,000 daltons through MALDI-TOF, besides the 30,000 peak you will also see a 15,000 dalton peak. Why?

    (a) MALDI will probably cleave the protein in half.
    (b) There is likely to be an impurity with MW=15,000.
    (c) Some of the protein will be doubly charged and will appear at 15,000.

22. Tandem mass spectrometry can provide enough information to give the sequence of peptides. But all you get from mass spectrometry is a list of numbers, the molecular weight of the fragments. How do you decode that into amino acids?

Proteins Can Be Synthesized by Automated Solid-Phase Methods

23. The amino acid sequence of a protein is known and strong antigenic determinants have been predicted from the sequence; however, you do not have enough of the pure protein to prepare antibodies. How could you circumvent this problem, using knowledge of peptide synthesis?

24. Which of the following is commonly used as a protecting group during peptide synthesis?

    (a) tert-butyloxycarbonyl
    (b) dicyclohexylcarbodiimide
    (c) dicyclohexylurea
    (d) hydrogen fluoride
    (e) phenyl isothiocyanate

25. The following reagents are often used in protein chemistry:

    (1) CNBr
    (2) urea
    (3) β-mercaptoethanol
    (4) trypsin
    (5) dicyclohexylcarbodiimide
    (6) 6 N HCl, cation exchanger
    (7) phenyl isothiocyanate
    (8) chymotrypsin
    (9) dilute F_3CCOOH

    Which of these reagents are best suited for the following tasks?

    (a) determination of the amino acid sequence of a small peptide
    (b) reversible denaturation of a protein devoid of disulfide bonds
    (c) hydrolysis of peptide bonds on the carboxyl side of aromatic residues
    (d) cleavage of peptide bonds on the carboxyl side of methionine
    (e) hydrolysis of peptide bonds on the carboxyl side of lysine and arginine residues
    (f) reversible denaturation of a protein that contains disulfide bonds (two reagents are needed)
    (g) activation of carboxyl groups during peptide synthesis
    (h) determination of the amino acid composition of a small peptide
    (i) removal of t-Boc protecting group during peptide synthesis

26. Solid-phase synthesis would be best for synthesis of what useful compound?
    (a) Vasopressin or Oxytocin (9 amino acid residues)
    (b) Insulin (51 amino acid residues)
    (c) Ribonuclease (124 amino acid residues)
    (d) All of the above

Three-Dimensional Protein Structure Can Be Determined by X-Ray Crystallography and NMR Spectroscopy

27. Must we be able to crystallize a protein in order to learn the three-dimensional structure?

28. Which of the following statements concerning x-ray crystallography is NOT true?
    (a) Only crystallized proteins can be analyzed.
    (b) The x-ray beam is scattered by the protein sample.
    (c) All atoms scatter x-rays equally.
    (d) The basic experimental data are relative intensities and positions of scattered electrons.
    (e) The electron-density maps are obtained by applying the Fourier transform to the scattered electron intensities.
    (f) The resolution limit for proteins is about 2 Å.

29. How can x-ray crystallography provide information about the interaction of an enzyme with its substrate?

ANSWERS TO SELF-TEST

1. In any given cell, at any given time, it is quite likely that many genes capable of producing protein are not being expressed. Single-celled organisms tend to respond to their environment, producing enzymes to deal with the nutrients and conditions in the area. Multicellular organisms need different proteins for different parts of the body. Humans have very different needs in the retina, the liver, and muscle cells. This is why the proteome is a useful concept, it is a description of the proteins actually present in a functioning cell.

2. Top $\dfrac{\text{e d a c b}}{}$ Bottom

3. High pH (–) $\dfrac{\text{b c e a d}}{}$ Low pH (+)

4. a, c. Regardless of the pH, a protein is never devoid of charges; at the pI, the sum of all the charges is zero.

5. b, e

6.

	pH 2.0	4.0	6.0	11.0
(a)	C	C	C	A
(b)	C	C	O	A
(c)	C	C	A	A
(d)	C	O	A	A
(e)	C	C	C	A

For example, peptide b carries a net charge of +1.5 at pH 2.0 (Lys side chain, +1; α-amino group, +1; Glu side chain, 0; and terminal carboxyl, –0.5, since the pH coincides with its pK value). At pH 4.0, the net charge is +0.5; the Glu side chain is half ionized (−0.5), but the terminal carboxyl is almost completely ionized (−1). At pH 6.0, the net charge is 0 due to a +2 charge contributed by the Lys residue and a −2 charge contributed by the Glu residue. At pH 11.0, the α-amino group is deprotonated (charge of 0) and the Lys side chain is half-protonated (charge of +0.5); thus, the net charge is −1.5. The same answer for peptide b can be given graphically (see Figure 3.1):

FIGURE 3.1

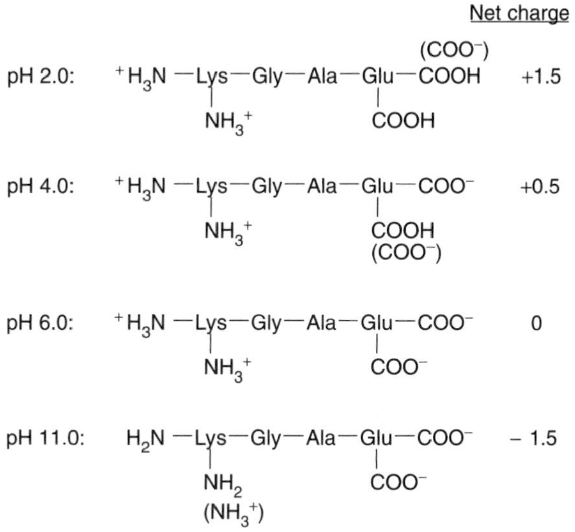

7. More time would be required for the separation, since the velocity of movement of the compounds would be slowed because of the increased viscosity of the solution. The velocity at which a molecule moves during electrophoresis is inversely dependent on the frictional coefficient, which is, itself, directly proportional to the viscosity of the solution.

8. Total activity drops as material is lost in each step of purification. In a good purification, total protein drops much faster, so that specific activity goes up dramatically. Not all proteins can be purified with affinity chromatography. When a protein has a unique substrate and works this well with affinity chromatography, it may be a good idea to leave out the ion exchange and molecular exclusion steps, and go straight to the "home run" technique. A standard source on protein purification states that "One-step purifications of 1,000-fold with nearly 100% recovery have been reported" with this technique. (R. K. Scopes. [1994]. *Protein purification, principles and practice,* 3rd ed. New York: Springer-Verlag.)

9. Determinations of mass by SDS-polyacrylamide gel electrophoresis are carried out on proteins that have been denatured by the detergent in a reducing medium; the reducing agent in the medium disrupts disulfide bonds. Therefore, to determine the molecular weight of a native protein containing subunits with disulfide bridges, you must use the sedimentation equilibrium method. Another nondenaturing method, gel-filtration chromatography, can be used to obtain approximate native molecular weights.

10. b. Students encountering amino acid analysis may underestimate the violence of the procedure. Water boils at 100° so you need a sealed glass vial (or pressure cooker) just to achieve that temperature. And 6M HCl is dangerous even at room temperature. Plus 24 hours is a long reaction time. Besides destroying all peptide bonds, the technique actually destroys three of the amino acids in the protein, glutamine, asparagine, and tryptophan.

11. a, b, c, d

12. b, c, d

13. The amino-terminal residue of the hemoglobin fragment can be determined by analyzing the intact fragment by the Edman degradation method; this shows that the amino-terminal residue is Val. Trypsin digestion, separation of peptides, and Edman degradation give

Val-Leu-Ser-Pro-Ala-Lys
Thr-Asn-Val-Lys
Ala-Ala-Trp-Gly-Lys
Val-Gly-Ala-His-Ala-Gly-Glu-Tyr-Gly-Ala-Glu-Ala-Thr-Glu
Chymotrypsin digestion, separation of peptides, and Edman degradation give
Val-Leu-Ser-Pro-Ala-Lys-Thr-Asn-Val-Lys-Ala-Ala-Trp
Gly-Lys-Val-Gly-Ala-His-Ala-Gly-Glu-Tyr
Gly-Ala-Glu-Ala-Thr-Glu

14. c, e. The performic acid oxidizes the disulfide bonds to SO_3^- groups and releases new peptides.

15. a, b, c, e, f. Answer (d) may be correct in some cases where homologous proteins are compared in terms of amino acid sequences and known three-dimensional structures.

16. The amino acid sequence derived from the DNA sequence is that of the nascent polypeptide chain before any posttranslational modifications. Since the function of a protein depends on its mature structure, it is often necessary to analyze the protein itself to determine if any changes have occurred after translation.

17. (a) 2   (b) 4, 8   (c) 1, 5, 7   (d) 1, 3, 6

18. d, e. Immunoelectron microscopy provides more precise localization than does fluorescence microscopy.

19. Because the interaction of an antibody with its antigen is highly specific, recognition and binding can occur in the presence of many other substances. If the antibody is coupled to a radioactive or fluorescent group or an enzyme whose activity can be detected in situ, then a sensitive method is available for the detection and quantitation of the antigen-antibody complex.

20. Mass spectrometry—Electrospray or MALDI-TOF—could easily distinguish a protein of this approximate mass (101 amino acids × approximately 110 d per amino acid = 12 kd) that contained an extra 41 atomic mass units as a result of the substitution of a valine for a glycine residue.

21. c. See the example of Lactoglobulin MALDI-TOF mass spec in the text chapter.

22. Subtract each MW from the next higher MW. This will give you a series of MW for each amino acid in sequence. Amino acids tend to have unique MW (except for leucine and isoleucine) so you can convert the MW into amino acids by consulting a table of amino acid molecular weights.

23. You could synthesize peptides containing the putative antigenic determinants, couple these peptides to an antigenic macromolecule, and prepare antibodies against the synthetic peptides. If the same antigenic determinants are present in the protein of interest and are not occluded by the structure of the protein, then the antibodies prepared against the synthetic peptides should also react with the protein.

24. a

25. (a) 7   (b) 2   (c) 8   (d) 1   (e) 4   (f) 2, 3   (g) 5   (h) 6   (i) 9

26. a. Solid phase synthesis of peptides works best with short sequences. It might be possible to synthesize an entire polypeptide but there are other ways to do that which work better.

27. No. Crystals are necessary for x-ray crystallography, but not for NMR spectroscopy. If a highly concentrated solution (1 mM) of a pure protein can be obtained, then significant information can be derived about the three-dimensional shape.

28. c, d, e. In x-ray crystallography, x-rays, not electrons, are scattered and detected.

29. If an enzyme can be crystallized with and without its substrate and the three-dimensional structures of both are obtained using x-ray crystallography, the difference between the two structures should reveal how the substrate fits in its binding site and which atoms and what kind of bonds are involved in the interaction.

PROBLEMS

1. The proteome is described as a dynamic collection of all proteins in use in a cell at a given time, including their modifications and associations. How could the techniques described in this chapter be used to obtain information about the proteome in certain cells?

2. How can a protein be assayed if it is not an enzyme?

3. Many of the methods described in Chapter 3 are used to purify enzymes in their native state. Why would the use of SDS-polyacrylamide gel electrophoresis be unlikely to lead to the successful purification of an active enzyme? What experiments would you conduct to determine whether salting out with ammonium sulfate would be useful in enzyme purification?

4. Of the techniques for analyzing proteins discussed in Chapter 3 of the text, which one would be the easiest to use for accurately determining the molecular weight of a small monomeric protein? Comment on the standards you would wish to use in this technique. What types of proteins might not be analyzed accurately by your suggested method?

5. In ultracentrifugation, the equation for the sedimentation coefficient is $s = m(1 - \nu\rho)/f$, where m is the mass of the particle, ν ("nu") is the partial specific volume, ρ ("rho") is the density of the medium, and f is the frictional coefficient. For proteins, $\nu = 0.72$ cm^3/g, and for DNA, $\nu = 0.5$ cm^3/g.

   (a) Which is denser, protein or DNA?
   (b) What are the dimensions (units) of $(1 - \nu\rho)$?
   (c) Given that "s" has units of "seconds" derive the units of "f."

6. A glutamine residue that is the amino-terminal residue of a peptide often undergoes spontaneous cyclization to form a heterocyclic ring; the cyclization is accompanied by the release of ammonium ion. Diagram the structure of the ring, showing it linked to an adjacent amino acid residue. How would the formation of the ring affect attempts to use the Edman procedure for sequence analysis? A similar heterocyclic ring is formed during the biosynthesis of proline; in this case, glutamate is the precursor. Can you propose a pathway for the synthesis of proline from glutamate?

7. A peptide composed of 12 amino acids does not react with phenyl isothiocyanate or

with dabsyl chloride (an end group reagent that reacts with the free amino end of peptides). Cleavage with cyanogen bromide yields a peptide with a carboxyl-terminal homoserine lactone residue, which is readily hydrolyzed, in turn yielding a peptide whose sequence is determined by the Edman procedure to be

E-H-F-W-D-D-G-G-A-V-L

On the other hand, cleavage with staphylococcal protease (see Table 3.3 in the text) yields an equivalent of aspartate and two peptides. Use of the Edman procedure gives the following sequences for these peptides:

G-G-A-V-L-M-E and H-F-W-D

Why does the untreated peptide fail to react with phenyl isothiocyanate?

8. A hexapeptide that is part of a mouse polypeptide hormone is analyzed by a number of chemical and enzymatic methods. When the hexapeptide is hydrolyzed and analyzed by ion-exchange chromatography, the following amino acids are detected:

Tyr Cys Glu

Ile Lys Met

Two cycles of Edman degradation of the intact hexapeptide released the following PTH-amino acids (see Figure 3.2):

FIGURE 3.2

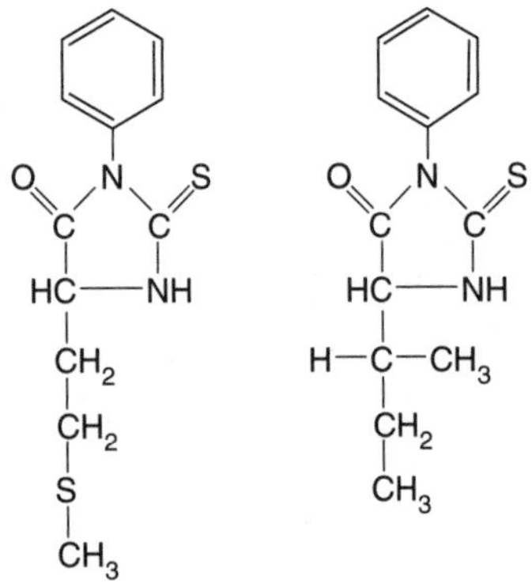

Cleavage of the intact protein with cyanogen bromide yields methionine and a pentapeptide. Treating the intact hexapeptide with trypsin yields a dipeptide, which contains tyrosine and glutamate, and a tetrapeptide. When the intact hexapeptide is treated with carboxypeptidase A, a tyrosine residue and a pentapeptide are produced. Bearing in mind that the hexapeptide is isolated from a mouse, write its amino acid sequence, using both three-letter and one-letter abbreviations.

9. The production of a small acidic protein, HCG or human chorionic gonadotropin, during pregnancy is the basis of most pregnancy test kits. What method makes the most sense for detecting a known protein like this?

10. A laboratory group wishes to prepare a monoclonal antibody that can be used to react with a specific viral coat protein in a Western blotting procedure. Why would it be a good idea to treat the viral coat protein with SDS before attempting to elicit monoclonal antibodies?

11. Mass spectrometry is often used for the sequence analysis of peptides. The procedure requires only microgram quantities of protein and is very sensitive; cationic fragments are identified by their charge-to-mass ratio. In one procedure, peptides are treated

with triethylamine and then with acetic anhydride. What will such a procedure do to amino groups? Next, the modified peptide is incubated with a strong base and then with methyl iodide. What groups will be methylated? Which two amino acids cannot be distinguished by mass spectrometry?

12. A map of the electron density is necessary for the determination of the three-dimensional structure of a protein, but other information is also needed. Hydrogen atoms have one electron and cannot be visualized by x-ray analyses of proteins. Bearing this in mind, compare the structures of amino acids like valine, threonine, and isoleucine and then describe what additional information would be needed along with an electron-density map.

13. Some crystalline forms of enzymes are catalytically active, and thus are able to carry out the same chemical reactions as they can in solution. Why are these observations reassuring to those who are concerned about whether crystallographic determinations reveal the normal structure of a protein?

14. In some ways it is easier to obtain protein structures by NMR spectroscopy. Proteins need not be crystallized, just purified and dissolved. Would it be desirable to use NMR to learn about the structure of the active site of an enzyme to design an inhibitor? Why or why not?

15. The text describes the use of fluorescence labeled antibodies to visualize the locations of certain proteins in the cell. Suppose you know that a certain protein exists with and without a certain phosphorylation, could you detect the different forms in this way?

16. Suppose you are investigating the genetic basis of a rare disease that causes degeneration of neurons. You have brain tissue from normal individuals and from a few unrelated patients who died from the disease. Should you look at differences between DNA or proteins to search for the cause of the disease? What sort of techniques would be likely to work?

ANSWERS TO PROBLEMS

1. Given the current state of knowledge about genome sequences, and the current availability of powerful computers, some of the older techniques are able to afford much more information than they used to. Matching mass spectrometry with a complete database of all possible peptides produced by digestion with trypsin in a genome makes identification of unknown proteins easy because a computer can generate a table of all expected molecular weights of all such peptides and proteins coded in the DNA. See text Fig. 3.37 for this method.

   A more modern method would add tandem MS to the tryptic digest to obtain sequence information about peptide fragments. Computers can be set to control which peaks are selected for successive stages of MS. Having actual sequence information makes matching known genes in a genome much easier. See text Fig. 3.36 for tandem MS.

   LCMS, or Liquid Chromatography-Mass Spectrometry, adds an HPLC column prior to the ionization step. See text Fig 3.6 for HPLC. Often a very thin capillary column is set to supply an electrospray ionizer, and a series of pure proteins or peptides can be analyzed as they emerge from the column. To obtain the maximum information, these proteins might be analyzed with tandem or higher multiples of MS (say MS-MS-MS). With a computer controlling selection of fragments for further analysis, a protein can be analyzed, and partially sequenced, in a matter of seconds.

An example of this sort of analysis is found in "Precision Mapping of an In Vivo N-Glycoproteome Reveals Rigid Topological and Sequence Constraints" [Zielinska et al., *Cell* 141 (2010) 897]. In Matthias Mann's laboratory, proteins with sugars attached were isolated using lectins (which bind to the sugars). These sugars were attached to the N of the side chain of asparagine. When the sugars were removed by an enzyme, the asparagine was converted into aspartate, and the difference of about one Dalton between $-NH_2$ and $-OH$ showed where the sugars had been attached. Tandem MS revealed that each protein had a MW one unit larger than it should be, and the sequencing afforded by tandem MS showed exactly where the sugars had been linked to the protein. Using a heavier isotope of oxygen (^{18}O-water) for the enzymatic deglycolation provided a 3 Dalton difference and allowed discrimination between spontaneous deamidation of asparagine and the intentional deglycolation step. What does this have to do with analysis of the proteome? These powerful techniques allowed for identification of more than 6000 N-glycosylation sites on 2352 different mouse proteins. So you have a simultaneous snapshot of essentially all N-glycoproteins in use at a given time.

2. This is a rather serious problem. Some proteins have a slight catalytic activity that can be utilized in an assay although they are not enzymes. Or perhaps the protein will serve as a substrate for a reaction. If a protein has no enzyme activity, but the molecular weight and/or pI is known, it can be detected by gel electrophoresis by looking for protein concentration at the right spot on the gel. Some proteins actually fluoresce, like the GFP (green fluorescent protein) produced by jellyfish. In these cases, the intensity of the fluorescence could serve as the basis for an assay.

If the gene is known, then there are ways to "fish" out the protein in very high yield by modifying the sequence. This bypasses the need for an assay. One common procedure is "his-tagging" in which six tandem histidines are added to the sequence of the gene. The protein expressed can then be purified in one step on a nickel-containing column, and eluted with imidazole. This method is mentioned in the text as a special case of affinity chromatography. (A good review of various "Affinity Fusion Strategies" [Nilsson et al., *Prot. Exp. Purif.* 11(1997):1].)

3. SDS disrupts nearly all noncovalent interactions in a native protein, so the renaturation of a purified protein, which is necessary to restore enzyme activity, could be difficult or impossible. You should therefore conduct small-scale pilot tests to determine whether enzyme activity would be lost upon SDS denaturation. Similarly, when salting out with ammonium sulfate is considered, pilot experiments should be conducted. In many instances, concentrations of ammonium sulfate can be chosen such that the active enzyme remains in solution while other proteins are precipitated, thereby affording easy and rapid purification.

4. SDS-polyacrylamide gel electrophoresis, a sensitive and rapid technique which takes only a few hours and which has a high degree of resolution, is probably the easiest and most rapid method for providing an estimate of molecular weight. Small samples (as low as 0.02 mg) can be detected on the gel. For standards or markers on the gel, you should use two or more proteins whose molecular weights are higher than that of the protein to be analyzed, as well as two or more whose molecular weights are lower. The relative mobilities of these markers on the gel can then be plotted against the logarithms of their respective molecular weights (see Figure 3.10 in the text), providing a straight line that can be used to establish the molecular weight of the protein to be analyzed. Proteins that have

carbohydrate molecules covalently attached, or those that are embedded in membranes, often do not migrate according to the logarithm of their mass. The reasons for these anomalies are not clear; in the case of glycoproteins, or those with carbohydrate residues, the large heterocyclic rings of the carbohydrates may retard the movement of the proteins through the polyacrylamide gel. Membrane proteins often contain a high proportion of hydrophobic amino acid residues and may not be fully soluble in the gel system.

Modern variations of mass spectrometry can yield highly accurate MW values very quickly.

5. (a) The specific volume is the inverse of the density, so for proteins d = 1/0.72 = 1.39 and for DNA d = 1/0.5 = 2.0. DNA is denser.
    (b) Because v is the inverse of density (cm^3/g) and d is density (g/cm^3) the units cancel out, and $(1 - v\rho)$ is dimensionless.
    (c) For s to have dimensions of "seconds" then f must have units of grams/second. Multiplying the friction coefficient times the velocity of the molecule in the centrifuge gives units of force (gm cm/sec^2).

6. Glutamine cyclizes to form pyrrolidine carboxylic acid (shown in Figure 3.3):

FIGURE 3.3

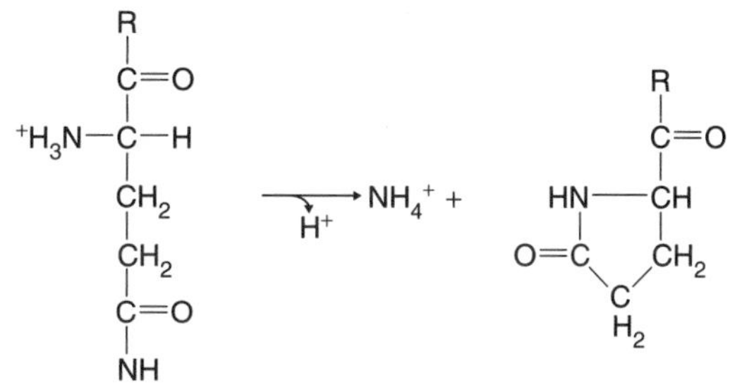

N-terminal
glutamine residue

Pyrrolidone carboxylate
residue

The Edman procedure begins with the reaction of phenyl isothiocyanate with the terminal α-amino group of the peptide. In the pyrrolidine ring, that group is not available. Therefore, the cyclized residue must be removed enzymatically before the Edman procedure can be used. During the biosynthesis of proline, glutamate undergoes reduction to form glutamate γ-semialdehyde; this compound cyclizes, with the loss of water, to form Δ1-pyrroline-5-carboxylate, which is then reduced to form proline (see Figure 3.4).

FIGURE 3.4

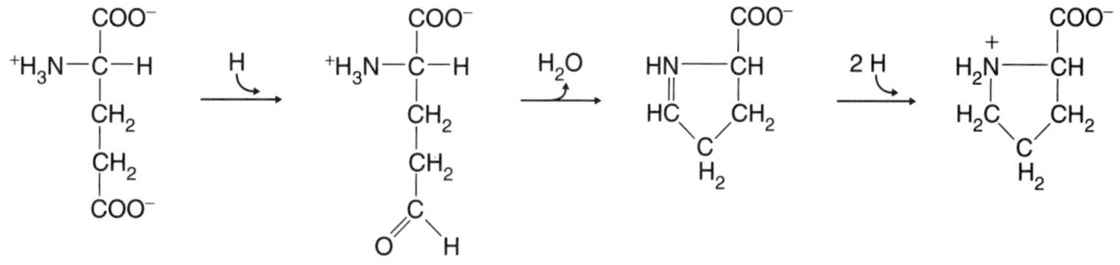

Glutamate

Glutamic
γ-semialdehyde

Δ^1-Pyrroline-
5-carboxylate

Proline

7. The sequences of the peptides produced by the two cleavage methods are circular permutations of each other. Thus, the peptide is circular, so it has no free α-amino group that can react with phenyl isothiocyanate (see Figure 3.5).

FIGURE 3.5

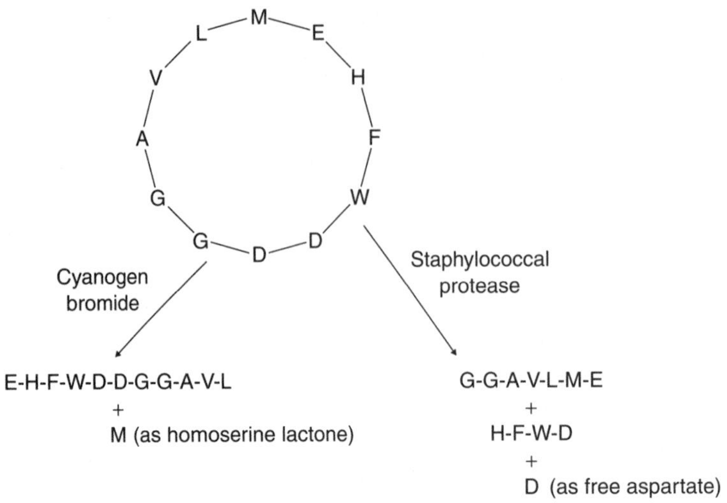

8. The sequence of the mouse hexapeptide is Met-Ile-Cys-Lys-Glu-Tyr, or MICKEY. Cyanogen bromide treatment cleaves methionine from one end, and the PTH-Met derivative places Met at the N-terminal end, with Ile next in the sequence. Trypsin treatment cleaves on the carboxyl side of Lys, so that Lys is on the C-terminal end of the tetrapeptide, next to Cys. Tyrosine is located on the C-terminal end, as shown by the observation that it is released as a single amino acid when the intact hexapeptide is treated with carboxypeptidase A. Glutamate must therefore be located between Lys and Tyr.

9. Most blood proteins do not show up in the urine, but HCG does. And it is produced very soon after the egg is fertilized, and then in increasing amounts as the pregnancy progresses. Sandwich ELISA (see Figure 3.29 in the text) is the ideal method for complex biological fluids, and it is relatively easy to produce two different monoclonal antibodies to epitopes on opposite sides of the protein. All home pregnancy test kits are based on variations of this method.

    For a better understanding of the use of ELISA in home pregnancy tests, view the Animated Technique: Elisa Method for Detecting HCG at www.whfreeman.com/biochem5.

10. Samples for assay by Western blotting are separated by electrophoresis in SDS before blotting and antibody staining, so the reacting proteins are denatured. The use of an SDS-denatured antigen to generate the monoclonal antibody response to the viral coat protein could assure that similar specificities are achieved in the test.

11. Treatment with triethylamine and then with acetic anhydride will yield acetylated amino groups. The strong base removes protons from amino, carboxyl, and hydroxyl groups. These groups would then be methylated with methyl iodide. Leucine and isoleucine have identical molecular weights, so they cannot be distinguished by mass spectrometry.

12. Even though individual atoms can be delineated at a resolution of 1.5 Ångstroms, the structure of individual side chains that are similar in shape and size cannot be clearly established. The primary structure of the polypeptide chain must be available. The path of the polypeptide backbone can be traced and the positions of the side chains established. Those that are similar in size and shape can be distinguished by using the primary structure as a guide as they are fitted by eye to the electron-density map.

13. As later chapters will demonstrate, the functions (explanation: hemoglobin does *not* carry out a chemical reaction) carried out by proteins like hemoglobin and lysozyme depend on the precise orientation of atoms involved in binding and acting on sub-

strates. The fact that catalysis can occur in enzyme crystals argues that these same orientations are preserved and that the structure of the enzyme must be the same as that found in solution.

14. The great advantage of x-ray crystallography is the high resolution that can be obtained. While it is true that NMR is an easier method, the structures obtained are always approximate. Look at Figure 3.49 in the text. This is a family of approximations of a single structure-not a group of related structures. NMR generally gives fuzzy results like this. So determining the dimensions of an enzyme's active site would not work well with NMR; x-ray crystallography would be preferred. NMR is excellent for obtaining approximate structures of small proteins.

15. When proteins are phosphorylated the phosphate nearly always is "visible" on the surface. Antibodies binding the phosphorylated region are different from those binding the unphosphorylated form. Matsumura et al. [*J. Cell Bio.* 140 (1998):119] were able to label both unphosphorylated myosin, and the phosphorylated version of the myosin light chain. Because myosin is involved in contraction, it was interesting to observe different phosphorylation states in different parts of the cells during cell division and other cellular processes. Use of different fluorescent colors allows simultaneous observation of the different forms. Mouse antibodies were used to bind plain myosin, and rabbit antibodies were used for phosphorylated myosin. Then red rhodamine-linked goat anti-mouse antibodies were used to bind plain myosin, and green FITC-linked goat anti-rabbit antibodies were used to bind phosphorylated myosin. Color pictures in the article clearly showed which part of the cell was relaxed (unphosphorylated = red), and which was contracting (phosphorylated = green).

16. Genetic linkage approaches may be useful with diseases where a large patient population is available but that is not the case here. While a genetic disease would show up in the DNA, remember that the human genome is enormous and any two individuals will have many genetic differences. Furthermore each nucleated cell has a complete set of genes, so that neurons have the same DNA as skin cells, etc. A more productive approach may hinge on differential proteomics, that is, looking at what proteins are present in the two tissue samples. If a difference in the expressed proteins can be found, it may be related to the disease. In one example of this approach, Sleat et al. [*Science* 277 (1997):1802] identified the defect in late infantile neuronal ceroid lipofuscinosis by selecting a subset of proteins which had been modified in the cell by addition of mannose-6-phosphate (a sugar label that sends the proteins to a certain compartment in the cell). They used two-dimensional electrophoresis (described in the text) and found one protein missing from diseased nerves which was present in normal patient samples. When this protein was purified and identified, this led to discovery of the genetic cause of the disease.

Another possible approach to differential proteomic analysis would be "shotgun" mass spectrometry in which many proteins are fragmented and analyzed together.

Chapter 4

DNA, RNA, and the Flow of Genetic Information

C hapters 2 and 3 introduced you to proteins. The authors now expand on the introductory material in Chapter 1 by turning back to a second class of macromolecules, the nucleic acids, that serve as the storage forms of genetic information and are critical participants in the elaboration of this information into functional forms. First, they describe in more detail the structures of the nucleoside building blocks of DNA and the phosphodiester bond that links them together. Following this, the Watson-Crick DNA double helix is presented again, an overview of how the strands of DNA separate for replication is given, and some of the various conformations and structures that nucleic acids can assume are described. The polymerases that form DNA chains are introduced next. The section describing the molecules that store genetic information ends by providing two examples of viruses in which the genetic material is not duplex DNA but rather single-strand RNA. The authors next describe the way in which RNA viruses replicate through double-strand nucleic-acid intermediates whose formation is also directed by specific base pairing.

How the information stored in DNA or RNA directs the formation of the proteins of a cell is discussed next. The authors start with descriptions of the basic structures and kinds of RNA and provide an explanation of the central roles of RNAs in the overall flow of genetic information and the regulation of its expression. They then present the specific functions of messenger RNA, transfer RNA, and ribosomal RNA in protein synthesis, along with a description of the polymerase that synthesizes all cellular RNAs. The genetic code, which relates the nucleotide sequence of RNA to the amino acid sequence of proteins, is described. The collinear relationship between the sequences of nucleotides in the DNA and the amino acids of the encoded protein is compared in prokaryotes and in eukaryotes where some genes are interrupted by noncoding sequences (introns). The authors next describe the process by which these intron sequences are removed from the initial transcript to form functional messenger RNA and the biological consequences of such splicing.

LEARNING OBJECTIVES

When you have mastered this chapter, you should be able to accomplish the following objectives.

Nucleic Acids (Text Section 4.1)

1. Locate the structural components of DNA, namely, the *nitrogenous bases*, the *sugar*, and the *phosphate* group. Know the various conventions used to represent these components and the structure of DNA. Know how the carbon atoms are numbered.

2. Differentiate *purines, pyrimidines, ribonucleosides, deoxyribonucleosides, ribonucleotides,* and *deoxyribonucleotides.*

3. Recognize the *deoxyadenosine, deoxycytidine, deoxyguanosine,* and *deoxythymidine* constituents of DNA, and describe the *phosphodiester* bond that joins them together to form DNA.

4. Compare the *phosphodiester backbones* of RNA and DNA. Contrast the composition and structures of RNA and DNA. Distinguish *thymine* from *uracil* and *2′-deoxyribose* from *ribose*. Understand the effect that a 2′-OH has on structure and base lability.

5. Relate the *polarity of the DNA* chain (5′→3′) to the convention for writing a single-letter DNA sequence abbreviation.

6. Compare the lengths of the DNA molecules in polyoma virus, the bacterium *E. coli,* and the average human chromosome.

Double Helices (Text Section 4.2)

7. List the important features of the *Watson-Crick DNA double helix*. Relate the *base pairing* of *adenine* with *thymine* and of *cytosine* with *guanine* to the duplex structure of DNA and to the replication of the helix. Explain the molecular determinants of the specific base pairs in DNA. Know the dimensions of the Watson-Crick double helix. Understand the forces that stabilize the helix.

8. Describe *supercoiling* and state its biological consequences.

9. Appreciate the variety of structures that double stranded and *single-stranded nucleic acids* can assume.

Transmission of Information (Text Section 4.3)

10. Know why hypochromism occurs. Know the wavelength where DNA maximally absorbs UV light. Outline the *Meselson-Stahl experiment* and relate it to *semiconservative replication*. Define *the melting temperature* (T_m) for DNA and relate it to the separation of the strands of duplex DNA. Describe *annealing*.

DNA Polymerases (Text Section 4.4)

11. List the *substrates* and the important enzymatic properties of *DNA polymerases* as they relate to *replication* and *repair* of DNA. Distinguish between a *primer* and a *template* and describe their functions.

12. Define *virus* and appreciate that RNA is the genome of some viruses.

13. Relate the catalytic activity of *reverse transcriptase* (an RNA-directed DNA polymerase) to the replication of *retroviruses*. Provide an overview of *retroviral replication*.

Gene Expression (Text Section 4.5)

14. State the role of DNA in protein synthesis. Outline the *flow of genetic information* during gene expression.

15. Define the terms *transcription* and *translation* and relate these processes to the flow of genetic information.

16. Name the *three major classes of RNA* found in *E. coli* and explain their functions. Compare their sizes and their relative amounts in the cell. List other classes of RNA molecules and describe their functions.

17. List the substrates and important enzymatic properties of *RNA polymerases (DNA-dependent RNA polymerases)*. Explain the roles of the *DNA template, promoter, enhancer sequences,* and *terminator* in transcription. Know how eukaryote MRNA is modified after transcription.

18. Describe the *transcription* of duplex DNA to form single-strand RNA. Relate the sequence of *mRNA* to that of the sequence of the DNA template from which it is transcribed.

19. Describe the role of *tRNA* as the *adaptor molecule* acting between mRNA and amino acids during protein synthesis. Outline how specific amino acids are covalently attached to specific tRNA molecules. Explain the relationship of the *codon* and *anticodon* to the specific interaction between mRNA and tRNA.

Amino Acids and the Genetic Code (Text Section 4.6)

20. Explain what the *genetic code* is and list its major characteristics. Define the terms *triplet code (codon), nonoverlapping, degenerate, synonym, triplet,* and *reading frame* as they apply to the genetic code. Recognize the *initiation* and *termination codons*.

21. Using the genetic code, predict the sequence of amino acids in a peptide encoded by a template DNA or mRNA sequence.

22. Discuss the universality of the genetic code.

Introns and Exons (Text Section 4.7)

23. Describe the function and composition of *spliceosomes*.

24. Contrast the *linear relationship* between the sequence of DNA in a gene and the sequence of the amino acids in the protein it encodes in bacteria and in higher eukaryotes. Apply the terms *intron* and *exons* to these relationships.

25. Relate exons to *protein domains* and list the advantages the existence of introns and exons confer upon eukaryotes.

26. Outline *RNA processing* in eukaryotes. Name the alterations made to the RNA after it is initially formed by RNA polymerase.

27. Recount a hypothesis relating exons and *functional domains* to the generation and evolution of protein diversity. Differentiate between nucleotide sequence rearrangements by *genetic recombination* at the DNA level and by *RNA splicing*.

SELF-TEST

Nucleic Acids

FIGURE 4.1

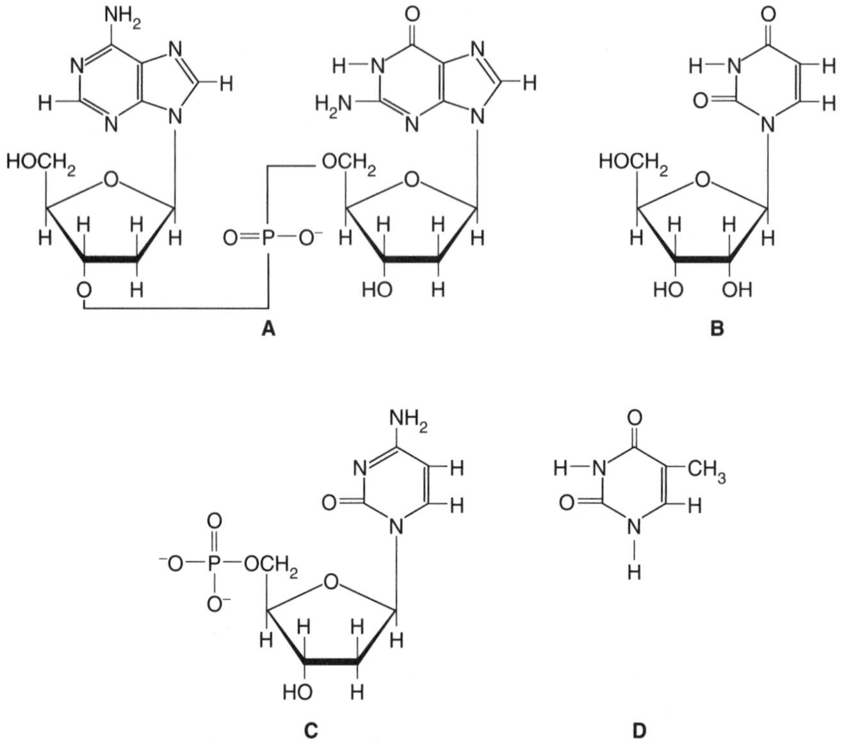

1. Which of the preceding structures in Figure 4.1
   (a) contains ribose?
   (b) contains deoxyribose?
   (c) contains a purine?
   (d) contains a pyrimidine?
   (e) contains guanine?
   (f) contains a phosphate monoester?
   (g) contains a phosphodiester?
   (h) is a nucleoside?
   (i) is a nucleotide?
   (j) would be found in RNA?
   (k) would be found in DNA?

Double Helices

2. Which of the following are characteristics of the Watson-Crick DNA double helix?
   (a) The two polynucleotide chains are coiled about one another and about a common axis.
   (b) Hydrogen bonds between A and C and between G and T help hold the two chains together.
   (c) The helix makes one complete turn every 34Å because each base pair is rotated by 36° with respect to adjacent base pairs and is separated by 3.4Å from them along the helix axis.

    (d) The purines and pyrimidines are on the inside of the helix and the phosphodi-ester-linked backbones are on the outside.

    (e) Base composition analyses of DNA duplexes isolated from many organisms show that the amounts of A and T are equal as are the amounts of G and C.

    (f) The sequence in one strand of the helix varies independently of that in the other strand.

3. If a region of one strand of a Watson-Crick DNA double helix has the sequence ACGTAACC, what is the sequence of the complementary region of the other strand?

4. Explain why A · T and G · C are the only base pairs possible in normal double-strand DNA.

5. Match the appropriate characteristics in the right column with the structures of double-strand or single-strand DNA.

    (a) double-strand DNA   (1) is a rigid rod

    (b) single-strand DNA   (2) shows a greater hyperchromic effect upon heating

                                    (3) contains equal amounts of A and T bases

                                    (4) may contain different amounts of C and G bases

                                    (5) contains U rather than T bases

                                    (6) may contain stem-loop structures

6. Haploid human DNA has 3×10^6 kilobase pairs (a kilobase pair, abbreviated kb, is 1000 base pairs). What is the total length of human haploid DNA in centimeters?

7. Outline the basic process by which a Watson-Crick duplex replicates to give two identical daughter duplexes. Explain the molecular reasons for the accuracy of the process.

8. The DNA in a bacterium is uniformly labeled with ^{15}N, and the organism shifted to a growth medium containing ^{14}N-labeled DNA precursors. After two generations of growth, the DNA is isolated and is subjected to density-gradient equilibrium sedimentation. What proportion of light-density DNA to intermediate-density DNA would you expect to find?

9. Purified duplex DNA molecules can be

    (a) linear.

    (b) circular and supercoiled.

    (c) linear and supercoiled.

    (d) circular and relaxed, that is, not supercoiled.

10. You are given two solutions containing different purified DNAs. One is from the bacterium *P. aeruginosa* and has a G + C composition of 68%, whereas the other is from a mammal and has a G + C composition of 42.5%.

    (a) You measure the absorbance of ultraviolet light of each solution as a function of increasing temperature. Which solution will yield the higher T_m value and why?

    (b) After melting the two solutions, mixing them together, and allowing them to cool, what would you expect to happen?

    (c) Would appreciable amounts of bacterial DNA be found associated in a helix with mammalian DNA? Explain.

DNA Polymerases

11. DNA polymerase activity requires

    (a) a template.
    (b) a primer with a free 5'-hydroxyl group.
    (c) dATP, dCTP, dGTP, and dTTP (which is the same as TTP).
    (d) ATP.
    (e) Mg^{2+}.

12. Derive the polarity of the synthesis of a DNA strand by DNA polymerase from the mechanism for the formation of the phosphodiester bond.

13. You are provided with a long, single-strand DNA molecule having a base composition of C = 24.1%, G = 18.5%, T = 24.6%, and A = 32.8%; DNA polymerase; $[\alpha\text{-}^{32}P]dATP$ (dATP with the innermost phosphate labeled), dCTP, dGTP, and dTTP; a short primer that is complementary to the single-strand DNA; and a buffer solution with Mg^{2+}. What is the base composition of the radiolabeled product DNA after the completion of one round of synthesis?

14. For the virus in the left column, indicate the appropriate characteristics from the right column.

    (a) tobacco mosaic virus
    (b) AIDS (HIV-1) virus

    (1) linear genome
    (2) genome contains U rather than T
    (3) single-strand nucleic acid genome
    (4) DNA intermediates are involved in replication
    (5) uses RNA-directed RNA polymerase to replicate
    (6) uses RNA-directed DNA polymerase to replicate

15. Propose how a single-strand DNA virus could replicate by incorporating semiconservative replication into the process.

16. From the following nucleic acids, select those that appear during the infection of a cell with a retrovirus, for example, the AIDS virus, and place them in the order in which genetic information flows during the process of forming a new progeny virus.

    (a) double-strand DNA-RNA helix in the cell
    (b) single-strand RNA in the virus
    (c) single-strand RNA in the cell
    (d) double-strand DNA in the cell
    (e) double-strand RNA in the virus
    (f) double-strand RNA in the cell

Gene Expression

17. Transcription is directly involved in which of the following possible steps in the flow of genetic information?

    (a) DNA to RNA
    (b) RNA to DNA
    (c) DNA to DNA
    (d) RNA to protein
    (e) protein to RNA

18. Translation is involved in which of the following possible steps in the flow of genetic information?
    (a) DNA to RNA
    (b) RNA to DNA
    (c) DNA to DNA
    (d) RNA to protein
    (e) protein to RNA

19. Answer the following questions about RNA.
    (a) What is the name of the bond joining the ribonucleoside components of RNA to one another?
    (b) Is this bond between the 2'- or the 3'-hydroxyl group of one ribose and the 5'-hydroxyl of the next?
    (c) Intramolecular base pairs form what kinds of structures in RNA molecules?
    (d) What bases pair with one another in RNA?
    (e) What are the three major classes of RNA in a cell and which is most abundant?

20. If you have samples of pure RNA and duplex DNA, how can you tell whether they have any complementary nucleotide sequences?

21. If all the RNA referred to in Question 20 turns out to have sequences that were complementary to the DNA, will its percentage of G and C be identical to that of the DNA? Explain.

22. If each of the three major classes of RNA found in a cell were hybridized to denatured DNA from the same cell and the presence of RNA-DNA hybrids were tested, which of the classes would be retained on the filter?
    (a) mRNA
    (b) rRNA
    (c) tRNA

RNA Polymerases and Transcription

23. Which of the following are required for the DNA-dependent RNA polymerase reaction to produce a unique RNA transcript?
    (a) ATP   (g) RNA
    (b) CTP   (h) Mg^{2+}
    (c) GTP   (i) promoter sequence
    (d) dTTP   (j) operator sequence
    (e) UTP   (k) terminator sequence
    (f) DNA

24. What is the sequence of the mRNA that will be synthesized from a template strand of DNA having the following sequence?

    . . .ACGTTACCTAGTTGC. . .

25. Describe the mechanism of chain growth during RNA synthesis. What is the polarity of synthesis and how is it related to the polarity of the template strand of DNA?

The Genetic Code and Protein Synthesis

26. Which of the following are characteristics or functions of tRNA?
    (a) It contains a codon.
    (b) It contains an anticodon.
    (c) It can become covalently attached to an amino acid.
    (d) It interacts with mRNA to stimulate transcription.
    (e) It can have any of a number of different sequences.
    (f) It serves as an adaptor between the information in mRNA and an individual amino acid.

27. What is the minimum number of contiguous nucleotides in mRNA that can serve as a codon? Explain.

28. What is the sequence of the polypeptide that would be encoded by the DNA sequence given in Question 24? Assume that the reading frame starts with the 5′ nucleotide given. The genetic code is given on page 129 of the text.

29. The following is a partial list of mRNA codons and the amino acids they encode:

    AGU = serine   AGC = serine
    AAU = asparagine   AAC = asparagine
    AUG = methionine   AUA = isoleucine

    From this list, which of the following statements are correct?

    (a) The genetic code is degenerate.
    (b) The alteration of a single nucleotide in the DNA directing the synthesis of these codons could lead to the substitution of a serine for an asparagine in a polypeptide.
    (c) The alteration of a single nucleotide in the DNA directing the synthesis of these codons would necessarily lead to an amino acid substitution in the encoded polypeptide.
    (d) A tRNA with the anticodon ACU would be bound by a ribosome in the presence of one of these codons.

30. Explain how mitochondria and some organisms can use a genetic code that is different from the standard code used in the nucleus.

Introns and Splicing

31. Explain how genetic techniques and amino acid sequence analyses could be used to show the collinear relationship of a prokaryotic gene and the protein it encodes.

32. Answer the following questions about what was revealed when DNA encoding the gene for the β-chain of hemoglobin and the mRNA for the β-chain were compared.
    (a) What was the major finding when the nucleotide sequence of the gene and the amino acid sequence of the β-chain were compared?
    (b) What did hybridization between the partially denatured DNA and the mRNA for β-globin show?
    (c) What must happen to the primary transcript from the β-globin gene before it can serve as an mRNA for protein synthesis?

33. How might the fact that some exons encode discrete functional domains in proteins be related to the evolution of new proteins?

34. Spliceosomes
    (a) recombine DNA sequences in a process called exon shuffling.
    (b) are composed of RNA and proteins.
    (c) recognize RNA sequences that signal for the removal of introns.
    (d) can produce different mRNA molecules by splicing at alternative sites.

ANSWERS TO SELF-TEST

1. (a) B   (b) A, C   (c) A   (d) B, C, D   (e) A   (f) C   (g) A   (h) B   (i) C; strictly speaking, A is called a dinucleotide, not a nucleotide   (j) B   (k) A, C, D

2. a, c, d, and e. Answer (b) is not correct because A pairs with T and G pairs with C. Answer (f) is not correct because the sequence of one strand determines the sequence of the other by base pairing.

3. GGTTACGT. The convention for indicating polarity is that the 5′-end of the sequence is written to the left. The two chains of the Watson-Crick double helix are antiparallel, so the correct complementary sequence is not TGCATTGG.

4. The space between the two deoxyribose-phosphodiester strands is precisely defined. This distance is not large enough for two purines to hydrogen-bond. Conversely, two pyrimidine bases would not be close enough to form stable hydrogen bonds. Furthermore, in the double-strand structure the hydrogen-bond donor and acceptor groups are not properly aligned to form stable G · T or A · C base pairs.

5. (a) 1, 2, 3   (b) 4, 6. Answer (5) does not apply because DNA does not contain uracil.

6. 102 cm. The math is as follows:

$$(3 \times 10^6 \text{ kb} \times 10^3 \text{ bases/kb} \times 3.4 \text{ Å/base} \times 10^{-8} \text{ cm/Å}) = 102 \text{ cm}$$

7. When replication occurs, the two strands of the Watson-Crick double helix must separate so that each can serve as a template for the synthesis of its complement. Since the two strands are complementary to one another, each bears a definite sequence relationship to the other. When one strand acts as a template, it directs the synthesis of its complement. The product of the synthesis directed by each template strand is therefore a duplex molecule that is identical to the starting duplex. The process is accurate because of the specificity of base pairing and because the protein apparatus that catalyzes the replication can remove mismatched bases.

8. After two generations, you should expect to find equal amounts of light-density DNA, in which both strands of each duplex were synthesized from ^{14}N precursors, and intermediate-density DNA, in which each duplex consists of a heavy ^{15}N strand paired with a light ^{14}N strand.

9. a, b, and d. Answer (d) is correct because, if at least one discontinuity exists in the phosphodiester backbone of either chain of a circular duplex molecule, the chains are free to rotate about one another to assume the relaxed circular form. Answer (c) is incorrect because supercoiling requires closed circular molecules. In a linear molecule, the ends of each strand are not constrained with respect to rotation about the helical axis; therefore, the molecule cannot be supercoiled.

10. (a) The bacterial DNA solution has the higher T_m value because it has the higher G + C content and is therefore more stable to the thermal-induced separation of its strands because G · C base pairs are more stable than A · T base pairs.

   (b) The complementary DNA strands from each species will anneal to form Watson-Crick double helices as the solution cools.

   (c) No; each strand will find its partner because the perfect match between the linear arrays of the bases of complementary strands is far more stable than the mostly imperfect matches in duplexes composed of one strand of bacterial and one strand of mammalian DNA would be.

11. a, c, and e. Answer (b) is not correct because, although the enzyme requires a primer, the nature of its 5'-end is irrelevant since dNMP residues are added to its 3'-end. A primer with a 3-OH is required. Answer (d) is not correct because the enzyme uses dNTP and not NTP molecules, where N means A, C, G, T, or U. Note that dTTP is used interchangeably with TTP.

12. The 3'-hydroxyl of the terminal nucleotide of the primer makes a nucleophilic attack on the innermost phosphorus atom of the incoming dNTP that is appropriate for Watson-Crick base pairing to the template strand to form the phosphodiester bond. As a result, a dNMP residue is added onto the 3'-end of the primer with the concomitant release of PP_i, and the chain grows in the 5' to 3' direction.

13. After the completion of one round of synthesis, the template strand will have directed the polymerization of a complement in which C = 18.5%, G = 24.1%, T = 32.8%, and A = 24.6%. Since the primer is short with respect to the template, its contribution to the composition of the product strand can be neglected.

14. (a) 1, 2, 3, 5   (b) 1, 2, 3, 4, 5, 6

15. The single-strand DNA penetrates the cell, where it has converted by enzymes to a duplex replicative form through Watson-Crick base pairing. The replicative form is then reproduced by a mechanism similar to that used for the semiconservative replication of the duplex chromosome of the host cell. Finally, after this stage of replication, the mechanism shifts to one in which the replicative form serves as a template to produce copies of the single-strand DNA found in the mature virus.

16. Starting with the single-strand RNA in the virus and ending with the single-strand RNA in the progeny viruses, the order in which genetic information flows during the infection of a cell with a retrovirus is: b, c, a, d, c, b. Retroviruses use the enzyme reverse transcriptase to convert their single-strand genomes into a DNA-RNA replicative form that is subsequently converted into a duplex DNA replicative form prior to insertion into the host chromosome and ultimate reconversion into the single-strand viral RNA by a DNA-dependent RNA polymerase.

17. a and b. Answer (b) is correct because the reverse transcription of RNA sequences into DNA sequences occurs during the replication of retroviruses. The term transcription is usually used to describe the formation of RNA from a DNA duplex by RNA polymerase.

18. d

19. (a) The bond is called the phosphodiester bond.

   (b) The bond joins the 3'-hydroxyl to the 5'-hydroxyl to form a 3' $\longrightarrow$ 5' phosphodiester bond.

   (c) Hairpin loops are formed when the RNA chain folds back upon itself and some of the bases become hydrogen bonded to form an antiparallel duplex stem with unpaired bases forming a loop at one end.

(d) A pairs with U, and G pairs with C; G can also pair with U, but the association is weaker than that of the G · C base pair.

(e) The three major classes of RNA found in a cell are mRNA, rRNA, and tRNA; the most abundant is rRNA.

20. You could sequence the RNA and DNA and compare the sequences of each to see if the two are complementary; this method provides definitive evidence of identity. An easier but less precise way would be to use hybridization. You would mix the samples, heat the mixture to melt the double-strand DNA and RNA hairpins, slowly cool the solution, and then examine it to see if it contains double-strand DNA-RNA hybrids. Such hybrids would indicate that the RNA and DNA sequences are complementary.

21. Not necessarily; RNA synthesis is asymmetric, and generally only one strand of any region of the DNA serves as a template. This can lead to RNA with a G + C composition different from that of the duplex DNA.

22. a, b, c. All cellular RNA is encoded by the DNA of the cell.

23. a, b, c, e, f, h, i, and k. Answer (f) is correct because DNA is needed to serve as the template. Answers (k) and (i) are correct because the promoter and terminator sequences are needed to specify the precise start and stop points, respectively, for the transcription.

24. The mRNA sequence will be ...GCAACUAGGUAACGU..., written in the 5′ to 3′ direction.

25. The 3′-hydroxyl terminus of the growing RNA chain makes a nucleophilic attack on the α-phosphate (the innermost phosphate) of the ribonucleoside triphosphate that has been selected by base pairing to the template strand of the duplex DNA. RNA polymerase catalyzes the reaction. A ribonucleoside monophosphate residue is added to the chain as a result, and the chain has grown in the 5′ to 3′ direction; that is, the chain has grown at its 3′ end. As with all Watson-Crick base pairing, the strands are antiparallel; that is, the RNA chain is assembled in the 3′ to 5′ direction with respect to the polarity of the template strand of the DNA.

26. b, c, e, and f. Answer (d) is incorrect because the interaction of tRNA with mRNA takes place during translation, not transcription.

27. Three contiguous nucleotides is the minimum that can serve as a codon. There are four kinds of nucleotides in mRNA. A codon consisting of only two nucleotides (either of which could be any of the four possible nucleotides) allows only 16 possible combinations ($4 \times 4 = 16$). This would not be sufficient to specify all 20 of the amino acids. A codon consisting of three nucleotides, however, allows 64 possible combinations ($4 \times 4 \times 4 = 64$), more than enough to specify the 20 amino acids.

28. The sequence of the polypeptide would be Ala-Thr-Arg. The reading frame is set by the nucleotide at the 5′ end of the mRNA transcript; the fourth codon of the mRNA transcript is UAA, which is a translation termination codon.

29. a, b, and d. Answer (a) is correct because both AGU and AGC specify serine; since more than one codon can specify the same amino acids, the genetic code is said to be degenerate. Answer (b) is correct because the alteration of a single nucleotide in the DNA could change a codon on the mRNA transcript from AGU, which specifies serine, to AAU, which specifies asparagine. Answer (d) is correct because the anticodon ACU would base-pair with the codon AGU. Answer (c) is *not* correct because the alteration of a single nucleotide in the DNA could result in another codon that specifies the same amino acid; for example, a codon changed from AGU to AGC would continue to specify serine.

30. Mitochondria and some organisms can use a genetic code that differs from the standard code because mitochondrial DNA encodes a distinct set of tRNAs that are matched to the genetic code used in their mRNAs.

31. The mutations in a given gene of *E. coli* could be mapped by recombination analysis. The proteins encoded by the wild-type and the mutant genes could then be sequenced, and the location and nature of the amino acid substitution for each mutation identified. The result would be that the order of the mutations on the genetic map is the same as the order of the corresponding changes in the amino acid sequence of the polypeptide produced by the gene; these experiments established that genes and their polypeptide products are collinear in prokaryotes.

32. (a) The number of nucleotides in the gene was significantly greater than three times the number of amino acids in the protein. There were two stretches of extra nucleotides between the exon sequences that encode the amino acids in the β-chain.
    (b) The mRNA hydridized to the DNA under conditions where DNA-RNA hybrids are more stable than DNA-DNA hybrids, but there were sections of duplex DNA between the hybrid regions. This indicated that there are intron sequences in the DNA that have no corresponding sequences in the mRNA. (See Figure 4.36 in the text.)
    (c) The intervening sequences (introns) in the nascent or primary transcript, which are complementary to the template strand of the DNA of the gene but do not encode amino acids in the protein, must be removed by splicing to generate the mRNA that functions in translation.

33. The shuffling of exons that encode discrete functional domains, such as catalytic sites, binding sites, or structural elements, preserves the functional units but allows them to interact in new ways, thereby generating new kinds of proteins.

34. b, c, and d. (a) is incorrect because exon shuffling takes place at the DNA level through breakage and rejoining of DNA, not RNA.

PROBLEMS

1. The genome of the mammalian virus SV40 is a circular DNA double helix containing 5243 base pairs. When a solution containing intact DNA molecules is heated, one observes an increase in the absorbance of ultraviolet light at 260 nm. When the solution is then cooled slowly, a decrease in absorbance is observed. If one or more breaks are made in the sugar-phosphate backbones of the SV40 double-strand circles, heating causes a similar hyperchromic effect. However, when the solution of nicked molecules is cooled, the reduction in absorbance is much slower than that observed in the solution containing intact molecules. Why do the two types of molecules behave differently when they are cooled after heating?

2. A number of factors influence the behavior of a linear, double-strand DNA molecule in a 0.25M sodium chloride solution. Considering this, explain each of the following observations.
   (a) The T_m increases in proportion to length of the molecule.
   (b) As the concentration of sodium chloride decreases, the T_m decreases.
   (c) Renaturation of single strands to form double strands occurs more rapidly when the DNA concentration is increased.
   (d) The T_m value is reduced when urea is added to the solution.

3. (a) Many proteins that interact with double-strand DNA bind to specific sequences in the molecule. Why is it unlikely that these enzymes operate by sensing differences in the diameter of the helix?

   (b) What other features of the double-strand helix might be recognized by the protein?

4. You have a double-strand linear DNA molecule, the appropriate primers, all the enzymes required for DNA replication, four ^{32}P-labeled deoxyribonucleoside triphosphates, Mg^{2+} ion, and the means to detect newly synthesized radioactive DNA. Why is this system not sufficient to distinguish between conservative and semiconservative replication of the DNA molecule?

5. Certain deoxyribonucleases cleave any sequence of single-strand DNA to yield nucleoside monophosphates; these enzymes do not hydrolyze base-paired DNA sequences. What products would you expect when you incubate a solution containing a single-strand specific deoxyribonuclease and the following oligodeoxyribonucleotide?

   5'-ApGpTpCpGpTpApTpCpCpTpCpTpApCpGpApCpTp-3'

6. Formaldehyde reacts with amino groups to form hydroxymethyl derivatives. Would you expect formaldehyde to react with bases in DNA? Suppose you have a solution that contains separated complementary strands of DNA. How would the addition of formaldehyde to the solution affect reassociation of the strands?

7. When double-strand DNA is placed in a solution containing tritiated water (^3H_2O), hydrogens associated with the bases readily exchange with protons in the solution. The greater the percentage of AT base pairs in the DNA, the greater the rate of exchange. Why?

8. While many experiments were suggesting that DNA in chromosomes is very long and continuous, it was established that DNA polymerase adds deoxyribonucleotides to the 3'-hydroxyl terminus of a primer chain and that a DNA template is essential. Why did investigators interested in DNA replication initially focus a great deal of attention on determining whether chromosomal DNA contained breaks in the sugar-phosphate backbone?

9. The value of the T_m for DNA in degrees Celsius can be calculated using the formula, $T_m = 69.3 + 0.41(G + C)$, where G + C is the mole percentage of guanine plus cytosine.

   (a) A sample of DNA from *E. coli* contains 50 mole percent G + C. At what temperature would you expect this DNA molecule to melt?

   (b) The melting curves for most naturally occurring DNA molecules reveal that their T_m values are normally greater than 65°C. Why is this important for most organisms?

   (c) What problem concerning replication does a T_m value > 65°C imply?

10. During early studies of the denaturation of double-strand DNA, it was not known whether the two strands unwind and completely separate from each other. Suppose that you have double-strand DNA in which one strand is labeled with ^{14}N and the other is labeled with ^{15}N. If density-gradient equilibrium sedimentation can be used to distinguish between both double- and single-strand molecules of different densities, how can you determine whether DNA strands separate completely after denaturation?

11. Under strongly acidic conditions, several atoms of DNA bases are protonated; these include the N-1 of adenine, the N-3 of cytosine, and the O-4 of thymine. Predict the effects of such protonations occurring at low pH on the stability of double-strand DNA.

12. The microbiologist Sol Spiegelman found that some types of single-strand RNA can associate with single-strand DNA to form double-strand molecules. What is the most important condition that must be satisfied in order to allow the formation of these hybrid molecules?

13. Many cells can synthesize deoxyuridine 5′-triphosphate (dUTP). Can dUTP be used as a substrate for DNA polymerase? If so, with which base will uracil pair in newly replicated DNA?

14. The DNA of bacteriophage λ is a linear double-strand molecule that has complementary single-strand ends. These molecules can form closed-circular molecules when two "cohesive" ends on the same molecule join, and they can form linear dimers, trimers, or longer molecules when sites on different molecules are joined.

    (a) What conditions should be chosen *in vitro* to ensure that λ phage DNA molecules form closed-circular monomers?

    (b) Under certain conditions, λ phage DNA molecules are infective. When a very low concentration of λ phage DNA is incubated with DNA polymerase I and the four deoxyribonucleoside triphosphates, the infectious activity of λ phage DNA is destroyed. Brief treatment of λ phage DNA with bacterial exonuclease III, an enzyme that removes 5′-mononucleotides from the 3′-ends of double-strand DNA molecules also destroys infectivity, but subsequent treatment of the DNA with DNA polymerase I and dNTP substrates can restore infectivity. Describe more completely the structure of λ phage DNA, and provide an interpretation of the action of the two enzymes on the molecule.

15. The isolation of viral DNA from animal cells that have been infected with adenovirus yields linear double-strand molecules that, when denatured and allowed to reassociate under conditions favoring intramolecular annealing, form single-strand circles. Although circular molecules can be detected using the electron microscope, resolution is not sufficient to visualize the ends of the molecule. Other analyses of the single-strand molecule show that each end has a sequence that allows the structure shown in Figure 4.2 to form.

FIGURE 4.2   A single-strand circle formed by intra-molecular annealing of adenovirus DNA.

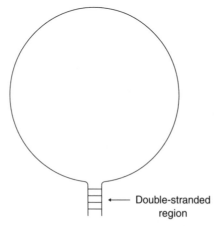

Double-stranded region

    (a) Suppose that the base sequence at one end of the single-strand molecule is 5′-ACTACGTA.... What is the corresponding sequence at the other end? Show

how these sequences would allow full-length, double-strand linear molecules to be formed.

(b) An alternate suggestion for the formation of the single-strand molecules was also proposed; it is shown in Figure 4.3. Why is this proposed pairing scheme unlikely?

FIGURE 4.3 Another proposal for formation of single-stranded molecules of adenovirus DNA.

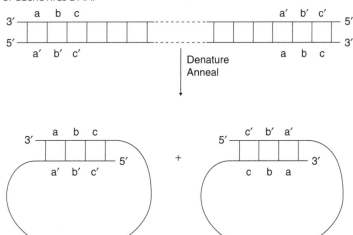

16. Thermoacidophilic bacteria can grow in volcanic sulfur springs at pH 2 and at temperatures as high as 85°C. DNA polymerase purified from the thermophile *Sufolobus acidocaldarius* has an optimal activity at 70°C and is stable at 80°C. When incubated with a circular DNA template at 100°C, the isolated polymerase can extend a 20-nucleotide primer by more than 100 nucleotides. These experiments require that enzyme-to-primer concentration be at least 1:1. The T_m value for the double-strand DNA used in the experiment is about 60°C. Unlike DNA polymerase I from *E. coli*, DNA polymerase from *S. acidocaldarius* has no demonstrable exonuclease activity to correct mistakes in DNA by removing mismatched nucleotides (such an enzyme activity is often referred to as *proofreading*).

(a) Why should the ratio of enzyme to primer be 1 in order for primer extension to take place at 100°C?

(b) Would you expect to find an auxiliary proofreading enzyme in *S. acidocaldarius*? Why?

(c) Would you expect DNA from the genome of *S. acidocaldarius* to have a G + C content higher or lower than that from a bacterium that grows at a more normal temperature? Why?

17. Terminal deoxynucleotidyl transferase (TdT), an enzyme found in bone marrow and thymus tissue, can extend a DNA primer by 5′ to 3′ polymerization using deoxyribonucleoside triphosphates as substrates. The primer must be at least three nucleotides in length and must have a free 3′-OH end. The enzyme does not require a template nor does it copy one.

(a) Compare TdT with DNA polymerase I.

(b) Would TdT be useful for synthesizing DNA molecules that carry genetic information? Why?

18. The 2′,3′-dideoxynucleosides can be used as reagents to inhibit DNA replication. These analogs must be converted to dideoxynucleoside triphosphates in order to have a measurable effect on DNA synthesis. When incorporated into a growing DNA chain, a single dideoxyribonucleoside residue can effectively block subsequent chain extension.

    (a) Why must a 2′,3′-dideoxyribonucleoside be converted to a dideoxyribonucleoside triphosphate to be incorporated into DNA?

    (b) What feature of a 2′,3′-dideoxynucleoside is most likely to account for inhibition of DNA chain extension?

FIGURE 4.4

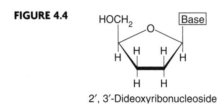

2′, 3′-Dideoxyribonucleoside

19. In each chain-elongation reaction catalyzed by DNA polymerase, a phosphodiester bond is formed and pyrophosphate is concomitantly released. Hydrolysis of pyrophosphate to two molecules of inorganic phosphate occurs rapidly because most cells have a potent pyrophosphorylase. By removing one of the products of the chain-elongation reaction, pyrophosphate cleavage in the cell is partially responsible for the forward progress of polymerization. However, isolated DNA polymerases can efficiently carry out chain extension in the absence of pyrophosphate cleavage, as long as the double-strand helix is allowed to form during elongation. What forces resulting from DNA helix formation might contribute to driving polymerization forward?

20. In his studies of DNA in the late 1940s, Erwin Chargaff established that DNA from all organisms has equal numbers of adenine and thymine bases and equal numbers of guanine and cytosine bases. Considering that thymine and uracil are equivalent in their abilities to form hydrogen bonds with adenine, state whether you would expect similar constraints on base composition to be found in the following:

    (a) single-strand RNA from tobacco mosaic virus.

    (b) the DNA-RNA hybrid molecule synthesized by reverse transcriptase.

    (c) RNA from a virus in the reovirus family, which have large genomes composed of double-strand RNA molecules.

21. Certain DNA endonucleases degrade double-strand DNA to yield mononucleotides and dinucleotides, but these enzymes do not degrade those duplex sequences to which other proteins are tightly bound.

    (a) How can you use such a DNA endonuclease and RNA polymerase to locate a promoter site?

    (b) Why should this process be performed in the absence of ribonucleoside triphosphates?

22. The amino acid at position 102 in the primary sequence of a certain bacterial enzyme is valine, and the corresponding codon in the mRNA sequence for the enzyme is GUU. Suppose a mutation that alters the codon to GCU has no effect on the activity of the enzyme, but another mutation that changes the codon to GAU completely inactivates the enzyme. Briefly explain these observations.

23. It is essential for spliceosomes to remove introns precisely, that is, between the terminal nucleotide of an intron and the first nucleotide of an exon. To see why, suppose that the sequence at the normal junction in a pre-spliced mRNA between an intron and an exon is

...UUAG | GCUAACGG...

Intron   Exon

Suppose further that a spliceosome occasionally miscleaves the pre-mRNA transcript between the C and U residues in the exon sequence to yield the following two splicing intermediates:

...UUAGGC   UAACGG...

What would be the consequence of this cleavage?

24. Although nearly all the proteins synthesized by a bacterial cell after it has been infected with T2 bacteriophage are determined by the viral genome, some bacterial proteins are also required for successful infection. What bacterial enzyme is needed to initiate viral infection when T2 DNA first enters the cell?

25. (a) The genome of bacteriophage ΦX174 is a single strand of DNA containing 5386 nucleotides. If only one AUG in the genome were used as an initiation signal, how many amino acids could be encoded by the genome? If the average molecular weight of an amino acid is 112, what is the maximum molecular weight of protein encoded by the genome?

    (b) Studies have shown that the ΦX174 genome can encode a larger number of proteins than expected for its genome length. One reason for this increased encoding capacity is that some of the genes overlap each other. For example, the coding sequence for gene B is located entirely within the sequence that codes for gene A. However, the amino acid sequences of the two proteins specified by these genes are entirely different. How is this possible?

26. In contrast to DNA polymerase, RNA polymerase has no nuclease capability to excise mismatched nucleotides. Suggest why the two enzymes are different in this respect.

27. (a) In *E. coli*, a tRNA that carries tyrosine is composed of 85 nucleotides. However, transcription of the gene that codes for tyrosine tRNA yields an RNA molecule consisting of 350 nucleotides. At least three ribonuclease enzymes cooperate in removing a 41-base segment on the 5′ side of the tRNA sequence and a 224-base segment that extends from the 3′ terminus of the tRNA sequence. The tRNA sequence in the primary transcript is continuous, and no nucleotides are removed from that part of the transcript during processing. How does this type of RNA processing differ from splicing?

    (b) Another primary transcript that is synthesized in bacteria contains 6500 nucleotides, including sequences for the 23S, 16S, and 5S RNA molecules found in ribosomes. This primary transcript has sequences on either side of the set of rRNA sequences, as well as "spacer" sequences between each of them. Suggest a reason for the synthesis of a transcript containing all three rRNA sequences.

28. The codons UAA, UAG, and UGA are signals for chain termination in protein synthesis because none of these codons are read by tRNA molecules. These codons are normally found at the ends of coding sequences for proteins. However, single-base mutations in certain codons can also cause premature termination of the protein chain.

    (a) Which codons can be converted to the chain-termination codon UAA by a single base change?

(b) Suppose a mutation creates a UAA codon that is three codons away from the 3′ end of the normal mRNA coding sequence. Why might you assume that the prematurely terminated protein might still be functional?

(c) Revertants of chain-termination mutants include those in which a single-base substitution changes a termination codon to one that can again be read by a tRNA molecule. For example, a UAG codon can mutate to UCG. What amino acid would then be found at the corresponding position in the protein?

(d) Other revertants retain the original termination codon at the premature termination site, but an amino acid is inserted at the corresponding site in the protein so that the protein has the same length as the nonmutant protein would have. These revertants are due to another mutation in which the anticodon of a tRNA molecule is altered so that the tRNA molecule can read a termination codon. These tRNA molecules are called suppressor tRNAs because they suppress the effect of a chain-termination mutation. Suppose you have a chain-termination mutation that is due to the presence of a UAG codon in the normal coding sequence. If the effect of the UAG codon is suppressed by a tRNA mutation, which amino acids could be found at the site corresponding to the premature termination signal? Assume that a single base change occurs in each case.

29. Polynucleotide phosphorylase, which polymerizes ribonucleoside diphosphates (NDP) to form RNA and P_i, was used in the laboratory to synthesize polyribonucleotides that were useful in determining the genetic code. The reaction catalyzed by polynucleotide phosphorylase is: $(NMP)_n + NDP \leftrightarrow (NMP)_{n+1} + P_i$, where $(NMP)_n$ is RNA, a polymer of nucleoside monophosphates, and P_i is phosphate. Why is it unlikely that this enzyme synthesizes RNA in the cell? Suggest how the cell uses this enzyme.

30. (a) When an experiment was done to form hybrids between mRNA produced after bacteriophage T2 infection and the denatured T2 genomic DNA (mRNA was in molar excess over DNA strands), significantly less than 100% of the DNA could form a DNA-RNA hybrid. What did this suggest about whether transcription takes place on one or both of the two DNA strands at any location on the chromosome?

(b) Later the principle of transcription on only one strand of DNA was established firmly by studies with DNA from the virus SP8, which infects the bacterium *Bacillus subtilis*. Because the two complementary strands of SP8 DNA have very different base compositions, they can be easily separated by density gradient centrifugation. How could you use these separated strands to show that the transcription of SP8 DNA occurred on one strand only?

31. You are studying the effects of amino acid replacements on the stabilbity of a particular α helix that is buried in the protein myoglobin. You carry out a series of replacements of a particular leucine residue located in the helix, using site-specific mutagenesis, a technique described in detail in Section 5.2 of the text. The replacements are as follows:

(i) leucine → arginine
(ii) leucine → valine
(iii) leucine → proline
(iv) leucine → glycine
(v) leucine → alanine

(a) For each replacement, predict whether the change would stabilize, destabilize, or have no effect on the structure of the α helix. Briefly explain each of your predictions.

(b) For each replacement, write the most likely mRNA codon required to code for the particular amino acid. Which of the replacements can be carried out by single-base changes? Which of the replacements can be produced only by altering two bases in the mRNA codon?

32. Cordycepin (3′-deoxyadenosine) is a compound that can block the synthesis of RNA, because a cordycepin residue in an RNA chain lacks the 3′-OH end needed for chain extension by RNA polymerase. The structure of cordycepin is shown below.

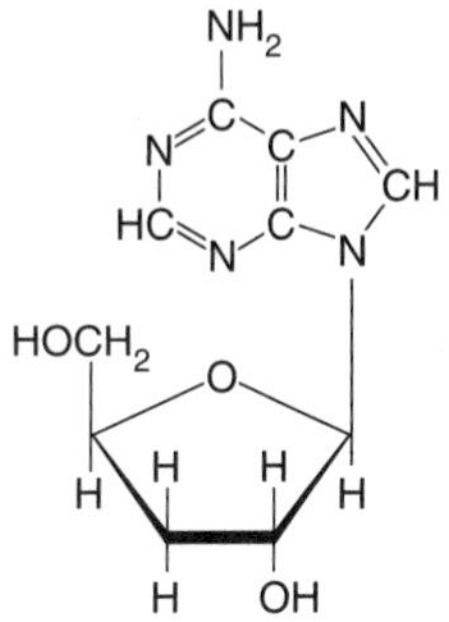

(a) Cordycepin does not inhibit the growth of bacteria, but it does inhibit growth and division of mammalian cells. Consider the reactions that are required for cordycepin to be converted into a substrate for RNA polymerase and then propose a reason for its ineffectiveness in bacteria.

(b) Would you expect cordycepin to block DNA synthesis as well? Why?

33. Raney nickel can convert cysteinyl-tRNA^{Cys} to alanyl-tRNA^{Cys}. When this altered aminoacyl-tRNA is used in a protein-synthesizing system *in vitro*, alanyl residues are placed in the position normally occupied by cysteinyl residues in the protein. What does this experiment tell you about the ability of the protein-synthesizing machinery to recognize an inappropriate aminoacyl-tRNA like alanyl-tRNA^{Cys}?

34. Many steps in the flow of genetic information are subject to regulation. Stringent control of the production of macromolecules limits expenditure of energy by the cell, permitting the synthesis of particular proteins only as they are required. Consider the steps in storage and transmission of genetic information, and describe which one, when regulated, makes it possible to achieve the greatest economy in energy expenditure by a mature cell.

35. In 1971, David Baltimore was investigating whether polymerase activities were contained in the Rauscher murine leukemia virus. This virus has an RNA genome and causes leukemia in mice. He disrupted purified virus particles and incubated the resulting mixture with Mg^{2+} and either the four dNTPs or the four NTPs in a buffered solution. One of the dNTPs or one of the NTPs was radiolabeled. After allowing time for a reaction to occur, the mixtures were treated with strong acid to precipitate nucleic acids while leaving unreacted nucleoside triphosphates in solution. By measuring the precipitated radioactivity, this assay allowed him to detect the formation of the product of a putative polymerase. He found the following: (1) NTPs were not incorporated into product; (2) dNTPs were incorporated into product; (3) the isolated, radiolabeled product was destroyed by DNase (an enzyme that hydrolyzes DNA) but not by RNase (an enzyme that hydrolyzes RNA); (4) the isolated product was not destroyed by NaOH; (5) pretreatment of the disrupted virus extract with DNase did not prevent the formation of product whereas pretreatment with RNase did.

Do these experiments suggest the presence of a polymerase? Why? What kind of polymerase is likely present? What is its template and what is the product formed? What did these experiments indicate, for the first time, about the flow of genetic information?

36. Radioisotopes have been critical for identifying specific molecules involved in biochemical processes. John Hershey and Martha Chase carried out an experiment in 1952 with bacteriophage T2 that had been radiolabeled by being grown in either $^{32}PO_4^{2-}$ or $^{35}SO_4^{2-}$-containing medium. Bacteriophage T2 has a DNA genome. After infecting the bacterial cells in separate cultures with the two different labeled virus preparations for a time short enough to ensure that newly made viruses did not develop to the point of lysing the cells, they put the culture of infected cells in a blender to strip off any part of the virus that did not enter the cell. They next collected the infected stripped cells by centrifugation and compared the amounts of radioisotope in the cells to that remaining in the supernatant. What do you think they observed and why? Why was this experiment important?

37. During their formation or processing, microRNA (miRNA) and small interfering RNA (siRNA) molecules, which have sequences determined by the DNA sequence of the organism, involve short duplex, self-complementary RNA structures. Propose two ways in which a self-complementary RNA duplex might be formed in the cell.

38. To gain an appreciation for the length of duplex DNA that is in an organism, calculate the distance the DNA of a human adult would cover were the DNA molecules (chromosomes) in each cell connected end-to-end. Use data in Chapter 4 of the text as well as the following possibly useful facts or approximations: 1) the distance from San Francisco, USA, to Chicago, USA, is 2000 miles; 2) the haploid human genome is 3×10^9 base pairs; 3) the distance from London, UK, to Sydney, Australia, is 10,500 miles; 4) 1 mile = 1.6 km; 5) each human cell is diploid; 6) the distance from the earth to the sun is 1.5×10^{11} m; 7) the Sears Tower building in Chicago is 1400 ft. high; 8) the distance from the Earth to the Moon is 384,000 km; 9) the human adult has 10^{14} cells (assume all have nuclei with chromosomes); 10) a typical plastic drinking water bottle is 8 inches tall; 11), the average man is about 6 ft tall; 12) each human cell has 23 pairs of chromosomes (neglect sex differences); and 13) one Ångstrom equals 10^{-8} cm.

39. Many biochemical techniques that replicate or detect DNA sequences rely on the binding of pieces of DNA or RNA called oligonucleotides. (Oligonucleotides are just short DNA or RNA single strands). These methods are known as hybridization-based techniques. When the oligonucleotide is mixed (hybridized) with the longer strand, the mixture is incubated at a temperature well below the T_m for that sequence. A "rule of thumb" is to incubate at 5°C below the T_m. Why incubate at this temperature?

40. A DNA-RNA hybrid double strand is placed in a solution of NaOH. What products result?

41. In cells, the DNA replication process begins with RNA polymerase generating a short piece of RNA that is complementary to the 3′-end of the template DNA strand. DNA polymerase then extends this primer to begin replication. Why is RNA polymerase used to make the DNA primer?

42. Extremophiles are organisms that live in environments that would normally not be conducive to life—hot springs and geysers, submarine vents in the ocean floor, the low-pH, high-sulfur run-off of metal mines, etc. When the DNA from extremophiles that live in high-temperature environments was analyzed, it was found to be GC-rich and more extensively positively supercoiled (overwound) than DNA from other organisms. What advantage do these particular adaptations bring the extremophiles?

ANSWERS TO PROBLEMS

1. When the intact double-strand circular DNA molecule is heated in solution, its base pairs are disrupted and an increase in the absorbance of light at 260nm is observed. However, the two resulting single-strand circles are so tangled about one another that they remain closely associated. When the molecules are cooled, the interlocked strands move relative to each other until their base sequences are properly aligned and a double-strand molecule is reformed. This molecule absorbs less ultraviolet light at 260nm (hypochromism) than does the pair of denatured single strands. Breaks in one or both strands of a double-strand DNA molecule allow the two strands to separate completely from one another during denaturation. In order to form a double-strand molecule, the separate strands collide randomly until at least a small number of correct base pairs is formed (nucleation); then the remaining base pairs form very rapidly to generate a completely double-strand molecule. Nucleation of a pair of separate strands in solution is slower than that of a pair of interlocked circles because the local concentration of strands is lower, so a corresponding difference in the reduction of absorbance will be observed.

2. (a) The longer the DNA molecule, the larger the number of base pairs it contains. As a result, more thermal energy is required to disrupt entirely the helical structure of the longer DNA molecule. Experiments show that such a relationship is true for molecules up to ~4000 base pairs in length.

    (b) Sodium ions neutralize the negative charges of the phosphate groups in both strands. As the concentration of NaCl decreases, repulsion between the negatively charged phosphate groups increases, making it easier to separate the two strands. The tendency for the strands to separate more easily means that dissociation occurs at a lower temperature, which is reflected in a lower T_m value of the molecule.

    (c) The reassociation of single strands begins when a short sequence of bases in one strand forms hydrogen bonds with a complementary sequence in another-a process called nucleation. Once a nucleation occurs, reassociation to form the longer double-strand molecule occurs rapidly. The higher the concentration of DNA, the greater the number of complementary sequences in the solution, and thus the quicker the complementary sequences will find and pair with each other.

    (d) Urea, which contains hydrogen bond donors ($-NH_2$) and hydrogen bond acceptors ($>C=O$), disrupts the hydrogen bonds between bases. Because hydrogen bonds are partly responsible for the stability of the double helix, the disruption of these bonds makes the structure more sensitive to denaturation by thermal energy and thereby reduces the T_m value. In addition to hydrogen bonding, the tendency of bases to stack also contributes significantly to the stability of the helix. Base stacking minimizes the contact of the relatively insoluble bases with water, and it also allows the sugar-phosphate chain to be located on the outside of the helix, where it can be highly solvated. Urea may also cause destabilization of the helix by allowing bases to associate more readily with water by disrupting its structure.

3. (a) The four base pairs found in the DNA double helix are almost identical in size and shape, so the diameter of the double helix is essentially uniform all along its length. It is therefore unlikely that a protein can identify a specific sequence by sensing differences in the diameter of the helix.

(b) Proteins that interact with specific sequences might do so by forming hydrogen bonds with the bases; in some cases, it might be necessary for the double strand to undergo local unwinding or melting in order for the bases to form hydrogen bonds with a protein. However, hydrogen-bond donors and acceptors are also found in the grooves of the intact helix. A protein could also bind to a specific location on DNA by forming hydrogen bonds with a particular group of atoms in one of the grooves of the helix. Hydrophobic interactions between amino acid side chains and the methyl group of thymine or the edges of the bases can also contribute to the specificity of the interaction.

4. Although the system described could yield ^{32}P-labeled daughter DNA molecules, chemical methods cannot distinguish DNA in which both strands are radioactively labeled from DNA in which one strand is labeled and one strand is unlabeled. In their experiments, Meselson and Stahl used a physical technique, density gradient equilibrium sedimentation, to separate the labeled molecules according to their content of ^{14}N and ^{15}N, which differ in their specific densities.

5. In solution, the oligodeoxyribonucleotide forms an interchain double-strand molecule with flush ends and a small single-strand loop containing the sequence 5′-pTpCpCpTpCp-3′. The deoxyribonuclease hydrolyzes the phophodiester bonds in this single-strand region to form nucleoside monophosphates, leaving a small double-strand linear molecule remnant containing seven base pairs.

6. Formaldehyde could react with the exocyclic amino groups on the C-6 carbon of adenine, the C-2 of guanine, and the C-4 of cytosine to form hydroxymethyl derivatives. Because these derivatives cannot form hydrogen bonds with complementary bases, formaldehyde-treated single strands would reassociate to a lesser extent than would untreated single strands. All the actual sites of the reaction of formaldehyde with DNA are not precisely known; these sites may also include the ring nitrogen atoms in pyrimidines.

7. The hydrogen bonds of base-paired regions of double-strand DNA may undergo reversible dissociation to form single-strand regions, often known as bubbles. The transient disruption of these hydrogen bonds allows the exchange of protons with the tritiated water. A · T pairs open more easily than G · C pairs. Thus, the greater the percentage of A · T pairs, the greater the rate of proton exchange.

8. A continuous, linear double-strand DNA molecule has only two 3′-OH groups available for the initiation of DNA synthesis by DNA polymerase; because each is located at opposite ends of the molecule, no template sequence is available. In order to construct a relatively simple mechanism for chromosomal replication, one could postulate that the enzyme initiates DNA replication at a number of breaks along the chromosome, with each of the breaks offering the 3′-OH group required for the initiation of the new DNA strand. The template required for replication would then be located on the strand opposite the break, thus ensuring that DNA synthesis could continue. It is now well established that DNA in chromosomes is very long and continuous. The fact that there are initially no breaks in the molecule makes the mechanism of replication complex. It involves a number of enzyme activities, as well as the use of RNA to prime the synthesis of DNA. For details, see page 823 of the text. Thus, the initial conjecture that the long molecules might have single-strand breaks where replication could initiate was not confirmed.

9. (a) The expected melting temperature for *E. coli* DNA containing 50% GC base pairs is

$$T_m = 69.3 + 0.41(G + C)$$
$$= 69.3 + 0.41(50)$$
$$= 69.3 + 20.5$$
$$= 89.8°C$$

(b) Most organisms live at temperatures that are considerably lower than 65°C. Because both the transmission and expression of genetic information depends on the integrity of the double-strand DNA molecule, it is important that the molecule not be disrupted by thermal energy.

(c) Some mechanism other than thermal denaturation must be involved in order to separate the strands for replication. Proteins that unwind and separate the strands will be described later.

10. First, you must determine the temperature at which the hydrogen bonds are disrupted and single strands are formed. You can do this by heating the double-strand DNA to various temperatures and measuring the extent of hyperchromicity. Once the DNA has been melted, centrifuge the sample using the density-gradient equilibrium sedimentation technique to attempt to separate the ^{14}N-labeled DNA strands from the ^{15}N-labeled DNA strands, which will be the denser of the two. If you are successful, this would suggest that the strands separate completely during thermal denaturation.

11. The protonation of the N-1 of adenine, the N-3 of cytosine, and the O-4 of thymine makes normal hydrogen bonding at these locations impossible because the atoms can no longer serve as hydrogen-bond acceptors. Therefore, at low pH, where proton concentrations are high and protonation of these atoms occurs, double-strand DNA is less stable than at neutral pH values. At high pH values (> 11) DNA is also denatured by deprotonation of other ring atoms.

12. The association of a molecule of RNA with a molecule of DNA to form a hybrid molecule depends primarily on the two molecules having complementary sequences of bases. The formation of hydrogen bonds between complementary bases will allow the formation of a double helix composed of RNA and DNA. Thus, an mRNA will anneal with the denatured DNA template from which it was made to form a DNA-RNA hybrid duplex.

13. The deoxyribonucleoside triphosphate dUTP can be used as a substrate for DNA polymerase during DNA replication because the structure and hydrogen-bonding properties of uracil are very similar to those of thymine. When incorporated into a double-strand DNA polymer, uracil pairs with adenine, as does thymine. For a discussion of the reasons uracil is not normally incorporated into DNA, see page 841 of the text.

14. (a) To ensure that λ phage DNA molecules form closed-circular monomers, the concentration of λ phage DNA should be relatively low so that the intrachain formation of hydrogen bonds is favored. At higher concentrations, the probability of interchain joining to form multimers is enhanced.

(b) The most reasonable model for the structure of the λ phage DNA molecule is a double-strand molecule having single-strand protrusions at the 5′-ends, as illustrated below. The 3′-ends have hydroxyl groups, which allow them to serve as primers for DNA synthesis catalyzed by DNA polymerase I. This enzyme fills in the single-strand regions of the molecule, producing a molecule with flush ends. Such a molecule no longer has cohesive ends that can form the required circular molecule needed for infectivity.

Molecules treated with exonuclease III have a longer single-strand sequence at both ends; they may not be infective because the newly exposed bases may not be fully complementary to each other, which would mean that the ends could no longer be joined. When the exonuclease-treated DNA is treated with DNA polymerase I, the single-strand regions are sufficiently filled in to reform a molecule that has protruding single strands that are approximately the same length as those in the native molecule. Hence, the molecule once again becomes infective. Further treatment of the molecule with DNA polymerase I will once again produce a molecule that has flush ends and is no longer infective.

15. (a) The sequence at the other end of the single-strand molecule must be composed of complementary bases. It must therefore be

$$. . .TACGTAGT\text{-}3'$$

The structure of the full-length, double-strand, linear molecule would be

$$5'\text{-}ACTACGTA\text{———}TACGTAGT\text{-}3'$$

$$3'\text{-}TGATGCAT\text{———}ATGCATCA\text{-}5'$$

Each single strand has a pair of inverted repeats.

(b) The formation of double-strand helical segments depends upon hydrogen bond formation between bases in nucleotide chains that are antiparallel, as follows:

$$5'...PuPyPuPyPu...3'$$

$$3'...PyPuPyPuPy...5'$$

where Py = pyrimidine and Pu = purine. When the suggested structure is labeled using this scheme, as in Figure 4.5, it can be seen that it would require the formation of base pairs between parallel chains, and such pairing cannot readily take place.

FIGURE 4.5　Base pairing between parallel nucleotide chains, required to form the circular structures shown in Figure 4.3, is unlikely in DNA.

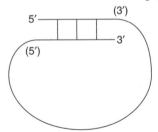

16. (a) Each enzyme molecule probably tightly clamps the primer as well as the extended, newly synthesized chain to the template, thereby protecting the helix from denaturation at high temperature. One primer would be required for each template strand. The association between primer and enzyme may protect the enzyme from thermal denaturation as well. There are many examples in biochemistry where substrate binding stabilizes protein structure.

   (b) During polymerization of a new DNA chain, the chance that base incorporation errors can occur will increase at high temperatures. Even though a proofreading or error-correcting activity is not found in the *S. acidocaldarius* polymerase polypeptide, you would expect an auxiliary enzyme to be present in the cells in order for the bacterial genome to be accurately replicated so that the genetic integrity of the organism is maintained.

   (c) The higher the G + C content of double-strand DNA, the higher the melting point at which the helix is denatured. Therefore, you might expect DNA from a thermophile to have a higher G + C composition. Surprisingly, DNA base ratios are sometimes not very different from those in bacteria living at lower temperatures, so that DNA isolated from these thermophilic bacteria melts at temperatures lower than those encountered in the hot springs where they grow. There must be proteins or other molecules in the bacterial cells that protect the genomic DNA of these thermophiles from thermal denaturation.

17. (a) Like DNA polymerase I, TdT can extend a DNA primer by using deoxynucleoside triphosphates as substrates. However, TdT does not use a template and cannot copy from one, so that the base composition of the newly synthesized single strand of DNA will depend solely on the relative concentrations of the deoxynucleoside triphosphate substrates. For chains synthesized by DNA polymerase I, the base composition will be complementary to that of the template strand. Although DNA polymerase has an exonuclease activity that removes mismatched bases from newly synthesized strands, TdT has no such activity and does not need one.

   (b) DNA molecules that carry genetic information must be synthesized as faithful copies of template strands; TdT cannot copy a template and would not be useful in genomic DNA synthesis. However, in cases where variability is desired, TdT is used to introduce sequence variation into DNA during antibody formation.

18. (a) DNA polymerase requires deoxyribonucleoside triphosphates as substrates for DNA chain extension. Nucleosides like the 2′,3′-dideoxy analogs must be converted to nucleoside triphosphates in order to serve as substrates for DNA polymerase. Studies on inhibition of DNA synthesis in living cells involve incubating those cells with the nucleoside forms of the analogs instead of their nucleoside triphosphate forms, because negatively charged phosphate anions cannot pass across the plasma membrane, while relatively neutral nucleosides can. Once inside the cell, nucleoside analogs are phosphorylated by cellular enzymes that normally function to "salvage" nucleosides generated by turnover of nucleotides from RNA and DNA.

   (b) Dideoxynucleosides lack a free 3′-hydroxyl group, which would normally serve as an acceptor for incorporation of the next nucleotide into the growing polynucleotide chain. The lack of a 3′-OH group also interferes with excision by the error-correcting exonuclease activity of some DNA polymerases, so that chain extension is blocked.

19. Noncovalent forces also contribute to driving the reaction forward. Hydrogen bonds form between opposing A and T bases and between G and C molecules in the antiparallel chains. There are also significant hydrophobic interactions between adjacent bases

on the same strand, and these stacking interactions may in fact contribute significantly to helix formation and stability. In vivo, it is likely that these noncovalent forces, along with the cleavage of the product, account for the forward progress of chain elongation.

20. (a) The ratios observed by Chargaff can be attributed to the requirement that in a double-strand polynucleotide, only certain bases can form hydrogen bonds with one another. Although a single-strand polynucleotide might form some hydrogen bonds between bases as it folds, the overall base ratios will not conform to Chargaff's established rules. This is also true for single-strand DNA in a virus like ΦX174.

    (b) Because hydrogen bonding between A and T (or U, in the case of RNA) and between G and C can occur in a duplex molecule formed by one strand of DNA and a complementary strand of RNA, you would expect to see base ratios like those observed by Chargaff.

    (c) Double-strand RNA molecules form hydrogen bonds between bases in a manner similar to those in DNA helices. Therefore, bearing in mind that U, not T, would normally be found in RNA, the number of uracil residues would equal the number of adenines, and the number of guanine bases would be expected to be the same as those of cytosine.

21. (a) To locate a promoter site, you would first incubate the double-strand DNA with RNA polymerase; the RNA polymerase will bind tightly to the promoter site. Next, you would add the DNA endonuclease, which will degrade the DNA that is not protected by the bound RNA polymerase. Electrophoresis can then be used to determine the size of the protected fragments of DNA, and the base sequence can be determined using methods discussed in Chapter 5 of the text.

    (b) Ribonucleoside triphosphates are substrates for RNA polymerase transcription. If present when this process is performed, they would allow the polymerase molecule to move from the promoter site to the site on the template where transcription begins, as well as beyond, as transcription progresses. As a result, the promoter site would no longer be protected from endonuclease degradation.

22. The GCU codon in the first mutation corresponds to a substitution of alanine for valine at position 102. Although the side chain of alanine is smaller than that of valine, both are aliphatic amino acids, so the alteration in the structure of the enzyme does not necessarily affect the enzyme activity. The GAU codon of the second mutation specifies the substitution of aspartate for valine at position 102. This substitution could have a detrimental effect because the side-chain carboxyl group of aspartate has a negative charge at neutral pH. The charged group could disrupt the native conformation of the enzyme, thereby inactivating it.

23. If the spliceosome cleaves the initial transcript within the normal exon sequence as shown, the exon coding sequence in the spliced mRNA will be altered because of the loss of two bases. Instead of beginning with the codon GCU in the normal exon, the reading frame will begin with the codon UAA in the altered exon. This codon is, in fact, a termination signal for protein synthesis, which means that the translation of the polypeptide specified by the spliced messenger RNA would be terminated prematurely.

24. Because only DNA and no protein from T2 enters the cell, the synthesis of viral-directed proteins cannot begin until T2 messenger RNA has been made. Transcription of the T2 DNA must therefore be carried out by bacterial RNA polymerase using ribonucleoside triphosphates synthesized by the bacterial cell.

25. (a) Three consecutive bases are required to encode an amino acid, so up to 1794 amino acids could be specified by the ΦX174 genome. The molecular weight of this much protein would be approximately 201,000

(b) Overlapping genes can yield proteins with different primary amino acid sequences only if each of the protein coding sequences is read in a different reading frame. The critical feature for the establishment of the proper reading frame is the location of the AUG initiation signal. As an example, consider the following mRNA sequence, which contains two AUG codons that are in overlapping but different reading frames:

When an initiator tRNA binds to the first AUG codon, reading frame 1 is established; similarly, reading frame 2 is established when an initiator tRNA binds to the second AUG codon. The polypeptides specified by the two different mRNA sequences will necessarily have different amino acid sequences.

26. DNA polymerase is responsible for the duplication of the DNA of chromosomes, which is the repository of the genetic information that is passed on to progeny cells. Any error that occurs in the copying of a DNA template will be transmitted not only to the duplicated chromosome but also to all the messenger RNA molecules transcribed from the miscopied DNA template. Therefore, it is crucial that DNA polymerase be able to correct errors that occur due to its incorporation during synthesis of mismatched nucleotides. RNA polymerase makes many copies of mRNA, but these molecules have relatively brief lives in the cell, and very few are passed on to progeny cells. Occasional errors in transcription can result in the production of defective proteins, but it appears that the cell can tolerate such errors provided that not too many occur.

27. (a) The processing of the primary tRNA transcript removes RNA on either side of the uninterrupted tRNA sequence of 85 nucleotides, whereas protein splicing operations remove RNA sequences that are located within regions of pre-mRNA that code for a continuous polypeptide; these sequences must be removed to ensure that the protein specified by the RNA will have the correct amino acid sequence. In some organisms, tRNA are produced by RNA splicing, which removes sequences from within the pre-tRNA transcript.

(b) The most obvious reason for transcribing all three rRNA sequences simultaneously is that it ensures that an equal number of ribosomal RNA molecules will be available for the assembly of ribosomes (the ratio of these three RNAs in a ribosome is 1/1/1). In addition, only one promoter, rather than three, is required for rRNA synthesis.

28. (a) The protein-encoding codons that could be mutated to UAA by a single base change are CAA, GAA, AAA, UCA, UUA, UAU, and UAC.

(b) Chain termination near the 3′ end of the normal coding sequence could allow the prematurely terminated protein to be functional because most of the polypeptide sequence would be intact. Removing a few amino acids from the C-terminal end of many, but not all, proteins does not appreciably affect their normal function.

(c) When a UAG codon reverts to UCG, a serine residue will be incorporated at the site in the protein that corresponds to the chain-termination site; it will be linked by a peptide bond to the next amino acid in the polypeptide.

(d) To determine which amino acids would be carried by suppressor tRNAs to a UAG codon site, you should identify all those tRNA molecules having an anticodon that, by a single base change, can read a UAG codon. Each such tRNA molecule would suppress the premature termination of the chain by inserting an amino acid at the corresponding site in the protein. The amino acids that could be found at the site (along with their codons) are Glu (GAG), Gln (CAG), Leu (UUG), Lys (AAG), Ser (UCG), Trp (UGG), and Tyr (UAC and UAU). When a cell contains a suppressor tRNA, proteins whose mRNA sequence normally ends with a single stop codon may not be efficiently terminated. Although the extension of such proteins could be lethal, most cells tolerate suppression. One explanation is that other proteins involved in chain termination may recognize a stop codon even though a tRNA that reads the codon is present. More work is needed to develop a full understanding of the reasons for toleration of suppression.

29. Equilibrium for the reaction catalyzed by polynucleotide phosphorylase lies toward the direction of RNA degradation rather than synthesis. High concentrations of ribonucleoside diphosphates are required to achieve the net synthesis of RNA; and it is likely that their concentrations in the cell are not sufficient to drive net polynucleotide synthesis. Also, polynucleotide phosphorylase does not use a template, so the polyribonucleotides it synthesizes contain random sequences, which makes them of no value for protein synthesis. The cell uses polynucleotide phosphorylase as a degradative enzyme in conjunction with other nucleases that regulate the lifetimes of RNA molecules, including mRNA. In bacteria mRNA lifetimes are relatively short.

30. (a) If transcription occurred simultaneously on both DNA strands, the excess, complementary mRNA molecules synthesized from each template strand could form double-strand structures with all the DNA strands. The fact that less than all the DNA could form hybrids indicated that some of the DNA strands lacked sequences complementary to the mRNA, that is, only one of the two strands at a given location along the DNA was being transcribed into DNA. In rare cases in some organisms over limited regions, RNA is synthesized from both strands of the template DNA.

    (b) To establish whether one or both strands of SP8 DNA are used for transcription, you can carry out hybridization experiments with the separate strands using radioactive RNA synthesized during the infection of *Bacillus subtilis* with SP8. The results show that such RNA hybridizes to only one of the two strands, which means that only one of the two strands of the DNA of the SP8 virus is transcribed. In most other organisms, different regions of each strand are used for transcription; SP8 virus is exceptional in that one strand is used exclusively for all mRNA synthesis.

31. (a) (i) Leucine → arginine destabilizes the helix because you substitute a hydrophilic residue for a hydrophobic residue (most likely in a hydrophobic region of the helix).

        (ii) Leucine → valine has no effect (a conservative replacement of one hydrophobic residue for another).

        (iii) Leucine → proline destabilizes the helix because proline does not allow rotation about its peptide bond and no hydrogen atom in this amino acid is available for hydrogen bonding.

        (iv) Leucine → glycine destabilizes because glycine is a very flexible residue and can act as a swivel, disrupting the helix (the frequency of glycines in a helix is as rare as the frequency of prolines).

        (v) Leucine → alanine has no effect (a conservative replacement of one hydrophobic residue for another).

    (b) (i) CUU → CGU   single-base change required
        (ii) CUU → GUU   single-base change required
        (iii) CUU → CCU   single-base change required

(iv) CUU → GGU   two-base alteration required
(v) CUU → GCU   two-base alteration required

32. (a) Cordycepin is a nucleoside, and it must be converted to the triphosphate nucleotide form before it can be incorporated (as cordycepin monophosphate) by RNA polymerase into a growing polynucleotide chain. The conversion of cordycepin to the triphosphate form is carried out by a number of kinase enzymes (see page 739 on nucleotide metabolism in the text) that utilize ATP as a phosphate donor. Bacteria probably cannot phosphorylate cordycepin efficiently, which makes them less susceptible to inhibition of RNA and DNA synthesis.

    (b) You would not expect cordycepin to inhibit DNA polymerase because, although a cordycepin residue that had been added to DNA would also lack a 3'-OH and act as a chain terminator, the presence of the 2'-OH on the ribose of the triphosphate form of cordycepin would be discriminated against by the DNA polymerase. You will learn later that DNA synthesis requires an RNA primer, and cordycepin might inhibit DNA synthesis by inhibiting RNA primer formation.

33. The ribosomal complex that carries out protein synthesis is unable to recognize alanyl-tRNA^{Cys} as an inappropriate or erroneous form of tRNA. An amino acid that is attached to a transfer RNA molecule will be transferred into a growing polypeptide chain solely on the basis of recognition between the anticodon in tRNA and the codon in the messenger RNA molecule. Once an aminoacyl-tRNA has been formed, accurate translation does not depend on recognition of the attached amino acid. This important point was established by Dintzis and von Ehrenstein, who carried out the incisive experiments using Raney nickel to reduce the cysteinyl residue on cysteinyl-tRNA^{Cys} to an alanyl residue, then analyzing the resulting protein using an hemoglobin-synthesizing system *in vitro*.

34. The step that would afford the maximum economy is probably transcription, through the control of the activity of RNA polymerase. Transcription by RNA polymerase to form RNA is the first step in the expression of genetic information. It follows that controlling messenger RNA production, by stimulating or inhibiting the activity of RNA polymerase, allows the cell to make particular types of mRNA and to synthesize the encoded proteins only when required. A cell that could not regulate RNA polymerase activity would produce unneeded mRNA molecules, and the energy required to produce those polynucleotides would be wasted, even if translation were stringently regulated.

35. Yes, the incorporation of nucleoside triphosphates into an acid-insoluble form is indicative of the presence of a polymerase. The polymerase is likely a DNA polymerase because dNTPs, and not NTPs, were used to form product. Further evidence for a DNA polymerase was that the radiolabeled product was destroyed by a nuclease, DNase, specific for hydrolyzing DNA, and not by one specific for RNA hydrolysis. Additionally, NaOH, which destroys RNA but not DNA, did not destroy the radiolabeled product. Pretreatment of the extract with the two hydrolytic enzymes demonstrated that the enzyme depends on an RNA and not a DNA template for its activity. Thus, this enzyme is an RNA-dependent DNA polymerase. No such enzyme had been observed previously in a cell, and this demonstration, along with similar findings by Howard Temin, of its existence in an RNA tumor virus caused a revision of Francis Crick's central dogma of molecular biology, which stated that information flowed from DNA to RNA to proteins. The demonstration of this RNA-dependent DNA polymerase suggested that in some cases information could flow from RNA to DNA. (This question was derived from D. Baltimore. Viral RNA-dependent DNA polymerase. *Nature* 226:[1971]1209–1213.)

36. Hershey and Chase observed that most of the ^{32}P was associated with the cells and most of the ^{35}S was in the supernatant. Since nucleic acids are rich in phosphorus and DNA does not contain sulfur, they concluded that DNA had entered the cell. The sulfur of sulfate is incorporated into the amino acids cysteine and methionine so that ^{35}S is a good marker for proteins. The experiment indicated that protein did not enter the cells. Recalling that bacteriophage T2 displays heredity, that is, passes genetic traits to its progeny, they concluded that DNA, not protein, is likely the genetic information because it entered the cells and was replicated. At the time these experiments were performed, they helped solidify the view that the genetic material was DNA, not protein. (This question was derived from A. D. Hershey and M. Chase. Independent functions of viral protein and nucleic acid in growth of bacteriophage. *J. Gen. Physiol.* 36:[1952]39–56.)

37. RNA duplexes might be formed as a result of an RNA polymerase transcribing both strands of a segment of DNA, and the resulting two product RNA chains annealing with each other to form a duplex. This would be contrary to the usual transcription to form mRNA, tRNA, or rRNA because RNA polymerase transcribes only one of the two strands at a given locus on the DNA. Alternatively, an RNA polymerase might produce a single RNA molecule by transcribing the DNA and this RNA might have self-complementary regions that anneal to one another to form a base-paired hairpin (Figure 4.31 in the text). For either duplex segment of RNA, hydrolytic enzymes could cut initial product to form a duplex of the needed size.

38. There are 3×10^9 base pairs of DNA/haploid set of human chromosomes/cell $\times$ 3.4 Ångstrom/base pair $\times 10^{-8}$ cm/Ångstrom $\times$ 2 haploid sets of chromosomes/diploid cell $\times 10^{-2}$ meter/cm = 2 meters of total DNA/cell; 2 meters DNA/cell $\times 10^{14}$ cells/adult = 2×10^{14} meters/1.5×10^{11} meters to sun = 1.3×10^3 distances to the sun or ~ 650 round trips to the Sun!

39. The T_m is the temperature at which the DNA is 50% denatured. To have complete annealing, it is crucial to be *below* this temperature. The quantity of five degrees is enough to ensure that the majority of the DNA is hybridized.

40. The products are single-stranded DNA and individual ribonucleotides. First, the high pH changes the ionization state of the bases, causing the helix to denature. Second, remember that RNA is base-labile. The NaOH causes the 2'-OH of the RNA to be deprotonated. The resulting 2'-alkoxide attacks the 3'-phosphate, resulting in hydrolysis of the phosphodiester bond and creating individual RNA nucleotides.

41. DNA polymerase requires a primer to initiate replication; RNA polymerase does not. So the RNA polymerase generates the primer for DNA polymerase.

42. The increased G-C content means that there are more base pairs with three hydrogen bonds rather than two. Thus, it takes more thermal energy to denature this DNA. Secondly, positive supercoiling (overwinding) stores energy in DNA, so the DNA has a higher T_m when in the more highly positively supercoiled state. Thus, both of these adaptations increase the thermal stability of DNA and make the organism able to withstand higher temperatures.

Exploring Genes and Genomes

The nature of hereditary material and the flow of information from DNA to protein by means of RNA were outlined in Chapter 4, which you should review in preparation for studying Chapter 5.

In this chapter, the authors present the methods and techniques used to analyze and manipulate DNA. They begin with an overview of recombinant DNA technology and the tools that make it possible. Of particular importance are the specificity of base pairing between nucleic acids and the enzymes that act on nucleic acids. Restriction endonucleases, the ability to immobilize nucleic acids onto solid supports, DNA sequencing, and chemical synthesis of oligodeoxyribonucleotides, plus the polymerase chain reaction (PCR) are introduced. A more detailed description of restriction enzymes and the joining of their products, specific DNA restriction fragments, by DNA ligase are described. They next present the major method of determining the sequence of DNA and an automated method for synthesizing oligodeoxyribonucleotides by chemical means. The authors then describe how specific fragments of genes can be amplified by PCR, a process that depends on specific hybridization of short oligodeoxyribonucleotide primers to a template strand followed by polymerase-catalyzed synthesis of DNA. A more detailed description of restriction enzymes and DNA ligase follows; these enzymes make possible the precise production and joining of DNA fragments. Next, various vectors, the self-replicating carriers of the target genes, are discussed and their roles in the cloning and expression of genes is described. The special role in recombinant DNA technology of complementary DNA (cDNA), which is produced from mRNA, is discussed. The authors describe how DNA chips can be used to monitor the pattern and level of gene expression in an organism. The methods for creating transgenic animals and plants are presented, and the information that can be obtained from them outlined. The authors describe how site-specific mutagenesis can be used with cloned genes to produce proteins having any desired amino acid at any position. An overview is provided to explain how the methods described allow the information in either protein or DNA to be manipulated. The authors then turn to describing the sequencing of entire genomes, including that of man, and to how such massive amounts of sequence information can be organized, analyzed, and exploited. The problems of locating specific genes in the genome and of inserting and expressing foreign genes in eukaryotes are also considered. The chapter closes with a description of RNA interference, introducing genes into plants, and the potential for gene therapy in medicine. Throughout the chapter, the authors use the example of amyotropic lateral sclerosis (ALS) to illustrate the effect that recombinant DNA technology has had on our knowledge of disease mechanisms.

LEARNING OBJECTIVES

When you have mastered this chapter, you should be able to accomplish the following objectives.

The Exploration of Genes Relies on Key Tools (Text Section 5.1)

1. List the basic tools of recombinant DNA technology and explain their applications.

2. Describe the reaction catalyzed by *restriction enzymes* and the characteristics of the *restriction sites* they recognize.

3. Explain why *gel electrophoresis* of DNA is essential to recombinant DNA technology. Describe how DNA *restriction fragments* can be detected in gels.

4. Contrast the *Southern, Northern, Western,* and *protein blotting techniques.*

5. Explain the process of *DNA sequencing* by controlled termination of DNA synthesis *in vitro* (Sanger dideoxy method), including the role for *fluorescence* in this process.

6. Describe how *oligodeoxyribonucleotides* are synthesized chemically. Indicate the roles of activated precursors, coupling, protecting groups, oxidation, and differential deprotection in the process. List some common experimental uses of oligonucleotides.

7. Describe the *polymerase chain reaction.* Explain the roles of the *primer* and *thermostable DNA polymerase* in amplifying the *target* DNA sequence.

8. Give examples of practical applications of PCR.

9. Outline five noteworthy features of the PCR technique.

10. Define RFLPs and describe how they have been used to obtain evidence for a disease-causing mutation in ALS. Identify the gene found on ALS-associated RFLPs.

Recombinant DNA Technology Has Revolutionized All Aspects of Biology
(Text Section 5.2)

11. Outline how restriction enzymes and *DNA ligase* have enabled recombinant DNA technology. Explain the roles of oligonucleotide *linkers* and *polynucleotide kinase* in creating recombinant molecules.

12. Name the substrates and describe the reaction catalyzed by *DNA ligase.* Draw the termini of the DNA fragments joined by DNA ligase. Distinguish between sticky and blunt ends of dsDNA.

13. List the desired characteristics of a *vector.* Distinguish *plasmids* and *viruses.* Outline the major steps in *cloning* a DNA molecule.

14. Name some common vectors used in prokaryotes and eukaryotes, and compare their properties and relative merits for cloning.

15. Distinguish between *screening* and *selection*. Explain *insertional inactivation*. Understand the role of *drug resistance genes* in cloning.

16. Outline how specific genes can be cloned from a digest of an organism to form a *genomic library*. Define genomic library.

17. Describe what a *probe* is and explain the biochemical basis of its specificity. Describe how probes are obtained.

18. Explain how to design a probe by converting an amino acid sequence into nucleotide sequences using the genetic code. Understand the complication introduced by the *degeneracy* of the code.

19. List the kinds of changes that can be introduced into genes by directed techniques and discuss how these changes can be employed to create new proteins with new functions (*protein engineering*).

20. Explain how the presence of *introns* in some eukaryotic genes complicates the expression of these genes in prokaryotes.

21. Define *cDNA*. Distinguish between genomic and *cDNA libraries*.

22. Describe the reactions catalyzed by *reverse transcriptase* and *terminal transferase*.

23. Explain the function of *expression vectors* and the use of *immunochemical screening* in isolating genes.

24. Explain the process of expression cloning. Give an example of a eukaryotic protein expressed in bacteria.

25. Describe how recombinant DNA technology has been used to study the potential effect of ALS-causing mutations on SODI structure and function.

Complete Genomes Have Been Sequenced and Analyzed (Text Section 5.3)

26. Provide examples of viruses and free-living organisms whose genomes have been sequenced and give the numbers of base pairs in each. Outline the *"shotgun"* approach to genome sequencing.

27. Describe functions for sequences of human DNA that do not encode for proteins.

28. List the ways in which one eukaryotic gene can encode more than one protein.

29. List some of the discoveries arising from *comparative genomics*.

Eukaryotic Genes Can Be Manipulated with Considerable Precision (Text Section 5.4)

30. Define and give examples of psuedogenes.

31. Explain how qPCR is used to determine copy number in cells.

32. Outline how *DNA microarrays (gene chips)* are used to monitor patterns of gene expression.

33. List the ways that exogenous DNA can be incorporated into eukaryotic cells.

34. Outline the role of *homologous recombination* in constructing *gene disruptions* and give an example of how it has been used to study development of muscle cells.

35. Explain how transgenetic mice were used to study the role of human superoxide dismustase in ALS.

36. Outline the mechanism of *RNA interference*.

37. Provide an overview of how *Ti plasmids, electroporation,* and *"gene guns"* can be used to introduce genes into plants. Discuss the first genetically modified organism to come to market.

38. Give an example of a human gene therapy target. Discuss some of the hurdles that must be overcome for gene therapy to become widely used.

SELF-TEST

The Exploration of Genes Relies on Key Tools

1. Which of the following portions of a longer duplex DNA segment are likely to be recognition sequences of a restriction enzyme?

   (a) 5′-AGTC-3′
       3′-TCAG-5′
   (b) 5′-ATCG-3′
       3′-TAGC-5′
   (c) 5′-ACCT-3′
       3′-TGGA-5′
   (d) 5′-ACGT-3′
       3′-TGCA-5′

2. Which of the following reagents would be useful for visualizing DNA restriction fragments that have been separated by electrophoresis in an agarose gel and remain in the wet gel?

   (a) $^{32}P_i$
   (b) $[\alpha\text{-}^{32}P]ATP$
   (c) diphenylamine
   (d) ethidium bromide
   (e) DNA polymerase
   (f) polynucleotide kinase

3. Which blotting technique is used for the detection of DNA that has been separated from a mixture of DNA restriction fragments by electrophoresis through an agarose gel and then transferred onto a nitrocellulose sheet?

   (a) Eastern blotting
   (b) Northern blotting
   (c) Southern blotting
   (d) Western blotting

4. Which of the following reagents would be useful for labeling the oligodeoxyribonucleotide d(GGATATCC)?

   (a) $[\gamma\text{-}^{32}P]ATP$
   (b) $^{32}P_i$
   (c) DNA-dependent RNA polymerase
   (d) polynucleotide kinase
   (e) DNA ligase

5. Complete the following statements about the Sanger dideoxy method of DNA sequencing.

   (a) The incorporation of a ddNMP onto a growing DNA chain stops the reaction because
   (b) The sequence of the DNA fragments emerging from the capillary is determined by
   (c) A sequencing primer does not have to anneal directly to the gene to be sequenced if
   (d) It is preferable to label the oligonucleotide primer with a fluorescent rather than a radioactive group because

6. Which of the following statements are correct? Chemically synthesized oligonucleotides can be used

   (a) to synthesize genes.
   (b) to construct linkers.
   (c) to introduce mutations into cloned DNA.
   (d) as primers for sequencing DNA.
   (e) as probes for hybridization.

7. Which of the following statements are correct? The efficient, successful chemical synthesis of oligonucleotides requires

   (a) high yields at each condensation step.
   (b) the protection of groups not intended for reaction.
   (c) a single treatment for the removal of all blocking groups.
   (d) methods for the removal of the blocking groups that do not rupture phosphodiester bonds.
   (e) a computer-controlled, automated "gene machine."

8. Match the conditions of the PCR reaction, in the left column, with the appropriate reaction in the right column.

   (a) cooling abruptly to 54°C   (1) DNA synthesis by Taq DNA polymerase
   (b) heating to 72°C, 30 s   (2) hybridization of primers
   (c) heating to 95°C, 15 s   (3) strand separation

9. Using the information in Question 8, give the sequence of the PCR reaction steps, and explain the rationale for each step.

10. Which of the following are possible applications of the PCR technique?

    (a) detection of very small amounts of bacteria and viruses
    (b) introduction of a normal gene into animals containing the corresponding defective gene
    (c) amplification of DNA in archaeological samples
    (d) monitoring of certain types of cancer chemotherapy
    (e) identification of matching DNA samples in forensic specimens

Recombinant DNA Technology Has Revolutionized All Aspects of Biology

11. Efficient covalent joining of two single-strand DNAs by DNA ligase requires

    (a) that the ends of the strands be juxtaposed so that a 3′-OH is adjacent to a 5′-OH.
    (b) a source of energy to form the phosphodiester bond.
    (c) a template or "splint" strand, which is complementary to the single-strand DNAs, to bring the ends to be joined into apposition.
    (d) the four dNTPs to fill any gap that may exist between the ends that are to be joined.

12. State whether each of the following is or is not a desired characteristic of vectors and explain why or why not.

    (a) autonomous replication   (d) small size
    (b) unique restriction sites   (e) circularity
    (c) genes that confer antibiotic resistance

13. Which of the following statements are correct? Very long DNA fragments (>100 kb) from eukaryotic genomes

    (a) can be efficiently packaged in a bacteriophage λ vector.
    (b) can be propagated in yeast artificial chromosomes.
    (c) must have cohesive ends for cloning.
    (d) can be analyzed by chromosome walking.
    (e) can be separated by standard polyacrylamide gel electrophoresis techniques.

14. You have been supplied with the linker oligonucleotide d(GGAATTCC) and an isolated and purified DNA restriction fragment that has been excised from a longer DNA molecule with a restriction endonuclease that produces blunt ends. Which of the

following reagents would you need to tailor the ends of the fragment so it could be inserted into an expression vector at a unique EcoRI cloning site?

(a) DNA polymerase

(d) ATP

(b) all four dNTPs

(e) DNA ligase

(c) EcoRI restriction endonuclease

(f) polynucleotide kinase

What would happen if the restriction fragment had an internal EcoRI site?

15. Inserting a long DNA fragment into the middle of a vector gene that specifies an enzyme that hydrolyzes an antibiotic and incorporating the altered vector into a bacterium

(a) leads to drug resistance transfer.

(b) is called insertional inactivation.

(c) renders the cell sensitive to that antibiotic.

(d) can be used to identify bacteria that contain the vector with the DNA fragment.

(e) is a method of destroying pathogenic bacteria.

16. Briefly describe genomic and cDNA libraries. Which library, from a given organism, has more clones?

17. Which of the following partial amino acid sequences from a protein whose gene you wish to clone would be most useful in designing an oligonucleotide probe to screen a cDNA library?

(a) Met-Leu-Arg-Leu

(b) Met-Trp-Cys-Trp

Explain why.

18. Explain how the presence of introns in eukaryotic genes complicates the production of the protein products they encode when expression is attempted in bacteria. How can this problem be circumvented?

19. Which of these reagents would be required to perform an immuno-chemical screen of a population of bacteria for the presence of a particular cloned gene if you have the pure protein encoded by the gene?

(a) $[\gamma\text{-}^{32}P]ATP$

(b) polynucleotide kinase

(c) DNA polymerase

(d) all four dNTPs

(e) a radioactive antibody to the protein encoded by the cloned gene

20. Reverse transcriptase requires the following for the conversion of a single-strand RNA into a double-strand DNA.

(a) all four NTPs

(d) an RNA template

(b) all four dNTPs

(e) a primer

(c) a DNA template

21. Which of the following statements are correct? Oligonucleotide-directed site-specific mutagenesis

(a) depends upon having an oligonucleotide with a sequence completely different from that of the target gene.

(b) can be used to produce deletion, insertion, and point mutations.

(c) is a good method for identifying the functional domains of an enzyme.

(d) is useful for determining the involvement of a particular amino acid in the catalytic mechanism of an enzyme.

(e) can involve the use of the same oligonucleotide to produce the mutant and to detect it.

22. Outline the steps necessary to synthesize a gene. Be explicit about the information and reagents, including enzymes, that you would need.

Complete Gemones Have Been Sequenced and Analyzed

23. Order the following genomes with respect to when their sequences were determined.

(a) *Saccharomyces cervisiae* (yeast)
(b) mitochondria of *Homo sapiens* (organelle)
(c) bacteriophage ΦX174 (virus)
(d) *Caenorhabditis elegans* (nematode)
(e) *Homo sapiens* (mammal)
(f) *Haemophilus influenzae* (eubacterium)

24. Which of the following discoveries is a result of comparative genomics?

(a) The human genome contains a large number of psuedogenes.
(b) The human genome contains a large amount of DNA that does not encode proteins.
(c) The human genome includes nearly one million SINES and LINES.
(d) Ninety percent of human genes have counterparts in the rat genome.

Eukaryotic Genes Can Be Manipulated with Considerable Precision

25. The gene for a eukaryotic polypeptide hormone was isolated, cloned, sequenced, and overexpressed in a bacterium. After the polypeptide was purified from the bacterium, it failed to function when it was subjected to a bioassay in the organism from which the gene was isolated. Speculate why the recombinant DNA product was inactive.

26. How might a gene that had a segment of DNA inserted into its interior by homologous recombination provide information about the product of the uninterrupted, normal gene?

27. Which of the following is true of C_T in qPCR?

(a) It is cycle number at which the fluorescence becomes detectable.
(b) It is directly proportional to the number of copies of the original template.
(c) It is the fluorescence detection threshold.
(d) All of the above are true.

28. Which of the following treatments of a yeast cell results in the largest changes in gene expression as monitored by microarry analysis?

(a) A heat shock treatment at 37°C
(b) Nitrogen depletion
(c) Amino acid starvation
(d) All three affect gene expression equally.

29. List some of the biochemical pathways that have been implicated in the cellular response to mutant, toxic forms of super oxide dismutase. Do these mutations alter the activity of the enzyme?

ANSWERS TO SELF-TEST

1. d. It has two-fold rotational symmetry; that is, the top strand, 5'-ACGT-3', has the same sequence as the bottom strand. Many restriction enzymes recognize and cut such palindromic sequences.

2. d. Ethidium bromide intercalates into the DNA double strand, and its quantum yield of fluorescence consequently increases. Upon uv irradiation, it fluoresces with an intense orange color wherever DNA is present in the gel.

3. c

4. a, d. The reaction of the oligonucleotide with these reagents would yield [5'-^{32}P] d(pGAATTCC).

5. (a) . . .the newly synthesized ddNMP terminus lacks the requisite 3'-hydroxyl onto which the next dNMP residue would add.
   (b) . . .the colors of the fluorescence tags on the chain terminators.
   (c) . . .it anneals to the vector that the gene has been cloned into.
   (d) . . .it avoids the use of radioisotopes and allows the automated detection of the terminated primers.

6. All are correct.

7. a, b, d, and e. Answer (c) is not correct, because the blocking groups must be removed differentially; for example, the dimethoxytrityl group must be removed from the 5'-hydroxyl (so that the next condensation with an incoming nucleotide can occur) without removing the blocking groups on the exocyclic amines of the bases. Answer (e) is correct because, although the synthesis can be carried out manually, it is slow and laborious.

8. (a) 2   (b) 1   (c) 3

9. The sequence of steps is 3, 2, and 1. Heating to 95°C completely separates all the double-strand DNA molecules. The subsequent rapid cooling to 54°C causes the excess primers to hybridize to the complementary parent DNA strands. Then at 72°C the Taq DNA polymerase (which retains its activity after the 95°C step) carries out DNA synthesis using the four dNTPs. Amplification of DNA is achieved by repeating these steps many times.

10. a, c, d, e.

11. b, c. (a) is incorrect because 5'-phosphate is required on one of the strands at the joining site. (d) is incorrect because the dNTPs are not substrates for DNA ligase. If the two strands, after annealing onto a complementary splint strand, had a gap between their ends, they could not join. If DNA polymerase and the dNTPs were added to this structure they could be used to convert the gapped, duplex DNA into a productive substrate for DNA ligase.

12. Answers (a), (b), (c), and (d) are desired characteristics. Autonomous replication allows amplification of the vector in the absence of extensive host cell growth. Unique restriction sites allow the cutting of vectors at single, specific sites for the insertion of the foreign DNA. Antibiotic resistance allows for the selection of those bacteria that carry the vector or for insertional inactivation. Small size allows the insertion of long pieces of foreign DNA without interfering with the introduction of the recombinant molecule into the host bacterium. Answer (e) is not an essential characteristic because vectors do not have to be circular to function effectively (e.g., bacteriophage λ).

13. b, d. Statements (a) and (e) are incorrect because very large DNA fragments do not fit into a λ capsid and are too big to be separated easily by standard polyacrylamide gel electrophoresis. Instead, large DNA fragments can be separated by pulsed field electrophoresis. Answer (c) is incorrect because linkers and adapters can be used in cloning DNAs with noncohesive ends.

14. c, d, e, f. The oligonucleotide must have a 5′-phosphate group to serve as a substrate for DNA ligase. Therefore, ATP and polynucleotide kinase would be used, as would bacteriophage T4 DNA ligase and ATP, to join the duplex form of the palindromic (self-complementary) oligonucleotide to the blunt-end fragment. Finally, the fragment with the linker covalently joined to it would be cut with EcoRI endonuclease to produce cohesive ends that match those of the cut vector. It is assumed that the fragment itself lacks EcoRI sites because, if one or more internal EcoRI sites were present, the fragment would be cut when it is treated with the enzyme to generate the cohesive ends. Such fragmentation of the DNA would complicate the joining reaction by forming product with various combinations of EcoRI-joined ends.

15. b, c, d. Insertion of the DNA fragment into the gene disrupts the production by the gene of the enzyme that confers drug resistance. A cell containing the altered vector is therefore sensitive to the antibiotic. When the vector contains a gene that confers resistance to a second drug, insertional inactivation can be incorporated into a selection scheme for isolating cells that contain vectors having the inserted fragment. Cells that remain resistant to the second antibiotic, while sensitive to the first, probably contain the vector bearing the foreign DNA.

16. A genomic library is composed of a collection of clones, each of which contains a fragment of DNA from the target organism. The entire collection should contain all the sequences present in the genome of the target organism. A cDNA library is composed of a collection of clones that contain the sequences present in the mRNA of the target organism from which the mRNA was isolated. A cDNA library contains far fewer clones than does a genomic library because only a small fraction of the genome is being transcribed into mRNA at any given time. The content of a cDNA library depends on the cells from which the mRNA was isolated. The type of cell, its state of development, and environmental factors influence the identity and quantity of its mRNA population.

17. b. This amino acid sequence is the better choice for reverse translation into a DNA sequence because it contains fewer amino acid residues having multiple codons; Trp and Met have one codon each and Cys has two. Thus, for the (b) sequence, Met-Trp-Cys-Trp, there are $1 \times 1 \times 2 \times 1 = 2$ different dodecameric oligonucleotide coding sequences. In contrast, Leu and Arg each have six codons, so for the (a) sequence, Met-Leu-Arg-Leu, there are $1 \times 6 \times 6 \times 6 = 216$ different coding sequences. Therefore, the probe for (b) would be simpler to construct and would be more likely to give unambiguous hybridization results.

18. In eukaryotes the introns are removed from the primary transcript by processing, to produce the mRNA that is translated. Prokaryotes lack the machinery to perform this processing; consequently, the translation product of the primary transcript would not be functional because it would encode amino acid sequences that are specified by the intron sequences. The problem can be circumvented by using cDNA prepared from the mRNA from the gene encoding the protein; the cDNA will contain only the sequences present in the processed RNA; that is, the intron sequences will have been removed.

19. e. An immunochemical screen could be performed by adding the radioactive antibody to lysed bacterial colonies and examining the population by autoradiography to see which colonies contain the antigen (protein) produced by the cloned gene.

20. b, d, e. Since reverse transcriptase makes DNA, it requires dNTPs not NTPs. RNA is required as a template to direct the synthesis of a complementary DNA strand. That DNA strand itself then serves as a template for the synthesis of its complement to form the duplex DNA product. All polymerases that form DNA need a primer to start the synthesis of a new DNA chain.

21. b, d, e. If the sequence of the oligonucleotide were completely different from the sequence of the target gene, it could not hybridize to the target gene and serve as a primer for DNA polymerase even under low-stringency hybridization conditions. Functional domains could be better identified by deletion mutagenesis, in which relatively large regions of the gene would be systematically removed and the resulting functional consequences tested. Although oligonucleotide-directed mutagenesis can be used to make deletions, it is not the method of choice for an initial survey to find functional domains because only a single, precisely defined deletion is produced with each oligonucleotide. For exploratory deletion analysis, nucleases are used to generate populations of deleted sequences for functional testing. Oligonucleotide-directed mutagenesis is better suited for changing specific regions when one wishes to test a specific model or hypothesis regarding the function of one or a few amino acids. At the correct conditions of hybridization stringency, the mutagenizing oligonucleotide will form a more stable hybrid with the newly produced mutant sequence than with the original unmodified sequence because it will form a perfect complement. Thus, it can be used to differentiate the mutant and the original sequences.

22. You would need to know the sequence of the gene you wish to synthesize. This could be derived from the amino acid sequence of the protein the gene encodes by reverse translation using the genetic code. You would also need to know what restriction sites you wish to build into the synthetic sequence for cloning the synthetic product. You would have to decide on the individual sequences of the different oligonucleotides that compose both strands of the gene. These sequences would be determined by the final desired sequence, the individual lengths (30 to 80 nucleotides long) that can be easily synthesized and purified, and the requirement for overlapping ends that will be necessary to allow unique joinings of the cohesive ends of the partially duplex segments. Self-complementary oligomers would be mixed together to form duplex fragments with cohesive ends. DNA ligase and ATP would be added to join these together to form the complete duplex. If appropriate ends have been designed into the synthesis, the product can then be ligated into a vector for cloning.

23. c, b, f, a, d, e. The order is from smallest genome to largest. The smaller genomes were sequenced first because they were easier to complete.

24. d. Although all statements are true, only the discovery that 90% of the human genome has a rat counterpart is a result of comparative genomics.

25. Omitting such an obvious explanation as the destruction of the polypeptide during the bioassay, it is possible that the polypeptide might not have undergone some posttranslational modification that is needed for it to function. For example, the polypeptide might need to be acetylated, methylated, or trimmed at the N- or C-terminus, or it might need to have a carbohydrate or lipid group attached to it. The bacterium in which it was produced would be unlikely to contain the enzymatic machinery necessary to carry out these modifications, or if it did, it might lack the ability to recognize the eukaryotic signals that direct these modifications. It is also possible that the bacterium might have contained a peptidase or protease that inactivated the peptide without destroying its antigenic properties.

26. Inserting a segment of DNA into the middle of a gene almost always disrupts the normal functioning of that gene. Examining the properties of an organism that survived with the disrupted or knocked-out gene might provide clues to what the normal gene did. For instance, if a cell contained a gene knockout, and one observed that the cell lacked a certain cell-surface protein, one could tentatively conclude that the gene had something to do with the production of that surface protein. Many other experiments would be needed to fully reveal the relationship of the disrupted gene to the protein.

27. a. C_T is determined by amplfying a standard and is the cycle number in which the fluorescence threshold is reached. It is inversely proportional to the number of copies of the original cDNA template.

28. b. In a microarray analysis, each gene in the sample is represented by an individual square. A black color indicates that expression of the gene was not affected by the conditions of the study relative to the control. Red indicates induction, or an increase in gene's expression relative to the control and green indicates repression, or a decrease. A large number of squares that are either red or green indicates a large number of genes have had their expression perturbed by the stressor. Using Figure 5.21 in the text, you can see that nitrogen depletion in yeast results in widespread increases (red) and decreases (green) in gene expression. Although there are some changes in the heat shock and amino acid starvation arrays, there are fewer colored squares in both relative to the nitrogen depletion sample.

29. Studies have implicated immunological activation, handling of oxidative stress and protein degradation in the cellular response to mutant, toxic forms of the SODI gene. The mutations did not lead to an alteration of the enzymatic activity of SOD, leading to one proposal that mutant SODI is prone to form toxic aggregates in the cytoplasm of neuronal cells.

PROBLEMS

1. You are studying a newly isolated bacterial restriction enzyme that cleaves double-strand circles of plasmid pBR322 once to yield unit-length, linear, double-strand molecules. After these molecules are denatured and are allowed to reanneal, all the double-strand molecules are unit-length linears. In another experiment, an enzyme that cleaves double-strand DNA at random sites is used at low concentration to cleave intact pBR322 molecules approximately once per molecule, again yielding unit-length, double-strand linear DNA molecules. Denaturation and renaturation yields some double-strand circles with a single, randomly located nick in each strand. How do these experiments show that the new restriction enzyme cleaves pBR322 DNA at a single specific site?

2. Before the development of modern methods for the analysis and manipulation of genes, many attempts were made to transform both prokaryotic and eukaryotic cells with DNA. Most of these experiments were unsuccessful. Suggest why these early efforts to transform cells largely failed.

3. Pseudogenes are composed of nonfunctional (unexpressed) DNA sequences that are related by sequence similarity to actively expressed genes. Some researchers have proposed that pseudogenes are copies of functional genes that have been inactivated during genome evolution. Suggest several ways that such genes could have become nonfunctional. Suppose you clone a number of closely related sequences, any of which may code for a particular protein. How can you tell which of the sequences is the functional gene, that is, which of the sequences codes for the protein?

4. The Sanger dideoxy method for determining DNA sequence is limited in that a stretch of only 1000 or fewer bases can be analyzed in one reaction. Suppose you wish to sequence a newly isolated double-strand DNA tumor virus that contains ~5000 base pairs. You decide to use the Sanger method on restriction fragments of the DNA for sequencing. You use an enzyme that makes a significant number of cuts to give, on average, fragments of ~275 nucleotides or less. Why might it be a good idea also to sequence a second set of fragments cleaved by another restriction enzyme?

5. The denaturation and reassociation of complementary DNA strands can be used as a tool for genetic analysis. Heating double-strand DNA in a dilute solution of sodium chloride or increasing the pH of the solution above 11 causes dissociation of the complementary strands. When the solution of single-strand molecules is cooled or when the pH is lowered, the complementary strands will reanneal as complementary base pairs reform. What causes base pairs to dissociate at pH 11 or higher?

   Both double- and single-strand DNA molecules can be visualized using electron microscopy in a technique called heteroduplex analysis. Suppose that two types of double-strand molecules, one type containing the sequence for a single gene and the other type containing the same sequence as well as an insertion of nonhomologous DNA, are mixed and used in a reannealing experiment. If the two types of molecules undergo denaturation and reannealing, what types of molecules would you expect to see?

6. Bacterial chromosome deletions of more than 50 base pairs can be detected by electron microscopy, using heteroduplex analysis as described in problem 5. When a heteroduplex is formed between a single-strand DNA molecule from a deletion strain and a single-strand molecule from a nondeletion, or wild-type, strain, a single-strand loop will be visible at the location of the deletion. Suppose you are studying a bacterial mutation, which appears to be a deletion of about 200 base pairs, located at a unique site on the bacterial chromosome, which contains over 3000 genes. Why would it be a good idea to clone DNA containing the site of the deletion, as well as the corresponding site in the wild-type strain, in order to study the deletion using heteroduplex analysis?

7. You wish to clone a yeast gene in λ phage. Why is it desirable to cleave both the yeast DNA and the λ-phage DNA with the same restriction enzyme?

8. Suppose you are studying the structure of a protein that contains a proline residue, and you wish to determine whether the substitution of a glycine residue will change the conformation of the polypeptide. You have cloned the gene for the protein, and you know the sequence of the protein. Using site-specific mutagenesis, what alterations would you make in the gene sequence in order to replace proline with glycine?

9. Cleavage of a double-strand DNA fragment that contains 500 bases with restriction enzyme A yields two unique fragments, one 100 bases and the other 400 bases in length. Cleavage of the DNA fragment with restriction enzyme B yields three fragments, two containing 150 nucleotides and one containing 200 nucleotides. When the 500-base fragment is incubated with both enzymes (this is called a double-digest), two fragments 100 bases in length and two 150 bases in length are found. Diagram the 500-base fragment, showing the cleavage sites of both enzymes. Now suppose you also have a double-strand DNA fragment that is identical with the original fragment, except that the first 75 base pairs at the left end are deleted. How can this fragment help you construct a cleavage map for the two enzymes?

10. Problem 2 of Chapter 5 in the text refers to the expression of a eukaryotic gene—chicken ovalbumin—in E. coli. To avoid transcribing and translating intron sequences, you should

use cDNA for protein expression. However, if you introduce only the chicken ovalbumin cDNA into bacteria, the level of expression of functional protein will likely be low. What other sequences are necessary in order to ensure optimal expression?

11. Because PCR can amplify DNA templates one millionfold or more, contaminating DNA must not be present in the sample to be used for amplification. To see why, consider a PCR procedure that begins with 1 μg DNA (about 10^6 templates) in a reaction mixture of 100 μL. This sample can be easily amplified about one millionfold in 20 cycles (an amplification of 2^{20}). Suppose that 0.1 μL of DNA from the initial amplification cycle is inadvertently introduced into another reaction mixture containing 1 μg of a different DNA. Could the contaminant cause problems with PCR analysis of the second sample? Why?

12. One method of analysis of evidence from cases of sexual assault often includes histocompatability locus antigen (HLA) type analysis using PCR. Samples collected from a victim may contain not only sperm but also epithelial cells from the victim. Such samples are first incubated in a protease-detergent mixture. The epithelial cells are lysed, while the sperm heads are not. The sperm heads are collected by centrifugation and then washed several times. They are then lysed in the presence of a reducing agent such as dithiothreitol, which makes sperm heads sensitive to the protease-detergent mixture. Lysis products are then used for PCR analysis. Why is it necessary to carry out separation of sperm and epithelial cells? Why are cells and sperm heads lysed before PCR analysis? Suppose that blood and hair samples are also found as evidence at the scene of the alleged crime. Why should precautions be taken to keep these samples isolated from each other?

13. Unlike DNA polymerase I from *E. coli,* DNA polymerase I from the bacterium *T. aquaticus* has no proofreading activity and is therefore unable to remove mismatched bases that are randomly incorporated into newly synthesized DNA strands. Under standard conditions used for the polymerase chain reaction, misincorporation of nucleotides occurs at a frequency of approximately 1 per 900 nucleotide residues in DNA.

    (a) Suppose that you are using PCR to detect copies of an oncogene in a tissue sample. You will challenge the amplified sample with a radioactive probe containing the oncogene sequence, using Southern blot analysis. Will low-frequency, random misincorporation of bases during amplification of the oncogene interfere with your analysis?

    (b) Suppose you are using PCR with a mutant primer, that is, a primer with a sequence differing from the wild-type sequence, to introduce a deliberate alteration in a eukaryotic gene. You plan to use the amplified mutant gene for cloning and expression in *E. coli* to determine how the directed change affects the expressed protein. How might misincorporation level in the amplification procedure interfere with your cloning and expression experiments? What could you do to solve the problem?

14. *E. coli* DNA polymerase I has a $5' \rightarrow 3'$ polymerase activity and also two other catalytic activities. One is a $3' \rightarrow 5'$ exonuclease activity whose function is to remove from the growing 3' end of the chain those mismatched bases occasionally incorporated erroneously by the polymerase activity. The other catalytic activity is a $5' \rightarrow 3'$ exonuclease, which removes both paired and mispaired DNA stretches ahead of the polymerase. If the $5' \rightarrow 3'$ exonuclease acts concomitantly with the polymerase in the same enzyme, new nucleotides are incorporated by the polymerase in place of the ones removed by the nuclease, and the nick is essentially "translated," that is, moved, in the $5' \rightarrow 3'$ direction. This nick-translating ability of DNA polymerase I has been exploited to create radioactive DNA probes for use in Southern blots and other techniques. Describe how nick translation could be used for such purposes.

15. Site-directed mutagenesis allows one to introduce virtually any desired mutation in a specific gene. One very useful application of the technique involves modifying a particular protein and evaluating the effect on biological or chemical activity. In the past, two methods have been used for producing modified proteins. One is to use chemical agents or ultraviolet light to induce mutations that result in changes in the amino acid sequence of proteins. The other is to modify certain residues in an isolated protein by treatment with chemical reagents; an example is the inactivation of a reactive serine in the active site of proteolytic enzymes like chymotrypsin, using diisopropylfluorophosphate.

   (a) Why is site-specific mutagenesis superior to the two older procedures described above?
   (b) In order to carry out the modification of a protein using site-specific mutagenesis in the most efficient way, what sort of information should you have about the protein you wish to modify?

16. Patients with a particular form of hemophilia (a deficiency in blood clotting) have a loss of an EcoRI restriction site within the gene for a coagulation factor protein. In one family with an affected son, PCR analysis was carried out on 200 μL blood samples from a male fetus and from several family members to determine whether the fetus also carried the mutation. Using appropriate primer oligonucleotides, DNA fragments 150 bp in length and spanning the EcoRI polymorphic site in intron 10 were synthesized. These fragments were incubated with EcoRI, and the resulting cleavage fragments were then separated by electrophoresis on a gel and stained with ethidium bromide. A diagram of the gel is shown in Figure 5.1, along with the source of the blood sample for each lane. Note that any 150-bp fragment that contains the EcoRI site will be cut by the enzyme into two fragments, 100 bp and 50 bp in length.

FIGURE 5.1

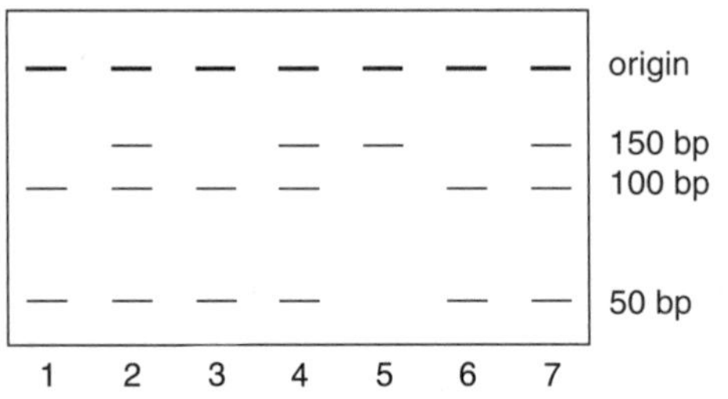

Lane	Source
1	Father
2	Mother
3	Fetus
4	Unaffected daughter
5	Affected son
6	Normal male control
7	Heterozygous control

   (a) Specify the genotype for each member of the family and for the controls. Does the fetus carry the mutation?
   (b) This analysis can also be done by using Southern blotting to detect single-copy sequences in genomic DNA. Why is the PCR analysis preferable?

17. Base pairing by hydrogen bond formation between complementary bases is a fundamental feature of many processes described in Chapter 4 of the text. Discuss the role of base pairing in the context of each of the following:

   (a) the fidelity of messenger RNA synthesis
   (b) synthesis of cDNA by reverse transcriptase
   (c) the use of primers in the polymerase chain reaction (PCR)
   (d) identifying a desired clone using a radioactive DNA probe
   (e) measuring the relatedness of two DNA species without sequencing them

18. DNA microarray or chip technology allows one to monitor simultaneously the level of mRNA production from every gene in a bacterium. Why might such an analysis of a microbe not give an accurate estimate of the levels of the proteins in the microbe?

19. The plasmid pBR322, a double-strand circular DNA molecule containing ~4.4 kilobase pairs, is commonly used in cloning experiments. A technician in a molecular biology laboratory needs to prepare a large quantity of pBR322 by growing a liter culture of *E. coli* containing the plasmid and then isolating the pBR322 DNA.

   (a) How many milligrams of plasmid DNA can be prepared from a liter of bacterial cells growing at a density of 10^8 cells per ml? Assume that each cell contains 100 plasmid molecules and that the molecular weight of the average base pair in the plasmid is ~660.

   (b) If the technician decides to use a nanogram of pBR322 as the template in a PCR experiment, how many templates will be present in the reaction mixture?

20. You have isolated cDNAs containing the genes encoding malarial proteins with the aim of developing an anti-malarial vaccine. How could you use these cDNAs to direct the efficient synthesis of their encoded proteins in an *in vitro* translation system in order to study their antigenic properties? Be sure to consider the entire information flow pathway.

21. You wish to use the restriction enzyme HhaI, which hydrolyzes the duplex sequence GCGC between the last G and C, to cut a large double-strand plasmid DNA (several kbp) at the single site operator site where a repressor protein binds very tightly to it. You know the site contains one HhaI site. Unfortunately, the rest of the DNA contains 31 HhaI site in its sequence. Considering what you learned about restriction endonucleases and modification methyl transferases (methylases), can you devise a method that would allow you to achieve the desired, unique cut in the DNA without fragmenting it elsewhere? You have at your disposal the DNA, repressor protein, and the HhaI restriction and modification enzymes. The HhaI DNA methylase adds a methyl group to the second C of the GCGC recognition sequence.

22. If you have access to the genomic library of an unkown eukaryote and you know the sequence from a segment of its genomic DNA, devise a method to determine the sequences of the adjacent genes.

23. If you wanted to make a human DNA library with the minimum number of clones required for complete coverage, why wouldn't you insert random human DNA fragments into the vector bacteriophage λ?

24. How would you use a synthetic oligodeoxyribonucleotide (ODN) to make an insertion of 20 bp in a protein-encoding bacterial gene of known sequence? Assume you make an ODN of 50 nucleotides long and that the gene is cloned in the vector of your choice.

25. You have sequenced the genome of a new, previously unknown, bacterium. How might you go about identifying the function of a particular open reading frame encoded by a gene in your bacterium? Reminder: an open reading frame is a gene sequence that contains an initiator codon and a terminator codon and likely encodes a protein product.

ANSWERS TO PROBLEMS

1. Cleavage of a circular molecule at one specific site, followed by denaturation, will yield single-strand DNA molecules with a specific end-to-end base sequence; that is, the molecules have base sequences that are perfectly complementary. Such molecules will anneal to form double-strand linears, rather than circles. Random single cleavages of the original intact molecules also yield double-strand linears with a variety of end-to-end (or permuted) sequences. Denaturation and renaturation allow the random association of these linears, which results in the formation of double-strand linears with overlapping, complementary ends. Such molecules then form circles as their overlapping ends anneal.

2. During the early years of such experiments, few ways were available to determine what happened to the DNA during transformation attempts, so specific remedies could not be sought. Consequently, the fate of the test DNA could not be determined. Among the reasons that these transformation attempts were not successful were the failure of the cells to take up the DNA, the rapid degradation of the DNA inside the cell (restriction enzymes in bacteria are a good example of a cause of this particular problem), the lack of accurate transcription or translation, and the inability of the host cells to replicate and maintain the foreign DNA as they divided.

3. Among the ways that a gene could be inactivated are the insertion of a stop codon in the sequence, which would prevent the complete translation of the protein; a mutation in the promoter region of the gene, which would prevent proper transcription; and other mutations that could prevent proper splicing or processing. To distinguish a functional gene from a pseudogene, you would have to determine the sequence of the protein and then compare it with the coding sequence for each of the gene sequences. These types of analyses remind us that protein sequencing remains a very necessary tool in molecular biology.

4. Whenever one attempts, using gel electrophoresis, to locate all the fragments produced by a particular enzyme, a chance exists that very small fragments generated by the cleavages may not be detected. Determining the sequences of a second set of fragments whose sequences extend across the junctions of the original set of fragments serves as a check on the overall assignment of sequence.

5. At high pH, protons dissociate from some of the bases, making them unable to participate in base pairing. One example is guanine, for which the pK_a of the proton on N-1 is 9.2. Removal of the hydrogen at this location disrupts the ability of guanine to pair with cytosine (see Chapter 1 in the text).

   If you mix the two types of double-strand molecules, you would expect to see linear molecules that are double-strand all along their length as well as some molecules that are only partially double-strand. These partially double-strand molecules will contain a single-strand loop that locates the position of the insertion; they are formed between one strand of the molecule containing the normal gene and one strand of the molecule containing the insertion.

6. Even if you were able to isolate intact, unbroken bacterial chromosomes, formation of intact heteroduplex molecules between the deletion and wild-type DNAs is difficult because the very long single strands become entangled as they pair with each other, making them impossible to analyze by electron microscopy. In addition, the time required for complete reassociation of the strands is very long. Generating shorter, randomly cleaved DNA fragments for heteroduplex analysis permits faster reassociation and easier analysis, but since the deletion is located at a single unique site in the chromosome, the probability of finding the desired molecule among the mixture of many heteroduplex molecules is rather low. Cloning DNA molecules containing the deletion or its corresponding wild-type sequence allows you to carry out reannealing experiments that yield a high concentration of heteroduplex molecules with the loop characteristic of deletion mutations.

7. To insert the yeast gene into the λ-phage vector, you must have complementary base pairs on the ends of each duplex in order for them to be joined efficiently by DNA ligase. Because each restriction enzyme cleaves at a unique sequence, the yeast and λ-phage molecules will have complementary ends if both have been cleaved with the same enzyme. Of course, you must also make sure that the sites of cleavage are in appropriate places so that the gene to be cloned is intact, and that the vector or fragment has not been fragmented by multiple cleavages.

8. The RNA codon for proline is 5′-CCN-3′ and the codon for glycine is 5′-GGN-3′, where N is any base. Suppose you determine that the proper codon for your protein is 5′-CCC-3′. Using the scheme outlined in Figure 5.20 of the text, you would prepare an oligodeoxyribonucleotide primer that is complementary to the region of the gene that specifies the proline residue, except that it would contain the DNA sequence 5′-CCC-3′ instead of 5′-GGG-3′. Elongation of the primer using DNA polymerase, followed by closure and replication, will yield progeny plasmids that will express a protein with a glycine substitution at the desired position.

9. The 500-base fragment has one site that is cleaved by enzyme A. This cleavage yields two fragments with two possible sets of products:

FIGURE 5.2

Enzyme B cleaves the 500-base molecule twice, so there are three possible cleavage patterns:

FIGURE 5.3

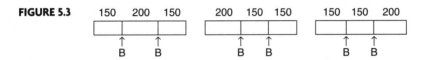

Since we cannot distinguish between ends of the molecule by this type of analysis, let us arbitrarily assume that enzyme A cuts the molecule of 100 nucleotides from the left end. We can then superimpose the possible cleavage patterns for enzyme B:

FIGURE 5.4

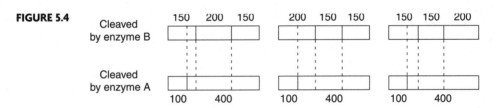

Only one of the patterns for enzyme B, that in which a cut occurs 200 bases from the left end, yields the results obtained when the fragment is incubated with both enzymes. The other patterns would yield at least one 50-base fragment. The correct pattern is therefore:

FIGURE 5.5

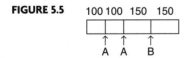

Because we still cannot distinguish between the right- and left-hand ends of the molecule, an alternative cleavage pattern can also be constructed from the analysis outlined above:

FIGURE 5.6

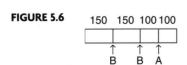

The deletion fragment serves as a marker for the left-hand end of the molecule, and using both enzymes in the double-digest technique allows us to establish which cleavage pattern is correct. For example, if the cleavage pattern shown below (Figure 5.7) on the left is correct, the cleaved deletion molecule will yield four fragments, including one only 25 nucleotides in length. Alternatively, the pattern shown below on the right means that double digestion of the deletion fragment will again yield four fragments, but the smallest will have a length of 75 nucleotides.

FIGURE 5.7

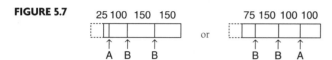

The complexity of cleavage patterns (known as restriction maps) increases greatly when additional cleavages are involved. Often the best way to use the double-digest technique is to isolate the fragments generated by one enzyme and then digest each of them with the other. This allows you to determine the location of different cleavage sites within a particular fragment. In restriction mapping, as in genetic mapping, it is important to remember that the sum of the fragment lengths generated by one enzyme must equal the sum of the fragment lengths generated by the other.

10. In order to obtain optimal expression in *E. coli*, you should have prokaryotic DNA sequences that include the appropriate transcriptional and translational signal elements. For example, in Chapter 4 of the text, promoter sites that determine where transcription begins are mentioned; these include the Pribnow box and the −35 region, both of which would be required to initiate efficient transcription of your cDNA clone by the bacterial RNA polymerase. You may also need the stem-loop and GC-rich terminator sequence at the 3′-end of your cDNA, in order to cause the nascent messenger RNA to terminate at the correct site. In addition, you should see that the Shine-Dalgarno ribosome recognition sequence and the proper start and stop signals for translation are also present, so that the mRNA code is read in the proper frame and that proper termination occurs. These signals ensure that the expressed protein has the proper amino acid sequence and is the correct length. The desired bacterial signals can be built into a vector so that only the cDNA itself need be cloned.

11. It is indeed possible that the contaminant could complicate the analysis of the second sample. If the 10^6 templates in the original sample are amplified one millionfold, then the concentration of templates at the completion of 20 cycles is 10^{12}/100 μL, or 10^{10} templates per μL. A contaminating volume of 0.1 μL will therefore contain 10^9 templates, compared with about 10^6 templates in the second sample. Such contamination could mask the identity of the DNA in the analyte sample. In practice, a number of precautions are taken to avoid introduction of foreign DNA into PCR reaction mixtures. These include use of sterile containers and reaction solutions; disposable gloves; laminar flow hoods; and separate work areas for preparing reaction mixtures, pipetting template samples, and analyzing the products. For the PCR reaction itself, it is important to run reactions that contain no added DNA in order to check for contamination with DNA from sources other than the solution containing the analyte.

12. In forensic PCR analysis, it is common to use cells from the victim as well as sperm from the alleged rapist to generate DNA templates for amplification and analysis. The differential lysis procedure allows the two to be efficiently separated, to avoid cross-contamination of DNA samples. Lysis is necessary so that the DNA templates will be accessible to DNA polymerase and substrates for amplification. During the gathering of evidence from crime scenes, it is necessary to keep samples that could contain large

amounts of DNA, such as bloodstains, separate from those that contain little DNA, like a single shed hair. As noted in Problem 11, the presence of contaminating DNA can confound PCR analyses.

13. (a) There should be little interference if you utilize amplified sequences with some misincorporated bases in Southern blot experiments. The results of the experiments depend on the complementarity of relatively long sequences of DNA, and occasional mismatches should not interfere with the ability of the amplified templates and the probe to anneal with each other. Remember that the primers, which are present in excess during the initial cycles of amplification, continue to initiate the synthesis of new DNA strands. If the incorporation of mismatches is random, the chances are that any particular strand will have few base changes and that it could still form complementary base pairs with the radioactive probe.

    (b) In these experiments, the low frequency of base misincorporation could interfere with your analysis, because each of your clones to be used for expression experiments will be derived from a single DNA molecule obtained in the amplification process. Cloning therefore amplifies any error introduced by the polymerase in the chain reaction, and random errors could alter the expressed protein in a non-controlled way. There are two ways to deal with this problem. You can use a thermostable DNA polymerase that has a proofreading activity and improved fidelity; several of these are commercially available. On the other hand, you can also prepare a number of clones and sequence them. Such a procedure will ensure that the clone you want to examine for altered expression has the sequence you wanted to generate during the amplification process. Even if you use the proofreading polymerase, you should sequence the product to see that no unintended changes were introduced.

14. Nick translation can be used with labeled nucleoside triphosphates to generate highly radioactive probes for use in Southern blot analysis as well as for other techniques that require such labeled DNA samples. The usual procedure includes isolating the DNA you wish to use for the probe, treating it with a nuclease that will create a small number of nicks or single-strand breaks per molecule, and then incubating the DNA with DNA polymerase I and α-^{32}P-labeled deoxyribonucleoside triphosphates (the α-phosphate is labeled because it is incorporated along with the deoxynucleoside when the DNA chain is extended by DNA polymerase). The DNA molecules then will contain stretches of radioactive sequences, and they can then be used for autoradiography in Southern or Northern blot experiments.

15. (a) Treatment of an organism with a chemical or radiation-inducing mutagen does not allow you to make changes soley in a particular region of a gene and its encoded protein because the changes occur at random. Much work would be required to find a mutant organism that had the desired alteration in a protein and to characterize that alteration. Treating a purified protein with a chemical agent may modify other amino acid residues in addition to the one or more specific residues of interest. Both the older approaches are relatively nonspecific, compared with the ability of site-specific mutagenesis to target a specific region of a particular gene.

    (b) One should have at a minimum the amino acid sequence for the protein and the exact DNA sequence coding for it. More information about the protein of interest would allow for more selective mutagenesis. Examples include the location of residues involved with the active site, with allosteric interactions, or with membrane association or those involved in protein-lipid or protein-nucleic acid interactions.

16. (a) The father and the normal male control have the EcoRI site, so that their amplified DNAs are cut into two fragments. The affected son has lost the site, and his DNA is not cut. The mother, daughter, and a heterozygous control have one copy of the normal gene (giving two fragments) and one copy of the mutant gene (giving one larger fragment). The amplified fetal DNA is cleaved completely by the enzyme, which shows that the male fetus is normal.

(b) Southern blotting, used to detect single-copy sequences in genomic DNA, requires relatively large amounts of material from which DNA is isolated, and it requires a highly radioactive probe to detect a particular sequence. Autoradiography, which is often used to detect annealing of the probe to the genomic sequence, can take a long time if the signal is weak. Analysis by PCR requires only nanogram amounts of DNA from very small amounts of tissue or blood, with minimal sample preparation. In addition, one can dispense altogether with radioactive probes, instead using a DNA staining reagent such as ethidium bromide to detect amplified sequences.

17. (a) Guided by a DNA template, hydrogen bonding mediates proper insertion of mononucleotides into the growing RNA chain, in the reaction catalyzed by RNA polymerase. Complementary base pairing is the guiding principle in determining the order of ribonucleotide assembly.

(b) Guided by an RNA template, hydrogen bonding mediates proper insertion of deoxymononucleotides into the growing DNA chain, in the reaction catalyzed by reverse transcriptase. Complementary base pairing guides the order of deoxyribonucleotide assembly. Reverse transcriptase can also use cDNA as a template for the creation of a duplex DNA molecule; again, hydrogen bonding of complementary base pairs establishes the order of deoxynucleotide addition.

(c) The PCR technique requires that DNA is denatured (breaking hydrogen bonds) and then annealed to a pair of primers, whose sequences are complementary to those flanking the target sequence in DNA. DNA polymerase then carries out chain extension by adding deoxynucleotides to those primers. After denaturation of the newly formed duplexes, the strands are reannealed with the excess primers. The process is carried out as many as 30 times. For each reaction sequence, specificity of the chain reaction for a particular DNA segment is mediated by specific base pairing between primers and templates.

(d) Radioactively labeled DNA probes are denatured and allowed to anneal to a mixture of DNA molecules. The extent of hydrogen bonding determines the homology between probe and DNA. Southern blotting uses this hybridization technique for DNAs resolved by gel electrophoresis. Location of a specific DNA is visualized by autoradiography, which locates the radioactive band in the gel.

(e) The two species are denatured and reannealed together. The more duplex DNA formed upon annealing, the greater the degree of relatedness. The extent of duplex DNA formation is determined by measuring the melting temperature of the DNA, which is a reflection of the number of hydrogen bonds between strands.

18. Levels of mRNA are not necessarily correlated with protein production. For instance, translational control might not allow an abundant mRNA to direct the synthesis of the protein it encodes. If one is interested in proteins, they have to be measured directly. Proteomics is the science of examining protein levels on a global level in an organism.

19. (a) First determine how many plasmid molecules are present in the bacterial cell culture: 10^8 cells/ml = 10^{11} cells/L, and 10^{11} cells/L × 100 plasmids = 10^{13} plas-

mid molecules in the culture. Then use the molecular weight of a base pair (~660 g/mol bp) in the plasmid, the length of the plasmid in base pairs (4.4×10^3 bp), and Avogadro's number to determine the mass of 10^{13} plasmid molecules.

$$\frac{10^{13}\,\text{plasmids} \times 4.4 \times 10^3\,\text{bp}/\text{plasmid} \times 6.6 \times 10^2\,\text{g}/\text{mol bp}}{6.023 \times 10^{23}\,\text{bp}/\text{mol}} = 4.8 \times 10^{-5}\,\text{g},$$

or 0.048 mg pBR322 DNA

(b) A nanogram of DNA equals 10^{-9} g. Divide this quantity by the molecular weight of a base pair to obtain the number of moles of DNA base pairs, and then multiply by Avogadro's number to determine how many molecules are present in the reaction mixture.

$$10^{-9}\,\text{g DNA}/6.6 \times 10^2\,\text{g/mol bp} = 1.51 \times 10^{-12}\,\text{mol DNA bp}$$

1.51×10^{-12} mol bp $\times$ 6.023×10^{23} molecules/mol $= 9.1 \times 10^{11}$ bp in the reaction mixture

$$\frac{9.1 \times 10^{11}\,\text{bp}}{4.4 \times 10^3\,\text{bp}/\text{plasmid}} = 2.1 \times 10^8\,\text{plasmids}$$

20. You could use PCR to isolate the DNA from the plasmids in which it had been cloned (or from the genome itself). To obtain maximal amounts of mRNA and optimal *in vitro* translation, you could design oligodeoxyribonucleotide primers (universal promoter primers) containing the signals necessary for efficient transcription and translation. Transcription of the amplified product would yield large amounts of mRNA customized for the chosen translation system. The primers could contain a promoter for bacteriophage T7 RNA polymerase, for which conditions have been developed that allow production of large amounts of transcript. In addition, an optimized upstream, untranslated region could be designed into the primers to produce an mRNA that contained optimal sequences for *in vitro* translation in the system of choice. Factors to consider would be the potential secondary structure of the mRNA (stem-loops inhibit translation), sequences preferred by the ribosomes, and the spacings between the various elements including the location of the start codon itself. This problem was derived from K. C. Kain, D. E. Lanar, and P. A. Orlandi. Universal promoter for gene expression without cloning. *Biotechniques* 10(1991):366–374.

21. You could take advantage of the fact that the tight binding of the repressor molecule to its operator DNA sequence prevents the action of enzymes at the sequence covered by the protein. Since two proteins cannot be in the same place on the DNA at the same time, one tightly bound protein prevents the binding of the other. Repressor binding thus would prevent HhaI endonuclease cleavage or methylase methylation. You could exploit this effect by binding the repressor to the DNA, treating the specific protein-DNA complex with HhaI methylase to methylate all the 30 remaining, uncovered GCGC sites. Then you would remove the repressor and treat the naked DNA with the endonuclease. The site covered by the repressor would be unmethylated and subject to cleavage. The result would be that the DNA would be cleaved uniquely at the one HhaI site that had been protected from methylation by having the repressor bound to it. This problem was based on M. Koob, E. Grimes, and W. Szybalski. Conferring operator specificity on restriction endonucleases. *Science* 241(1988):1084–1086.

22. You could synthesize oligodeoxyribonucleotide primers that were complementary to the sequences at the 3′-ends of the two strands of the known DNA. These could then be used as sequencing primers with subcloned pieces of DNA from a DNA library of the organism. Using the Sanger sequencing technique you could sequence the DNA that was downstream, in other words, 3′ to each of the strands of known sequence. You could repeat this procedure with new sequencing primers based on the newly determined sequence, and in that way, "walk" along the chomosome determining more and more sequence that was adjacent to the original DNA.

23. Bacteriophage λ, even those strains constructed especially to carry large amounts of DNA, can accept foreign DNA fragments of only ~45 kb in length. A YAC vector could take fragments ~1000 kb in length and thus ~20X fewer clones would be required to carry the human DNA. Furthermore, because of sampling problems and overlap, the actual number of clones required to have a >99% chance of covering the entire genome is much larger than the number derived by dividing the genome size by the fragment size carried by the vector.

24. You would design the sequence of the ODN so that sequences of 15 nt at its 3′ end and at its 5′ end were exactly complementary to the sequences immediately surrounding the site (15 nt on each side of the site) at which you wanted to make the insertion in the gene. The interior nucleotides (nt) of the ODN would form a loop when it was hybridized to the single strand of the gene that had been cloned into an M13 vector. Using M13 vector is important because it provides you with a single strand of the gene. When this circluar M13 strand with the ODN primer containing a looped-out interior bulge is extended with dNTPs and DNA polymerase in the presence of DNA ligase and ATP, the product would be a double-strand, covalently closed replicative form of the vector, one strand of which contained an extra 20 nucleotides. When this replicative form of the vector is transformed into a bacterium and replicated, daughter molecules containing the gene with an inserted 20 nucleotides will be produced. They will be mixed with wild-type replicated molecules lacking the insert, and would have to be purified from them. Drawing the single-strand circular M13 vector with the gene and the annealed ODN will aid in understanding how this insertion is accomplished.

25. You would use a computer and readily available software to compare the sequence of your gene to the known sequences from all genes from all the organisms and genes sequenced and placed into the sequence database. You might find that many organisms have sequences that are more or less similar to your new sequence. If you were lucky, earlier studies with one of the related organisms will have assigned a function to one of these evolutionary gene relatives by showing that it encodes a particular protein with a particular activity. You could then assay for the protein in your organism to see if the activity was present. If so, you could knock out the gene, and see if the activity was absent to further show that your gene encoded that protein. If you are unlucky, you face the arduous task of identifying the function of your gene. Consider how you might do this. How might doing a gene knockout help?

Chapter 6

Exploring Evolution and Bioinformatics

I n this chapter you will learn specific facts about what relationships can be seen between different genes and proteins and between different organisms. Related proteins (from related genes) are called homologous. Related proteins used for different tasks within an organism are paralogous, and related proteins used for similar tasks in different species are orthologous. Proteins can be shown to be related by shared sequences and by visible similarities in three-dimensional structure. Various tests are described for relatedness including sequence shuffling and substitution matrices. Truly related proteins generally have similar three-dimensional structures even if there is hardly any similarity in the sequences. Proteins cannot be considered to be related simply because they have similar functions or similar mechanisms. Convergent evolution can produce different proteins that function similarly. Sequence information can be used to construct evolutionary trees. It is also easy to detect repeated domains within a protein using sequence analysis. Modern methods including PCR can allow recovery of sequence information from certain fossils. Evolution of RNA sequences can be observed in vitro.

LEARNING OBJECTIVES

When you have mastered this chapter, you should be able to accomplish the following objectives.

Homologs Are Descended from a Common Ancestor (Text Section 6.1)

1. Define *homolog, ortholog,* and *paralog.*

Statistical Analysis of Sequence Alignments Can Detect Homology (Text Section 6.2)

2. Explain how "sliding" sequences and "gaps" are used in sequence alignment.

3. Distinguish sequence shuffling from sequence sliding.

4. Describe how the Blosum-62 (Block Sum) matrix is used.

5. Analyze the significance of a high score in the Blosum-62 matrix for a given pair of amino acids, for example L-M, E-Q, or R-K. Contrast the situation when the score is low, for example with N-W.

6. Know that 25% sequence identity (or higher) is considered to prove that two proteins are homologous and that the correspondence is not the result of chance. Less than 15% identity would lead to the conclusion that the relationship has not been demonstrated.

7. Understand that large amounts of sequence data are available online, and that newly obtained sequence data are routinely compared to data existing on the Internet.

Examination of Three-Dimensional Structure Enhances Our Understanding of Evolutionary Relationships (Text Section 6.3)

8. Ponder the fact that two proteins can have identical folds even if *all* the amino acids are different. If this situation were observed, the proteins would be considered to have a common ancestor.

9. Understand that proteins with similar shapes often do not have parallel functions in the cell, despite a generally close relationship between form and function.

10. Describe a sequence template. Explain how it can be used to compare protein sequences.

11. Know that even though key catalytic residues are aligned similarly in the active sites of two enzymes, they need not be related. The explanation would be convergent evolution.

12. Understand why analysis of self-pairing is important in comparing homologous RNAs.

Evolutionary Trees Can Be Constructed on the Basis of Sequence Information (Text Section 6.4)

13. Describe how sequence comparisons can be utilized to make an evolutionary tree. Explain how dates can be estimated for divergence of related genes.

Modern Techniques Make the Experimental Exploration of Evolution Possible (Text Section 6.5)

14. Dinosaurs lived more than 65 million years ago. Understand why it is unlikely that DNA from dinosaurs can be found and sequenced.

SELF-TEST

Homologs Are Descended from a Common Ancestor

1. Human myoglobin and human hemoglobin α are paralogs. Human myoglobin and chimpanzee myoglobin are orthologs. What are human myoglobin and chimpanzee hemoglobin α called?

Statistical Analysis of Sequence Alignments Can Detect Homology

2. High-scoring replacements for amino acids on the BLOSUM62 matrix (text Figure 6.9) are generally amino acids with similar properties. With that in mind, which amino acid would be the highest scoring replacement for F = phenylalanine?

   (a) A, alanine
   (b) Y, tyrosine
   (c) C, cysteine
   (d) D, aspartate
   (e) E, glutamate

3. What sort of results would be obtained if the mRNA or DNA gene sequences for the proteins in question were shuffled and scored?

4. In the introduction to this chapter, angiogenin and ribonuclease were described as 35% identical. Are they related?

5. If Protein A is homologous with B, and B is homologous with C, can we deduce that A must therefore be homologous with C?

Examination of Three-Dimensional Structure

6. Find a representation of the heavy chain of immunoglobin G (IgG) in Chapter 33 of your textbook. If you did a sliding alignment of the heavy chain of IgG with itself, what would you find?

7. The mitochondrial enzyme malate dehydrogenase and the cytoplasmic version of malate dehydrogenase catalyze the same reaction (and hence have the same name), but their sequences and three-dimensional structures show no relationship. Explain how this can be.

8. RNA is analyzed for the location of hairpin folds. Which of the following sequences could form a mini-hairpin?

   (a) AGGUUUCCU
   (b) AGGUUUGGA
   (c) AGGUUUAGG
   (d) AAAAAAAAA
   (e) none of the above

Evolutionary Trees Can Be Constructed

9. In this chapter (Figure 6.21 in the text), an evolutionary tree is shown using various mostly paralogous globin genes. What could we learn from a similar comparison of orthologous genes (from different species)?

10. Many animals, vertebrate and invertebrate, use globin proteins to carry oxygen. What animal mentioned in the text is probably our closest relative without tetrameric hemoglobin in its blood?

    (a) starfish
    (b) horseshoe crab
    (c) lamprey
    (d) shark
    (e) lemur

Modern Techniques Make the Experimental Exploration of Evolution Possible

11. Why is detailed knowledge of the fossil record important in determining when genes diverged?

12. Mitochondrial DNA was extracted from a molar of a 9000-year-old skeleton found in Cheddar Gorge, England. Would you expect this DNA to be comparable to modern human DNA?

ANSWERS TO SELF-TEST

1. The working definitions for the terms ortholog and paralog state that orthologs arose by speciation and paralogs arose by gene duplication. Thus, these are paralogs because they originally diverged by gene duplication within a species. The common ancestor of myoglobin and hemoglobin would have existed hundreds of millions of years ago. The common ancestor of humans and chimpanzees existed about 10 million years ago. Even though the proteins occur within two different species, that is not the cause of the separation. The text is a bit unclear on this point.

2. Looking at the text Figure 6.9, you can see that the amino acids form clusters, and the group represented in blue letters to the right of the diagram are the hydrophobic amino acids FILMYVW (a mnemonic for this grouping is that gasoline, or octane, is hydrophobic, and that is what I use to "Fill my VW" when it is low on fuel). So the best replacement for one member of that group would be another member of that group, hence F is replaced by Y with a score of +3. If you know the structures of those two amino acids it is also striking that they differ only by one phenolic hydroxyl, and so are very similar.

3. Shuffling would not work well with DNA or RNA because there are only four kinds of "letters" or nucleotides. This would make the shuffled score artificially high compared to proteins, which have twenty different "letters" or amino acids.

4. Yes, two proteins with 35% identity would definitely be considered homologous. Anything over 25% sequence identity "proves" the relationship.

5. Yes. The example in the chapter is myoglobin, which is homologous with the alpha chain of hemoglobin, which is homologous with leghemoglobin. It is easy to understand this if one remembers that most homologous proteins have visibly similar shapes.

6. The heavy chain of IgG would have three very vivid repeats representing the C_H domains and a sketchier repeat representing the V_H domain.

7. Oxidation of malate is a problem that evolution had to solve. There were evidently two different solutions, resulting in convergent evolution. While the structure at the active site of enzymes that do similar jobs is often similar, the protein folds are quite different.

8. a. Answer (a) is correct because AGG pairs with the antiparallel CCU.

9. The tree that would be drawn from orthologous gene data would reveal relationships between the various species studied rather than the divergence of similar genes within a species. Relationships between species is known as taxonomy. The shark and lemur would have tetrameric hemoglobin like ours.

10. (c) The lamprey is a jawless fish. The horseshoe crab uses huge copper-containing proteins to carry oxygen, so its blood is blue rather than red.

11. To construct a "molecular clock" we need an approximate date for the divergence of various species. Consider *Escherichia coli* and *Salmonella*. These are similar organisms, which appear to be related, but *Salmonella* lives in the gut of reptiles and birds, and *E. coli* in the gut of various mammals. So when did the two microorganisms diverge? To answer this question, we have to look for the divergence of reptiles and mammals in the fossil record and find a reasonably accurate date.

12. Members of the human species, even members of our species from thousands of years ago, have quite similar DNA. In fact, the fossil DNA was found to match the mitochondrial DNA of a local schoolteacher who lived a few miles from where the fossil was found.

PROBLEMS

1. Scientists have found that relationships between species can appear quite different depending on which common gene is being studied. For example, comparisons of one eukaryote's gene might make the organism appear close to the prokaryotes, and another gene might make the same eukaryote appear closer to the archaea. Why?

2. Name the amino acids forming high-scoring pairs on the Blosum-62 matrix. Now find the highest scoring pairs on the codon chart. What do you notice about them? Are all "close" codons high scoring? Look up PH, PL, and PT on the codon chart and the Blosum-62 matrix, and comment.

3. Here are two amino acid sequences. One is from a form of glucokinase from *Pyrococcus furiosus,* an archaeal organism that lives in very hot water. The other is from the fruit fly genome (eukaryotic). What percentage identity do you see in the sequences as aligned here? How many of the nonidentical amino acids have positive scores on the Blosum-62 grid in the chapter?

   Pyrococcus glucokinase   `SVGLNEVELASIMEIL`

   Drosophila CG6650 gene   `SLGMNEQELSNLQQVL`

4. Some researchers are using genome databases to identify "COGs," or clusters of orthologous genes. What can one learn from finding these related proteins?

5. In the text, actin and Hsp-70 are shown to be homologous on the basis of their shared three-dimensional structures. Actin is found in essentially all eukaryotes, often as part of the contractile apparatus with myosin. Hsp-70 is found in eukaryotes, prokaryotes, and archaea as a chaperone for protein folding. What can we deduce from this distribution?

6. In 1977, Carl Woese published an article that showed that the Archaea were a separate kingdom from the Bacteria. Microbiologists had assumed that all single-celled organisms without a nucleus were rather closely related. Woese had to pick something that would be in every organism no matter how exotic. What would you choose to compare that would be present in all living organisms?

7. There were reports in the literature several years ago that DNA had been recovered from dinosaur fossils and amplified and sequenced. But the genes obtained appeared to be human. What is the explanation?

8. Compare and contrast the molecular evolution experiment described in the last section of the chapter (6.5) with the following phage Qβ evolution experiment: The

Qβ experiment started with one naturally occurring RNA, the genome of bacteriophage Qβ. Replication of Qβ was accomplished with a single enzyme, an RNA replicase. Selection in the Qβ experiment was fairly simple, that is, shortening the available time for reproduction. Compare the sort of RNA that was used at the start of each experiment. How was the RNA reproduced in each experiment? How was selection applied to the population of molecules? Which process would be closer to what happens in nature?

9. This is an exercise to show you how to use BLAST, which stands for "Basic Local Alignment Search Tool." To do an alignment you must have two sequences. We will look at proteins, which means we will be using BLASTP (BLAST for Proteins). But first we need to have a sequence to compare. Here is how to obtain one:

   a. Go to http://www.ncbi.nlm.nih.gov; while you are there make it a "favorite" or "bookmark." At the top right is a window labeled "Popular Resources." Click on BLAST. This will take you to http://blast.ncbi.nlm.nih.gov. Scroll down to Basic BLAST and click on "protein blast." The large window at the top is labeled "Enter Query Sequence-Enter accession number, gi, or FASTA sequence."

   b. Look at Figure 2.53 in the text. This shows the sequence of Bovine Pancreatic Ribonuclease. We want to retrieve this sequence, so in the large box we type in the beginning letters, KETAAAKFRQ. Scroll to the bottom of the page and hit the button labeled BLAST. There will be a pause of a few seconds. Then you will see an entire screen full of 100% matches to your sequence. Scroll down to where the sequences are named, and look for Bovine Pancreatic Ribonuclease. At this writing, one of the first clearly labeled sequences has the accession code "3DIC_A." Click on that and you will find a lot of information about the protein, and at the bottom you should see the complete sequence. Compare this to the figure in the text.

   c. Now go back to the original Protein Blast screen. In the large window at the top, you could either paste in the entire sequence of amino acids, or simply type in "3DIC_A," so type in the accession code, which amounts to the same thing. Scroll down to "Choose Search Set" and type in "chicken" beside "Organism." We are going to compare the bovine (cow) protein to the database of chicken proteins. Scroll down and push the BLAST button and wait.

   d. How many similar proteins were found? What are their names? How similar are they? How do the results reinforce the text chapter's discussion of homologous proteins?

10. Using BLASTP as described in the previous problem, identify the following unknowns, all of them human. This exercise simulates the real experimental situation where one has obtained a partial sequence of a protein by Edman degradation or some other technique and needs to find the entire sequence. Don't forget to set the organism to "human" before pressing BLAST. After you have learned the name of each protein you might vary the species as in question 9 to see how much homology you get.

   (a) pkkyipgtkm ifvgikkkee radliaylkk
   (b) alegslqkrg   iveqcctsic slyqlenycn
   (c) pgdfgadaqg amnkalelfr kdmasnykel
   (d) sqraglqfpv   grihrhlksr   ttshgrvgat

11. In the 1970's, when the first edition of the Stryer textbook came out, only a few globin sequences were known, mostly hemoglobin and myoglobin from a variety of vertebrate animals. Attempts to understand the significance of globin structure tended to emphasize a handful of "invariant residues," which were the same in all known globins. Now that several hundred globin sequences are known, including several

from invertebrates, the only truly "invariant" residue is the histidine, which attaches to the iron of the heme group (see Chapter 7 of the text for more discussion of that). Two of the invariant residues were a proline and a glycine. Look at the BLOSUM62 matrix, Figure 6.9, and suggest why a set of 11 vertebrate proteins might all share a proline and a glycine.

ANSWERS TO PROBLEMS

1. Now that several complete genomes have been sequenced, it is obvious that horizontal gene transfer, also called lateral gene transfer, is very common in nature. This means that a gene can move from one species to another, even if the species are very different. Hence in the human genome we have some "bacterial" genes as well as "archaeal" genes, and using one of those to construct a tree of relationships would have very strange results.

2. The highest scoring replacements are F-Y, I-V, V-I, and Y-F, all with scores of +3. Looking at the codon chart, F or Phe is UUU or UUC, and Y or Tyr is UAU or UAC. So the only difference is in the second position. I or Ile has three codons, AUU, AUC, and AUA. V or Val has four codons, but the closest in sequence would be GUU, GUC, and GUA, or a change in the first position. So are the scores high because the replacements are "easy"? Evidently not, because other "easy" replacements have much lower scores. The question mentions P, or proline, changing to H, L, or T. These all have negative scores: T = −2, H = −3, and L = −5, even though the codon structures are close. The difference is the similarity in amino acid structure. Having similar codons *permits* the mutation to occur. Having similar amino acids, pairs like valine/isoleucine or phenylalanine/tyrosine, makes the protein have a very similar functionality. So, in general, very similar amino acids will have high Blosum-62 replacement scores.

3. Of the 16 amino acids shown from each sequence, seven, or 44%, are identical. Another seven have positive scores on the Blosum-62 matrix:

Pyrococcus glucokinase	SVGLNEVELASIMEIL
	I+I+II II+++ ++I
Drosophila CG6650	SLGMNEQELSNLQQVL

VL	+1	IL	+2
LM	+3	EQ	+3
AS	+2	IV	+4
SN	+1		

Based on these data, the two proteins clearly appear to be homologous despite being from very distantly related species. (*Pyrococcus* enzyme from *J. Biochem.* 128[2000]:1079–1085.)

4. For one thing, if a single member of a COG has been crystallized so that its three-dimensional structure could be solved, then the shapes of other members can be inferred. Also, although it isn't always the case, often orthologous proteins will have either the same or very similar functions. There are many possible uses for these data sets.

5. The universal distribution of Hsp-70 implies that the earliest use of this protein fold was as a chaperone for protein folding. This would also be consistent with the theory that the earliest cells lived in a hot environment. Hsp stands for "heat shock protein,"

and proteins can have difficulty folding in high heat. Even though actin is highly conserved and found in all eukaryotes, it is not found in the other kingdoms and thus is probably a later development. It is interesting that another well-known protein, mammalian hexokinase-I, belongs to this family of related proteins, even though its use is quite different. There is also a bacterial protein that forms actinlike filaments in some but not all bacteria, called MreB. It is close in structure to actin and has the same fold as the other proteins mentioned in this problem. (*Nature*, 413[2001]:39–44.)

6. An enzyme would be a risky choice because many of the Archaea live in very extreme environments—high salt, strong acid, water at 100°C or higher—and they might need very different enzymes. But all organisms have ribosomes. So Carl Woese picked the RNA found in the smaller ribosomal subunit, the 16S rRNA. Sequencing methods were primitive back in the early 1970s, so Woese found a way to fragment the 16S rRNA and sequence just the small fragments. To everyone's surprise, the methanogens he was studying, and later most of the "funny bugs" that live in extreme environments, turned out to be quite different from the "regular" bacteria that live in easier locations. Hence Woese split them off into a new kingdom, now called the Archaea. (*Proc. Natl. Acad. Sci. USA*, 74[1997]:5088–5095.)

7. Modern methods such as PCR are so powerful that they can amplify the tiniest trace of DNA to an amount that can be sequenced. This technology does indeed allow for sequencing of some DNA from fossils, usually only a few thousand years old. But it also allows errors to be made. It appears that a fingerprint or a tiny flake of skin or something containing the researchers' DNA got into the fossil sample. And since all the actual dinosaur DNA had degraded over the millions of years since the animal died, this was the only DNA available as a starting material for PCR. (Debunking article: *Trends. Ecol. Evol.*, 12[1997]:303–306.)

8. Both experiments aim to simulate evolution by (a) generating a diverse population, (b) selecting "fit" population members based on a predetermined criterion, and (c) allowing the selected members to reproduce, thereby producing successive generations of increasingly "fit" populations. For step (a), the textbook's experiment begins with an assortment of artificially constructed RNA molecules, which are replicated using an elaborate process involving several enzymes—reverse transcriptase, DNA polymerase, and RNA polymerase—to go from RNA to DNA and back to RNA. In contrast, the phage experiment begins with just one naturally occurring RNA, and this is replicated using a single enzyme, RNA replicase. In both cases, errors in replication increase diversity in the population. For step (b), in the textbook's experiment, the selection criterion is somewhat contrived (testing for binding to ATP on an affinity column), while in the phage experiment it is fairly simple (shortening reproduction time). For step (c), in both cases, reproduction is effected through repeated rounds of steps (a) and (b).

   Oddly enough, the complicated process in this chapter is modeled more closely on "real" cellular processes. For example, retroviruses have RNA genomes, but reproduction of these viruses involves reverse transcription into DNA.

9. The chicken genome, viewed as expressed proteins, codes for several proteins that are homologous with ribonuclease from cow. One is chicken ribonuclease (an ortholog, separated from cow RNAse by speciation and used for the same purpose), and another is angiogenin (a paralog, separated by gene duplication and used for very different purposes). Bovine ribonuclease and angiogenin are clearly related and share the same fold, as seen in text Figures 6.1 and 6.2. You will also see other proteins with lesser

homology including MHC proteins—"Major Histocompatibility Complex" proteins are used for tissue typing (see text Chapter 33, "The Immune System"). If you want to continue "playing" try again with the organism set to iguana or fruit fly or whatever your favorite animal is.

10. (a) cytochrome c   (b) proinsulin   (c) myoglobin   (d) histone

11. The argument made in the first edition of the Stryer text was that where proline was an invariant residue, it enforced a necessary "elbow" bend in the alpha-helical structure of the globin fold, and where glycine was invariant, any other amino acid would have been too large for the helices to pass quite so closely to each other. These arguments make a lot of sense, and still have validity. But the existence of globins with replacements at those positions shows that there are other ways to solve those problems. And, in fact, looking at Figure 6.9 shows us that proline, glycine, and cysteine are the three "irreplaceable" amino acid residues in protein structure. In general, there is no positive scoring replacement for any of them because of their unique properties. So it could be argued that it was inevitable that in a small set of a dozen or so examples of orthologous proteins, you would find P, G, or C, which would be kept the same throughout the entire set.

Chapter 7

Hemoglobin: Portrait of a Protein in Action

At this point in the text, students should have enough information to understand what proteins are and how they are investigated in the laboratory. Homologous proteins should also be a familiar concept. In this chapter, the authors describe the structure and function relationships of two very important, well-studied proteins: hemoglobin and myoglobin. These homologous proteins have the important jobs of carrying oxygen in the blood and storing oxygen in the muscles. They also provide a clear example of the difference between allosteric (sigmoidal) binding and simpler (hyperbolic) single-ligand binding. The cooperative binding of oxygen to hemoglobin is critical to its ability to transport oxygen efficiently in blood and release it to myoglobin in tissues. Because it is also regulated by H^+, CO_2, and 2,3 BPG, hemoglobin also provides an excellent example of allosteric regulation of proteins. Finally the chapter describes the genetic diseases sickle-cell anemia and thalassemia, which are caused by abnormal hemoglobin genes.

LEARNING OBJECTIVES

When you have mastered this chapter, you should be able to accomplish the following objectives.

Introduction

1. Describe the physiologic roles of *myoglobin* and *hemoglobin* in vertebrates.

2. Compare myoglobin and hemoglobin, and explain how they can have such similar structural folding and such different properties.

Myoglobin and Hemoglobin Bind Oxygen at Iron Atoms in Heme (Text Section 7.1)

3. Describe the structure of the *heme prosthetic group* and its properties when free and when bound to the *globins*.

4. Define the terms *superoxide* and *metmyoglobin*.

5. Define fMRI, and know it uses changes in hemoglobin to allow noninvasive imaging of the brain.

6. Discuss the three-dimensional structures of myoglobin and hemoglobin and how the heme groups attach to each one.

7. Describe how the distal histidine in myoglobin prevents release of superoxide anion and formation of metmyoglobin.

Hemoglobin Binds Oxygen Cooperatively (Text Section 7.2)

8. Contrast the *oxygen-binding* properties of myoglobin and hemoglobin. Define the *cooperative binding* of oxygen by hemoglobin and summarize how it makes hemoglobin a better oxygen transporter.

9. Explain the significance of the differences in *oxygen dissociation curves,* in which the *fractional saturation (Y)* of the oxygen-binding sites is plotted for myoglobin and hemoglobin as a function of the *partial pressure of oxygen (pO_2).*

10. State the major structural differences between the *oxygenated* and *deoxygenated* forms of hemoglobin.

11. Distinguish between the *sequential model* and the *concerted model* for cooperative interactions between the subunits of proteins.

12. Explain the effect of *2,3-bisphosphoglycerate (BPG)* (also known as 2,3-diphosphoglycerate) on the affinity of hemoglobin for oxygen.

13. Summarize the general properties of an allosteric protein, as exemplified by hemoglobin.

14. Rationalize the existence of *fetal hemoglobin.*

15. Explain how carbon monoxide disrupts oxygen transport by hemoglobin

Hydrogen Ions and Carbon Dioxide Promote the Release of Oxygen: The Bohr Effect (Text Section 7.3)

16. Explain the effects of CO_2 and H^+ (the *Bohr effect*) and 2,3-bisphosphoglycerate (BPG) on the binding of oxygen by hemoglobin. Describe the structural bases for the effects of these molecules on the binding of oxygen by hemoglobin. Explain the consequences of the metabolic production of CO_2 and H^+ on the oxygen affinity of hemoglobin.

Mutations in Genes Encoding Hemoglobin Subunits Can Result in Disease
(Text Section 7.4)

17. Describe *sickle-cell anemia* as a genetically transmitted *molecular disease*.

18. Contrast the biochemical and structural properties of deoxygenated *sickle-cell hemoglobin (HbS)* with those of *hemoglobin A (HbA)*.

19. Correlate the clinical observations of patients with *sickle-cell anemia* and the *sickle-cell trait* to their hemoglobins.

20. Explain how deoxyhemoglobin S forms *fibrous precipitates*.

21. Outline the possible functional consequences of *amino acid substitutions* in *mutant hemoglobins*.

22. Define *thalassemias*, and explain why β-thalassemia (deficiency of the β chain of hemoglobin) is more common than α-thalassemia (deficiency of the α chain).

Binding Models Can Be Formulated in Quantitative Terms: The Hill Plot and the Concerted Model (Appendix)

23. Starting with the equilibrium expression for oxygen binding by myoglobin, derive the equation

$$Y = \frac{pO_2}{(pO_2 + P_{50})}$$

24. Relate the empirical expression for the fractional saturation of hemoglobin

$$\frac{(pO_2)}{(pO_2)^n + (P_{50})^n}$$

with the equation

$$\frac{Y}{(1 - Y)} = \frac{(pO_2)^n}{P_{50}}$$

with the *Hill Plot*. Explain the significance of the *Hill coefficient*.

SELF-TEST

Introduction

1. Where are hemoglobin and myoglobin located in the body?

2. Hemoglobin differs from myoglobin in that it has four binding sites for oxygen (compared to myoglobin's single binding site) and is cooperative. Are all molecules with multiple binding sites cooperative?

Myoglobin and Hemoglobin Bind Oxygen at Iron Atoms in Heme

3. Which of the following statements about heme structure is true?

   (a) Heme contains a tetrapyrrole ring with four methyl, four vinyl, and four propionate side chains.

(b) The iron atom in heme may be present in the ferrous or the ferric state.

(c) The iron atom is coplanar with the tetrapyrrole ring in deoxymyoglobin.

(d) The axial coordination positions of heme are occupied by tyrosine residues in myoglobin.

4. Match the forms of myoglobin in the left column with their corresponding properties in the right column.

(a) Metmyoglobin _____

(b) Oxymyoglobin _____

(c) Deoxymyoglobin _____

(1) Iron in the +2 oxidation state

(2) Iron in the +3 oxidation state

(3) Oxygen bound to the sixth coordination position of iron

(4) Empty sixth coordination position

(5) Water bound to sixth coordination position

(6) Histidine at fifth coordination position

5. How does the structure of myoglobin help to prevent bound oxygen from being released as superoxide (O_2^-)?

6. One way to observe whether a heme group has bound to oxygen (allowing the iron to shift position into the heme plane) is to use fMRI as described in the text. Can you think of another way?

Hemoglobin Binds Oxygen Cooperatively

7. Which of the following statements are false?

(a) The oxygen dissociation curve of myoglobin is sigmoidal, whereas that of hemoglobin is hyperbolic.

(b) The affinity of hemoglobin for O_2 is regulated by organic phosphates, whereas the affinity of myoglobin for O_2 is not.

(c) Hemoglobin has a higher affinity for O_2 than does myoglobin.

(d) The affinity of both myoglobin and hemoglobin for O_2 is independent of pH.

8. Hemoglobin is a tetrameric protein consisting of two α and two β polypeptide subunits. The structures of the α and β subunits are remarkably similar to that of myoglobin. However, at a number of positions, hydrophilic residues in myoglobin have been replaced by hydrophobic residues in hemoglobin.

(a) How can this observation be reconciled with the generalization that hydrophobic residues fold into the interior of proteins?

(b) In this regard, what can you say about the nature of the interactions that determine the quaternary structure of hemoglobin?

9. Much of the information we have concerning hemoglobin structure comes from the study of crystals. Which of the following lines of evidence support the notion that crystallized hemoglobin has a structure similar to that of hemoglobin in solution?

(a) The amino acid sequence of crystallized hemoglobin is the same as that found in hemoglobin in solution.

(b) The visible absorption spectrum of crystallized hemoglobin is virtually the same as that of hemoglobin in solution.

(c) Crystallized hemoglobin is functionally active.

(d) The α helix content of crystallized hemoglobin is similar to that found in hemoglobin in solution.

10. The book describes one structural difference between fetal hemoglobin (HbF) and adult hemoglobin (HbA). What is it?

11. One effect of carbon monoxide binding to hemoglobin is that it increases hemoglobin's affinity for oxygen. Why isn't that a good thing?

Hydrogen Ions and Carbon Dioxide Promote the Release of Oxygen: The Bohr Effect

12. The oxygen dissociation curve for hemoglobin reflects allosteric effects that result from the interaction of hemoglobin with O_2, CO_2, H^+, and BPG. Which of the following structural changes occur in the hemoglobin molecule when O_2, CO_2, H^+, or BPG bind?

    (a) The binding of O_2 pulls the iron into the plane of the heme and causes a change in the interaction of all four globin subunits, mediated through the proximal His.
    (b) BPG binds at a single site between the four globin subunits in deoxyhemoglobin and stabilizes the deoxyhemoglobin form by cross-linking the β subunits.
    (c) The deoxy form of hemoglobin has a greater affinity for H^+ because the molecular environment of His and the α-NH_2 groups of the a chains changes, rendering these groups less acidic when O_2 is released.

13. Several oxygen dissociation curves are shown in Figure 7.1. Assuming that curve 3 corresponds to isolated hemoglobin placed in a solution containing physiologic concentrations of CO_2 and BPG at a pH of 7.0, indicate which of the curves reflects the following changes in conditions:

    (a) decreased CO_2 concentration
    (b) increased BPG concentration
    (c) increased pH
    (d) dissociation of hemoglobin into subunits

FIGURE 7.1   Oxygen dissociation curves.

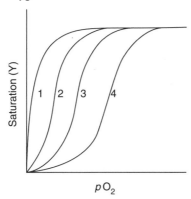

14. Which of the following statements concerning the Bohr effect are true?

    (a) Lowering the pH shifts the oxygen dissociation curve of hemoglobin to the right.
    (b) The acidic environment of an exercising muscle allows hemoglobin to bind O_2 more strongly.
    (c) The affinity of hemoglobin for O_2 is diminished by high concentrations of CO_2.
    (d) In the lung, the presence of higher concentrations of H^+ and CO_2 allows hemoglobin to become oxygenated.
    (e) In the lung, the presence of higher concentrations of O_2 promotes the release of CO_2 and H^+.

15. The structure of deoxyhemoglobin is stabilized by each of the following interactions *except*
    (a) BPG binding.
    (b) salt bridges between acidic and basic side chains.
    (c) coordination of the hemes with the distal histidine.
    (d) hydrophobic interactions.
    (e) salt bridges involving *N*-terminal carbamates.

16. In the transition of hemoglobin from the oxy to the deoxy form, an aspartate residue is brought to the vicinity of His 146. This increases the affinity of this histidine for protons. Explain why.

17. Explain why fetal hemoglobin has a higher affinity for oxygen than does maternal hemoglobin and why this is a necessary adaptation.

Mutations in Genes Encoding Hemoglobin Subunits Can Result in Disease

18. The *N*-terminal tryptic peptides from the β chains of HbA and HbS are as follows:

    HbA: Val-His-Leu-Thr-Pro-Glu-Glu-Lys (or VHLTPEEK)

    HbS: Val-His-Leu-Thr-Pro-Val-Glu-Lys (or VHLTPVEK)

    (a) Would these peptides separate from each other in an electric field at pH 7.0?
    (b) What is the approximate net charge on each peptide at pH 7.0?

19. Which of the following answers are true? Hemoglobin S forms fibrous precipitates
    (a) because the valine at position 6 of the β chain forms a sticky hydrophobic patch on the surface of the protein.
    (b) that are reversible upon oxygenation.
    (c) that distort the shape of cells.
    (d) only in the deoxy form.
    (e) only in homozygotes.

Binding Models Can Be Formulated in Quantitative Terms: The Hill Plot and the Concerted Model

20. Match the parameters in the left column with the appropriate definitions in the right column.

    (a) Hill coefficient (n) _____
    (b) $(pO_2/P_{50})n$ _____
    (c) P_{50} _____
    (d) Y _____

    (1) The fractional occupancy of the oxygen-binding sites
    (2) The pO_2 at half-saturation of the hemes with oxygen
    (3) The ratio of oxyheme to deoxyheme
    (4) The cooperativity of oxygen binding

21. BPG (2,3-bisphosphoglycerate) plays a role in high-altitude adaptation. What is the effect of an increase in the level of BPG on the amount of oxygen transported to muscle in a person living at 10,000 feet compared with a person living at sea level? Assume that the increase in BPG level has shifted P_{50} from 26 torr to 35 torr and that alveolar pO_2 is 67 torr and pO_2 in muscle capillaries is 20 torr at 10,000 feet. At sea level the alveolar pO_2 is 100 torr and pO_2 in muscle capillaries is still 20 torr. Assume that the Hill coefficient is 2.8.

22. Which of the following answers completes the sentence *incorrectly*? Hemoglobin differs from myoglobin in that

   (a) hemoglobin is multimeric whereas myoglobin is monomeric.
   (b) myoglobin binds O_2 more tightly than does hemoglobin at any given O_2 concentration.
   (c) hemoglobin binds CO_2 more effectively than does myoglobin.
   (d) the Hill coefficient for O_2 binding is smaller for hemoglobin than it is for myoglobin.
   (e) the binding of O_2 by hemoglobin depends on the concentrations of CO_2, H^+, and BPG, whereas the binding of O_2 by myoglobin does not.

ANSWERS TO SELF-TEST

1. Hemoglobin is found in red blood cells and myoglobin in muscles. "Haem" is a Greek prefix meaning blood, and "myo" is a Greek prefix meaning muscle.

2. No. A simple example would be the phosphate molecule, PO_3^{-3} which can bind one, two, or three protons. The binding of one proton does not change the affinity of the molecule for a second or third proton. Hence the whole population of phosphate molecules will have one, two, or three protons depending on the pH. In contrast, binding one oxygen molecule "opens" hemoglobin to binding a second, third, and fourth molecule (because of its cooperative nature) so that only the deoxy and tetra-oxy forms can be observed. The mono-, di-, and tri-oxy forms are essentially transition states.

3. b

4. (a) 2, 5, 6 (b) 1, 3, 6 (c) 1, 4, 6

5. The distal histidine binds to the bound oxygen in a way that favors binding as superoxide but release as dioxygen. It also aids the reversibility of oxygen binding—without it, oxygen could bind much more tightly.

6. The color difference in arterial and venous blood reveals the state of the heme groups. Arterial blood is a bright crimson red, which is the color of the heme when oxygen is bound. Venous blood is a bluish purple, showing the deoxy-heme group.

7. a, c, d

8. (a) Hydrophobic patches occur on the surface of the hemoglobin subunits where the α and β chains fit together. As a result, these patches are not on the surface of the multimeric protein.
   (b) Hydrophobic interactions play an important role in stabilizing the tetrameric subunit structure of hemoglobin.

9. b, c, d. Regarding answer (b), the visible absorption spectrum of heme varies with its oxidation state and its environment. The fact that both crystallized hemoglobin and hemoglobin in solution have the same spectra is indicative of a similar structure of the heme pocket.

10. Fetal hemoglobin has a serine residue replacing a histidine residue in its γ chains (there are 2 γ and 2 α chains in HbF). This makes the central cavity where 2,3 BPG binds have less of a positive charge, and thus it binds less and has slightly more affinity for oxygen than HbA. See text Figure 7.18.

11. The higher hemoglobin's affinity for oxygen is, the closer its loading curve is to the loading curve of myoglobin. The amount of oxygen released to tissues is represented by the space between those two curves. See text Figure 7.8. Both carbon monoxide's tight binding to hemoglobin, and the change it causes in the affinity for oxygen, lead to oxygen starvation of the tissues.

12. a, b, c

13. (a) 2   (b) 4   (c) 2   (d) 1

14. a, c, e

15. c

16. The pK values of ionizable groups are sensitive to their environment. The change in the environment of His 146 in deoxyhemoglobin increases its affinity for protons as a result of the electrostatic attraction between the negative charge of the aspartate and the proton.

17. Fetal hemoglobin is composed of different subunits than adult hemoglobin and binds BPG less strongly. As a result, the affinity of fetal hemoglobin for oxygen is higher, and the fetus can extract the O_2 that is transported in maternal blood.

18. (a) Yes, the peptides will separate.
    (b) The net charge for HbS is between -0.9 and -1 and that for HbA is between 0 and $+0.1$. Since the pK of His is approximately 6.0, this residue will be partially ionized with a charge of less than $+0.1$ at pH 7.0, which leads to the nonintegral answers.

19. a, b, c, d

20. (a) 4   (b) 3   (c) 2   (d) 1

21. Using the equation $Y = (pO_2)^n / [(pO_2)^n + (P_{50})^n]$, the increased level of BPG at 10,000 feet changes P_{50} and yields a ΔY (the difference between the fractional saturation of hemoglobin with oxygen in the lungs versus that in the tissues) of 0.69. Compare this with a ΔY of 0.65 at sea level. Thus, the increase in BPG results in a similar delivery of oxygen to the tissues even when alveolar pO_2 is significantly decreased. In general, the larger the value of ΔY, the higher the delivery of O_2 to the tissues.

22. (d)

PROBLEMS

1. The dense, rich color of human blood tells us that there are very many erythrocytes in every drop of blood. About how many red blood cells are in the adult human body? About what percent of human cells would this be? Considering that red blood cells have a half-life of only about 120 days, the supply of erythrocytes must be replaced about every 4 months.

2. Glutathione is a tripeptide consisting of glutamic acid, cysteine, and glycine; it is abundant in human erythrocytes as well as in many types of tissue. Reduced glutathione (GSH) acts as a reducing agent in tissues because its side-chain —SH group can be readily oxidized to form disulfide bonds. Unless glutathione is maintained substantially in its reduced form as opposed to its oxidized form in erythrocytes, they lose their ability to transport oxygen effectively. Why do you think this is so?

3. The heme prosthetic group found in hemoglobin and myoglobin is also found in cytochromes b and c, which are proteins that transfer electrons in the electron transport chain. How can the same prosthetic group serve such different functions as oxygen binding and electron transport?

4. The iron in hemoglobin must be in the ferrous ($+2$) state to bind oxygen. If the iron of hemoglobin becomes oxidized to the ferric ($+3$) state, the corresponding hemoglobin, called methemoglobin, cannot bind oxygen. To maintain iron in the ferrous state, red blood cells contain a reducing system to convert any methemoglobin back

to hemoglobin. Some variant forms of hemoglobin, such as hemoglobin M, cannot be reduced by that system. What would be the expected clinical presentation of a patient having HbM?

5. If fMRI shows what is happening inside someone's brain, do you think it would provide the basis of the best possible lie detector? What other uses might there be for fMRI?

6. Superoxide anion could be generated by heme groups in a variety of ways, either from hemoglobin/myoglobin or by cytochromes in the mitochondria. As the text states, superoxide can damage biological materials, so there are safeguards to avoid large scale production of this toxic ion, as well as the enzyme superoxide dismutase, which removes it. Still, over a lifetime lived in an oxygen atmosphere, there will be harmful effects. What do you imagine some of the specific damages would include?

7. One molecule of 2,3-bisphosphoglycerate binds to one molecule of hemoglobin in a central cavity of the hemoglobin molecule. Is the interaction between BPG and hemoglobin stronger or weaker than it would be if BPG bound to the surface of the protein instead? Explain your answer.

8. An effective respiratory carrier must be able to pick up oxygen from the lungs and deliver it to peripheral tissues. Oxygen dissociation curves for substances A and B are shown in Figure 7.2. What would be the disadvantage of each of these substances as a respiratory carrier? Where would the curve for an effective carrier appear in the figure?

FIGURE 7.2   Oxygen dissociation curves for substances A and B.

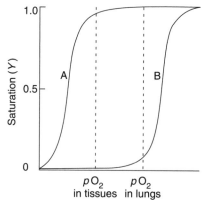

9. A portion of the lungs of patients suffering from pneumonia are filled with fluid, and therefore the lung surface area available for oxygen exchange is reduced. Standard hospital treatment of these patients involves placing them on a ventilating machine set to deliver enough oxygen to keep their hemoglobin approximately 92% saturated. Why is this value selected rather than one lower or higher?

10. What major differences exist between the sequential and concerted models for allostery in accounting for hemoglobin that is partially saturated with oxygen?

11. Predict whether each of the following manipulations will increase or decrease the tendency of HbS molecules to polymerize in vitro. Give a brief rationale for each answer.
    (a) increase in temperature
    (b) increase in the partial pressure of oxygen
    (c) stripping the HbS molecules of BPG
    (d) increase in the pH

12. One avenue of approach to therapy for sickle-cell anemia involves finding a way of turning on the synthesis of HbF in afflicted adults. Briefly explain why such a ma-

nipulation might be beneficial.

13. One approach to the management of sickle-cell anemia involves a search for osmotically active agents that would expand the volume of erythrocytes.

    (a) Give a brief rationale for this approach.
    (b) Suppose that the volume of an erythrocyte is increased by 10%. Calculate the rate of HbS polymer formation in the enlarged cell as a fraction of its rate in a normal-sized cell.

14. Prenatal diagnosis of sickle-cell anemia can be carried out by the treatment of fetal DNA with the restriction endonuclease MSTII, which recognizes the sequence CCTNAGG, where N is any nucleotide. Digestion of the β^S gene with MSTII yields a 1.3-kb fragment, whereas digestion of the β^A gene yields a 1.1-kb fragment.

    (a) Using your knowledge of the amino acid substitution that occurs in HbS, give the identity of nucleotide N on both the β^A and β^S genes. Also identify the mutational change at the DNA level that leads to the amino acid substitution. Explain your answer. (Refer to the genetic code in your text.) Note that the (+) strand of DNA is given so you need to look for the complementary antiparallel RNA sequence.
    (b) Give the identity of the amino acid at position 5 in the β chain using the information provided here and the genetic code.

15. The cooperative nature of hemoglobin can be understood using either the sequential model, in which each of the four subunits changes from T to R individually upon binding oxygen, or the MWC symmetry model, in which the whole tetramer shifts at once from T to R. What structural features of hemoglobin support each model?

16. Why is endemic malaria such a powerful selective force for the sickle-cell trait? Why can't malaria be treated with antibiotic drugs?

ANSWERS TO PROBLEMS

1. According to this Web site, written for a popular audience—
   http://www.madsci.org/posts/archives/feb2001/981770369.An.r.html
   —there are about 30 trillion RBC (red blood cells) in the adult human body, or 3×10^{13} cells (a trillion = 10^{12}). Of the roughly 90 trillion cells in the average human body, about 40 trillion are bacteria in the gut and other places. So the RBC make up about half of the "human" cells in the body.

2. Reduced glutathione helps keep the iron of hemoglobin in the ferrous (+2) valence state. When the iron of heme is oxidized to the ferric (+3) state to form methemoglobin, it can no longer combine reversibly with oxygen.

3. The oxidation state and the binding properties of heme vary markedly with its environment; therefore, the different environments caused by the different amino acids in these two classes of proteins change the functions of the heme.

4. An individual with HbM would have a bluish appearance, a condition called *cyanosis*, because of the lower concentration of ferrous hemoglobin ++ and thus less O_2 in RBC's.

5. Use of fMRI as a lie detector occurred to several people around the world, and there are some who claim to have systems that work. The central problem appears to be that

the method works too well, in a sense. Some people will consider lying before they actually tell the truth. The same area of your brain "lights up" if you actually lie, or if you just think about lying. Still the research potential of fMRI seems limitless, and one could study decision making, or what part of the brain is active during religious meditation or prayer, and other things which are otherwise almost impossible to visualize.

6. The "Free Radical Theory of Aging" holds that the changes observed as the body ages are largely due to inappropriate cross-linking of biomolecules. Thus we get atherosclerosis as blood vessels lose their flexibility, and wrinkling of the skin. So while oxygen, and an aerobic style of metabolism, rewards us with vast amounts of energy compared to anaerobic cells, the price we pay is that we must age. Even the phenomenon of graying hair has been linked to a lessening in superoxide dismutase levels during ageing. (I. Emerit et al., *Photochemistry and Photobiology* 80[2004]:579–582)

7. BPG binds to hemoglobin by electrostatic interactions. These interactions between the negatively charged phosphates of BPG and the positively charged residues of hemoglobin are much stronger in the interior, hydrophobic environment than they would be on the surface, where water would compete and weaken the interaction by binding both to BPG and to the positively charged residues. Remember that the force of electrostatic interactions, given by Coulomb's law (in Chapter 1 of the text), is inversely proportional to the dielectric constant of the medium. The dielectric constant in the interior of a protein may be as low as 2. Hence, electrostatic interactions there are much more stable than those on the surface, where the dielectric constant is approximately 80.

8. Substance A would never unload oxygen to peripheral tissues. Substance B would never load oxygen in the lungs. An effective carrier would have an oxygen dissociation curve between the curves depicted for substance A and substance B. It would be relatively saturated with oxygen in the lungs and relatively unsaturated in the peripheral tissues.

9. Look at the oxygen saturation curve for hemoglobin given in Figure 7.7 of the text. Administering enough oxygen to give saturation levels greater than approximately 92% would be wasteful of oxygen, because one reaches the point of diminishing returns. Administering oxygen in amounts less than that required for 92% saturation runs the risk of compromising oxygen delivery to the tissues.

10. According to the concerted model, hemoglobin partially saturated with oxygen is composed of a mixture of fully oxygenated molecules, with all subunits in the R form, and fully deoxygenated molecules, with all subunits in the T form. According to the sequential model, individual molecules would have some subunits that are oxygenated (in the R form) and some that are deoxygenated (in the T form).

11. (a) An increase in temperature will favor the polymerization of HbS. The interaction between Hb molecules in polymer formation is hydrophobic in nature. Hydrophobic interactions have negative temperature coefficients; that is, they become more stable with increasing temperature.
    (b) An increase in the partial pressure of oxygen will inhibit polymerization because only deoxyhemoglobin S polymerizes.
    (c) BPG stabilizes deoxyhemoglobin S. Since only the deoxy form polymerizes, the removal of BPG from HbS would inhibit polymer formation.
    (d) Increasing the pH (decreasing the acidity) stabilizes oxyhemoglobin S. Since only the deoxy form polymerizes, this would inhibit formation.

12. HbF is devoid of β chains, having γ chains instead, so it does not polymerize. If adults with sickle-cell anemia could synthesize HbF, each erythrocyte would contain a mix-

ture of HbS and HbF, which would reduce the degree of polymerization of HbS.

13. (a) The rate of polymerization of HbS is proportional to the tenth power of its concentration. Therefore, increasing the cell volume would decrease the HbS concentration, which would slow the rate of HbS polymerization.

    (b) If the cell has been expanded by 10%, the HbS concentration has been decreased to 10/11 or 90.9% of its original value. The rate of polymerization under these conditions is $(0.909)^{10} = 0.386$ of the rate in an unexpanded cell.

14. (a) Nucleotide N must be C on both the β^A and β^S genes. In HbA, Glu is present at position 6; it is encoded by GAG (on mRNA). Therefore, the sequence CTC must be present on the informational strand of β^A DNA. The mutation that leads to the substitution of Val for Glu at position 6 is T ->A.

    (b) The amino acid at position 5 in the β chain is Pro, which is encoded by CCU on RNA (or AGG on DNA).

15. The sequential model is supported by the fact that as each globin subunit binds oxygen, the iron moves into the heme plane pulling on the proximal histidine. This alters the position of the attached alpha helix and directly produces a conformational change. The MWC symmetry model is supported by the fact that each tetramer contains only one molecule of 2,3-BPG. The fact that 2,3-BPG must be either present or absent supports the concept that the tetramer should either be tense (in the presence of BPG) or relaxed (in its absence). In fact both models produce the same sigmoidal curve and both are useful ways of looking at hemoglobin.

16. Even though people who are heterozygotic for sickle cell (and have both HbS and HbA in every erythrocyte) are unlikely to have serious sickling attacks, there is still some formation of HbS fibers. This is enough to shorten the life of erythrocytes, which normally have a half-life in the body of 120 days. When the cells break down, the *Plasmodium falciparum* (malaria) parasites inside are exposed to the body's immune system and destroyed. So individuals with the sickle-cell trait survive while many with normal HbA die from the disease. Antibiotics are generally effective only against prokaryotes (bacteria). Malaria and several other tropical diseases are caused by eukaryotes.

Chapter 8

Enzymes: Basic Concepts and Kinetics

Enzymes catalyze almost all chemical reactions in a cell and are also involved in the transformations of one form of energy into another. Most enzymes are proteins, but RNA also catalyzes physiologically important reactions. The authors begin this chapter with a brief overview of the catalytic power and specificity of enzymes. They point out that many enzymes require small molecule partners (cofactors) to effect catalysis. They then explain how the thermodynamic concepts of free energy change and free energy of activation are used to determine whether or not chemical reactions can occur and the rate at which they will occur, respectively. They explain how enzyme binding to the transition state of a reaction provides the chemical basis for catalysis. They explain how the velocity of enzyme-catalyzed reactions is analyzed, and they describe enzyme inhibitors and their analysis. This chapter draws on your knowledge of protein structure (Chapter 2) and the interactions between biomolecules (Chapter 1). It sets the stage for the majority of the remaining chapters of the text that deal with biochemical reactions.

LEARNING OBJECTIVES

When you have mastered this chapter, you should be able to accomplish the following objectives.

Enzymes Are Powerful and Highly Specific Catalysts (Text Section 8.1)

1. Explain why *enzymes* are versatile *biological catalysts*.

2. Appreciate that *catalytic power* and *specificity* are critical characteristics of enzymes. Give examples of the rate enhancements of enzymes and the substrate selectivity they display.

3. Realize that both *protein* and *RNA molecules* are enzymes.

4. Define *substrate, cofactor, prosthetic group, apoenzyme,* and *holoenzyme*. Relate *vitamins* to cofactors.

5. Provide examples of *proteases* with diverse *substrate specificity,* and explain how substrate specificity arises from precise interactions of the enzyme with the substrate.

6. Provide examples of enzymes that transduce one form of energy into another.

Free Energy Is a Useful Thermodynamic Function for Understanding Enzymes (Text Section 8.2)

7. Describe how ΔG can be used to predict whether a reaction can occur spontaneously.

8. Write the equation for the ΔG of a chemical reaction. Define the *standard free-energy change* ($\Delta G°$); define $\Delta G'$ and $\Delta G°'$. Interconvert *kilojoules* and *kilocalories*.

9. Derive the relationship between $\Delta G°'$ and the *equilibrium constant* (K'_{eq}) of a reaction. Relate each tenfold change in K'_{eq} to the change in $\Delta G°'$ in kilojoules per mole (kJ/mol) or kcal/mol.

10. Relate the concentrations of reactants and products to $\Delta G'$. Define *endergonic* and *exergonic*.

11. Explain why enzymes do not alter the *equilibrium* of chemical reactions but change only their *rates*.

Enzymes Accelerate Reactions by Facilitating the Formation of the Transition State (Text Section 8.3)

12. Define the *transition state* and the *free energy of activation* ($\Delta G^{\ddagger}$), and describe the effect of enzymes on $\Delta G^{\ddagger}$.

13. Describe the formation of *enzyme-substrate (ES) complexes* and discuss their properties.

14. Summarize the common features of the *active sites* of enzymes, and relate them to the specificity of binding of the substrate.

15. Define the *binding energy* for the association of an enzyme and its substrate in terms of free energy change and describe the chemical bases for the interaction.

16. Explain why enzyme-catalyzed reaction rates reach a maximum value.

The Michaelis-Menten Model Accounts for the Kinetic Properties of Many Enzymes (Text Section 8.4)

17. Review the fundamental terms and equations of the kinetics of chemical reactions. Define *first-order, second-order,* and *zero-order* reactions.

18. Outline the *Michaelis-Menten model of enzyme kinetics* and describe the molecular nature of each of its components.

19. Reproduce the derivation of the *Michaelis-Menten equation* in the text. Relate the Michaelis-Menten equation to experimentally derived plots of *velocity (V)* versus *substrate concentration [S]*. List the assumptions underlying the derivation.

20. Define V_{max} and K_M, and explain how these parameters can be obtained from a plot of V versus [S] or a plot of $1/V$ versus $1/[S]$ (a *Lineweaver-Burk plot*).

21. Explain the significance of V_{max}, K_M, k_2, k_{cat}, and k_{cat}/K_M. Define *kinetic perfection* as it pertains to enzyme catalysis.

22. Distinguish *sequential displacement* and *double displacement* in reactions involving multiple substrates. Provide examples of enzymes using each mechanism.

23. Contrast the *kinetics of allosteric enzymes* with those displaying simple Michaelis-Menten kinetics. Describe the molecular basis of allostery.

Enzymes Can Be Inhibited by Specific Molecules (Text Section 8.5)

24. Describe the functions and uses of *enzyme inhibitors*. Contrast *reversible* and *irreversible* inhibitors.

25. Describe the effects of *competitive, uncompetitive,* and *noncompetitive inhibitors* on the kinetics of enzyme reactions. Apply kinetic measurements and analysis to determine the nature of an inhibitor. Know how inhibitors change double-reciprocal plots.

26. Explain how irreversible inhibitors are used to learn about the active sites of enzymes. Provide examples of *group-specific, reactive substrate-analog, suicide, and transition-state* inhibitors. Know how DIPF reacts with chymotrypsin.

27. Contrast the properties of substrates and *transition-state analogs*.

28. Describe the formation of *catalytic antibodies* and recognize their uses.

29. Outline the mechanism of action of the antibiotic *penicillin*.

SELF-TEST

Enzymes Are Powerful and Highly Specific Catalysts

1. Which of the following are NOT true of enzymes?
   (a) Enzymes are proteins.
   (b) Enzymes have great catalytic power.
   (c) Enzymes bind substrates with high specificity.
   (d) Enzymes use hydrophobic interactions exclusively in binding substrates.
   (e) The catalytic activity of enzymes is often regulated.

2. Enzymes catalyze reactions by
   (a) binding regulatory proteins.
   (b) covalently modifying active-site residues.
   (c) binding substrates with great affinity.
   (d) selectively binding the transition state of a reaction with high affinity.

3. The combination of an apoenzyme with a cofactor forms what? What are the two types of cofactors? What distinguishes a prosthetic group from a cosubstrate?

4. Name a process that converts the energy of light into the energy of chemical bonds.

Free Energy Is a Useful Thermodynamic Function for Understanding Enzymes

5. Which of the following statements is correct? The free energy change of a reaction

   (a) if negative, enables the reaction to occur spontaneously.
   (b) if positive, enables the reaction to occur spontaneously.
   (c) is greater than zero when the reaction is at equilibrium.
   (d) determines the rate at which a reaction will attain equilibrium.

6. Explain why the thermodynamic parameter ΔS cannot be used to predict the direction in which a reaction will proceed.

7. If the standard free-energy change ($\Delta G°$) for a reaction is zero, which of the following statements about the reaction are true?

   (a) The entropy ($\Delta S°$) of the reaction is zero.
   (b) The enthalpy ($\Delta H°$) of the reaction is zero.
   (c) The equilibrium constant for the reaction is 1.0.
   (d) The reaction is at equilibrium.
   (e) The concentrations of the reactants and products are all 1 M at equilibrium.

8. The enzyme triose phosphate isomerase catalyzes the following reaction:

   $$\text{dihydroxyacetone phosphate} \underset{k_{-1}}{\overset{k_1}{\rightleftharpoons}} \text{glyceraldehyde 3-phosphate}$$

   The $\Delta G°'$ for this reaction is 7.66 kJ/mol (1.83 kcal/mol). In light of this information, which of the following statements are correct?

   (a) The reaction would proceed spontaneously from left to right under standard conditions.
   (b) The rate of the reaction in the reverse direction is higher than the rate in the forward direction at equilibrium.
   (c) The equilibrium constant under standard conditions favors the synthesis of the compound on the left, dihydroxyacetone phosphate.
   (d) The data given are sufficient to calculate the equilibrium constant of the reaction.
   (e) The data given are sufficient to calculate the left-to-right rate constant (k_1).

9. Glycogen phosphorylase, an enzyme involved in the metabolism of the carbohydrate polymer glycogen, catalyzes the reaction:

   $$\text{Glycogen}_n + \text{phosphate} \rightleftharpoons \text{glucose 1-phosphate} + \text{glycogen}_{n-1}$$

   $$K'_{eq} = \frac{[\text{glucose 1} - \text{phosphate}][\text{glycogen}_{n-1}]}{[\text{phosphate}][\text{glycogen}_n]} = 0.088$$

   Based on these data, which of the following statements are correct?

   (a) Because glycogen phosphorylase normally *degrades* glycogen in cellular metabolism, there is a paradox in that the equilibrium constant favors synthesis.
   (b) The $\Delta G°'$ for this reaction at 25°C is 5.98 kJ/mol (1.43 kcal/mol).
   (c) The phosphorolytic cleavage of glycogen consumes energy, that is, it is endergonic.
   (d) If the ratio of phosphate to glucose 1-phosphate in cells is high enough, phosphorylase will degrade glycogen.

10. The reaction of the hydrolysis of glucose 6-phosphate to give glucose and phosphate has a $\Delta G°' = -13.8$ kJ/mol (-3.3 kcal/mol). The reaction takes place at 25°C. Initially, the concentration of glucose 6-phosphate is 10^{-5} M, that of glucose is 10^{-1} M, and that of phosphate is 10^{-1} M. Which of the following statements pertaining to this reaction are correct?

(a) The equilibrium constant for the reaction is 260.

(b) The equilibrium constant cannot be calculated because standard conditions do not prevail initially.

(c) The $\Delta G'$ for this reaction under the initial conditions is –3.26 kJ/mol (–0.78 kcal/mol).

(d) Under the initial conditions, the synthesis of glucose 6-phosphate will take place rather than hydrolysis.

(e) Under standard conditions, the hydrolysis of glucose 6-phosphate will proceed spontaneously.

Enzymes Accelerate Reactions by Facilitating the Formation of the Transition State

11. The transition state of an enzyme-catalyzed reaction that converts a substrate to a product

    (a) is a transient intermediate formed along the reaction coordinate of the reaction.

    (b) has higher free energy than either the substrates or products.

    (c) is the most populated species along the reaction coordinate.

    (d) is increased in concentration because the enzyme binds tightly to it.

    (e) determines the velocity of the reaction.

12. Explain briefly how enzymes accelerate the rate of reactions.

13. Which of the following statements is true? Enzyme catalysis of a chemical reaction

    (a) decreases $\Delta G'$ so that the reaction can proceed spontaneously.

    (b) increases the energy of the transition state.

    (c) does not change $\Delta G^{\circ\prime}$, but rather changes the ratio of products to reactants at equilibrium.

    (d) decreases the entropy of the reaction.

    (e) increases the forward and reverse reaction rates.

14. Which of the following statements regarding an enzyme-substrate complex (ES) is true?

    (a) The heat stability of an enzyme frequently changes upon the binding of a substrate.

    (b) At sufficiently high concentrations of substrate, the catalytic sites of the enzyme become filled and the reaction rate reaches a maximum.

    (c) An enzyme-substrate complex can usually be isolated.

    (d) Enzyme-substrate complexes can usually be visualized by x-ray crystallography.

    (e) Spectroscopic changes in the substrate or the enzyme can be used to detect the formation of an enzyme-substrate complex.

15. Why is there a high degree of stereospecificity in the interaction of enzymes with their substrates?

16. Explain why the forces that bind a substrate at the active site of an enzyme are usually weak.

The Michaelis-Menten Model Accounts for the Kinetic Properties of Many Enzymes

17. Which of the following statements regarding simple Michaelis-Menten enzyme kinetics are correct?

    (a) The maximal velocity V_{max} is related to the maximal number of substrate molecules that can be "turned over" in unit time by a molecule of enzyme.

(b) K_M is expressed in terms of a reaction velocity (e.g., mol s^{-1}).

(c) K_M is the dissociation constant of the enzyme-substrate complex.

(d) K_M is the concentration of substrate required to achieve half of V_{max}.

(e) K_M is the concentration of substrate required to convert half the total enzyme into the enzyme-substrate complex.

18. Explain the relationship between K_M and the dissociation constant of the enzyme-substrate complex K_{ES}.

19. Myoglobin binds and releases O_2 in muscle cells: myoglobin + O_2 $\longrightarrow$ myoglobin·O_2. The fraction of myoglobin (Y) saturated with O_2 is given by the equation

$$Y = \frac{pO_2}{pO_2 + P_{50}}$$

where p is the partial pressure of the O_2 and P_{50} is the pressure of O_2 at which 50% of the myoglobin is saturated with O_2. (This value reflects the equilibrium constant for the reaction.) Note the similarity between this equation and the Michaelis-Menten equation

$$\frac{V}{V_{max}} = \frac{[S]}{[S] + K_m}.$$

Explain the relationships between the two equations.

20. From the plot of velocity versus substrate concentration shown in Figure 8.1, obtain the following parameters. (The amount of enzyme in the reaction mixture is 10^{-3} mmol.)

(a) K_M

(b) V_{max}

(c) k_2/K_M

(d) Turnover number

FIGURE 8.1   Plot of reaction velocity versus substrate concentration.

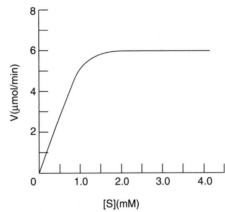

21. What is the significance of k_{cat}/K_M?

22. Which of the following statements is correct? The turnover number for chymotrypsin is 100 s^{-1}, and for DNA polymerase it is 15 s^{-1}. This means that

    (a) chymotrypsin binds its substrate with higher affinity than does DNA polymerase.
    (b) the velocity of the chymotrypsin reaction is always greater than that of the DNA polymerase reaction.
    (c) the velocity of the chymotrypsin reaction at a particular enzyme concentration and saturating substrate levels is lower than that of the DNA polymerase reaction under the same concentration conditions.
    (d) the velocities of the reactions catalyzed by both enzymes at saturating substrate levels could be made equal if 6.7 times more DNA polymerase than chymotrypsin were used.

Enzymes Can Be Inhibited by Specific Molecules

23. Which of the following statements about the different types of enzyme inhibition are correct?

    (a) Competitive inhibition occurs when a substrate competes with an enzyme for binding to an inhibitor protein.
    (b) Competitive inhibition occurs when the substrate and the inhibitor compete for the same active site on the enzyme.
    (c) Uncompetitive inhibition of an enzyme cannot be overcome by adding large amounts of substrate.
    (d) Competitive inhibitors are often similar in chemical structure to the substrates of the inhibited enzyme.
    (e) Noncompetitive inhibitors often bind to the enzyme irreversibly.

24. If the K_M of an enzyme for its substrate remains constant as the concentration of the inhibitor increases, what can be said about the mode of inhibition?

25. The kinetic data for an enzymatic reaction in the presence and absence of inhibitors are plotted in Figure 8.2. Identify the curve that corresponds to each of the following:

    (a) no inhibitor
    (b) noncompetitive inhibitor
    (c) competitive inhibitor
    (d) mixed inhibitor

FIGURE 8.2   Effects of inhibitors on a plot of V versus [S].

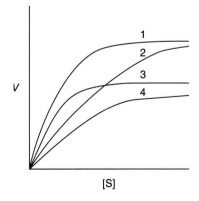

26. Draw approximate Lineweaver-Burk plots for each of the inhibitor types in question 25.

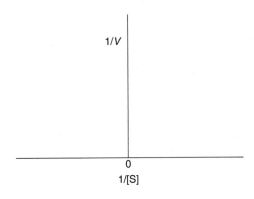

27. Which statements are NOT true about a transition state analog?
    (a) It fits better in the active site than the substrate.
    (b) It increases the rate of product formation.
    (c) It can be used as a hapten to produce catalytic antibodies.
    (d) It is usually a distorted or strained molecule.
    (e) It is a potent inhibitor of the enzyme.

28. The inhibition of bacterial cell wall synthesis by penicillin is a classic example of a medically significant inhibition of an enzymatic reaction. Which of the following statements about the inhibition of glycopeptide transpeptidase by penicillin is true?
    (a) The inhibition is noncompetitive.
    (b) Penicillin binds irreversibly to an allosteric site of the enzyme.
    (c) Penicillin inhibits bacterial cell wall synthesis by incorrectly cross-linking the peptides of the proteoglycan.
    (d) The penicilloyl-enzyme intermediate may be dissociated by high concentrations of D-alanine.
    (e) Penicillin resembles acyl-D-Ala-D-Ala, one of the substrates of the transpeptidase.

ANSWERS TO SELF-TEST

1. d. (a) is incorrect because some enzymes are RNA.

2. d. (c) is incorrect because, although tight binding to the substrates helps confer specificity on the reaction, it increases the activation barrier to reaction. Tight substrate binding makes binding to the transition state of the reaction more energetically costly, that is, it increases the free energy of activation of the reaction.

3. Holoenzyme. Cofactors may be metal ions or low molecular weight organic molecules. A prosthetic group is a tightly bound cofactor that seldom dissociates from the enzyme. Cofactors that are loosely bound behave like cosubstrates; they are easily bound and released from the enzyme.

4. Photosynthesis. The sun provides light energy that photosynthesis converts into chemical bond energy in the form of ATP. Other examples of energy transduction include the use of an ion gradient in mitochondria to drive the synthesis of chemical bonds, and the use of the energy in ATP to cause the movement of muscles.

5. a

6. The thermodynamic parameter ΔS for a chemical reaction is not easily measured. Even if it were easily determined, its value depends on changes that occur not only in the system under study but also in the surroundings (see Chapter 1). Intrinsically unfavorable reactions ($\Delta G^{o\prime} > 0$) can take place if a change in the surroundings compensates for a decrease in the entropy (negative ΔS) of the reaction.

7. c, e. $\Delta G^{o} = -RT \ln_{10} K_{eq}$. When $K_{eq} = 1$, $\Delta G^{o} = 0$ because the natural log of $1 = 0$. (e) is correct by definition.

8. c, d

9. All of the statements are correct.

   (a) The paradox is that although glycogen normally degrades glycogen to form glucose 1-phosphate, the standard free energy change of the reaction is positive, that is, the reaction is endergonic. See the answer to (d) for a resolution of the paradox.

   (b) Using K'_{eq}, one can calculate the $\Delta G^{o\prime}$ for the phosphorylase reaction:

   $$\Delta G^{o\prime} = -RT \ln K'_{eq}$$

   $$= -8.31 \frac{J}{mol\,^{\circ}K} \times 298\,^{\circ}K \times \ln(0.088)$$

   $$= -2476\ J/mol \times -2.43$$

   $$= 6017\ J/mol = 6.02\ kJ/mol\ (1.44\ kcal/mol)$$

   (c) In part (b) the $\Delta G^{o\prime}$ for the phosphorylase reaction of 6.02 kJ/mol was calculated; therefore, energy is consumed rather than released by this reaction.

   (d) In cells, the ratio of phosphate to glucose 1-phosphate is so large that phosphorylase is mainly involved with glycogen degradation.

10. a, d, e

   (a) $$\Delta G^{o\prime} = -RT \ln K'_{eq}$$

   $$-13.8\ kJ/mol = -8.31\ J/mol \times 298jK \times \ln K'_{eq}$$

   $$\ln K'_{eq} = 5.56$$

   $$K'_{eq} = 260$$

   (b) Incorrect. K'_{eq} is a constant; it is independent of the initial concentrations.

   (c) Incorrect.

   $$\Delta G' = \Delta G^{o\prime} + RT\ \ln \frac{[glucose][phosphate]}{[glucose\ 6\text{-}phosphate]}$$

   $$= -13.8\ kJ/mol + \left(2.48\ kJ/mol \times \ln \frac{10^{-1} \times 10^{-1}}{10^{-5}}\right)$$

   $$= -13.8\ kJ/mol + \left(2.48\ kJ/mol \times \ln \frac{10^{-2}}{10^{-5}}\right)$$

   $$= -13.8\ kJ/mol + (2.48\ kJ/mol \times 6.91)$$

   $$= +3.3\ kJ/mol\ (+0.79\ kcal/mol)$$

(d) Correct. Under the initial conditions, $\Delta G'$ is positive; therefore, the reaction will proceed toward the formation of glucose 6-phosphate.

(e) Correct. The negative $\Delta G^{o\prime}$ value (at standard conditions) indicates that the reaction will proceed spontaneously toward the hydrolysis of glucose 6-phosphate.

11. a, b, d, e. (c) is incorrect because it has the most energy and is therefore hardest to form. The velocity of the reaction is directly proportional to the concentration of the transition state.

12. Enzymes have evolved to bind tightly the transition state of the reaction they catalyze. By binding the transition state with high affinity, they facilitate its formation. Hydrogen bonds and ionic and hydrophobic interactions can be involved in binding the transition state. The more transition state formed, the faster the reaction.

13. e. The enzyme speeds up the rate of attainment of equilibrium.

14. a, b, e. Turnover of ES to form P usually makes isolating ES difficult. In reactions requiring two substrates, an enzyme-substrate complex of one of the substrates can be isolated in the absence of the other substrate if the complex is very stable. The absence of the cosubstrate precludes turnover of ES. The same consideration applies to ES complexes formed for x-ray crystallography.

15. The formation of an enzyme-substrate complex involves a close, complementary fitting of the atoms of the amino-acid-residue side chains that make up the active site of the enzyme with the atoms of the substrate. Since stereoisomers have different spatial arrangements of their atoms, only a single stereoisomer of the substrate usually fits into the active site in a form capable of being acted upon by the enzyme.

16. The enzyme-substrate and enzyme-product complexes must be reversible for catalysis to proceed; therefore, weak forces are involved in the binding of substrates to enzymes.

17. a, d. Answer (e) is correct only when $K_M = K_{ES}$. See Question 18.

18. K_M can be equal to K_{ES} when the rate constant $k_2 \ll k_{-1}$. Since $K_M = (k_2 + k_{-1})/k_1$, when k_2 is negligible relative to k_{-1}, K_M becomes equal to k_{-1}/k_1, which is the dissociation constant of the enzyme-substrate complex.

19. These equations are related because they express the occupancy of saturable binding sites as a function of either O_2 or substrate concentration. The fraction of active sites filled, as reflected in V/V_{max}, is analogous to Y, the degree of myoglobin saturation with oxygen; [S] and pO_2 are the concentrations of substrate and O_2, respectively; and K_M and P_{50} are substrate or O_2 concentrations at half-maximal saturation.

20. (a) $K_M = 5 \times 10^{-4}$ M. The value of the asymptote in Figure 8.1 is 6.0 μmol/min. K_M is equal to [S] at $1/2\ V_{max}$. Note that the units of [S] are mM.

(b) $V_{max} = 6$ μmol/min. V_{max} is obtained from Figure 8.1; it is the maximum velocity.

(c) $k_3/K_M = 2 \times 10^5\ \text{s}^{-1}\ \text{M}^{-1}$. In order to calculate this ratio, k_2 must be known. Since $V_{max} = k_2[E_T]$, $k_2 = V_{max}/[E_T]$. Thus

$$k_2 = \frac{6\,\mu\text{mol}\,/\,\text{min}}{10^{-3}\,\mu\text{mol}}$$

$$= 6 \times 10^3\ \text{min}^{-1}$$

$$= 100\ \text{s}^{-1}$$

Using K_M from part (a),

$$\frac{K_2}{K_M} = \frac{100 \text{ s}^{-1}}{5 \times 10^{-4} \text{ M}} = 2 \times 10^5 \text{ s}^{-1} \text{ M}^{-1}$$

(d) The turnover number is 100 s^{-1}, equal to k_2, which was calculated in part (c).

21. Since $V_0 = (k_{cat}/K_M)$ [S] [ET], k_{cat}/K_M represents the second-order rate constant for the encounter of S with E. The ratio k_{cat}/K_M thus allows one to estimate the catalytic efficiency of an enzyme. The upper limit for k_{cat}/K_M, 10^8 to 10^9 M^{-1} s^{-1}, is set by the rate of diffusion of the substrate in the solution, which limits the rate at which it encounters the enzyme. If an enzyme has a k_{cat}/K_M in this range, its catalytic velocity is restricted only by the rate at which the substrate can reach the enzyme, which means that the enzymatic catalysis has attained kinetic perfection.

22. d. $V_{max} = k_2[E_T]$; thus, if 6.7 times more DNA polymerase than chymotrypsin is used, V_{max} for both enzymes is the same:

$$100 \text{ s}^{-1} = 6.7 \times 15 \text{ s}^{-1}$$

Answer (a) is incorrect because the affinity of substrate for the enzyme is given by $K_{ES} = k_{-1}/k_1$. Answer (b) is incorrect because the velocity of the enzymatic reactions is a function of K_M, V_{max}, and substrate concentration. Answer (c) is incorrect because for the same enzyme concentration, $V_{max} = k_2[E_T]$ is greater for chymotrypsin than for DNA polymerase.

23. b, c, d

24. The inhibition is noncompetitive because the proportion of bound substrate remains the same as the concentration of the inhibitor increases.

25. (a) 1 (b) 3 (c) 2 (d) 4

26. See Figure 8.3. Plots 1 and 2 have the same 1/V intercept; plots 1 and 3 have the same 1/[S] intercept; and plots 1 and 4 have different 1/V and 1/[S] intercepts.

FIGURE 8.3   Lineweaver-Burk plots for competitive (2), noncompetitive (3), and mixed (4) inhibition, relative to the enzymatic reaction in the absence of inhibitors (1).

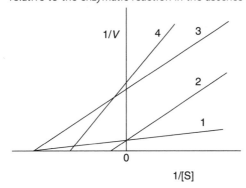

27. b, d. Answer (b) is incorrect because transition state analogs are inhibitors of the corresponding enzymes. Therefore, they decrease rather than increase enzyme reaction rates. Answer (d) is incorrect because transition-state analogs are not necessarily strained or distorted; rather, they mimic the shape of the transition state, which may itself be strained or distorted.

28. e

PROBLEMS

1. Calculate the values for $\Delta G^{o'}$ that correspond to the following values of K'_{eq}. Assume that the temperature is 25°C.

    (a) 1.5×10^4
    (b) 1.5
    (c) 0.15
    (d) 1.5×10^{-4}

2. Calculate the values for K'_{eq} that correspond to the following values of $\Delta G^{o'}$. Assume that the temperature is 25°C.

    (a) -41.84 kJ/mol (-10 kcal/mol)
    (b) -4.18 kJ/mol (-1 kcal/mol)
    (c) $+4.18$ kJ/mol ($+1$ kcal/mol)
    (d) $+41.84$ kJ/mol ($+10$ kcal/mol)

3. The enzyme aldolase catalyzes the following reaction:

    $$\text{Fructose 1,6-bisphosphate} \rightleftharpoons$$

    $$\text{dihydroxyacetone phosphate} + \text{glyceraldehyde 3-phosphate}$$

    For this reaction, $\Delta G^{o'} = +23.8$ kJ/mol ($+5.7$ kcal/mol).

    (a) Calculate the change in free energy $\Delta G'$ for this reaction under typical intracellular conditions using the following concentrations: fructose 1,6-bisphosphate, 0.15 mM; dihydroxyacetone phosphate, 4.3×10^{-6} M; and glyceraldehyde 3-phosphate, 9.6×10^{-5} M. Assume that the temperature is 25°C.
    (b) Explain why the aldolase reaction occurs in cells in the direction written despite the fact that it has a positive free-energy change under standard conditions.

4. The text states (page 235) that a decrease of 5.69 kJ/mol (1.36 kcal/mol) in the free energy of activation of an enzyme-catalyzed reaction has the effect of increasing the rate of conversion of substrate to product by a factor of 10. What effect would this decrease of 5.69 kJ/mol in the free energy of activation have on the reverse reaction, the conversion of product to substrate? Explain.

5. What is the ratio of [S] to K_M when the velocity of an enzyme-catalyzed reaction is 80% of V_{max}?

6. The simple Michaelis-Menten model (equation 12 in the text, page 240) applies only to the initial velocity of an enzyme-catalyzed reaction, that is, to the velocity when no appreciable amount of product has accumulated. What feature of the model is consistent with this constraint? Explain.

7. Two first-order rate constants, k_{-1} and k_2, and one second-order rate constant, k_1, define K_M by the relationship

    $$K_M = \frac{k_{-1} + k_2}{k_1}$$

    By substituting the appropriate units for the rate constants in this expression, show that K_M must be expressed in terms of concentration.

8. Suppose that two tissues, tissue A and tissue B, are assayed for the activity of enzyme X. The activity of enzyme X, expressed as the number of moles of substrate converted to product per gram of tissue, is found to be five times greater in tissue A than in tissue B under a variety of circumstances. What is the simplest explanation for this observation?

9. Sketch the appropriate plots on the following axes. Assume that simple Michaelis-Menten kinetics apply, and that the pre-steady state occurs so rapidly that it need not be considered (see Section 8.4).

FIGURE 8.10

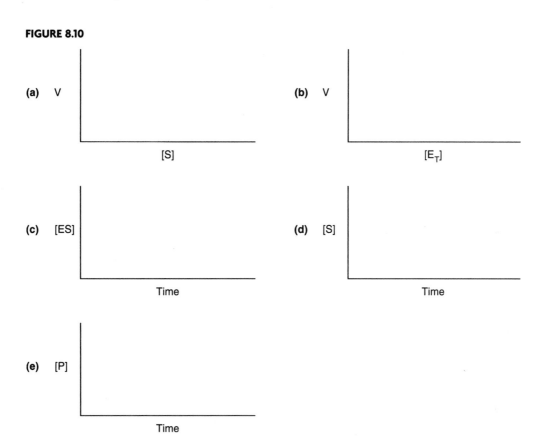

10. Suppose that the data shown below are obtained for an enzyme-catalyzed reaction.

[S](mM)	V (mmol ml^{-1} min^{-1})
0.1	3.33
0.2	5.00
0.5	7.14
0.8	8.00
1.0	8.33
2.0	9.09

(a) From a double-reciprocal plot of the data, determine K_M and V_{max}.

(b) Assuming that the enzyme present in the system had a concentration of 10^{-6} M, calculate its turnover number.

11. Suppose that the data shown below are obtained for an enzyme-catalyzed reaction in the presence and absence of inhibitor X.

	V (mmol ml^{-1} min^{-1})	
[S](mM)	Without X	With X
0.2	5.0	3.0
0.4	7.5	5.0
0.8	10.0	7.5
1.0	10.7	8.3
2.0	12.5	10.7
4.0	13.6	12.5

(a) Using double-reciprocal plots of the data, determine the type of inhibition that has occurred.
(b) Does inhibitor X combine with E, with ES, or with both? Explain.
(c) Calculate the inhibitor constant K_i for substance X, assuming that the final concentration of X in the reaction mixture was 0.2 mM.

12. Suppose that the data shown below are obtained for an enzyme-catalyzed reaction in the presence and absence of inhibitor Y.

	V (mmol ml^{-1} min^{-1})	
[S](mM)	Without Y	With Y
0.2	5.0	2.0
0.4	7.5	3.0
0.8	10.0	4.0
1.0	10.7	4.3
2.0	12.5	5.0
4.0	13.6	5.5

(a) Using double-reciprocal plots of the data, determine the type of inhibition that has occurred.
(b) Does inhibitor Y combine with E, with ES, or with both? Explain.
(c) Calculate the inhibitor constant K_i for substance Y, assuming that the final concentration of Y in the reaction mixture was 0.3 mM.

13. Although the double-reciprocal plot is the most widely used plotting form for enzyme kinetic data, it suffers from a major disadvantage. If linear increments of substrate concentration are used, thereby minimizing measurement errors in the laboratory, data points will be obtained that cluster near the vertical axis. Thus the intercept on the ordinate can be determined with great accuracy, but the slope of the line will be subject to considerable error, because the least reliable data points, those obtained at low substrate concentrations, have greater weight in establishing the slope. (Remember that many enzymes are protected against denaturation by the presence of their substrates at high concentrations.)

Because of the limitation of double-reciprocal plots described above, other linear plotting forms have been devised. One of these, the Eadie plot, graphs V versus V/[S]. Another, the Hanes-Woolf plot, ([S]/V versus [S]) is perhaps the most useful in minimizing the difficulties of the double-reciprocal plot.

(a) Rearrange the Michaelis-Menten equation to give [S]/V as a function of [S].

(b) What is the significance of the slope, the vertical intercept, and the horizontal intercept in a plot of [S]/V versus [S]?

(c) Data shown below were obtained for the hydrolysis of *o*-nitrophenyl-β-D-galactoside (ONPG) by *E. coli* β-galactosidase. Use both double-reciprocal and Hanes-Woolf plots to analyze these data, and calculate values for K_M and V_{max} from both plots. (We suggest that you use a graphing program to generate a scatterplot, and then fit the data using a linear curve-fitting algorithm.)

[S](mM)	V (μmol ml^{-1} min^{-1})
0.5	8.93
1.0	14.29
1.5	16.52
2.0	19.20
2.5	19.64

(d) Make a sketch of a plot [S]/V versus [S] in the absence of an inhibitor as in the presence of a competitive inhibitor and in the presence of a noncompetitive inhibitor.

14. Suppose that a modifier Q is added to an enzyme-catalyzed reaction with the results depicted in Figure 8.4. What role does Q have? Does it combine with E, with ES, or with both E and ES?

FIGURE 8.4   Effects of modifier Q on a plot of 1/V versus 1/[S].

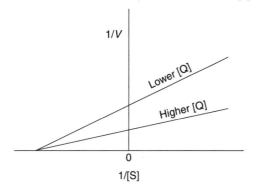

15. The enzyme DNA ligase catalyzes the formation of a phosphodiester bond at a break (nick) in the phosphodiester backbone of a duplex DNA molecule. The enzyme from bacteriophage T4 uses the free energy of hydrolysis ATP as the energy source for the formation of the phosphodiester bond. A covalently modified form of the enzyme in which AMP is bound to a lysine side chain is an intermediate in the reaction. The intermediate is formed by the reaction of E + ATP to form E-AMP + PP_i. In the next step, the AMP is transferred from the enzyme to a phosphate on the DNA to form a pyrophosphate-linked DNA-AMP. In the last step of the reaction, the phosphodiester bond is formed by the free enzyme to seal the nick in the DNA and AMP is released.

(a) Write chemical equations that show the individual steps that occur over the course of the overall reaction.

(b) Does this enzyme catalyze a double-displacement reaction?

(c) Do you think that if DNA were omitted from the reaction mixture, the enzyme would catalyze a partial reaction? If so, what reaction might it catalyze?

16. If you were studying an enzyme that catalyzed the reaction of ATP and fructose 1-phosphate to form fructose 1,6-bisphosphate and ADP and discovered that a plot of the initial velocity of formation of fructose 1,6-bisphosphate versus ATP concentration was not hyperbolic, but rather sigmoid, what would you suspect?

17. Assume that you have allowed the alcohol dehydrogenase reaction illustrated on page 242 of this chapter in the text to come to equilibrium, *in vitro,* so that the concentrations of reactant ethanol and the product acetaldehyde were not changing. What would happen to the equilibrium ratio of products if you added more ethanol and acetaldehyde (assume 5% of the equilibrium concentrations of each) to the reaction mixture? If you added a small amount of high-specific-activity ^{14}C-labelled ethanol (not enough to shift the equilibrium position) to the reaction mixture, waited, and then isolated the ethanol and acetaldehyde from the mixture, where would you expect to find the radioisotopic label? Why?

18. Converting between calories and joules is a skill that scientists frequently need. A rule of thumb is to remember that a calorie is about one-fourth of a joule. Specifically, 0.239 cal = 1 J. The following problem puts this conversion in context. In Europe, food energy content is expressed in kJ rather than kcal. (Food "calories" are really kcal.) If a cup of yogurt has 1370 kJ of energy and you are trying to keep your lunch to a maximum of 350 kcal (food calories), could you eat the entire container of yogurt? How many kcal are represented by 1370 kJ?

19. The enzyme cyclooxygenase (COX) is part of a larger enzyme called prostaglandin H2 synthase. This enzyme converts the 20-carbon fatty acid arachidonic acid to lipid messengers called prostaglandins. The conversion of arachidonate to prostaglandin H2 is shown below. Prostaglandins are responsible for many physiological processes, including the contraction and relaxation of smooth muscle, and they play a role in mediating inflammation.

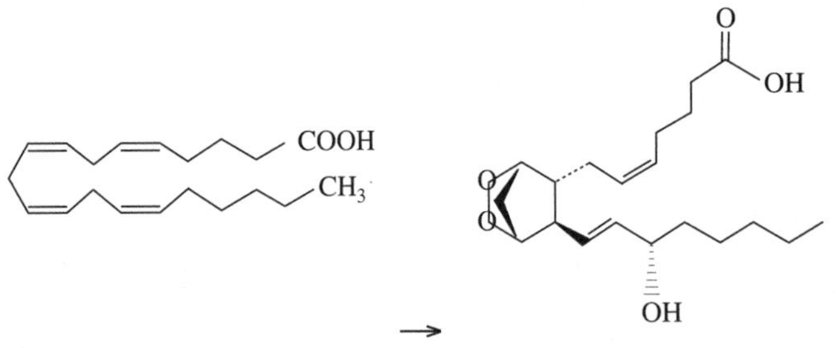

Arachidonic Acid → Prostaglandin H$_2$

(a) Aspirin (shown below) acetylates a specific serine residue of prostaglandin H$_2$ synthase, preventing arachidonate from reaching the cyclooxygenase active site. What kind of an inhibitor is aspirin? Explain.

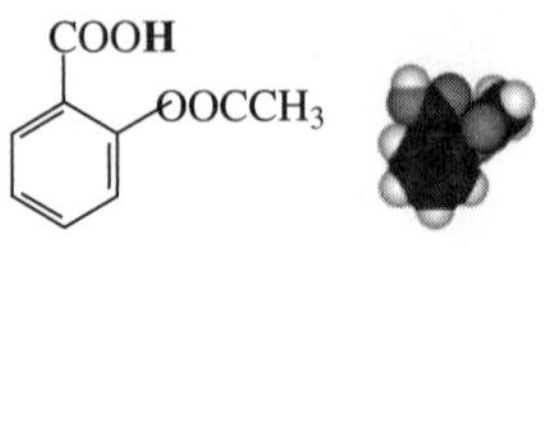

(b) Ibuprofen (shown below), the active ingredient in Motrin™ and Advil™, binds noncovalently to the enzyme in its active site. How would you expect the K_M and V_{max} of the enzyme to be affected by Ibuprofen? Draw a Lineweaver-Burk plot for this enzyme in the presence and absence of Ibuprofen and label each line.

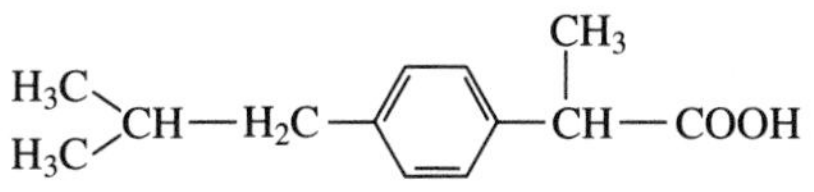

$$H_3C \diagdown CH-H_2C-\!\!\langle\ \rangle\!\!-\overset{\overset{\displaystyle CH_3}{|}}{CH}-COOH$$

(c) How could one overcome the inhibition of the enzyme caused by Ibuprofen?

20. It is important to "have a feel" for what a K_m means in a physical sense. One way to attain this intuition is to compare different enzymes. For example, two K_m values can be reported for the enzyme hexokinase with two different substrates. Hexokinase catalyzes the phosphorylation of six-carbon sugars, trapping them in cells. The K_m for hexokinase with glucose is 0.15 mM. The K_m for hexokinase with fructose is 1.5 mM. Which substrate does hexokinase bind more strongly?

21. Proteins such as cell surface receptors are often described by a value called $K_{0.5}$. This value is analogous to a K_m in that it reflects tightness of binding. It is defined as the concentration of ligand required to reach 50% saturation. Brain cells have glucose-binding proteins called GLUT 3 receptors on their membranes. Muscle and fat cells have similar receptors called GLUT 4 receptors. Will the $K_{0.5}$ of brain GLUT 3 receptors be higher or lower than the $K_{0.5}$ of muscle/fat GLUT 4 receptors? What is the physiological significance of this difference?

ANSWERS TO PROBLEMS

1. The values for $\Delta G^{o\prime}$ are found by substituting the values for K'_{eq} into equation 3 on page 233 of the text.

   (a) $\Delta G^{o\prime} = -RT \ln K'_{eq}$

   $$= -8.31 \times 298 \ln (1.5 \times 10^4)$$

   $$= -20.4 \text{ kJ} / \text{mol} (-5.7 \text{ kcal} / \text{mol})$$

   (b) −1.00 kJ/mol (−0.24 kcal/mol)
   (c) +4.60 kJ/mol (+1.1 kcal/mol)
   (d) +21.76 kJ/mol(+5.2 kcal/mol)

2. Equation 8 in Section 8.2 is used to find the answers.

   (a) $$K'_e = e^{\frac{-\Delta G'}{RT}} = e^{\frac{-\Delta G}{2.48}}$$

   $$= e^{\frac{41.84}{2.48}} = e^{16.87}$$

   $$= 2.12$$

   (b) 5.42
   (c) 0.18
   (d) 4.71×10^{-8}

3. (a) The applicable relationship is the same equation used in 3a:

$$\Delta G' = \Delta G^{\circ\prime} + RT \ln \frac{[C][D]}{[A][B]}$$

$$= \Delta G^{\circ\prime} + RT \ln \frac{[DHAP][G3P]}{[FBP]}$$

$$= +2.38 \text{ kJ}/\text{mol} + (248)$$

$$\times \ln \frac{\left(4.3 \times 10^{-6}\right) \times \left(9.6 \times 10^{-5}\right)}{0.15 \times 10^{-3}}$$

$$= +23.8 \text{ kJ}/\text{mol} - 31.7 \text{ kJ}/\text{mol}$$

$$= -7.9 \text{ kJ}/\text{mol} \ (-1.9 \text{ kcal}/\text{mol})$$

(b) The reaction occurs in the direction written because of the effects of the concentrations on the free-energy change. The concentration term in the equation is much smaller than 1.0, which is its value under standard conditions. Removal of G3P by a subsequent reaction keeps its concentration low.

4. The rate of the reverse reaction must also increase by a factor of 10. Enzymes do not alter the equilibria of processes; they affect the rate at which equilibrium is attained. Since the equilibrium constant K_{eq} is the quotient of the rate constants for the forward and reverse reactions, both rate constants must be altered by the same factor. If the rate of the forward reaction is increased by a factor of 10, the rate of the reverse reaction must also increase by the same factor.

5. Start with the Michaelis-Menten equation, equation 26 on page 242 of the text:

$$V = V_{max} \frac{[S]}{[S] + K_M}$$

Substituting $0.8 \, V_{max}$ for V yields

$$0.8 \, V_{max} = V_{max} \frac{[S]}{[S] + K_M}$$

$$0.8[S] + 0.8 K_M = [S]$$

$$0.8 K_M = 0.2[S]$$

$$[S] = 4 K_M$$

$$\frac{[S]}{K_M} = 4$$

Thus, a substrate concentration four times greater than the Michaelis constant yields a velocity that is 80% of maximal velocity.

6. Equation 12 on page 240 of the text shows the k_2 step as being irreversible. This is true in practice at the initial stage of the reaction because P and E cannot recombine

to give ES at an appreciable rate if negligible P is present. Note that the equation reveals nothing about the relative magnitudes of k_2 and the reverse rate constant for this step, k_{-2}:

$$E + S \overset{k_1}{\underset{k_{-1}}{\leftrightarrow}} ES \overset{k_2}{\underset{k_{-2}}{\leftrightarrow}} E + P$$

The reverse constant k_{-2} may actually be quite large compared with k_2; nevertheless, the reverse reaction will not occur when little product is present, since the rate of the k_{-2} step depends on the concentrations of P and E as well as on the magnitude of its rate constant.

7. The first-order rate constants have the dimensions t^{-1}, whereas the second-order constant has the dimension $conc^{-1} \, t^{-1}$. Thus, we can carry out the following dimensional analysis:

$$K_M = \frac{k_{-1} + k_2}{k_1}$$

$$= \frac{t^{-1} + t^{-1}}{conc^{-1}t^{-1}}$$

$$= conc$$

8. For the activity of enzyme X to be five times greater in tissue A than in tissue B, tissue A must have five times the amount of enzyme X as does tissue B. Enzyme activity is directly proportional to enzyme concentration.

9. The sketches should resemble the following:

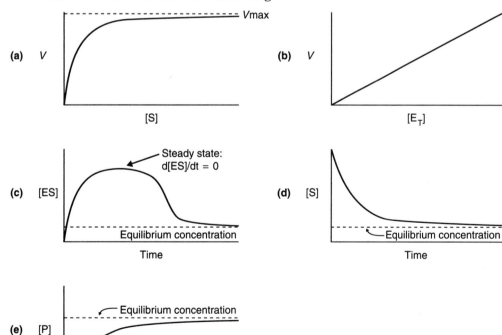

10. (a) See the graph, Figure 8.5. $V_{max} = 1/0.1 = 10$ mmol ml^{-1} min^{-1}.

$$\text{Slope} = \frac{0.3 - 0.1}{10} = 0.02$$

$$\text{Slope} = \frac{K_M}{V_{max}}$$

$K_M = 0.02 \times 10 = 0.2$ mM

FIGURE 8.5 A double-reciprocal plot of data for problem 11.

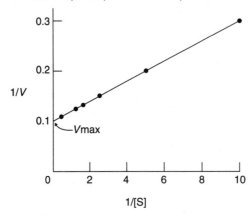

(b) The turnover number is equal to the rate constant k_2 in equation 25 on page 240 of the text. Rearrangement of the equation gives

$$k_2 = \frac{V_{max}}{[E_T]}$$

$$= \frac{10 \text{ mmol ml}^{-1} \text{ min}^{-1}}{10^{-6} \text{ mol liter}^{-1}}$$

$$= \frac{10 \text{ mol liter}^{-1} \text{ min}^{-1}}{10^{-6} \text{ mol liter}^{-1}}$$

$$= 10^7 \text{ min}^{-1} \quad \text{or} \quad 1.7 \times 10^5 \text{ s}^{-1}$$

11. (a) See Figure 8.6. The double-reciprocal plots intersect on the y-axis, so the inhibition is competitive.

FIGURE 8.6 A double-reciprocal plot of data for problem 12 showing the effects of an inhibitor X.

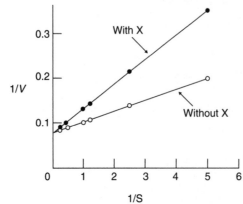

(b) The inhibitor combines only with E, the free enzyme. A competitive inhibitor cannot combine with ES because the inhibitor and the substrate compete for the same binding site on the enzyme.

(c) An inhibitor increases the slope of a double-reciprocal plot by a factor of $1 + [I]/K_i$:

$$\text{Slope}_{\text{inhib}} = \text{slope}_{\text{uninhib}}\left(1 + \frac{[I]}{K_i}\right)$$

The slope with X is

$$\text{Slope}_{\text{inhib}} = \frac{0.333 - 0.067}{5} = 0.0532$$

The slope without X is

$$\text{Slope}_{\text{uninhib}} = \frac{0.200 - 0.067}{5} = 0.0266$$

Substituting in these values yields

$$0.0532 = 0.0266\left(1 + \frac{0.2 \text{ mM}}{K_i}\right)$$

$$K_i = 0.2 \text{ mM}$$

12. (a) See Figure 8.7. The inhibition was noncompetitive, as indicated by the fact that the double-reciprocal plots intersect to the left of the y-axis.

FIGURE 8.7   A double-reciprocal plot of data for problem 13 showing the effects of an inhibitor Y.

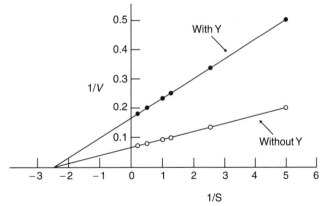

(b) A noncompetitive inhibitor combines at a site other than the substrate binding site. Thus, it may combine with both E and ES. In the case illustrated, the inhibitor has equal affinity for E and ES, which is shown by the fact that the plots intersect on the x-axis.

(c) Again, the slope increases by a factor of $1 + [I]/K_i$ in the presence of an inhibitor.

$$\text{Slope}_{\text{inhib}} = \text{Slope}_{\text{uninhib}}\left(1 + \frac{[I]}{K_i}\right)$$

The slope with Y is

$$\text{Slope}_{\text{inhib}} = \frac{0.500}{5.0 - (-2.5)} = 0.0667$$

The slope without Y is

$$\text{Slope}_{\text{uninhib}} = \frac{0.200}{5.0 - (-2.5)} = 0.0267$$

Substituting in these values yields

$$0.0667 = 0.0267 \left(1 + \frac{0.3 \text{ mM}}{K_i}\right)$$

$$K_i = 0.2 \text{ mM}$$

13. (a) We start with the Michaelis-Menten equation:

$$V = V_{\text{max}} [S]/(K_M + [S])$$

Cross multiplying yields

$$V(K_M + [S]) = V_{\text{max}}[S]$$

Division of both sides by V/V_{max} gives

$$[S]/V = (K_M + [S])/V_{\text{max}}$$

$$[S]/V = K_M/V_{\text{max}} + [S]/V_{\text{max}}$$

$$[S]/V = (1/V_{\text{max}})[S] + K_M/V_{\text{max}}$$

(b) The linear equation above is in the form, $y = mx + b$, where m is the slope, and b the y-intercept. Therefore, the slope of a Hanes-Woolf plot is $(1/V_{\text{max}})$, and the intercept on the y-axis is K_M/V_{max}. The plot will intercept the x-axis when $[S]/V$ is zero. Then

$$0 = (1/V_{\text{max}})[S] + K_M/V_{\text{max}}$$

$$-K_M/V_{\text{max}} = (1/V_{\text{max}})[S]$$

$$[S] = -K_M$$

(c) See Figure 8.8. The y-intercept of the double-reciprocal plot is $1/V_{\text{max}}$. Therefore $V_{\text{max}} = 1/0.034 = 29.4 \ \mu\text{mol l}^{-1} \ \text{min}^{-1}$. The slope of the double-reciprocal plot is K_M/V_{max}. Therefore,

$$0.039 = K_M/29.4$$

$$K_M = 1.15 \text{ mM}$$

FIGURE 8.8 A double-reciprocal plot of data for problem 14.

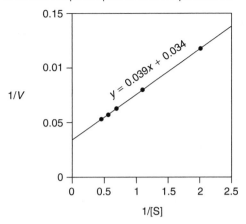

See Figure 8.9. The slope of the Hanes-Woolf plot is $1/V_{max}$. Therefore $V_{max} = 1/0.035 = 28.6$ μmol l^{-1} min^{-1}. The y-intercept of the Hanes-Woolf plot is K_M/V_{max}. Therefore,

$$0.037 = K_M/28.6$$

$$K_M = 1.06 \text{ mM}.$$

FIGURE 8.9 Hanes-Woolf plot of data for problem 14.

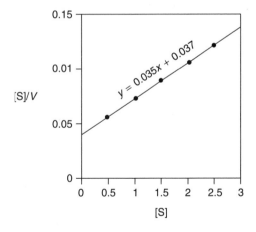

In this instance, both plots give good fits of the data, and the values derived from each for K_M and V_{max} do not differ significantly. We can conclude that the measurements at low substrate concentration are reliable.

(d) See Figure 8.10.

FIGURE 8.10 Hanes-Woolf plots depicting effects of competitive and noncompetitive inhibitors.

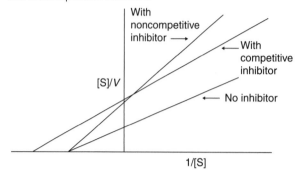

14. Q increases the rate of reaction, so it is an activator, or perhaps a second substrate. It combines with both E and ES.

15. (a) The overall reaction proceeds as follows:

(1) E + ATP $\longleftrightarrow$ E-AMP + PP$_i$
(2) E-AMP + nicked DNA $\longleftrightarrow$ nicked DNA-AMP + E
(3) nicked DNA-AMP + E $\longleftrightarrow$ sealed DNA + E

Σ ATP + nicked DNA $\longleftrightarrow$ sealed DNA + AMP + PP$_i$

(b) Yes, a substituted enzyme intermediate (E-AMP) is formed.

(c) In the absence of DNA, the enzyme catalyzes the partial reaction of the formation of the E-AMP with the release of PP$_i$. DNA is not involved in the first part of the double-displacement reaction. (This problem is derived from B. Weiss, and C. C. Richardson. Enzymatic breakage and joining of deoxyribonucleic acid. 3. An enzyme-adenylate intermediate in the polynucleotide ligase reaction. *J. Biol. Chem.* 243[1964]:4556–4563. See also I. R. Lehman. DNA ligase: Structure, mechanism, and function. *Science* 186[1974]:790–797, for a complete review.)

16. In the absence of additional information, you would suspect that the enzyme had allosteric properties; its initial velocity was being influenced by binding of one of the substrates to a site different from the active site.

17. Because you added ethanol and acetaldehyde in amounts that did not change their concentration ratios, you would expect no change in the equilibrium position of the reaction. Recall that the equilibrium constant of a reaction is proportional to the ratio of the products to reactants. When you add such a small amount of highly radioactive ethanol that you do not change the ratio of products and reactants, you will not change the equilibrium significantly. However, because the reaction continues, even at equilibrium, you would expect to find radioactivity in the acetaldehyde after incubation. Given enough time the ratio of radioactivity in the product and reactant would equal their mass ratio.

18. $1370 \text{ kJ} \left(\dfrac{0.239 \text{ kcal}}{\text{kJ}} \right) = 327$ kcal or 327 food calories Yes, you could eat the entire container of yogurt and keep your food calories under 350.

19. (a) Since aspirin makes a covalent bond with the enzyme, aspirin is an irreversible inhibitor.

(b) Since Ibuprofen binds to the active site noncovalently and it somewhat resembles the substrate, Ibuprofen is a competitive inhibitor.

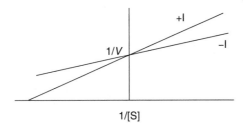

(c) Flooding the system with substrate can result in reversal of competitive inhibition. Therefore the inhibitory effect of Ibuprofen could be overcome by sufficiently increasing the presence of arachidonate, the natural substrate of the enzyme.

20. Hexokinase binds more strongly to glucose. It takes less substrate to reach half the V_{max}, so the K_m is lower and the binding is tighter. On the other hand, a higher concentration of fructose is required to reach half the V_{max}, so fructose shows more dissociation from the enzyme. Remember that the K_m can be thought of like a K_d.

21. The $K_{0.5}$ of brain cell GLUT 3 will be lower. This means that GLUT 3 will bind glucose more tightly and will reach saturation at a lower concentration of glucose than is seen with the muscle/fat GLUT 4 receptors. This lower affinity of the GLUT 4 receptors for glucose helps to ensure that the organ that most needs glucose, that is, the brain, receives it preferentially. In conditions of low blood sugar, brain glucose receptors will still be available to bind glucose.

Catalytic Strategies

In the previous chapter, you learned that the catalytic activity of enzymes is based on their ability to stabilize the transition states of chemical reactions and thereby decrease the energy-activation barrier to reactivity. In Chapter 9, the authors describe in detail the structures, active-site configurations, binding of substrates, and catalytic mechanisms of four well-understood enzymes: chymotrypsin, carbonic anhydrase, restriction endonuclease EcoRV, and the ATPase domain of myosin. Using these specific enzymes as models, fundamental principles of enzyme catalysis are exemplified: specific binding of substrates, induced fit of enzyme–substrate complexes, covalent catalysis, general acid–base catalysis by active-site residues, catalysis by propinquity and by metal ions, formation and stabilization of transition states, and reversibility of catalytic steps. The principles employed by these enzymes illustrate how enzymes use basic chemistry to perform reactions at rapid rates and with high fidelity. Along the way, the authors describe how active sites have evolved to carry out specific reactions. Because the interactions of enzymes with substrates depends on the chemical properties of amino acid residues and on protein structure in general, a review of Chapter 2 would be helpful before reading this chapter. In addition, refresh your understanding of the basic concepts of enzyme action, thermodynamics, and kinetics presented in Chapter 8.

LEARNING OBJECTIVES

When you have mastered this chapter, you should be able to accomplish the following objectives.

A Few Basic Catalytic Principles Are Used by Many Enzymes

1. Define *binding energy* as it relates to enzyme–substrate interactions and explain how it can be used in *enzyme catalysis.*

2. List four strategies commonly employed by enzymes to effect catalysis.

Proteases Facilitate a Fundamentally Difficult Reaction (Text Section 9.1)

3. Define *proteolysis.* Draw the reaction for peptide-bond hydrolysis, and explain why peptide bonds are resistant to spontaneous hydrolysis.

4. Indicate the amino acid sequence specificity of the cleavage catalyzed by *chymotrypsin* and explain its molecular basis.

5. List the evidence that indicates that a *serine* hydroxyl serves as a nucleophile in the reaction catalyzed by chymotrypsin.

6. Explain why a *burst* of product appears when chymotrypsin reacts with a *chromogenic ester* substrate, and relate this phenomenon to *covalent catalysis.*

7. Indicate how *x-ray crystallography* was used to learn about the mechanism of the chymotrypsin reaction.

8. Describe the formation and stabilization of the transient *tetrahedral intermediate* produced from the scissile, planar peptide bond during hydrolysis.

9. Explain why Ser195 is especially reactive in chymotrypsin.

10. Summarize the roles of the *catalytic triad* in the mechanism of chymotrypsin and the relationship of the *oxyanion hole* to the tetrahedral intermediate of the reaction. Appreciate that these features are present in other *proteases, esterases,* and *lipases.*

11. Describe how *site-directed mutagenesis* was used to prove the role of the catalytic triad in *subtilisin* catalysis.

12. Explain how a *binding pocket* determines the specificity of chymotrypsin.

13. List other catalytic mechanisms by which peptide bonds can be hydrolyzed and provide examples for each.

14. Provide examples of *protease inhibitors* that serve as therapeutic agents.

Carbonic Anhydrases Make a Fast Reaction Faster (Text Section 9.2)

15. Outline the relationship of CO_2 to aerobic metabolism and indicate how most of the CO_2 generated by peripheral tissues is transported to the lungs.

16. Write the chemical equation for the *hydration of carbon dioxide,* and explain why *bicarbonate* is formed at physiologic pH values.

17. Indicate the physiologic requirement for catalysis of the reaction that hydrates CO_2.

18. Explain how *carbonic anhydrase* uses Zn^{2+} [Zn(II)] to activate a water molecule to attack CO_2.

19. Describe why a *buffer* must be present at high concentrations to allow carbonic anhydrase to function rapidly, and explain how a *proton shuttle* is involved in buffer action.

20. Using the carbonic anhydrases as examples, describe why *convergent evolution* is thought to have selected a common active-site structure.

Restriction Enzymes Perform Highly Specific DNA-Cleavage Reactions
(Text Section 9.3)

21. Write the reaction catalyzed by *restriction endonucleases* and explain why these enzymes must show very high substrate specificity to achieve their biological function.

22. Write the reaction and explain the biological role of *methylases (DNA methyltransferases)* in *restriction-modification* systems.

23. Draw a *phosphodiester bond*. Deduce how *phosphorothioates* and ^{18}O could be used to differentiate the *achiral oxygens* of a phosphodiester bond and to distinguish between *direct hydrolysis* of the bond and a mechanism involving a *covalent enzyme-DNA intermediate*.

24. Draw the *pentacoordinate, trigonal bipyramidal structure* of the transition state of a phosphodiester bond undergoing an *in-line displacement* reaction.

25. Compare the primary role of Mg^{2+} [Mg(II)] in the mechanism of restriction endonucleases with that of Zn^{2+} in the carbonic anhydrases.

26. Summarize the ways in which many restriction enzymes use binding energy to attain high substrate specificity. Consider the role of *DNA rotational symmetry* and *distortion* in achieving catalytic fidelity.

27. Contrast *substrate binding* and *catalytic activity* as mechanisms for attaining substrate specificity.

28. Provide evidence that restriction enzymes employed *horizontal gene transfer* to spread among bacteria.

Myosins Harness Changes in Enzyme Conformation to Couple ATP Hydrolysis to Mechanical Work (Text Section 9.4)

29. Write the reaction for ATP hydrolysis.

30. Know the role of divalent metals in forming the reaction substrate.

31. Know why the ATPase domain must undergo a conformational change in order to carry out ATP hydrolysis.

32. Know how ATP serves as a base to help deprotonate Ser-236 and facilitate ATP hydrolysis.

33. Know the role of Ser-236 in abstracting a proton from the attacking water and giving up a proton to the MG^{2+}-ATP complex.

34. Know the role of the 60-amino-acid carboxyl terminal region in communicating structural changes at the active site to other regions within the myosin.

35. Explain why ATP hydrolysis is highly favorable for free molecules in solution, yet is slow when catalyzed by the ATPase domain of myosin.

36. Know the significance of ATP's ability to re-form and persist on the enzyme in coupling to downstream processes.

37. Know the structural characteristics of a P-loop, some examples of other proteins where the motif is found, and what ability it confers to these proteins.

38. Understand why the K_{eq} for the enzyme-catalyzed reactions is much lower than the K_{eq} for the molecules free in solution.

SELF-TEST

A Few Basic Catalytic Principles Are Used by Many Enzymes

1. The free energy released when an enzyme binds a substrate

    (a) arises from many weak intermolecular interactions.
    (b) contributes to the catalytic efficiency of the enzyme.
    (c) is more negative when an incorrect substrate is bound.
    (d) becomes more positive as the transition state of the reaction develops.
    (e) becomes more negative the more tightly the enzyme binds the substrate.

2. Which of the following are used by enzymes to catalyze specific reactions?

    (a) metal ions   (d) general acid–base reactions
    (b) temperature changes   (e) covalent enzyme-substrate complexes
    (c) proximity between substrates

Proteases Facilitate a Fundamentally Difficult Reaction

3. Why is the peptide bond, which is thermodynamically unstable, resistant to spontaneous hydrolysis?

Chymotrypsin and Other Protelytic Enzymes

4. The alkoxide group on chymotrypsin that attacks the carbonyl oxygen of the peptide bond of the substrate arises from which amino acid side chain?

    (a) aspartate   (d) threonine
    (b) histidine   (e) tyrosine
    (c) serine

5. Which of the following experimental observations provide evidence for the formation of an acyl-enzyme intermediate during the chymotrypsin reaction?

    (a) A biphasic release of p-nitrophenol occurs during the hydrolysis of the p-nitrophenyl ester of N-acetyl-phenylalanine.
    (b) The active serine can be specifically labeled with organic fluorophosphates.
    (c) The pH dependence of the catalytic rate is bell shaped, with a maximum at pH 8.
    (d) A deep pocket on the enzyme can accommodate a large hydrophobic side chain of the recognized substrate.

6. Three essential amino acid residues in the active site of chymotrypsin form a catalytic triad. Which of the following are roles for these residues in catalysis?

    (a) The histidine residue facilitates the reaction by acting as an acid–base catalyst.
    (b) The aspartate residue orients the histine properly for reaction.
    (c) The serine residue acts as a nucleophile during the reaction with the substrate.
    (d) The aspartate residue acts as an electrophile during the reaction with the substrate.
    (e) The aspartate residue initiates the deacylation step by a nucleophilic attack on the carbonyl carbon of the acyl intermediate.
    (f) They make up the oxyanion hole.

7. Which of the following enzymes can be irreversibly inactivated with diisopropylphosphofluoridate (DIPF)?

    (a) carboxypeptidase II
    (b) trypsin

   (c) lysozyme

   (d) subtilisin

   (e) thrombin

8. The three enzymes trypsin, elastase, and chymotrypsin

   (a) probably evolved from a common ancestor.

   (b) have major similarities in their amino acid sequences and three-dimensional structures.

   (c) catalyze the same general reaction: the cleavage of a peptide bond.

   (d) catalyze reactions that proceed through a covalent intermediate.

   (e) have structural differences at their active sites.

9. Match the enzyme in the right column with the proteolytic-enzyme class to which it belongs in the left column.

   (a) metalloprotease   (1) papain

   (b) serine protease   (2) pepsin

   (c) thiol (cysteine) protease   (3) elastase

   (d) acid (aspartyl) proteases   (4) thermolysin

10. Why might inhibitors of specific proteases be useful therapeutic agents? Provide a specific example.

Carbonic Anhydrases Make a Fast Reaction Faster

11. Match the molecule in the first column with the appropriate item in the second column.

   (a) water   (1) $pK_a \sim 3.5$

   (b) bicarbonate   (2) $pK_a \sim 7$

   (c) carbonic acid   (3) $pK_a \sim 14$

   (d) water bound to Zn^{2+} (4) $pK_a \sim 10.3$

        in carbonic anhydrase II

12. Given the pK_a values of the compounds shown in Question 11, what is the significance of the water that is bound to the zinc ion in carbonic anhydrase?

13. Several different carbonic anhydrases coordinate Zn(II) in their active sites using the amino acid side chains of His exclusively or of His and of Cys. Rationalize how the binding of water to the coordinated Zn^{2+} lowers the pK_a value of the water.

Restriction Enzymes Perform Highly Specific DNA-Cleavage Reactions

14. Is the DNA sequence 5′-GAATTC-3′ palindromic when it is in a duplex? Why?

15. Bacteria use restriction enzymes to destroy invading, exogenous DNA, for instance, DNA injected during bacteriophage infection. How can the restriction enzyme hydrolyze the foreign DNA and not destroy the DNA of the bacterium in which it resides?

16. List all the substrates and cofactors used by type II restriction endonucleases and type II DNA methylases (DNA methyltransferases).

17. Which of the following DNA sequences are most likely to be cut by a restriction enzyme? Only one strand, written in the 5′ to 3′ orientation is shown, but you should assume that the opposite strand is present to form a duplex.

(a) TAGCAT
(b) CTGCAG
(c) CAGGAC
(d) GAATTC
(e) TCGA

18. Which of the following amino acids in the active site of a typical restriction enzyme would you expect to be involved in binding Mg^{2+} [Mg(II)]?

(a) D   (d) N
(b) Y   (e) E
(c) C

Myosins Harness Changes in Enzyme Conformation to Couple ATP Hydrolysis to Mechanical Work

19. Which of the following are roles for Mg^+ in myosin's reactions that catalyze the hydrolysis of ATP?

(a) binds to the enzyme and activates a water molecule
(b) binds to Ser-236 to activate it
(c) binds to ATP to create the substrate for the enzyme
(d) to organize the active site

20. The P-loops of NTPase domains

(a) contain several conserved G (glycine) residues.
(b) have a central β sheet flanked by α helices.
(c) have a loop between the first β-sheet and the first α-helix.
(d) move extensively upon ATP binding.
(e) all of the above.

ANSWERS TO SELF-TEST

1. a, b, e. The $\Delta G^{o\prime}$ of the reaction becomes more negative as the binding affinity of the enzyme for the substrate increases. Interactions between the substrate and the enzyme promote the reaction when they are fully formed during the development of the transition state of the reaction. Favorable interactions between the enzyme and the substrate in its ground state before development of the transition state can hinder the reaction by lowering the valley preceding the activation barrier in the reaction coordinate diagram if they do not also contribute to binding the transition state. For instance a substrate analog that is a good competitive inhibitor forms strong interactions with the enzyme, but cannot develop a transition state.

2. a, c, d, e

3. A peptide bond is stabilized by resonance, which gives the carbonyl–carbon-to-amide–nitrogen link partial double-bond character, making it more stable to hydrolysis. In addition, the carbonyl carbon of the peptide bond is linked to a partially negatively charged carbonyl oxygen that decreases the susceptibility of the carbon atom to nucleophilic attack by a hydroxyl ion.

4. c

5. a, b. The pH versus activity curve indicates only that some step in the mechanism is sensitive to the state of dissociation of a proton donor on the protein.

6. a, b, c. (d) is incorrect because the aspartic acid carboxylate is ionized, and bearing a negative charge, it is not an electrophile.

7. a, b, d, e. The mechanism of lysozyme does not involve the nucleophilic attack on the substrate by an activated hydroxyl of the enzyme. The other three enzymes do have such an activated serine hydroxyl and react to form a covalent, inactive complex with DIPF.

8. a, b, c, d, e. The enzymes differ in structure at the sites at which they interact with the amino acid side chains to determine their substrate specificity.

9. (a) 4   (b) 3   (c) 1   (d) 2

10. Proteases that are specific for particular amino acid sequences play important roles in normal and pathological physiology in humans. For example, a protease, the angiotensin-converting enzyme (ACE), is involved in blood-pressure regulation. A specific inhibitor would prevent the hypertension that arises from overactivity of ACE. Similarly, a specific protease is necessary for human immunodeficiency virus maturation after infection. Inhibition of this protease could limit HIV infection.

11. (a) 3   (b) 4   (c) 1   (d) 2. The CO_2 buffering system is unusual because one of its components, dissolved CO_2, is a volatile gas in equilibrium with atmospheric CO_2. By convention, $[H_2CO_3]$ is used to represent the total concentration of dissolved CO_2 + H_2CO_3. Carbonic acid that dissociates to form H^+ + bicarbonate is immediately replaced by the reaction of CO_2 with water. The observed pK_a value of carbonic acid in a solution in equilibrium with gaseous CO_2 in the lungs is approximately 6.1, not 3.5, because of its equilibrium with dissolved CO_2, which exceeds it in concentration by approximately 1000-fold. Thus, although the "true" pK_a value of carbonic acid is 3.5, it behaves in gas transport in mammals as if the value were approximately 6.1. The pK_a value for the dissociation of bicarbonate is ~10.3.

12. With the pK_a value of water lowered to near physiological pH values, an appreciable amount of zinc-bound hydroxyl ion will be formed by dissociation of a proton from the zinc-bound water. The hydroxyl ion is the nucleophile that attacks the carbonyl carbon of CO_2 to form the bicarbonate ion. Thus, the enzyme generates a reactive substrate by binding water to Zn^{2+}, thereby facilitating its dissociation to form the reactive substrate.

13. The positive charge on the zinc ion withdraws electrons from the oxygen of the bound water, weakens the bonds to its hydrogen atoms, and promotes the dissociation of a proton to form an enzyme-bound hydroxyl.

14. Yes, because the complementary strand is identical, namely, 5′-GAATTC-3′. Remember, the strands of duplex DNA have opposite polarity. A palindromic sequence has two-fold rotational symmetry. If you rotate the duplex molecule 180° about an axis located perpendicular to its long axis and piercing between the two strands between the AT sequences in each strand, you will generate the starting configuration of atoms.

15. A restriction enzyme recognizes and hydrolyzes a particular DNA sequence. The same sequence is recognized and methylated by the partner DNA methylase of the restriction enzyme. A methylated restriction site is immune to cleavage by the restriction enzyme. The methylase keeps the host DNA methylated and thus protected. The invading DNA, if unmethylated itself, will be cleaved by the restriction enzyme and subsequently destroyed by less specific nucleases.

16. Restriction enzymes require only target DNA, Mg^{2+}, and water. DNA methylases require only unmethylated target DNA and S-adenosylmethionine.

17. b, d, e. Each of these sequences has an identical complementary strand.

18. a, e. The carboxyl groups of Asp and Glu can bind Mg^{2+} effectively.

19. c. The Mg^{2+}-ATP complex is the substrate. Mg^{2+} or Mn^{2+} with NTPs form the substrates for virtually all NTP-dependent enzymes.

20. e. All of the above characteristics are common to P-loops.

PROBLEMS

1. Why is histidine a particularly versatile amino acid residue in its involvement in enzymatic reaction mechanisms?

2. Although chymotrypsin is a proteolytic enzyme, it is quite resistant to digesting itself. How would you explain its resistance to self-proteolysis?

3. For each enzyme in the left column, indicate the appropriate transition state or chemical entity in the right column that has a postulated involvement in its catalytic mechanism.

    (a) carbonic anhydrase   (1) mixed anhydride
    (b) nucleoside monophosphate kinase   (2) oxyanion hole
    (c) restriction endonuclease EcoRV   (3) pentacovalent phosphorus
    (d) chymotrypsin   (4) carbonium ion
         (5) tetrahedral carbon intermediate

4. A pH-enzyme activity curve is shown in Figure 9.1. Which of the following pairs of amino acids would be likely candidates as catalytic groups? (See Table 2.1 in the text for the pK_a values of amino acid residues.)

    (a) glutamic acid and lysine   (d) histidine and histidine
    (b) aspartic acid and histidine   (e) histidine and lysine
    (c) histidine and cysteine

FIGURE 9.1

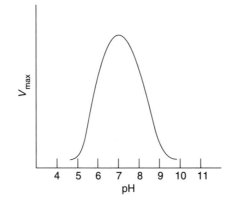

5. Consider the fact that DNA methylases put methyl groups only onto preformed DNA, that is, after DNA has been synthesized from unmethylated dNTPs. Because DNA replication is semiconservative, what would be the methylation state of a restriction site immediately after replication in a bacterium containing a restriction modification system? What can you conclude about the number of methyl groups needed per restriction site to render the DNA refractory to cleavage by the cognate restriction endonuclease?

6. On average, how many EcoRV restriction sites would you expect in the genome of E. coli? The genome is 4.6×10^6 base pairs and its composition is approximately 50% G + C.

7. Trypsin, chymotrypsin, and carboxypeptidase A fail to cleave peptide bonds involving proline. Trypsin, for example, will not cleave a peptide at a Lys-Pro junction. Why do you think this is the case?

8. Place slash marks at the sites where you would expect chymotrypsin to cleave the following peptide:

Lys-Gly-Phe-Thr-Tyr-Pro-Asn-Trp-Ser-Tyr-Phe

9. Many enzymes can be protected against thermal denaturation during purification procedures by the addition of substrate. Propose an explanation for this phenomenon.

10. What are the main structural features of an enzyme that determine its substrate specificity?

11. DNA methyltransferases (DNA methylases) use S-adenosylmethionine (AdoMet) as the methyl donor in a reaction that methylates a specific base at a specific sequence in DNA and releases the AdoMet remnant S-adenosylhomocysteine (AdoHcy). The DNA methyltransferase RsrI catalyzes the reaction DNA + AdoMet $\rightarrow$ methylated DNA + AdoHcy where the methyl group is deposited on the exocyclic amino group of the second A in the recognition sequence GAATTC. A burst of incorporation of methyl groups into DNA occurred in an experiment in which the enzyme was saturated with AdoMet radiolabeled with ^{14}C in its activated methyl group. The enzyme was first saturated with an excess of [^{14}C]AdoMet and then a saturating excess of unmethylated DNA containing the target sequence was added along with more radiolabeled AdoMet to maintain its original concentration and specific activity. The incorporation of isotope into the DNA was monitored. A rapid incorporation of methyl groups occurred (burst) on addition of the DNA + [^{14}C]AdoMet. The burst was followed by a slower, steady-state rate of DNA methylation. A plot of the formation of labeled DNA as a function of time is shown in Figure 9.2. The mol of methyl groups incorporated into DNA/mol of enzyme is plotted on the ordinate, and time in seconds is plotted on the abcissa. When the steady-state phase of the reaction curve was extrapolated back to the ordinate (y-axis), the value obtained was 0.94 mol methyl group deposited on DNA/mol of enzyme.

a. What can you conclude from this experiment about (1) the mechanism of the reaction and (2) the proportion of molecules of enzyme that were active?

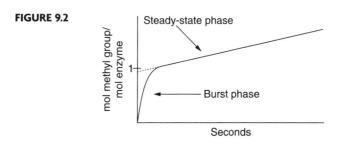

FIGURE 9.2

b. In a second experiment with RsrI methylase, the same protocol described for the burst experiment was followed except that when the excess DNA was added to the enzyme that had been saturated with radiolabeled AdoMet, the solution containing the DNA also contained the initial 50-fold concentration excess of *unlabeled* AdoMet. The incorporation of the radiolabeled methyl groups of the prebound AdoMet into DNA was again followed. In this case, a smaller burst, approximately 10% of that observed in the first experiment, was detected before the steady-state rate of reaction began. Explain what the smaller burst implies about the order of addition of the substrates, DNA and AdoMet, to the enzyme.

12. At physiological pH values, the dissociation of water cannot supply a sufficient concentration of protons to support the full catalytic potential of carbonic anhydrase. As a result, the enzyme has evolved to use buffers as acid-base catalysts to increase local proton concentrations in the active site. Some of these buffers have molecular dimensions too large to allow them to penetrate into the active site and gain proximity to the protein-bound Zn(II). Despite their exclusion by steric factors from the active site, the buffers support efficient catalysis by the enzyme. In addition, ionizable groups in the active site of the enzyme are involved in the delivery and removal of the required protons. Site-specific mutagenesis that substituted some of these residues with other ionizable amino acids having different pK_a values failed to inactivate the enzyme. What can you conclude about the molecular mechanisms by which protons are shuttled into the active site of carbonic anhydrase?

13. When the aspartate at postion 32 in subtilisin was converted to alanine by site-directed mutagenesis, the k_{cat} value for the hydrolysis catalyzed by the enzyme fell to 0.005% of its wild-type value (see pages 250–251 of the text). Because Asp32 is a crucial part of the catalytic triad of this enzyme, provide a hypothesis to explain the residual enzyme activity after its conversion to alanine.

14. X-ray crystallography of some restriction enzymes in complexes with their target DNA sequences has been very useful in explaining, in part, how these remarkably sequence-specific enzymes distinguish their targets from the sea of surrounding nonspecific DNA sequences. Why has this technique been of less value in determining the role of Mg(II) in the reaction? What can be done to learn about the role of the divalent cations in the reaction by using X-ray crystallography? Why has trying to determine the basis for the catalytic specificity been particularly problematic with the EcoRV restriction endonuclease?

15. Why is it thought that almost all NTPs (ribonucleoside triphosphates) in the cell are present as their complexes with Mg(II)?

16. What are two ways to make a reactive group in an enzyme into a more potent nucleophile? Provide examples from the enzymes studied in this unit.

17. Acetylchoinesterase is a serine protease found at the synapses between nerve cells. It cleaves the neurotransmitter acetylcholine. Organic fluorophosphates such as DIFP (Diisopropylphosphofluoridate, shown below) inhibit acetylcholinesterase. Suggest how this enzyme might be inhibited by DIFP.

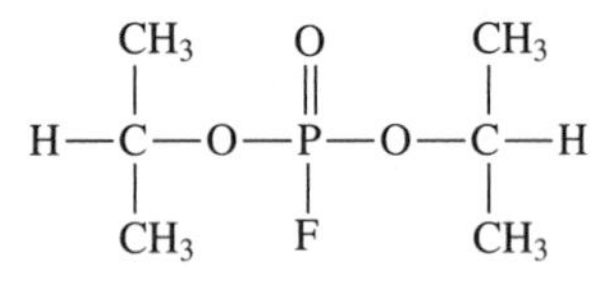

$$
\begin{array}{c}
\text{CH}_3 \quad\quad \text{O} \quad\quad \text{CH}_3 \\
| \quad\quad\quad || \quad\quad\quad | \\
\text{H—C—O—P—O—C—H} \\
| \quad\quad\quad | \quad\quad\quad | \\
\text{CH}_3 \quad\quad \text{F} \quad\quad \text{CH}_3
\end{array}
$$

18. What are three things that a Zn^{2+} ion can do in an enzyme to speed up a reaction?

19. When an enzyme is subjected to x-ray crystallography, sometimes the active site is obvious. A cleft or pocket may be present of the correct size, shape, and amino acid sequence to bind the substrate. Sometimes, however, the active site is not at all obvious from the structure. Name something that scientists can do to help identify the active site in crystallography experiments.

ANSWERS TO PROBLEMS

1. The imidazole ring of histidine can act as an acid–base catalyst, a nucleophile, or a chelator (coordinator) of metal ions. The first and third functions were illustrated by chymotrypsin and carbonic anhydrase, respectively.

2. Chymotrypsin specifically cleaves peptide bonds whose C-terminal amino acid is adjacent to nonpolar aromatic amino acid residues or the bulky, hydrophobic methionine. Because these residues are often buried in the interior of proteins, including chymotrypsin, the self-hydrolysis of native, folded chymotrypsin is very inefficient. In fact, during digestion, chymotrypsin acts most effectively on partially degraded and denatured (unfolded) proteins.

3. (a) 1   (b) 1, 3   (c) 3   (d) 2, 5

4. c

5. Both strands of the restriction sites in a parent DNA molecule in a cell with a restriction-modification (R-M) system would be methylated. On semiconservative replication, the newly synthesized daughter strand would be transiently unmethylated. Because the cell survives during this time, you can conclude that only one methyl group on a restriction site can stop the endonuclease from cutting. This fact has been verified by experiments in vitro with purified restriction enzymes and unmethylated, hemimethylated (one methyl group on one strand), and fully methylated DNAs.

6. For DNA that has equal proportions of A, C, G, and T, each base has a 0.25 probability of appearing at any position in the sequence. Since the EcoRV site, GATATC, is six bases long, $(0.25)^6 \times 4.6 \times 10^6 = 976$. We would thus expect ~1000 EcoRV sites in the genome of *E. coli*.

7. Because of its ring structure, the imino acid proline cannot be accommodated in the substrate binding sites of trypsin, chymotrypsin, or carboxypeptidase A. Therefore these proteases fail to cleave peptide bonds involving proline.

8. Chymotrypsin would produce the following four fragments:

   Lys-Gly-Phe, Thr-Tyr-Pro-Asn-Trp, Ser-Tyr, and Phe

9. When the substrate occupies the active site in the enzyme, the weak bonds that it forms with groups on the enzyme help to stabilize the tertiary structure of the enzyme and protect it against thermal denaturation.

10. The enzyme must have functional groups in the active site that can interact specifically with the substrate to distinguish it from other similar molecules and position it properly for a productive reaction. Usually, the enzyme must also have catalytic residues that react with a specific chemical bond of the substrate during the development of the transition state. Both ground-state interactions with the substrate by specific binding and the ability to catalyze the chemistry of the reaction determine the ability of an enzyme to convert a substrate to a product.

11. a. Unlike chymotrypsin, the existence of the burst with RsrI DNA methylase is not due to a covalent enzyme–substrate complex. The appearance of a burst in an enzyme reaction reveals only that some step after the chemistry occurring during bond making and breaking is limiting the overall rate. The RsrI methylase reaction is known to proceed without a covalent enzyme–substrate intermediate. With this methylase, the burst indicates that some step subsequent to the

addition of the methyl group onto the DNA is the rate-limiting step of the reaction. The rate-limiting step is likely to be the release of product from the enzyme. The fact that 0.94 mol of DNA was methylated per mol of enzyme indicates that at least 94% of the enzyme molecules were active. The experiment measured the initial reaction of all the enzyme molecules present because the enzyme was preloaded with AdoMet and then given DNA at a concentration that would also saturate it with the methyl acceptor. No excess free enzyme existed in the solution, and the initial reaction observed (the burst) measured a single turnover.

b. The protocol in problem 13b is an isotope-partitioning experiment. That a burst was detected when the enzyme was preloaded with labeled AdoMet before being mixed with excess DNA and a 50-fold excess of unlabeled AdoMet means that AdoMet bound to the enzyme before the DNA binds can be catalytically competent. If the radiolabeled AdoMet had dissociated from the enzyme before reacting, its specific activity would have been decreased 50-fold by the unlabeled AdoMet in the solution, and the maximum incorporation would have been 2% of that seen in the burst experiment (problem 13a). This result does not prove that the reaction is ordered with the order of binding being AdoMet first and DNA second. It shows only that AdoMet can be bound first and used in the reaction after DNA binds. The order of addition of the substrates to the enzyme might be random with either AdoMet or DNA binding first. Further experiments would be needed to resolve this question. (Both these questions were derived from S. S. Szegedi, N. O. Reich, and R. I. Gumport. Substrate binding *in vitro* and kinetics of *RsrI* [N6-adenine] DNA methyltransferase. *Nucleic Acids Res.* 28[2000]:3962–3971.)

12. The buffers must donate and accept protons at some distance from the active center of the enzyme because they are too large to access it. The protons supplied by these buffers reach the reaction center by being transported or shuttled through a network of proton carriers that comprises ionizable groups on the protein and water molecules. The fact that active site residues with pK_a values different from those of the wild-type enzyme function in the reaction suggests that the precise location and strength of the ionizable groups are not critical to the functioning of the shuttle. The malleability of the positioning of the active site, ionizing amino acid side chains, probably results from the formation of different networks of variable numbers of hydrogen-bonded water molecules. These networks form in various shapes to accommodate the altered positions of the variant amino acid side chains. (This question was derived from M. Qian, J. N. Earnhardt, M. Qian, C. Tu, P. J. Laipis, and D. N. Silverman. Intramolecular proton transfer from multiple sites in catalysis by murine carbonic anhydrase V. *Biochemistry* 37[1998]:7649–7655.)

13. As stated in the text, the mutated enzyme retains much of its catalytic ability because the remaining two residues of the catalytic triad can continue to generate a nucleophile powerful enough to attack the scissile peptide bond of the protein substrate. The enzyme can still bind the two substrates (water and protein), which brings them into proximity with each other. Approximation of the interacting molecules, which is one of the general means by which enzymes catalyze reactions, remains in effect despite the mutation, and the local concentration of the two substrates is far higher in the presence of the enzyme, even if mutated, than in its absence.

14. Recall that restriction enzymes hydrolyze their target DNA in the presence of Mg(II). Thus, crystals cannot be formed in the presence of this divalent metal ion because the substrate sequence would be cut. To circumvent this difficulty, cocrystals of the enzyme and its target sequence can be formed in the absence of Mg(II) and the divalent cation subsequently diffused into the crystal. The reaction then may occur and the

metal ion can be located with respect to the cleaved product and thereby information gained about its possible role in the cleavage. Another approach uses a divalent cation, for example, Ca(II) that was found to bind to the complex but not support catalysis by the enzyme. In this case the Ca(II) surrogate could provide information about where the Mg(II) would be with the intact substrate and enzyme. Finally, enzymes with mutations that deactivate cleavage can be crystallized with the target and Mg(II), and attempts made to locate the metal ion. EcoRV is particularly problematic because it binds both its target sequence and nontarget sequences with high affinity in the absence of Mg(II) and it assembles a catalytically competent complex only when the metal is bound. It enforces its catalytic fidelity at the level of catalysis rather than at the level of DNA substrate binding. The precatalytic complex lacks many of the crucial contacts that affect the catalysis, and these come into play only when the metal is bound and the target DNA distorted. It is interesting that, after many crystal structures have been solved and many biochemical experiments performed, precisely how many (one, two, or three) Mg(II) ions are involved in restriction enzyme-catalyzed hydrolysis of DNA remains unclear.

15. Total Mg(II) concentrations in the cell are typically in the mM range, as are the concentrations of the NTPs. Because dNTPs are typically present at 1/10 or lower concentrations than the NTPs, they can be neglected in this problem. Given that the equilibrium dissociation constant for the NTP-Mg(II) complex is approximately 0.1 mM, almost all the NTPs are present as their complexes with Mg(II).

16. As in chymotrypsin, the nucleophile could be close to a base such as histidine, which could itself be made a stronger base by a nearby ionized carboxylic acid such as aspartate. The histidine could pull the H^+ of any alcohol toward itself, making an effective alkoxide, which is a stronger nucleophile.

    Another method is used by carbonic anhydrase. The enzyme contains Zn^{2+}, which is bound to a water molecule. Because the zinc(II) pulls electron density toward itself, the bound water loses its hold on H^+ and the pKa of the water is lowered considerably. Thus, the water turns into a hydroxide, which is a more potent nucleophile.

17. DIPF is an affinity label. The serine O^- attacks the phosphate group and displaces the F^-. But there is no leaving group for any additional chemistry, so the DIPF stays bound and prevents acetylcholine from entering the active site.

18. Zn^{2+} could make a bound water a more powerful nucleophile by drawing electron density toward itself and facilitating deprotonation of the water.

    Zn^{2+} could stabilize an anionic intermediate.

    Zn^{2+} could orient substrates for reaction.

19. Scientists can crystallize the enzyme bound to a substrate analog inhibitor. This structure is called a cocrystal. The substrate analog inhibitor will bind in the active site and will thus identify the active site.

Regulatory Strategies

The theme of Chapter 10 is the regulation of protein function. Four major types of regulatory mechanisms are discussed in detail: allosteric control, isozymes, reversible covalent modification, and proteolytic activation. The authors use specific examples to illustrate the general structure-function relationships involved in these control mechanisms. To illuminate allosteric control, the authors discuss *E. coli* aspartate transcarbamoylase (ATCase), one of the best understood allosterically regulated proteins. ATCase is the enzyme that catalyzes the condensation of carbamoyl phosphate and aspartate in the first step of pyrimidine biosynthesis. Its activity is regulated both positively and negatively and provides a classic example of feedback inhibition of enzymes in multistep biosynthetic pathways. After the section on allosteric control, the authors illustrate the use of isozymes to regulate enzymes in a developmental and/or tissue-specific manner using lactate dehydrogenase as an example. Next the authors discuss the regulation of enzymes by covalent modifications such as phosphorylation, acetylation, lipidation, and ubiquination. The authors focus on reversible phosphorylation as a control mechanism and use cAMP-dependent protein kinase (PKA) as an example of how phosphorylation of target proteins can be regulated. The authors then turn to the activation of enzymes by proteolytic cleavage. They describe the proteolytic steps and conformational rearrangements that produce the active forms of chymotrypsin and trypsin from their inactive zymogens. The mechanisms of action of these enzymes were presented in Chapter 9. The authors conclude Chapter 10 with a discussion of the blood clotting cascade—the series of proteolytic activations of clotting factors that lead to the formation of fibrin clots. Several specific stimulating and inhibiting proteins are described in connection with the proteolytic enzymes.

LEARNING OBJECTIVES

When you have mastered this chapter, you should be able to accomplish the following objectives.

Introduction

1. List the major regulatory mechanisms discussed in this chapter that control enzyme activity and give examples of each.

Aspartate Transcarbamoylase Is Allosterically Inhibited by the End Product of Its Pathway (Text Section 10.1)

2. Describe the reaction catalyzed by *aspartate transcarbamoylase (ATCase)*, the regulation of ATCase by CTP and ATP, and the biological significance of this regulation.

3. Define feedback inhibition and describe how ATCase uses it as a regulatory mechanism.

4. Explain the differences in the kinetic curves of an enzyme following Michaelis-Menten kinetics and an allosterically regulated enzyme.

5. Describe the composition and arrangement of the subunits of ATCase and the major features of its active site as revealed by the binding of N-*(phosphonacetyl)-L-aspartate (PALA)* and treatment with p-hydroxymercuribenzoate. Explain the effects of subunit dissociation on the *allosteric behavior* of the enzyme.

6. Outline the structural effects of binding of CTP and PALA to ATCase.

7. Differentiate between *concerted* and *sequential* mechanisms of allosteric regulation and describe the experimental evidence for a concerted allosteric transition during the binding of substrate analogs to ATCase.

8. Outline the effects of *heterotropic* and *homotropic* allosteric interactions on the equilibrium between the T and R forms of ATCase.

9. Identify similarities between cooperativity in ATCase and hemoglobin.

Isozymes Provide a Means of Regulation Specific to Distinct Tissues and Developmental Stages (Text Section 10.2)

9. Define *isozyme*. Give examples of ways in which isozymes of a given enzyme can be differentiated from each other.

10. Explain the purpose of isozymes in metabolism.

11. Explain the role of isozymes in the tissue-specific regulation of lactate dehydrogenase.

Covalent Modification Is a Means of Regulating Enzyme Activity (Text Section 10.3)

12. List the common covalent modifications used to regulate protein activity.

13. Write the basic reactions catalyzed by *protein kinases* and *protein phosphatases*.

14. List the reasons why phosphorylation is such an effective control mechanism.

15. Discuss the determinants of specificity of protein kinases.

16. Describe the activation of *protein kinase A (PKA)* by cyclic *AMP (cAMP)* and the mode of interaction of PKA with its *pseudosubstrate*.

Many Enzymes Are Activated by Specific Proteolytic Cleavage (Text Section 10.4)

17. Define *zymogen*. Give examples of enzymes and proteins that are derived from zymogens and the biological processes they mediate.

18. Describe the formation of the substrate-binding site of chymotrypsin by proteolysis of chymotrypsinogen.

19. Summarize the enzymes and conditions required for the activation of all the *digestive enzymes*.

20. Explain how *trypsin* is inhibited by the *pancreatic trypsin inhibitor*.

21. Compare the extrinisic and intrinsic pathways of blood clotting.

22. Explain the role of *thrombin* in the activation of fibrinogen into fibrin and describe the structure of fibrin arrays.

23. Discuss the requirement for *vitamin K* in the synthesis of *prothrombin*. Outline the mechanism of prothrombin activation.

24. Explain the genetic defect in *hemophilia*.

25. State the general mechanisms for the control of clotting and explain the specific role of *antithrombin III* in the clotting cascade. Note the effect of *heparin* on antithrombin III.

26. Describe the *lysis* of fibrin clots by *plasmin* and the activation of *plasminogen* by *tissue-type plasminogen activator (TPA)*.

SELF-TEST

Aspartate Transcarbamoylase Is Allosterically Inhibited by the End Product of Its Pathway

1. The dependence of the reaction velocity on the substrate concentration for an allosteric enzyme is shown in Figure 10.1 as curve A. Which of the following would be expected to cause a shift to curve B?

   (a) addition of an irreversible inhibitor.
   (b) addition of an allosteric activator.
   (c) addition of an allosteric inhibitor.
   (d) dissociation of the enzyme into subunits.

FIGURE 10.1   Reaction velocity versus substrate concentration for an allosteric enzyme.

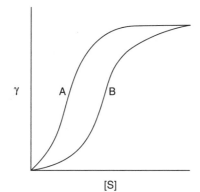

2. In *E. coli*, ATCase is inhibited by CTP and is activated by ATP. Explain the biological significance of these effects.

3. Which of the following statements regarding the structure of ATCase in *E. coli* are incorrect?

   (a) ATCase consists of two kinds of subunits and a total of 12 polypeptide chains.
   (b) Reaction with mercurials dissociates each ATCase into three r_2 and two c_3 subunits.
   (c) The substrate and allosteric effectors bind at different sites.
   (d) The active sites of ATCase are located at the interface between c and r subunits.

4. Which of the following is true for a concerted mechanism of allosteric regulation?

   (a) All of the subunits are in either T state or R state.
   (b) Binding of ligand to one site can affect neighboring sites without causing all of the subunits to undergo the T to R transition.
   (c) The regulation of ATCase cannot be explained using a concerted model.
   (d) Binding of CTP to the enzyme shifts the equilibrium of all subunits simultaneously to the R state.
   (e) Most allosteric effectors act via a concerted mechanism.

5. Which of the following best describes the allosteric effect of CTP on ATCase?

   (a) homotropic activation
   (b) homotropic inhibition
   (c) heterotropic activation
   (d) heterotropic inhibition

6. In which of the following ways are ATCase and Hemoglobin similar?

   (a) Both are tetramers, consisting of two different types of subunits.
   (b) Both are described best by a sequential model of allostery.
   (c) Both exhibit cooperativity in binding ligands.
   (d) Both exist in a either a T or an R state.

Isozymes Provide a Means of Regulation Specific to Distinct Tissues and Developmental Stages

7. Which of the following is NOT true about isozymes?

   (a) They can often be distinguished from one another by biochemical properties such as electrophoretic mobility.
   (b) Isozymes usually display identical kinetic parameters.
   (c) They are encoded by different genetic loci.
   (d) They permit fine-tuning of metabolism.

8. The isozymes of human Lactate Dehydrogenase

   (a) are expressed only in the heart.
   (b) are called M_4 and H_4; M_4 has a higher affinity for substrates than does H_4.
   (c) are allosterically inhibited by pyruvate.
   (d) catalyze different chemical reactions.
   (e) function optimally under different oxygen conditions.

Covalent Modification Is a Means of Regulating Enzyme Activity

9. Protein kinases

   (a) transfer a phosphoryl group from one protein to another.
   (b) use AMP as a substrate.

   (c) use Thr, Ser, or Tyr as the acceptor groups for phosphoryl transfer.

   (d) transfer the phosphorus atom of ATP.

   (e) are located on the external surface of cells.

10. Explain how a phosphoryl group can change the conformation of a protein.

11. Protein kinase A

   (a) is activated by ATP.

   (b) consists of two catalytic (c) and two regulatory (r) subunits in the absence of the activator.

   (c) upon binding the activator dissociates into one c_2 and two r subunits.

   (d) contains a pseudosubstrate sequence in the c subunits.

12. Which of the following are examples of irreversible protein modifications?
   (i) Phosphorylation   (ii) Myristoylation   (iii) Farnesylation

   (a) i and ii

   (b) i and iii

   (c) ii and iii

   (d) All of the above.

Many Enzymes Are Activated by Specific Proteolytic Cleavage

13. Give five examples of how specific proteolysis activates enzymes and other proteins in biological systems.

14. Activation of chymotrypsinogen requires

   (a) the cleavage of at least two peptide bonds by trypsin.

   (b) structural rearrangements that complete the formation of the substrate cavity and the oxyanion hole.

   (c) structural rearrangements of the entire protein molecule.

   (d) the concerted proteolytic action of trypsin and pepsin to give α-chymotrypsin.

15. The pancreas is the source of the proteolytic enzyme trypsin. Which of the following are reasons trypsin does not digest the tissue in which it is produced?

   (a) It is synthesized in the form of an inactive precursor that requires activation.

   (b) It is stored in zymogen granules that are enclosed by a membrane.

   (c) It is active only at the pH of the intestine, not at the pH of the pancreatic cells.

   (d) It requires a specific noncatalytic modifier protein in order to become active.

16. Explain why the new carboxyl-terminal residues of the polypeptide chains produced during the activation of pancreatic zymogens are usually Arg or Lys.

17. The inactivation of trypsin by pancreatic trypsin inhibitor involves

   (a) an allosteric inhibition.

   (b) the covalent binding of a phosphate to the active site serine.

   (c) the facilitated self-digestion of the enzyme.

   (d) denaturation at the alkaline pH of the duodenum.

   (e) the nearly irreversible binding of the protein inhibitor at the active site.

18. Match fibrinogen and fibrin with the appropriate properties in the right column.

   (a) fibrinogen   (1) is soluble in blood

   (b) fibrin   (2) is insoluble in blood

                                (3) forms ordered fibrous arrays

                                (4) contains α-helical coiled coils

                                (5) may be cross-linked by transamidase

19. Which of the following statements about prothrombin are incorrect?
    (a) It requires vitamin K for its synthesis.
    (b) It can be converted to thrombin by the decarboxylation of γ-carboxyglutamate residues.
    (c) It is activated by Factor IX_a and Factor VIII.
    (d) It is anchored to platelet phospholipid membranes through Ca^{2+} bridges.
    (e) It is part of the common pathway of clotting.

20. Explain the role of the α-carboxyglutamate residues found in clotting factors.

21. Which of the following mechanisms is not involved in the control of the clotting process?
    (a) the specific inhibition of fibrin formation by antielastase
    (b) the degradation of Factors V_a and VIII_a by protein C, which is in turn switched on by thrombin
    (c) the dilution of clotting factors in the blood and their removal by the liver
    (d) the specific inhibition of thrombin by antithrombin III

22. Which of the following statements about plasmin are true?
    (a) It is a serine protease.
    (b) It diffuses into clots.
    (c) It cleaves fibrin at connector rod regions.
    (d) It is inactivated by $α_1$-antitrypsin.
    (e) It contains a "kringle" region in its structure for binding to clots.

ANSWERS TO SELF-TEST

1. c

2. The activation of ATCase by ATP occurs when metabolic energy is available for DNA replication and the synthesis of pyrimidine nucleotides. Feedback inhibition by CTP prevents the overproduction of pyrimidine nucleotides and the waste of precursors.

3. d

4. a

5. d

6. c

7. b

8. e

9. c

10. A phosphoryl group introduces two negative charges that can affect the electrostatic interactions within the protein. In addition, a phosphoryl group can form three highly directional hydrogen bonds to adjacent H-bond partners in the protein. These local effects can be transmitted to more distant parts of the protein in a manner similar to allosteric effects.

11. b

12. c

13. The following five examples are given in the text: 1) The *digestive enzymes* hydrolyze proteins that are synthesized as zymogens in the stomach and pancreas to activate them; 2) *Blood clotting* is mediated by a cascade of proteolytic activations that ensures a rapid and amplified response to trauma; 3) Some protein hormones are synthesized as inactive

precursors; 4) The fibrous protein *collagen*, the major constituent of skin and bone, is derived from proteolysis of *procollagen*, a soluble precursor; 5) *Programmed cell death*, or *apoptosis*, is mediated by proteolytic enzymes called *caspases*, which are synthesized in precursor form as *procaspases*. When activated by various signals, caspases function to cause cell death in most organisms, ranging from *C. elegans* to human beings.

14. c

15. a

16. Because trypsin is the common activator of the pancreatic zymogens, its specificity for Arg-X and Lys-X peptide bonds will produce Arg and Lys carboxyl-terminal residues.

17. e

18. (a) 1, 4   (b) 2, 3, 4, 5

19. a, b, c

20. The α-carboxyglutamate residues are effective chelators of Ca^{2+}. This Ca^{2+} is the electrostatic anchor that binds the protein to a phospholipid membrane, thereby bringing interdependent clotting factors into close proximity.

21. a and c

22. a, b, c

PROBLEMS

1. What would be the kinetic consequences if a substrate were to have exactly equal affinities for the R form and the T form of an allosteric enzyme?

2. Aspartate transcarbamoylase catalyzes the first step in the biosynthetic pathway leading to the synthesis of cytidine triphosphate (CTP). CTP serves as an allosteric inhibitor of aspartate transcarbamoylase that shuts off the biosynthetic pathway when the cell has an ample supply of CTP. Although the first step in a pathway may often be the principal regulatory step, such is not always the case. Figure 10.2 shows a hypothetical degradative metabolic pathway in which step 3 is the principal regulatory step. In this pathway, what advantage does regulation at step 3 have over regulation at step 1 or 2?

FIGURE 10.2   A hypothetical metabolic pathway.

3. Explain why the reagent *N*-(phosphonacetyl)-L-aspartate (PALA) has been especially useful in the investigation of the properties of ATCase.

4. Explain how PALA can act as both an activator and an inhibitor of ATCase.

5. Predict whether each of the following peptide sequences is likely to be phosphorylated by protein kinase A. Briefly explain your answers, and indicate which residue would be phosphorylated.

   (a) Ala-Arg-Arg-Ala-Ser-Leu
   (b) Ala-Arg-Arg-Ala-His-Leu
   (c) Val-Arg-Arg-Trp-Thr-Leu
   (d) Ala-Arg-Arg-Gly-Ser-Asp
   (e) Gly-Arg-Arg-Ala-Thr-Ile

6. Consider the hypothetical metabolic sequence shown in Figure 10.3. Suppose it is known that protein kinase A phosphorylates both enzyme 1 and enzyme 2, and that an increase in intracellular cAMP levels increases the steady-state [B]/[A] ratio. In order for a given increment in cAMP concentration to result in the largest change in the steady-state [B]/[A] ratio, what should be the effect of phosphorylation on the activities of enzymes 1 and 2?

FIGURE 10.3   Hypothetical metabolic sequence.

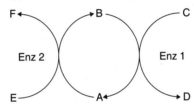

7. Figure 10.4 shows time courses for the activation of two zymogens, I and II. Which of the time courses more resembles that of the activation of trypsinogen, and which corresponds to the activation of chymotrypsinogen? Explain.

FIGURE 10.4   Time courses for the activation of two zymogens, I and II.

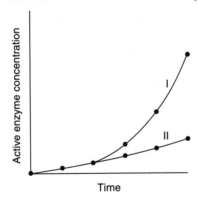

8. Trypsin has 13 lysine and 2 arginine residues in its primary structure. Why does trypsin not cleave itself into 16 smaller peptides?

9. Although thrombin has many properties in common with trypsin, the conversion of prothrombin to thrombin is not autocatalytic whereas the conversion of trypsinogen to trypsin is autocatalytic. Why is the conversion of prothrombin to thrombin not autocatalytic?

10. Because many clotting factors are present in blood in small concentrations, direct chemical measurements often cannot be used to determine whether the factors are within normal concentration ranges or are deficient. Once a deficiency has been established,

however, plasma from the affected person can be used to screen for the presence of the deficiency in other people. A rare deficiency in Factor XII leads to a prolongation of clotting time. Assuming that you have plasma from someone in which this deficiency has been established, design a test that might help determine whether another person has a Factor XII deficiency.

11. In general, regulatory enzymes catalyze reactions that are irreversible in cells, that is, reactions that are far from equilibrium. Why must this be the case?

12. Amplification cascades, such as the one involved in blood clotting, are important in a number of regulatory processes. Figure 10.5 shows a hypothetical cascade involving conversions between inactive and active forms of enzymes. Active enzyme A serves as a catalyst for the activation of enzyme B. Active B in turn activates C, and so forth. Assume that each enzyme in the pathway has a turnover number of 10^3. How many molecules of enzyme D will be activated per unit time when one molecule of active enzyme A is produced per unit time?

FIGURE 10.5 A hypothetical regulatory cascade.

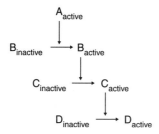

13. Thrombin and trypsin are both serine proteases that are capable of cleaving the peptide bond on the carboxyl side of arginine; thrombin, however, is specific for Arg-Gly bonds. Describe briefly the similarities and differences in the active sites of these two enzymes.

ANSWERS TO PROBLEMS

1. If a substrate were to have equal affinities for the R and T forms, the forms would be indistinguishable kinetically and the system would behave as if all the enzyme were present in a single form. Thus, Michaelis-Menten kinetics would apply, and a plot of the reaction velocity versus the substrate concentration would be hyperbolic.

2. The pathway shown in Figure 10.2 is branched. If regulation were to occur at step 1 only, there would be no control over the production of X from B. If only step 2 were regulated, there would be no regulation over the production of X from A. Regulation at step 3 provides control of the amount of X produced from both A and B. In branched pathways, the principal regulatory step is usually after the branch point.

3. PALA is a bisubstrate analog; that is, it resembles a combination of both substrates, and it is a transition state analog for the carbamoyl phosphate-aspartate complex during catalysis by ATCase. X-ray diffraction analysis of ATCase with bound PALA has revealed the location of the active site and interactions that occur within it. In addition, comparisons of structures with and without PALA have indicated the large structural changes that ATCase undergoes upon binding substrates.

4. PALA is an analog of both substrates of ATCase; therefore, it binds to the active site and acts as a potent inhibitor. At low concentrations, however, binding of PALA shifts the distribution of ATCase molecules to the R conformation. This increases binding of substrates and enzymatic activity.

5. The consensus motif recognized by protein kinase A is Arg-Arg-X-Ser-Z, where X is a small residue and Z is a large hydrophobic residue. The site of phosphorylation is either Ser or Thr.

   (a) Ser would be phosphorylated.
   (b) There would be no phosphorylation because neither Ser nor Thr is present.
   (c) There would be no phosphorylation because Trp is a bulky group and residue X must be small.
   (d) There would be no phosphorylation because Asp is polar and charged and residue Z must be hydrophobic.
   (e) Thr would be phosphorylated.

6. Phosphorylation should increase the activity of enzyme 2 and decrease the activity of enzyme 1. That being the case, an increase in intracellular cAMP levels could greatly increase the steady-state ratio of [B]/[A]. Such coordinated, reciprocal control of opposing metabolic sequences is observed frequently in cells.

7. Curve I corresponds to the activation of trypsinogen, a process that is autocatalytic. As the process occurs, trypsin is produced, which can then cleave yet more trypsinogen. Curve II corresponds to the activation of chymotrypsinogen. The activation of chymotrypsinogen is not autocatalytic. Rather, tryspin catalyzes the conversion of chymotrypsinogen to active α-chymotrypsin. Therefore, its time course is initially linear.

8. The lysine and arginine residues must be partially buried and inaccessible to the active site of trypsin.

9. Thrombin specifically cleaves Arg-Gly bonds. The two bonds that are broken when prothrombin is converted to thrombin are Arg-Thr and Arg-Ile. Therefore, the conversion cannot be autocatalytic.

10. Prepare two samples of blood from the person to be tested. Add normal plasma to one sample and Factor XII-deficient plasma to another. If clotting time is restored to normal in both samples, Factor XII deficiency is probably not involved. If the addition of normal plasma restores normal clotting time but the addition of Factor XII-deficient plasma does not, then a Factor XII deficiency must be suspected.

11. Suppose that a reaction is at equilibrium. If the enzyme catalyzing that reaction were made more active, nothing would happen. The reaction would still be at equilibrium. If, on the other hand, the reaction is displaced far from equilibrium and the enzyme catalyzing the reaction is made more active, more product will be produced. Thus, a regulatory enzyme must catalyze an irreversible step if it is to increase the flux rate through a pathway when it is allosterically activated.

12. One molecule of active A will lead to the activation of 10^9 molecules of enzyme D per unit time. Active A will produce 10^3 molecules of active B. Each of the 10^3 molecules of active B will activate 10^3 molecules of C per unit time. Since there are 10^3 molecules of B, this gives a total of 10^6 molecules of active C. Similar reasoning leads to the answer of 10^9 molecules of active D.

13. Because both thrombin and trypsin are serine proteases, they both have an oxyanion hole and a catalytic triad at the active site. Also, the substrate-specificity sites of both have a similar, negatively charged pocket capable of binding Arg. However, thrombin probably has just enough space to accommodate a Gly residue next to the Arg binding site in contrast to trypsin, which has no restrictions as to the amino acid residue that can be accommodated at the corresponding position.

Carbohydrates

Carbohydrates are one of the four major classes of biomolecules; the others are proteins, nucleic acids, and lipids. In Chapter 11, the authors describe the chemical nature of carbohydrates and summarize their principal biological roles. First, they introduce monosaccharides, the simplest carbohydrates, and describe their chemical properties. Since these sections assume familiarity with the properties of aldehydes, ketones, alcohols, and stereoisomers, students with a limited background in organic chemistry should review these topics in any standard organic chemistry text. Next, the chapter discusses simple derivatives of monosaccharides, including sugar phosphates and disaccharides. *Sugar* is the common name for monosaccharides and their derivatives. You have already seen some monosaccharide derivatives in the structures of nucleic acids in Chapter 4 and nucleotides in Chapter 9. Then the text discusses polysaccharides and oligosaccharides as storage and structural polymers and as components of proteoglycans and glycoproteins.

Glycoproteins are proteins with carbohydrates attached, generally as oligosaccharides. The attachment of sugars takes place either in the lumen of the endoplasmic reticulum or in the Golgi complex. One reason for attachment of sugars is the targeting of specific proteins to specific sites. For example, attachment of mannose 6-phosphate sends proteins from the Golgi complex to the lysosomes. A eukaryotic cell has many different subcellular compartments, each of which has to have a certain array of enzymes and proteins. The Golgi complex functions as the "post office" for the cell, and the attached oligosaccharides function as the "ZIP codes." Attached sugars can also function as signals for proper folding, or as sites of interaction between cells. Lectins and selectins are proteins that bind specific oligosaccharide clusters on the cell surface. The A, B, and O blood group antigens are examples of cell-surface oligosaccharides. Hemagglutinin allows the influenza virus to bind to sialic acid and thus attach to cells before invading them.

LEARNING OBJECTIVES

When you have mastered this chapter, you should be able to accomplish the following objectives.

Introduction

1. List the main roles of *carbohydrates* in nature.

Monosaccharides Are the Simplest Carbohydrates (Text Section 11.1)

2. Define *carbohydrate* and *monosaccharide* in chemical terms.

3. Relate the absolute configuration of monosaccharide D or L *stereoisomers* to those of *glyceraldehyde*.

4. Associate the following monosaccharide class names with their corresponding structures: *aldose* and *ketose*; *triose, tetrose, pentose, hexose,* and *heptose; pyranose* and *furanose*.

5. Distinguish among *enantiomers, diastereoisomers,* and *epimers* of monosaccharides.

6. Draw the *Fischer* (open-chain) *structures* and the most common *Haworth* (ring) *structures* of D-*glucose*, D-*fructose*, DD-*galactose*, and D-*ribose*.

7. Explain how ring structures arise through the formation of *hemiacetal* and *hemiketal* bonds. Draw a ring structure, given a Fischer formula.

8. Distinguish between α and β *anomers* of monosaccharides.

9. Compare the *chair, boat,* and *envelope conformations* of monosaccharides.

10. Define *O-glycosidic* and *N-glycosidic* bonds in terms of *acetal* and *ketal* bonds. Draw the bonds indicated by such symbols as α-1,6 or β-1,4.

11. Explain why *phosphorylated* sugars are important components of living cells.

12. Explain what makes a sugar a reducing sugar.

13. Explain how Advanced Glycation End-Products (AGEs) arise.

Monosaccharides Are Linked to Form Complex Carbohydrates (Text Section 11.2)

14. Explain the role of O-glycosidic bonds in the formation of monosaccharide derivatives, *disaccharides,* and *polysaccharides*.

15. Draw the structures of *sucrose, lactose,* and *maltose*. Give the natural sources of these common disaccharides.

16. Describe the structures and biological roles of *glycogen, starch, amylose, amylopectin,* and *cellulose*.

17. Give examples of enzymes involved in the digestion of carbohydrates in humans.

18. List the major kinds of *glycosaminoglycans* and name their sugar components.

19. Describe *proteoglycans* and explain their importance in cartilage.

20. Explain the differences between the oligosaccharide antigens for A, B, and O blood types.

21. Explain why humans cannot digest cellulose.

Carbohydrates Can Be Attached to Proteins to Form Glycoproteins (Text Section 11.3)

22. Name the amino acid residues that are used for attachment of carbohydrates to glycoproteins.

23. Discuss *erythropoietin* (EPO) structure and function.

24. Describe the use of the lipid *dolichol phosphate* as an intermediary in attachment of oligosaccharides to proteins in the endoplasmic reticulum (ER).

25. State the molecular basis of *I-cell disease*. Explain how this disorder revealed the molecular signal that directs hydrolytic enzymes to the lysosome.

26. Explain briefly how biological oligosaccharides can be "sequenced" using mass spectrometry methods in conjunction with specific enzyme cleavage.

Lectins Are Specific Carbohydrate-Binding Proteins (Text Section 11.4)

27. Give examples of *lectins* and outline their functions and uses.

28. Describe the functions of *selectins* in the human body.

29. Explain why the influenza virus would have two proteins, hemagglutinin and neuraminidase, which perform diametrically opposite tasks.

SELF-TEST

Introduction

1. Which of the following are roles of carbohydrates in nature? Carbohydrates
   (a) serve as energy stores in plants and animals.
   (b) are major structural components of mammalian tissues.
   (c) are constituents of nucleic acids.
   (d) are conjugated to many proteins and lipids.
   (e) are found in the structures of all the coenzymes.

2. In the human diet, carbohydrates constitute approximately half the total caloric intake, yet only 1% of tissue weight is carbohydrate. Explain this fact.

Monosaccharides Are Aldehydes or Ketones with Multiple Hydroxyl Groups

3. Examine the five sugar structures in Figure 11.1:

FIGURE 11.1

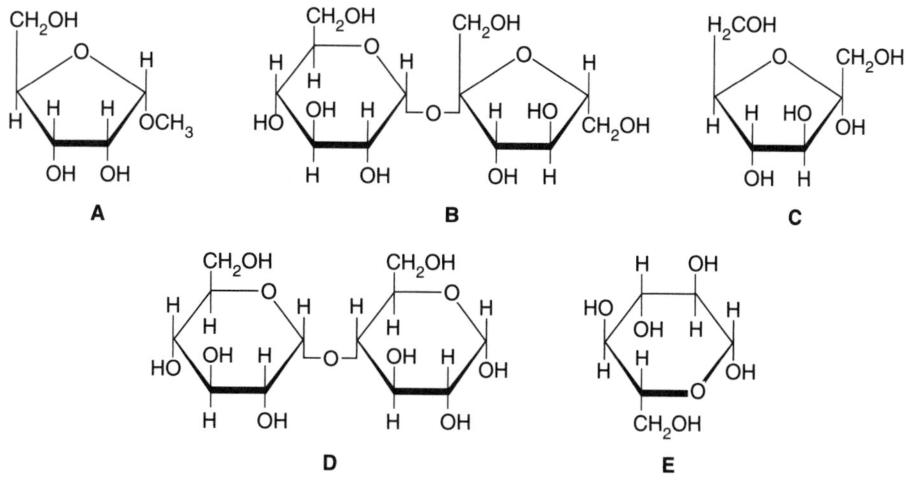

Which of these sugars

(a) contain or are pentoses?
(b) contain or are ketoses?
(c) contain the same monosaccharides? Name the monosaccharides.
(d) will yield different sugars after chemical or enzymatic hydrolysis of glycosidic bonds?
(e) are reducing sugars?
(f) contain a β-anomeric carbon?
(g) is sucrose?
(h) are released upon the digestion of starch?

4. Consider the aldopentoses in Figure 11.2.

FIGURE 11.2

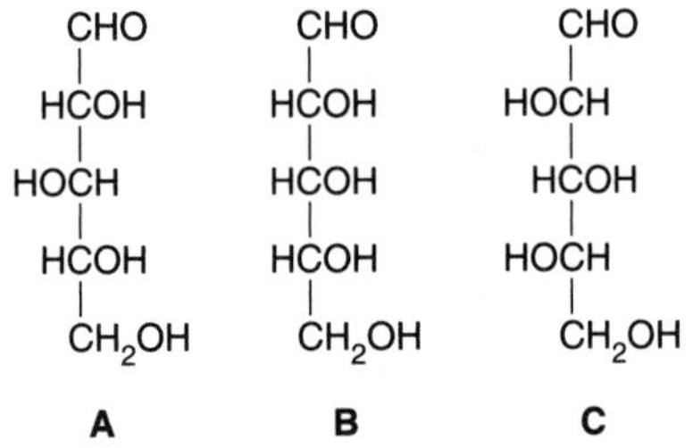

$$
\begin{array}{ccc}
\text{CHO} & \text{CHO} & \text{CHO} \\
| & | & | \\
\text{HCOH} & \text{HCOH} & \text{HOCH} \\
| & | & | \\
\text{HOCH} & \text{HCOH} & \text{HCOH} \\
| & | & | \\
\text{HCOH} & \text{HCOH} & \text{HOCH} \\
| & | & | \\
\text{CH}_2\text{OH} & \text{CH}_2\text{OH} & \text{CH}_2\text{OH} \\
\text{A} & \text{B} & \text{C}
\end{array}
$$

Aldopentoses

(a) Name the types of stereoisomers represented by each pair.
     A and B are
     B and C are
     A and C are
(b) Name sugar B.
(c) Draw the α-anomeric form of the furanose Haworth ring structure for sugar A.

5. Identify the properties common to D-glucose and D-ribose. Both monosaccharides

(a) are reducing sugars.
(b) form intramolecular hemiacetal bonds.
(c) have functional groups that can form glycosidic linkages.
(d) occur in hexose form.
(e) are major constituents of glycoproteins.

6. Referring to the structure of ATP in Figure 11.3, which of the statements are true?

FIGURE 11.3

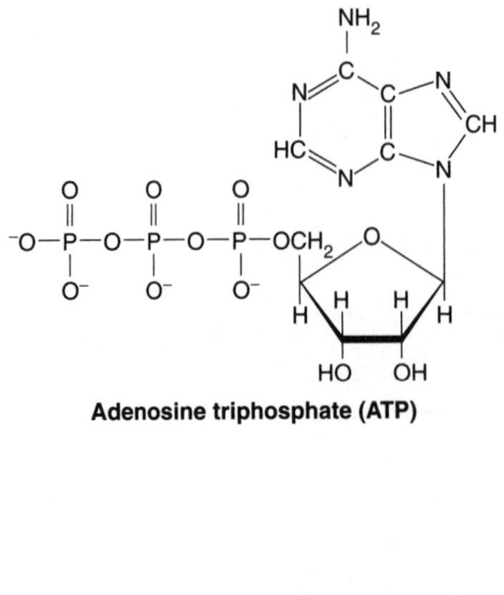

Adenosine triphosphate (ATP)

The structure of ATP

(a) contains a β-N-glycosidic linkage.

(b) contains a pyranose ring.

(c) exists in equilibrium with the open Fischer structure of the sugar.

(d) preferentially adopts a chair conformation.

(e) contains a ketose sugar.

Complex Carbohydrates Are Formed by Linkage of Monosaccharides

7. Draw the structure of the disaccharide glucosyl α-1,6-galactose in the β-anomeric form.

8. If one carries out the partial mild acid hydrolysis of glycogen or starch and then isolates from the product oligosaccharides all the trisaccharides present, how many different kinds of trisaccharides would one expect to find? Disregard α or β anomers.

(a) 1

(b) 2

(c) 3

(d) 4

(e) 5

9. A sample of bread gives a faint positive color with Nelson's reagent for reducing sugars. After an equivalent bread sample has been masticated, the test becomes markedly positive. Explain this result.

10. Why does cellulose form dense linear fibrils, whereas amylose forms open helices?

11. For the polysaccharides in the left column, indicate all the descriptions in the right column that are appropriate.

(a) amylose

(b) cellulose

(c) dextran

(d) glycogen

(e) starch

(1) contains α-1,6 glucosidic bonds

(2) is a storage polysaccharide in yeasts and bacteria

(3) can be effectively digested by humans

(4) contains β-1,4 glucosidic bonds

(5) is a branched polysaccharide

(6) is a storage polysaccharide in humans

(7) is a component of starch

12. α-Amylase

(a) removes glucose residues sequentially from the reducing end of starch.

(b) breaks the internal α-1,6 glycosidic bonds of starch.

(c) breaks the internal α-1,4 glycosidic bonds of starch.

(d) cleaves the α-1,4 glycosidic bond of lactose.

(e) can hydrolyze cellulose in the presence of an isomerase.

13. Which of the following statements about glycosaminoglycans are true?

(a) They contain derivatives of either glucosamine or galactosamine.

(b) They constitute 5% of the weight of proteoglycans.

(c) They contain positively charged substituent groups.

(d) They include heparin, chondroitin sulfate, and keratan sulfate.

(e) They have repeating units of four sugar groups.

14. Look at Figure 11.26 in the text, which shows the structures of the A, B, and O blood antigens. Based on the structures of the three antigens, can you suggest why type O blood is the "universal donor" and can be transfused into people with type A or type B without provoking an immune response?

Carbohydrates Can Be Attached to Proteins to Form Glycoproteins

15. Glycoproteins
    (a) contain oligosaccharides linked to the side chain of lysine or histidine residues.
    (b) contain oligosaccharides linked to the side chain of asparagine, serine, or threonine residues.
    (c) contain linear oligosaccharides with a terminal glucose residue.
    (d) bind to liver cell-surface receptors that recognize sialic acid residues.
    (e) are mostly cytoplasmic proteins.

16. Translocated proteins may undergo which of the following modifications in the lumen of the ER?
    (a) signal sequence cleavage
    (b) the attachment of dolichol phosphate to form a lipid anchor
    (c) folding, disulfide-bond formation and isomerization, and cis-trans isomerization of X-pro peptide bonds
    (d) the addition of oligosaccharides to their asparagine residues to form N-linked derivatives
    (e) the addition of oligosaccharides to their tyrosine residues to form O-linked derivatives

17. Which of the following statements about dolichol phosphate is false?
    (a) It serves as an acceptor of monosaccharides.
    (b) It serves as a donor of both monosaccharides and oligosaccharides.
    (c) It acts as a lipid carrier to facilitate the transfer of sugar residues from the cytosol to the lumen of the ER.
    (d) It helps transfer O-linked oligosaccharides onto serine residues

18. Which of the following statements about I-cell disease are correct?
    (a) It results from the inability of lysosomes to hydrolyze glycosaminoglycans and glycolipids.
    (b) It results from a chromosomal deletion of the genes specifying at least eight acid hydrolases ordinarily found in the lysosomes.
    (c) It arises from a deficiency in an enzyme that transfers mannose 6-phosphate onto a core oligosaccharide that is normally found on lysosomal enzymes.
    (d) It arises from the absence of a mannose 6-phosphate receptor in the *trans* Golgi complex.

Lectins Are Specific Carbohydrate-Binding Proteins

19. Which of the following statements are true? Lectins
    (a) are produced by plants and bacteria.
    (b) contain only a single binding site for carbohydrate.
    (c) are glycosaminoglycans.
    (d) recognize specific oligosaccharide patterns.
    (e) mediate cell-to-cell recognition.

20. Which of the following statements are true? Selectins
    (a) circulate in blood as free proteins.
    (b) are cell-surface receptor proteins.
    (c) are carbohydrate-binding adhesive proteins.
    (d) recognize and bind collagen in the extracellular matrix.
    (e) mediate the binding of immune cells to sites of injury during the inflammation process.

ANSWERS TO SELF-TEST

1. a, c, d

2. Most of the carbohydrates in the human diet are used as fuel to supply the energy requirements of the organism. Although some carbohydrate is stored in the form of glycogen, the mass stored is relatively small compared with adipose tissue and muscle mass. The carbohydrate present in nucleic acids, glycoproteins, glycolipids, and cofactors, although functionally essential, contributes relatively little to the weight of the body.

3. (a) A
   (b) B, C
   (c) B and C contain fructose; B, D, and E contain or are glucose. Note that glucose is in the α-anomer form in sugars B and D and is in the β-anomer form in sugar E.
   (d) A, B, D
   (e) C, D, E
   (f) B and E. In structure B, the fructose ring is flipped over, and E is D-glucose filpped over.
   (g) B
   (h) D and E. Although E is in the β-anomer form, recall that in solution it can "mutarotate" or change back to the α-anomer.

4. (a) A and B are 3-epimers. B and C are diastereoisomers. A and C are enantiomers.
   (b) D-ribose
   (c) See Figure 11.4.

FIGURE 11.4

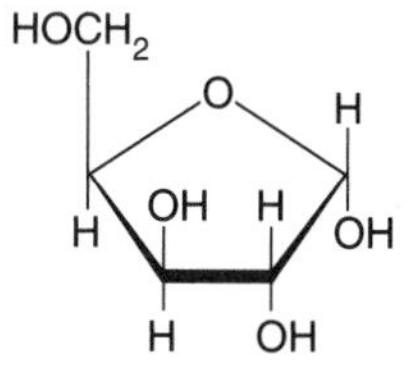

Sugar A
(α-anomeric form)

5. a, b, c. Note that *glycosidic* refers to bonds involving any sugars; however, *glucosidic* and *galactosidic* refer specifically to bonds involving the anomeric (reducing) carbons of glucose and galactose, respectively.

6. a

7. See Figure 11.5.

FIGURE 11.5

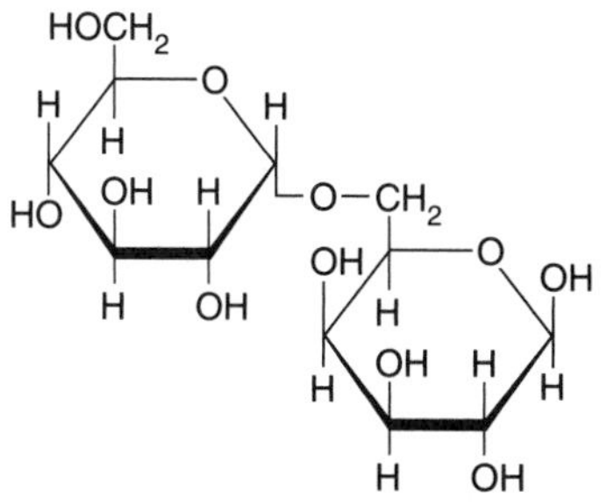

Glucose α (1–6)-galactose
(β-anomeric form)

8. Both c and d are correct. Since there are two glucosidic bonds in each trisaccharide and each bond can be α-1,4 or α-1,6, the total number of possible kinds of trisaccharides is four. However, two consecutive α-1,6 bonds would be very rare in glycogen or starch; therefore, one would be more likely to find three kinds.

9. The carbohydrate in bread is mostly starch, which is a polysaccharide mixture containing D-glucose residues linked by glucosidic bonds. All the aldehyde groups in each polysaccharide, except one at the free end, are involved in acetal bonds and do not react with Nelson's reagent. During mastication (chewing), α-amylase in saliva breaks many of the internal α-1,4 glucosidic bonds and exposes reactive aldehyde groups (reducing groups). Note: Nelson's reagent consists of copper sulfate in a hot alkaline solution; a reducing sugar, such as glucose, reduces the copper, which in turn reduces the arsenomolybdate in the reagent, producing a blue complex.

10. Both cellulose and amylose are linear polymers of D-glucose, but the glucosidic linkages of cellulose are β-1,4 whereas those of amylose are α-1,4. The different configuration at the anomeric carbons determines a different spatial orientation of consecutive glucose residues. Thus, cellulose is capable of forming a linear, hydrogen-bonded structure, whereas amylose forms an open helical structure (see Figure 11.14 in the text).

11. (a) 3, 7 (b) 4 (c) 1, 2, 5 (d) 1, 3, 5, 6 (e) 1, 3, 5, and, if you wish, 7.

12. c

13. a, d

14. The O antigen lacks the extra galactose or N-acetylgalactosamine that the other antigens have. Antibodies will react to the presence of an unfamiliar "bump" in the shape of an oligosaccharide but will evidently not react to the lack of a sugar. It is also possible that individuals with Type A or Type B blood have a small amount of O antigen because of inefficient transfer of the final galactose or perhaps hydrolysis of the galactose. This would prevent the immune system from seeing the O antigen as "foreign."

15. b

16. a, c, d. Answer (e) is incorrect because threonine and serine provide hydroxyls for the formation of *O*-linked oligosaccharides. Answer (b) is incorrect because dolichol phosphate is attached to an oligosaccharide, not a protein.

17. d. Sugar-substituted dolichol phosphates serve both as acceptors of monosaccharides from nucleotide sugars and other dolichol phosphate sugars and as donors of monosaccharides and oligosaccharides to other dolichol phosphate sugar derivatives and proteins. But dolichol phosphates are involved in production of only *N*-linked oligosaccharides in the ER. *O*-linked oligosaccharides are produced in the Golgi apparatus.

18. c. The disease results from a deficiency in a sugar phosphotransferase that initiates a two-step sequence leading to the formation of a mannose 6-phosphate terminus on an oligosaccharide substituent of the eight or more affected lysosomal hydrolases. The phosphotransferase attaches a GlcNAc phosphate to a mannose residue of the oligosaccharide. Removal of the GlcNAc leaves the phosphate on the mannose. The enzymes lacking this mannose 6-phosphate "address" label are erroneously exported from the cell rather than being directed to the lysosomes. (See Figure 11.31 of the text.)

19. a, d, e

20. b, c, e. Collagen is a fibrous protein that is bound by proteins called "integrins."

PROBLEMS

1. Glucose and other dietary monosaccharides like fructose and galactose are very soluble in water at neutral pH. For example, over 150 g of glucose can be dissolved in 100 ml g water at 25°C.

    (a) What features of the chemical structure of glucose make it so soluble in water?

    (b) What features of the proteoglycans found in cartilage make them so highly hydrated and contribute to their ability to spring back after deformation?

2. Indicate whether the following pairs of molecules are enantiomers, epimers, diastereoisomers, or anomers.

    (a) D-xylose and D-lyxose

    (b) α-D-galactose and β-D-galactose

    (c) D-allose and D-talose

    (d) L-arabinose and D-arabinose

3. What is the name of the compound that is the mirror image of α-D-glucose?

4. Compound X, an aldose, is enzymatically reduced using NADPH as an electron donor, yielding D-sorbitol (Figure 11.6). This sugar alcohol is then oxidized at the C-2 position with NAD^+ as the electron acceptor; the products are NADH and a ketose, compound Y.

FIGURE 11.6

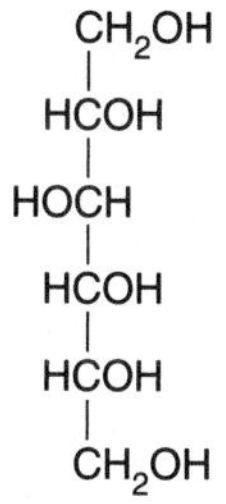

$$CH_2OH$$
$$|$$
$$HCOH$$
$$|$$
$$HOCH$$
$$|$$
$$HCOH$$
$$|$$
$$HCOH$$
$$|$$
$$CH_2OH$$

D-Sorbitol

    (a) Name compound X and write its structure.

    (b) Will sorbitol form a furanose or pyranose ring? Why?

    (c) Name compound Y and write its structure.

5. In Section 11.1 of the text, reducing sugars are defined as those with a free aldehyde or keto group that can reduce cupric ion to the cuprous form. The reactive species in the reducing sugar reaction is the open-chain form of the aldose or ketose. The reaction can be used to estimate the total amount of glucose in a solution such as blood plasma. An aqueous solution of glucose contains only a small amount of the open-chain form. How can the reaction be used to provide a *quantitative* estimate of glucose concentration?

6. Compare the number of dimers that can be prepared from a pair of alanine molecules and from a pair of D-galactose molecules, each of which is present as a pyranose ring. For the galactose molecules, pairs may be made using the α or β anomers.

7. Storage polysaccharides, like starch and glycogen, often contain over a million glucose units. The energetic cost of synthesizing polysaccharides is high (about one high-energy phosphate bond per sugar residue added). Suppose that in a liver cell, the glucosyl residues in large numbers of glycogen molecules were replaced with an equivalent number of molecules of free glucose. What problems would this cause for the liver cell?

8. You have a sample of glycogen that you wish to analyze using exhaustive methylation and acid hydrolysis. You incubate a sample of 0.4 g glycogen with methyl iodide, which methylates all *free* primary or secondary alcohol groups on sugars. Then you subject the sample to acid hydrolysis, which cleaves glycosidic linkages between adjacent glucose residues. You then determine the yield of 2,3-dimethylglucose in your sample.

   (a) Why is a 2,3-dimethylglucose residue produced from a branch point in glycogen?
   (b) The yield of 2,3-dimethylglucose is 0.247 mmol. What fraction of the total residues in each sample are branch points? The molecular weight of a glucosyl residue in glycogen is 162.
   (c) Could you use this technique to determine the anomeric nature of the glycogen branch? Why?

9. Shown in Figure 11.7 is one example of the storage oligosaccharides that account in part for the flatulence caused by eating beans, peas, and other legumes. These oligosaccharides cannot be digested by enzymes in the small intestine, but they can be metabolized by anaerobic microorganisms in the large intestine. There, they undergo oxidation, with the production of large quantities of carbon dioxide, hydrogen sulfide, and other gases. Solutions are now on the market containing one or more enzymes that, when ingested with the offending legumes at mealtime, convert the oligosaccharides to digestible products.

FIGURE 11.7

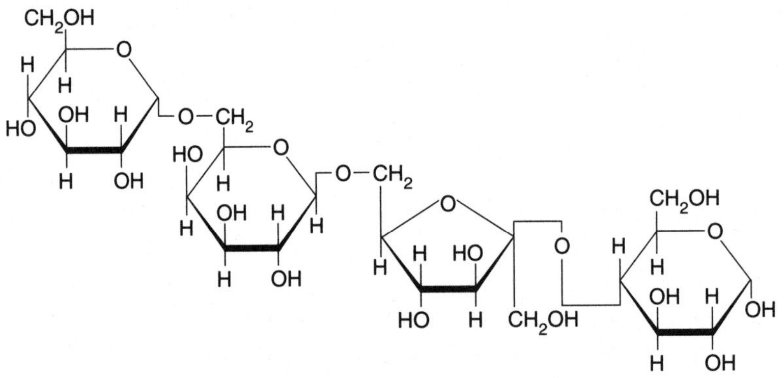

   (a) Name the oligosaccharide shown in the figure.
   (b) Given that free hexoses can pass easily through intestinal cells into the blood, what types of enzymes do you think are included in the commercial products that aid in legume oligosaccharide digestion?
   (c) The concentration of oligosaccharides in beans can be reduced by cooking or by sprouting. What happens to the oligosaccharides in cooking? When the beans sprout before cooking or eating?
   (d) When small amounts of cellulose are ingested purposely or accidentally (for example, by pets or young children), there is usually no gas production. In fact, the primary concern about paper ingestion by pets or small children is intestinal blockage. Why?

10. Explain the roles of (a) the phosphate group and (b) the long lipid chain of dolichol phosphate in the transport of polysaccharides across membranes.

12. MALDI–TOF MS stands for Matrix Assisted Laser Desorption/Ionization–Time of Flight Mass Spectrometry. It is a highly sophisticated technique (also used for proteins, see text Section 3.5), but it can't solve oligosaccharide structures without input from other techniques. Why not?

13. Why is the structural analysis of an oligosaccharide containing eight monosaccharide residues more complicated than a similar analysis for an octanucleotide or an octapeptide? This is not a quantitative question; a qualitative description will do.

14. In the 1950s, Morgan and Watkins showed that N-acetylgalactosamine and its α-methylglycoside inhibit the agglutination of type A erythrocytes by type A-specific lectins, whereas other sugars had little effect. What did this information reveal about the structure of the glycoprotein on the surface of type A cells?

15. N-Acetyl neuraminic acid (or sialic acid, see Fig. 11.9 of the text) is a particularly interesting compound because it is one of the few molecules that are unique to human metabolism. Other higher animals produce a hydroxylated version, as in Figure 11.8.

FIGURE 11.8

$$\text{Human Neu} \quad \text{5Ac} \quad \underset{\underset{H}{|}}{\overset{\overset{O}{\|}}{CH_3\text{-}C\text{-}N}}\text{-R} \qquad \text{Other Neu} \quad \text{5Gc} \quad \underset{\underset{H}{|}}{\overset{\overset{O}{\|}}{HO\text{-}CH_2\text{-}C\text{-}N}}\text{-R}$$

Sialic acid is an important cell surface antigen. As an example, at the end of the chapter in the text we learn that the influenza virus uses hemagglutinin to bind sialic acid and then neuraminidase to cleave and remove it. Now, mouse cells are often mixed with human stem cells in culture. Can you imagine what problems might arise with this practice?

16. Because red blood cells (RBC) carry oxygen, athletes have long been interested in finding ways to increase the numbers of RBC in their bodies. Honest athletes go through physical training, which increases their blood volume. Thirty years ago, some dishonest athletes hit on the idea of "donating" blood that could be centrifuged so that their own RBC could be reinjected before an athletic event (blood doping). More recently some athletes in endurance events (like cycling) have been caught using EPO, a hormone that stimulates production of RBC. As described in the text, EPO is a protein that is rather easy to produce by modern techniques. In the year 2000 tests were developed that could reveal the presence of EPO based on differences in glycosylation. These differences were there because human genes had been expressed in hamster cells. In 2004, a new form of EPO called dynEPO was produced using human cell lines. The glycosylation appears to be identical because this is a human protein expressed by human cells. Can you think of any way to detect athletes who cheat using this highly engineered product?

17. Exercise is fueled primarily by the oxidation of glucose (glycolysis) and fat (a process called β-oxidation). A typical runner might oxidize carbohydrate at the rate of 4 g/min and fat at the rate of 0.5 g/minute. The glucose that is being oxidized can come from the breakdown of liver and muscle glycogen. What advantages does the structure of glycogen provide for a runner?

18. Blood glucose is carefully regulated so that its normal concentration (the level between meals) is approximately 4.4–6.1 mM (82–110 mg/dL). This concentration represents about four teaspoons of dissolved glucose. Many mechanisms exist for blood glucose homeostasis (keeping blood glucose levels as constant as possible). Why is glucose homeostasis so important? (That is, what are the negative effects of high blood glucose levels?)

19. If you have ever ripped a piece of newspaper, then you may have noticed that if you tear the paper in one direction, you get a smooth tear. If you tear it in the perpendicular direction, the paper is more difficult to tear smoothly and you get a jagged edge. Given that newsprint is made from cellulose, explain this phenomenon.

20. Alcoholic beverages are made from the fermentation of glucose. A source of glucose can be the amylose and amylopectin of potatoes (for vodka), barley (for beer), or rice (for sake). The starch is treated with amylases that are similar to the α-amylase secreted by the salivary glands and pancreas. Malted barley (barley that has begun to germinate) is a source of these enzymes. This treatment hydrolyzes amylose to glucose and maltose and a few short oligosaccharides. The same products appear from amylopectin but with the addition of a class of products called "limit dextrins." What would you expect would be the structure of these limit dextrins?

ANSWERS TO PROBLEMS

1. (a) Glucose and other hexose monosaccharides have five hydroxyl groups and an oxygen in the heterocyclic ring that can all form hydrogen bonds with water. The ability to form these hydrogen bonds with water and other polar molecules enables hexoses and other carbohydrates to dissolve easily in aqueous solution.

   (b) In addition to hydrogen bonding of water to hydroxyl groups and oxygen atoms in the repeating disaccharide units of cartilaginous proteoglycans like keratan sulfate and chondroitin sulfate, these molecules also contain charged sulfate and carboxylate groups that can also interact with water. Compression of these large hydrated polyanions can drive some water out of the cavities between them, but the high degree of hydration of the molecules, as well as charge repulsion between the sulfate and carboxylate groups, contributes to the tendency of these compounds to resume their normal conformations after deformation.

2. (a) D-Xylose and D-lyxose differ in configuration at a single asymmetric center; they are epimers.

   (b) α-D-Galactose and β-D-galactose have differing configurations at the C-1, or anomeric, carbon; they are anomers.

   (c) D-Allose and D-talose are diastereoisomers because they have opposite configurations at one or more chiral centers, but they are not complete mirror images.

   (d) L-arabinose and D-arabinose are mirror images of each other and are therefore enantiomers.

3. Although the mirror image of a D compound is an L compound, the mirror image of an α compound is an α compound. (An α compound has a 1-hydroxyl group in the α position.) Thus, α-L-glucose is the compound that is the mirror image, or enantiomer, of α-D-glucose.

4. (a) Compound X is D-glucose; it is the only D-aldose whose reduction will yield a hexitol with the same conformation as that of D-sorbitol. The less common sugar L-glucose would also yield the same result. L-sorbitol is commonly used for applications like sweetening toothpaste (without promoting decay). With sugar alcohols, there is no most-oxidized carbon, so it is hard to define which end is "carbon one," but the use of an enzyme greatly favors D-aldose as the starting material. L-Sorbitol probably originates from D-glucose. Turn the molecule upside down to see the relationship.

   (b) Sorbitol cannot form a hemiacetal because it has no aldehyde or ketone group. Therefore, neither type of ring can be formed by sorbitol.

   (c) Compound Y is D-fructose, a ketose that is produced by the oxidation of sorbitol. Enzymes would be very unlikely to produce an L-ketose, so this is the only expected result.

5. In water, an equilibrium exists among three forms of glucose. Two-thirds is present as the β anomer, one-third as the α anomer, and less than 1% as the open-chain form. When excess cupric ion reacts with the open-chain form, glucose is oxidized to gluconic acid. Through the law of mass action, the α and β anomers of glucose are then converted to the open-chain aldose form. Continued production of gluconic acid from the open-chain form leads to the ultimate conversion of all glucopyranoses to the open-chain form, which reacts quantitatively with cupric ion. Thus the total amount of glucose in a known volume of blood plasma or other solution can be determined.

6. Only one dimer, alanylalanine, can be made from two alanine molecules linked via a peptide bond. However, the presence of several hydroxyl groups and the aldehydic function at the C-1 position of each D-galactose molecule provides an opportunity to make a larger number of dimers. Both the α and β forms of one molecule can form glycosidic linkages with the C-2, C-3, C-4, or C-6 hydroxyl groups of the other. Recall that the C-5 position is not available, because it participates in the formation of the pyranose ring. To these eight dimers can be added dimers formed through glycosidic linkages involving the αα, αβ, or ββ configurations. Thus, 11 possible dimers exist. If one is allowed to use L forms, then the number of possible dimers increases greatly. This variety of linkages makes the sugars very versatile molecules and yields many different structures that may be useful in biology. However, this variety has also made the systematic study of the chemistry of polysaccharides very difficult.

7. The primary consequence of a high concentration of free glucose molecules in the cell would be a dramatic and probably catastrophic increase in osmotic pressure. In aqueous solutions, colligative properties like boiling and freezing points, vapor pressure, and osmotic pressure depend primarily on the number of molecules in the solution. Thus a glycogen molecule containing a million glucose residues exerts one-millionth the osmotic pressure of a million molecules of free glucose. Osmotic pressure exerted by high glucose concentration would induce entry of water into the cell in an attempt to equalize pressure inside and outside the cell. Unlike bacterial or plant cells which have a rigid cell wall that can help resist high pressures, animal cells have a comparatively fragile plasma membrane, which will burst when osmotic pressures are too high.

8. (a) A glucosyl residue at a branch point has three of its five carbons linked to other glucose residues; these are carbons 1, 4, and 6. Only C-2 and C-3 of a branch point residue will have alcohol or hydroxyl groups that are free and therefore available for methylation. Thus residues at a branch point are converted to 2,3-dimethylglucose after methylation and hydrolysis. The glucosyl residues not at a branch point would be converted to 2,3,6-trimethylglucose by the same procedure, except for the single residue at the reducing end, which could be converted to 1,2,3,6-tetramethylglucose.

   (b) The original sample of 0.4 g corresponds to 0.4 g ÷ 162 g/mole, or 2.47×10^{-3} mole, or 2.47 mmol glucose residues, which is 10% of the total sample. Thus 10% of the glucosyl residues are at branch points.

   (c) The analysis using methylation and acid hydrolysis does not allow determination of the anomeric linkage. Acid hydrolysis cleaves both α- and β-anomeric linkages and does not allow distinctions between them.

9. (a) Glu α-1,6   Gal α-1,6   Fru β-1,4   Glu.

   (b) The solution must contain enzymes that hydrolyze the glycosidic linkages between the monosaccharides. For example, an activity that would be required for the oligosaccharide shown would be a type of α-1,6-glycosidase, which would cleave the α-1,6 linkage between glucose and galactose. Another would be the β-1,4-fructosidase, a different glycosidase. The glycosidases are needed to convert the oligosaccharides to free hexoses, which then pass easily into the circulation. The three common sugars found in the oligosaccharide shown in this problem are easily metabolized by the liver and other cells.

(c) Cooking by heating in water probably hydrolyzes some of the glycosidic linkages found in the oligosaccharides. Sprouting or germinating beans undergo a reduction in oligosaccharide concentration because hydrolase proteins induced during germination produce free hexoses, which can be used in the developing plant tissues as a source of carbon for biosynthesis.

(d) Because cellulose is an unbranched polymer of glucose residues joined by β-1,4 linkages, the molecule is resistant to hydrolysis even by anaerobic bacteria in the human intestine. Small amounts of cellulose and other indigestible complex carbohydrates are virtually unaltered as they pass through the digestive system. Thus no gases from carbohydrate breakdown are generated in the large intestine. Intestinal blockage may result from ingestion of large quantities of cellulose because there are no enzymes available to cleave the glycosidic linkages. Organisms that use cellulose as an energy source (for example, cows and termites) have gut flora that make cellulase and can provide the service of breaking these β-1,4 bonds.

10. (a) The phosphate group serves as the site for the covalent attachment of sugar residues to the carrier.

(b) The long lipid chain renders the carrier highly hydrophobic and thus membrane-permeable.

12. MALDI–TOF MS provides a very accurate molecular weight for only an oligosaccharide or other complex molecule. If you have, say, ten sugars, they can be rearranged in many different isomeric forms that all would have the same molecular weight. Enzymes that can cleave only certain sugars in certain positions provide extra information that is critical to the "sequencing" of an oligosaccharide.

13. In oligosaccharides, there are a number of different types of potential glycosidic linkages that can be formed among eight residues, because each free hydroxyl group as well as the anomeric carbon on a particular monosaccharide could be linked to similar groups on adjacent residues. An octo-oligosaccharide could be linear or branched and could be composed of as many as eight different monosaccharides, each of which could require additional steps to analyze completely. Analysis of an oligonucleotide is somewhat less complicated, because usually only four different bases are found during the analysis, and the linkage between adjacent nucleotides is almost always $3' \rightarrow 5'$; in addition, the oligonucleotide molecule is not likely to be branched. Although there may be as many as eight different amino acid residues in an octapeptide, all 20 different amino acids found in most proteins are relatively easy to characterize and the octapeptide is unlikely to be branched.

14. The observations of Morgan and Watkins suggested that the sugar N-acetylgalactosamine in a linkage is the determinant of blood group A specificity. The galactose derivative binds to type A lectins, occupying the sites that would otherwise bind to glycoproteins having N-acetylgalactosamine end groups on the surfaces of type A cells. The papers establishing the structures of the blood group oligosaccharides were among the first of Winifred M. Watkins's long and distinguished career. The fields of biochemistry and molecular biology have provided several early female role models including such important scientists as Maud Menten (who collaborated with L. Michaelis to study enzymology) and Rosalind Franklin (who determined the structure of the A-form and worked on the B-form of double-helical DNA). Dr. Watkins was elected as a Fellow of the Royal Society in 1998. [W. M. Watkins & W. T. J. Morgan, *Nature* 178[1956]:1289, and other papers.]

15. As you might imagine, culturing human stem cells with other cells can result in cells that have N-glycolyl neuraminic acid (Neu5Gc). Attempting to treat human diseases with stem cells prepared in this way can raise antibodies, because there are likely to be nonhuman cell surface antigens in the resulting culture. (For one article on this phenomenon, see M. J. Martin et al., *Nat. Med.* 11[2005]:228.) On a related topic, one reason why bird flu can't normally infect humans is that the hemagglutinin in bird flu viruses normally bind Neu5Gc. A mutation could allow binding to Neu5Ac, and the virus would infect humans. Viral designations like "H5N1" indicate the varieties of hemagglutinin ("H") and neuraminidase ("N") that are found in the virus.

16. Assuming that dynEPO is identical to normal human EPO, it could be impossible to detect as a "foreign substance" because it is not "foreign." It should be possible to have some indication that a person is using illegal drugs by following their hematocrit (which shows levels of RBC in the blood) and other secondary effects. In other words, if you can't see the drug, you still should be able to see the effect of the drug. Here is an article about the difficulties of testing:

    http://hum-molgen.org/NewsGen/08-2004/msg19.html

    and an article about dynEPO and the sport of cycling:
    http://www.bike-zone.com/features/?id=EPOv2

17. Glycogen consists of a homopolymer of glucose molecules linked via $\alpha 1 \rightarrow 4$ linkages with branches of the same attached via $\alpha 1 \rightarrow 6$ linkages every 10 residues or so. Glycogen is broken down into monomers of glucose-1-phosphate by the enzyme glycogen phosphorylase. One advantage of the glycogen structure lies in its branches. Each branch has a non-reducing end and because glycogen is broken down from these ends, the branches simply increase the number of ends that are accessible to glycogen phosphorylase. The branches thus enable glucose 1-phosphate to be released at a faster rate than it would be without these branches. The process provides the body with more glucose to meet its energy demands.

18. Glucose homeostasis is important because glucose is so reactive in its linear form. It will react with the primary amines found in proteins (N-termini, Lys residues, etc.) to form Advanced Glycation Endproducts (AGEs). Many enzymes and proteins do not work as well when they are glycosylated. Second, as you will learn later in this course, elevated blood glucose levels can result in increased insulin secretion, and insulin is a hormone that directs biochemical processes toward fat storage and away from fat oxidation. Elevated insulin levels are associated with obesity and type 2 diabetes.

19. Cellulose is made of glucose monomers with $\beta 1 \rightarrow 4$ linkages. These linkages result in a rather straight chain. Thus, these chains are able to line up side-by-side as well as above and below, and they hydrogen bond with each other. When you rip the newsprint between these hydrogen-bonded chains, the newsprint rips smoothly. When you rip the newsprint in the perpendicular direction, you are ripping it at natural breaks in the cellulose chain. That is, you are ripping it where there is no $\beta(1 \rightarrow 4)$ covalent bond. The jagged-edged tear shows the random distribution of these breaks in the cellulose polymers.

20. Limit dextrins are the branched structures that remain when only the $\alpha 1 \rightarrow 4$ linkages of amylopectin are hydrolyzed. Unlike amylose, which is made entirely of $\alpha(1 \rightarrow 4)$ linkages, amylopectin has branches that are attached via $\alpha(1 \rightarrow 6)$ linkages. These linkages are not hydrolyzed by the α-amylase enzymes. What remains after amylase treatment are short oligosaccharides with branches. Additional enzymes (sometimes from mold) must be added to hydrolyze the $\alpha(1 \rightarrow 6)$ bonds at these branches.

Lipids and Cell Membranes

Chapter 12

In this chapter, the authors describe the composition, structural organization, and general functions of biological membranes. After outlining the common features of membranes and discussing the structure of fatty acids, a new class of biomolecules, the lipids, is introduced in the context of their role as membrane components. The authors focus on the three main kinds of membrane lipids—phospholipids, glycolipids, and cholesterol. The amphipathic nature of membrane lipids and their ability to organize into vesicles and bilayers in water are then described. An important functional feature of membranes is their selective permeability to molecules, in particular the inability of ions and most polar molecules to cross membrane bilayers. This aspect of membrane function is discussed next and will be revisited when the mechanisms for transport of ions and polar molecules across membranes are discussed in Chapter 13.

Next, the authors turn to membrane proteins, the major functional constituents of biological membranes. The arrangement of proteins and lipids in membranes is described and the asymmetric, fluid nature of membranes is stressed. The important differentiation between integral and peripheral membrane proteins is discussed as well as the chemical forces that bind them to the membrane. The high-resolution analyses of the structures of selected membrane proteins are discussed, including structure prediction of membrane-spanning proteins. The chapter concludes with a discussion of internal membranes within eukaryotic cells and the mechanisms by which proteins are targeted to specific compartments within cells.

LEARNING OBJECTIVES

When you have mastered this chapter, you should be able to accomplish the following objectives.

Introduction

1. List the functions of *biological membranes*.
2. Describe the common features of biological membranes.

Fatty Acids Are Key Constituents of Lipids (Text Section 12.1)

3. Draw the general chemical formula of a *fatty acid* and be able to use standard notation for representing the number of carbons and double bonds in a fatty acid chain.
4. Distinguish between *saturated* and *unsaturated* fatty acids.
5. Explain the relationship between fatty acid *chain length* and *degree of saturation* and the physical property of *melting point*.

There Are Three Common Types of Membrane Lipids (Text Section 12.2)

6. Define *lipid* and list the major kinds of *membrane lipids*.
7. Recognize the structures and the constituent parts of *phospholipids (phosphoglycerides and sphingomyelin)*, *glycolipids*, and *cholesterol*.
8. Describe the general properties of the *fatty acid chains* found in phospholipids and glycolipids.
9. Draw the general chemical formula of a phosphoglyceride, and recognize the most common *alcohol moieties* of phosphoglycerides (for example, *choline, ethanolamine*, and *glycerol*).
10. Distinguish between membranes of *archaea* and those of eukaryotes and bacteria.
11. Describe the composition of *glycolipids*. Note the location of the carbohydrate components of membranes.
12. Recognize the structure of *cholesterol*.
13. Describe the properties of an *amphipathic molecule*.

Phospholipids and Glycolipids Readily Form Bimolecular Sheets in Aqueous Media (Text Section 12.3)

14. Distinguish among *oriented monolayers, micelles*, and *lipid bilayers*.
15. Describe the *self-assembly process* for the formation of lipid bilayers. Note the stabilizing intermolecular forces.
16. Outline the methods used to prepare *lipid vesicles (liposomes)* and *planar bilayer membranes*. Point out some applications of these systems.
17. Explain the relationship between the *permeability coefficients* of small molecules and ions and their *solubility* in a nonpolar solvent relative to their solubility in water.

Proteins Carry Out Most Membrane Processes (Text Section 12.4)

18. Distinguish between *peripheral* and *integral membrane proteins*.

19. Give an example of α-helical and β-sheet membrane-bound proteins.

20. Describe the structure of *glycophorin*. Explain how *transmembrane α helices* can be predicted from *hydropathy plots*.

Lipids and Many Membrane Proteins Diffuse Rapidly in the Plane of the Membrane (Text Section 12.5)

21. Describe the evidence for the *lateral diffusion* of membrane lipids and proteins. Contrast the rates for *lateral diffusion* with those for *transverse diffusion*.

22. Describe the features of the *fluid mosaic model* of biological membranes.

23. Explain the roles of the fatty acid chains of membrane lipids and cholesterol in controlling the *fluidity of membranes*.

24. Describe the components of lipid rafts and discuss their effect on *membrane fluidity*.

25. Discuss the origin and the significance of membrane asymmetry.

Eukaryotic Cells Contain Compartments Bounded by Internal Membranes (Text Section 12.6)

25. Give examples of the compositional and functional varieties of biological membranes.

26. Discuss the role of *targeting sequences* in eukaryotic proteins.

27. Describe the recognition of a nuclear localization signal by α *karyopherin*.

28. Describe the process of *receptor-mediated endocytosis* of *transferrin*.

30. Define SNARE and describe their role in *membrane fusion*.

SELF-TEST

Introduction

1. Which of the following statements about biological membranes are true?
   (a) They constitute selectively permeable boundaries between cells and their environment and between intracellular compartments.
   (b) They are formed primarily of lipid and carbohydrate.
   (c) They are involved in information transduction.
   (d) Targeting across them requires specific systems.
   (e) They are dynamic structures.

2. Which of the following statements about biological membranes is NOT true?
   (a) They contain carbohydrates that are covalently bound to proteins and lipids.
   (b) They are very large, sheetlike structures with closed boundaries.
   (c) They are symmetric because of the symmetric nature of lipid bilayers.
   (d) They can be regarded as two-dimensional solutions of oriented proteins and lipids.
   (e) They contain specific proteins that mediate their distinctive functions.

Fatty Acids Are Key Constituents of Lipids

3. Which of the following fatty acids is polyunsaturated?
   (a) arachididic
   (b) arachidonic
   (c) oleic
   (d) palmitic
   (e) stearic

There Are Three Common Types of Membrane Lipids

4. Which of the following substances are membrane lipids?
   (a) cholesterol
   (b) glycerol
   (c) phosphoglycerides
   (d) choline
   (e) cerebrosides

5. The phosphoinositol portion of the phosphatidyl inositol molecule is called the
   (a) amphipathic moiety.
   (b) hydrophobic moiety.
   (c) hydrophilic moiety.
   (d) micelle.
   (e) polar head group.

6. Acid hydrolysis will break all ester, amide, and acetal chemical linkages. Which of the following statements about the acid hydrolysis of various lipids is NOT correct?
   (a) A cerebroside releases two fatty acids and one monosaccharide per mole of cerebroside.
   (b) Phosphatidylcholine releases two fatty acids and one glycerol molecule per mole of phosphatidylcholine.
   (c) Sphingomyelin and phosphatidylcholine releases equivalent molar amounts of choline and phosphoric acid.
   (d) Cerebrosides and sphingomyelin each release one mole of sphingosine.

7. After examining the structural formulas of the four lipids in Figure 12.1, answer the following questions.
   (a) Which formulas are phosphoglycerides?
   (b) Which is a glycolipid?
   (c) Which contain sphingosine?
   (e) Which contain glycerol?
   (d) Which contain choline?
   (f) Name the lipids.

FIGURE 12.1 Membrane lipids R$_1$ and R$_2$ represent hydrocarbon chains.

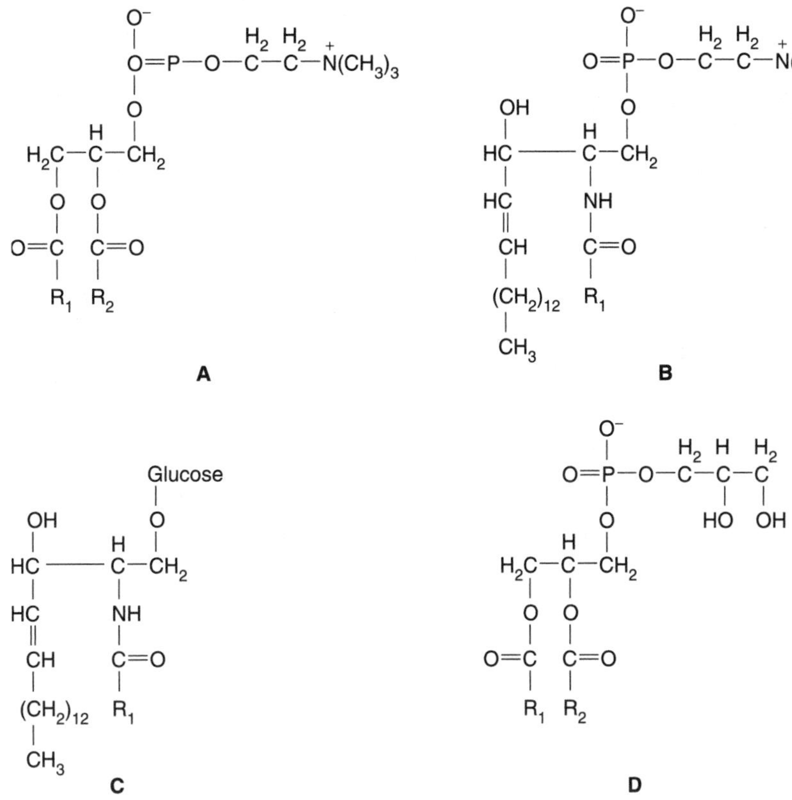

Phospholipids and Glycolipids Readily Form Bimolecular Sheets in Aqueous Media

8. Which of the following statements about a micelle and a lipid bilayer are NOT true?

   (a) Both assemble spontaneously in water.
   (b) Both are made up of amphipathic molecules.
   (c) Both are very large, sheetlike structures.
   (d) Both have the thickness of two constituent molecules in one of their dimensions.
   (e) Both are stabilized by hydrophobic interactions, van der Waals forces, hydrogen bonds, and electrostatic interactions.

9. A triglyceride (triacylglycerol) is a glycerol derivative that is similar to a phosphoglyceride except that all three of its glycerol hydroxyl groups are esterified to fatty acid chains. Would you expect a triglyceride to form a lipid bilayer? Explain.

10. What is the volume of the inner water compartment of a liposome that has a diameter of 500 Å and a bilayer that is 40 Å thick?

    (a) 5.6×10^5 Å^3
    (b) 7.3×10^6 Å^3
    (c) 3.9×10^7 Å^3
    (d) 7.3×10^7 Å^3
    (e) 3.9×10^8 Å^3

11. Arrange the following substances in the order of decreasing permeability through a lipid bilayer.

    (a) urea
    (b) tryptophan
    (c) H_2O
    (d) Na$^+$
    (e) glucose

Proteins Carry Out Most Membrane Processes

12. Why is an α-helix the preferred structure for transmembrane protein segments?

13. Show which of the properties listed at the right are characteristics of peripheral membrane proteins and which are characteristics of integral membrane proteins.

    (a) peripheral
    (b) integral

    (1) require detergents or organic solvent treatment for dissociation from the membrane
    (2) require mild salt or pH treatment for dissociation from the membrane
    (3) bind to the surface of membranes
    (4) have transmembrane domains

14. Which of the following sequences would target a protein to the nucleus?

    (a) -SKL-COO$^-$
    (b) -KKLK-
    (c) -KDEL-COO$^-$
    (d) -KKLK-COO$^-$

Lipids and Many Membrane Proteins Diffuse Rapidly in the Plane of the Membrane

15. Which of the following statements about the diffusion of lipids and proteins in membranes is NOT true?

    (a) Many membrane proteins can diffuse rapidly in the plane of the membrane.
    (b) In general, lipids show a faster lateral diffusion than do proteins.
    (c) Membrane proteins do not diffuse across membranes at measurable rates.
    (d) Lipids diffuse across and in the plane of the membrane at equal rates.

16. Which of the following statements about the asymmetry of membranes are true?

    (a) It is absolute for glycoproteins.
    (b) It is absolute for phospholipids, but only partial for glycolipids.
    (c) It arises during biosynthesis.
    (d) It is structural but not functional.

17. If phosphoglyceride A has a higher T_m than phosphoglyceride B, which of the following differences between A and B may exist? (In each case only one parameter—either chain length or double bonds—is compared.)

    (a) A has shorter fatty acid chains than B.
    (b) A has longer fatty acid chains than B.
    (c) A has more unsaturated fatty acid chains than B.
    (d) A has more saturated fatty acid chains than B.
    (e) A has trans unsaturated fatty acid chains, whereas B has cis unsaturated fatty acid chains.

18. Explain how the presence of lipid rafts modifies the original fluid mosaic model for biological membranes.

Eukaryotic Cells Contain Compartments Bounded by Internal Membranes

19. Which of the following statements about receptor-mediated endocytosis of transferrin is false?
    (a) Transferrin binds to a cell receptor, which is an integral membrane protein.
    (b) The transferrin receptor is recycled back to the membrane once it has transported iron to the cytoplasm.
    (c) The iron-free transferrin complex is recycled to the plasma membrane, where transferrin is released into the bloodstream.
    (d) Proton pumps within the vesicle membrane raise the lumenal pH, which lowers the affinity of iron ions for transferrin, releasing the iron ions.
    (e) All of the above are true.

ANSWERS TO SELF-TEST

1. a, c, d, e
2. c
3. b
4. a, c, e
5. c, e
6. a
7. (a) A, D (b) C (c) B, C (d) A, B (e) A, D (f) A is phosphatidyl choline, B is sphingomyelin, C is cerebroside, and D is phosphatidyl glycerol.
8. c
9. No. Although a triglyceride has hydrophobic fatty acyl chains attached to a glycerol backbone, it lacks a polar head group; therefore, it is not an amphipathic molecule and is incapable of forming a bilayer.
10. The correct answer is (c). The volume of a sphere is $4/3\, \pi\, r^3$, so we just need the radius of the inner compartment to do the calculation. Using Figure 12.2 to represent the liposome, we can calculate the diameter of the inner water compartment by subtracting the width of the bilayer from the left and right sides of the liposome from the diameter of the outer compartment.

FIGURE 12.2

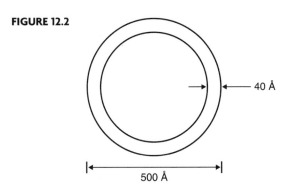

40 Å

500 Å

Diameter of inner water compartment = 500 Å − (2 × 40 Å) = 420 Å

Since the radius of a circle is half the diameter, the radius of the inner compartment $r = 1/2(420\ \text{Å}) = 210\ \text{Å}$.

Therefore the volume of inner water compartment in the liposome 4/3 π (210 Å)^3 = 3.9×10^7 Å^3.

11. c, a, b, e, d

12. Transmembrane protein segments usually consist of nonpolar amino acids. The main-chain peptide CO and NH groups, however, are polar and tend to form hydrogen bonds with water. In an α helix, these groups hydrogen-bond to each other, thereby decreasing their overall polarity and facilitating the insertion of the protein segment into the lipid bilayer.

13. (a) 2, 3   (b) 1, 4

14. b

15. d

16. a, c

17. b, d, e. Trans unsaturated fatty acid chains have a straighter conformation than do cis unsaturated chains; the packing of trans chains in bilayers is therefore more highly ordered, so they require higher temperatures to melt.

18. Lipid rafts are formed from cholesterol in specific complexes with lipids that contain the sphingosine backbone and with GPI-anchored proteins. Lipid rafts moderate membrane fluidity (making them less fluid) and at the same time make them less subject to phase transitions. The fluid mosaic model says that membrane fluidity is controlled by fatty acid composition and cholesterol content. Normally the addition of cholesterol will increase membrane fluidity by disrupting packing between fatty acid chains, but when in a lipid raft complex, may actually decrease membrane fluidity.

19. d

PROBLEMS

1. The ability of bacteria, yeasts, and fungi to convert aliphatic hydrocarbons to carbon dioxide and water has been studied intensively over the past decade because of concerns about the effects of crude oil spills on the environment. Microorganisms cannot survive when they are placed in high concentrations of crude oil or any of its components. However, they can utilize hydrocarbons very efficiently when they are placed in a medium in which an extensive lipid–water interface is created by agitation and aeration. Why?

2. Phytol, a long-chain alcohol, appears as an ester in plant chlorophyll. When consumed as part of the diet, phytol is converted to phytanic acid (see Figure 12.3).

FIGURE 12.3

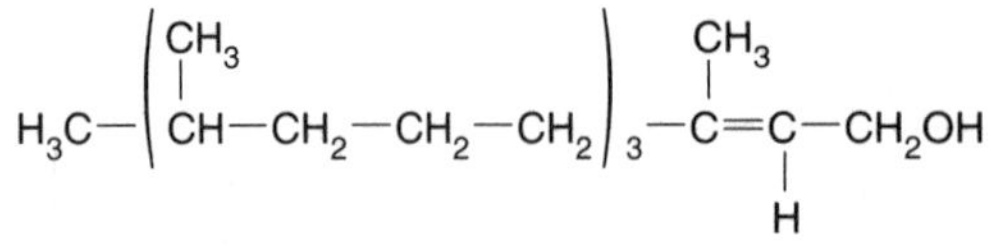

$$H_3C-\left(\underset{\substack{|\\ CH}}{\overset{CH_3}{\underset{}{}}}-CH_2-CH_2-CH_2\right)_3-\underset{\underset{H}{|}}{C}=\overset{CH_3}{\underset{}{C}}-CH_2OH$$

Phytol

$$H_3C-\left(\underset{\substack{|\\ CH}}{\overset{CH_3}{\underset{}{}}}-CH_2-CH_2-CH_2\right)_3-\overset{CH_3}{\underset{|}{CH}}-CH_2-C\overset{\diagup O}{\diagdown O^-}$$

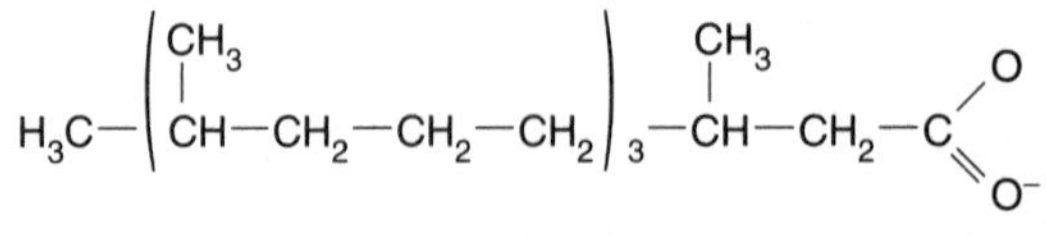

Phytanic acid

People who cannot oxidize phytanic acid suffer from a number of neurological disorders that together are known as *Refsum's disease*. The symptoms may be related to the fact that phytanic acid accumulates in the membranes of nerve cells. What general effects of phytanic acid on these membranes would be observed?

3. Bacterial mutants that are unable to synthesize fatty acids will incorporate them into their membranes when fatty acids are supplied in their growth medium. Suppose that each of two cultures contains a mixture of several types of straight-chain fatty acids, some saturated and some unsaturated, ranging in chain length from 10 to 20 carbon atoms. If one culture is maintained at 18°C and the other is maintained at 40°C over several generations, what differences in the composition of the cell membranes of the two cultures would you expect to observe?

4. Given two bilayer systems, one composed of phospholipids having saturated acyl chains 20 carbons in length and the other having acyl chains of the same length but with cis double bonds at C-5, C-8, C-11, and C-14, compare the effect of the acyl chains on T_m for each system.

5. Hopanoids are pentacyclic molecules that are found in bacteria and in some plants. A typical bacterial hopanoid, bacteriohopanetetrol, is shown in Figure 12.4. Compare the structure of this compound with that of cholesterol. What effect would you expect a hopanoid to have on a bacterial membrane?

FIGURE 12.4

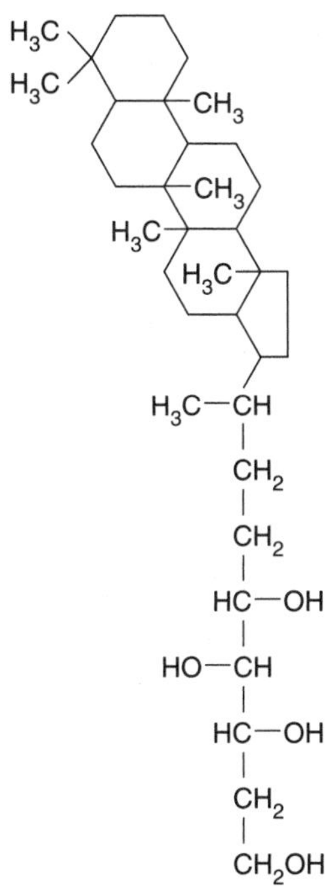

Bacteriohopanetetrol

6. As early as 1972, it was known that many biological membranes are asymmetric in the distribution of phospholipids between the inner and outer leaflets of the bilayer. Once such asymmetry is established, what factors act to preserve it?

7. As discussed in the text, the length and degree of saturation of the fatty acyl chains in membrane bilayers can affect the melting temperature T_m.

   (a) The value of T_m for a pure sample of phosphatidyl choline that contains two 12-carbon fatty acyl chains is $-1°C$. Values for phosphatidyl choline species with longer acyl chains increase by about 20°C for each two-carbon unit added. Why?

   (b) Suppose you have a phosphatidyl choline species that has one palmitoyl group esterified to C-1 of the glycerol moiety, as well as an oleoyl group esterified at C-

2 of glycerol. How would T_m for this species compare with that of dipalmitoylphosphatidyl choline, which contains two esterified palmitoyl groups?

(c) Suppose you have a sample of sphingomyelin that has palmitate esterified to the sphingosine backbone. Compare the T_m for this phospholipid with that of dipalmitoylphosphatidyl choline.

(d) The transition temperature for dipalmitoylphosphatidyl ethanolamine is 63°C. Suppose you have a sample of this phospholipid in excess water at 50°C, and you add cholesterol until it constitutes about 50% of the total lipid, by weight, in the sample. What would you expect when you attempt to determine the transition temperature for the mixture?

8. At least two segments of the polypeptide chain of a particular glycoprotein span the membrane of an erythrocyte. All the sugars in the glycoprotein are O-linked.

(a) Which amino acids might be found in the portion of the chain that is buried in the lipid bilayer?

(b) Why would you expect to find serine or threonine residues in the glycoprotein?

9. (a) Many integral membrane proteins are composed of a number of membrane-spanning segments, which form bundles of a helices packed closely together, often forming a membrane channel or pore. Each membrane-spanning sequence of most integral membrane proteins is an α helix composed of 18 to 20 amino acids. What is the width of the hydrocarbon core of the membrane?

(b) The sequence of one of the α helices in a particular integral membrane protein is shown in Figure 12.5, and the 19 residues in this helix are plotted in a helical wheel plot. Such a plot projects the side chains of the amino acid residues along the axis of the α helix (z-axis) onto an x-y plane. In an α helix, a full turn occurs every 3.6 residues, so each successive residue is 100° apart on the helix wheel. Compare the location of hydrophobic side chains on the helix surface with those that are polar or hydrophilic. Where are the hydrophobic side chains, and how are they accommodated in the membrane? Where are the polar side chains? How are they accommodated in the protein-membrane complex?

Helix sequence:

Ser Val Tyr Asp Ile Leu Glu Arg Phe Asn Glu Thr Met Asn His Ala Val Ser Gly

FIGURE 12.5

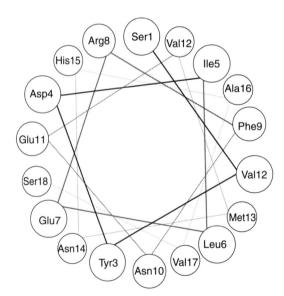

10. A series of experiments that shed some light on the movement of lipids in membranes was conducted by Rothman and Kennedy, using a gram-positive bacterium. They used 2,4,6-trinitrobenzenesulfonic acid (TNBS), which reacts with amino groups in phosphoethanolamine residues. Note that TNBS, shown in Figure 12.6, is charged at physiologic pH and cannot penetrate intact membrane vesicles. Incubation of TNBS with intact bacterial cells and with disrupted cells revealed that about two-thirds of the phosphoethanolamine molecules are located on the outside of the membrane, with the remaining residues on the inside. Rothman and Kennedy then incubated growing cells with a pulse of radioactive inorganic phosphate to label newly synthesized phosphoethanolamine molecules in the membrane. Using TNBS once again to distinguish between residues on the two sides of the membrane, they determined that immediately after the radioactive pulse, all newly synthesized phosphoethanolamine residues were located on the inner face of the membrane. After 30 minutes, however, the original distribution of phosphoethanolamine residues on the inner and outer faces of the bacterial cell membrane was restored. What do these observations suggest about the movement of phospholipids in membranes?

FIGURE 12.6

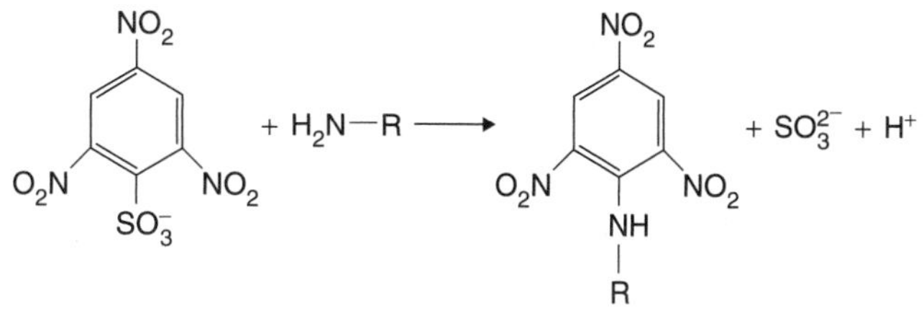

2,4,6-Trinitrobenzenesulfonic acid (TBNS)

11. Mycoplasma cells can be grown under conditions such that their plasma membrane contains one type of glycolipid, such as mono- or diglucosylated sphingosine molecules.

   (a) Membranes prepared from mycoplasma cells undergo a phase transition when heated. Suppose that sample A is isolated from cells whose glycolipids contain a very high percentage of unsaturated fatty acyl chains, whereas sample B is isolated from cells whose glycolipids contain a high percentage of saturated fatty acyl chains of the same length. When heated, which sample will exhibit a higher melting temperature? Why?

   (b) Glycolipids from samples A and B are analyzed for the carbohydrate content of their polar head groups. Those from sample A have a higher percentage of diglucosyl residues than those from sample B, which have mostly monoglucosyl residues. Explain how this observation is consistent with the lipid content of the two samples.

12. In mammals, lysophosphoglycerides (1-monoacylglycerol-3-phosphates) are generated in small quantities to trigger physiologic responses. Hydrolysis of a fatty acyl group from the C-2 position of a glycerophospholipid yields a lysophosphoglyceride. The reaction is catalyzed by phospholipase A_2, whose activity is strictly regulated. However, large quantities of phospholipase A_2 are found in snake venom, and the active venom enzyme can generate high concentrations of lysophosphoglycerides from membranes of snakebite victims. Lysophosphoglycerides are so named because in high concentrations they can disrupt membrane structure. Why?

13. What features of liposomes make them potentially useful as a delivery system for transporting water-soluble drugs to target cells? Suggest how one could prepare a liposome that is specific for a particular type of cell.

14. Explain the role of cholesterol in cell membranes.

15. During the solubilization of membranes, the purification of integral membrane proteins, and the reconstitution of membranes, gentler detergents, such as octyl glucoside, are used in preference to sodium dodecyl sulfate (SDS). Explain why.

16. Why do membrane proteins not diffuse, that is, flip-flop, across membranes?

17. In a membrane, an integral membrane protein diffuses laterally an average distance of 4×10^{-6} m in 1 minute, whereas a phospholipid molecule diffuses an average distance of 2 μm in 1 second.

    (a) Calculate the ratio of the diffusion rates for these two membrane components.
    (b) Provide reasons for the difference between the two rates.

ANSWERS TO PROBLEMS

1. All organisms require water for many biochemical reactions, and thus organisms can live only in an aqueous environment. Although a bacterial cell placed in a solution of crude oil might at least survive, it would not have enough water for growth and division. Microorganisms can best utilize crude oil or its hydrocarbon components when the microorganisms are present at a boundary layer between water and lipid. Aeration and agitation increase the effective area of such a layer. Some microorganisms that degrade hydrocarbons have a glycolipid-rich cell wall in which those compounds are soluble. After being solubilized, the compounds are transferred to the cytoplasmic membrane, where water-requiring reactions that initiate hydrocarbon degradation occur.

2. The four methyl side chains of each phytanic acid molecule interfere with the ordered association of fatty acyl chains; thus, they increase the fluidity of nerve cell membranes. This increase in fluidity could interfere with myelin function or ion transport, but the actual molecular basis for the symptoms is not yet known. Many of the symptoms of Refsum's disease can be eliminated by adopting diets that are free of phytol. The primary source of phytol in the human diet is from dairy products and other fats from ruminants. Cows, for example, consume large quantities of chlorophyll as they ingest grasses and plant materials. The symbiotic bacteria that inhabit the bovine rumen readily degrade chlorophyll, releasing free phytol, which is then converted to phytanic acid. Up to 10% of the fatty acids in bovine blood plasma are found as phytanic acid, which can then be incorporated into cell membranes and milk. For those who have Refsum's disease, it is therefore necessary to restrict consumption of beef as well as dairy products like milk and butter. Because humans do not degrade chlorophyll extensively during digestion, restriction of green plants in the diet is usually unnecessary.

3. You would expect to find that the bacteria grown at the higher temperature has incorporated a higher number of the longer fatty acids and a greater proportion of the saturated fatty acids. The membranes of bacteria grown at 18°C have more short-chain fatty acids and more that are unsaturated. These cells select fatty acids that will remain fluid at a lower temperature in order to prevent their membranes from becoming too rigid. The cells grown at the higher temperatures can select fatty acids that pack more closely. Cells in both cultures thus employ strategies designed to achieve optimal membrane fluidity.

4. The higher the number of cis double bonds, the less ordered the bilayer structure and the more fluid the membrane system. You would therefore expect T_m for the bilayer system containing the acyl chains with four unsaturated bonds to be much lower than that for the system containing the saturated fatty acid chains.

5. Like cholesterol, bacteriohopanetetrol is a pentacyclic molecule with a rigid, plate-like, hydrophobic ring structure; it has a hydrophilic region as well, although that region is on the opposite end of the molecule when compared with cholesterol. In bacterial membranes, hopanoids may have a function similar to that of cholesterol in mammalian membranes; that is, they may moderate bacterial membrane fluidity by blocking the motion of fatty acyl chains and by preventing their crystallization.

6. Phospholipids have polar head groups, so their transfer across the hydrophobic interior of the bilayer as well as their dissociation from water at the bilayer surface would require a positive change in free energy. Without the input of free energy to make the process a spontaneous one, the transfer of the polar head group is very unlikely, so the asymmetric distribution of the phospholipids is preserved.

7. (a) The longer the acyl groups, the larger the number of noncovalent interactions that can form among the hydrocarbon chains. Higher temperatures are therefore required to disrupt the interactions of phospholipid species that have longer fatty acyl groups.

   (b) The cis double bond in oleate produces a bend in the hydrocarbon chain, interfering with the formation of noncovalent bonds between the acyl chains. Less heat energy is therefore required to cause a phase transition; in fact, the melting temperature for phosphatidyl choline with a palmitoyl and an oleoyl unit is $-5°C$, and T_m for dipalmitoylphosphatidyl choline is $41°C$.

   (c) As shown in Figures 12.5 and 12.6 of the text, the structures of phosphatidyl choline and of sphingomyelin are very similar to each other; both contain phosphoryl choline and both have a pair of hydrocarbon chains. Given similar chain lengths in palmitoylspingomyelin and in dipalmitoylphosphatidyl choline, you would expect that values of T_m for the two molecules are similar. Both species in fact exhibit a phase transition at $43°C$.

   (d) At $50°C$, you should expect cholesterol to diminish or even to abolish the transition, by preventing the close packing of the fatty acyl chains that impart rigidity to the molecular assembly. At higher temperatures, cholesterol in the mixture also prevents larger motions of fatty acyl chains, making the assembly less fluid. Studies show that in mixtures containing 30 to 35 mol % cholesterol, phase transitions are extinguished.

8. (a) You should expect to find nonpolar amino acid residues in the portions of the glycoprotein chain that are buried in the membrane. Because the core portion of the membrane is 30 Å wide, up to 20 amino acids could be included in the buried segments, assuming that the amino acids are part of an α helix, in which the translation distance for each amino acid is 1.5 Å.

   (b) The glycoprotein has O-linked carbohydrate residues, and in most such proteins the sugars are attached to the side chains of serine or threonine residues. Were the sugars N-linked instead, you would expect to find one or more asparagine residues in the glycoprotein.

9. (a) In an α helix, each amino acid residue extends 1.5 Å (1.5×10^{-1} nm) along the helix axis. Therefore a span of 20 amino acids in an α helix will be about 30 Å in length, corresponding to the width of the hydrophobic core of the membrane.

   (b) The plot clearly shows that hydrophobic amino acids are concentrated along one side of the surface of the helix. The side chains of those residues are likely to face the hydrophobic core of the membrane. Polar side chains are located on the opposite side of the helical surface; they are likely to be on the side of the chain that faces other α-helical bundles. They could form hydrogen or ionic bonds with polar residues in other bundles.

10. In model membrane systems, the transfer of phospholipid head groups from one side of the bilayer to the other is very slow, presumably because of the energy required to move the polar head group through the hydrophobic bilayer. The experiments carried out by Rothman and Kennedy indicate that a process that mediates the flip-flop of membrane lipids is operating in bacterial cells. Phospholipid synthesis takes place on the cytosolic face (the inner leaflet) of the membrane, and some of the newly synthesized lipids are moved through the bilayer to the outside surface of the membrane bilayer. While aminophospholipid translocases that can move polar lipids across membranes have been found in eukaryotes, it is not yet known how such a process occurs in bacteria.

11. (a) You would expect sample B, with a higher percentage of saturated fatty acids, to have a higher melting point. The saturated chains will aggregate more closely with each other, requiring more thermal energy to disrupt that aggregation. The acyl chains of unsaturated fatty acids are kinked and therefore cannot aggregate in regular arrays like saturated acyl chains of the same length. They are therefore disrupted at a lower temperature.

    (b) In the membrane, the cross-sectional area occupied by unsaturated fatty acids is larger than that occupied by saturated chains because of kinks in the hydrocarbon chains due to double bonds. A diglucosyl head group is larger (has a larger cross-section size) than that of a monoglucosyl derivative, so the larger head group would match the increase in cross-sectional area in the interior of the bilayer composed of unsaturated fatty acyl chains.

12. Glycerophospholipids contain two fatty acyl groups esterified to glycerol, to which a polar head group is also attached at the C-3 carbon. When phospholipase A_2 removes one of the fatty acyl chains, the polar head group is too large in relation to the single hydrocarbon chain to allow optimal packing in the bilayer. The regular association of the hydrocarbon tails is disrupted, and the plasma membrane dissolves.

13. Liposomes are essentially impermeable to water-soluble molecules. Therefore, water-soluble drugs could be trapped inside the liposomes and then be delivered into the target cells by fusing the liposomes with the cell membrane. To make a liposome specific for a particular type of cell, antibodies that have been prepared against a surface protein of the target cell could be attached to the liposome via a covalent bond with a bilayer lipid, for example, phosphatidyl ethanolamine. This would enable the liposome to recognize the target cells. Of course, strategies would also have to be devised to prevent the premature, nonspecific fusion of the liposome with other cells.

14. Cholesterol modulates the fluidity of membranes. By inserting itself between the fatty acid chains, cholesterol prevents their "crystallization" at temperatures below T_m and sterically blocks large motions of the fatty acid chains at temperatures above T_m. In fact, high concentrations of cholesterol abolish phase transitions of bilayers. This modulating effect of cholesterol maintains the fluidity of membranes in the range required for biological function.

15. Although sodium dodecyl sulfate (SDS) is a very effective detergent for solubilizing membrane components, the strong electrostatic interactions of its polar head groups with charged groups on the membrane proteins disrupt protein structure. A detergent such as octyl glucoside, which has an uncharged head group, allows the proteins to retain their three-dimensional structures while it interacts with their hydrophobic domains.

16. Membrane proteins are very bulky molecules that contain numerous charged amino acid residues and polar sugar groups (in the case of glycoproteins) that are highly hydrated. Such molecules do not diffuse through the hydrophobic interior of the lipid bilayer.

17. (a)   Rate of protein diffusion:

$$4 \times 10^{-6} \text{ m/min} = \frac{1 \text{ min}}{60 \text{ s}} = 6.7 \times 10^{-8} \text{ m/s}$$

Rate of phospholipid diffusion: 2 μm/s = 2×10^{-6} m/s

$$\frac{2 \times 10^{6} \text{ m/s}}{6.7 \times 10^{-8} \text{ m/s}} = 30$$

Ratio of phospholipid diffusion rate to protein diffusion rate:

(b)   The difference in diffusion rates is due primarily to the difference in mass between phospholipids, which have a molecular weight of approximately 800, and proteins, which have a molecular weight greater than 10,000. In addition, integral membrane proteins may associate with peripheral proteins, which would further decrease their lateral diffusion.

Membrane Channels and Pumps

Three classes of transmembrane proteins—pumps, carriers, and channels—can circumvent the intrinsic impermeability of the lipid bilayer to polar molecules and ions. This chapter describes some of the structural and functional features of these proteins. The authors first differentiate between active transport (used by pumps) and passive transport (used by channels) of molecules across a membrane and discuss how to quantitate the free energy stored in concentration gradients. The authors then discuss three types of active transport systems: (1) the P-type ATPases and (2) the ATP-binding cassette (ABC) pumps, both of which use ATP hydrolysis to drive the transport of ions across the membrane, and (3) the secondary transporters, which couple the thermodynamically uphill flow of one molecule with the downhill flow of another. The well-studied Na^+-K^+ ATPase and sarcoplasmic reticulum Ca^{2+} ATPase are used as examples of P-type ATPases, which have many common structural and mechanistic features. The authors then look at the more recently identified family of ABC pumps, including the multidrug resistance protein. The discussion of active transport is concluded with an examination of the mechanism of secondary transporters, using the bacterial lactose permease as an example, which uses the proton-motive force to drive the uptake of lactose against a concentration gradient. The mechanistic similarities between the symporters and the ATPase pumps are pointed out.

In addition to active transport, ions can be transported across membranes by passive methods, such as through ion channels. The powerful patch-clamp technique is described, which allows researchers to measure the activity of a single ion channel. The authors differentiate between voltage-gated and ligand-gated channels and discuss the key properties of all channels. The high-resolution structure of the voltage-gated K^+ channel is discussed, as is the structural basis for ion specificity. The acetylcholine receptor, a ligand-gated ion channel important in the propagation of nerve impulses, is examined in detail as well. The coordinated action of ligand-gated and ion-gated channels is examined through an analysis of action potentials. The chapter concludes with a discussion of gap junctions, which act as cell-to-cell channels and allow all polar molecules with a molecular mass of less than 1 kDa to pass through, and aquaporins, which are specific water channels.

LEARNING OBJECTIVES

When you master this chapter, you should be able to accomplish the following objectives.

Introduction

1. Distinguish between *channels, carriers,* and *pumps.*
2. List the forms of energy that can drive *active transport.* Use examples to distinguish between primary and secondary active transport.
3. Discuss how expression of transporters determines the metabolic activities of a cell type.

The Transport of Molecules Across a Membrane May Be Active or Passive
(Text Section 13.1)

4. List the two factors that determine whether a molecule will cross a membrane.
5. Distinguish between *simple* and *facilitated diffusion.*
6. Use the concepts of free-energy change (ΔG) and electrochemical potential to predict active or passive transport.

Two Families of Membrane Proteins Use ATP Hydrolysis to Pump Ions and Molecules Across Membranes (Text Section 13.2)

7. Describe the defining features of *P-type ATPases.*
8. Describe the functions of the Na^+–K^+ *ATPase* or Na^+–K^+ *pump.*
9. Describe the structure and the functional sites of the sarcoplasmic Ca^{2+} *ATPase.*
10. Outline the reaction cycle of the Ca^{2+} *ATPase.*
11. Compare the sarcoplasmic Ca^{2+} *ATPase* and the Na^+–K^+ *pump* in terms of functional sites and reaction cycles.
12. Discuss the inhibition of the Na^+–K^+ *pump* by *cardiotonic steroids.*
13. Define *multidrug resistance.*
14. Describe the function of the multi-drug resistance (MDR) protein and under what circumstances it is expressed.
15. Describe the architecture of the ABC transporter family of proteins.
16. Outline the mechanism of the ABC transporter.

Lactose Permease is an Archetype of Secondary Transporters That Use One Concentration Gradient to Power the Formation of Another (Text Section 13.3)

17. Define *symporter, antiporter,* and *uniporter.*
18. Compare the structure and mechanism of the lactose permease with that of the *P-type ATPases* and the ABC transporters.

Specific Channels Can Rapidly Transport Ions Across Membranes
(Text Section 13.4)

19. List the key properties of *ion channels*. Compare the rates of active transport with rates of transport through channels.

20. Define *action potential* and explain its mechanism in terms of the transient changes in Na^+ and K^+ permeability of the plasma membrane of a neuron.

21. Outline the *patch-clamp technique* and note its use in electrical measurements of membranes.

22. Explain the effects of *tetrodotoxin* on the *sodium channel*.

23. Outline the possible role of the *S4 segments* of sodium channels as *voltage sensors*. List the sequence of steps in the cycling of sodium channels during an action potential.

24. Outline the possible role of the *S5* and *S6 segments* of sodium, potassium, and calcium channels as a key region of the *ion channel pore*.

25. Describe the *selectivity filter* of potassium channels.

26. Relate the structure of the potassium channel to its rapid rate of transport.

27. Describe models for voltage gating and inactivation of ion channels.

28. Distinguish between *ligand-gated* and *voltage-gated* channels.

29. Explain the function of *neurotransmitters* and *ligand-gated channels* in the transmission of nerve impulses across *synapses*. Outline the effects of *acetylcholine* on the *postsynaptic membrane*.

30. Describe the subunit structure, ligand binding sites, and channel architecture of the *acetylcholine receptor* from *Torpedo marmorata*.

31. Discuss the role of action potentials in coordinating the effects of ion-gated and ligand-gated ion channels.

32. Explain the features and molecular basis of the disorder Long QT syndrome (LQTS).

33. Discuss side-effects induced by drugs that propagate the cardiac action potential.

Gap Junctions Allow Ions and Small Molecules to Flow Between Communicating Cells (Text Section 13.5)

34. Distinguish between *gap junctions* and other membrane channels. Give examples of molecules that can pass through gap junctions.

35. Describe the structure of a gap junction and the role of connexin in the formation of the structure.

Specific Channels Increase the Permeability of Some Membranes to Water
(Text Section 13.6)

36. Discuss the role of aquaporins in water transport in cells.

SELF-TEST

Introduction

1. Which of these statements about membrane channels, carriers, and pumps are true?

   (a) All are integral, transmembrane proteins.
   (b) All can be ligand- or voltage-gated.
   (c) All contain multiple subunits or domains.
   (d) All carry out active transport of ions and polar molecules.
   (e) All allow bidirectional flux of the transported molecule.

2. Which of the following are used as forms of energy that can drive active transport?

   (a) Light absorption
   (b) ATP hydrolysis
   (c) Ion gradient
   (d) a and b
   (e) All of the above

The Transport of Molecules Across a Membrane May Be Active or Passive

3. What will be the free-energy change generated by transport of one mole of Na^+ from a concentration of 10 mM to 150 mM with a membrane potential of -25 mV at 37° C? Would the transport need to be active or passive?

4. Which of the following is an example of simple diffusion?

   (a) Passage of steroid hormones through a membrane
   (b) Movement of sodium ions across a membrane
   (c) Transport of glucose across a membrane
   (d) Transport of lactose across a membrane

A Family of Membrane Proteins Uses ATP Hydrolysis to Pump Ions Across Membranes

5. The orientation of the Na^+–K^+ pump in cell membranes determines the side of the membrane where the various processes involved in the transport of Na^+ and K^+ will take place. Assign each of the steps or processes in the right column to the intracellular or extracellular side of the pump.

   (a) intracellular side
   (b) extracellular side
       of the pump

   (1) binding of cardiotonic steroids
   (2) hydrolysis of ATP and phosphorylation
   (3) binding of K^+
   (4) binding of Na^+

6. The proposed model for the mechanism of the Ca^{2+} ATPase is based on the existence of four conformational states of this enzyme. Match each conformational state in the left column with the appropriate descriptions in the right column.

   (a) E_1
   (b) E_1-P
   (c) E_2
   (d) E_2-P

   (1) low affinity for Ca^{2+}
   (2) high affinity for Ca^{2+}
   (3) is phosphorylated by ATP on Ca^{2+} binding
   (4) ion-binding sites open to the cytosol
   (5) ion-binding sites open to the luminal side of the membrane
   (6) is dephosphorylated upon the release of Ca^{2+}

7. Explain why an electric current is generated during the transport of Na^+ and K^+ by the Na^+–K^+ pump.

8. Which of the following statements describe properties that are common to the Ca^{2+} ATPase of the sarcoplasmic reticulum and the Na^+–K^+ pump?

    (a) Both are abundant membrane proteins in the sarcoplasmic reticulum.
    (b) Both have homologous N-terminal subunits containing numerous transmembrane helices.
    (c) Both contain an aspartate residue that is phosphorylated by ATP.
    (d) Both translocate the same number of ions per transport cycle.
    (e) Both probably have four major conformational states.

9. Which of the following statements about proteins containing ABC domains is NOT correct?

    (a) They are members of the P-loop NTPase superfamily.
    (b) They usually consist of two membrane spanning domains and two ATP-binding domains.
    (c) The membrane spanning domains and ATP-binding domains are always on separate polypeptide chains.
    (d) The ABC domains undergo conformational changes upon ATP binding and hydrolysis.

10. Explain how the MDR protein functions in the phenomenon of multidrug resistance.

Lactose Permease is an Archetype of Secondary Transporters That Use One Concentration Gradient to Power Formation of Another

11. Which of the following statements about the sodium-calcium exchanger is NOT correct?

    (a) It is a symporter for sodium and calcium transport.
    (b) It is an antiporter for sodium and calcium transport.
    (c) It is driven by the Na^+ gradient generated by the Na^+–K^+ pump.
    (d) It has a lower affinity for Ca^{2+} than the Ca^{2+} ATPase.
    (e) It has a higher transport rate for Ca^{2+} than the Ca^{2+} ATPase.

12. Which of the following statements about the lactose permease of E. coli under physiologic conditions are correct?

    (a) It derives energy for transport from an Na^+ gradient.
    (b) It derives energy for transport from an H^+ gradient.
    (c) It derives energy for transport from a Ca^{2+} gradient.
    (d) It is an antiporter for lactose and H^+.
    (e) It is a symporter for lactose and Na^+.

Specific Channels Can Rapidly Transport Ions Across Membranes

13. Ascribe the characteristics in the right column either to active transport or to transport of ions or molecules through channels.

    (a) active transport
    (b) transport through channels

    (1) flux $\sim 10^7$ s^{-1}
    (2) flux 30 to 2000 s^{-1}
    (3) Ions can flow from either side of the membrane.
    (4) flux in a specific direction

14. List the events in the transmission of nerve impulses in synapses in their proper sequence.
    (a) binding of acetylcholine to acetylcholine receptor
    (b) depolarization of the postsynaptic membrane
    (c) release of acetylcholine from synaptic vesicles into the synaptic cleft
    (d) increase in postsynaptic membrane permeability to Na^+ and K^+
    (e) increase of acetylcholine concentration in the synaptic cleft from ~10 nM to 500 mM

15. Explain the use of cobratoxin in the purification of acetylcholine receptor.

16. Which of the following is not a characteristic of the structure of the acetylcholine receptor of *Torpedo*?
    (a) It has five subunits of four different kinds.
    (b) Acetylcholine binds between the α–λ and α–δ interfaces.
    (c) It has a uniform channel 20 Å in diameter.
    (d) The genes for the subunits arose by the duplication and divergence of a common ancestral gene.
    (e) It has pentagonal symmetry.

17. Which of the following statements about the plasma membrane of a neuron are correct?
    (a) In the resting state, the membrane is more permeable to Na^+ than to K^+.
    (b) In the resting state, the membrane potential is approximately $+30$ mV.
    (c) The Na^+ and K^+ gradients across the membrane are maintained by the Na^+-K^+ pump.
    (d) The equilibrium potential for K^+ across the membrane is near -75 mV.
    (e) During the action potential, the membrane potential varies between the limits of $+30$ mV and -75 mV.

18. Place the following events of the action potential in their correct sequence.
    (a) spontaneous closing of sodium channels
    (b) membrane potential of -75 mV
    (c) depolarization of the plasma membrane to approximately -40 mV
    (d) opening of the potassium channels
    (e) opening of sodium channels
    (f) membrane potential of $+30$ mV
    (g) membrane potential of -60 mV

19. Which of the following statements about the sodium channel, purified and reconstituted in lipid bilayers, are correct?
    (a) It is about 10 times more permeable to Na^+ than to K^+.
    (b) It is sensitive to voltage.
    (c) It is inhibited by cobratoxin.
    (d) It becomes inactivated spontaneously.
    (e) It consists of seven hydrophobic transmembrane segments.

20. Match the sodium (eel electric organ) and potassium (Shaker) channels with the corresponding properties listed in the right column.
    (a) sodium channel
    (b) potassium channel

    (1) four 70-kd subunits
    (2) single 260-kd polypeptide chain
    (3) fourfold symmetry
    (4) tetrodotoxin binding site
    (5) positively charged, voltage-sensing S4 helical segment
    (6) ball-and-chain inactivation mechanism
    (7) 3-Å-diameter channel

21. Describe the physical manifestations and molecular causes of Long QT syndrome (LQTS).

Gap Junctions Allow Ions and Small Molecules to Flow Between Communicating Cells

22. Which of the following statements about gap junctions between cells are NOT true?
    (a) They allow the exchange of ions and metabolites between cells.
    (b) They allow the exchange of cytoplasmic proteins.
    (c) They are essential for the nourishment of cells that are distant from blood vessels.
    (d) They are made up of 10 molecules of connexin.
    (e) Ca^{2+} and H^+ concentrations in cells control them.

ANSWERS TO SELF-TEST

1. a, c

2. e

3. Using the following equation,

$$\Delta G = RT \ln \frac{c_2}{c_1} + ZF\Delta V,$$

$$\Delta G = (8.314 J \cdot mol^{-1} \cdot K^{-1})(310K) \ln \frac{150mM}{5mM} + (1)(96.45kJ \cdot mol^{-1} \cdot V)(-0.025V)$$

$$\Delta G = +6.36 kJ/mol$$

   Since the sign of ΔG is positive, the transport is unfavorable in the direction indicated, and therefore active transport must be used.

4. (a) Simple diffusion involves the passage of lipophilic molecules across a membrane according to their concentrations gradient.

5. (a) 2, 4   (b) 1, 3

6. (a) 2, 3, 4   (b) 2, 4   (c) 1, 5, 6   (d) 1, 5

7. Since three Na^+ ions are transported out for every two K^+ ions that are transported in, there is a net efflux of one positively charged ion. The net movement of ions sets up an electric current.

8. b, c, e

9. The answer is (c). While prokaryotic ABC proteins are often multisubunit proteins, eukaryotic ABC proteins usually contain both membrane spanning and ATP-binding domains on the same polypeptide.

10. The multidrug resistance (MDR) protein is a 170kd P-glycoprotein that is expressed in cells displaying multidrug resistance. It acts as an ATP-dependent pump that extrudes a wide range of small molecules from cells that express it. When cells expressing the MDR protein are exposed to a drug, they are pumped out of the cell before the drug can exert its effects.

11. Answer (a) is incorrect. The sodium-calcium exchanger transports these cations in the opposite direction; therefore, it is an antiporter, not a symporter.

12. b

13. (a) 2, 4   (b) 1, 3

14. c, e, a, d, b

15. Cobratoxin binds specifically and with very high affinity to the acetylcholine receptor; therefore, a column with covalently attached cobratoxin can be used in the affinity purification of the receptor from a mixture of macromolecules in the postsynaptic membrane that has been solubilized by adding nonionic detergents.

16. Answer (c) is incorrect. The channel is not of uniform diameter.

17. c, d, e

18. g, c, e, f, a, d, b, g

19. a, b, d

20. (a) 2, 3, 4, 5, 6   (b) 1, 3, 5, 6, 7

21. LQTS is a genetic disorder in which the recovery of the action potential from its peak potential to the resting equilibrium potential is delayed. The physical manifestations of LQTS can range from brief losses of consciousness to disruption of normal cardiac rhythm and sudden death. The molecular basis most often appears to be mutations that inactivate K^+ channels or prevent the propert trafficking of these channels to the plasma membrane. This leads to a loss in potassium permeability which slows the repolarization of the membrane and delays the induction of subsequent cardiac contraction.

22. b, d

PROBLEMS

1. Calculate the free-energy change for the transport of an uncharged species from a concentration of 5 mM outside a cell to a concentration of 150 mM inside. Assume that the temperature is 25°C. Now repeat the calculation for an ion with a charge of +1 that is crossing a membrane with a potential of −60 mV, with the interior negative with respect to the exterior. Would the transport of an ion with a −1 charge be more or less favorable?

2. In dog skeletal muscle, the extracellular and intracellular concentrations of Na^+ are 150 mM and 12 mM, and those of K^+ are 2.7 mM and 140 mM, respectively.
   (a) Calculate the free-energy change as three Na^+ are transported out and two K^+ are transported in by the Na^+–K^+ pump. Assume that the temperature is 25°C and that the membrane potential is −60 mV.
   (b) Does the hydrolysis of a single ATP provide sufficient energy for the process in part (a)? Explain.

3. An uncharged molecule is transported from side 1 to side 2 of a membrane.
   (a) If its concentration is 10^{-3} M on side 1 and 10^{-6} M on side 2, will the transport be an active or a passive process? Explain your answer.
   (b) If the concentration is 10^{-1} M on side 1 and 10^{-4} M on side 2, how will the free-energy change compare with that in part (a)? Explain.
   (c) How will the rate of transport in (a) and (b) compare? Explain your answer.

4. In addition to the Na^+–K^+ ATPase, eukaryotic cells contain other ATP-driven pumps. One such pump is the H^+–K^+ ATPase, in which a hydrogen ion is extruded from the cytoplasm in exchange for a potassium ion at the expense of ATP hydrolysis. Given that the interior of most animal cells is electrically negative with respect to the exte-

rior, explain why the Na^+–K^+ ATPase can contribute to the membrane potential but the H^+–K^+ ATPase cannot.

5. What is the molecular basis for the phenomenon of multidrug resistance?

6. In experiments to investigate the mechanism of transport of two substances, X and Y, across cell membranes, cells were incubated in media containing various concentrations of X and Y, and the initial rate of transport of each of the substances into the cell was determined. The results that were obtained are depicted in Figure 13.1. What conclusion is suggested by the results? Explain. It may be helpful to refer to the discussion of enzyme kinetics in the text.

FIGURE 13.1   Initial velocity of transport versus concentration for substances X and Y.

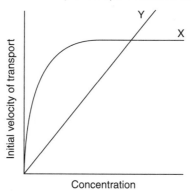

7. Figure 13.2 depicts a typical action potential that might be measured in an isolated axon, such as the giant axon of a squid. Give the events that are responsible for (a) the rising phase of the action potential, and (b) the falling phase. Specify in each case whether ion flow occurs with or against concentration gradients, electrical gradients, or both.

FIGURE 13.2   Action potential.

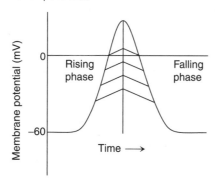

8. When the sciatic nerve is removed from a frog, placed in an isotonic salt solution, and stimulated electrically, it generates action potentials that can be measured by an electrode placed at some distance from the site of stimulation. When metabolic poisons are added to the preparation, the nerve retains the capability of generating action potentials even though the supply of ATP to drive its Na^+–K^+ pump has been depleted and it is thus incapable of carrying out active transport. Explain how this can be the case.

9. Suppose that a Glu residue is present in the narrow region of the sodium channel. A mutant protein is found in which that Glu is replaced by Val.

    (a) Compare the Na^+ conductance of the mutant as opposed to the normal channel. Explain.

(b) Compare the sodium permeability as a function of pH in each case.

(c) Compare the sensitivity of the normal and the mutant channels to tetrodotoxin.

(d) Compare the magnitude of an action potential in nerves containing sodium channels of the mutant type, as opposed to the normal type.

10. Acetylcholine opens a single kind of cation channel that has a very similar permeability to Na^+ and K^+, yet the influx of Na^+ is much larger than the efflux of K^+. Explain this fact.

11. The K^+ channel is over 100 times more permeable to K^+ than to Na^+. Explain the molecular mechanism for this selectivity.

12. Explain the experimental evidence supporting the *ball-and-chain* model for channel inactivation.

ANSWERS TO PROBLEMS

1. We use the following equation from the text:

$$\Delta G = RT \ln \frac{c_2}{c_1}$$

$$\Delta G = (8.314 J \cdot mol^{-1} \cdot K^{-1})(298K) \ln \frac{150 mM}{5 mM} + 8.43 kJ \cdot mol^{-1}$$

Now for the ion with a $+1$ charge:

$$\Delta G = RT \ln \frac{c_2}{c_1}$$

$$\Delta G = 8.43 kJ \cdot mol^{-1} + (1)(96.45 kJ \cdot V^{-1} \cdot mol^{-1})(-0.060V)$$

$$\Delta G = +2.64 \, kJ / mol.$$

Note that the membrane potential favors the entry of a positively charged ion and overcomes the unfavorable concentration gradient. If the calculation is repeated for a negatively charged molecule, the free energy change will be more positive and therefore less favorable.

2. (a) To solve this problem, we first calculate the free-energy change for transporting three Na^+ and then that for two K^+. The total free-energy change will be the sum of the two.

Again we use the following equation from the text:

$$\Delta G = RT \ln \frac{c_2}{c_1} + ZF \Delta V$$

Substituting the values for Na^+ yields

$$\Delta G_{Na^+} = (8.314 J \cdot mol^{-1} \cdot K^{-1})(298K)\ln\frac{150mM}{12mM} + (1)(96.45 kJ \cdot mol^{-1} \cdot V)(0.0606)$$

$$\Delta G_{Na^+} = +6.26 kJ/mol + 5.79 kJ/mol = +12.1 kJ/mol$$

Note that when Na^+ is transported out of the cell, work must be done against both a concentration gradient and an electrical gradient.

Now we carry out the corresponding calculation for the K^+ ion.

$$\Delta G_{K^+} = (8.314 J \cdot mol^{-1} \cdot K^{-1})(298K)\ln\frac{140mM}{2.7mM} + (1)(96.45 kJ \cdot mol^{-1} \cdot V)(0.0606)$$

$$\Delta G_{K^+} = +9.78\ kJ/mol - 5.79\ kJ/mol = +3.99\ kJ/mol$$

Note that potassium ion is being transported against a concentration gradient but with an electrical gradient. Accordingly, the sign for the electrical term in the equation is negative.

To get the total energy expenditure, we must account for the stoichiometry of transport by summing the energy required for the transport of Na^+ and that required for the transport of K^+.

$$\Delta G = 3\Delta G_{Na^+} + 2\Delta G_{K^+}$$

$$\Delta G = 3(12.1\ kJ/mol) + 2(3.99\ kJ/mol) = +44.3\ kJ/mol$$

(b) The free-energy change for the ATP concentrations that exist in typical cells is approximately -50 kJ/mol. Thus, the energy furnished by the hydrolysis of a single ATP is sufficient.

3. An uncharged molecule is transported from side 1 to side 2 of a membrane.

(a) The transport will be a passive process. Because the concentration on side 1 is higher than that on side 2, the molecule will move spontaneously down its concentration gradient. One can use the following expression to show that ΔG has a negative value.

$$\Delta G = RT \ln \frac{c_2}{c_1}$$

$$\Delta G = RT \ln \frac{10^{-6}}{10^{-3}} = (8.314 kJ \cdot mol^{-1} \cdot K^{-1})(298K) = -17.1 kJ \cdot mol^{-1}$$

A negative ΔG value in the direction of movement is the hallmark of passive transport.

(b) Since the ratios of the concentrations in (a) and (b) are equal ($c_2/c_1 = 10^{-3}$), the free-energy change for the transport is the same in both cases.

(c) The rate of a chemical process is always equal to a rate constant multiplied by the concentration of the chemical species undergoing the change. Thus, the rate of transport will be equal to $k(c_1)$. Since the concentration of c_1 is greater in (b) than in (a), the rate of transport will be greater in (b).

4. The difference between the two ATPase systems lies in the stoichiometry of their exchange of ions. In the case of the Na^+–K^+ ATPase, three Na^+ ions are extruded and two K^+ ions are taken up during each pump cycle, making the interior of the cell more negative (less positive) for each pump cycle. The H^+–K^+ ATPase, in contrast, extrudes one H^+ ion for each K^+ taken up, so its operation is electrically neutral.

5. Multidrug resistance is said to occur when resistance to one drug makes cells less sensitive to a range of other drugs. The development of multidrug resistance is correlated with expression and activity of a 170 kDa protein called multidrug resistance protein (MDR). MDR contains an ATP-binding cassette (ABC) domain and pumps drugs out of cells before the drugs can exert their effects.

6. The curve for X shows saturation, which would be expected if some membrane carrier is involved in the transport of substance X. The curve for Y shows no saturation, which is consistent with the notion that substance Y diffuses through the membrane without a carrier. Such behavior is shown by lipid soluble substances, which dissolve in the hydrophobic tails of membrane phospholipids and can thus enter cells without a carrier.

7. Nerve cells, like most animal cells, have a higher concentration of K^+ inside than outside and a higher concentration of Na^+ outside than inside. In addition, there is a membrane potential; that is, the inside of the cell is negative (in this case -60 mV) with respect to the outside.

    (a) The rising phase of the action potential is due to the influx of Na^+ ions down a concentration gradient and an electrical gradient.

    (b) The falling phase is due to the efflux of K^+ ions down a concentration gradient but against an electrical gradient.

8. For each action potential that is generated in an axon, only a very few Na^+ ions enter and a very few K^+ ions depart from the cell. Thus, in a poisoned nerve cell, many tens of thousands of impulses may be conducted before ionic equilibrium across the membrane is achieved. The active transport of Na^+ and K^+ across the membrane may be best viewed as necessary in the long run but not in the short run.

9. If a Glu residue is present in the narrow region of the sodium channel and a mutant protein is found in which that Glu is replaced by Val:

    (a) The mutant sodium channel would have decreased Na^+ conductance. Charge attraction between Na^+ ions and carboxylate anions is important in drawing Na^+ into the channel. In the mutant channel, Val, with its uncharged side chain, will not attract Na^+ ions.

    (b) In the normal channel, Na^+ conductance decreases as pH is lowered below 5.4 and the negatively charged carboxylate side chains are titrated to uncharged carboxyl groups. The mutant channel would show no such sensitivity to pH, because no charged groups are present in the side chain of Val.

    (c) The normal channel would be sensitive to tetrodotoxin, whereas the mutant channel would be less sensitive. Tetrodotoxin contains a positively charged guanido group (pg. 384 of the text) that presumably interacts electrostatically with a negatively charged group in the sodium channel.

    (d) The magnitude of the action potential would be reduced in nerves containing the mutant channel. The action potential is generated by sodium flowing from the outside of the cell to the inside. Decreased sodium conductance in the mutant would reduce the amount of sodium influx, and hence reduce the magnitude of the action potential.

10. The electrochemical gradient for Na^+ influx is steeper than that for K^+ efflux. The concentration gradients across the membrane are similar for both ions, but Na^+ moves from the positive to the negative side of the membrane, whereas K^+ moves from the negative to the positive side; that is, the membrane potential favors Na^+ influx.

11. At the entrance of the pore of the potassium channel there is a glutamate residue that binds cations. The pore is about 3 Å in diameter at its narrowest point, so that only dehydrated small cations can fit; however, the energy required to dehydrate Na^+ and smaller cations is too large and is not compensated by favorable polar interactions that occur in the case of K^+.

12. There are two pieces of experimental evidence given in the text in support of the ball-and-chain model of channel inactivation. The first is that treatment of the cytoplasmic side of either the Na^+ or K^+ channel with trypsin yields a trimmed channel that stays open after depolarization. The second is that N-terminal splice variants of the potassium channel have altered inactivation kinetics. A deletion of 42 amino acids at the N-terminus of the Shaker channel causes the channel to open upon depolarization but not inactivate. Addition of a synthetic peptide corresponding to the deleted amino acids restores inactivation to the channel.

Chapter 14

Signal-Transduction Pathways: An Introduction to Information Metabolism

In Chapter 12 you learned how biological membranes serve as semipermeable boundaries that isolate the cell from its surroundings and separate intracellular compartments from one another. Chapter 13 also described how selective, controlled breaching of the membrane barrier generates ion gradients across the bilayer, thereby producing electrical signals. In this chapter you will learn how molecules external to the cell bind to integral membrane protein receptors to initiate specific responses within the cell. The text describes how these binding and transmission mechanisms lead to an amplification of the initial signal and to specific effects that adapt the cell to its environment through effects on intracellular enzymes and regulatory proteins. The text also describes how disorders in these pathways of information flow can lead to diseases.

The authors use three examples of signal transduction pathways as examples. The first is the epinephrine-initiated pathway, the next is the pathway activated by insulin, and the last is the pathway activated by the epidermal growth factor. After a brief overview of signal transduction, the text describes the structure of the seven-helix transmembrane β-adrenergic receptor and indicates how it transmits to the intracellular side of the plasma membrane a signal arising from binding the hormone epinephrine on the extracellular surface of the cell. The common features of the G proteins are presented next. The description of the information-transmission pathway from hormone stimulus to G proteins to adenylate cyclase is completed by a discussion of how cAMP activates specific protein kinases to modulate the activities of the phosphorylated target proteins. A small number of hormone molecules outside the cell results in an amplified response because each activated enzyme in the triggered cascade forms numerous products. There are many distinct seven-helix transmembrane hormone receptors.

The text next describes an analogous hormone-stimulated system—the phospho-ionositide cascade. In this system, the hormone activates, by means of G proteins, a specific phospholipase (phospholipase C) that cleaves a plasma membrane phospholipid, phosphatidyl inositol 4,5-bisphosphate (PIP_2), to form two second messengers. The inositol phosphate derivative, inositol 1,4,5-trisphosphate (IP_3), which is short-lived, triggers the opening of ion channels so that the Ca^{+2} concentration in the cytosol is increased. The remnant of the PIP_2 molecule, diacylglycerol (DAG), is also a second messenger that activates protein kinase C. The increased Ca^{+2} levels and the activated protein kinase C affect a variety of biochemical reactions. The authors then describe the structure of Ca^{+2}-binding proteins, focusing on calmodulin, and explain how the binding of the ion is highly specific and leads to a large conformational change in the protein—qualities desirable in molecules serving as Ca^{2+} sensors and signal transducers.

The text then introduces another class of receptors, the transmembrane receptor tyrosine kinases that are often activated by a ligand-induced dimerization. The activated dimers phosphorylate some of their own tyrosine residues to provide docking sites for effector proteins on the cytosolic side of the membrane. Once bound, these effector enzymes are themselves phosphorylated and thereby activated by the tyrosine receptor kinase. The insulin and EGF receptors are used to discuss this class of receptors. A description of the susceptibility of signal transduction pathways to malfunctions that produce cancer follows, and the roles of oncogenes and their normal cellular counterparts (proto-oncogenes) in cell growth and differentiation are presented last.

LEARNING OBJECTIVES

When you have mastered this chapter, you should be able to accomplish the following objectives.

Introduction

1. List the components of *signal-transduction cascades*.
2. Summarize the physiological triggers of, and responses to, release of *ephinephrine, insulin,* and *epidermal growth factor.*
3. Draw a generalized *molecular circuit* based on a signal-transduction cascade. Outline the roles of *membrane receptors, ligands, primary messengers, second messengers, effectors,* and *cross talk* in the process.
4. Explain how a small number of *hormone* molecules outside the cell can effect a change involving many molecules within the cell, and consider that the cascade must be curtailed after being initiated.

Heterotrimeric G Proteins Transmit Signals and Reset Themselves (Text Section 14.1)

5. Describe the structure of the *seven-transmembrane-helix (7TM)* class of cell surface receptors. List some of the functions of members of this family and recognize their importance as drug targets.
6. Compare the structure of human *β-andrenergic receptor (β₂-AR)*, which binds epinephrine, with that of *rhodopsin.*

7. Identify the cellular localization of *guanyl nucleotide-binding proteins* (*G proteins*) and describe their structures, catalytic characteristics, and molecular mechanisms of activation and inactivation. Describe the roles of G proteins in coupling a *hormone-receptor complex* to *adenylate cyclase* and in amplifying the stimulus.

8. Understand that families of G proteins enable diverse hormones to effect a variety of physiologic functions.

9. Recognize the structure of *cAMP*, and write the reaction catalyzed by *Adenylate Cyclase* that forms it.

10. State the role of the hormone bound receptor in activation of *heterotrimeric G proteins* and describe the structural changes involved in the activation.

11. Describe the mechanism by which cAMP modulates the activity of *protein kinase A* (*PKA*).

12. List the steps in a G protein-*cAMP cascade* that contribute to the amplification of the hormonal stimulus, and explain how the amplified response is achieved. Use the epinephrine pathway as a specific example.

13. Write the reaction catalyzed by *phospholipase C* to produce the second messengers *inositol 1,4,5-trisphosphate* (IP_3) and *diacylglycerol* (*DAG*). Note that there are several forms of mammalian phospholipase C.

14. Describe the effects of IP_3 on the *IP_3-gated channel*. Describe the effects of Ca^{2+} released from endoplasmic reticulum of smooth muscle, and list some biochemical processes affected by an increased intracellular Ca^{2+} concentration.

15. Outline the features that suit Ca^{2+} in its role as a eukaryotic signaling ion.

16. Describe the structure of *calmodulin* and its biochemical function. Relate calmodulin to the *calmodulin-dependent protein kinase* (*CaM Kinase*). Note the value of calcium ionophores, calcium buffers, and fluorescent indicators in studying the functions of Ca^{2+} in cells.

17. Describe the *EF hand structural motif* of *calcium-binding proteins*, explain how it binds Ca^{2+}, and describe how ion binding affects its structure.

Insulin Signaling: Phosphorylation Cascades Are Central to Many Signal-Transduction Processes (Text Section 14.2)

18. Describe the structure of the *insulin receptor* (*IR*) including the location of the insulin-binding site. Recognize the significance of the insulin-dependent dimerization of the receptor.

19. Describe the general structures of the *receptor tyrosine kinases* and outline the process that converts them from inactive proteins to active enzymes. List some of the hormones that activate tyrosine kinases.

20. Outline the changes in the *activation loop* that occur on binding of insulin to the insulin receptor.

21. List the domains present in the IR and *insulin-receptor substrates* (*IRS*) and describe how they mediate protein–protein interactions in insulin signaling.

22. Summarize the series of events that occur after binding of insulin to the IR.

23. Describe the mechanism of termination of the insulin signal transduction cascade.

EGF Signaling: Signal-Transduction Systems Are Poised to Respond (Text Section 14.3)

24. Describe how the quaternary structure (*monomer-dimer equilibrium*) of the *human growth hormone receptor* changes on binding *human growth hormone*.

25. Discuss how the structure of the *epidermal growth factor receptor (EGFR)* provides evidence for why it is a monomer in the absence of the *epidermal growth factor (EGF)*, but dimerizes in its presence. Explain how the structure and activity of the Her2 receptor supports this model.

26. Define *cross-phosphorylation* and give examples of it in signal transduction.

27. Explain the role of *adaptor proteins* in signal transduction and give examples of their use.

28. Name some members of the *small G protein* family, and distinguish their structures from those of the heterotrimeric G proteins.

29. Appreciate the essential roles of the receptor tyrosine kinases and small G proteins in controlling cell growth and differentiation.

30. Outline the EGF signaling pathway including the mechanism for signal transmission between steps.

Many Elements Recur with Variation in Different Signal-Transduction Pathways (Text Section 14.4)

31. List and give examples of three common themes that underlie many singaling pathways.

Defects in Signal-Transduction Pathways Can Lead to Cancer and Other Diseases (Text Section 14.5)

32. Define *cancer* in terms of cell growth.

33. Describe the effect of *Rous sarcoma* virus gene product (*v-src*) on cell growth and note the relationship of *v-src* to its cellular counterpart, *c-src*. Describe the relationship between *proto-oncogene* and *oncogene*.

34. Explain the role of *SH2* and *SH3 domains* in tyrosine kinase function. Appreciate that proto-oncogenes, which provide normal, essential functions in cell growth and proliferation, can give rise to cancer on mutation to oncogenes.

35. Outline the biochemical mechanism of *v-ras protein*–induced cancer and note the role of the normal (noncarcinogenic) *c-ras* protein in cellular growth. Appreciate that a diminished GTPase activity, in this case, leads to cancer.

36. Explain how an inhibitor of a specific protein kinase might be an effective anticancer drug. Compare it with the mechanism of *monoclonal antibodies* in treatment of cancer.

37. Explain the molecular mechanisms causing *cholera* and *pertussis*.

SELF-TEST

Introduction

1. Identify the receptor and the physiological response to release of epinephrine, insulin, and the epidermal growth factor.

2. Signal transduction cascades are produced by molecular assemblies of which of the following components?

(a) enzymes
(b) regulatory proteins
(c) receptors
(d) transmembrane channels
(e) nuclear pores

3. Match the process in the left column with the function it performs in the right column.

(a) second messenger
(b) effectors
(c) membrane receptors
(d) signal terminator

(1) carries a signal outside the cell across the membrane into the cell
(2) returns the signal-transduction system to its original state
(3) Directly alter the physiological response
(4) relays information from the membrane receptor

Heterotrimeric G Proteins Transmit Signals and Reset Themselves

4. Describe an essential structural property of seven-transmembrane helix receptors (7TM) that allows them to respond to stimuli.

5. Guanyl nucleotide-binding proteins (G proteins) have which of the following properties?

(a) bind GMP in their inactivated state
(b) act as intermediates in 7TM receptor-initiated signal transductions
(c) are heterodimers
(d) have an intrinsic GTPase activity
(e) can be activated in large numbers by a single activated membrane receptor

6. The β-adrenergic receptor is a member of which of the following receptor families?

(a) G-protein coupled receptors (GPCR)
(b) seven-transmembrane receptors (7TM)
(c) serpentine receptors
(d) all of the above
(5) none of the above

7. Which nucleotides are bound to G proteins in their unactivated state and activated state, respectively? How does the 7TM activate a G protein? How is the activated state of a G protein returned to the unactivated form?

8. Which of the following statements about GTP and its role in the cAMP-mediated hormone response system are correct?

(a) GTP is associated with the α subunit of a guanyl nucleotide–binding protein (G protein).
(b) GTP reduces the magnitude of the hormone response because it is converted to cGMP—a compound that antagonizes the effects of cAMP.
(c) GTP maintains the steady-state level of cAMP by rephosphorylating AMP to ATP in a nucleotide kinase-catalyzed reaction.
(d) GTP activates G protein so that its G_{α}−GTP subunit interacts with adenylate cyclase.
(e) GTP couples the stimulus from a hormone-receptor complex or an activated receptor to a system that produces an allosteric effector.
(f) The effect of GTP on hormone response is antagonized by the GTPase activity of the G_{α} subunit of the G protein.

(g) A single GTP-binding event with a stimulatory G protein leads to the formation of one cAMP molecule.

9. Which of the following are correct statements about G proteins and their functioning in cAMP-mediated hormonal systems?

    (a) G proteins bind hormones.
    (b) G proteins are integral membrane proteins.
    (c) G proteins are heterotrimers.
    (d) G proteins bind adenylate cyclase.
    (e) In their GDP form and in the absence of hormone, G proteins bind to hormone receptors and are converted to their GTP forms.
    (f) When G protein in the GDP form binds to a hormone-receptor complex, GTP exchanges with GDP.
    (g) The α subunit of G proteins is a GTPase.

10. If cells with β-adrenergic receptors are exposed for extended times to epinephrine, a hormone that causes activation of adenylate cyclase, the G protein fails to carry out efficiently the GDP–GTP exchange reaction and adenylate cyclase is no longer activated. What is this phenomenon called, what is its biological function, and how does it occur?

11. Both cAMP and AMP contain one adenine base, one ribose, and one phosphorus atom. How are they different?

12. Which of the following statements about cAMP and its functioning in hormone action are correct?

    (a) Most effects of cAMP in eukaryotic cells are exerted through the activation of protein kinase A (PKA).
    (b) Cyclic AMP binds the catalytic subunits of PKA and activates the enzyme allosterically.
    (c) Cyclic AMP binds the regulatory subunits of PKA and activates the enzyme by releasing the catalytic subunits.
    (d) Cyclic AMP is bound by the activated hormone receptor and PKA simultaneously to convey the hormonal signal in order to activate the kinase.

13. During cAMP-mediated hormone activation, what are the three steps at which amplification occurs?

14. Which of the following statements about cAMP and the second-messenger mechanism of hormone function are correct?

    (a) The hormonal stimulus leads to increased amounts of adenylate cyclase.
    (b) The formation of a hormone-receptor complex leads to the activation of adenylate cyclase.
    (c) Cyclic AMP acts as an allosteric modulator to affect the activities of specific protein kinases.
    (d) Cyclic AMP interacts with a hormone-receptor complex to dissociate the hormone.
    (e) The hormone-receptor complex enters the cell and affects the activities of target enzymes.

15. Why do you think the cells of one kind of tissue respond to a given hormone, whereas cells of another tissue may NOT do so?

16. Suppose a patient is suffering from a disorder in which adenylate cyclase is impaired and, as a result, cAMP levels are not readily increased by hormones. Explain why the infusion of cAMP probably will not remedy the problem.

17. Which of the following statements about the phosphoinositide cascade are correct?

(a) The phosphoinositide cascade depends on the hydrolysis of a phospholipid component of the plasma membrane.

(b) A polypeptide hormone interacts with a G_{M1} ganglioside on the cell surface to trigger the phosphoinositide cascade.

(c) In some cases, a G-protein system acts to transduce the stimulus from the receptor to the phosphoinositidase.

(d) At least four kinds of phospholipase C play a crucial role in the phosphoinositide cascade.

(e) The phosphoinositide cascade directly produces a unique second-messenger molecule.

18. Which of the following are the second messengers that are produced by the phosphoinositide cascade?

(a) phosphatidyl inositol 4,5-bisphosphate (PIP_2)

(b) inositol 1,4,5-trisphosphate (IP_3)

(c) inositol 4-phosphate

(d) inositol 1,3,4,5-tetrakisphosphate

(e) inositol 1,3,4-trisphosphate

(f) diacylglycerol (DAG)

19. Which of the following statements about inositol 1,4,5-trisphosphate (IP_3) are correct?

(a) IP_3 leads to the uptake of Ca^{2+} by the endoplasmic reticulum and the sarcoplasmic reticulum.

(b) IP_3 may be rapidly inactivated by either a phosphatase or a kinase.

(c) IP_3 opens calcium ion channels in the membranes of the endoplasmic reticulum and the sarcoplasmic reticulum.

(d) IP_3 reacts with CTP to form CDP-inositol phosphate, a precursor of PIP_2.

(e) IP_3 acts by altering the intracellular-to-extracellular Na^+-to-K^+ ratio, thereby altering the transmembrane potential.

20. Which of the following statements about the actions or targets of the second messengers of the phosphoinositide cascade are correct?

(a) Diacylglycerol (DAG) activates protein kinase C (PKC).

(b) Most of the effects of IP_3 and DAG are antagonistic.

(c) DAG increases the affinity of PKC for Ca^{2+}.

(d) PKC requires Ca^{2+} for its activity.

21. Which of the following statements about Ca^{2+} and its roles in the regulation of cellular metabolism are correct?

(a) The solubility product of calcium phosphate is small; therefore, low Ca^{2+} levels must be maintained in the cell to avoid its precipitation.

(b) Intracellular Ca^{2+} is maintained at concentrations that are several orders of magnitude smaller than the extracellular concentration by ATP-dependent Ca^{2+} pumps.

(c) The transient opening of ion channels in the plasma membrane or endoplasmic reticulum can rapidly raise cytosolic Ca^{2+} levels.

(d) The binding of Ca^{2+} by a protein can induce a large conformational change because the ion simultaneously coordinates to several anionic groups within the protein.

(e) Ca^{2+} is bound by a family of regulatory proteins that have a characteristic EF hand, helix-loop-helix structure.

(f) When calmodulin binds Ca^{2+} at its low-affinity site, it undergoes a conformational change that allows the complex to interact with target proteins.

22. Explain how a Ca^{2+} ionophore could mimic the effects of a hormone.

23. If it were incubated with cells *in vitro*, why would EGTA prevent either a Ca^{2+}-triggering hormone or a Ca^{2+} ionophore from acting?

24. Which of the following answers complete the sentence correctly? Calmodulin

    (a) is a member of the EF hand family of calcium-binding proteins.
    (b) activates target molecules by recognizing negatively charged β sheets.
    (c) serves as a calcium sensor in most eukaryotic cells.
    (d) is activated when intracellular Ca^{2+} concentrations rise above 0.5 μM.
    (e) activates CAM kinase II, which then phosphorylates many different proteins.
    (f) undergoes a large conformational change on binding Ca^{2+} ions.

Insulin Signaling: Phosphorylation Cascades are Central to Many Signal-Transduction Processes

25. Which of the following statements about the protein kinase domain of the insulin receptor are true?

    (a) It is a tyrosine kinase.
    (b) The kinase domain is inactive when covalently modified.
    (c) The protein kinase domain is a dual-specificity kinase in that is it phosphory-lates Ser, Thr, and Tyr.
    (d) An activation loop is responsible for inactivation of the kinase domain in the IR.
    (e) The kinase recognizes the sequence Tyr-X-X-Met in substrates.

26. Place the following events in the insulin signal transduction domain in order of occurrence.

    (a) activation of Akt
    (b) phosphorylation of IRS
    (c) dimerization of the insulin receptor
    (d) binding of IRS to the insulin receptor
    (e) phosphorylation of the insulin receptor
    (f) binding of IRS to Phosphatidyl-inositol-3 kinase
    (g) activation of PDK1

27. Which of the following answers complete the sentence correctly? Receptor tyrosine kinases

    (a) are seven-transmembrane-helix receptors.
    (b) are integral membrane enzymes.
    (c) activate their targets via the G-protein cascade.
    (d) are often activated by ligand-induced dimerization.
    (e) can phosphorylate themselves on their cytoplasmic domains when activated.
    (f) that have been activated by hormone binding are recognized by target proteins having SH2 (src protein homology region 2) sequences.
    (g) are so named because they contain extraordinarily high amounts of tyrosine.

EGF Signaling: Signal-Transduction Systems Are Posted to Respond

28. Which of the following statements about the tyrosine kinases or hormones that affect them are correct?

    (a) Epidermal growth factor (EGF) stimulates epidermal and epithelial cells to divide.
    (b) EGF is a protein kinase that phosphorylates tyrosine residues.
    (c) EGF and insulin share the common mechanism of dimerization for signal trans-duction across the plasma membrane.

(d) Receptors for EGF and insulin are integral membrane proteins.

(e) Some oncogenes encode tyrosine kinases.

(f) Specialized adaptor proteins link the phosphorylation of the EGF receptor to the stimulation of cell growth.

29. Match each compound in the left column with its characteristic in the right column.

(a) cyclic AMP
(b) GTP
(c) G proteins
(d) adenylate cyclase
(e) PIP_2
(f) Ca^{2+}
(g) IP_3
(h) DAG
(i) insulin receptor
(j) Ras
(k) β-Adrenergic receptor
(l) arachidonic acid

(1) is cleaved by phospholipase C
(2) binds the regulatory subunits of specific protein kinases
(3) exchanges with GDP on G_α subunits
(4) is a downstream hormone product of PIP_2 catabolism
(5) binds epinephrine
(6) is a second messenger arising from PIP_2
(7) is a small G protein GTPase
(8) transduces hormone stimulus from an activated 7TM membrane receptor to adenylate cyclase
(9) has inducible tyrosine kinase activity.
(10) is activated by G_α-GTP
(11) has its intracellular concentration increased by IP_3
(12) activates protein kinase C

Many Elements Recur with Variation in Different Signal-Transduction Pathways

30. Which of the following signaling components directly amplify the signal?
(a) protein kinases
(b) receptor tyrosine kinases
(c) primary messengers
(d) second messengers
(e) signaling domains

Defects in Signaling Pathways Can Lead to Cancer and Other Diseases

31. Which of the following answers complete the sentence correctly? A mammalian protein, src,
(a) has a viral counterpart, v-src, that is oncogenic.
(b) is a proto-oncogene.
(c) is a component of a signaling pathway for cell growth and differentiation.
(d) can be converted to an oncogene by the alteration of some of its C-terminal amino acids.
(e) is a protein tyrosine kinase.

32. Which of the following statements about hormones in mammals are correct?
(a) Hormones are enzymes.
(b) Hormones are synthesized in specific tissues.
(c) Hormones are secreted into the blood.
(d) Hormones alter one or more activities in the cells to which they are targeted.
(e) Hormones display specificity toward the tissues with which they interact.
(f) Hormones are involved in biochemical amplification systems.

33. Cholera toxin (choleragen)
    (a) inactivates a G protein by locking it in the off state (inactivates the GTPase).
    (b) A subunit enters the cell and ADP-ribosylates the $G_{\alpha S}$ subunit of a G protein.
    (c) B subunit interacts with a GM_1 ganglioside on the target-cell surface.
    (d) causes the activation of protein kinase A, which opens a membrane channel and inhibits a $Na^+–H^+$ exchanger.
    (e) causes the retention of Cl^- in the cell.

ANSWERS TO SELF-TEST

1. The receptor for epinephrine is the β-adrenergic receptor and it triggers energy-store mobilization. The receptor for insulin is the insulin receptor and it triggers increased glucose uptake in addition to many other physiological effects. EGF binds to the EGF receptor and results in expression of growth-promoting genes.

2. a, b, c, d

3. (a) 4   (b) 3   (c) 1   (d) 2

4. A signal, in the form of a molecule or a photon interacts with a part of a 7TM on the outside surface of the cell. This interaction causes a conformational change in the protein that is transmitted to the inside of the cell.

5. b, d, e. G proteins are heterotrimers and alternate between states in which GTP or GDP is bound.

6. d.

7. GDP is bound to G proteins when they are inactive and GTP when they are activated. The 7TM receptor, when activated by binding its cognate signaling molecule outside the cell, catalyzes the exchange on a G protein of GDP by GTP inside the cell to activate the G protein. An intrinsic GTPase of the G protein converts GTP to GDP to cause inactivation.

8. a, d, e, f. Answer (g) is incorrect because the GTP form of the G protein activates adenylate cyclase and it forms many cAMP molecules; that is, an amplification occurs.

9. c, d, f, g. Answers (a) and (b) are incorrect because G proteins are peripheral membrane proteins inside cells. They do bind not the hormone but rather the activated hormone-receptor complex, and they carry the signal to adenylate cyclase. Answer (e) is incorrect because the hormone receptor must have the hormone bound to it or it must have been activated by hormone binding before the G protein will bind.

10. The phenomenon is called *desensitization* or *adaptation*. It allows the system to adapt to a given level of hormone so that it can respond to changes in hormone concentrations rather than to absolute amounts. Desensitization is effected by phosphorylation by β-adrenergic receptor kinase at multiple seine and threonine sites on the carboxyl-terminal region of the β-adrenergic receptor when it has epinephrine bound to it. These covalent modifications of the hormone-receptor complex allow β-arrestin to bind it and further inhibit but not completely prevent the GDP–GTP exchange. These events thereby decrease the activation of adenylate cyclase. However, the desensitized receptor can still respond to an increase in epinephrine concentrations. Ultimately, a phosphatase reverses the effects of the modification and resensitizes the receptor.

11. AMP has a single phosphomonoester attached to the 5′-hydroxyl of the adenosine moiety. cAMP has its single phosphate group attached to both the 5′ and 3′ hydroxyls of its adenosine to form a phosphodiester bond.

12. a, c

13. A single hormone molecule combines with a single receptor to form several stimulatory G_α–GTP molecules. Each of these stimulatory molecules activates an adenylate cyclase molecule to form many cAMP molecules. The cAMP molecules activate protein kinases, mainly PKA molecules, each of which can phosphorylate many target enzymes.

14. b, c. Answer (a) is incorrect because the hormone leads to an increase in the activity of adenylate cyclase, not an increase in the amount of the enzyme. Answer (e) is incorrect because the hormone need not enter the cell to carry out its action.

15. The simplest explanation for the tissue specificity of hormones is the presence or absence of receptors for particular hormones on the extracellular surfaces of the tissues. Whether or not a given cell type has a given hormone receptor depends upon which genes have been expressed within it.

16. Aside from the likelihood that serum phosphodiesterases might destroy it, cAMP is a polar molecule that does not readily traverse the plasma membrane. Even if a more hydrophobic derivative, such as dibutyryl-cAMP, were used to overcome the permeability problem, there would be no tissue specificity, and all cells would have increased cAMP levels, leading to a massive, nonspecific response.

17. a, c, d. Answer (e) is incorrect because two messengers are formed.

18. b, f. Answer (a) is incorrect because PIP_2 is the precursor of the second messengers. The other incorrect choices are all downstream products of IP_3 metabolism.

19. b, c. Answer (a) is incorrect because IP_3 causes the release, not the uptake, of Ca^{2+}. Answer (b) is correct because not only does a phosphatase act on IP_3 but a specific kinase phosphorylates it to form the inactive tetrakisphosphate derivative. Answer (d) is incorrect because free inositol reacts with CDP-diacylglycerol to form phosphatidyl inositol, which is then phosphorylated to form PIP_2.

20. a, c, d. Answer (b) is incorrect because most of the effects of IP_3 and Ca^{2+} are synergistic, not antagonistic.

21. a, b, c, d, e, f

22. The ionophore allows Ca^{2+} to enter cells by rendering the membrane permeable to the ion. Since the extracellular Ca^{2+} concentration is higher than the intracellular concentration, the ion enters the cell and the cytosolic level increases. Because some hormones act to raise intracellular Ca^{2+} levels to carry out their physiological roles, the ionophore could lead to the same response.

23. EGTA is a specific Ca^{2+} chelator. It would bind tightly to the ion and markedly lower Ca^{2+} concentration in the extracellular medium. Consequently, when a hormone or a Ca^{2+} ionophore acted to allow Ca^{2+} influx, none could occur because the concentration gradient of Ca^{2+} would be insufficient.

24. a, d, e. Phosphorylation (covalent modification) of the activation loop leads to activation of the receptor kinase. So b is incorrect and d is correct. The specificity is for Tyr residues so a is correct and c is incorrect.

25. The correct order is c, e, d, b, f, g, a.

26. a, c, d, e, f. Answer (b) is incorrect because activated calmodulin recognizes complementary positively charged amphipathic α helices on target proteins. Complementary hydrophobic interactions also contribute to the recognition.

27. b, d, e, f. Answer (g) is incorrect because the name arises from the amino acid that they phosphorylate in their target proteins.

28. a, d, e, f. Answer (b) is incorrect because the hormone itself does not have tyrosine

kinase activity; only the activated receptor is an active tyrosine kinase. Answer (c) is incorrect because the insulin receptor exists as a dimer and merely requires binding insulin to activate its intrinsic tyrosine kinase activity.

29. (a) 2 (b) 3 (c) 8 (d) 10 (e) 1 (f) 11 (g) 6 (h) 12 (i) 9 (j) 7 (k) 5 (l) 4

30. a, b, d. Answer (c) is incorrect because it is the signal that intiates the cascade Answer (e) is incorrect because it passes along the signal but generally does not directly lead to amplification. Signaling domains do mediate specifically in the pathway and help control cross talk

31. a, b, c, d, e

32. b, c, d, e, f

33. b, c, d. Choleragen stabilizes the GTP form of the G protein to keep it in the activated state, resulting in a loss of Cl^- and H_2O from the cell.

PROBLEMS

1. Based on the material so far covered in the text and your general understanding of regulation and signal transduction, list properties that a substance should have for it to be classified as a hormone.

2. Bee venom is particularly rich in phospholipase A_2, an enzyme that hydrolytically removes the fatty acyl residue at position 2 of phospholipids. The action of phospholipase A_2 on phosphatidyl choline is shown in Figure 14.1. One of the mediators of the inflammatory response following a bee sting (swelling, redness, pain, heat, and loss of function) is lysophosphatidyl choline, the remainder of the phospholipid following the hydrolysis of the fatty acyl residue at position 2. Lysophosphatidyl choline stimulates mast cells to release histamine, which triggers the inflammatory response.

FIGURE 14.1   Action of phospholipase A2 on phosphatidyl choline.

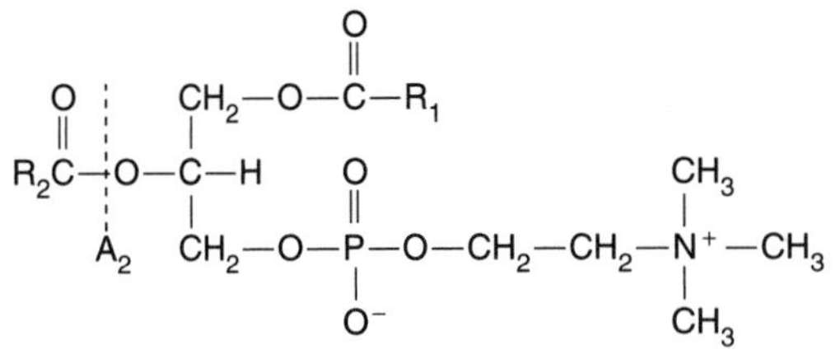

(a) Explain the major point of similarity between the system described here and one (phospholipase C hydrolysis of PIP_2) described in Section 14.1 in the text.

(b) Suppose that the hydrolysis product of phosphatidyl choline that is important as a mediator of the inflammatory response were unknown. Suggest an experiment that might help establish the identity of the active agent.

3. Suppose that epinephrine stimulates the conversion of compound A to compound B in liver cells by means of a regulatory cascade involving a G protein (page 406 of the text), cAMP, protein kinase A, and enzymes E_1 and E_2, as shown in Figure 14.2. Assume that each catalytically active enzyme subunit in the regulatory cascade has a turnover number of 1000 s^{-1}. Assume further that 10 G-GTP are formed for each molecule of epinephrine bound to receptor. Calculate the theoretical number of molecules

of A that would be converted to molecules of B per second as a result of the interaction of one molecule of epinephrine with its receptor on a liver cell membrane.

FIGURE 14.2 Hypothetical regulatory cascade for problem 3.

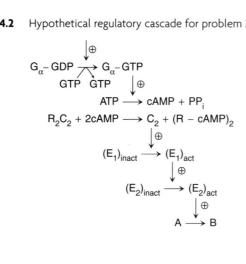

4. In the early days of research on insulin action, it was not known whether insulin might enter cells and directly mediate intracellular effects or whether it might act through a second messenger. In a classic experiment, Pedro Cuatrecasas attached insulin covalently to Sepharose beads many times the size of fat cells and showed that the addition of the insulin-Sepharose complexes to isolated fat cells gave the same stimulation of glucose oxidation as did addition of insulin alone.

   (a) What conclusion might follow from this experiment? Explain.
   (b) What assumptions have you made about the effects of adding sepharose without attached insulin to fat cells and about the attachment of the insulin to the Sepharose bead?

5. In kinetic studies on the interaction of human growth hormone with its receptor, each functional receptor dimer was found to bind one hormone molecule, and the monomer receptors needed to dimerize in order to transduce the signal from the hormone. Explain how this occurs.

6. To be effective, intracellular signals must be readily inactivated when their effects are no longer needed. Give a method of inactivation for each of the following classes of intracellular messengers;

   (a) G proteins
   (b) cyclic nucleotides
   (c) phosphoproteins
   (d) calcium ion
   (e) inositol 1,4,5-trisphosphate (IP_3)
   (f) diacylglycerol

7. A tissue is known to increase cyclic AMP production on stimulation by a certain hormone. Addition of an analog of GTP in which the terminal phosphate group is replaced by a sulfate to a homogenate of the tissue results in sustained production of cyclic AMP. Propose an explanation for this observation.

8. What properties of Ca^{2+} render it so useful as a messenger in cells? What protein is often used in cells to "sense" Ca^{2+}? How does the cell overcome the problem of the low-solubility product of Ca^{2+} with P_i, phosphorylated compounds, and carboxyl groups?

9. What signal-transduction functions do SH2 domains in proteins serve?

10. After the insulin–insulin-receptor complex autophosphorylates itself, a series of downstream events carries the signal to molecules directly involved in promoting, among

other things, the entry of glucose into muscle and adipose cells. Insulin thus promotes a lowering of the blood glucose (hypoglycemia). When both copies of the gene (*Akt2*) for a particular serine–threonine kinase (a protein kinase B isoform) were ablated in a strain of mice, the "knockout" mice could no longer lower their blood glucose by taking it into muscle cells upon administration of insulin. (Isoenzymes are sometimes called isoforms.(See Section 10.2 of the text to see how lactate dehydrogenase exemplifies isoenzymes.) What conclusions can you draw about the role of the Akt2 isoform of protein kinase B in glucose homeostasis? Can you think of alternative explanations for the observation with the knockout mice?

11. List three characteristics of many signaling pathways and give an example of each.

ANSWERS TO PROBLEMS

1. The major criteria for classifying a substance as a hormone are:
   (1) In order to carry messages from one tissue to another, it should be produced by one type of cell and have effects on another type of cell.
   (2) Its effects should involve the chemical amplification of the original signal.
   (3) It should be produced in response to a stimulus, and its production should cease upon cessation of the stimulus.
   (4) It should be selectively destroyed following cessation of the stimulus.
   (5) The addition of the purified substance to tissues should mimic physiologic responses produced in vivo.
   (6) Specific inhibitors of the physiologic response should also abolish the response elicited by the addition of the purified substance to tissues.
   (7) Specific receptors for the hormone should exist and should be more abundant in tissues that are more sensitive to the hormone.

2. (a) The bee venom system resembles the phosphoinositide cascade discussed in section 14.1 of the text. In that system, a membrane phospholipid is also converted into an active mediator of the response of several hormones.

   (b) One could inject each of the hydrolysis products—lysophosphatidyl choline and the fatty acid—into tissues separately to see which might elicit the inflammatory response.

3. The theoretical number of molecules of A converted to B per second would be 10^{13}. One molecule of epinephrine would result in the production of 10 G-GTP. Each activated α subunit would stimulate adenylate cyclase to produce 1000 cAMP molecules for a total of 10,000 molecules of cAMP. Each of these cAMP molecules would activate one catalytic subunit of protein kinase. (Remember that a molecule of protein kinase exists as an R_2C_2 complex. Two cAMP molecules combine with two R subunits to give two catalytically active C subunits.) Each of the 10,000 active C subunits would result in the production of 1000 molecules of active E_1 for a total of 10^7 molecules of active E_1. Each molecule of active E_1 would in turn activate 1000 molecules of E_2 for a total of 10^{10} molecules of active E_2. Since each molecule of active E_2 would convert 1000 molecules of A to B per second, the total would be $1000 \times 10^{10} = 10^{13}$ per second. (*Note:* This is a greatly oversimplified example, but it illustrates the profound chemical amplification that can occur in systems under hormonal control.)

4. (a) A reasonable conclusion is that insulin need not enter the cell to have an effect. The results are consistent with the notion that insulin affects cells by combining

with a membrane receptor site outside the cell thereby causing some second messenger to be formed within the cell that mediates the effects. Note that the experiment does not prove that insulin fails to enter cells.

(b) You probably assumed that the addition of Sepharose alone gave no stimulation of glucose oxidation. You also have to assume that the covalent attachment of the insulin to the Sepharose is stable so that free insulin is not formed during the course of the experiment.

5. A given molecule of growth hormone contains two domains, each of which binds a receptor monomer. Thus, a single molecule of growth hormone could be bound by two receptors, bringing them together to form the activated hormone-receptor dimer complex.

6. (a) G proteins are active while GTP is bound, but they become inactive as GTP is hydrolyzed to GDP and phosphate.

(b) Cyclic nucleotides are converted by phosphodiesterases to $5'$-mononucleotides.

(c) Phosphates are cleaved from phosphoproteins by protein phosphatases.

(d) Calcium ions are pumped from the cell interior into the extracellular fluid or intracellular storage organelles, for example, the endoplasmic reticulum.

(e) Inositol 1,4,5-trisphosphate can be degraded to inositol and inorganic phosphate by the sequential action of phosphatases or it can be phosphorylated by a kinase to form inositol 1,3,4,5-tetrakisphosphate.

(f) Diacylglycerol may be converted to phosphatidate or hydrolyzed to glycerol and fatty acids.

7. The observation could be explained if the sulfate-containing analog of GTP bound to G protein, stimulating cyclic AMP production, could not be hydrolyzed to GDP and sulfate by the GTPase activity of G. Thus the production of cyclic AMP would persist.

8. Energy-requiring molecular pumps maintain a steep concentration gradient of Ca^{2+} across the plasma membrane between the outside and inside of the cell and across the membrane between intracellular organelles and the cytoplasm. When the membrane, for instance, is rendered permeable to Ca^{2+} as a result of the opening of a Ca^{2+} channel, a flux of ions passes through the membrane raising the cytoplasmic Ca^{2+} concentration. Such a sudden increase in Ca^{2+} can act as a signal to Ca^{2+}-sensing proteins within the cell. Calmodulin binds Ca^{2+} and interacts with several proteins and enzymes as a consequence of the binding. Because Ca^{2+} can interact simultaneously with several anionic amino acid side chains, the carbonyls of the peptide backbone, or the carbonyls of Gln and Asn, it can cause large conformational changes in the protein to which it binds. Conformational changes in response to binding a ligand are the hallmarks of a molecular switch. Thus, the ability to rapidly change its concentration and to effect large conformational changes renders Ca^{2+} an effective intracellular messenger. The cell avoids precipitating the Ca^{2+} salts of its intracellular components by maintaining the Ca^{2+} concentration below the solubility product for various compounds. Endergonic pumps and exchangers maintain the low intracellular Ca^{2+} concentrations.

9. SH2 domains bind to peptides or sections of proteins that contain phosphotyrosine residues in particular sequence contexts. The formation of phosphorylated tyrosine residues in a receptor often results from hormone activation of the receptor. The phosphorylated tyrosine-containing peptides in the receptor can be recognized and bound by other proteins that have SH2 domains. The SH2 domain allows different proteins to respond to and be affected by the phosphorylated tyrosines that arise in proteins as a result of a signal transduction event.

10. The simplest interpretation of the observation is that this particular protein kinase B isoform is directly involved in mediating the ability of insulin to lower blood glucose concentrations by facilitating its entry into muscle cells. The kinase presumably acts by phosphorylating a target molecule that, in turn, facilitates the movement of a glucose transporter (GULT4) to the surface of the cell. An alternative explanation could be that the lack of the Akt2 kinase during the growth of the knockout mouse led to the failure to synthesize a molecule that was, itself, the active component in the insulin-signaling pathway. The mutation-induced lack of the protein in some other tissue could also have caused the effect if that tissue normally supplied a compound needed in the muscle cells for the insulin response. For instance, adipose tissue is known to affect glucose uptake by muscle cells. The gene deletion could have affected the ability of adipose tissue to make that compound. Further experiments would be required to verify the simplest conclusion. (This problem is based on H. Cho, J. Mu, J. K. Kim, J. L. Thorvaldsen *et al*. Insulin resistance and a diabetes mellitus-like syndrome in mice lacking the protein kinase Akt2 (PKBβ). *Science* 292[2001]:1728–1731.)

11. The text lists three common themes in signal transduction. The first is that protein kinases are central to many pathways. The kinases can be either part of receptors or cytoplasmic. Although not discussed in the text, kinases with a specificity for serine and threonine phosphyorylation are actually more common than tyrosine kinases. All three signal-transduction pathways discussed in the text contain protein kinases, either as tyrosine-kinase receptors (IR, EGFR) or as cytoplasmic kinases (PKA, Raf, MEK, ERK). The second theme is that second messengers participate in many signal-transduction pathways. The text discusses several, including cAMP, Ca^{2+}, IP_3, and DAG. These messengers are mobile so they can move between signaling components and can be either soluble (cAMP, Ca^{2+}, and IP_3) or membrane-associated (DAG). The last theme is that specialized domains that mediate specific interactions are present in many signaling proteins. Protein-protein interactions lie at the heart of cellular signaling. These are mediated in large part through protein domains which repeat in a large number of signaling components and may even be present in multiple copies within a particular protein. Some examples given in the text are SH2 and SH3 domains, which bind phosphotryosines and polyprolines respectively and PH which interact with the lipid PIP_3.

Metabolism: Basic Concepts and Design

This chapter is an introduction to the next two parts of the text, which are devoted to metabolism. Metabolism is the interconnected, integrated ensemble of chemical reaction cells used to extract energy and reducing power from their environments, synthesize the building blocks of their macromolecules, and carry out all the other processes that are required to sustain life. Basic thermodynamic postulates were presented in Chapter 8 and should now be reviewed for a better understanding of the principles of metabolism presented in this chapter. Because energy is an essential concept in understanding metabolism, the authors begin this chapter with a review of free-energy changes in the context of a description of coupled reactions. They explain how an energetically unfavorable reaction can occur if it is coupled to one that occurs spontaneously. The most important molecules for storing and carrying energy in metabolic processes, including ATP, the universal currency of energy in biological systems, are described next. The role of creatine phosphate, the specialized energy storage molecule of vertebrate muscle, is also given.

The authors present a broad outline of energy metabolism. The energy for ATP synthesis comes from the oxidation of carbon compounds, and the pathways that perform these oxidations can be classified into three stages. All living cells draw on a spectrum of a few activated carriers to help run these reactions, including the electron carriers NAD^+ and FAD, plus CoA, thiamine, biotin, and several others. Most metabolic reactions fall into a handful of predictable types of reactions. The student would do well to pay particular attention to this discussion because these reactions form a major part of the subsequent 350 pages of the text, and learning them now can save time and energy later. The chapter concludes with a cogent discussion of regulation of metabolic pathways, followed by speculation about the origin of nucleotide-containing cofactors.

LEARNING OBJECTIVES

When you have mastered this chapter, you should be able to accomplish the following objectives.

Metabolism Is Composed of Many Coupled, Interconnecting Reactions
(Text Section 15.1)

1. Define *metabolism*.

2. List the three major purposes of living organisms that require a constant input of free energy.

3. Distinguish between *phototrophs* and *chemotrophs*.

4. Define *catabolism* and *anabolism*.

5. State the significance of the *free-energy change (ΔG)* of reactions and the relationship of ΔG to ΔG°′, the *equilibrium constant*, and the *concentrations of reactants* and *products* of the reaction.

6. Describe the *additivity* of ΔG values for *coupled reactions* and explain the ability of a thermodynamically favorable (exergonic) reaction to drive an energetically unfavorable (endergonic) one.

ATP is the Universal Currency of Free Energy in Biological Systems
(Text Section 15.2)

7. Give the structure of *adenosine triphosphate*. Describe the role of ATP as the major energy-coupling agent *(energy currency)* in metabolism.

8. Explain how coupling a reaction with the hydrolysis of ATP can change the equilibrium ratio of the concentrations of the products to the concentrations of the reactants by a factor of 10^8. Know the ΔG for ATP hydrolysis under typical celllular conditions.

9. Describe the structural and electronic bases for the *high-phosphoryl group-transfer potential* of ATP, and give the free energy liberated by the hydrolysis of ATP under standard and cellular conditions.

10. Recognize that compounds that have a *high group-transfer potential*, that is, compounds that release large amounts of free energy on hydrolysis or oxidation.

11. Describe how *creatine phosphate* serves as a "high-energy" buffer in vertebrate muscle.

The Oxidation of Carbon Fuels Is an Important Source of Cellular Energy
(Text Section 15.3)

12. Describe the *ATP–ADP cycle* of energy exchange in biological systems.

13. Explain in general terms how oxidation of carbon compounds can drive the formation of ATP.

14. Define *proton gradient,* and explain how a proton gradient can couple unfavorable reactions to favorable ones.

15. Describe the three major *stages in the extraction of energy* from foodstuffs.

Metabolic Pathways Contain Many Recurring Motifs (Text Section 15.4)

16. Recognize the structures of *nicotinamide adenine dinucleotide (NAD⁺)* and *flavin adenine dinucleotide (FAD)*, describe their reduction to NADH and FADH₂, and explain their roles in metabolism.

17. Contrast the metabolic roles of *NADPH* and *NADH*.

18. Explain the fact that most *high-energy compounds* and *reduced electron carriers* are kinetically stable, despite their high-energy status, and that they require enzymes for their reactions.

19. Describe the structure of *coenzyme A (CoA)* and its role as a *carrier of acetyl* or *acyl groups*.

20. List the major *activated carriers*, both of electrons and activated groups, in metabolic reactions.

21. Describe the six basic types of chemical reactions encountered in cellular metabolism.

22. Discuss the three major mechanisms for the *regulation of metabolism*.

23. Define *energy charge* and compare it with the *phosphorylation potential*.

24. Describe the common structural features of NAD, FAD, and ATP.

SELF-TEST

Metabolism Is Composed of Many Coupled, Interconnecting Reactions

1. Which of the following functions is not a purpose of metabolism?
   (a) extract chemical energy from substances obtained from the external environment
   (b) form and degrade the biomolecules of the cell
   (c) convert exogenous foodstuffs into building blocks and precursors of macromolecules
   (d) equilibrate extracellular substances and the biomolecules of the cell
   (e) assemble the building-block molecules into macromolecules

2. If the ΔG of the reaction A $\longrightarrow$ B is -12.5 kJ/mol (-3.0 kcal/mol), which of the following statements are correct?
   (a) The reaction will proceed spontaneously from left to right at the given conditions.
   (b) The reaction will proceed spontaneously from right to left at standard conditions.
   (c) The equilibrium constant favors the formation of B over the formation of A.
   (d) The equilibrium constant could be calculated if the initial concentrations of A and B were known.
   (e) The value of $\Delta G^{o\prime}$ is also negative.

ATP is the Universal Currency of Free Energy in Biological Systems

3. The text compares ATP to currency. How is ATP similar to money?

4. Glucose 1-phosphate is converted to fructose 6-phosphate in two successive reactions:

   Glucose-1-phosphate $\longrightarrow$ glucose-6-P $\Delta G^\prime = -7.1$ kJ/mol (-1.7 kcal/mol)

   Glucose-6-phosphate $\longrightarrow$ fructose-6-P $\Delta G^{o\prime} = -1.7$ kJ/mol (-0.4 kcal/mol)

What is the $\Delta G^{o\prime}$ for the overall reaction?

(a) −8.8 kJ/mol (−2.1 kcal/mol)   (d) 5.4 kJ/mol (1.3 kcal/mol)
(b) −7.1 kJ/mol (−1.7 kcal/mol)   (e) 8.8 kJ/mol (2.1 kcal/mol)
(c) −5.4 kJ/mol (−1.3 kcal/mol)

5. The reaction

phosphoenolpyruvate + ADP + H^+ ⟶ pyruvate + ATP

has a $\Delta G^{o\prime} = -31.4$ kJ/mol (−7.5 kcal/mol). Calculate $\Delta G^{o\prime}$ for the hydrolysis of PEP.

6. Inside cells, the ΔG value for the hydrolysis of ATP to ADP + P_i is approximately −50 kJ/mol (−12 kcal/mol). Calculate the approximate ratio of [ATP] to [ADP][P_i] found in cells at 37°C.

(a) 5000/1   (d) 1000/1
(b) 4000/1   (e) 200/1
(c) 2000/1

7. Which of the following processes are ways by which two reactions can be coupled energetically to each other?

(a) As common intracellular components of a compartment, two reactions become automatically coupled.
(b) An ionic gradient across a membrane that is formed by one reaction can drive another reaction that uses the gradient to render it exergonic.
(c) A shared, common intermediate can couple two reactions.
(d) A protein that is activated by binding another molecule or by being covalently modified can provide energy to drive another reaction.

8. Which of the following statements about the structure of ATP are correct?

(a) It contains three phosphoanhydride bonds.
(b) It contains two phosphate ester bonds.
(c) The sugar moiety is linked to the triphosphate by a phosphate ester bond.
(d) The nitrogenous base is called *adenosine*.
(e) The active form is usually in a complex with Mg^{2+} or Mn^{2+}.

9. Which of the following factors contributes to the high-phosphate group-transfer potential of ATP?

(a) greater resonance stabilization of ADP and P_i than of ATP
(b) increase in the electrostatic repulsion of oxygens on hydrolysis of ATP
(c) interaction of the terminal phosphoryl group with the ribose group in ADP
(d) formation of a salt bridge between the base amino group and the negative charges of the phosphate oxygens in ATP

10. Which of the following are high-energy compounds?

(a) glycerol 3-phosphate   (d) 1,3-bisphosphoglycerate
(b) adenosine diphosphate   (e) fructose 6-phosphate
(c) glucose 1-phosphate

11. ATP falls in the middle of the list of compounds having high phosphate group-transfer potentials. Explain why this is advantageous for energy coupling during metabolism.

12. Which of the following statements about the phosphoryl transfer potential of skeletal muscle are correct?

(a) The ATP of muscle can sustain contraction for less than a second.
(b) Creatine phosphate serves as a phosphoryl reservoir that replenishes the ATP pool.

(c) Creatine phosphate can support contraction for up to 4 minutes.

(d) The phosphoguanidino group of creatine phosphate has a large negative standard free energy of hydrolysis.

(e) Creatine phosphate is formed by a reaction between creatine and ATP.

The Oxidation of Carbon Fuels Is an Important Source of Cellular Energy

13. Which of the following features are part of the ATP–ADP cycle in biological systems?

    (a) ATP hydrolysis is used to drive reactions that require an input of free energy.

    (b) The oxidation of fuel molecules forms $ADP + P_i$ from ATP.

    (c) The oxidation of fuel molecules forms ATP from $ADP + P_i$.

    (d) Light energy drives ATP hydrolysis.

    (e) A transmembrane proton-motive force drives ATP synthesis.

14. Which of the following statements about the third of the three stages of metabolism that generate energy from foodstuffs are correct?

    (a) It is common to the oxidation of all fuel molecules.

    (b) It involves the breakdown of the macromolecular components of food into smaller units, such as amino acids, sugars, and fatty acids.

    (c) It releases relatively little energy compared with the second stage.

    (d) It involves the conversion of sugars, fatty acids, and amino acids into a few common metabolites.

    (e) It produces most of the ATP and CO_2 in cells.

Metabolic Pathways Contain Many Recurring Motifs

15. Which of the following answers complete the sentence correctly? NAD^+

    (a) is a flavin nucleotide.

    (b) is the major electron acceptor used in fuel metabolism.

    (c) contains a nicotinamide ring that accepts a hydride ion during reduction.

    (d) loses a plus charge upon reduction.

    (e) contains ATP as a part of its structure.

16. Which of the following answers complete the sentence correctly? During the reduction of FAD,

    (a) a flavin group is transferred.

    (b) an equivalent of a hydride ion is transferred.

    (c) the isoalloxazine ring becomes charged.

    (d) two hydrogen atoms are added to the isoalloxazine ring.

    (e) the adenine ring opens.

17. Match the four cofactors in the left column with the appropriate structural features and properties from the right column.

    (a) ATP   (1) nicotinamide ring
    (b) FAD   (2) adenine group
    (c) NAD^+ (3) phosphoanydride bond
    (d) CoA   (4) sulfur atom
                            (5) isoalloxazine ring
                            (6) ribose group
                            (7) acyl group transfer
                            (8) electron transfer
                            (9) phosphate transfer

18. ATP and NADH release large amounts of free energy on the transfer of the phosphate group to H_2O and electrons to O_2, respectively. However, both molecules are relatively stable in the presence of H_2O or O_2. Explain why.

19. Which of the following pairs correctly matches a coenzyme with the group transferred by the coenzyme?

    (a) CoA, electrons
    (b) biotin, CO_2
    (c) ATP, one-carbon unit
    (d) NADPH, phosphoryl group
    (e) thiamine pyrophosphate, acyl group

20. Which of the following water-soluble vitamins forms part of the structure of CoA?

    (a) pantothenate   (d) pyridoxine
    (b) thiamine   (e) folate
    (c) riboflavin

21. In Table 15.5 in the text, six types of chemical reactions are listed. In metabolic pathways, these reactions would be catalyzed by enzymes. Enzymes are given functional names. For example, "isomerization" is catalyzed by an "isomerase." What would be the enzyme name, ending in "-ase," for the following: oxidation, ligation, group transfer, hydrolysis?

22. Which of the following statements are reasons the biochemical pathway for the catabolism of a molecule is almost never the same as the pathway for the biosynthesis of that molecule?

    (a) It would be extremely difficult to regulate the pathway if it served both functions.
    (b) The free-energy change would be unfavorable in one direction.
    (c) The reactions never take place in the same type of cell.
    (d) Enzyme-catalyzed reactions are irreversible.
    (e) Biochemical systems are usually at equilibrium.

23. Which of the following phrases are ways by which metabolism is regulated?

    (a) accessibility of substrates   (d) control of enzyme activities
    (b) pressure fluxes   (e) temperature cycles
    (c) amounts of enzymes

24. Which of the following statements about the energy charge are correct?

    (a) It can have a value between 0 and 1.
    (b) It is around 0.1 in energy-consuming cells, such as the muscle cells.
    (c) It can regulate the rates of reactions in energy-consuming and energy-producing pathways.
    (d) It is also called the *phosphorylation potential*.
    (e) It is buffered in the sense that its value is maintained within narrow limits.

ANSWERS TO SELF-TEST

1. d. The intracellular and extracellular concentrations of most substances are not at equilibrium, and one of the functions of metabolism is to maintain these nonequilibrium concentrations.

2. a, d. The expression for ΔG contains two variables: $\Delta G^{o\prime}$ (a derivative of K_{eq}) and the ratio of the product concentrations to the reactant concentrations. Therefore, ΔG alone cannot provide information about $\Delta G^{o\prime}$ or K_{eq}. Answer (d) is correct because $\Delta G^{o\prime}$ and K_{eq} can be calculated when ΔG and the reactant and product concentrations are known.

3. The way ATP is used in the cell is remarkably similar to the way money is used in society. The ATP is "earned" by oxidizing food molecules and "spent" to build "expensive" molecules. It is useful to put energetic calculations into these terms—"I am spending $30.50 worth of ATP to buy $9.20 worth of glycerol-3-phosphate, so my change will be $21.30." In other words, there will be -21.3 kJ/mol left over to drive the reaction far to the right. Energy values mentioned here are given on page 434 of the text.

4. a

5. The answer is -61.9 kJ/mol (-14.8 kcal/mol).

   The overall reaction can be separated into two steps:

   (1) PEP $\longrightarrow$ Pyruvate + P_i unknown $\Delta G^{o\prime}$
   (2) ADP + $P_i \longrightarrow$ ATP $\Delta G^{o\prime} = -30.5$ kJ/mol ($+7.3$ kcal/mol)

   The sum of standard free energies for the two steps is -31.4 kJ/mol (-7.5 kcal/mol). Combining what we know,

   $$\Delta G^{o\prime} = -31.4 \text{ kJ/mol} - (+30.5 \text{ kJ/mol}) = -61.9 \text{ kJ/mol}$$

   or in kcal: -7.5 kcal/mol $- (+7.3$ kcal/mol$) = -14.8$ kcal/mol

6. (c) ATP $\longrightarrow$ ADP + P_i ^{2} $\Delta G^{o\prime} = -30.5$ kJ/mol (-7.3 kcal/mol)

   $$\text{Using } \Delta G = \Delta G^{o\prime} + RT \ln \frac{[\text{ADP}][\text{P}_i]}{[\text{ATP}]}$$

   $$-50 \text{ kJ/mol} = -30.5 \text{ kJ.mol} + 0.0083 \text{ kJ/mol K} \times 310 \text{ K} \times \ln \frac{[\text{ADP}][\text{P}_i]}{[\text{ATP}]}$$

   $$-50 + 30.5 = -19.5 \text{ kJ/mol}$$

   $$0.0083 \text{ kJ/mol K} \times 310 \text{ K} = 2.57 \text{ kJ/mol}$$

   $$-19.5 = 2.57 \ln \frac{[\text{ADP}][\text{P}_i]}{[\text{ATP}]}$$

   $$\frac{[\text{ADP}]}{[\text{ATP}][\text{P}_i]} = \text{anti} \ln (-7.6) = e^{-7.6} = 5 \times 10^{-4}$$

   $$\text{so } \frac{[\text{ATP}]}{[\text{ADP}][\text{P}_i]} = \frac{1}{5 \times 10^{-4}} = 2000$$

   Alternatively, in calories,

   $$-12 \text{ kcal/mol} = -7.3 \text{ kcal/mol} + 2.303 \times 1.98 \text{ cal/mol K} \times 310 \text{ K} \times \log_{10} \frac{[\text{ADP}][\text{P}_i]}{[\text{ATP}]}$$

   $$-12 + 7.3 = -4.7 \text{ kcal/mol}$$

   $$2.30 \times 0.00198 \text{ kcal/mol K} \times 310 \text{ K} = 1.41 \text{ kcal/mol}$$

   $$-4.7 = 1.414 \log_{10} \frac{[\text{ADP}][\text{P}_i]}{[\text{ATP}]}$$

   $$\frac{[\text{ADP}][\text{P}_i]}{[\text{ATP}]} = antilog (-3.32) = 10^{-3.32} = 4.79 \times 10^{-4}$$

   $$\text{so } \frac{[\text{ATP}][\text{P}_i]}{[\text{ADP}]} = \frac{1}{4.79 \times 10^{-4}} = 2089$$

Answers slightly different due to approximation. 12 kcal/mol is more like 50.15 kJ/mol.

7. b, c, d. Being in the same compartment does not necessarily couple two reactions.

8. c, e

9. a

10. b, d

11. The intermediate phosphate group-transfer potential of ATP means that, although ATP hydrolysis can drive a very large number of thermodynamically unfavorable biochemical reactions in metabolic pathways, it can itself be regenerated by coupling with other reactions that release more free energy than −30.5 kJ/mol (−7.3 kcal/mol). Thus, ATP can act as an effective carrier, since it both accepts and donates phosphoryl groups.

12. a, b, d, e. Answer (c) is incorrect because the amount of creatine phosphate is sufficient to maintain only a few seconds of intense contraction. The ATP for the creatine kinase-catalyzed reaction given in (d) is generated by glycolysis in anaerobic muscle or by respiration in muscle with sufficient oxygen.

13. a, c, e

14. a, e

15. b, c, d. Answer (e) is incorrect because ADP, not ATP, forms a part of the structure of NAD$^+$.

16. d

17. (a) 2, 3, 6, 9   (b) 2, 3, 5, 6, 8   (c) 1, 2, 3, 6, 8   (d) 2, 3, 4, 6, 7

18. Although the transfer reactions of the cofactors ATP and NADH have large negative free-energy changes, there are high activation-energy barriers that greatly slow spontaneous reactions with H_2O or O_2, respectively. In other words, cofactors with high group-transfer potentials and fuel molecules are thermodynamically unstable yet kinetically stable. Consequently, specific enzymes are required to catalyze their reactions.

19. b

20. a

21. Oxidase, ligase, transferase, and hydrolase. In fact, the name "oxidase" is reserved for oxidations where a molecule of oxygen is present. Most oxidations encountered in metabolic pathways are catalyzed by "dehydrogenase" enzymes, which utilize cofactors such as FAD or NAD$^+$ as the oxidant. Removal of functional groups without hydrolysis is done by "lyase" enzymes. The classification of enzymes is discussed in Chapter 8.

22. a, b

23. a, c, d

24. a, c, e

PROBLEMS

1. Under standard conditions, the free energy of hydrolysis of L-glycerol phosphate is −9.2 kJ/mol (−2.2 kcal/mol), and for ATP hydrolysis it is −30.5 kJ/mol (−7.3 kcal/mol). Show that when ATP is used as a phosphoryl donor for the formation of L-glycerol phosphate, the value of the equilibrium constant is altered by a factor of over 10^5.

2. When a hexose phosphate is hydrolyzed to free hexose and inorganic phosphate, the ratio of the concentration of hexose to the concentration of hexose phosphate at equilibrium is 99 to 1. What is the free-energy change for the reaction under standard conditions?

3. Sucrose phosphorylase catalyzes the phosphorolytic cleavage of sucrose in certain microorganisms.

$$\text{sucrose} + P_i \longrightarrow \text{glucose-1-phosphate} + \text{fructose}$$

(a) Use the following information to calculate the standard free-energy change for the phosphorolysis of sucrose.

$\text{sucrose} + H_2O \longrightarrow \text{glucose} + \text{fructose} \qquad \Delta G^{o\prime} = -29.3$ kJ/mol $(-7.0$ kcal/mol$)$

$\text{glucose 1-phosphate} + H_2O \longrightarrow \text{glucose} + P_i \qquad \Delta G^{o\prime} = -20.9$ kJ/mol $(-5.0$ kcal/mol$)$

(b) Calculate the equilibrium constant for the phosphorolysis of glucose at 25°C.

4. Phosphocreatine can be used as a phosphoryl donor for the synthesis of ATP in a reaction catalyzed by creatine kinase. Refer to Section 15.2 of the text for free energies of hydrolysis of ATP and creatine phosphate.

(a) What effect does creatine kinase have on the value of $\Delta G^{o\prime}$ for the reaction?
(b) From the typical concentrations of ATP, ADP, creatine phosphate, and creatine cited in Section 15.2 of the text, calculate ΔG for the reaction in a resting muscle cell at 25°C.
(c) Suppose that during muscle contraction the concentration of creatine phosphate drops to 1 mM, and ATP concentration drops to 3.9 mM. At these concentrations, will creatine phosphate serve as a donor of phosphoryl groups to ADP at 25°C?

5. Chemotrophs derive free energy from the oxidation of fuel molecules, such as glucose and fatty acids. Which compound, glucose or a saturated fatty acid containing 18 carbons, would yield more free energy per carbon atom when subjected to oxidation in the cell? See Figure 15.10 in the text for a comparison.

6. The process of catabolism releases free energy, some of which is stored as ATP and some of which is lost as heat to the surroundings. Explain how these observations are consistent with the fact that catabolic pathways are essentially irreversible.

7. It is well known that putting sugar in the gas tank of an internal combustion engine will disable the engine. But sugar is mostly carbon, and a sugar cube burns readily. Explain why it causes a problem on a molecular level, and relate the problem to this chapter.

8. Figure 15.12 of the textbook shows that stage III of catabolism requires oxygen. What can we deduce about metabolism in obligate anaerobes? Many prokaryotes (Eubacteria and Archaea) are anaerobic.

9. In a typical cell, the concentrations of pyridine nucleotides ($NAD^+/NADP^+$) and flavins (FAD/FMN) are relatively low compared with the number of substrate molecules that must be oxidized. What does this observation suggest about the rate of oxidation and reduction of these electron carriers?

10. The flavins FAD and FMN (flavin mononucleotide) are both bright yellow compounds; in fact, the name *flavin* was taken from *flavus,* the Latin word for "yellow." The corresponding reduced compounds, $FADH_2$ and $FMNH_2$, are nearly colorless. What portion of a flavin molecule accounts for these color changes? The enzyme glucose oxidase, found in fungi, catalyzes the conversion of free glucose to gluconic acid. Glucose oxidase utilizes two molecules of FAD as cofactors. How could you use the light-absorbing properties of flavins as a means of monitoring glucose oxidase activity?

11. The flow of electrons from reduced pyridine nucleotides, such as NADH, provides energy that can drive the formation of ATP. Why must the reaction

$$NADH \rightarrow NAD^+ + H^+ + 2\,e^-$$

have a negative value for $\Delta G^{o\prime}$?

12. Refer to problem 5 in the homework problems of the text. The formation of a number of other important compounds in biosynthetic reactions involves the generation of pyrophosphate and its subsequent hydrolysis to two molecules of P_i. For example, the formation of UDP-glucose from UTP and glucose 1-phosphate yields PP_i, which is then cleaved. What does this tell you about the group transfer potential of UDP-glucose?

13. Many important reactions are "driven to completion" by formation of pyrophosphate in the reaction ATP $\rightarrow$ AMP + PP_i followed by $PP_i \rightarrow 2\,P_i$. Examples would include DNA polymerase, attachment of amino acids to transfer RNA, synthesis of NAD^+, synthesis of acyl CoA derivatives, and many other reactions. Yet in text Table 15.1, the standard free energy of hydrolysis of pyrophosphate is given as -19.3 kJ/mol (-4.6 kcal/mol), a rather low value. Why should a reaction with such a small standard free-energy change be utilized in such critically important processes?

14. In Table 15.2 in the text, pantothenate, part of coenzyme A, is listed as a vitamin precursor. Coenzyme A also contains AMP and mercaptoethylamine. Why do you think these components are not listed as vitamins?

15. The coenzyme biotin acts as a carrier of carbon dioxide molecules in carboxylase enzymes. During catalytic cycles biotin undergoes successive carboxylation and decarboxylation, but the coenzyme itself is not chemically altered at the end of each cycle. Yet a small amount of biotin is required on a daily basis in the diet. Why?

16. Look in the index of the textbook and find leucine catabolism. The degradation of the amino acid leucine produces isovaleryl CoA, which is then further catabolized. What features of this catabolism are familiar from Section 15.4?

17. Why is it desirable for a cell to regulate the *first* reaction in a biosynthetic pathway?

18. The text discusses the fact that many cofactors contain nucleotides as evidence for the former existence of an RNA world. Are there other "fossil" nucleotide-containing carriers not mentioned in Section 15.4? Do all of these "fossils" prove that RNA came before protein in living cells?

19. Approximately 4% of cellular enzymes use coenzyme A. Understanding the role of this cofactor is critical to understanding many metabolic processes such as glucose oxidation and fatty acid oxidation. What is coenzyme A? What is its role (or the role of Acetyl-CoA) in metabolism? What advantages does this cofactor impart?

20. It will be critical later on to know which forms of the nicotinamide adenine dinucleotides and mononucleotides are the reduced forms and which are the oxidized forms. The following exercises will help to reinforce this knowledge. Using the abbreviated structure below, fill in the blanks for the reactions for the reduction of NAD^+.

21. The sale of nutritional supplements has become an enormously profitable industry. The following are supplements that are available for over-the-counter purchase. Identify the role of these substances in metabolism.

    (a) Pantothenic acid (vitamin B_5)
    (b) Biotin
    (c) Niacin or Nicotinate
    (d) Creatine
    (e) Riboflavin (vitamin B_2)

22. NAD^+ and FAD are both oxidizing agents, but they participate in different types of reactions. What are these two reaction types?

ANSWERS TO PROBLEMS

1. For the synthesis of L-glycerol phosphate from glycerol and phosphate, the value of $\Delta G^{\circ\prime}$ is $+9.2$ kJ/mol ($+2.2$ kcal/mol). Under standard conditions,

$$\Delta G^{\circ\prime} = -RT \ \ln \ K'_{eq}$$

$$+9.2 \ \ kJ/\ mol = (.0083 \ kJ/\ ^{\circ}Kmol)(298K) \ln K'_{eq}$$

$$\ln \ K'_{eq} = \frac{+9.2}{(-2.47)}$$

$$= -3.72$$

$$K'_{eq} = \ e^{-3.72} = 0.024 = 2.40 \times 10^{-2}$$

[In Chapter 8 of the text, we saw that $RT\ln K_{eq}$ could be calculated using standard conditions ($T = 298$) and base 10 logs so that $RT^*2.303 = 1.36$ assuming that R is in cal

$$\Delta G^{\circ\prime} = 1.36 \ log_{10} \ K'_{eq} \ (using \ calories)$$

$$log_{10} \ K'_{eq} = \frac{(+2.2)}{(-1.36)}$$

$$= -1.62$$

$$K'_{eq} = anti\,log \ (-1.62)$$

$$= 2.40 \times 10^{-2} \ (the \ same \ answer.)]$$

The overall value of $\Delta G^{\circ\prime}$ for the formation of L-glycerol phosphate using ATP as a phosphoryl donor is equal to the sum of the free-energy values for the two individual reactions:

$$= (+9.2) + (-30.5) = -21.3 \ kJ/mol$$

The equilibrium constant for the overall reaction is

$$\ln K'_{eq} = \frac{-21.3}{(-2.47)} \text{ (as in the calculations above)}$$

$$= +8.62$$

$$K'_{eq} = 5.6 \times 10^3$$

[Using calories, and 1.36 as described above, we get

$$= (+2.2) + (-7.3) = -5.1 \, kcal / mol$$

$$log_{10} K'_{eq} = \frac{(-5.1)}{(-1.36)}$$

$$= +3.75$$

$$K'_{eq} = antilog \, 3.75$$

$$= 5.6 \times 10^3 \, (the \, same \, anwer.)]$$

The ratio of the two equilibrium constants is $5.6 \times 10^3 \div 2.4 \times 10^{-2} = 2.3 \times 10^5$.

2. The reaction is

hexose phosphate $+ H_2O \longrightarrow$ hexose $+ P_i$

The expression for the equilibrium constant is

$$K'_{eq} = \frac{[\text{hexose}][P_i]}{[\text{hexose phosphate}]} = \frac{99}{1}$$

Thus, $log_{10} K'_{eq}$ is approximately equal to 2 (note that $\ln 99 = 4.6$). Using the free-energy equation,

$$\Delta G^{\circ\prime} = RT \ln K'_{eq} = -(0.0083)(298) \ln 99 = -(2.47)(4.6) = -11.4 \, kJ / mol$$

[$\Delta G^{\circ\prime} = -1.36 \times 2$ (in calories)

$$= -2.72 \, kcal/mol$$

For the origin of the "1.36" term, see problem 1.]

3. (a)   To calculate the standard free-energy change for the phosphorolysis of sucrose, add the standard free-energy changes of the two reactions written so that their sum yields the required net reaction:

H$_2$O + sucrose $\longrightarrow$ glucose + fructose $\Delta G^{\circ\prime} = -29.3$ kJ/mol

$$\text{glucose} + P_i \longrightarrow \text{glucose 1-phosphate} + H_2O \qquad \Delta G^{o'} = +20.9 \text{ kJ/mol}$$

Net:

$$\text{sucrose} + P_i \longrightarrow \text{glucose 1-phosphate} + \text{fructose} \qquad \Delta G^{o'} = -8.4 \text{ kJ/mol}$$

$[H_2O + \textit{sucrose} \longrightarrow \textit{glucose} + \textit{fructose} \qquad \Delta G^{o'} = -7.0 \textit{ kcal/mol}$

$\textit{glucose} + P_i \longrightarrow \textit{glucose 1-phosphate} + H_2O \qquad \Delta G^{o'} = +5.0 \textit{ kcal/mol}$

Net:

$\textit{sucrose} + P_i \longrightarrow \textit{glucose 1-phosphate} + \textit{fructose} \qquad \Delta G^{o'} = -2.0 \textit{ kcal/mol}]$

(b) The equation that describes the relationship between the standard free energy and the equilibrium constant is

$$\Delta G^{o'} = -RT \ln K'_{eq} = -(0.0083)(298) \ln K'_{eq} = -2.47 \ln K'_{eq}$$

Solving for K'_{eq},

$$K'_{eq} = \text{antiln}(-\Delta G^{o'} / (2.47)) = \text{antiln}(-(-8.4)/ (2.47)) = e^{+3.4} = 30$$

Alternatively,

$$[\Delta G^{o'} = -1.36 \log_{10} K'_{eq} \text{ at } 25°C \text{ in calories}).$$

Solving for K'_{eq},

$$\log_{10} K'_{eq} = -\Delta G^{o'}/1.36$$

$$= -(-2.0/1.36) = +1.47$$

$$K'_{eq} = \textit{antilog } 1.47 = 30$$

For the origin of the "1.36" term see problem 1.]

4. (a) Although the enzyme controls the rate at which equilibrium is attained, it has no effect on the equilibrium constant, K'_{eq}. Because $\Delta G^{o'}$ is a function of the equilibrium constant, the action of the enzyme has no effect on its value.

(b) Under the conditions described in the text we know that

$$ADP + P_i + H^+ \longrightarrow ATP + H_2O \qquad \Delta G^{o'} = +30.5 \text{ kJ/mol}$$
$$\text{creatine phosphate} + H_2O \longrightarrow \text{creatine} + P_i + H^+ + \qquad \Delta G^{o} = -43.1 \text{ kJ/mol}$$

so adding these two reactions we know that

$$\text{creatine phosphate} + ADP \longrightarrow \text{creatine} + ATP \qquad \Delta G^{o'} = -12.6 \text{ kJ/mol}$$

$$\text{So } \Delta G = -12.6 \text{ kJ/mol} + RT \ln \frac{[\text{ATP}][\text{creatine}]}{[\text{ADP}][\text{creatine phosphate}]}$$

$$\Delta G = -12.6 \text{ kJ/mol} + (0.0083) \times (298) \ln \frac{[4 \text{ mM}][13 \text{ mM}]}{[0.03 \text{ mM}][25 \text{ mM}]}$$

Note that the brackets imply molar concentrations. We can use millimolar concentrations because the expression has the form $A \times B / C \times D$ and the concentrations cancel out. With expressions of the form $A \times B / C$ it is important to do the work in moles per liter and not mM.

$$\Delta G = -12.6 \text{ kJ/mol} + (2.47) \ln (160) = -12.6 + 12.5$$

$$\Delta G = -0.1 \text{ kJ/mol, practically zero}$$

At these concentrations, there is no free energy released in the reaction, and there is no net formation of ATP.

Alternatively,

$[ADP + P_i + H^+ \rightarrow ATP + H_2O$ ⠀⠀⠀⠀⠀⠀⠀ $\Delta G^{o\prime} = +7.3 \text{ kcal/mol}$

$creatine\ phosphate + H_2O \rightarrow creatine + P_i + H^+ +$ ⠀⠀ $\Delta G^o = -10.3 \text{ kcal/mol}$

so adding these two reactions we know that

$creatine\ phosphate + ADP \rightarrow creatine + ATP$ ⠀⠀⠀⠀⠀ $\Delta G^{o\prime} = -3.0 \text{ kcal/mol}$

So $\Delta G = -3.0 + RT\ 2.303 \log_{10} \dfrac{[ATP][creatine]}{[ADP][creatine\ phosphate]}$

$\Delta G = -3.0 \text{ kcal/mol} + 1.36 \log_{10} \dfrac{[4\ mM][13\ mM]}{[0.03\ mM][25\ mM]}$

$\Delta G = -3.0 + 1.36 \log_{10}(160) = -3.0 + 1.36(2.2)$

$\Delta G = -0.01 \text{ kcal/mol, practically zero}]$

(c) Under these conditions, because the sum of (creatine + creatine phosphate) remains the same, the concentration of creatine is 37 mM, and by the same logic the concentration of ADP increases to about 0.113 mM. (AMP levels will be constant and low enough to ignore here.) Using the free-energy equation,

$$\Delta G = -12.6 \text{ kJ/mol} + (0.0083)(298) \dfrac{[3.9\ mM][37\ mM]}{[0.113\ mM][1\ mM]}$$

$$\Delta G = -12.6 \text{ kJ/mol} + 2.47 \ln (1277) = -12.6 + 2.47 (7.15)$$

$$\Delta G = -12.6 + 17.7 = +5.1 \text{ kJ/mol}$$

Alternatively,

$[\Delta G = -3.0 \text{ kcal/mol} + 1.36 \log_{10} \dfrac{[3.9\ mM][37\ mM]}{[0.113\ mM][1\ mM]}$

$\Delta G = -3.0 \text{ kcal/mol} + 1.36 \log_{10}(1277) = -3.0 + 1.36(3.105)$

$\Delta G = -3 + 4.2 = +1.2 \text{ kcal/mol}]$

The positive value of ΔG shows that the reaction will not proceed toward the net formation of ATP, so that under these conditions, creatine phosphate does not serve as a donor of phosphoryl groups to ATP. Instead, the reaction proceeds toward the net formation of creatine phosphate, with phosphoryl groups donated from ATP.

5. Of the 18 carbon atoms in a saturated fatty acid, 17 are saturated (as –CH$_2$–groups) and are more reduced than the partially oxidized carbon atoms in glucose. In glucose, five of the six carbons are partially oxidized to the hydroxymethyl level, and the sixth is at the more oxidized aldehyde level. A greater number of electrons per carbon are available in the fatty acid, so more metabolic energy is available from it than from glucose.

6. The heat that is lost contributes to an increase in the entropy of the surroundings. A positive change in entropy means that the free energy for a catabolic process is more likely to be negative. Reactions with negative free-energy values are irreversible in that they require an input of energy to proceed in the opposite direction.

7. Look at Figure 15.10 in the text. Gasoline is largely composed of octane, which would be similar to the structure of the fatty acid shown—truncated to eight carbons and missing the carboxyl group on the left. "Sugar" could be thought of as the glucose structure shown. Octane is much more highly reduced than sugar, which has many hydroxyl groups. Sugar, or syrup, can't provide enough power to propel a vehicle. Octane also burns cleaner—sugars tend to form "caramel" when heated and oxidized, and this would mean that pistons would jam and valves would stick.

8. Anaerobic organisms cannot perform "stage III" of catabolism as shown in Figure 15.12. As a result, they obtain much less energy from each molecule of glucose ingested or synthesized. They utilize fuel very inefficiently compared to aerobic organisms. One molecule of glucose metabolized anaerobically yields only 2 ATP, whereas aerobic catabolism of glucose yields about 30 ATP. Obviously this disadvantage doesn't slow down the "germs"—they simply eat more food to stay alive.

9. Pyridine nucleotides, such as NADH, serve as acceptors and donors of electrons in many metabolic reactions, including those that generate energy for the cell. Because the absolute number of pyridine nucleotides in the cell is low, the cycle of oxidation and reduction for these compounds must occur rapidly for the oxidation of fuel molecules to proceed at a sufficient rate.

10. The conjugated π system in the isoalloxazine ring of oxidized flavins like FAD and FMN accounts for their intense yellow color. The reduced molecules are partially saturated, and their remaining double bonds are not conjugated, making them nearly colorless. To monitor glucose oxidase activity, one could use spectrophotometry to determine the wavelength of maximum absorption for FAD and then monitor the oxidation of the flavin in the enzyme by observing changes in absorption. As glucose is oxidized and FAD is reduced to FADH$_2$, one would observe a corresponding decrease in absorption.

11. For the synthesis of ATP to proceed spontaneously, the overall value of ΔG$^{o'}$ must be negative. The value of ΔG$^{o'}$ for ATP synthesis is positive, so a negative value would be expected for the oxidation of NADH to NAD$^+$.

12. The text's answer to this problem shows that the cleavage of pyrophosphate ensures that the coupled reactions will proceed toward the net formation of desired product; that is, the overall reaction will have a rather large negative free-energy value. The fact that PP$_i$ is formed during the synthesis of UDP-glucose suggests that the free energy released by the coupled reactions for the formation of UDP-glucose and the hydrolysis of UTP is

small. Therefore, the free energy of the hydrolysis of UDP-glucose would be similar to that of UTP. Thus, you should surmise that the group transfer potential of glucose from UDP-glucose would be high. This is the case, as UDP-glucose serves as a donor of glucose residues for the synthesis of glycogen.

13. If you look at older biochemistry textbooks, you will see that many of them list the $\Delta G^{o\prime}$ for pyrophosphate hydrolysis as a much more negative number, between -29 and -33 kJ/mol (-7 and -8 kcal/mol). A recent paper by Perry A. Frey (*Biochem.* 34[1995]:11307) shows that the real driving force in such reactions is the high energy of the α, β phosphoanhydride bond. In other words, it is the first reaction ATP $\rightarrow$ AMP + PP_i with a $\Delta G^{o\prime}$ of about -64.9 kJ/mol (-15.5 kcal/mol), which provides the driving force for such reactions. The subsequent pyrophosphatase step $PP_i \rightarrow 2P_i$ (-19.2 kJ/mol)(-4.6 kcal/mol) makes only a relatively minor contribution to functional irreversibility.

14. AMP and mercaptoethylamine are not listed as vitamins because they can be synthesized de novo from other precursors in cells. Pantothenate is required in the diet because one or more of the biochemical steps needed to synthesize it are deficient in higher organisms. All three components of coenzyme A can be put together by a cell to form the required cofactor.

15. In cells, there is constant synthesis and degradation of enzymes in response to the need for enzymatic activity. Enzyme degradation through proteolysis will often release coenzymes like biotin. Although some biotin molecules can be incorporated into newly synthesized proteins, others are carried by the blood to the kidney, where they are then excreted. Daily excretion of coenzymes leads to a requirement for their replenishment in the diet.

16. The point here is that most steps in this pathway fall into a familiar pattern. The oxidation of isovaleryl CoA produces a C=C bond, and the cofactor is FAD just as in reaction 1 on page 434 of the text. The subsequent carboxylation with ATP as a cofactor resembles reaction 3. Browsing through the many chapters on metabolic pathways would reveal many similar examples.

17. Regulation of the first reaction in a biosynthetic pathway ensures that the intermediates in the pathway will be synthesized only when the ultimate product is required. In this way the cell can conserve energy as well as precursors of all intermediates. Such a regulatory scheme will be found when none of the intermediates are utilized in other pathways.

18. In Table 15.2 in the text, we also see uridine diphosphate glucose and cytidine diphosphate diacylglycerol. So besides the ADP in common redox cofactors and CoA, we also have CDP and UDP. These "fossils" make a convincing case that RNA played a greatly expanded role in the distant past. But just because we find dinosaur bones, and they resemble chicken bones, we can't conclude that dinosaurs (or chickens) were the first form of life on Earth. There is positive evidence for an RNA world, but there isn't really negative evidence showing a complete lack of protein during or before this phase of evolution.

19. Coenzyme A is a thiol that reacts with carboxylic acids (such as pyruvate from glycolysis and fatty acids, as we shall see) to form a thioester. It serves as a carrier of acyl groups and it activates carboxylic acids for further reaction. Commonly, this acyl group is an acetyl group and the molecule formed is Acetyl-CoA. Two important roles for Acetyl CoA that will emerge are its shuttling carbon atoms from glycolysis or from fatty acids from the cytoplasm to the mitochondria to be oxidized. One advantage of

esterifying acyl groups to coenzyme A is that the hydrolysis of a thioester has a large negative $\Delta G°'$ value (-31.4 kJ/mol or -7.5 kcal/mol). This exergonic reaction can then be used to drive other reactions forward.

20.

21. (a) Panthothenic acid (vitamin B_5)—part of coenzyme A. Coenzyme A is a cofactor that carries two-carbon acetyl groups, activating them as a thioester.

(b) Biotin—used to carry CO_2 groups.

(c) Niacin or Nicotinate—precursor to the nicotinamide ring that stores electrons in NADH and NADPH.

(d) Creatine-used to create creatine phosphate, a molecule with a high phosphoryl potential that serves as a reservoir of phosphoryl groups to transfer to ADP in muscle during strenuous exercise.

(e) Riboflavin (Vitamin B_2)—precursor to the isoalloxazine ring of Flavin Adenine Dinucleotides (FAD, $FADH_2$) and Flavine Mononucleotides (FMN, $FMNH_2$), used to carry electrons.

22. NAD^+ tends to participate in reactions in which it oxidizes alcohols to carboxylic acids, producing NADH and H^+. FAD participates in the oxidation of alkanes to alkenes, producing $FADH_2$.

Glycolysis and Gluconeogenesis

Chapter 16 examines one of the most well-studied metabolic pathways—the metabolism of carbohydrates via the glycolytic and gluconeogenic pathways. Glycolysis is a series of reactions that converts glucose into pyruvate with the concomitant trapping of a portion of the energy as ATP. Gluconeogenesis, on the other hand, is a biosynthetic pathway that generates glucose from noncarbohydrate precursors. The chapter begins with glycolysis, a classic metabolic pathway whose study ushered in biochemistry as a discipline separate from chemistry. The glycolytic pathway can be broken down into three distinct stages: (1) the conversion of glucose into fructose 1,6-bisphosphate; (2) cleavage of fructose-1,6-biphosphate into triose phosphate intermediates; and (3) the oxidation of the three-carbon fragments into pyruvate, leading to the formation of ATP. The authors discuss the individual reactions within each stage, along with some of the reaction mechanisms and enzyme structures of particular interest.

After summarizing the energetics of glycolysis, the authors discuss the various fates of pyruvate (conversion to ethanol, lactate or acetyl CoA), which differs depending on the organism, cell type, and metabolic state. In addition to glucose, fructose and galactose can also be oxidized by enzymes in the glycolytic pathway, and their mode of entry into glycolysis is described, as are the physiological results of defects in lactose and galactose metabolism. The regulation of glycolysis by the enzymes that catalyze the irreversible reactions in the pathway is discussed next. Phosphofructokinase, the most prominent regulatory enzyme in glycolysis, is examined in detail. Hexokinase and pyruvate kinase, two other important glycolytic regulatory enzymes, are also discussed. The discussion of glycolysis concludes with a description of the family of glucose transporters as examples of the ability of isoforms of proteins to perform diverse and specialized functions.

The chapter concludes by discussing the process of gluconeogenesis, or the synthesis of glucose from noncarbohydrate precursors such as lactate, amino acids, and glycerol. Gluconeogenesis is not simply a reversal of glycolysis, due to the fact that the equilibrium of glycolysis lies far on the side of pyruvate formation. The steps of glycolysis that lie near equilibrium are used in gluconeogenesis, and three new steps are substituted for those that are essentially irreversible. The authors discuss these three new steps in detail, in which (1) phosphoenolpyruvate is produced from pyruvate, in a two-step reaction with ox-aloacetate as an intermediate; (2) fructose-6-phosphate is synthesized from fructose-1,6-biphosphate; and finally (3) glucose is produced from glucose-6-phosphate. The authors emphasize the reciprocal regulation of these two pathways, ensuring that cells respond quickly to the need for energy.

LEARNING OBJECTIVES

When you have mastered this chapter, you should be able to accomplish the following objectives.

Introduction

1. Define *glycolysis* and explain its role in the generation of *metabolic energy*.

2. List the alternative end points of the glycolytic degradation of *glucose*.

3. Define *gluconeogenesis* and explain how it differs from a simple reversal of glycolysis.

4. Outline the early work in delineating the glycolytic pathway.

5. Discuss reasons why glucose is such an important fuel for most organisms.

Glycolysis Is an Energy-Conversion Pathway in Many Organisms (Text Section 16.1)

6. Outline the three stages of glycolysis.

7. Discuss the *induced-fit rearrangements* that occur in hexokinase upon glucose binding and list two important consequences of this step in glycolysis.

8. Describe the steps in the conversion of glucose to *fructose 1,6-bisphosphate,* including all the intermediates and enzymes. Note the steps where *ATP* is consumed.

9. List the reactions that convert fructose 1,6-bisphosphate, a hexose, into the triose *glyceraldehyde 3-phosphate*. Summarize the most important features of the catalytic mechanism of *triosephosphate isomerase I*.

10. Outline the steps in glycolysis between glyceraldehyde 3-phosphate and *pyruvate*. Recognize all the intermediates and enzymes and the cofactors that participate in the ATP-generating reactions. Summarize the most important features of the catalytic mechanism of *glyceraldehyde 3-phosphate dehydrogenase*.

11. Explain the role of 2,3-BPG in the interconversion of 3-phosphoglycerate and *2-phosphoglycerate*.

12. Explain the role of the *enol to ketone conversion* in the *phosphoryl transfer* catalyzed by *pyruvate kinase*.

13. Write the net reaction for the transformation of glucose into pyruvate and enumerate the ATP and NADH molecules formed.

14. Outline the reactions for the conversion of pyruvate into *ethanol, lactate,* or *acetyl CoA*. Explain the role of *alcoholic fermentation* and lactate formation in the regeneration of NAD$^+$.

15. Describe the structure of the NAD⁺-binding region common to many *NAD⁺-linked dehydrogenases.*

16. Outline the pathways for the conversion of *fructose* and *galactose* into *glyceraldehyde 3-phosphate* and *glucose-6-phosphate*, respectively. Note the role of *UDP-activated sugars.*

17. Describe the biochemical defects in *lactose intolerance* and *galactosemia.*

The Glycolytic Pathway Is Tightly Controlled (Text Section 16.2)

18. Identify the features of a regulated enzyme in a metabolic pathway and list the regulated enzymes in glycolysis.

19. Describe the allosteric regulation of *phosphofructokinase* including the reason AMP is used as a positive regulator rather than ADP. Explain the role of *fructose 2,6-bisphosphate* in its regulation.

20. Discuss the regulation of *hexokinase.* Contrast the properties and physiologic roles of hexokinase and *glucokinase.*

21. Describe the regulation of the isozymes of *pyruvate kinase.*

22. Compare the regulation of glycolysis in skeletal muscle cells to that in liver cells.

23. Describe the features of the five different isozymes of *glucose transporters.*

24. Define HIF-1 and describe the role it plays in some tumors and in physical exercise training.

Glucose Can Be Synthesized from Noncarbohydrate Precursors (Text Section 16.3)

25. Describe the physiologic significance of *gluconeogenesis.* List the primary precursors of gluconeogenesis.

26. Describe the enzymatic steps in the conversion of *pyruvate* to *phosphoenolpyruvate.* Name the enzymes, intermediates, and cofactors involved in these reactions.

27. Name the major organs that carry out gluconeogenesis. Locate the various enzymes of gluconeogenesis in cell compartments.

28. Explain the role of *biotin* as a carrier for *activated CO_2* in the *pyruvate carboxylase* reaction. Describe the control of pyruvate carboxylase by *acetyl CoA* and its role in maintaining the level of citric acid cycle intermediates.

29. Calculate the number of high-energy phosphate bonds consumed during gluconeogenesis and compare it with the number formed during glycolysis.

Gluconeogenesis and Glycolysis Are Reciprocally Regulated (Text Section 16.4)

30. Describe the coordinated control of the enzymes in glycolysis and gluconeogenesis. Include a discussion of the effects of the hormones *insulin* and *glucagon.*

31. Describe the *fused-domain structure* of *phosphofructokinase 2 (PFK2)/fructose bisphosphatase 2* that forms and degrades fructose 2,6-bisphosphate. Describe the reciprocal regulation of the enzyme.

32. Explain how *substrate cycles* may *amplify metabolic signals* or *produce heat.*

33. Outline the *Cori cycle* and explain its biological significance.

34. Contrast the properties and roles of the *H* and *M isozymes* of *lactate dehydrogenase.*

SELF-TEST

Introduction

1. Which of the following are reasons why glucose is so prominent (relative to other monosaccharides) as a metabolic fuel?

    (a) It has a relatively low tendency to nonenzymatically glycosylate proteins.
    (b) It can be formed from formaldehyde under prebiotic conditions.
    (c) It has a strong tendency to stay in the ring formation.
    (d) Its oxidation yields more energy than other monosaccharides.

2. What are the three primary fates of pyruvate?

Glycolysis Is an Energy-Conversion Pathway in Many Organisms

3. For each of the following types of chemical reactions, give one example of a glycolytic enzyme that carries out such a reaction.

    (a) aldol cleavage
    (b) dehydration
    (c) phosphoryl transfer
    (d) phosphoryl shift
    (e) isomerization
    (f) phosphorylation coupled to oxidation

4. Which of the following answers completes the sentence correctly? Hexokinase

    (a) catalyzes the conversion of glucose 6-phosphate into fructose 1,6-bisphosphate.
    (b) requires Ca^{2+} for activity.
    (c) uses inorganic phosphate to form glucose 6-phosphate.
    (d) catalyzes the transfer of a phosphoryl group to a variety of hexoses.
    (e) catalyzes a phosphoryl shift reaction.

5. During the phosphoglucose isomerase reaction, the pyranose structure of glucose 6-phosphate is converted into the furanose ring structure of fructose 6-phosphate. Does this conversion require an additional enzyme? Explain.

6. The steps of glycolysis between glyceraldehyde 3-phosphate and 3-phosphoglycerate involve all of the following except

    (a) ATP synthesis.
    (b) utilization of P_i.
    (c) oxidation of NADH to NAD^+.
    (d) formation of 1,3-bisphosphoglycerate.
    (e) catalysis by phosphoglycerate kinase.

7. In the mechanism of G3PDH, glyceraldehyde-3-phosphate forms a covalent bond with which amino acid residue of the enzyme?

    (a) lysine
    (b) serine
    (c) cysteine
    (d) glutamate
    (e) histidine

8. Why are there bubbles in beer and champagne? Why does bread "rise"?

9. Which of the following answers complete the sentence correctly? The phosphofruc-tokinase and the pyruvate kinase reactions are similar in that
   (a) both generate ATP.
   (b) both involve a "high-energy" sugar derivative.
   (c) both involve three-carbon compounds.
   (d) both are essentially irreversible.
   (e) both enzymes undergo induced-fit rearrangements after binding of the substrate.

10. The reaction phosphoenolpyruvate + ADP + H^+ (pyruvate + ATP has a $\Delta G^{o\prime} = -31.4$ kJ/mol and a $\Delta G' = -16.7$ kJ/mol under physiologic conditions. Explain what these free-energy values reveal about this reaction.

11. If the C-1 carbon of glucose were labeled with ^{14}C, which of the carbon atoms in pyruvate would be labeled after glycolysis?
   (a) the carboxylate carbon
   (b) the carbonyl carbon
   (c) the methyl carbon

12. Starting with fructose 6-phosphate and proceeding to pyruvate, what is the net yield of ATP molecules?
   (a) 1   (d) 4
   (b) 2   (e) 5
   (c) 3

13. Which of the following statements about triosphosphate isomerase (TIM) is NOT true?
   (a) The mechanism of action of TIM involves a conformational change in the structure of the enzyme that prevents escape of an activated intermediate.
   (b) The rate-limiting step in the reaction catalyzed by TIM is the release of the product glyceraldehyde 3-phosphate.
   (c) The $kcat/K_M$ ratio for the reaction catalyzed by TIM is close to the diffusion-controlled limit for a bimolecular reaction.
   (d) The isomerization of a hydrogen atom from one carbon atom to another in the TIM-catalyzed reaction is assisted by a base (the γ-carboxyl of a glutamate residue) in the enzyme.
   (e) TIM catalyzes an intramolecular oxidation-reduction reaction.

14. Since lactate is a "dead-end" product of metabolism in the sense that its sole fate is to be reconverted into pyruvate, what is the purpose of its formation?

15. Galactose metabolism involves the following reactions: (1) galactose + ATP $\longrightarrow$ galactose 1-phosphate + ADP + H^+; (2) ?; (3) UDP-galactose $\longrightarrow$ UDP-glucose.
   (a) Write the reaction for step 2.
   (b) Which step is defective in galactosemia?
   (c) Which enzymes catalyze steps 1, 2, and 3?

The Glycolytic Pathway Is Tightly Controlled

16. The essentially irreversible reactions that control the rate of glycolysis are catalyzed by which of the following enzymes?
   (a) pyruvate kinase
   (b) aldolase
   (c) glyceraldehyde 3-phosphate
   (d) phosphofructokinase
   (e) hexokinase
   (f) phosphoglycerate kinase dehydrogenase

17. Both phosphofructokinase and pyruvate kinase are inhibited by high levels of ATP. Why is this logical?

18. In which of the following is the enzyme correctly paired with its allosteric effector?

    (a) hexokinase: ATP
    (b) phosphofructokinase: glucose 6-phosphate
    (c) pyruvate kinase: alanine
    (d) phosphofructokinase: AMP
    (e) glucokinase: fructose 2,6-bisphosphate

19. Match hexokinase and glucokinase with the descriptions from the right column that are appropriate.

    (a) hexokinase
    (b) glucokinase

    (1) is found in the liver
    (2) is found in nonhepatic tissues
    (3) is specific for glucose
    (4) has a broad specificity for hexoses
    (5) requires ATP for reaction
    (6) has a high K_M for glucose.
    (7) is inhibited by glucose 6-phosphate

20. Which of the following statements about glucose transporters is NOT true?

    (a) They are transmembrane proteins.
    (b) They accomplish the movement of glucose across animal cell plasma membranes.
    (c) Their tissue distribution and concentration can depend on the tissue type and metabolic state of the organism.
    (d) Their glucose binding site is moved from one side of the membrane to the other by rotation of the entire protein.
    (e) They constitute a family of five isoforms of a protein.

Glucose Can Be Synthesized from Noncarbohydrate Precursors

21. Which of the following statements about gluconeogenesis are true?

    (a) It occurs actively in the muscle during periods of exercise.
    (b) It occurs actively in the liver during periods of exercise or fasting.
    (c) It occurs actively in adipose tissue during feeding.
    (d) It occurs actively in the kidney during periods of fasting.
    (e) It occurs actively in the brain during periods of fasting.

22. Glucose can be synthesized from which of the following noncarbohydrate precursors?

    (a) adenine
    (b) alanine
    (c) lactate
    (d) palmitic acid
    (e) glycerol

23. Which statement about glucose 6-phosphatase is true? Glucose 6-phosphatase

    (a) is bound to the inner mitochondrial membrane.
    (b) requires an associated Ca^{2+}-binding protein for activity.
    (c) is directly associated with a glucose transporter.
    (d) produces glucose and phosphate in a reaction that consumes energy.
    (e) has an identical active site with hexokinase.

24. Figure 16.1 below shows the sequence of reactions of gluconeogenesis from pyruvate to phosphoenolpyruvate. Match the capital letters indicating the reactions of the gluoneogenic pathway with the following statements:

FIGURE 16.1 Reactions of gluconeogenesis from pyruvate to phosphoenolpyruvate.

Pyruvate → oxaloacetate → malate → oxaloacetate → phosphoenolpyruvate
　　　　　A　　　　　B　　　　C　　　　　　D

(a) occurs in the mitochondria
(b) occurs in the cytosol
(c) produces CO_2
(d) consumes CO_2

(e) requires ATP
(f) requires GTP
(g) is regulated by acetyl CoA
(h) requires a biotin cofactor

25. How many "high-energy" bonds are required to convert oxaloacetate to glucose?

(a) 2
(b) 3
(c) 4

(d) 5
(e) 6

Gluconeogenesis and Glycolysis Are Reciprocally Regulated

26. Which of the following statements correctly describe what happens when acetyl CoA is abundant?

(a) Pyruvate carboxylase is activated.
(b) Phosphoenolpyruvate carboxykinase is activated.
(c) Phosphofructokinase is activated.
(d) If ATP levels are high, oxaloacetate is diverted to gluconeogenesis.
(e) If ATP levels are low, oxaloacetate is diverted to gluconeogenesis.

27. When blood glucose levels are low, glucagon is secreted. Which of the following are the effects of increased glucagon levels on glycolysis and related reactions in liver?

(a) Phosphorylation of phosphofructokinase 2 and fructose bisphosphatase 2 occurs.
(b) Dephosphorylation of phosphofructokinase 2 and fructose bisphosphatase 2 occurs.
(c) Phosphofructokinase is activated.
(d) Phosphofructokinase is inhibited.
(e) Glycolysis is accelerated.
(f) Glycolysis is slowed down.

28. In the coordinated control of phosphofructokinase (PFK) and fructose 1,6-bisphosphatase (F-1,6-BPase),

(a) citrate inhibits PFK and stimulates F-1,6-BPase.
(b) fructose 2,6-bisphosphate inhibits PFK and stimulates F-1,6-BPase.
(c) acetyl-CoA inhibits PFK and stimulates F-1,6-BPase.
(d) AMP inhibits PFK and stimulates F-1,6-BPase.
(e) NADPH inhibits PFK and stimulates F-1,6-BPase.

29. Indicate which of the conditions listed in the right column *increase* the activity of the glycolysis or gluconeogenesis pathways.

(a) glycolysis
(b) gluconeogenesis

(1) increase in ATP
(2) increase in AMP
(3) increase in F-2,6-BP
(4) increase in citrate
(5) increase in acetyl-CoA
(6) increase in insulin
(7) increase in glucagon
(8) starvation
(9) fed state

30. Which of the following statements about the Cori cycle and its physiologic consequences are true?
    (a) It involves the synthesis of glucose in muscle.
    (b) It involves the release of lactate by muscle.
    (c) It involves lactate synthesis in the liver.
    (d) It involves ATP synthesis in muscle.
    (e) It involves the release of glucose by the liver.

31. Which part of glycolysis appears to be evolutionarily "oldest": the hexose portion or the triose portion?

ANSWERS TO SELF-TEST

1. a, b, c

2. ethanol, lactate, and CO_2/water

3. (a) aldolase
   (b) enolase
   (c) hexokinase, phosphofructokinase, phosphoglycerate kinase, or pyruvate kinase
   (d) phosphoglycerate mutase
   (e) phosphoglucose isomerase, triosephosphate isomerase
   (f) glyeraldehyde 3-phosphate dehydrogenase

4. d

5. No. The reaction catalyzed by phosphoglucose isomerase is a simple isomerization between an aldose and a ketose and involves the open-chain structures of both sugars. Since glucose 6-phosphate and fructose 6-phosphate are both reducing sugars, their Haworth ring structures are in equilibrium with their open-chain forms. This equilibration is very rapid and does not require an additional enzyme. Note that this isomerization reaction is of the same type as that catalyzed by triosephosphate isomerase.

6. c

7. (c) cysteine. The resulting thioester bond is important catalytically.

8. See text Figure 16.9. Anaerobic glycolysis in yeast produces ethanol and carbon dioxide gas. Louis Pasteur's studies on this alcoholic fermentation gave rise to the entire science of biochemistry. So capping a bottle before the fermentation is finished gives rise to carbonation, or bubbles. The bubbles in champagne are generally "real" carbonation produced directly by yeast. Beer is generally fermented to completion and then artificially carbonated, but the bubbles originated in the same way. In bread dough, the alcohol cooks off when the bread is baked but it is there and it gives a sweet smell to the rising dough. The bubbles that make the bread rise are CO_2, just as in beer and champagne.

9. d

10. The large negative $\Delta G^{o'}$ value indicates that equilibrium favors product formation by a very large margin. The -16.7 kJ/mol value for $\Delta G'$ means that under physiologic conditions the reaction will also proceed toward product formation essentially irreversibly. The fact that $\Delta G'$ has a smaller negative value than $\Delta G^{o'}$ indicates that under physiologic conditions the ratio of the concentrations of products over reactants is considerably smaller than in the standard state.

11. c

12. c

13. b. The rate-limiting step of the reaction, which by definition can be no faster than the rate at which the product appears, is the diffusion-controlled encounter of the substrate with the enzyme, not the release of product.

14. The reduction of pyruvate to lactate converts NADH to NAD^+, which is required in the glyceraldehyde 3-phosphate dehydrogenase reaction. This prevents glycolysis from stopping owing to too low a concentration of NAD^+ and allows continued production of ATP.

15. (a) Galactose 1-phosphate + UDP-glucose $\longrightarrow$ glucose 1-phosphate + UDP-galactose
    (b) Step 2 is defective in galactosemia.
    (c) Galactokinase catalyzes step 1; galactose 1-phosphate uridyl transferase catalyzes step 2; and UDP-galactose 4-epimerase catalyzes step 3.

16. a, d, e

17. Control points are generally the irreversible steps in a pathway, and for glycolysis that means the kinase enzymes. Catabolic pathways produce ATP by breaking down food metabolites, so it is appropriate for any catabolic pathway to be inhibited by a high energy charge, or high ATP. If you have enough ATP you should stop "burning" your fuel supply; the concept is like a thermostat controlling the temperature in a house. If your house is at 80°F, turn off the furnace!

18. a, c, d

19. (a) 1, 2, 4, 5, 7   (b) 1, 3, 5, 6

20. d

21. b, d

22. b, c, e

23. b. The other answers are incorrect because glucose 6-phosphatase is bound to the luminal side of the endoplasmic reticulum membrane. It is associated with the glucose 6-phosphate and phosphate transporters, but not with the glucose transporter. The hydrolysis of glucose 6-phosphate is an exergonic reaction. The active site of hexokinase is distinct from that of the phosphatase, since the hexokinase binds ATP.

24. (a) A, B   (b) C, D   (c) D   (d) A   (e) A   (f) D   (g) A   (h) A

25. c. The two steps in gluconeogenesis that consume GTP or ATP are:

    Oxaloacetate + GTP $\longrightarrow$ phosphoenolpyruvate + GDP + CO_2

    3-Phosphoglycerate + ATP $\longrightarrow$ 1,3-bisphosphoglycerate + ADP

    Since two oxaloacetate molecules are required to synthesize one glucose molecule, a total of four "high-energy" bonds are required.

26. a, d

27. a, d, f

28. a

29. (a) 2, 3, 6, 9   (b) 1, 4, 5, 7, 8

30. b, d, e

31. According to the last paragraph of the textbook chapter, the triose portion of glycolysis is absolutely universal, unlike the hexose portion, which is missing in some species, particularly among the archaea. The fact that the triose portion is universal means that it must have been present in LUCA, the Last Universal Common Ancestor, and hence it is probably at least 3 billion years old.

PROBLEMS

1. Triose Phosphate Isomerase (TPI) as shown in the text Figure 16.4 has an eight-fold alpha-beta barrel supporting the active site. According to the text, what special properties does this enzyme have? How much can be linked to the structure shown?

2. The text states that lactose intolerance in adult humans is the "normal" state, and that the mutation that allows for lactose tolerance should not be older than 10,000 years because that is when the practice of dairying started. Can you think of a way that this can be proven, that a certain mutation is relatively "young"? Hint, it involves DNA and chromosomes.

3. Inorganic phosphate labeled with ^{32}P is added with glucose to a glycogen-free extract from liver, and the mixture is then incubated in the absence of oxygen. After a short time, 1,3-bisphosphoglycerate (1,3-BPG) is isolated from the mixture. On which carbons would you expect to find radioactive phosphate? If you allow the incubation to continue for a longer period, will you find any change in the labeling pattern? Why?

4. Mannose, the 2-epimer of glucose (Figure 11.2 in text), and mannitol, a sugar alcohol, are widely used as dietetic sweeteners. Both compounds are transported only slowly across plasma membranes, but they can be metabolized by the liver. Propose a scheme by which mannitol and mannose can be converted into intermediates of the glycolytic pathway. You may wish to take advantage of the fact that hexokinase is relatively nonspecific. Why should such sugars be brought into glycolysis as early in the sequence as possible?

5. The value of $\Delta G^{o\prime}$ for the hydrolysis of sucrose to glucose and fructose is -31.4 kJ/mol. You have a solution that is 0.10 M in glucose and that contains sufficient sucrase enzyme to bring the reaction rapidly to equilibrium.
   (a) What concentration of fructose would be required to yield sucrose at an equilibrium concentration of 0.01 M, at 25°C?
   (b) The solubility limit for fructose is about 3.0 M. How might this limit affect your experiment?

6. In 1905, Harden and Young, two English chemists, studied the fermentation of glucose using cell-free extracts of yeast. They monitored the conversion of glucose to ethanol by measuring the evolution of carbon dioxide from the reaction vessel. In one set of experiments, Harden and Young observed the evolution of CO_2 when inorganic phosphate (P_i) was added to a yeast extract containing glucose. In the graph in Figure 16.2, curve A shows what happens when no P_i is added. Curve B shows the effect of adding P_i in a separate experiment. As the evolution of CO_2 slows with time, more P_i is added to stimulate the reactions; this is shown in curve C.
   (a) Why is glucose fermentation dependent on P_i?
   (b) During fermentation, what is the ratio of P_i consumed to CO_2 evolved?
   (c) How does the formation of ethanol ensure that the fermentation process is in redox balance?
   (d) Harden and Young found that they could recover phosphate from the reaction mixture, but it was not precipitable by magnesium citrate, as is P_i. Name at least three organic compounds that would be phosphorylated when P_i is added to the fermenting mixture.
   (e) As the rate of CO_2 evolution decreased, Harden and Young found that an unusual compound accumulated in the reaction mixture. In 1907, Young identified the compound as a hexose bisphosphate. Name the compound and explain why it might accumulate when P_i becomes limiting.

(f) Later, Meyerhof showed that the addition of adenosine triphosphatase (ATPase, an enzyme that hydrolyzes ATP to yield ADP and P_i) to the reaction mixture stimulates the evolution of CO_2. Explain this result.

FIGURE 16.2 Evolution of CO_2 in the Harden-Young experiment.

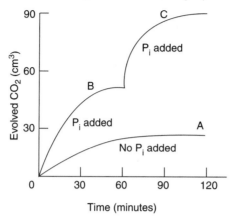

7. Ahlfors and Mansour studied the activity of purified sheep phosphofructokinase (PFK) as a function of the concentration of ATP in experiments that were carried out at a constant concentration of fructose 6-phosphate. Typical results are shown in Figure 16.3. Explain these results, and relate them to the role of PFK in the glycolytic pathway.

FIGURE 16.3 The effects of ATP concentration on sheep PFK.

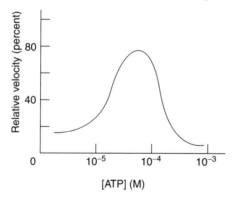

6. Several researchers have mutated Cys149 in glyceraldehyde 3-phosphate dehydrogenase (GAPDH) to determine the role of the side chain in the GAPDH reaction. Upon mutation of Cys149 to serine, the catalytic activity of GAPDH decreases over 10^4-fold. Muller and Branlant (*Arch. Biochem. Biophys.* 363[1999]:259–266) looked at how the dehydrogenase activity changed with pH. They found that base titration of the wild-type enzyme resulted in an increase in activity with a single pK_a near 7.2, while the activity of a Cys149 to serine mutant increased linearly with pH, and did not show a titratable proton below pH 10.

(a) What can you conclude about the wild-type protein from the pH study?
(b) Since serine acts as a nucleophile in the serine protease mechanism (see text Section 9.1), why is it unable to catalyze the dehydrogenase reaction?
(c) What effect would you expect such a mutation to have on glycolysis in living cells?

9. Hexokinase catalyzes the formation of glucose 6-phosphate at a maximum velocity of 2×10^{-5} mol/min, whereas V_{max} for the formation of fructose 6-phosphate is 3×10^{-5} mol/min. The value of K_M for glucose is 10^{-5} M, whereas K_M for fructose is 10^{-3} M. Suppose that in a particular cell the observed rates of phosphorylation are 1.0×10^{-8} mol/min for glucose and 1.5×10^{-5} mol/min for fructose.

   (a) Estimate the concentrations of glucose and fructose in the cell.
   (b) Which of these hexoses is more important in generating energy for this cell?

10. Galactose is an important component of glycoproteins. Explain why withholding galactose from the diet of galactosemic patients has no effect on their synthesis of glycoproteins.

11. For the glyceraldehyde 3-phosphate dehydrogenase reaction, explain how the oxidation of the aldehyde group ultimately gives rise to an acyl phosphate product.

12. Explain the high phosphoryl group-transfer potential of phosphoenolpyruvate as it is displayed in the pyruvate kinase reaction.

13. Glycerol can enter the glycolytic pathway through phosphorylation to glycerol 3-phosphate, catalyzed by glycerol kinase, and then by oxidation to dihydroxyacetone phosphate, catalyzed by glycerol 3-phosphate dehydrogenase. Glyceraldehyde is funneled into the glycolytic pathway through phosphorylation to glyceraldehyde 3-phosphate, catalyzed by triose kinase. Glycerate enters the glycolytic pathway when it is phosphorylated to 3-phosphoglycerate by glycerate kinase. Lactate-forming bacteria can metabolize glycerol, glyceraldehyde, or glycerate in the presence of oxygen, but only one of these substrates can be converted to lactate under anaerobic conditions. Which one, and why?

14. Hexokinase and glucokinase are isozymes, which is to say that they are related and they both catalyze the phosphorylation of glucose at carbon 6. Hexokinase is the first enzyme of glycolysis and is thus more or less universal in living cells. Why do we need a different isozyme, found only in certain organs (liver and pancreas)?

15. The textbook points out that both certain kinds of cancer and anaerobic exercise produce similar responses. Both result in *hypoxia-inducible transcription factor* (HIF-1). How does this make sense? Can you see a parallel logic for development of a tumor and development of stronger muscles?

16. Aminotransferases are enzymes that catalyze the removal of amino groups from amino acids to yield α-keto acids. How could the action of such enzymes contribute to gluconeogenesis? Consider the utilization of alanine, aspartate, and glutamate in your answer.

17. Even if the concentration of lactate and other precursors were high, why is it unlikely that liver cells would be carrying out gluconeogenesis under anaerobic conditions?

18. Explain why a CO_2 is added to pyruvate in the pyruvate carboxylase reaction only to be subsequently removed by the phosphoenolpyruvate carboxykinase reaction. Identify the high-energy intermediate in the carboxylation reaction.

19. In liver, V_{max} for fructose bisphosphatase is three to four times higher than V_{max} for phosphofructokinase, whereas in muscle it is only about 10 percent of that of phosphofructokinase. Explain this difference.

20. In muscle, lactate dehydrogenase produces lactate from pyruvate, whereas in the heart it preferentially synthesizes pyruvate from lactate. Explain how this is possible.

21. Briefly explain what the purpose is of each of the following steps in glycolysis? In other words, why are these steps necessary or useful?

a) Glucose → Glucose-6-P
b) Glucose-6-P → Fructose-6-phosphate
c) Dihydroxyacetone-phosphate → Glyceraldehyde-3-phosphate
d) 2-Phosphoglycerate → Phosphoenolpyruvate

ANSWERS TO PROBLEMS

1. Besides looking nice, the alpha-beta barrel confers some stability on the residues of the active site, which probably contributes to the speed and efficiency of the enzyme. The text says it is a "kinetically perfect enzyme," and having the active site residues in the correct position must help the speed of the reaction. Furthermore the loop shown in red in Figure 16.4 serves as a "door" to the active site, which prevents release of the harmful side product methyl glyoxal. Form and function are always very closely linked in biochemistry.

2. Even students who have not studied DNA in detail might know that there is a steady level of recombination in chromosomes. A part of one chromosome is swapped for part of another chromosome. So in the very first generation with a helpful gene (like lactose tolerance in adults)—say a woman has this mutation, and several of her children have it. At that time, everyone with the gene will have the entire chromosome, the whole chromosome will be 100% the same. Over the centuries, bits of that chromosome will be swapped away, but the portion that contains the useful gene will often cause the survival of that individual (during a famine) and thus be passed on. The region passed on with the gene becomes smaller and smaller over time. What is interesting about the lactose tolerance mutation is that the northern European version of the gene is surrounded by a block of about a *million* base pairs. This is interpreted to imply that the mutation could have occurred as recently as 2000 years ago (Bersaglieri, T. et al., *Am. J. Hum. Genet.* 74 (2004) p. 1111). This matches the time frame mentioned in the text (less than 10,000 years) very nicely.

3. After a short incubation time, labeled phosphate will be found on C-1 of 1,3-BPG. Inorganic phosphate enters the glycolytic pathway at the step catalyzed by glyceraldehyde 3-phosphate dehydrogenase. However, after a longer incubation time, the radioactive label will be found on both C-1 and C-3 of 1,3-BPG because the step subsequent to the formation of 1,3-BPG involves the phosphorylation of ADP to form ATP, which will be radioactively labeled in the γ-phosphoryl group. In other glycolytic reactions, the radioactively labeled ATP can phosphorylate at C-1 of fructose 6-phosphate and C-6 of glucose, both of which are equivalent to C-3 in 1,3-BPG. Thus, after prolonged incubation, both labeled inorganic phosphate and labeled ATP will be present in the mixture, and 1,3-BPG with a radioactive label at both C-1 and C-3 will be present in the extract. One must assume that a small amount of unlabeled ATP is available at the start to initiate hexose phosphorylation.

4. The first step is the conversion of mannitol to mannose. This requires oxidation at the C-1 of mannitol, using a dehydrogenase enzyme with NAD^+ or $NADP^+$ as an electron acceptor. One could then propose a number of schemes, using nucleotide derivatives with isomerase or epimerase activities in combination with one or more phosphorylated intermediates. An established pathway uses hexokinase and ATP for the synthesis of mannose 6-phosphate; this is then converted by mannose phosphate isomerase to form fructose 6-phosphate, an intermediate of the glycolytic pathway. Bringing such sugars into the glycolytic pathway as soon

as possible means that already existing enzymes can be used to process the inter-mediates derived from each of a number of different sugars. Otherwise, a separate battery of enzymes would be needed to obtain energy from each of the sugars found in the diet.

5. The reaction you are concerned with is glucose + fructose ⟶ sucrose.

   (a) For the reaction in this direction, $\Delta G^{\circ\prime}$ = +31.4 kJ/mol. The equilibrium constant K'_{eq} is

   $$K'_{eq} = \frac{[\text{sucrose}]}{[\text{glucose}][\text{fructose}]}$$

   At the start of the reaction, [sucrose] = 0, [glucose] = 0.10 M, and [fructose] = x. At equilibrium, [sucrose] = 0.01 M, [glucose] = 0.09 M, and [fructose] = x – 0.01 M. First, calculate K'_{eq} for the reaction at equilibrium:

   $$\Delta G^{\circ\prime} = -RT\ln K_{eq}{}' = (-8314\,\text{JK}^{-1}\text{mol}^{-1})(298\text{K})\ln K_{eq}{}' = +31.4\,\text{kJ}/\text{mol}$$

   $$\ln K_{eq}{}' = \frac{-31.4\,\text{kJ}/\text{mol}}{(8.314\,\text{J K}^{-1})(298K)} = -12.7$$

   $$K_{eq}{}' = 3.05\times10^{-6}\,\text{M}$$

   Then find the concentration of fructose that satisfies the conditions at equilibrium. Assume that the unknown concentration of fructose at equilibrium (x – 0.01 M) is approximately equal to x.

   $$x = \frac{0.01\text{M}}{(0.09\text{m})(3.05\times10^{-6}\text{M})} = 3.64\times10^{4}\text{M}$$

   (b) The concentration of fructose required to generate sucrose at a concentration of 0.01 M exceeds the solubility limit for fructose; it is therefore impossible to establish such conditions in solution.

6. Harden-Young Experiment

   (a) Inorganic phosphate is required for one of the reactions of the glycolytic pathway: the phosphorylation by which glyceraldehyde 3-phosphate is converted to 1,3-bisphosphoglycerate.

   (b) The P_i/CO_2 ratio is 1.0, with one P_i being consumed for each pyruvate that undergoes decarboxylation.

   (c) In most cells, the absolute concentration of NAD^+ and NADH is low. Successive and continuous reduction and oxidation of NAD^+ and NADH, respectively, is necessary for them to continue to serve as donors and acceptors of electrons. In this case, NAD^+ must be constantly available for the continued activity of glyceraldehyde 3-phosphate dehydrogenase in the glycolytic pathway. The NADH generated during the oxidation of glyceraldehyde 3-phosphate is reoxidized to NAD^+ when acetaldehyde is reduced to ethanol.

   (d) Initially, glyceraldehyde 3-phosphate is phosphorylated when P_i is added to the fermenting mixture. ADP is phosphorylated when 1,3-bisphosphoglycerate donates a phosphoryl group to the nucleotide and is itself converted to 3-phosphoglycerate. The ATP formed in this reaction can be used in two earlier reactions of glycolysis, the phosphorylations of glucose and of fructose 6-phosphate.

   (e) The hexose bisphosphate is fructose 1,6-bisphosphate, the only such intermediate in the glycolytic pathway. This compound accumulates when glycolytic flux is blocked at the glyceraldehyde 3-phosphate dehydrogenase step by limited Pi availability. As the phosphorylation of glucose continues, intermediates from the steps preceding the formation of 1,3-bisphosphoglycerate build up.

(f) The hydrolysis of ATP to ADP and P_i makes more P_i available for the phosphorylation of glyceraldehyde 3-phosphate. Under such conditions, glycolytic activity and CO_2 production are stimulated.

7. The rate of the reaction catalyzed by PFK initially increases with the ATP concentration because ATP is a substrate for the reaction; it binds at the active site of PFK with fructose 6-phosphate and serves as a phosphoryl donor. At higher concentrations, ATP binds not only at the active site but also at the allosteric site; this alters the conformation of the enzyme and decreases the level of its activity. The effects of ATP on PFK are consistent with the role of PFK as a control element for the glycolytic pathway. When concentrations of ATP are relatively low, the activity of PFK is stimulated so that additional fructose 1,6-bisphosphate is made available for subsequent energy-generating reactions; when concentrations of ATP are higher and the demand of the cell for energy is lower, ATP inhibits PFK activity, thereby allowing glucose and other substrates to be utilized in other pathways. In many cells, ATP concentration is maintained at relatively high and constant levels, so that PFK is always subject to inhibition by ATP. Inhibition can be relieved by fructose 2,6-bisphosphate, which is synthesized when glucose is readily available. This allows cells to carry out glycolysis even when ATP levels are high, permitting the synthesis of building blocks from glucose.

8. Glyceraldehyde-3-phosphate dehydrogenase

(a) According to the mechanism presented in the text, Cys149 must be deprotonated to attack the aldehyde of GAP. That means that the active site of the enzyme must be designed to lower the normally high pK_a of the cysteine (see the discussion on cysteine proteases in the text, pages 263–264). One explanation for the pH results is that the increase in activity in the wild-type protein is due to deprotonation of an activated Cys 149 upon increasing pH.

(b) Since the pK_a of the serine does not appear to be near neutral pH (based on the pH study), it is not activated as are the serines in serine proteases, such as chymotrypsin. One explanation for this is that the catalytic triad present in the serine proteases is not present in GAPDH. The active site is presumably designed to activate a cysteine residue rather than a serine. The serine cannot act as a nucleophile at physiological pH and is unable to catalyze the dehydrogenase reaction.

(c) You would expect such a mutation to shut down glycolysis in living cells since they would be unable to convert GAP to 1,3-BPG and therefore enter into stage 3 of glycolysis.

9. Hexokinase kinetics

(a) The values of V_{max}, K_M, and V, the measured velocity, are given for glucose and fructose. You can use the Michaelis-Menten equation by solving for [S], the substrate concentration:

$$V = V_{max} \frac{[S]}{[S] + K_M}$$

$$V_{max}[S] = V[S] + VK_M$$

$$[S](V_{max} - V) = VK_M$$

$$[S] = \frac{VK_M}{V_{max} - V}$$

Then you can use the values provided to calculate the concentrations of the two sugars in the cell. For glucose, [S] = 5 × 10^{-9} M; whereas for fructose, [S] = 1 × 10^{-3} M.

(b) The phosphorylation of a hexose such as glucose or fructose is the initial step in the oxidation of the sugar, a process in which energy in the form of ATP is generated. Fructose is more important in the provision of energy for the cell because V, the observed rate of formation, is 1500 times faster for fructose 6-phosphate than it is for glucose 6-phosphate.

10. Because their epimerase activity is normal, galactosemic patients are able to synthesize UDP-galactose from UDP-glucose. The UDP-galactose is then used in the synthesis of glycoproteins.

11. The oxidation of the aldehyde group by NAD$^+$ is an energetically favorable reaction that leads to the formation of a high-energy thioester bond between the substrate and the thiol group of a cysteine residue of the enzyme. Inorganic phosphate then attacks the thioester bond, which gives rise to an acyl phosphate product, 1,3-bisphosphoglycerate.

12. When pyruvate kinase transfers the phosphoryl group from phosphoenolpyruvate to ADP, the remaining enediol remnant, enolpyruvate, is much more unstable than its ketone tautomer, pyruvate. This enol-ketone tautomerization drives the overall reaction toward ATP formation by removing the enolpyruvate by converting it to pyruvate.

13. Only glyceraldehyde can be converted to lactate under anaerobic conditions. The pathway for glyceraldehyde to lactate produces net formation of one ATP with no net oxidation per molecule metabolized. Every glycerol molecule converted to lactate under anaerobic conditions generates 2 NADH, one produced during the conversion of glycerol 3-phosphate into DHAP and another during the formation of 1,3-BPG from glyceraldehyde 3-phosphate. Because there is only one step, catalyzed by lactate dehydrogenase, that regenerates an NAD$^+$ molecule for every glyceraldehyde molecule metabolized, NADH accumulates. The glycolytic pathway is interrupted because there is no NAD$^+$ available to accept electrons from glycerol 3-phosphate or glyceraldehyde 3-phosphate. Glycerate cannot be metabolized under anaerobic conditions, because during its conversion to lactate there is no net formation of ATP. In addition, the pathway from glycerate to lactate has no pathway for generation of NADH, which would be required to balance the generation of NAD$^+$ during the reduction of pyruvate to form lactate.

14. Hexokinase phosphorylates glucose even at low concentrations. This is important in muscle and other tissues because phosphorylating newly absorbed glucose prevents it from leaving the cell. But remember that the liver must be able to export glucose, so allowing hexokinase to put a phosphate on right after the liver has gone to some trouble to allow glucose-6-phosphatase to take a phosphate *off* would make no sense. Thus both in liver and pancreas, glucokinase is used because it only works at high glucose concentrations. When glucose is particularly high it allows glycogen synthesis and helps to signal for secretion of the hormone insulin.

15. A malignant tumor is likely to grow. As new tissue is added, it won't have enough oxygen unless new blood vessels form. Thus it will be hypoxic, which will cause formation of HIF-1. This will stimulate a variety of responses (see Table 16.5 and textbook Figure 16.23) including growth of new blood vessels. Exercising enough to strengthen muscles also is likely to add new tissue as muscles get larger, producing the same situation. HIF-1 is produced and new blood vessels will form. As the text points out, one promising focus of chemotherapy is prevention of new blood vessel formation in tumors.

16. Examination of the structures of the α-keto acid analogs of alanine, aspartate, and glutamate shows that each can be used for gluconeogenesis. Specific amino transferases convert alanine to pyruvate, aspartate to oxaloacetate, and glutamate to α-ketoglutarate. These amino acids, along with others whose carbon skeletons can be used for the synthesis of glucose, are termed glucogenic amino acids.

17. Gluconeogenesis requires six high-energy phosphate bonds for every two molecules of pyruvate converted to glucose. These phosphate molecules come from ATP, most of which is generated in the liver by oxidative phosphorylation in the presence of oxygen. Under anaerobic conditions, the only source of ATP is glycolysis, but only two molecules of ATP are produced per glucose converted to pyruvate. The extra price of generating glucose from pyruvate would lead to a deficit in the supply of ATP. The balance between gluconeogenesis and glycolysis is stringently controlled; therefore, it seems unlikely that ATP generation via glycolysis would occur if cellular conditions favored gluconeogenesis.

18. The carboxylation reaction produces an activated carboxyl group in the form of a high-energy carboxybiotin intermediate. The cleavage of this bond and release of CO_2 in the phosphoenolpyruvate carboxykinase reaction or the transfer of the CO_2 to acceptors in other reactions in which biotin participates allows endergonic reactions to proceed. Thus, the formation of phosphoenolpyruvate from oxaloacetate is driven by the release of CO_2 ($\Delta G^{o\prime}$ = −19.7 kJ/mol) and the hydrolysis of GTP ($\Delta G^{o\prime}$ = −30.6 kJ/mol).

19. In contrast to muscle tissue, which oxidizes glucose to yield energy, liver tissue generates glucose primarily for export to other tissues. Thus, one would expect the rate of gluconeogenesis in the liver to be greater than the rate of glycolysis. Therefore, the relative catalytic capacity (as measured by V_{max}) of fructose bisphosphatase, a key enzyme in gluconeogenesis, should be expected to exceed that of phosphofructokinase, which is a regulatory enzyme of the glycolytic pathway.

20. Muscle and heart have distinct lactate dehydrogenase isozymes. Heart lactate dehydrogenase contains mostly H-type subunits. This enzyme has higher affinity for substrates and is inhibited by high concentrations of pyruvate; that is, it is designed to form pyruvate from lactate. In contrast, muscle lactate dehydrogenase, which consists of M-type subunits, is more effective for forming lactate from pyruvate.

21. (a) The negatively charged phosphate keeps glucose from diffusing out of the cell.
    (b) The isomerization reaction sets up the sugar for a second phosphorylation at C-1 and symmetric cleavage by aldolase.
    (c) The isomerization allows both 3-C sugars to be oxidized by the same pathway.
    (d) The isomerization converts 2-phosphoglycerate (a low energy compound) into phosphoenolpyruvate whose hydrolysis can be coupled to ATP synthesis.

The Citric Acid Cycle

The citric acid cycle, also known as the *tricarboxylic acid cycle* or the *Krebs cycle,* is the final oxidative pathway for carbohydrates, lipids, and amino acids. It is also a source of precursors for biosynthesis. The authors begin Chapter 17 with a detailed discussion of the reaction mechanisms of the pyruvate dehydrogenase complex, followed by a description of the reactions of the citric acid cycle. This description includes details of mechanism and stereospecificity of some of the reactions and homologies of the enzymes to other proteins. In the following sections, they describe the stoichiometry of the pathway including the energy yield (ATP and GTP) and then describe control mechanisms. They conclude the chapter with a summary of the biosynthetic roles of the citric acid cycle and its relationship to the glyoxylate cycle found in bacteria and plants.

The chapters on enzymes (Chapters 8 through 10), the introduction to metabolism (Chapter 15), and the chapter on glycolysis (Chapter 16) contain essential background material for this chapter.

LEARNING OBJECTIVES

When you have mastered this chapter, you should be able to accomplish the following objectives.

Introduction

1. Outline the role of the *citric acid cycle* in aerobic metabolism.
2. Locate the enzymes of the cycle in eukaryotic cells.

Pyruvate Dehydrogenase Links Glycolysis to the Citric Acid Cycle (Text Section 17.1)

3. Account for the origins of acetyl CoA from various metabolic sources.
4. Describe *pyruvate dehydrogenase* as a *multienzyme complex.*
5. List the *cofactors* that participate in the pyruvate dehydrogenase complex reactions and discuss the roles they play in the overall reaction.

The Citric Acid Cycle Oxidizes Two-Carbon Units (Text Section 17.2)

6. Outline the enzymatic mechanism of *citrate synthase.*
7. Explain the importance of the *induced-fit* structural rearrangements in citrate synthase during catalysis.
8. Describe the role of iron in the enzyme aconitase.
9. Compare the reaction catalyzed by the α-*ketoglutarate dehydrogenase complex* to the reaction catalyzed by the *pyruvate dehydrogenase complex.*
10. Name all the *intermediates* of the citric acid cycle and draw their structures.
11. List the enzymatic reactions of the citric acid cycle in their appropriate sequence. Name all the enzymes.
12. Give examples of *condensation, dehydration, hydration, decarboxylation, oxidation,* and *substrate-level phosphorylation* reactions.
13. Indicate the steps of the cycle that yield CO_2, *NADH*, $FADH_2$, and *GTP*. Note the biological roles of GTP.
14. Calculate the *yield of ATP* from the complete oxidation of pyruvate or of acetyl CoA.

Entry into the Citric Acid Cycle and Metabolism Through It Are Controlled (Text Section 17.3)

15. Summarize the *regulation* of the pyruvate dehydrogenase complex through reversible *phosphorylation.* List the major *activators* and *inhibitors* of the kinase and phosphatase.
16. Indicate the *control points* of the citric acid cycle and note the activators and inhibitors.
17. Understand how prolyl hydroxylase 2 and HIF-1 interact with certain enzymes of the citric acid cycle in some forms of cancer.
18. Explain how cancer could be considered a "metabolic disease"?

The Citric Acid Cycle Is a Source of Biosynthetic Precursors (Text Section 17.4)

19. Indicate the citric acid cycle intermediates that may be used as *biosynthetic precursors.*

20. Describe the role of *anaplerotic reactions* and discuss the *pyruvate carboxylase* reaction.

21. Describe the consequences and the biochemical basis of *thiamine deficiency.* Compare the effects of heavy-metal poisoning with mercury or arsenite.

The Glyoxylate Cycle Enables Plants and Bacteria to Grow on Acetate
(Text Section 17.5)

22. Compare the reactions of the *glyoxylate cycle* and those of the citric acid cycle. List the reactions that are unique to the glyoxylate cycle.

SELF-TEST

Introduction

1. If a eukaryotic cell were broken open and the subcellular organelles were separated by zonal ultracentrifugation on a sucrose gradient, in which of the following organs would the citric acid cycle enzymes be found?
   (a) nucleus
   (b) lysosomes
   (c) Golgi complex
   (d) mitochondria
   (e) endoplasmic reticulum

Pyruvate Dehydrogenase Links Glycolysis to the Citric Acid Cycle

2. What are the potential advantages of a multienzyme complex with respect to the isolated enzyme components? Explain.

3. Match the cofactors of the pyruvate dehydrogenase complex in the left column with their corresponding enzyme components and with their roles in the enzymatic steps in the right column.
   (a) coenzyme A
   (b) NAD^+
   (c) thiamine pyrophosphate
   (d) FAD
   (e) lipoamide

   (1) pyruvate dehydrogenase component
   (2) dihydrolipoyl dehydrogenase
   (3) dihydrolipoyl transacetylase
   (4) oxidizes the hydroxyethyl group
   (5) decarboxylates pyruvate
   (6) oxidizes dihydrolipoamide
   (7) accepts the acetyl group from acetyl-lipoamide
   (8) provides a long, flexible arm that conveys intermediates to different enzyme components
   (9) oxidizes $FADH_2$

The Citric Acid Cycle Oxidizes Two-Carbon Units

4. Which of the following statements concerning the enzymatic mechanism of citrate synthase is correct?
   (a) Citrate synthase uses an NAD^+ cofactor.
   (b) Acetyl CoA binds to citrate synthase before oxaloacetate.
   (c) The histidine residues at the active site of citrate synthase participate in the hydrolysis of acetyl CoA.
   (d) After citryl CoA is formed, additional structural changes occur in the enzyme.
   (e) Each of the citrate synthase subunits binds one of the substrates and brings the substrates into close proximity to each other.

5. Citrate synthase binds acetyl CoA, condenses it with oxaloacetate to form citryl CoA, and then hydrolyzes the thioester bond of this intermediate. Why doesn't citrate synthase hydrolyze acetyl CoA?

6. Which of the following answers complete the sentence correctly? Succinate dehydrogenase

   (a) is an iron–sulfur protein like aconitase.
   (b) contains FAD and NAD^+ cofactors like pyruvate dehydrogenase.
   (c) is an integral membrane protein unlike the other enzymes of the citric acid cycle.
   (d) carries out an oxidative decarboxylation like isocitrate dehydrogenase.

7. The conversion of malate to oxaloacetate has a $\Delta G^{o'} = +29.7$ kJ/mol (7.1 kcal/mol), yet in the citric acid cycle the reaction proceeds from malate to oxaloacetate. Explain how this is possible.

FIGURE 17.1 Citric acid cycle and the pyruvate dehydrogenase reaction.

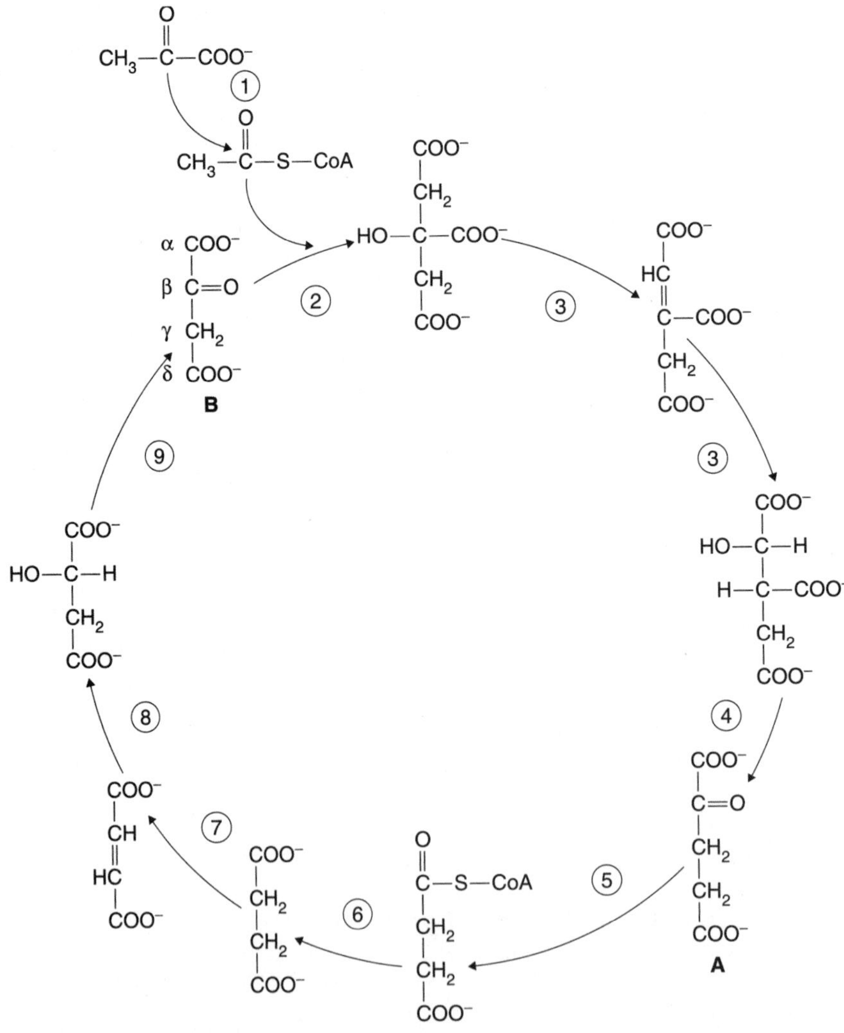

8. Given the biochemical intermediates of the pyruvate dehydrogenase reaction and the citric acid cycle (Figure 17.1), answer the following questions.

   (a) Name the intermediates **A** and **B**.
   (b) Draw the structure of isocitrate and show those atoms that come from acetyl CoA in bold letters.
   (c) Which reaction is catalyzed by α-ketoglutarate dehydrogenase?

(d) Which enzyme catalyzes step 2?

(e) Which reactions are oxidations? Name the enzyme that catalyzes each of them.

(f) At which reaction does a substrate-level phosphorylation occur? Name the enzyme and the products of this reaction.

(g) Which of the reactions require an FAD cofactor? Name the enzymes.

(h) Indicate the decarboxylation reactions and name the enzymes.

9. If the methyl carbon atom of pyruvate is labeled with ^{14}C, which of the carbon atoms of oxaloacetate would be labeled after one turn of the citric acid cycle? (See the lettering scheme for oxaloacetate in Figure 17.1 in this book.) Note that the "new" acetate carbons are the two shown at the bottom of the first few structures in the cycle, because aconitase reacts stereospecifically.

(a) None. The label will be lost in CO_2.

(b) α

(c) β

(d) γ

(e) δ

10. Considering the citric acid cycle steps between α-ketoglutarate and malate, how many high-energy phosphate bonds, or net ATP molecules, can be generated?

(a) 4   (d) 10

(b) 5   (e) 12

(c) 7

11. The standard free-energy change (in terms of net ATP production) when glucose is converted to 6 CO_2 and 6 H_2O is about how many times as great as the free-energy change when glucose is converted to two lactate molecules?

(a) 2   (c) 15

(b) 7   (d) 28

Entry into the Citric Acid Cycle and Metabolism Through It Are Controlled

12. Although O_2 does not participate directly in the reactions of the citric acid cycle, the cycle operates only under aerobic conditions. Explain this fact.

13. Which of the following answers complete the sentence correctly? The pyruvate dehydrogenase complex is activated by

(a) phosphorylation of the pyruvate dehydrogenase component (E_1).

(b) stimulation of a specific phosphatase by Ca^{2+}.

(c) inhibition of a specific kinase by pyruvate.

(d) decrease of the $NADH/NAD^+$ ratio.

(e) decreased levels of insulin.

14. First select the enzymes in the left column that regulate the citric acid cycle. Then match them with the appropriate control mechanisms in the right column.

(a) citrate synthase

(b) aconitase

(c) isocitrate dehydrogenase

(d) α-ketoglutarate dehydrogenase

(e) succinyl CoA synthetase

(f) succinate dehydrogenase

(g) fumarase

(h) malate dehydrogenase

(1) feedback inhibited by succinyl CoA

(2) allosterically activated by ADP

(3) inhibited by NADH

(4) regulated by the availability of acetyl CoA and oxaloacetate

(5) inhibited by ATP

15. Although the ATP/ADP ratio and the availability of substrates and cycle intermediates are very important factors affecting the rate of the citric acid cycle, the $NADH/NAD^+$ ratio is of paramount importance. Explain why.

16. Prolyl hydroxylase 2 (PHD-2) regulates Hypoxia-Inducible Factor 1 (HIF-1). How does the regulation work? What is the cofactor for PHD-2? Have you seen a similar reaction with the same cofactor before?

17. Which specific enzymes of the citric acid cycle appear to be related to cancer, as described in the textbook?

The Citric Acid Cycle Is a Source of Biosynthetic Precursors

18. Which of the following answers complete the statement correctlly? The citric acid cycle
    (a) does not exist as such in plants and bacteria because its functions are performed by the glyoxylate cycle.
    (b) oxidizes acetyl CoA derived from fatty acid degradation.
    (c) produces most of the CO_2 in anaerobic organisms.
    (d) provides succinyl CoA for the synthesis of carbohydrates.
    (e) provides precursors for the synthesis of glutamic and aspartic acids.

19. Match the intermediates of the citric acid cycle in the left column with their biosynthetic products in mammals, in the right column.

    (a) isocitrate   (1) aspartic acid
    (b) α-ketoglutarate   (2) glutamic acid
    (c) succinyl CoA   (3) cholesterol
    (d) *cis*-aconitate   (4) porphyrins
    (e) oxaloacetate   (5) none

20. Which of the following answers complete the sentence correctly? Anaplerotic reactions
    (a) are necessary because the biosynthesis of certain amino acids requires citric acid cycle intermediates as precursors.
    (b) can convert acetyl CoA to oxaloacetate in mammals.
    (c) can convert pyruvate into oxaloacetate in mammals.
    (d) are not required in mammals because mammals have an active glyoxylate cycle.
    (e) include the pyruvate dehydrogenase reaction operating in reverse.

21. Which of the following answers complete the sentence correctly? Pyruvate carboxylase
    (a) catalyzes the reversible decarboxylation of oxaloacetate.
    (b) requires thiamine pyrophosphate as a cofactor.
    (c) is allosterically activated by NADH.
    (d) requires ATP.
    (e) is found in the cytoplasm of eukaryotic cells.

22. Which of the following enzymes have impaired activity in vitamin B_1 deficiency?
    (a) succinate dehydrogenase
    (b) pyruvate dehydrogenase
    (c) isocitrate dehydrogenase
    (d) α-ketoglutarate dehydrogenase
    (e) dihydrolipoyl transacetylase
    (f) transketolase

The Glyoxylate Cycle Enables Plants and Bacteria to Grow on Acetate

23. Malate synthase, an enzyme of the glyoxylate cycle, catalyzes the condensation of glyoxylate with acetyl CoA. Which enzyme of the citric acid cycle carries out a similar reaction? Would you expect the binding of glyoxylate and acetyl CoA to malate synthase to be sequential? Why?

24. All organisms require three- and four-carbon precursor molecules for biosynthesis, yet bacteria can grow on acetate whereas mammals cannot. Explain why this is so.

25. Starting with acetyl CoA, what is the approximate yield of high-energy phosphate bonds (net ATP formed) via the glyoxylate cycle?
    (a) 3
    (b) 6
    (c) 9
    (d) 12
    (e) 15

ANSWERS TO SELF-TEST

1. d

2. A multienzyme complex can carry out the coordinated catalysis of a complex reaction. The intermediates in the reaction remain bound to the complex and are passed from one enzyme component to the next, which increases the overall reaction rate and minimizes side reactions. In the case of isolated enzymes, the reaction intermediates would have to diffuse randomly between enzymes.

3. (a) 3, 7 (b) 2, 9 (c) 1, 5 (d) 2, 6 (e) 3, 4, 8

4. d

5. Citrate synthase binds acetyl CoA only after oxaloacetate has been bound and the enzyme structure is rearranged to create a binding site for acetyl CoA. After citryl CoA is formed, there are further structural changes that bring an aspartate residue and a water molecule into the vicinity of the thioester bond for the hydrolysis step. Thus, acetyl CoA is protected from hydrolysis.

6. a, c

7. Although this step is energetically unfavorable in standard conditions, in mitochondria the concentrations of malate and NAD^+ are relatively high and the concentrations of the products, oxaloacetate and NADH, are quite low, so the overall ΔG for this reaction is negative.

8. (a) **A**: α-ketoglutarate; **B**: oxaloacetate
   (b) See the structure of isocitrate in the margin. The text doesn't go into detail about the stereochemistry of the enzyme aconitase, but the enzyme always puts the double bond and then the hydroxyl on the side of the molecule away from the "new" carbons introduced from Acetyl CoA.

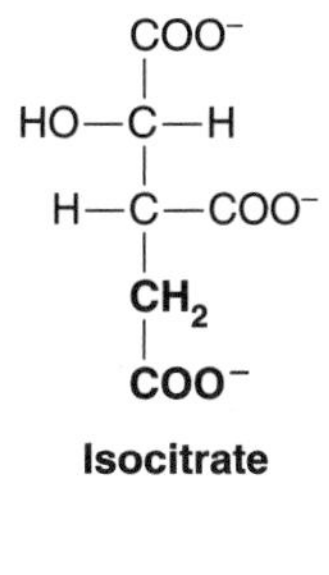

```
            COO⁻
            |
     HO—C—H
            |
     H—C—COO⁻
            |
           CH₂
            |
           COO⁻
```
Isocitrate

    (c)   reaction 5

    (d)   citrate synthase

    (e)   step 1, pyruvate dehydrogenase; step 4, isocitrate dehydrogenase; step 5, α-ketoglutarate dehydrogenase; step 7, succinate dehydrogenase; step 9, malate dehydrogenase

    (f)   step 6; the enzyme is succinyl CoA synthetase; the products of the reaction are succinate, CoA, and GTP.

    (g)   step 1, dihydrolipoyl dehydrogenase component of the pyruvate dehydrogenase complex; step 5, dihydrolipoyl dehydrogenase component of the α-ketoglutarate dehydrogenase complex; step 7, succinate dehydrogenase.

    (h)   step 1, pyruvate dehydrogenase; step 4, isocitrate dehydrogenase; step 5, α-ketoglutarate dehydrogenase.

9. c and d. Both the middle carbons of oxaloacetate will be labeled because succinate is a symmetrical molecule.

10. b

11. c. From glucose to lactate, two ATP are formed; from glucose to CO_2 and H_2O, about 30 ATP are formed.

12. The citric acid cycle requires the oxidized cofactors NAD^+ and FAD for its oxidation–reduction reactions. The oxidized cofactors are regenerated by transfer of electrons through the electron transport chain to O_2 to give H_2O (see Chapter 18).

13. b, c, d

14. Regulators are. (a) 4, 5   (c) 2, 3, 5   (d) 1, 3, 5. The inhibition of citrate synthase by ATP is species specific (found in certain bacteria), as the text points out. Citrate synthase is quite sensitive to the levels of available oxaloacetate and acetyl CoA in all organisms.

15. The oxidized cofactors NAD^+ and FAD are absolutely required as electron acceptors in the various dehydrogenation reactions of the citric acid cycle. When these oxidized cofactors are not available, as when their reoxidation stops in the absence of O_2 or respiration, the citric acid cycle also stops.

16. Normally PHD-2 hydroxylates proline residues on HIF-1, which causes it to be destroyed in the proteasome. The cofactor for the reaction is ascorbate, or vitamin C, although the enzyme also requires oxygen and α-ketoglutarate. Near the end of Chapter 2 in the textbook there is a discussion of the disease scurvy, which is caused by a lack of vitamin C. Specifically scurvy is a lack of mature collagen. One of the main differences between pro-collagen and collagen is that many of the proline residues have to be hydroxylated. So that is another prolyl hydroxylase that requires ascorbate as its cofactor.

17. The book mentions succinate dehydrogenase, fumarase, and pyruvate dehydrogenase kinase. The enzymes that are directly part of the citric acid cycle (succinate DH and fumarase) can be defective or lacking, in which case succinate and fumarate build up and spill out into the cytoplasm from the mitochondrial matrix. These are competitive inhibitors of PHD-2.

18. b, e

19. (a) 5   (b) 2   (c) 4   (d) 5   (e) 1

20. a, c

21. a, d

22. b, d, f

23. The condensation of glyoxylate and acetyl CoA carried out by malate synthase in the glyoxylate cycle is similar to the condensation of oxaloacetate and acetyl CoA carried out by citrate synthase in the citric acid cycle. The initial binding of glyoxylate, which induces structural changes in the enzyme that allow the subsequent binding of acetyl CoA, would be expected in order to prevent the premature hydrolysis of acetyl CoA. See Question 5.

24. Bacteria are capable, via the glyoxylate cycle, of synthesizing four-carbon precursor molecules for biosynthesis (for example, malate) from acetate or acetyl CoA. Mammals do not have an analogous mechanism; in the citric acid cycle, the carbon atoms from acetyl CoA are released as CO_2, and there is no net synthesis of four-carbon molecules.

25. a. One NADH is formed that can yield approximately 2.5 molecules of ATP.

PROBLEMS

1. In addition to its role in the action of pyruvate dehydrogenase, thiamine pyrophosphate (TPP) serves as a cofactor for other enzymes, such as pyruvate decarboxylase, which catalyzes the *nonoxidative* decarboxylation of pyruvate. Propose a mechanism for the reaction catalyzed by pyruvate decarboxylase. What product would you expect? Why, in contrast to pyruvate dehydrogenase, are lipoamide and FAD not needed as cofactors for pyruvate decarboxylase?

2. Sodium fluoroacetate is a controversial poison also known as *compound 1080*. When an isolated rat heart is perfused with sodium fluoroacetate, the rate of glycolysis decreases and hexose monophosphates accumulate. In cardiac cells, fluoroacetate is condensed with oxaloacetate to give fluorocitrate. Under these conditions, cellular citrate concentrations increase while the levels of other citric acid cycle components decrease. What enzyme is inhibited by fluorocitrate? How can you account for the decrease in glycolysis and the buildup of hexose monophosphates?

3. The conversion of citrate to isocitrate in the citric acid cycle actually occurs by a dehydration-rehydration reaction with aconitate as an isolatable intermediate. A single enzyme, aconitase, catalyzes the conversion of citrate to aconitate and aconitate to isocitrate. An equilibrium mixture of citrate, aconitate, and isocitrate contains about 90%, 4%, and 6% of the three acids, respectively.

   (a) Why must citrate be converted to isocitrate before oxidation takes place in the citric acid cycle?

   (b) What are the respective equilibrium constants and standard free-energy changes for each of the two steps (citrate ⇌ aconitate; aconitate ⇌ isocitrate)? For the overall process at 25°C?

   (c) Could the citric acid cycle proceed under standard conditions? Why or why not?

   (d) Given the thermodynamic data you have gathered about the reactions catalyzed by aconitase, how can the citric acid cycle proceed under cellular conditions?

4. Lipoic acid and FAD serve as prosthetic groups in the enzyme isocitrate dehydrogenase. Describe their possible roles in the reaction catalyzed by the enzyme.

5. Malonate anion is a potent competitive inhibitor of succinate dehydrogenase, which catalyzes the conversion of succinate to malate.

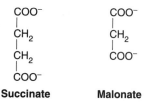

Succinate   Malonate

(a) Why is malonate unreactive?

(b) In work that led to the elucidation of the citric acid cycle, Hans Krebs employed malonate as an inhibitor of succinate dehydrogenase. Earlier studies by Martius and Knoop had shown that in animal tissues there is a pathway from citrate to succinate. Krebs had also noticed that citrate catalytically enhances respiration in minced muscle tissues. Knowing that malonate reduces the rate of respiration in animal cells, he then added citrate to malonate-poisoned muscle. In another experiment, Krebs added fumarate to malonate-poisoned muscle. What changes in succinate concentration did Krebs observe in each of the experiments with malonate-treated muscle, and what was the significance of each finding?

(c) Krebs carried out a final set of crucial studies by showing that citrate can be formed in muscle suspensions if oxaloacetate is added. What is the significance of this experiment, and how did it provide a coherent scheme for terminal oxidation of carbon atoms?

6. Recent studies suggest that succinate dehydrogenase activity is affected by oxaloacetate. Would you expect the enzyme activity to be enhanced or inhibited by oxaloacetate?

7. Winemakers have to understand some biochemistry to know what is happening as crushed grapes turn to wine. The major pathway involved is glycolysis, leading to ethanol and CO_2. Early bottling can lead to sparkling wine as more CO_2 is produced. A secondary fermentation is allowed to take place in many wines, both red and white, called "malolactic fermentation." This fermentation is classically produced by bacteria that have an enzyme that binds L-malic acid and decarboxylates it to form L-lactate. This process alters the flavor, making the wine more complex and less acidic. The secondary fermentation is so desirable that biotechnologists inserted the gene for this enzyme into *Saccharomyces cerevisiae,* the yeast used to ferment wine or beer. Initial experiments failed to produce malolactic fermentation using only yeast, but after some thought, researchers inserted another gene into the yeast and the process succeeded.

(a) Why does wine taste less acidic when malate is converted into lactate?

(b) What was the second gene that researchers had to insert to make the process work?

8. Oysters and some other mollusks live their adult lives permanently cemented to a support on the seafloor. The local environment can occasionally become anaerobic. This means that these higher animals have to function as *facultative anaerobes.* When oysters are deprived of oxygen, they accumulate succinate. Even though the citric acid cycle cannot be run as a cycle in the absence of oxygen, the reactions can be exploited in a way that maintains redox balance. The "four-carbon" reactions are run backward, from oxaloacetate to succinate, which produces reduced NAD^+ and FAD. Simultaneously, the cycle runs forward from citrate to succinate, which produces two molecules of NADH. Assuming that the oysters manage a steady supply of oxaloacetate to run these reactions, how much energy would they derive from this process?

9. In the early 1900s, Thunberg proposed a cyclic pathway for the oxidation of acetate. In his scheme, two molecules of acetate are condensed, with reduction, to form succinate, which in turn is oxidized to yield oxaloacetate. The decarboxylation of oxaloacetate to pyruvate and the oxidative decarboxylation of pyruvate to acetate complete the cycle. Assuming that electron carriers like NAD^+ and FAD would be part of the scheme, compare the energy liberated by the Thunberg scheme with that liberated by the now-established citric acid cycle. Which of the steps in Thunberg's scheme was not found in subsequent studies?

10. (a) A cell is deficient in pyruvate dehydrogenase phosphate phosphatase. How would such a deficiency affect cellular metabolism?

    (b) A cell has a metabolic defect in the citric acid cycle that causes inhibition of prolyl dehydrogenase kinase 2, and consequent activation of HIF-1, hypoxia-inducible factor 1. Compare the metabolic effects of this situation to part "a" of this problem.

11. ATP is an important source of energy for muscle contraction. Pyruvate dehydrogenase phosphate phosphatase is activated by calcium ion, which increases greatly in concentration during exercise. Why is activation of the phosphatase consistent with the metabolic requirements of muscle during contraction?

12. As the textbook points out, the symptoms of Beriberi are very similar to the symptoms of arsenite poisoning. At first glance this seems impossible because two very different cofactors are affected; Beriberi is a deficiency of thiamine, and arsenite inactivates lipoamide. Explain how the two are linked, and explain why the symptoms are neurological.

13. The fermentation process for beer or Scotch whiskey begins with barley. The grains are moistened and allowed to swell and almost sprout (the process is called "malting") and then they are roasted in an oven (an "oast") before they are fermented. Why couldn't one simply grind up the grains, mix with water, and ferment without bothering with the malting process? What pathway is involved (one that humans do not have)?

14. In addition to the carboxylation of pyruvate, there are other anaplerotic reactions that help to maintain appropriate levels of oxaloacetate. For example, the respective amino groups of glutamate and aspartate can be removed to yield the corresponding α-keto acids. How can these α-keto acids be used to replenish oxaloacetate levels?

15. The oxidation of a fatty acid with an even number of carbon atoms yields a number of molecules of acetyl CoA, whereas the oxidation of an odd-numbered fatty acid yields molecules of not only acetyl CoA but also propionyl CoA, which then gives rise to succinyl CoA. Why does only the oxidation of odd-numbered fatty acids lead to the *net* synthesis of oxaloacetate?

16. Some microorganisms can grow using ethanol as their sole carbon source. Propose a pathway for the utilization of this two-carbon compound. The pathway should convert ethanol into one or more molecules that can be used for energy generation and as biosynthetic precursors.

17. The citric acid cycle provides most of the energy for eukaryotes. It is how we "make a living" biochemically. But there are biochemical strategies that are completely unrelated. Bacteria called *methanophiles* (or *methanotrophs*) can use methane as a fuel. Propose an energy-conserving reaction sequence for converting methane into CO_2. What is the likely yield of ATP of this pathway?

18. Günter Wächtershäuser has proposed several theories about prebiotic metabolism and the origin of life. He thinks that the earliest cells originated in the vicinity of hot, deep sea volcanic vents, and that the process which propelled the chemistry was an inorganic reaction between iron sulfide (FeS) and hydrogen sulfide (H_2S). Normally we think of metal sulfides as being insoluble, but the waters inside a volcanic vent are much hotter than the boiling point of water (this is possible because they are deep in the ocean and thus under a lot of pressure). The vents are called "black smokers" because most of the sulfides precipitate when the hot water meets the cold surrounding seawater. The reaction is $FeS + H_2S \rightarrow FeS_2 + H_2$. FeS_2 is pyrite or fool's gold. The pathway he proposes is a "backward" citric acid cycle. What would be the overall results of running the citric acid cycle backward?

ANSWERS TO PROBLEMS

1. The mechanism is similar to that shown on page 500 of the text, in which the C-2 carbanion of TPP attacks the α-keto group of pyruvate. The subsequent decarboxylation of pyruvate is enhanced by the delocalization of electrons in the ring nitrogen of TPP. The initial product is hydroxyethyl-TPP, which is cleaved on protonation to yield acetaldehyde and TPP. In contrast to the reaction catalyzed by pyruvate dehydrogenase, no net oxidation occurs, so lipoamide and FAD, which serve as electron acceptors, are not needed.

2. The accumulation of citrate and the decrease in the levels of other citric acid cycle intermediates suggest that aconitase is inhibited by fluorocitrate. Excess citrate inhibits phosphofructokinase, causing a decrease in the rate of glycolysis and an accumulation of hexose monophosphates such as glucose 6-phosphate and fructose 1,6-bisphosphate. The controversy over compound 1080 is between environmentalists, who want it banned from outdoor use, and farmers and ranchers, who find it useful against rodents and predators. Because of the way it acts on cells, there is no antidote and it produces a slow and painful death.

3. (a) The oxidation of isocitrate involves oxidation of a secondary alcohol. Citrate has an alcohol function, but it is a tertiary alcohol, which is much more difficult to oxidize. Isomerization of citrate to isocitrate provides an easier route to oxidative decarboxylation.

   (b) For the citrate–aconitate pair, the equilibrium constant is equal to the ratio of product and substrate concentration. Because $K_{eq} = (4/90) = 0.0444$, the value of the standard free-energy change for the citrate-aconitate pair would be

$$\Delta G^{o\prime} = -RT \ln K_{eq} = -0.0083\ (298) \ln (4.44 \times 10^{-2})$$
$$= (-2.47) \times (-3.11) = +7.68\ \text{kJ/mol}$$

   Similar calculations for the aconitate-isocitrate pair give $K_{eq} = 6/4 = 1.5$ and thus for the aconitate isocitrate pair we would have

$$= -0.0083(298) \ln (1.5) =$$
$$(-2.47) \times (0.405) = -1.00\ \text{kJ/mol}$$

   Summing the two numbers, we get

$$= 7.68 - 1.00 = +6.68\ \text{kJ/mol}$$

   [In calories, $\Delta G^{o\prime}$ is $-1.36 \log_{10} (4.44 \times 10^{-2}) = -1.36(-1.35) = +1.84$ kcal/mol for the citrate–aconitate pair. Similar calculations for the aconitate–isocitrate pair give $K_{eq} = 6/4 = 1.5$ and of $\Delta G^{o\prime} = -0.24$ kcal/mol.

   The overall standard free-energy value for the conversion of citrate to isocitrate is the sum of the two values for the individual reactions:

$$\Delta G^{o\prime} = 1.84 + (-0.24) = +1.60\ \text{kcal/mol.}]$$

   (c) Under standard conditions, the citric acid cycle could not proceed because the positive free-energy value for the reaction indicates that it would proceed toward net formation of citrate. Note that under standard conditions, everything would be present at 1 molar concentration.

   (d) The net conversion of citrate to isocitrate can occur in the mitochondrion if the isocitrate produced is then converted to α-ketoglutarate. This would lower the concentration of isocitrate, pulling the reaction toward net formation of that molecule. Concentrations of citrate could also be increased, driving the reaction once again toward the formation of isocitrate. Although accurate concentrations of metabolites in mitochondria are difficult to establish, it appears that both mechanisms may operate to ensure net isocitrate synthesis.

4. Lipoic acid contains a sulfhydryl group that could act as an acceptor for electrons from isocitrate. Those electrons could then be transferred to NAD^+ via FAD. The roles of the prosthetic groups would be similar to those they play in the reaction catalyzed by pyruvate dehydrogenase.

5. (a) Succinate, which has two methylene groups, loses two hydrogens during its oxidation by succinate dehydrogenase. Malonate, which has only one methylene group, cannot be dehydrogenated and is therefore unreactive.

 (b) In both experiments, Krebs observed an increase in the concentration of succinate. We know now that both citrate and fumarate can be viewed as precursors of succinate or other components of the citric acid cycle. A malonate-induced block in the conversion of succinate to fumarate would cause an increase in succinate concentration. The first experiment, in which citrate addition caused an increase in succinate concentration, showed that the pathway from citrate to succinate is physiologically significant and is related to the process of respiration using carbohydrates as a fuel. The second experiment with fumarate suggested that a pathway from fumarate to succinate exists that is separate from the reaction catalyzed by succinate dehydrogenase. Krebs realized that a cyclic pathway could account for all these observations. The piece of the puzzle that was left was to learn how citrate might be generated from pyruvate or acetate. The results from those experiments are described in (c).

 (c) The generation of citrate from oxaloacetate enabled Krebs to devise a scheme that incorporated two-carbon molecules from acetate or pyruvate into citrate, with oxaloacetate serving as an acceptor of the carbon atoms. He then was able to use the results of his experiments and those of others to show how a cyclic pathway could function to carry out oxidation of carbon molecules while regenerating oxaloacetate. Krebs was prepared for the development of the cyclic scheme because he had shown earlier that the ornithine cycle, which is used for urea synthesis, is also a cyclic metabolic pathway. Krebs's famous paper, which describes the entire cycle as well as the malonate inhibition study, appeared in *Enzymologia* 4(1937):148.

6. Oxaloacetate is derived from succinate by the sequential action of succinate dehydrogenase, fumarase, and malate dehydrogenase. When levels of oxaloacetate are high, one would expect the activity of the enzyme to be reduced. Low levels of oxaloacetate would call for an increase in succinate production.

7. (a) Malic acid has two carboxyl groups. Lactic acid has only one. The pH of the wine changes significantly as malate is converted into lactate and carbon dioxide.

 (b) The yeast cells with the extra gene for the malolactic enzyme were quite capable of running the reaction, but they had no transport system for malate. The researchers realized that they could also insert a gene for malate permease, which would allow malate to enter the yeast cells and be metabolized. With both genes, the system started working well. An example of a malate permease is shown in Figure 18.37 on page 553 of the text (*Nature Biotech.* 15[1997]:224, 253).

8. One high-energy bond is generated by succinyl CoA synthetase, which produces GTP. And notice that while the NADH used up cancels out the NADH produced, a second NADH can provide electrons through the electron transport chain to reduce FAD (see Chapter 18). The passage of a pair of electrons forward through complex I (NADH-Q reductase) and backward through complex II (succinate-Q reductase) should produce

enough of a proton gradient to form another ATP. The details of ATP production in mitochondria will be discussed in Chapter 18, but the basic facts are described in Chapter 17 of the text. Metabolism in facultative anaerobes is discussed in *J. Biol. Chem.* 251(1976):3599.

9. As shown in Figure 17.2, there are at least four steps that generate reduced electron carriers. For each acetate group consumed, 3 NADH and 1 $FADH_2$ are generated, and their subsequent reoxidation in the electron transport chain provides energy for the generation of nine molecules of ATP. The same number of reduced electron carriers is generated through the action of pyruvate dehydrogenase and the enzymes of the citric acid cycle, so that the energy liberated by both schemes is the same. Each of the reactions shown in Thunberg's scheme is known to occur, except for the condensation of two acetyl groups to form succinate.

FIGURE 17.2 Thunberg's cycle.

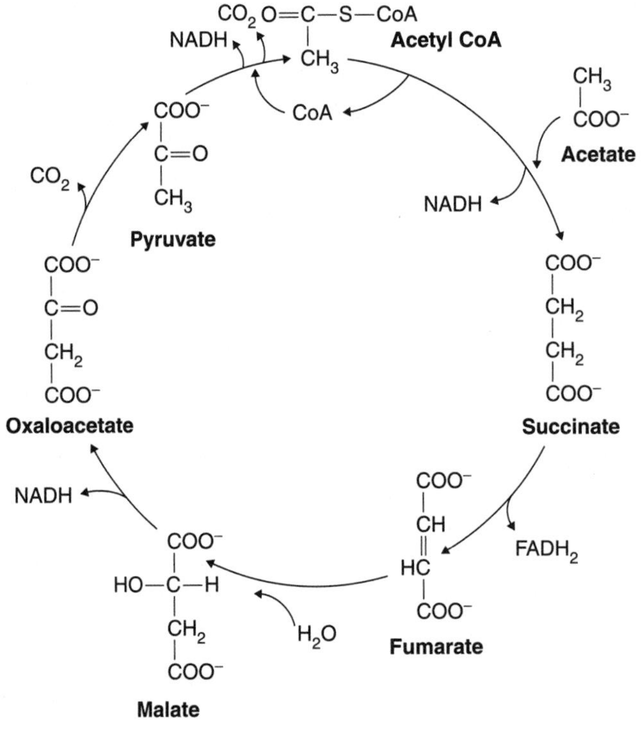

10. (a) Pyruvate dehydrogenase phosphate phosphatase removes a phosphoryl group from pyruvate dehydrogenase, activating the enzyme complex and accelerating the rate of synthesis of acetyl CoA. Cells deficient in phosphatase activity cannot activate pyruvate dehydrogenase, so that the rate of entry of acetyl groups into the citric acid cycle will decrease, as will aerobic production of ATP. Under such conditions, stimulation of glycolytic activity and a subsequent increase in lactate production would be expected as the cell responds to a continued requirement for ATP synthesis. See the clinical note on page 514 of the text (*phosphatase deficiency*).

(b) As described in the second clinical note on page 537 (*development of cancer*), the metabolic state of the cell is rather similar to the situation in part "a" of this problem. But instead of a deficiency in pyruvate dehydrogenase phosphatase, we see an increase in pyruvate dehydrogenase kinase. So in both cases the PDH is phosphorylated and hence "off" or inactive. But under the circumstances, with high HIF-1, just as described in the previous chapter of the textbook, aerobic glycolysis is favored so that the accumulating pyruvate will be channeled toward formation of lactate. This appears to contribute to the development of certain cancers.

11. As discussed in the previous problem, the phosphatase activates pyruvate dehydrogenase, stimulating the rate of both glycolysis and the citric acid cycle. Calcium-mediated activation of pyruvate dehydrogenase therefore promotes increased production of ATP, which is then available for muscle contraction.

12. Both thiamine and lipoamide are cofactors of the pyruvate dehydrogenase complex. So a deficiency of either cofactor will cause the complex to slow down or shut down. As described in the textbook, if this enzyme is not working then pyruvate will accumulate and glycolysis will be blocked, not for want of glucose, but through accumulation of product. Because nerve tissue is particularly dependent on glycolysis, this will have a deleterious effect on nerve functions.

13. Fermentation starts with sugar. All that is inside un-malted barley (or many other grains) is starch and oil. The pathway involved here is the glyoxylate cycle, which allows many plants to convert stored supplies of oil into acetyl CoA, and then build the acetyl units up into molecules of glucose. By moistening the barley we convince the seeds that it is springtime and so they make a lot of glucose in preparation for turning it into cellulose for sprouts and leaves and roots. Roasting the malt kills the seeds and arrests the process at a stage where there is maximum sugar. At this point you have something that is easily fermented. As a side note, the smoke from the roasting process is what gives some Scotch whiskeys a smoky aroma.

14. Examination of the structures of the α-keto acid analogs of glutamate and aspartate shows that they are in fact both citric acid cycle intermediates, α-ketoglutarate and oxaloacetate. Aspartate, when it is deaminated, thus contributes directly to the insertion of additional molecules of oxaloacetate. Glutamate produces α-ketoglutarate, which, as a component of the citric acid cycle, is a precursor of oxaloacetate.

15. The entry of acetyl groups from acetyl CoA into the citric acid cycle does not contribute to the net synthesis of oxaloacetate, because two carbons are lost as CO_2 in the pathway from citrate to oxaloacetate. Only the entry of compounds with three or more carbons, like succinate, can increase the relative number of carbon atoms in the pathway. Thus, although odd-numbered fatty acids contribute to the net synthesis of oxaloacetate, compounds with an even number of fatty acids do not.

16. The microorganism first converts ethanol to acetic acid, or acetate, by carrying out two successive oxidations, with acetaldehyde as an intermediate. Two molecules of a reduced electron carrier such as NADH are also produced. Next, acetate is activated through the action of acetyl CoA synthetase to form acetyl CoA, and then the acetyl group is transferred to oxaloacetate to form citrate. After citrate is converted to isocitrate, two enzymes from the glyoxylate cycle, isocitrate lyase and malate synthase,

assist in the net formation of oxaloacetate from isocitrate and another molecule of acetyl CoA, as discussed in the text on page 519. Oxaloacetate can then be used for generation of energy as well as production of biosynthetic intermediates. Note that a small amount of oxaloacetate and other intermediates of the citric acid cycle must be present initially for acetate to enter the pathway.

17. Methane is first oxidized by a monooxygenase to methanol; NADH, the reductant, is converted to NAD^+, and methanol and water are generated. Methanol is then oxidized to formaldehyde; PQQ, a novel quinone, is the electron acceptor in this step. Formaldehyde is oxidized to formic acid, which in turn is oxidized to CO_2. NADH is formed in each of these two steps. About four ATP are formed (2.5 ATP from NADH and 1.5 from $PQQH_2$). See G. Gottschalk (1986), *Bacterial Metabolism* (2nd ed., pp. 155, 163), Springer-Verlag.

18. Running the "modern" citric acid cycle backward would take electrons from NADH and $FADH_2$ and use them to incorporate CO_2 into organic molecules. This sort of "carbon fixation" is an attractive feature for reactions near the origin of life. There is no guarantee that cofactors like NADH and $FADH_2$ would have been available, but it is assumed that hydrogen could have contributed to the functioning of the cycle in some way. The reaction that forms pyrite is energetically favorable, so there is an unending supply of hydrogen for reactions like this. There is no way to prove that Wächtershäuser's controversial ideas are right. But the presence in our cells of so many iron sulfur clusters seems to point back to a time when life was developing in an environment that had plenty of iron and sulfur. Wächtershäuser's papers are summarized in an interesting article by B. Edward H. Maden, *Trends in Biochemical Sciences* 20[1995]:337.

Oxidative Phosphorylation

The reduced coenzymes NADH and $FADH_2$ that are formed during glycolysis (Chapter 16) and as a result of the functioning of the citric acid cycle (Chapter 17) ultimately transfer their electrons via a series of carriers to oxygen and thereby release a large amount of useful energy. The energy from this electron transfer is coupled to the formation of ATP by means of a transmembrane proton–motive force composed of a pH gradient and an electric potential. This process, known as *oxidative phosphorylation,* produces the bulk of the ATP in aerobic organisms. After introducing the mitochondria, the text begins with the concept of reduction potential as another measure of the free energy of chemical reactions (see Chapters 8 and 15) involving electron transfers. The text then lists the three proton pumps and the electron carriers they bear, as well as two mobile electron carriers. The location of these electron transfer chain constituents in the mitochondria of eukaryotes and the sequence of electron transfer reactions that occur between the reduced cofactors and the final electron acceptor, O_2, are also outlined. The authors also discuss the potentially damaging free radicals that are produced as a side reaction of oxidative metabolism. Next, the text discusses the generation of a proton gradient as a consequence of the electron flow and the use of the resulting proton–motive force by ATP synthase to form ATP from ADP and P_i by ATP synthase. Because ATP synthesis occurs inside mitochondria, whereas most of the reactions that utilize ATP take place in the cytosol, the membrane shuttle systems for ATP–ADP and other cofactors and biomolecules are described. The text then discusses the reasons for the variation in the estimation of the number of ATP molecules formed per glucose molecule oxidized. The control of oxidative phosphorylation and the mechanisms that uncouple electron transfer and oxidative phosphorylation as well as disease states resulting from mitochondiral misfunction are disucssed. The chapter ends by stressing the central role of proton gradients in interconverting free energy in cells.

LEARNING OBJECTIVES

INTRODUCTION

When you have mastered this chapter, you should be able to accomplish the following objectives.

1. Compare the amount of ATP required for a typical day's worth of activity to that stored in a sedentary male. Recognize the implications of the disparity and how a human body compensates for it.

2. Write the overall, balanced reaction for the electron-transport chain along with the value of $\Delta G^{o\prime}$.

3. Appreciate that proton gradients are an interconvertible form of free energy in the cell.

4. Define *oxidative phosphorylation* and *respiration*.

5. Realize that *coupling* of *oxidation* to *phosphorylation* by a *proton gradient (proton–motive force)* forms *ATP*.

Eukaryotic Oxidative Phosphorylation Takes Place in Mitochondria
(Text Section 18.1)

6. Describe the *compartments* and *membranes of mitochondria* and locate the *respiratory assemblies* and the *N* and *P sides* of the inner membrane.

7. Provide a hypothesis for the evolutionary origin of mitochondria.

Oxidative Phosphorylation Depends on Electron Transfer (Text Section 18.2)

8. Relate quantitatively *redox potential* ($\Delta E_0\prime$) and *free-energy change* ($\Delta G^{o\prime}$).

9. Describe the meaning and the measurement of the redox potential ($E_0\prime$) for a *redox couple* relative to the *standard reference half-cell*.

10. Explain the meaning of $E_0\prime$ in electron transfer.

11. Calculate ΔG° for *oxidation–reduction reactions* from the redox potentials for individual redox couples.

12. Identify the driving force of oxidative phosphorylation and be able to calculate the amount of chemical work that can be coupled to the reduction of O_2 with NADH.

13. Calculate the free energy associated with a proton gradient.

The Respiratory Chain Consists of Four Complexes: Three Proton Pumps and a Physical Link to the Citric Acid Cycle (Text Section 18.3)

14. List the components of the *respiratory chain* and the *electron-carrying molecules*.

15. Describe the entry of electrons from *NADH* into *NADH-Q oxidoreductase (Complex I)* and trace their path through this *proton pump*. State the roles of *flavin mononucleotide (FMN)*, *iron-sulfur clusters*, and *coenzyme* Q.

16. Distinguish among the *quinone, semiquinone,* and *ubiquinol* forms of coenzyme Q. Explain how reduction of a quinone can consume two protons.

17. Discuss the role of coenzyme Q as a mobile electron carrier between NADH-Q *oxidoreductase* and *cytochrome reductase (Complex III)*.

18. Describe the entry of electrons into the respiratory chain at the *succinate-Q reductase complex (Complex II)* from *flavoproteins* such as *succinate dehydrogenase (a component*

of Complex II), *glycerol phosphate dehydrogenase*, and *fatty acyl CoA dehydrogenase* by way of *FADH$_2$*. Appreciate that Complex II is not a proton pump.

19. Describe the prosthetic group and the functions of the *cytochromes* and contrast the features of cytochromes b, c_1, c, a, and a_3.

20. List the components of the *Q-cytochrome c oxidoreductase complex,* and explain how *ubiquinol* transfers its electrons to cytochromes c_1 and b and ultimately to cytochrome c. Describe the roles of *heme* in these processes.

21. Explain the origin of the nomenclature used in cytochrome b (heme b_L and heme b_H). Give a physical explanation for the difference between redox potentials of the two hemes.

22. Explain how a two-electron carrier, ubiquinol, can interact with a one-electron carrier, the *Fe-S cluster*. Describe the steps of the Q cycle and write a balanced reaction for the cycle.

23. Describe the *cytochrome oxidase proton pump (Complex IV)* and its electron-carrying groups. Write a balanced reaction for the process catalyzed by cytochrome c oxidase.

24. Outline the mechanism for the reduction of O$_2$ to H$_2$O on cytochrome oxidase. Describe the path of the electrons from the *heme* a–*CuA* cluster to the *heme* a$_3$–*CuB* cluster and state the changes in the oxidation states of Fe and O. Note the formation of a *superoxide anion* intermediate.

25. Describe the salient features of the three-dimensional structure of cytochrome c and relate them to its interaction with cytochrome reductase and cytochrome oxidase.

26. List the *reactive oxygen species* that are generated during electron transport. Explain why oxygen is a potentially toxic substance. Summarize the reactions and the biological roles of *superoxide dismutase, catalase,* and the *peroxidases*.

27. Discuss the functional implications of the conservation of the structure of cytochrome c in evolution.

A Proton Gradient Powers the Synthesis of ATP (Text Section 18.4)

28. Describe the *chemiosmotic model* of oxidative phosphorylation and relate experimental evidence that only the proton–motive force links the respiratory chain and ATP synthesis.

29. Describe the mitochondrial location, the subunit structure, and the function of eukaryotic *ATP synthase*.

30. Outline the proposed mechanism of ATP synthesis by ATP synthase during proton flow. Relate *catalytic cooperativity* to the *binding-change mechanism*.

31. Describe evidence for *rotational catalysis* in the ATP synthase.

32. Relate the structure and mechanism of ATP synthase with that of the G proteins, discussed in Chapter 14.

Many Shuttles Allow Movement Across the Mitochondrial Membranes
(Text Section 18.5)

33. Explain the roles of the *glycerol phosphate* and *malate–aspartate shuttles* in carrying the electrons of cytoplasmic NADH into the mitochondrion. Estimate the ATP yields in each case.

34. Describe the *ATP–ADP translocase* mechanism and state its cost relative to the energy yield from electron transfer by the respiratory chain.

35. List the most important *mitochondrial transport systems* for ions and metabolites and describe the common feature of their structures.

The Regulation of Cellular Respiration Is Governed Primarily by the Need for ATP (Text Section 18.6)

36. Estimate the net yield of ATP from the complete oxidation of glucose, taking into account the different shuttles for the cytoplasmic reducing equivalents of NADH. Discuss the sources of uncertainty in the estimation.

37. Describe *respiratory control* and relate it to *energy charge*.

38. List compounds that specifically block electron transport and locate their sites of action.

39. Explain the effect of *uncouplers* on oxidative phosphorylation. Note how regulated uncoupling can be used for *thermogenesis*. Give examples of thermogenesis in mammals and describe the physiological consequences for its absence.

40. Relate disorders in energy generation in mitochondria to human disease. Describe *apoptosis*.

41. List examples of *energy conversions* by proton gradients and appreciate their central roles in free-energy interconversions.

42. Explain the difference between oxidative phosphorylation and the substrate-level phosphorylation occurring in glycolysis and the citric acid cycle.

SELF-TEST

INTRODUCTION

1. For a sedentary male weighing 70 kg, how much more ATP is needed each day than than is present in the body?

   (a) 83 kg
   (b) 250 g
   (c) 82.75 kg
   (d) 82.75 g
   (e) None of the above. The body contains the amount needed.

2. Which of the following constitute cellular respiration?

   (a) biosynthesis of glycogen in the liver and muscles
   (b) conversion of an electron–motive force into a proton–motive force
   (c) formation of compounds with high electron transfer potential
   (d) conversion of a proton–motive force into a phosphoryl-transfer force

Oxidative Phosphorylation in Eukaryotes Takes Place in Mitochondria

3. Which of the following statements regarding mitochondria and their components are correct?

   (a) Mitochondria are approximately 20 nm in diameter.
   (b) The matrix compartment contains the enzymes of glycolysis.
   (c) Mitochondria are bounded by two membrane systems: an inner membrane and an outer membrane.

(d) The inner membrane contains pores and is readily permeable to most small metabolites.

(e) The inner membrane has a large surface area because it is highly folded.

4. Which of the following answers complete the sentence correctly? Mitochondria

(a) are found in all kingdoms of life.

(b) are semiautonomous organelles.

(c) all contain proteins encoded by their own genes and encoded by the nuclear genome.

(d) likely arose from the engulfment of a virus by a bacterium.

Oxidative Phosphorylation Depends on Electron Transfer

5. Which of the following statements about the redox potential for a reaction are correct?

(a) It is used to describe phosphate group transfers.

(b) It is unrelated to the free energy of the reaction.

(c) It can be used to predict whether a given compound can reduce another.

(d) It can be used to predict whether a given oxidation will provide sufficient energy for the formation of ATP from ADP and P_i.

(e) It can be used to predict the rate of O_2 uptake upon the oxidation of a given substrate.

6. The equation for the reduction of cytochrome a by cytochrome c is

$$\text{Cyt } a \ (+3) + \text{Cyt } c \ (+2) \rightarrow \text{Cyt } a \ (+2) + \text{Cyt } c \ (+3)$$

$$\text{where Cyt } a \ (+3) + e^- \rightarrow \text{Cyt } a \ (+2): E_0' = 027 \text{ V}$$

$$\text{Cyt } c \ (+3) + e^- \rightarrow \text{Cyt } c \ (+2): E_0' = +0.22 \text{ V}$$

Which of the following answers completes the sentence correctly? Under standard conditions ([products] = [reactants] = [1M]; pH = 7), the reaction

(a) proceeds spontaneously.

(b) yields sufficient energy for ATP synthesis.

(c) does not alter the absorption spectra of the cytochromes.

(d) involves the transfer of two electrons.

7. What is the $\Delta G^{o\prime}$ value for the following reaction? Use Table 18.1 in the text for values of $E^{o\prime}$. Note that 1 kcal = 4.187 kJ.

$$\text{succinate} + \text{FAD} \rightarrow \text{fumarate} + \text{FADH}_2$$

(a) +5.79 kJ/mol

(b) −2.89 kJ/mol

(c) +0.59 kJ/mol

(d) +2.89 kJ/mol

(e) −5.79 kJ/mol

8. The parameters $\Delta G^{o\prime}$ and $\Delta E_0'$ can be used to predict the direction of chemical reactions in standard conditions. On the other hand, $\Delta G'$ can be used for any concentration of reactants and products to predict in what direction a chemical reaction will proceed. Using the expressions

$$\Delta G' = \Delta G^{o\prime} + RT \ln \frac{[\text{products}]}{[\text{reactants}]} \quad \text{and} \quad \Delta G' = -nF\Delta E_0'$$

derive an expression for $\Delta E'$. Explain the significance of this redox potential.

9. Obtain $\Delta G'$ and $\Delta E'$ for the reaction given in question 7 when the succinate concentration is 2×10^{-3} M, the fumarate concentration is 0.5×10^{-3} M, the FAD concentration is 2×10^{-3} M, the $FADH_2$ concentration is 0.2×10^{-3} M, and the temperature is 37°C. ($R = 8.314$ J/mol K)

10. For each proton transported out of the matrix across the inner membrane and into the inner membrane space of a mitochondrion, how much free-energy potential is generated across the inner membrane?

    (a) -220.2 kJ/mol
    (b) -30.6 kJ/mol
    (c) -2.18 kJ/mol
    (d) -21.8 kJ/mol

The Respiratory Chain Consists of Four Complexes: Three Proton Pumps and a Physical Link to the Citric Acid Cycle

11. Place the following respiratory-chain components in their proper sequence. Also, indicate which are mobile carriers of electrons.

    (a) cytochrome c
    (b) NADH-Q oxidoreductase
    (c) cytochrome c oxidase
    (d) ubiquinone
    (e) Q-cytochrome c oxidoreductase

12. Match the enzyme complexes of the respiratory chain in the left column with the appropriate electron-carrying groups in the right column.

    (a) cytochrome c oxidase   (1) heme c_1
    (b) Q-cytochrome c oxidoreductase   (2) FAD
    (c) NADH-Q oxidoreductase   (3) heme a_3
    (d) succinate-Q reductase   (4) heme b_L
                                           (5) iron-sulfur complexes
                                           (6) Cu_A and Cu_B
                                           (7) FMN
                                           (8) heme a
                                           (9) heme b_H

13. Which of the following statements about the enzyme complexes of the electron transport system are correct?

    (a) They are located in the mitochondrial matrix.
    (b) They cannot be isolated from one another in functional form.
    (c) They have very similar visible spectra.
    (d) They are integral membrane proteins located in the inner mitochondrial membrane.
    (e) They transfer electrons to one another by means of mobile electron carriers.

14. Which of the following statements about ubiquinol are correct?

    (a) It is the mobile electron carrier between cytochrome c oxidoreductase and cytochrome c oxidase.
    (b) It is an integral membrane protein.
    (c) Its oxidation involves the simultaneous transfer of two electrons to the Fe-S center of cytochrome reductase.
    (d) It is oxidized to ubiquinone by way of a semiquinone intermediate.
    (e) It is a lipid-soluble molecule.

15. Which cytochrome has a protoporphyrin IX heme that is not covalently bound to protein?

    (a) cytochrome a

    (b) cytochrome a_3

    (c) cytochrome b

    (d) cytochrome c

    (e) cytochrome c_1

16. Explain the roles of cytochrome c_1 and the b cytochromes (b_L and b_H) in the oxidation of ubiquinol to ubiquinone. Are protons pumped across the inner mitochondrial membrane during these reactions?

17. In the reduction of O_2 to H_2O by cytochrome oxidase, four electrons and four protons are used. How can this occur when a single electron at a time is transferred by heme iron and by copper?

18. Which of the following statements about the structure and properties of cytochrome c are NOT correct?

    (a) It is an integral membrane protein.

    (b) It contains very little α helix or β-pleated sheet secondary structure.

    (c) It lacks a heme.

    (d) It has retained a highly conserved conformation throughout evolution.

    (e) It is soluble in water.

19. Which of the following answers correctly complete the sentance? Reactive oxygen species (ROS)

    (a) serve as substrates for enzymes that render them less reactive.

    (b) arise from intermediates generated in electron transport.

    (c) are transported out of the cell on specialized carriers.

    (d) include $OH\cdot$, H_2O_2, O_2^{2-}.

20. How can the $FADH_2$ generated by the succinate-Q-reductase complex participate in electron transport if it is not free to diffuse from the enzyme complex? Does the oxidation of succinate transport protons?

21. Which of the following statements about an aerated, functional mitochondrial preparation in which the reduced substrate is succinate are correct?

    (a) Approximately 1.5 ATP molecules will be formed per succinate oxidized to fumarate.

    (b) Approximately two protons will be pumped across the inner membrane by the succinate-Q reductase complex.

    (c) The addition of CN^- will result in the synthesis of only one ATP per succinate.

    (d) Reduction of NADH-Q oxidoreductase will occur.

    (e) Reduction of Q-cytochrome oxidoreductase will occur.

A Proton Gradient Powers the Synthesis of ATP

22. Which of the following experimental observations provide evidence that supports the chemiosmotic model of oxidative phosphorylation?

    (a) A closed membrane or vesicle compartment is required for oxidative phosphorylation.

    (b) A system of bacteriorhodopsin and ATPase can produce ATP in synthetic vesicles when light causes proton pumping.

    (c) A proton gradient is generated across the inner membrane of mitochondria during electron-transport.

    (d) ATP is synthesized when a proton gradient is imposed on mitochondria.

23. Explain why one cannot precisely predict the sites in the electron transport chain where the coupling of oxidation to phosphorylation occurs on the basis of the redox potentials of the electron-transport chain components.

24. Which of the following statements about the mitochondrial ATP-synthesizing complex are correct?
    (a) It contains more than 10 subunits.
    (b) It is located in the intermembrane space of mitochondria.
    (c) It contains a subassembly that constitutes the proton channel.
    (d) It is sensitive to oligomycin inhibition.
    (e) It translocates ATP through the mitochondrial membranes.

25. Match the major units of the ATP-synthesizing system in the left column with the appropriate components and functions in the right column.
    (a) F_0
    (b) F_1

    (1) contains the proton channel
    (2) contains the catalytic sites for ATP synthesis
    (3) contains α, β, γ, δ, and ε subunits
    (4) contains sequences homologous to P-loop NTPase family members
    (5) spans the inner mitochondrial membrane
    (6) is mostly in the matrix

26. Which of the following statements about the proposed mechanism for ATP synthesis by ATP synthase are correct?
    (a) ATP synthase forms ATP only when protons flow through the complex.
    (b) ATP synthase contains sites that change in their affinity for ATP as protons flow through the complex.
    (c) ATP synthase binds ATP more tightly when protons flow through the complex.
    (d) ATP synthase has two active sites per complex.
    (e) ATP synthase has active sites that are not functionally equivalent at a given time.

27. ATP synthase can form ATP in the absence of a proton gradient when it is mixed with ADP and P_i. Explain how this can happen when the formation of ATP from ADP and P_i requires 30.6 kJ/mol of free energy.

28. Identify the similarities in the mechanisms of G proteins and ATP synthase.

Many Shuttles Allow Movement Across the Mitochondrial Membranes

29. The inner mitochondrial membrane contains translocases—that is, specific transport proteins—for which pairs of substances?
    (a) NAD^+ and NADH
    (b) glycerol 3-phosphate and dihydroxyacetone phosphate
    (c) AMP and ADP
    (d) citrate and pyruvate
    (e) glutamate and aspartate

30. Which of the following pairs of molecules are transported in opposition to each other across the mitochondrial membrane in the malate-aspartate shuttle?
    (a) Malate and aspartate
    (b) Aspartate and glutamate
    (c) Malate and α-ketoglutarate
    (d) Malate and glutamate
    (e) Oxaloacetate and glutamate

31. Match each mitochondrial transporter with the molecules it transports.
    (a) Dicarboxylate carrier
    (b) Tricarboxylate carrier
    (c) Pyruvate carrier
    (d) Phosphate carrier

    (1) Phosphate
    (2) Hydroxide ion
    (3) Malate
    (4) Citrate
    (5) Pyruvate
    (6) Protons

32. Explain why the rate of eversion of the binding site from the matrix to the cytosolic side is more rapid for ATP than for ADP when the ATP–ADP translocase functions in the presence of a proton gradient.

The Regulation of Oxidative Phosphorylation Is Governed by the Need for ATP

33. Approximately how many ATP are formed for each extramitochondrial NADH that is oxidized to NAD^+ by O_2 via the electron transport chain. Assume that the glycerol phosphate shuttle is operating.

(a) 1.0   (c) 2.5
(b) 1.5   (d) 3.0

34. How many ATP molecules are generated during the complete oxidative degradation of each of the following to CO_2 and H_2O? Assume that the glycerol phosphate shuttle is operating.

(a) acetyl CoA
(b) phosphoenolpyruvate
(c) glyceraldehyde 3-phosphate

35. What is meant by the term *respiratory control*?

36. Which of the following answers completes the sentence more correctly? The rate of flow of electrons through the electron transport chain is most directly regulated by

(a) the ATP:ADP ratio.
(b) the concentration of acetyl CoA.
(c) the rate of oxidative phosphorylation.
(d) feedback inhibition by H_2O.
(e) the catalytic rate of cytochrome oxidase.

37. Which of the following answers completes the sentence correctly? Uncouplers, such as dinitrophenol (DNP) or thermogenin, uncouple electron transport and phosphorylation by

(a) inhibiting cytochrome reductase.
(b) dissociating the F_0 and F_1 units of ATP synthase.
(c) blocking electron transport.
(d) dissipating the proton gradient.
(e) blocking the ATP-ADP translocase.

38. Which of the following are the products of the reaction of superoxide dismutase?

(a) O_2^- (d) O_2
(b) H_2O (e) H_3O^+
(c) H_2O_2

39. Match each inhibitor in the left column with its *primary effect* in the right column.

(a) azide   (1) inhibition of electron transport
(b) atractyloside   (2) uncoupling of electron transport and
(c) rotenone   (3) oxidative phosphorylation
(d) dinitrophenol   (3) inhibition of ADP–ATP translocation
(e) carbon monoxide (4) inhibition of ATP synthase
(f) oligomycin
(g) antimycin A

40. Which of the following are directly involved in apoptosis?
    (a) cytochrome oxidase
    (b) caspases
    (c) nuclear RNases
    (d) RNA polymerase II
    (e) cytochrome *c*
    (f) caspase-activated DNases

41. Proton gradients are used for which of the following?
    (a) generating heat
    (b) free-energy storage
    (c) ATP generation
    (d) active transport
    (e) mechanical movement

ANSWERS TO SELF-TEST

1. c. Approximately 83 kg is needed per day but only 250 g (0.250 kg) is present. That means there is a 83 kg − 0.25 kg = 82.75 kg deficit.

2. b, c, d. Although glycogen is a molecule that stores energy, answer (a) is incorrect because respiration is defined as the collection of reactions that use the reductive power of $NADH^+$ and $FADH_2$ to form ultimately ATP.

3. c, e. Answer (a) is incorrect because mitochrondia are ~500 nm in diameter. They are roughly the size of bacteria.

4. b, c. Answer (d) is incorrect because a bacterium, not a virus, was probably engulfed.

5. c, d

6. a. The $\Delta E_0'$ for the reaction is 0.05 V. Calculating $\Delta G^{o\prime}$,

$$\Delta G^{o\prime} = -nF\Delta E_0'$$

where *n*, the number of electrons transferred, is 1, and *F* is 96.49 kJ/V·mol.

$$\Delta G^{o\prime} = -1(96.49 \text{ kJ/V mol})(0.05 \text{ V}) = -4.82 \text{ kJ/mol}$$

Therefore, the reaction will proceed spontaneously. However, when considered by itself, it is insufficiently exergonic to drive ATP synthesis, which requires −30.6 kJ/mol under standard conditions. In the cell, this comparison is relatively meaningless because ATP is not synthesized during oxidative phosphorylation by direct chemical coupling of redox reactions to ATP formation, but rather by being coupled to a proton–motive force. In addition, the concentrations of the reactants can alter the actual free-energy change observed in the reaction. Answer (c) is incorrect because the state of oxidation of a cytochrome alters its absorption spectrum. The heme group contributes significantly to the adsorption spectrum of cytochomes, and the state of oxidation of the heme affects its adsorption spectrum.

7. e. The $\Delta E_0'$ for this reaction is −0.03 V. Calculating $\Delta G^{o\prime}$,

$$\Delta G^{o\prime} = -nF\Delta E_0' = -2(96.49 \text{ kJ/V mol})(-0.03 \text{ V}) = -5.79 \text{ kJ/mol}$$

8. The same proportionality constants that relate $\Delta G^{o\prime}$ and $\Delta E_0'$ can be used to relate $\Delta G'$ and $\Delta E'$. Substituting $\Delta G^{o\prime} = -nF\Delta E_0'$ and $\Delta G' = -nF\Delta E'$ into the expression for $\Delta G'$,

$$-nF\Delta E' = -nF\Delta E_0' + RT \ln \frac{[\text{products}]}{[\text{reactants}]}$$

$$\Delta E' = \Delta E_0' + \frac{-RT}{nF} \ln \frac{[\text{products}]}{[\text{reactants}]}$$

$\Delta E'$ is a measure of the direction in which an oxidation–reduction reaction will proceed for any given concentration of reactants and products. If $\Delta E'$ is positive the reaction is exergonic in the direction written.

9

$$\Delta G' = \Delta G^{\circ\prime} + RT \ln \frac{[\text{Fumarate}][\text{FADH}_2]}{[\text{Succinate}][\text{FAD}]}$$

$$\Delta G' = -5.79 \text{ kJ/mol} + (8.314 \text{ J/mol K})(310\text{K}) \ln\left[\frac{(0.5 \times 10^{-3}\text{M})(0.2 \times 10^{-3}\text{M})}{(2 \times 10^{-3}\text{M})(2 \times 10^{-3}\text{M})}\right]$$

$$\Delta G' = -5.79 \text{ kJ/mol} + -9.51 \text{ kJ/mol} = -15.30 \text{ kJ/mol}$$

10. d. Answers (a) and (b) are incorrect because -220.2 kJ/mol is the free energy released by the oxidation of an NADH by $\frac{1}{2}$ O_2, and -30.6 kJ/mol is the free energy released by the hydrolysis of ATP to ADP + P_i.

11. The proper sequence is b, d, e, a, and c. The mobile carriers are (a) and (d).

12. (a) 3, 6, 8   (b) 1, 4, 5, 9   (c) 5, 7   (d) 2, 5

13. d, e. Answer (c) is incorrect because each enzyme complex has a unique absorption spectrum that reflects the environment of its electron carriers and its oxidation state.

14. d, e

15. c

16. See the Q cycle in Figure 18.12 in the text. Ubiquinol transfers one electron to cytochrome c_1 through a Rieske Fe-S cluster in cytochrome oxidoreductase. The semiquinone derived from the Q in this process donates an electron to cytochrome b_L, giving rise to ubiquinone. In turn, the electron from cytochrome b_L is transferred to cytochrome b_H, which then reduces another semiquinone to ubiquinol. Thus, the b cytochromes act as a recycling device that allows ubiquinol, a two-electron carrier, to transfer its electrons, one at a time, to the Fe-S cluster of cytochrome oxidoreductase. Cytochrome c_1 accepts the electrons (one electron/cytochome c_1) from the Fe-S cluster and transfers them to cytochrome c through the b cytochromes. Protons pumping across the mitochondrial membrane is tightly coupled to the oxidation of ubiquinol. The full oxidation of one QH_2 yields two reduced cytochome c molecules and removes two protons from the matrix.

17. See Figure 18.14 in the text. Molecular O_2 is bound between the Fe^{2+} and Cu^+ ions of the heme a_3-Cu_B center of cytochrome oxidase. The oxygen remains bound while four electrons and four protons are sequentially added to its various intermediates, resulting in the net release of two H_2O. The heme a-Cu_A center supplies the electrons for this process. Although four electrons are used in the reduction of O_2 to 2 H_2O, the individual steps of the reaction cycle involve single electron transfers.

18. a, b, c

19. a, b, d

20. The $FADH_2$ generated by the succinate-Q-reductase complex upon oxidation of succinate transfers it electrons to iron-sulfur center and finally to ubiquinone. No, this system does not transport protons across the inner mitochondrial membrane.

21. a, e. Answer (b) is incorrect; no protons are pumped by Complex II.

22. a, b, c, d. Answer (d) is true, although not mentioned in the text.

23. The free-energy change ($\Delta G^{\circ\prime}$) and for each electron transfer step of the respiratory chain can be calculated from the redox-potential change ($\Delta E_0'$) for that step, using the equation $\Delta G^{\circ\prime} = -nF\Delta E_0'$. Since $\Delta G^{\circ\prime}$ for the synthesis of ATP from ADP is

+30.5 kJ/mol, $\Delta G^{o'}$ for an electron transfer reaction in which the coupling of oxidation to phosphorylation could occur must be more negative than +30.6 kJ/mol. However, we cannot conclude that there is a direct, quantitative relationship between a given redox reaction and the phosphorylation of ADP in the cell because the free-energy coupling is by means of a spatially delocalized proton–motive force.

24. a, c, d

25. (a) 1, 5   (b) 2, 3, 4, 6

26. b, e

27. The β subunits of F_1, when in the T form, can bind ATP so tightly that they will condense ADP and P_i to form ATP with the release of water. The binding energy between the protein and the substrates is used to form the chemical bond. The K_{eq} of a reaction can be markedly different when the substrates are bound by an enzyme compared with when they are free in solution. In such cases, the protein is a stoichiometric part of the reaction and the K_{eq} is for a reaction different from that of the substrates themselves. The proton gradient is used to move, by directional rotation, the β subunits into different environments where their affinities for the substrates and products are different, thereby leading to net synthesis of ATP.

28. Both G proteins and the ATP synthase bind different nucleotides based on interaction with different proteins. The G proteins have a high affinity for GDP until they are activated by interactions with a receptor or an effector protein which can trigger exchange of GDP for GTP. ATP synthase will bind either ADP or ATP depending on which of the three different faces of the γ subunit they interact with.

29. b, e. Answer (a) is incorrect because only the electrons of NADH are transported, not the entire molecule. Answers (c) and (d) are incorrect because these pairs of compounds are not transported by the same translocase.

30. b and c. While all of the molecules listed except oxaloacetate are transported in the shuttle, only certain pairs are transported through the same protein in opposition to each other.

31. (a) 1,3   (b) 3,4,6   (c) 2, 5   (d) 1, 2

32. The proton gradient and the membrane potential make the cytosolic side of the inner mitochondrial membrane more positive than the matrix side; therefore, ATP, which has one more negative charge than ADP, is more attracted to the cytosolic side than is ADP.

33. b. The oxidation of NADH by the electron transport chain leads to the synthesis of approximately 2.5 ATP. However, when the reducing equivalents of an extramitochondrial NADH enter the mitochondrial matrix via the glycerol phosphate shuttle, they give rise to $FADH_2$, which yields 1.5 ATP.

34. (a) About 10 ATP: citric acid cycle (3 NADH → 2.5, 1 $FADH_2$ → 1.5, and 1 GTP)
    (b) About 13.5 ATP: citric acid cycle (10 ATP), pyruvate dehydrogenase reaction (1 NADH, intramitochondrial → 2.5), pyruvate kinase reaction (1 ATP)
    (c) About 16 ATP: citric acid cycle (10 ATP), pyruvate dehydrogenase reaction (1 NADH → 2.5), pyruvate kinase reaction (1 ATP), phosphoglycerate kinase reaction (1 ATP), glyceraldehyde 3-phosphate dehydrogenase reaction (1 NADH, extramitochondrial, which yields 1.5 ATP by the glycerol phosphate shuttle)

35. The regulation of the rate of oxidative phosphorylation by the availability of ADP is referred to as *respiratory control*.

36. a

37. d

38. c, d

39. (a) 1 (b) 3 (c) 1 (d) 2 (e) 1 (f) 4 (g) 1

40. b, e, f

41. a, b, c, d, e. Answer (b) is correct, but you should realize that free-energy storage by a proton gradient across a membrane is more transient than storage in a molecule such as glucose or NADH.

42. Oxidative phosphorylation is the process by which ATP is formed as electrons are transferred from NADH or $FADH_2$ to O_2. The coupling of the release of free energy from the reduced cofactors to the consumption of free energy in synthesizing the ATP occurs through a proton–motive gradient. In substrate level phosphorylation, the syntheses of ATP or GTP are coupled through chemical intermediates shared with other chemical reactions that release sufficient energy to drive the formation of ATP or GTP from ADP and GDP, respectively. Substrate-level phosphorylations have defined stoichiometries because they use a shared, defined intermediate to couple reactions. Oxidative phosphorylation has a less precise stoichiometry because its driving force is a delocalized energy gradient (proton– motive force) composed of a proton concentration and electric charge difference across a membrane.

PROBLEMS

1. The reduction potential for methylene blue is such that it can be reduced by components of the electron transport chain and can itself, when reduced, reduce O_2 to H_2O. Suggest why massive doses of methylene blue might serve to counteract cyanide poisoning.

2. What is the effect on the proton–motive potential across the inner membrane when an ATP is moved from the matrix to the cytosolic side?

3. Predict the relative oxidation–reduction states of NAD^+, NADH-Q reductase, ubiquinone, cytochrome c_1, cytochrome c, and cytochrome a in liver mitochondria that are amply supplied with isocitrate as substrate, P_i, ADP, and oxygen but are inhibited by

   (a) rotenone.
   (b) antimycin A.
   (c) cyanide.

4. Nitrite (NO_2^-) is toxic to many microorganisms. It is therefore often used as a preservative in processed foods. However, members of the genus *Nitrobacter* oxidize nitrite to nitrate (NO_3^-), using the energy released by the transfer of electrons to oxygen to drive ATP synthesis. Given the following E_0' values, calculate the maximum ATP yield per mole of nitrate oxidized.

$$NO_3^- + 2H^+ + 2\,e^- \longrightarrow NO_2^- + H_2O \qquad E_0' = +0.42\ V$$

$$\tfrac{1}{2}\,O_2 + 2\,H^+ + 2\,e^- \longrightarrow H_2O \qquad E_0' = -0.82\ V$$

5. How are mitochondria thought to have arisen? What evidence suggests a particular bacterium as the origin of the mitochondria in all eukaryotes?

6. A newly discovered compound called *coenzyme U* is isolated from mitochondria.

   (a) Several lines of evidence are presented in advancing the claim that coenzyme U is a previously unrecognized carrier in the electron transport chain.

   (1) When added to a mitochondrial suspension, coenzyme U is readily taken up by mitochondria.

   (2) Removal of coenzyme U from mitochondria results in a decreased rate of oxygen consumption.

   (3) Alternate oxidation and reduction of coenzyme U when it is bound to the mitochondrial membrane can be easily demonstrated.

   (4) The rate of oxidation and reduction of coenzyme U in mitochondria is the same as the overall rate of electron transport.

   Which of the lines of evidence do you find the most convincing? Why?

   Which are the least convincing? Why?

   (b) In addition to the evidence cited in (a), the following observations were recorded when coenzyme U was incubated with a suspension of submitochondrial particles.

   (1) The addition of NADH caused a rapid reduction of coenzyme U.

   (2) Reduced coenzyme U caused a rapid reduction of added cytochrome *c*.

   (3) In the presence of antimycin A, the reduction of coenzyme U by added NADH took place as rapidly as in the absence of antimycin A. However, the reduction of cytochrome *c* by reduced coenzyme U was blocked in the presence of the inhibitor.

   (4) The addition of succinate caused a rapid reduction of coenzyme U.

   Assign a tentative position for coenzyme U in the electron-transport chain.

7. Coenzyme Q can be selectively removed from mitochondria using lipid solvents. If these mitochondria are then incubated in the presence of oxygen with an electron donor that is capable of reducing NAD^+, what will be the redox state of each of the carriers in the electron transport chain?

8. The treatment of submitochondrial particles, which are prepared from the inner mitochondrial membrane by sonication and which have the orientation of their membrane reversed (see Figure 18.23 in the text, which shows a structure analogous to a submitochondrial particle), with urea removes F_1 subunits. When these treated particles are incubated in air with an oxidizable substrate and calcium ion, the concentration of calcium inside the particles increases.

   (a) What do these observations tell you about the source of free energy required for accumulation of calcium ion?

   (b) Would you expect the accumulation of calcium ion to be more sensitive to DNP or to oligomycin? Why?

9. Analysis of the electron-transport pathway in a pathogenic gram-negative bacterium reveals the presence of five electron-transport molecules with the redox potentials listed in Table 18.1.

Table 18.1 Reduction potentials for pathogenic gram-negative bacterium

Oxidant	Reductant	Electrons transferred	E'_0 (V)
NAD^+	NADH	2	−0.32
Flavoprotein b (ox)	Flavoprotein b (red)	2	−0.62
Cytochrome c (+3)	Cytochrome c (+2)	1	+0.22
Ferroprotein (ox)	Ferroprotein (red)	2	+0.85
Flavoprotein a (ox)	Flavoprotein a (red)	2	+0.77

(a) Predict the sequence of the carriers in the electron-transport chain.

(b) How many molecules of ATP can be generated under standard conditions when a pair of electrons is transported along the pathway?

(c) Why is it unlikely that oxygen is the terminal electron acceptor?

10. Calculate the minimum value of $\Delta E'_0$ that must be generated by a pair of electron carriers to provide sufficient energy for ATP synthesis. Assume that a pair of electrons is transferred.

11. The value of $\Delta E'_0$ for the reduction of $NADP^+$ is $+0.32$ V.

(a) Calculate the equilibrium constant for the reaction catalyzed by NADPH dehydrogenase:

$$NADP^+ + NADH \rightarrow NADPH + NAD^+$$

(b) What function could NADPH dehydrogenase serve in the cell?

12. Why is it important for the value $\Delta E'_0$ for the NAD^+:NADH redox couple to be less negative than those for the redox couples of oxidizable compounds that are components of the glycolytic pathway and the citric acid cycle?

13. Arsenate, AsO_4^{3-}, is an uncoupling reagent for oxidative phosphorylation, but unlike DNP it does not transport protons across the inner mitochondrial membrane. How might arsenate function as an uncoupler?

14. A newly isolated soil bacterium grows without oxygen but requires ferric ion in the growth medium. Succinate suffices as a carbon source, but neither hexoses nor pyruvate can be utilized. The bacteria require riboflavin as a growth supplement. Neither niacin nor thiamin is required, and neither the substances nor compounds derived from them can be found in the cells. The electron carriers found in the bacteria are cytochrome b, cytochrome c, FAD, and coenzyme Q.

(a) Propose a reasonable electron-transport chain that takes these observations, including the requirement for ferric ion, into account.

(b) Which of the other observations help to explain why the bacterium cannot utilize pyruvate or hexoses?

(c) Why is riboflavin required for the growth of the bacterium?

15. Yeast can grow both aerobically and anaerobically on glucose. Explain why the rate of glucose consumption decreases when yeast cells that have been maintained under anaerobic conditions are exposed to oxygen.

16. The addition of the drug dicyclohexylcarbodiimide (DCCD) to mitochondria markedly decreases both the rate of electron transfer from NADH to O_2 and the rate of ATP formation. The subsequent addition of 2,4-dinitrophenol leads to an increase in the rate of electron transfer without changing the rate of ATP formation. What does DCCD likely inhibit?

17. Mitochondria isolated from the liver of a particular patient will oxidize NADH at a relatively high rate even if ADP is absent. The P:O ratio for oxidative phosphorylation (the ratio of the number of P_i molecules incorporated into organic molecules per atom of oxygen consumed) by these mitochondria is less than normal. Predict the likely symptoms of this disorder.

18. Acidic aromatic compounds like 2,4-dinitrophenol (DNP) act as uncouplers of electron transport and oxidative phosphorylation because they carry protons across the inner mitochondrial membrane, disrupting the proton gradient. The structure of the neutral, protonated form of DNP is shown in Figure 18.1.

FIGURE 18.1

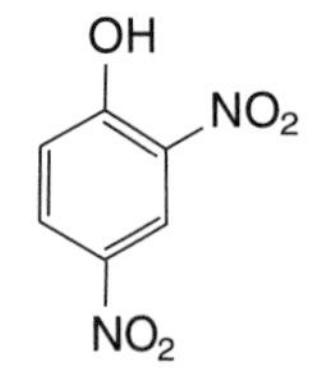

2, 4-Dinitrophenol
(DNP)

(a) Although you might expect that 2,4-dinitrophenylate anion (pK_a = 4.0) would be unable to cross the inner mitochondrial membrane, the deprotonated form of DNP is membrane-soluble. Explain why, by drawing resonance forms of 2,4-dinitrophenolate anion, showing how negative charge is distributed over the phenyl ring structure.

(b) Suppose you are studying the effects of DNP on proton transport in an artificial phospholipid membrane system. You observe that the rate of proton transport by DNP increases at temperatures above the transition temperature for the membrane. Explain.

19. Explain why the K_{eq} of the reaction ADP + P_i + H^+ $\rightleftharpoons$ ATP is ~1 on the surface of ATP synthase considering that you have been told repeatedly that the $\Delta G^{o\prime}$ for the formation of ATP is +30.6 kJ/mol.

20. The hemes in cytochrome bc1 have different redox potentials because they are in different polypeptide environments. Speculate on what physical properties of the protein could lead to a higher or lower redox potential in the hemes.

ANSWERS TO PROBLEMS

1. Cyanide blocks the transfer of electrons from cytochrome oxidase to O_2. Therefore, all the respiratory-chain components become reduced and electron transport ceases; consequently, oxidative phosphorylation stops. An artificial electron acceptor with an appropriate redox potential, such as methylene blue, can reoxidize some components of the respiratory chain, reestablish a proton gradient, and thereby restore ATP synthesis. The methylene blue takes the place of cytochrome oxidase as a means of transferring electrons to O_2, which remains the terminal electron acceptor.

2. The ATP-ADP translocase (ANT) allows the adenosine diphosphate and triphosphates to cross the inner mitochondrial membrane and in so doing brings one ADP in for each exiting ATP. In addition, since ATP bears one more negative charge than does ADP, one negative charge is removed from the matrix by an exchange thereby reducing the membrane potential. Approximately a quarter of the energy yield of electron transport through the respiratory chain is "lost" through the ATP-ADP exchange.

3. Upstream from the inhibitor, reduced respiratory-chain components will accumulate; downstream, oxidized components will be present. The point of inhibition is the crossover point.

(a) Rotenone inhibits the step at NADH-Q reductase; therefore, NADH and NADH-Q reductase will be more reduced, and ubiquinone, cytochrome c_1, cytochrome c, and cytochrome a will be more oxidized.

(b) Antimycin A blocks electron flow between cytochromes b and c_1; therefore, NADH, NADH-Q reductase, and ubiquinol will be more reduced, and cytochrome c_1, cytochrome c, and cytochrome a will be more oxidized.

(c) Cyanide inhibits the transfer of electrons from cytochrome oxidase to O_2, so all the components of the respiratory chain will be more reduced.

4. The two reactions that generate nitrate are

$$NO_2^- + H_2O \longrightarrow NO_3^- + 2\,H^+ + 2\,e^- \qquad E'_0 = -0.42\;V$$

$$\tfrac{1}{2}\,O_2 + 2\,H^+ + 2\,e^- \longrightarrow H_2O \qquad E'_0 = +0.82\;V$$

The net reaction is

$$NO_2^- + \tfrac{1}{2}\,O_2 \longrightarrow NO_3^- \qquad E'_0 = +0.40\;V$$

The Nernst equation is used to calculate the free energy liberated by the oxidation of nitrite under standard conditions:

$$\Delta G^{\circ\prime} = -nF\Delta E'_0 = -2(96.49\;\text{kJ/V} \cdot \text{mol})(0.40\;V) = -77.2\;\text{kJ/mol}$$

Therefore, each mole of nitrite oxidized yields, in principle, energy sufficient to drive the formation of ~2.5 moles of ATP under standard conditions ($\Delta G^{\circ\prime} = -30.6$ kJ/mol).

5. Mitochondria probably arose when an ancient free-living organism capable of oxidative phosphorylation invaded another cell and formed a symbiont with enhanced survival capabilities. The sequences of mitochondrial DNAs from a number of eukaryotes indicate an evolutionary relationship. The existence of a common set of encoded proteins in all mitochondrial DNAs, each of which is in the genome of one bacterium, *Rickettsia prowazeki,* is strong evidence for this organism being in the initial symbiont that gave rise to the mitochondria in all eukaryotes.

6. (a) If coenzyme U is a component of the electron-transport chain, it should undergo successive reduction and oxidation in the mitochondrion, and its overall rate of electron transfer should be close to the overall rate. Observations 3 and 4 are therefore the most convincing. In addition to electron carriers, other compounds can be taken up by mitochondria, and some of them, such as pyruvate, can affect the rate of oxygen consumption because they are substrates that donate electrons to carriers. Therefore, observations 1 and 2 are less convincing.

   (b) The first two observations show that coenzyme U lies along the electron transport chain between NADH, which can reduce it, and cytochrome *c*, which is reduced by it. The fact that antimycin A blocks cytochrome *c* reduction by coenzyme U suggests that the carrier lies before cytochrome reductase. Succinate, which can transfer electrons to Q, can also transfer electrons to coenzyme U, so the position of coenzyme U in the chain is similar to that of Q.

7. The removal of ubiquinone from the electron transport chain means that no electrons can be transferred beyond Q in the pathway. You would therefore expect all carriers preceding Q to be more reduced and those beyond Q to be more oxidized.

8. (a) The treated submitochondrial particles can carry out electron transport and can establish a proton gradient, but because the F_1 subunits have been removed, they cannot synthesize ATP. The source of free energy for calcium accumulation must therefore be the proton–motive force generated by electron transport.

   (b) DNP disrupts the proton–motive force by carrying protons across the mitochondrial membrane, thereby dissipating the free energy required for calcium accumulation. Oligomycin inhibits ATP synthase in mitochondria. Because the treated particles do not depend on ATP as a source of energy for calcium accumulation, oligomycin will have little effect.

9. (a) The carrier with the most negative reduction potential has the weakest affinity for electrons and so transfers them most easily to an acceptor. The carrier with the most positive reduction potential will be the strongest oxidizing substance and will have the greatest affinity for electrons. A carrier should be able to pass

electrons to any carrier having a more positive reduction potential. Thus, the probable order of the carriers in the chain is flavoprotein b, NADH, cytochrome c, flavoprotein a, and ferroprotein.

(b) $\Delta E'_0 = +0.85 \text{ V} - (-0.62 \text{ V}) = 1.47 \text{ V}$

The total amount of free energy released by the transfer of two electrons is

$$\Delta G^{\circ\prime} = -nF\Delta E'_0 = -2(96.49 \text{ kJ/V} \cdot \text{mol})(1.47 \text{ V}) = -283.7 \text{ kJ/mol}$$

Because $+30.6$ kJ/mol is required to drive ATP synthesis under standard conditions, the number of molecules of ATP synthesized per pair of electrons is $283.7/30.6 = 9.3$.

(c) It is unlikely that oxygen is the terminal electron acceptor because the reduction potential for the ferroprotein is slightly more positive than that of oxygen, so under standard conditions, the ferroprotein could not transfer electrons to oxygen.

10. The minimum amount of free energy that is needed to drive ATP synthesis under standard conditions is $+30.6$ kJ/mol. The value of $\Delta E'_0$ needed to generate this amount of free energy can be determined using the equation

$$\Delta G^{\circ\prime} = -nF \, \Delta E'_0$$

$$\Delta E'_0 = -\frac{-\Delta G^{\circ\prime}}{nF}$$

If a pair of electrons is transferred, then

$$\Delta E'_0 = -\frac{-30.6 \text{ kJ/mol}}{2(96.49 \text{ kJ/Vmol})} = +0.158 \text{ V}$$

11. (a) The transfer of a pair of electrons from NADH to $NADP^+$ occurs with no release of free energy:

$$NAD^+ + H^+ + 2e^- \longrightarrow NADH \qquad E'_0 = -0.32 \text{ V}$$

$$NADP^+ + H^+ + 2e^- \longrightarrow NADPH \qquad E'_0 = -0.32 \text{ V}$$

For the overall reaction, $\Delta E'_0 = 0.00$ V. Because

$$\Delta G^{\circ\prime} = -nF \, \Delta E'_0,$$

$$\Delta G^{\circ\prime} = 0 = -RT \ln K'_{eq}$$

$$\ln K'_{eq} = 0$$

$$K'_{eq} = 1$$

$K'_{eq} = 1$ means that this reaction is at equilibrium when the ratio [NADPH][NAD^+]/ [$NADP^+$] [NADH] equals 1. Any combination of concentrations that give a ratio of 1 in this quotient represents an equilibrium condition.

(b) In the cell, NADPH dehydrogenase serves to replenish NADPH when the reduced cofactor is needed for biosynthetic reactions. On the other hand, metabolites such as isocitrate and glucose 6-phosphate are substrates for $NADP^+$-linked dehydrogenases. NAD^+ can accept reducing equivalents generated as NADPH through the action of these enzymes.

12. NADH is a primary source of electrons for the respiratory chain. Oxidizable substrates must have a more negative reduction potential to donate electrons to NAD^+. A more negative redox potential for the NAD^+:NADH couple would make it unsuitable as an electron acceptor.

13. Arsenate chemically resembles inorganic phosphate; therefore, it can enter into many of the same biochemical reactions as P_i. For example, it substitutes for phosphate in the glyceraldehydes 3-phosphate dehydrogenase reaction forming arseno-phosphoglycerate instead of the normally formed bisphosphoglycerate. The arsenate compound is labile to hydrolysis and the free energy resulting from the oxidation forming it is lost, thereby precluding phosphorylation. Arsenate can replace phosphate during oxidative phosphorylation also, presumably forming an arsenate anhydride with ADP. Such compounds are similarly unstable and are rapidly hydrolyzed, effectively causing the uncoupling of electron transport and oxidative phosphorylation.

14. (a) Because neither niacin nor NAD^+ is found in these cells, the transfer of electrons must proceed from succinate to FAD and then to those carriers that have successively more positive redox potentials. In order, these are coenzyme Q, cytochrome b, cytochrome c, and ferric ion, which serves as the terminal electron acceptor in the absence of oxygen.

    (b) To utilize hexoses in the glycolytic pathway, NAD^+ is required as a cofactor for glyceraldehyde 3-phosphate dehydrogenase. Because the bacterium has no NAD^+, the glycolytic pathway is not operating. Similarly, the fact that thiamin is neither required nor found in the cells means that pyruvate dehydrogenase cannot be used to convert pyruvate to acetyl CoA.

    (c) Riboflavin is a precursor of FAD, which is required as an acceptor of electrons from succinate.

15. The decrease in glucose consumption when oxygen is introduced is known as the *Pasteur effect*. Under anaerobic conditions, glucose cannot be oxidized completely to CO_2 and H_2O, because NADH and QH_2 generated in the citric acid cycle cannot be reoxidized in the absence of oxygen. In order to regenerate NAD^+, needed for continued operation of the glycolytic pathway, pyruvate is converted to ethanol and CO_2. There is a net production of only two molecules of ATP from each glucose molecule metabolized by the glycolytic pathway. This means that the pace of glycolysis must be relatively high to generate sufficient amounts of ATP for cell maintenance.

    When oxygen is introduced, reduced cofactors in the citric acid cycle can be reoxidized in the electron transport chain and oxidative phosphorylation occurs. Under these conditions the yeast cell can utilize glucose much more efficiently, producing ~30 molecules of ATP for each glucose molecule oxidized completely to CO_2 and H_2O. The rate of glucose consumption is greatly reduced under aerobic conditions because less glucose is needed to provide the amount of ATP needed to maintain the cell. Other factors that decelerate the pace of glycolysis include increases in concentrations of citrate and ATP under aerobic conditions. Both molecules are key regulators of phosphofructokinase 1, a rate-limiting glycolytic enzyme.

16. The fact that the rate of electron transport increases on addition of DNP without a concomitant increase in ATP synthesis suggests that DCCD inhibits ATP synthase. Experiments show that indeed DCCD blocks proton flow through the C subunit of F_0, inhibiting ATP synthase activity. When DCCD is added to respiring mitochondria, protons cannot move back into the mitochondrial matrix and the rate of ATP synthesis decreases. The flow of electrons through the respiratory chain slows as the need

to maintain the proton gradient decreases. The metabolic uncoupler 2,4-dinitrophenol is an effective ion carrier that dissipates the proton–motive force generated by electron transfer by allowing protons to freely cross the inner mitochondrial membrane. Although the rate of electron transport increases in response to the dissipation of the proton gradient, the rate of phosphorylation of ADP through oxidative phosphorylation does not increase because ATP synthase remains inhibited by DCCD.

17. From these observations it appears that electron transfer and ATP synthesis are uncoupled, so there is no way to control electron transport by limiting ADP availability. One would expect the patient to have a very high rate of metabolism, along with a possibly elevated temperature. The fact that much of the energy available from electron transport is not utilized means that the energy is released as heat rather than being utilized for the formation of ATP.

18. (a) See Figure 18.2 for structures. Resonance structures that include the two nitro groups show that the negative charge can be distributed among a number of forms. Therefore the dinitrophenylate anion is soluble in the membrane bilayer. This explains why DNP can rapidly carry protons across the inner mitochondrial membrane.

FIGURE 18.2

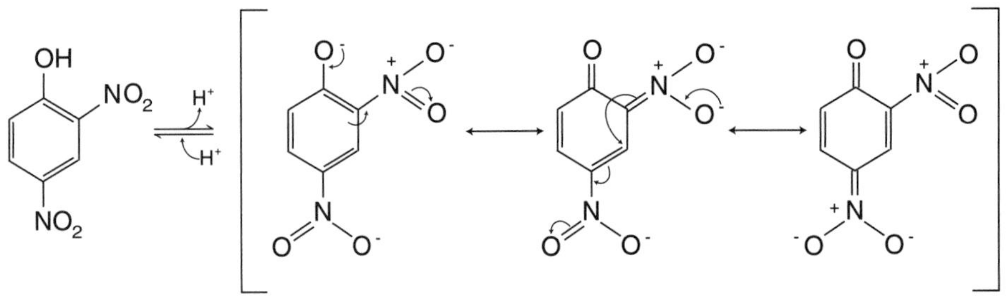

(b) DNP is a mobile proton carrier that is soluble in the membrane bilayer. Below the transition temperature for the phospholipid, the bilayer is in the gel state, where molecules like DNP may not be able to diffuse rapidly from one side of the bilayer to the other. At temperatures above the transition temperature, the bilayer is in the fluid state, in which DNP is more mobile, more rapidly transporting protons across the bilayer.

19. The ATP synthase binds ATP with such high affinity that the reaction is shifted toward synthesis when ADP and P_i are present. The binding is so tight that only rotation of the complex, which is powered by the proton gradient, can release the product ATP. Reactions on the surface of enzymes are not the same as those free in solution. The enzyme is a stoichiometric component of the reaction, and its concentration must be considered when calculating the free-energy change occurring.

20. The higher (more positive) the redox potential for a molecule is, the higher it's electron affinity. Reduction of cytochrome b converts the iron from the +3 state to +2, reducing the positive charge and along with the proprionates, giving the heme a net neutral charge. Any physical property that would tend to increase that electron affinity (stabilize the reduced state) would increase the redox potential for the molecule. Placing a heme in an environment with a positive charge nearby would be expected to increase the redox potential of the group by providing a stabilizing force for the additional electron. Conversely, placing the heme in an environment where there is a nearby negative charge should tend to decrease the redox potential. A polar environment would tend to favor the oxidized state (more charged) than the reduced state (more neutral).

Chapter 19

The Light Reactions of Photosynthesis

To this point, the authors have dealt with the mechanisms by which organisms obtain energy from their environment by oxidizing fuels to generate ATP and reducing power. In this chapter, they describe how light energy is transduced into the same forms of chemical energy, leading to conversion of CO_2 into carbohydrate by photosynthetic organisms. Carbon fixation and sugar synthesis (the dark reactions) will be covered in Chapter 20 of the text.

The authors begin with the basic equation of photosynthesis and an overview of the process. Next come descriptions of the chloroplast, chlorophyll, and the relatively simple reaction center of a photosynthetic bacterium. The authors then describe the overall structures, components, and reactions of photosystems II and I and the cytochrome *bf* complex, including the absorption of light, charge separation, electron-transport events, and the evolution of O_2. They explain how these light reactions lead to the formation of proton gradients and the synthesis of ATP and NADPH.

A review of the basic concepts of metabolism in Chapter 15 and mitochondrial structure, redox potentials, the proton-motive force, and free-energy changes in Chapter 18 will help you to understand this chapter.

LEARNING OBJECTIVES

When you have mastered this chapter, you should be able to accomplish the following objectives.

Introduction

1. Define *photosynthesis* and write the overall reaction of photosynthesis
2. Distinguish between the *light* and *dark reactions* of photosynthesis.

Photosynthesis Takes Place in Chloroplasts (Text Section 19.1)

3. Describe the structure of the *chloroplast.* Locate the *outer, inner,* and *thylakoid membranes,* the *intermembrane space,* the *thylakoid space,* the *granum,* and the *stroma.* Associate these structures with the functions they perform.
4. Describe the properties of the thylakoid membrane.
5. Discuss the probable origin of the chloroplast and compare it to theories of the origin of mitochondria.

Light Absorption by Chlorophyll Induces Electron Transfer (Text Section 19.2)

6. List the structural components of chlorophyll *a,* and explain why chlorophylls are effective *photoreceptors.*
7. Summarize the common features of diverse photosynthetic reaction centers, including bacterial, photosystem II, and photosystem I.
8. Distinguish between *bacteriochlorophyll,* chlorophyll, and *bacteriopheophytin.*
9. Explain the significance of the two plastoquinone binding sites, Q_A and Q_B, in the bacterial reaction center.

Two Photosystems Generate a Proton Gradient and NADPH in Oxygenic Photosynthesis (Text Section 19.3)

10. Diagram photosystem II and identify its major components. Describe the roles of P680, *pheophytin,* and *plastoquinone* in the absorption of light, *separation of charge,* and electron transfer in photosystem II.
11. Explain the function of the *manganese center* in the extraction of electrons from water.
12. Describe the composition and function of the *cytochrome bf complex,* and outline the roles of *plastoquinol, plastocyanin (Cu^{2+}),* and *Fe-S clusters* in the formation of a *transmembrane proton gradient.*
13. Compare and contrast the roles of plastocyanin in chloroplasts and cytochrome *c* in mitochondria.
14. Diagram photosystem I. Indicate the components and reactions of photosystem I, including the roles of P700, A_0, A_1, *ferredoxin, FAD, NADPH,* and plastocyanin(Cu^+) in these processes.

A Proton Gradient Across the Thylakoid Membrane Drives ATP Synthesis (Text Section 19.4)

15. Discuss the similarities and differences between the CF_1–CF_0 ATP synthase of chloroplasts and the F_1–F_0 synthase of mitochondria.

16. Explain how photosystem I can synthesize ATP without forming NADPH or O_2.

17. Contrast the formation of ATP by *cyclic photophosphorylation* and by *oxidative phosphorylation*.

18. Write the net reaction carried out by the combined actions of photosystem II, the cytochrome *bf* complex, and photosystem I.

Accessory Pigments Funnel Energy into Reaction Centers (Text Section 19.5)

19. Explain how the components of the *light-harvesting complexes* interact to funnel light to the reaction centers.

20. Relate the structure and color coding of Figures 19.28 and 19.30 in the text. Explain what is missing from Figure 19.30.

21. Rationalize the differences in the *photosynthetic assemblies* in the *stacked* and *unstacked* regions of the thylakoid membranes.

22. Explain the mechanisms of common herbicides that work by inhibiting the light reaction.

The Ability to Convert Light into Chemical Energy Is Ancient (Text Section 19.6)

23. List electron donors utilized by photosynthetic bacteria. Write the overall photosynthetic reaction when H_2S is the electron donor.

SELF-TEST

1. Write the basic reaction for photosynthesis in green plants.

Introduction

2. Assign each function or product in the right column to the appropriate structure or pathway in the left column.

   (a) chlorophyll
   (b) light-harvesting complex
   (c) photosystem I
   (d) photosystem II

   (1) O_2 generation
   (2) ATP synthesis
   (3) light collection
   (4) NADPH synthesis
   (5) separation of charge
   (6) light absorption
   (7) transmembrane proton gradient

Photosynthesis Takes Place in Chloroplasts

3. Thylakoid membranes contain which of the following?

   (a) light-harvesting complexes
   (b) reaction centers
   (c) ATP synthase
   (d) electron-transport chains
   (e) galactolipids
   (f) sulfolipids
   (g) phospholipids

4. The similarities between mitochondria and chloroplasts are obvious. In what ways are they opposite?

Light Absorption by Chlorophyll Induces Electron Transfer

5. Which of the following are constituents of chlorophylls?

   (a) substituted tetrapyrrole
   (b) plastoquinone
   (c) Mg^{2+}
   (d) Fe^{2+}
   (e) phytol
   (f) iron porphyrin

6. Why do chlorophylls absorb and transfer visible light efficiently?

7. Carefully read the description of the L, M, and H subunits of the bacterial reaction center and subunits D1 and D2 in photosystem II. How would you mark the locations of L, M, and H on the "box" structure of Figure 19.10 in the text? Where are D1 and D2 in text Figure 19.14? Note that D1 contains the "loose" plastoquinone.

Two Photosystems Generate a Proton Gradient and NADPH in Oxygenic Photosynthesis

8. Which of the following statements about photosystem II are correct?

   (a) It is a multimolecular transmembrane assembly containing several polypeptides, several chlorophyll molecules, a special chlorophyll (P680), pheophytin, and plastoquinones.
   (b) It transfers electrons to photosystem I via the cytochrome *bf* complex.
   (c) It uses light energy to create a separation of charge whose potential energy can be used to oxidize H_2O and to produce a reductant, plastoquinol.
   (d) It uses an Fe^{2+}-Cu^+ center as a charge accumulator to form O_2 without generating potentially harmful hydroxyl radicals, superoxide anions, or H_2O_2.

9. Which statement about the Mn center of photosystem II is NOT correct?

   (a) The Mn center has four possible oxidation states.
   (b) Electrons are transferred from the Mn center to $P680^+$.
   (c) A tyrosine residue on the D1 protein is an intermediate in electron transfer.
   (d) The O_2 released by the Mn center comes from the oxidation of water.
   (e) Each photon absorbed by the reaction center leads to the removal of an electron from the Mn cluster.

10. On the following diagram of photosystem II (Figure 19.2), identify the listed components, sites, and functions listed. The figure here is a more complex version of Figure 19.12 in the text. Note that D1 contains the "loose" plastoquinone.

FIGURE 19.1

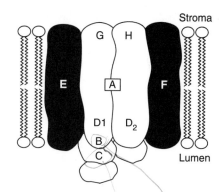

(a) Q_A plastoquinone site
(b) Q_B plastoquinone site
(c) chlorophyll P680
(d) Mn^{2+} site
(e) antennae chlorophylls
(f) photon absorption
(g) extraction of electrons from H_2O
(h) reaction center
(i) electron transfer chain
(j) tyrosyl radical

11. Match the photosystems of the purple sulfur bacterium or of green plants with the appropriate properties listed in the right column.

    (a) reaction center of
        Rhodopseudomonas viridis
    (b) photosystem II of green plants

    (1) contains an Mn center
    (2) contains two binding sites
        for plastoquinones
    (3) absorbs light of >900 nm
    (4) energy conserving event is separation
        of charge from a chlorophyll$^+$
        to pheophytin
    (5) transfers electrons from QH_2
        to a cytochrome
    (6) special-pair chlorophyll$^+$ reduced by
        electrons from H_2O
    (7) special-pair chlorophyll$^+$ reduced
        through a cytochrome with four
        covalently attached hemes

12. Explain how plastocyanin and plastoquinol are involved in ATP synthesis.

13. Write the net equation of the reaction catalyzed by photosystem I, and describe how NADPH is formed. What is the role of FAD in this process?

A Proton Gradient Across the Thylakoid Membrane Drives ATP Synthesis

14. Describe the experiment by which Jagendorf showed that chloroplasts could synthesize ATP in the dark when an artificial pH gradient was created across the thylakoid membrane.

15. Which of the following statements about cyclic photophosphorylation are correct?
    (a) It doesn't involve NADPH formation.
    (b) It uses electrons supplied by photosystem II.
    (c) It is activated when $NADP^+$ is limiting.
    (d) It does not generate O_2.
    (e) It leads to ATP production via the cytochrome bf complex.
    (f) It involves a substrate-level phosphorylation.

16. What is the overall stoichiometry of photosynthesis in chloroplasts? If eight photons are absorbed, the net yield is _____ O_2, _____ NADPH, and _____ ATP.

Accessory Pigments Funnel Energy into Reaction Centers

17. Which of the following statements about the light-harvesting complex are true?
    (a) It is a single chlorophyll molecule.
    (b) It collects light energy through the absorption of light by chlorophyll molecules.
    (c) It surrounds a reaction center with a specialized chlorophyll pair that contributes to the transduction of light energy into chemical energy.
    (d) It contains chlorophyll molecules that transfer energy from one to another by direct electromagnetic interactions.
    (e) It is the product of Planck's constant h and the frequency of the incident light ν.

18. Match the descriptions with the pigments

    A. tetrapyrrole   a. chlorophyll a
    B. polyene   b. β carotene
    C. contains Mg   c. pheophytin

19. Which of the following statements about the thylakoid membrane are correct?
    (a) It contains photosystem I and ATP synthase in the unstacked regions.
    (b) It contains the cytochrome bf complex in the unstacked regions only.
    (c) It contains photosystem II mostly in the stacked regions.
    (d) It facilitates communication between photosystems I and II by the circulation of plastoquinones and plastocyanins in the thylakoid space.
    (e) It allows direct interaction between P680* and P700* reaction centers through its differentiation into stacked and unstacked regions.

The Ability to Convert Light into Chemical Energy Is Ancient

20. *Chlorobium thiosulfatophilum* uses hydrogen sulfide as a source of electrons for photosynthesis. Write the basic equation for H_2S-based photosynthesis. Given the fact that the standard reduction potential for $S + 2 H^+ \rightarrow H_2S$ is +0.14 V, does it seem likely that two photosystems would be required for this process? Why or why not?

ANSWERS TO SELF-TEST

1. The basic reaction for photosynthesis in green plants is

$$H_2O + CO_2 \xrightarrow{\text{Light}} (CH_2O) + O_2$$

where (CH_2O) represents carbohydrate.

2. (a) $3, 5, 6$ (b) $3, 6$ (c) $2, 3, 4, 5, 6, 7$ (d) $1, 2, 3, 5, 6, 7$. Chlorophylls are involved in light absorption, light collection in the antennae, and reaction center chemistry. Photosystems I and II cooperate to generate a transmembrane proton-motive force that can synthesize ATP.

3. a, b, c, d, e, f, g

4. Chloroplasts *produce* oxygen from water; mitochondria *use* oxygen and *produce* water. The direction of the proton gradient and ATPase are reversed in the two organelles. Electrons travel only from higher to lower energy in mitochondria, but with the aid of photons, they can travel "uphill" in chloroplasts. Other differences (iron in heme versus. magnesium in chlorophyll; cytochrome *c* versus. plastocyanin) aren't "opposites."

5. a, c, e

6. The polyene structure (alternating single and double bonds) of chlorophylls causes them to have strong absorption bands in the visible region of the spectrum. Their peak molar absorption coefficients are higher than $10^5 \, M^{-1} \, cm^{-1}$. Also, although the fact is not emphasized in the text, iron porphyrins (heme groups) return to the ground state much more rapidly than excited magnesium tetrapyrroles. Thus chlorophyll has more time to transfer a high-energy electron before the excitation is dissipated as heat.

7. The H subunit would be outside the box, mainly underneath Figure 19.10 in the text. To delineate the L and M subunits, draw a vertical line dividing the box in half. We know that steps 1 and 2 show electron transfers in the L subunit, so it is the half on the left. The M subunit, on the right, contains Q_B, the loosely bound quinone. Figure 19.14 in the text is already divided into vertical halves. Notice that it is upside-down compared with Figure 19.10 because the special pair is shown at the bottom. The text doesn't emphasize the fact, but it is known that D1 contains the exchangeable plastoquinone Q_B. That means that D1 is parallel to M and is represented by the blue rectangle on the left of Figure 19.14. That means that D2, where the first electron transfers occur, is parallel to L.

8. a, b, c. Answer (d) is incorrect because a cluster of four manganese ions serves as a charge accumulator by interactions with the strong oxidant $P680^+$ and H_2O to form O_2.

9. a. Answer (a) is incorrect. The Mn center contains four Mn atoms, and can adopt five oxidation states (S_0–S_4). See Figure 19.16. Each manganese ion can exist in four oxidation states.

10. (a) H (b) G (c) A (d) B (e) E, F (f) A, E, F (g) B (h) D1, D2 (i) D1, D2 (j) D1

11. (a) $2, 3, 4, 5, 7$ (b) $1, 2, 4, 5, 6$

12. Two electrons from plastoquinol (QH_2) are transferred to two molecules plastocyanin (PC) in a reaction catalyzed by the transmembrane cytochrome *bf* complex; in the process, two protons are pumped across the thylakoid membrane to acidify the thylakoid space with respect to the stroma, and two more protons are contributed by QH_2 (Figure 19.18 in the text). The transmembrane proton gradient is used to synthesize ATP. This process closely resembles the mitochondrial Q cycle except that plastoquinone replaces ubiquinone (CoQ), plastocyanin replaces cytochrome *c*, and "inside" and "outside" are reversed.

13. The net reaction catalyzed by photosystem I is

$$PC(Cu^+) + ferredoxin_{oxidized} \rightarrow PC(Cu^{2+}) + ferredoxin_{reduced}$$

where PC is plastocyanin. Reduced ferredoxin is a powerful reductant. Two reduced ferredoxins reduce $NADP^+$ to form NADPH and two oxidized ferredoxins in a reaction catalyzed by ferredoxin-NADP$^+$ reductase. FAD is a prosthetic group on the enzyme that serves as an adapter to collect two electrons from two reduced ferredoxin molecules for their subsequent transfer to a single NADP$^+$ molecule.

14. In Jagendorf's experiment, chloroplasts were equilibrated with a buffer at pH 4 to acidify their thylakoid spaces. The suspension was then rapidly brought to pH 8, and ADP and P$_i$ were added. The pH of the stroma suddenly increased to 8, whereas that of the thylakoid space remained at 4, resulting in a pH gradient across the thylakoid membrane. Jagendorf observed that ATP was synthesized as the pH gradient dissipated and that the synthesis occurred in the dark.

15. a, c, d, e. Answer (b) is incorrect because photosystem I provides the electrons for photophosphorylation.

16. Eight photons would yield one O$_2$, two NADPH, and three ATP.

17. b, c, d

18. A: a,c; B: a,b,c; C: a.

19. a, c, d. Answer (b) is incorrect because the cytochrome *bf* complex is uniformly distributed throughout the thylakoid membrane. Answer (e) is incorrect because the differentiation into stacked and unstacked regions probably prevents direct interaction between the excited reaction center chlorophylls P680* and P700*.

20. The basic equation for this process is given in section 19.6 of the text:

$$CO_2 + 2\ H_2S \longrightarrow (CH_2O) + 2\ S + H_2O$$

Compare the standard reaction potential of S + 2 H$^+$ $\longrightarrow$ H$_2$S = +0.14 V with that of water, given in Section 18.2 of the text: 1/2 O$_2$ + 2 H$^+$ $\longrightarrow$ H$_2$O = +0.82 V. Hydrogen sulfide's electrons are much higher in energy than those from water, so it is easier to promote them to the level of NADPH (standard reduction potential = -0.32, same as NADH). If we reverse the calculation in Section 18.2, we see that the free energy change to move two electrons from oxygen to NADH would cost +220 kj (or +52.6 kcal) of free energy. Starting with hydrogen sulfide cuts this value roughly in half, so a single photosystem suffices.

PROBLEMS

1. NADP$^+$ and A$_0$ are two components in the electron-transport chain associated with photosystem I (see Figure 19.23 in the text). A$_0^-$ is a chlorophyll that carries a single electron, whereas NADPH carries two electrons. Write the overall reaction that occurs, and calculate $\Delta E'_0$ and $\Delta G^{o\prime}$ for the reduction of NADP$^+$ by A$_0^-$ using the fact that the standard reduction potential for A$_0$ is $\Delta E'_0 = -1.1$ V. See Chapter 18 for similar calculations.

2. Calculate the maximum free-energy change $\Delta G^{o\prime}$ that occurs as a pair of electrons is transferred from photosystem II to photosystem I, that is from P680* (excited) to P700 (unexcited). Estimate the E'_0 values from Figure 19.23 in the text. Then compare your answer with the free-energy change that occurs in mitochondria as a pair of electrons is transferred from NADH + H$^+$ to oxygen.

3. Explain the defect or defects in the hypothetical scheme for the light reactions of photosynthesis depicted in Figure 19.2.

FIGURE 19.2 Hypothetical scheme for photosynthesis.

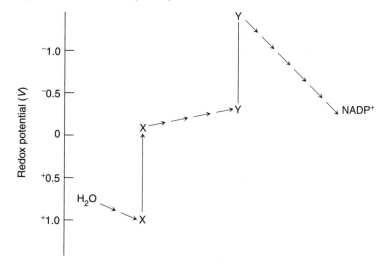

4. Explain why ATP synthesis requires a larger pH gradient across the thylakoid membrane of a chloroplast than across the inner membrane of a mitochondrion.

5. Would you expect oxygen to be evolved when $NADP^+$ is added to an illuminated suspension of isolated chloroplasts? Explain briefly.

6. Would your answer to problem 6 change if the chloroplasts were illuminated with extremely monochromatic light of 700 nm? Explain the basis for your answer.

7. Suppose you were designing spectrophotometric assays for chlorophyll *a* and chlorophyll *b*. What wavelengths would you use for the detection of each? (See Figure 19.29 in the text.) Explain your answer very briefly.

8. Green light has a wavelength of approximately 520 nm. Explain why solutions of chlorophyll appear to be green. (See Figure 19.29 in the text.)

9. If you were going to extract chlorophylls *a* and *b* from crushed spinach leaves, would you prefer to use acetone or water as a solvent? Explain your answer briefly.

10. The end of Section 19.5 of the text describes inhibitors of the light reaction that are used as herbicides. Inhibiting photosynthesis is a good way to produce compounds that can kill plants while keeping toxicity toward animals to a minimum. Can you think of another way to produce herbicides that would be relatively safe for animals?

11. Why is it considered likely that the photosystems found in chloroplasts evolved from earlier photosynthetic organisms? What is the minimum age for water-based, oxygen-producing photosynthesis?

12. What's so "special" about the special pair? Name three special pairs described in the text chapter, and tell what each one does.

ANSWERS TO PROBLEMS

1. A_0^- is the stronger reductant in the system because it has the more negative standard reducing potential. A_0^- will therefore reduce $NADP^+$ to NADPH under standard conditions. Following the convention for redox problems presented in text Chapter 18, we first write the partial reaction for the reduction involving the weaker reductant (the half-cell with the more positive standard reducing potential):

$$NADP^+ + H^+ + 2\,e^- \longrightarrow NADPH \qquad E'_0 = -0.32\text{ V} \qquad (1)$$

Next, we write, *again as a reduction,* the partial reaction involving the stronger reductant (the half-cell with the more negative standard reducing potential):

$$A_0 + e^- \longrightarrow A_0^- \qquad E'_0 = -1.1\text{ V} \qquad (2)$$

To get the overall reaction that occurs, we must equalize the number of electrons by multiplying equation 2 by 2. (We do *not*, however, multiply the half-cell potential by 2.)

$$2\,A_0 + 2\,e^- \longrightarrow 2\,A_0^- \qquad E'_0 = -1.1\text{ V} \qquad (3)$$

Then we *subtract* equation 3 from equation 1, which yields

$$NADP^+ + H^+ + 2\,A_0^- \longrightarrow NADPH + 2\,A_0 \qquad \Delta E'_0 = +0.78\text{ V}$$

In calculating $\Delta E'_0$ values, do not make the mistake of multiplying half-cell reduction potentials by factors used to equalize the number of electrons. Remember that $\Delta E'_0$ is a *potential difference* and hence, at least for our purposes, is independent of the *amount* of electron flow. For example, in a house with an adequate electrical power supply, the potential difference measured at the fuse box is approximately 117 V regardless of whether the house is in total darkness or all the lights are turned on.

To get the free-energy change for the overall reaction, we start with the relationship given in Section 18.2 of the text:

$$\Delta G^{o\prime} = -nF\,\Delta E'_0$$

Substitution yields

$$\Delta G^{o\prime} = -2 \times 96.48 \times 0.78 = -151\text{ kJ/mol}$$

or $\qquad \Delta G^{o\prime} = -2 \times 23.06 \times 0.78 = -36\text{ kcal/mol}$

2. In this process, electrons are transferred down an electron-transport chain from P680* to P700. The E'_0 value for P680* is approximately -0.8 V, and that for P700 is approximately 0.4 V; $\Delta E'_0$ is therefore $+1.2$ V. The free-energy change is calculated from the relationship given on in section 18.2 of the text.

$$\Delta G^{o\prime} = -nF\,\Delta E'_0$$

$$= -2 \times 96.48 \times 1.2 = -232\text{ kJ/mol}$$

or $\qquad = -2 \times 23.06 \times 1.2 = -55\text{ kcal/mol}$

In the mitochondrial electron-transport chain, the free-energy change as a pair of electrons is transferred from NADH + H^+ to oxygen is -52.6 kcal/mol (see

Section 18.2 of the text). In both cases the large "span" of free energy is used to drive the formation of ATP.

3. In the scheme in Figure 19.3 in the text, electrons are shown flowing "uphill" from X* to Y as ATP is being formed. This is a thermodynamic impossibility. For electrons to flow spontaneously from X* to Y, the redox potential of X* must be more negative than that of Y. For ATP to be formed as electron transfer occurs, the free-energy change must be of sufficient magnitude to allow for ATP biosynthesis. In order to make electrons flow from X* to Y as depicted in the hypothetical scheme, ATP would be consumed, not generated.

4. The synthesis of ATP in both the chloroplast and the mitochondrion is driven by the proton-motive force across the membrane. In mitochondria, a membrane potential of 0.14 V is established during electron transport. In chloroplasts, the light-induced potential is close to 0. Therefore, there must be a greater pH gradient in the chloroplast to give the same free-energy yield (see Chapter 18 and Section 19.4 of the text).

5. Oxygen would be evolved. $NADP^+$ is the final electron acceptor for photosynthesis; see the summary in Figure 19.25 in the text. Adding $NADP^+$ will drive the process to the right.

6. Yes. Little oxygen would be evolved when 700-nm light is used. Oxygen is evolved by photosystem II, which contains P680 and is therefore not maximally excited by 700-nm light.

7. You would use 430 nm for chlorophyll a and 455 nm for chlorophyll b. These are the wavelengths of maximum absorbance, so they would provide the most sensitive spectrophotometric assays.

8. Chlorophyll appears to be green because it has no significant absorption in the green region of the spectrum and therefore transmits green light.

9. Acetone is the preferred solvent. Because of the hydrophobic porphyrin ring and the very hydrophobic phytol tail of the chlorophylls, they are soluble in organic solvents like acetone but are insoluble in water.

10. Higher animals tend to have limited biosynthetic abilities because their diet contains plants and sometimes other animals. Thus there are many amino acids that animals can't synthesize. One of the most popular herbicides in use today is glyphosate (Roundup®), which inhibits the synthesis of phenylalanine. See Chapter 24 in the text for more discussion. There are several herbicides that block amino acid biosynthesis. They are not very toxic to animals because they block enzymes we lack.

11. The fact that it requires two separate photosystems to promote electrons from water to NADPH implies that the system evolved from something simpler with a single photosystem, which is why the apparatus from *Rhodopseudomonas viridis* is discussed in the chapter. Also, the 4-manganese center that interacts with water (Figure 19.16 in the text) is quite complex and might relate to a 2-manganese center found in catalase. There is evidence for some O_2 appearing in the atmosphere a little more than 2 billion years ago. Water-based photosynthesis cannot be younger than this, and could be significantly older, assuming the oxygen evolved was scavenged up locally. Stromatolites resembling those found today in Shark Bay, Australia (which use oxygen-producing photosynthesis), can be found in layers dated 3.2 billion years old. One interesting theory about early photosynthesis postulates that a major source of electrons could have been the Ferrous (Fe^{+2}) ions that were abundant in the Earth's oceans before oxygen precipitated most of the iron as banded iron formations (*Trends in Biochem. Sci.* 23[1998]:94).

12 "Special pairs" are so named because they are fundamental to photosynthesis. Photosynthesizing bacteria have one such pair, P960. Green plants have two photosystems, each with a special pair: P680 in photosystem II and P700 in photosystem I. Each pair is named for the wavelength of light it best absorbs; P stands for pigment.

Each special pair consists of two adjacent chlorophyll molecules which absorb light and, when excited, emit an electron. This electron is captured by nearby electron carriers and then sent through a photosynthetic electron transport chain. The departure of an electron leaves the special pair positively charged, producing an initial charge separation without which photosynthesis could not take place. The special pairs are then replenished: P680 gets a replacement electron from a nearby cytochrome, P680 takes its electron from H_2O (generating the oxygen in the Earth's atmosphere), and P700 takes its electron from plastocyanin, which in some cases has been reduced by Photosystem II.

The Calvin Cycle and the Pentose Phosphate Pathway

Chapter 16 introduced the glycolytic and gluconeogenic pathways in which glucose was either broken down into or synthesized from pyruvate. These pathways were in many ways mirror images of each other in which many of the same enzymes were used. This chapter introduces two pathways that, like the glycolytic and gluconeogenic pathways, are mirror images of each other. The Calvin cycle (sometimes referred to as the reductive pentose phosphate pathway) uses NADPH to convert carbon dioxide into hexoses, and the pentose phosphate pathway breaks down glucose into carbon dioxide to produce NADPH. The Calvin cycle constitutes the dark reactions of photosynthesis. The light reactions were discussed in Chapter 19; they transform light energy into ATP and biosynthetic reducing power, nicotinamide adenine dinucleotide phosphate (NADPH). While the dark reactions do not directly require light, they do depend on the ATP and NADPH that are produced by the light reactions. The Calvin cycle synthesizes hexoses from carbon dioxide and water in three stages: (1) fixation of CO_2 by ribulose-5-phosphate to form two molecules of 3-phosphoglycerate, (2) reduction of 3-phosphoglycerate to form hexose sugars, and (3) regeneration of ribulose-5-phosphate so that more CO_2 can be fixed. After a discussion of the reactions of the Calvin cycle, the authors proceed to the regulation of the cycle. Carbon dioxide assimilation by the Calvin cycle operates during the day, and carbohydrate degradation to yield energy occurs at night. The discussion of the Calvin cycle concludes with two environmentally dependent modifications to the pathway used by tropical plants and succulents to respond to high temperatures and drought.

The authors next turn their attention to the pentose phosphate pathway, which is common to all organisms. The role of the pentose phosphate pathway is to produce NADPH, which is the currency of reducing power utilized for most reductive biosyntheses. In addition, this pathway generates ribose 5-phosphate needed for DNA synthesis and can produce various size sugars for other uses. The pathway can be separated into oxidative steps in which glucose-6-phosphate and $NADP^+$ are converted into ribulose-5-phosphate, CO_2, and NADPH, and nonoxidative steps in which ribulose-5-phosphate is converted into three-, four-, five-, six-, and seven-carbon sugars. The pentose phosphate pathway is linked to glycolysis (Chapter 16) by the common intermediates glucose-6-phosphate, fructose-6-phosphate, and glyceraldehyde-3-phosphate. The authors discuss the mechanisms of the two enzymes that catalyze the conversion of ribose-5-phosphate into glyceraldehyde-3-phosphate and fructose 6-phosphate, transketolase and transaldolase, respectively. The regulation of the pentose phosphate pathway and the ways in which its activity is coordinated with glycolysis are discussed. The chapter concludes with the role of glucose-6-phosphate dehydrogenase in protection against reactive oxygen species and the physiological consequences of deficiencies in the enzyme.

LEARNING OBJECTIVES

When you have mastered this chapter, you should be able to accomplish the following objectives.

Introduction

1. Distinguish between the *dark* and *light reactions* of photosynthesis.

2. Explain the functions of the *Calvin cycle* and the *pentose phosphate pathway*.

3. Discuss the mirror image nature of the Calvin cycle and the *pentose phosphate pathway*.

The Calvin Cycle Synthesizes Hexoses from Carbon Dioxide and Water
(Text Section 20.1)

4. Differentiate between *heterotrophs* and *autotrophs*.

5. Outline the three stages of the Calvin cycle.

6. Describe the formation of *3-phosphoglycerate* by *ribulose 1,5-bisphosphate carboxylase* (*rubisco*). Note the two different roles of CO_2 and the role of Mg^{2+} in the reaction.

7. Describe the structure of rubisco. Relate the large amount of rubisco present in plants to its slow catalytic rate.

8. Outline the formation of *phosphoglycolate* by the *oxygenase reaction* of rubisco, and follow its subsequent metabolism. Define *photorespiration*.

9. Give two explanations for why it does not seem possible to remove the *oxygenase* activity from rubisco.

10. Outline the conversion of 3-phosphoglycerate into fructose 6-phosphate and the *regeneration of ribulose 1,5-bisphosphate*.

11. Write a balanced equation for the Calvin cycle, and account for the ATP and NADPH expended to form a hexose molecule.

12. Explain the formation of *starch* and *sucrose*.

The Activity of the Calvin Cycle Depends on Environmental Conditions
(Text Section 20.2)

13. List the four light-dependent changes in the stroma that regulate the Calvin cycle.

14. Outline the role of rubisco and *thioredoxin* in coordinating the light and dark reactions of photosynthesis.

15. Describe the C_4 *pathway* and its adaptive value to tropical plants. Explain how CO_2 transport suppresses the oxygenase reaction of rubisco.

16. Explain the use of *crassulacean acid metabolism* by plants growing in arid climates.

The Pentose Phosphate Pathway Generates NADPH and Synthesizes Five-Carbon Sugars (Text Section 20.3)

17. List the two phases of the pentose phosphate pathway. List the biochemical pathways that require NADPH from the pentose phosphate pathway.

18. Describe the reactions of the *oxidative branch* of the pentose phosphate pathway and the regulation of *glucose 6-phosphate dehydrogenase* by $NADP^+$ levels.

19. Explain how the pentose phosphate pathway and the glycolytic pathway are linked through reactions catalyzed by *transaldolase* and *transketolase*.

20. Outline the sugar interconversions of the *nonoxidative branch* of the pentose phosphate pathway.

21. Compare the role of *thiamine pyrophosphate (TPP)* in transketolase with its role in *pyruvate dehydrogenase* and α-*ketoglutarate dehydrogenase*. Outline the enzymatic mechanisms of transketolase and transaldolase.

The Metabolism of Glucose-6-Phosphate by the Pentose Phosphate Pathway Is Coordinated with Glycolysis (Text Section 20.4)

22. State the different product stoichiometries obtained from the pentose phosphate pathway under conditions in which (1) more ribose 5-phosphate than NADPH is needed, (2) there is a balanced requirement for both, (3) more NADPH than ribose 5-phosphate is needed, and (4) both NADPH and ATP are required.

23. Describe the ways in which the Calvin cycle and the pentose phosphate pathway are mirror images of each other.

Glucose-6-Phosphate Dehydrogenase Plays a Key Role in Protection Against Reactive Oxygen Species (Text Section 20.5)

24. Explain why red blood cells are especially senstive to oxidative stress.

25. Discuss the effects of *glucose 6-phosphate dehydrogenase deficiency* on red cells in drug-induced hemolytic anemia, and relate them to the biological roles of *glutathione*.

26. Discuss the oxidation and reduction of glutathione by *glutathione reductase* and *glutathione peroxidase* respectively.

27. Explain the role of reduced glutathione maintaining the normal structure of red blood cells.

28. Give an example of a benefit in a deficiency of glucose-6-phosphate dehydrogenase.

SELF-TEST

The Calvin Cycle Synthesizes Hexoses from Carbon Dioxide and Water

1. Which of the following statements about ribulose 1,5-bisphosphate carboxylase (rubisco) are correct?

   (a) It is present at low concentrations in the chloroplast.

   (b) It is activated by the addition of CO_2 to the ε-amino group of a specific lysine to form a carbamate that then binds a divalent metal cation.

   (c) It catalyzes, as one part of its reaction sequence, an extremely exergonic reaction, the cleavage of a six-carbon diol derivative of arabinitol to form two three-carbon compounds.

   (d) It catalyzes a reaction between ribulose 1,5-bisphosphate and O_2 that decreases the efficiency of photosynthesis.

   (e) It catalyzes the carboxylase reaction more efficiently and the oxygenase reaction less efficiently as the temperature increases.

2. The rubisco-catalyzed reaction of O_2 with ribulose 1,5-bisphosphate forms which of the following substances?

   (a) 3-phosphoglycerate

   (b) 2-phosphoacetate

   (c) phosphoglycolate

   (d) glycolate

   (e) glyoxylate

3. Which of the following statements about 3-phosphoglycerate (3-PG) produced in the Calvin cycle is NOT true?

   (a) It can be used to produce glucose-1-phosphate, glucose-6-phosphate, and fructose-6-phosphate.

   (b) It is converted to hexose phosphates in a series of reactions that are identical to those in the gluconeogenic pathway.

   (c) It produces glyceraldehyde 3-phosphate, which can be transported to the cytosol for glucose synthesis.

   (d) The conversion of 3-PG into hexose phosphates produces energy and reducing equivalents.

   (e) Although both glyceraldehyde 3-phosphate (GAP) and dihydroxyacetone phosphate (DHAP) can be produced from 3-PG, only GAP can be used in further sugar-producing reactions.

4. Place the following sugar conversions in the correct order used to regenerate starting material for the Calvin cycle, and name the enzyme that catalyzes each reaction.

   (a) C_7-ketose + C_3-aldose $\longrightarrow$ C_5-ketose + C_5-aldose

   (b) C_6-ketose + C_3-aldose $\longrightarrow$ C_4-aldose + C_5-ketose

   (c) C_4-aldose + C_3-ketose $\longrightarrow$ C_7-ketose

5. Match the two major storage forms of carbohydrates, starch and sucrose, with the appropriate properties listed in the right column.

   (a) starch
   (b) sucrose

   (1) contains glucose
   (2) contains fructose
   (3) is a polymer
   (4) is synthesized in the cytosol
   (5) is synthesized in chloroplasts
   (6) is synthesized from UDP-glucose

6. Scientists have been unable to create a recombinant rubisco that does not have oxygenase activity. Which of the following are possible explanations as to why?

    (a) The enzyme cannot discriminate between oxygen and carbon dioxide.
    (b) The oxygenase activity is biochemically important for an unknown reason and cannot be removed.
    (c) Because the enzyme is so fast, any diatomic gas will react with it.
    (d) Photorespiration is critically important to plants and without the oxygenase activity, it would not be possible.

The Activity of the Calvin Cycle Depends on Environmental Conditions

7. Which of the following statements about the Calvin cycle are true?

    (a) It regenerates the ribulose 1,5-bisphosphate consumed by the rubisco reaction.
    (b) It forms glyceraldehyde 3-phosphate, which can be converted to fructose 6-phosphate.
    (c) It requires ATP and NADPH.
    (d) It is exergonic because light energy absorbed by the chlorophylls is transferred to rubisco.
    (e) It consists of enzymes, several of which can be activated through reduction of disulfide bridges by reduced thioredoxin.
    (f) It is controlled, in part, by the rate of the rubisco reaction.
    (g) Its rate decreases as the level of illumination increases because both the pH and the level of Mg^{2+} of the stroma decrease.

8. Which of these statements about thioredoxin is correct?

    (a) It contains a heme that cycles between two oxidation states.
    (b) Its oxidized form predominates while light absorption is taking place.
    (c) It activates some biosynthetic enzymes by reducing disulfide bridges.
    (d) It activates some degradative enzymes by reducing disulfide bridges.
    (e) Oxidized thioredoxin is reduced by plastoquinol.

9. Answer the following questions about the C_4 pathway in tropical plants.

    (a) What is the three-carbon CO_2 acceptor in mesophyll cells?
    (b) What is the four-carbon CO_2 donor in bundle-sheath cells?
    (c) What is the net reaction for the C_4 pathway?
    (d) Is the C_4 pathway a type of active or passive transport?

10. Plants growing in hot environments have developed adaptations in CO_2 storage and use. Contrast the adaptive differences between plants growing in tropical and arid environments.

The Pentose Phosphate Pathway Generates NADPH and Synthesizes Five-Carbon Sugars

11. Which of the following compounds is NOT a product of the pentose phosphate pathway?

    (a) NADPH
    (b) glycerate 3-phosphate
    (c) CO_2
    (d) ribulose 5-phosphate
    (e) sedoheptulose 7-phosphate

12. Figure 20.1 shows the first four reactions of the pentose phosphate pathway. Use it to answer the questions.

FIGURE 20.1   Oxidative reactions of the pentose phosphate pathway.

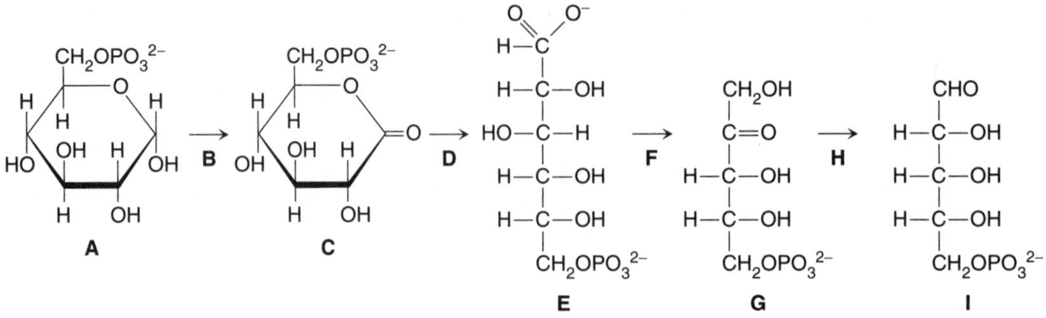

(a) Which reactions produce NADPH?
(b) Which reaction produces CO_2?
(c) Which compound is ribose 5-phosphate?
(d) Which compound is 6-phosphoglucono-δ-lactone?
(e) Which compound is 6-phosphogluconate?
(f) Which reaction is catalyzed by phosphopentose isomerase?
(g) Which enzyme is deficient in drug-induced hemolytic anemia?
(h) Which compound can be a group acceptor in the transketolase reaction?

13. Which of the following statements about glucose 6-phosphate dehydrogenase are correct?

(a) It catalyzes the committed step in the pentose phosphate pathway.
(b) It is regulated by the availability of NAD^+.
(c) One of its products is 6-phosphogluconate.
(d) It contains thiamine pyrophosphate as a cofactor.
(e) It is important in the metabolism of glutathione in erythrocytes.

14. The nonoxidative branch of the pentose phosphate pathway does NOT include which of the following reactions?

(a) Ribulose 5-P ⟶ ribose 5-P
(b) Xylulose 5-P + ribose 5-P ⟶ sedoheptulose 7-P ⟶ glyceraldehyde 3-P
(c) Ribulose 5-P + glyceraldehyde 3-P ⟶ sedoheptulose 7-P
(d) Sedoheptulose 7-P + glyceraldehyde 3-P ⟶ fructose 6-P + erythrose 4-P
(e) Ribulose 5-P ⟶ xylulose 5-P

15. Liver synthesizes fatty acids and lipids for export to other tissues. Would you expect the pentose phosphate pathway to have a low or a high activity in this organ? Explain your answer.

16. Transaldolase and transketolase have which of the following similarities?

(a) require thiamine pyrophosphate
(b) form a Schiff base with the substrate
(c) use an aldose as a group donor
(d) use a ketose as a group donor
(e) form a covalent addition compound with the donor substrate

The Metabolism of Glucose-6-Phosphate by the Pentose Phosphate Pathway Is Coordinated with Glycolysis

17. Which enzyme catalyzes the rate limiting step in the pentose phosphate pathway?

(a) lactonase
(b) transaldolase
(c) transketolase

    (d) glucose 6-phosphate dehydrogenase
    (e) 6-phosphogluconate dehydrogenase

18. Which of the following conversions take place in a metabolic situation that requires much more NADPH than ribose 5-phosphate, as well as complete oxidation of glucose 6-phosphate to CO_2? The arrows represent one or more enzymatic steps.
    (a) glucose 6-phosphate $\rightarrow$ ribulose 5-phosphate
    (b) fructose 6-phosphate $\rightarrow$ glyceraldehyde 3-phosphate $\rightarrow$ ribose 5-phosphate
    (c) ribose 5-phosphate $\rightarrow$ fructose 6-phosphate $\rightarrow$ glyceraldehyde 3-phosphate
    (d) glyceraldehyde 3-phosphate $\rightarrow$ pyruvate
    (e) fructose 6-phosphate $\rightarrow$ glucose 6-phosphate

19. List two ways in which the Calvin cycle and the pentose phosphate pathway are mirror images of each other.

Glucose-6-Phosphate Dehydrogenase Plays a Key Role in Protection Against Reactive Oxygen Species

20. Which of the following statements about reduced glutathione is NOT true?
    (a) It contains one γ-carboxyglutamate, one cysteine, and one glycine residue.
    (b) It keeps the cysteine residues of proteins in their reduced states.
    (c) It is regenerated from oxidized glutathione by glutathione reductase.
    (d) It reacts with hydrogen peroxide and organic peroxides.
    (e) It is decreased relative to oxidized glutathione in glucose 6-phosphate dehydrogenase deficiency.

21. Suggest reasons why glucose 6-phosphate dehydrogenase deficiency may be manifested in red blood cells but not in adipocytes, which also require NADPH for their metabolism.

ANSWERS TO SELF-TEST

1. b, c, d. Answer (e) is incorrect because the rate of the oxygenase reaction increases relative to that of the carboxylase reaction as the temperature increases; the altered ratio of the two reaction rates decreases the efficiency of photosynthesis as the temperature increases.

2. a, c.

3. The incorrect statements are b, d, and e. Statement (b) is incorrect because the gluconeogenic pathway uses NADH, not NADPH; (d) is incorrect because the conversion requires both ATP and NADPH; e is incorrect because both GAP and DHAP can be used to produce larger sugars.

4. The correct order is b, c, a. Transketolase catalyzes reactions (a) and (b); aldolase catalyzes reaction (c).

5. (a) 1, 3, 5   (b) 1, 2, 4, 6

6. a, b. Answer (c) is incorrect because the enzyme is very slow. Answer (d) is incorrect because photorespiration is harmful to plants, and ideally would be avoided.

7. a, b, c, e, f

8. c. Thioredoxin contains cysteine residues that cycle between two oxidation states. It is reduced by ferredoxin while the light reactions are proceeding. It activates biosynthetic enzymes and inhibits degradative enzymes by reducing their disulfide bridges.

9. (a) Phosphoenolpyruvate is the three-carbon CO_2 acceptor in mesophyll cells.
   (b) Malate is the four-carbon CO_2 donor in bundle-sheath cells.
   (c) CO_2 (in mesophyll cell) + ATP + H_2O → CO_2 (in bundle-sheath cell) + AMP + $2P_i$ + H^+
   (d) It is a type of active transport because it requires ATP to function.

10. Plants in arid environments use crassulacean acid metabolism to store CO_2 absorbed at night in the form of malate until it can be used during the day. Plants in tropical environments use the C4 pathway to increase the concentration of CO_2 in bundle-sheath cells thus accelerating the carboxylase reaction relative to the oxygenase reaction and minimizing photorespiration. In contrast with C4 plants, CAM plants separate CO_2 accumulation from CO_2 utilization temporally rather than spatially.

11. b

12. (a) B, F   (b) F   (c) I   (d) C   (e) E   (f) H   (g) B   (h) I

13. a, e

14. c

15. The activity of the pentose phosphate pathway in the liver is high. The biosynthesis of fatty acids and lipids requires reducing equivalents in the form of NADPH. In all organs that carry out reductive biosyntheses, the pentose phosphate pathway supplies a large proportion of the required NADPH.

16. d, e

17. d

18. a, c, e. Glucose 6-phosphate is converted to ribulose 5-phosphate, producing CO_2 and NADPH in the process. Then ribulose 5-phosphate, via ribose 5-phosphate, is transformed into fructose 6-phosphate and glyceraldehyde 3-phosphate. These two glycolytic intermediates are converted back to glucose 6-phosphate, and the cycle is repeated until the equivalent of six carbon atoms from glucose 6-phosphate are converted to CO_2.

19. (1) Calvin cycle fixes CO_2 and utilizes NADPH to form sugars while the PPP oxidizes a sugar to form CO_2 and generates NADPH. (2) Calvin cycle converts C6 and C3 molecules into a C5 molecule while the PPP converts a C5 molecule into C6 and C3 molecules.

20. a. Answer (a) is incorrect because the glutamate residue in glutathione is not γ-carboxyglutamate; rather, the glutamate in glutathione forms a peptide bond with the adjacent cysteine residue via its γ-carboxyl group.

21. The glucose 6-phosphate dehydrogenase in erythrocytes and that in adipocytes are specified by distinct genes; they have the same function but different structures—that is, they are isozymes. Furthermore, NADPH synthesis by the pentose phosphate pathway may not be as critical in the cells of other tissues as it is in erythrocytes because other tissues have other sources of NADPH.

PROBLEMS

1. Outline the synthesis of fructose 6-phosphate from 3-phosphoglycerate.

2. How many moles of ATP and NADPH are required to convert 6 moles of CO_2 to fructose 6-phosphate?

3. Describe photorespiration, and explain why it decreases the efficiency of photosynthesis.

4. It is said that the C_4 pathway increases the efficiency of photosynthesis. What is the justification for this statement when more than 1.6 times as much ATP is required to convert 6 moles of CO_2 to a hexose when this pathway is used in contrast with the pathway used by plants lacking the C_4 apparatus? Account for the extra ATP molecules used in the C_4 pathway.

5. In addition to the well-understood ferredoxin-thioredoxin couple, NADPH can regulate Calvin cycle enzymes. The text gives the example of a recently discovered assembly protein CP12 (page 599), which binds to and inhibits phosphoribulose kinase (PRK) and glyceraldehyde 3-phosphate dehydrogenase (GAPDH) in the dark and releases them in the light (Wedel et al. *Proc. Nat. Acad. Sci.* 94 [1997]:10479–10484). The authors of the paper found that NADPH triggers the release of PRK and GADPH from CP12 and is also necessary for PRK activity after its release. They also noted that PRK is rapidly oxidized in the absence of reduced thioredoxin; it remains reduced when bound to CP12.

   (a) Why would PRK require NADPH for full activity given that is does not catalyze a reduction reaction?

   (b) Given this information, what is a possible role of PRK binding to CP12?

6. The conversion of glucose 6-phosphate to ribose 5-phosphate via the enzymes of the pentose phosphate pathway and glycolysis can be summarized as follows:

$$5 \text{ glucose 6-phosphate} + \text{ATP} \longrightarrow 6 \text{ ribose 5-phosphate} + \text{ADP} + \text{H}^+$$

   Which enzyme uses the molecule of ATP shown in the equation?

7. Liver and other organ tissues contain relatively large quantities of nucleic acids. During digestion, nucleases hydrolyze RNA and DNA, and among the products is ribose 5-phosphate.

   (a) How can this molecule be used as a metabolic fuel?

   (b) Another product formed by the degradation of nucleic acids is 2-deoxyribose 5-phosphate. Can this molecule be converted to glycolytic intermediates through the action of the pentose phosphate pathway? Explain your answer.

8. You have glucose that is radioactively labeled with ^{14}C at C-1, and you have an extract that contains the enzymes that catalyze the reactions of the glycolytic and the pentose phosphate pathways, along with all the intermediates of the pathways.

   (a) If the enzymes of the *oxidative* branch of the pentose phosphate pathway are *not* active in your extract, is it possible to obtain labeled sedoheptulose 7-phosphate using glucose labeled with ^{14}C at C-1? Explain.

   (b) Suppose that in a second experiment *all* the enzymes of both the oxidative branch and the nonoxidative branch of the pentose phosphate pathway are active. Will the labeling pattern of sedoheptulose 7-phosphate be different? Explain.

   (c) Can sedoheptulose 7-phosphate form a heterocyclic ring?

9. Why is the pentose phosphate pathway more active in cells that are dividing than in cells that are not?

10. A bacterium isolated from a soil culture can utilize ribose as a sole source of carbon when grown anaerobically. Experiments show that in the anaerobic pathways leading to ATP production, three molecules of ribose are converted to five molecules of CO_2

and five molecules of ethanol. These organisms also use ribose for the production of NADPH. The assimilation of ribose begins with its conversion to ribose 5-phosphate, with ATP serving as a phosphoryl donor.

(a) Explain how ribose can be converted to CO_2 and ethanol under anaerobic conditions. Write the overall reaction, showing how much ATP can be produced per pentose utilized.

(b) Write an equation for the generation of NADPH using ribose as a sole source of carbon.

11. Mature erythrocytes, which lack mitochondria, metabolize glucose at a high rate. In response to the increased availability of glucose, erythrocytes generate lactate and also evolve carbon dioxide.

(a) Why is generation of lactate necessary to ensure the continued utilization of glucose?

(b) In erythrocytes, what pathway is likely to be used for the generation of carbon dioxide from glucose? Can glucose be completely oxidized to CO_2 in erythrocytes? Explain.

12. A biochemist needs to determine whether a particular tissue homogenate has a high level of pentose phosphate pathway activity. She incubates one sample with ^{14}C-1 glucose and another with ^{14}C-6 glucose. Then she measures the specific activity of radioactive CO_2 generated by each sample. Her measurements show that the specific activity of CO_2 from the experiment using glucose labeled at C-1 is much higher than that from the sample in which glucose labeled at C-6 was used. What is her conclusion?

13. Even if glucose 6-phosphate dehydrogenase is deficient, the synthesis of ribose 5-phosphate from glucose 6-phosphate can proceed normally. Explain how this is possible.

ANSWERS TO PROBLEMS

1. Phosphoglycerate kinase converts 3-phosphoglycerate, the initial product of photosynthesis, to the glycolytic intermediate 1,3-bisphosphoglycerate, which is then converted to glyceraldehyde 3-phosphate (G-3-P) by an NADPH-dependent G-3-P dehydrogenase in the chloroplast. Triosephosphate isomerase converts G-3-P to dihydroxyacetone phosphate, which aldolase can condense with another G-3-P to form fructose 1,6-bisphosphate. The phosphate ester at C-1 is hydrolyzed to give fructose 6-phosphate. The result of this pathway, which is functionally equivalent to the gluconeogenic pathway, is the conversion of the CO_2 fixed by photosynthesis into a hexose.

2. Eighteen moles of ATP and twelve moles of NADPH are required to fix six moles of CO_2. Two moles of ATP are used by phosphoglycerate kinase to form two moles of 1,3-bisphosphoglycerate, and one mole of ATP is used by ribulose 5-phosphate kinase to form one mole of ribulose 1,5-bisphosphate per mole of CO_2 fixed. Two moles of NADPH are used by G-3-P dehydrogenase to form two moles of G-3-P per mole of CO_2 incorporated. Therefore, three moles of ATP and two moles of NADPH are used for each mole of CO_2 fixed.

3. The oxygenase reaction of rubisco and the salvage reactions that convert two resulting phosphoglycolate molecules into serine are called *photorespiration* because CO_2 is released and O_2 is consumed in the process. Unlike genuine respiration, no ATP or NADPH is produced by photorespiration. Ordinarily, no CO_2 is released during

photosynthesis, and all the fixed CO_2 can be used to form hexoses. During photorespiration, no CO_2 is fixed, and the products into which ribulose 1,5-bisphosphate is converted by the oxygenase reaction of rubisco cannot be completely recycled into carbohydrate because of the loss of CO_2 in the phosphoglycolate salvage reactions.

4. Plants lacking the C_4 pathway cannot compensate for the relative increase in the rate of the oxygenase reaction of rubisco with respect to the rate of the carboxylase reaction that occurs as the temperature rises. Plants with the C_4 pathway increase the concentration of CO_2 in the bundle-sheath cell, where the Calvin cycle occurs, thereby increasing the ability of CO_2 to compete with O_2 as a substrate for rubisco. As a result, more CO_2 is fixed and less ribulose 1,5-bisphosphate is degraded into phosphoglycolate, which cannot be efficiently converted into carbohydrate. Thus, the Calvin cycle functions more efficiently in these specialized plants under conditions of high illumination and at higher temperatures than it would otherwise. The concentration of CO_2 is increased by an expenditure of ATP. The collection of one CO_2 molecule and its transport on C_4 compounds from the mesophyll cell into the bundle-sheath cell is brought about by the conversion of one ATP to AMP and PP_i in a reaction in which pyruvate is phosphorylated to PEP. The PP_i is hydrolyzed, and two ATP are required to resynthesize ATP from AMP. Thus, an *extra* ATP/CO_2 × 6 CO_2/hexose = 12 ATP/hexose are used by the C_4 pathway.

5. (a) Since the purpose of PRK is to regenerate ribulose 1,5-bisphosphate for use in the Calvin cycle, it does not make sense to have it active when there is not enough NADPH to run the cycle. The PRK reaction requires ATP and would be wasteful if ribulose 1,5-bisphosphate were not needed.

   (b) In the absence of CP12 complex formation, PRK is rapidly oxidized and becomes inactive. In conditions of low NADPH, if complex formation did not occur, PRK would reoxidze and become inactive before producing ribulose 1,5-bisphosphate. The light energy used to reduce thioredoxin would therefore be wasted. By keeping thioredoxin-reduced PRK bound to CP12 until enough NADPH is present, the light energy is not wasted.

6. Phosphofructokinase uses ATP to convert fructose 6-phosphate to fructose 1,6-bisphosphate, which is then cleaved by aldolase to yield dihydroxyacetone phosphate (DHAP) and glyceraldehyde 3-phosphate. The conversion of DHAP to a second molecule of glyceraldehyde 3-phosphate provides the molecules that are needed for the synthesis of ribose 5-phosphate.

7. (a) The most direct route for the oxidative degradation of ribose 5-phosphate is its conversion to glycolytic intermediates by the nonoxidative enzymes of the pentose phosphate pathway. The overall reaction is

   3 ribose 5-phosphate $\rightarrow$ 2 fructose 6-phosphate $\rightarrow$ glyceraldehyde 3-phosphate

   (b) The formation of glycolytic intermediates from 2-deoxyribose 5-phosphate is not possible, because unlike ribose 5-phosphate, 2-deoxyribose 5-phosphate lacks a hydroxyl group at C-2. It is therefore not a substrate for phosphopentose isomerase, whose action is required to convert ketopentose phosphates to substrates that can be utilized by other enzymes of the pentose phosphate pathway. Most deoxyribose phosphate molecules are used in salvage pathways to form deoxynucleotides.

8. (a) Yes. The most direct route would be the conversion of glucose to fructose 6-phosphate, followed by the condensation of fructose 6-phosphate with erythrose 4-phosphate to form sedoheptulose 7-phosphate and glyceraldehyde 3-phosphate.

The labeled carbon of glucose becomes the C-1 of fructose 6-phosphate and C-1 of sedoheptulose 7-phosphate.

(b) The labeling pattern will be the same, although the amount of labeled carbon incorporated into the heptose will be reduced. In the oxidative branch of the pentose phosphate pathway, the labeled glucose is converted to glucose 6-phosphate with the ^{14}C label on C-1. Glucose 6-phosphate then undergoes successive oxidations and decarboxylation to form ribulose 5-phosphate. The label is lost when the C-1 carbon is removed during decarboxylation.

(c) Sedoheptulose 7-phosphate is a ketose and can form a heterocyclic ring through a hemiketal linkage. The most likely link would be between the keto group at C-2 and the hydroxyl group at C-6.

9. Cells have a high rate of nucleic acid biosynthesis when they grow and divide. Among the precursors needed is ribose 5-phosphate, which is synthesized through the action of the enzymes of the glycolytic and the pentose phosphate pathways. Biosynthetic reactions requiring NADPH occur at a high rate in growing and dividing cells. For these reasons, the enzymes of the pentose phosphate pathway will be extremely active in dividing cells.

10. (a) To generate ATP, ethanol, and CO_2, ribose must first be converted to ribose 5-phosphate, with ATP serving as a phosphate donor. Then, in the nonoxidative branch of the pentose phosphate pathway, three molecules of ribose 5-phosphate are converted to two molecules of fructose 6-phosphate and one molecule of glyceraldehyde 3-phosphate. Two molecules of ATP are required for the production of fructose 1,6-bisphosphate from fructose 6-phosphate. The formation of a total of five molecules of glyceraldehyde 3-phosphate is achieved through the action of aldolase and triose phosphate isomerase. These five molecules are converted to five molecules of pyruvate, yielding ten ATP molecules and five NADH molecules. To keep the anaerobic cell in redox balance, the pyruvate molecules are converted to five molecules of ethanol, with the production of five CO_2 molecules and five NAD^+. The overall reaction is 3 ribose + 5 ADP + P_i → 5 ethanol + 5 CO_2 + 5 ADP.

(b) Ribose 5-phosphate molecules must first be converted to glucose 6-phosphate for the oxidative enzymes of the pentose pathway to generate NADPH. The stoichiometry of the reactions is

6 ribose 5-phosphate → 4 fructose 6-phosphate + 2 glyceraldehyde 3-phosphate

4 fructose 6-phosphate → 4 glucose 6-phosphate

2 glyceraldehyde 3-phosphate → glucose 6-phosphate + P_i

5 glucose 6-phosphate + 10 $NADP^+$ + 5 H_2O → 5 ribose 5-phosphate + 10 NADPH + 10 H^+ + 5 CO_2

The net reaction is

ribose 5-phosphate + 10 $NADP^+$ + 5 H_2O → 10 NADPH + 10 H^+ + 5 CO_2 + P_i

11. (a) Because erythrocytes lack mitochondria, they cannot use the citric acid cycle to regenerate the NAD^+ needed to sustain glycolysis. Instead, they regenerate NAD^+ by reducing pyruvate through the action of lactate dehydrogenase; NAD^+ is then reduced in the reaction catalyzed by glyceraldehyde 3-phosphate dehydrogenase during glycolysis. Failure to oxidize the NADH generated in the glycolytic pathway will cause a reduction in the rate of glucose breakdown.

(b) In erythrocytes, the pentose phosphate pathway is the only route available to yield CO_2 from glucose. Glucose can be completely oxidized by first entering the oxidative branch of the pathway, generating NADPH and ribose 5-phosphate. Transaldolase and transketolase then convert the pentose phosphates to fructose 6-phosphate and glyceraldehyde 3-phosphate. Part of the gluconeogenic pathway is used to convert both the products to glucose 6-phosphate. The net reaction is

$$\text{glucose 6-P} + 12 \text{ NADP}^+ + 7 \text{ H}_2\text{O} \longrightarrow 6 \text{ CO}_2 + 12 \text{ NADPH} + 12 \text{ H}^+ + \text{P}_i$$

12. The experiments show that the activity of the pentose phosphate pathway is high. In the pentose phosphate pathway, glucose labeled at C-1 is decarboxylated, while glucose labeled at C-6 is not. On the other hand, both C-1- and C-6-labeled glucose are decarboxylated to the same extent by the combined action of the glycolytic pathway and the citric acid cycle. Because in these experiments, the specific activity (ratio of labeled CO_2 to total CO_2) is higher for C-1-labeled glucose, much of the glucose in the experiment must be moving through the pentose phosphate pathway.

13. Ribose 5-phosphate can be synthesized from fructose 6-phosphate and glyceraldehyde 3-phosphate, both of which are glycolytic products of glucose 6-phosphate. These reactions are carried out by transketolase and transaldolase in a reversal of the nonoxidative branch of the pentose phosphate pathway and do not involve glucose 6-phosphate dehydrogenase.

Glycogen Metabolism

The topic of carbohydrate metabolism presented in Chapters 16 and 20 is further developed in this chapter with a detailed discussion of the metabolism of glycogen, the intracellular storage form of glucose. Glycogen is important in the metabolism of higher animals because its glucose residues can be easily mobilized by the liver to maintain blood glucose levels and used by muscle to satisfy its energy needs during bursts of contraction. The text first reviews briefly the structure and the physiologic roles of glycogen and provides an overview of its metabolism. You were introduced briefly to the structure of glycogen, a polymer of glucose, in Section 11.2 of Chapter 11. Next, with glucose as the ending and starting points, the text presents first the enzymatic reactions of glycogen degradation and then of synthesis. The control of these catabolic and anabolic reactions by allosteric mechanisms and the phosphorylation and dephosphorylation of the key enzymes in response to hormonal signals is discussed. AMP, ATP, glucose, and glucose 6-phosphate act as allosteric effectors, and the hormones insulin, glucagon, and epinephrine function as signals in transduction pathways that control critical enzyme phosphorylations and dephosphorylations. The text describes relevant structures and control mechanisms for phosphorylase, phosphorylase kinase, glycogen synthase, the branching enzyme, and protein phosphatase 1. The differences in glycogen metabolism in muscle and liver are related to the distinct physiologic functions these tissues perform. The text concludes the chapter with a discussion of the biochemical basis of several glycogen storage diseases in humans.

LEARNING OBJECTIVES

When you have mastered this chapter, you should be able to accomplish the following objectives.

Introduction

1. Describe the structure of *glycogen* and its roles in the *liver* and *muscle*. Distinguish between *α-1,4 glycosidic linkages* and *α-1,6 glycosidic linkages*.

2. Describe the three steps of *glycogen catabolism* and the three fates of its product, *glucose 6-phosphate*.

3. Describe the precursors of *glycogen anabolism*.

4. Explain the role of *allosteric responses* and *hormones* in regulating glycogen metabolism.

Glycogen Breakdown Requires the Interplay of Several Enzymes (Text Section 21.1)

5. Write the phosphorolysis reaction catalyzed by *glycogen phosphorylase*.

6. Explain the advantage of the *phosphorolytic cleavage* of glycogen over *hydrolytic cleavage*.

7. Describe the roles of *pyridoxal 5′-phosphate, orthophosphate, general acid–base catalysis, Schiff-base formation,* and the *carbonium ion intermediate* in the mechanism of action of *glycogen phosphorylase.*

8. Define *processivity* as it relates to glycogen phosphorylase activity.

9 Outline the steps in the degradation of glycogen, and relate them to the action of *phosphorylase, transferase,* and *α-1,6-glucosidase,* which is also known as the *debranching enzyme.* Explain why the glycogen molecule must be remodeled during its degradation.

10. Compare the reaction mechanisms of *phosphoglucomutase* and *phosphoglycerate mutase* and describe their common mechanistic features.

11. Explain the importance of *glucose 6-phosphatase* in the release of glucose by the liver. Note the absence of this enzyme in the brain and muscle and rationalize this tissue distribution.

Phosphorylase is Regulated by Allosteric Interactions and Reversible Phosphorylation (Text Section 21.2)

12. Appreciate that the two primary regulatory mechanisms for glycogen phosphorylase are interactions with *allosteric effectors* and *reversible covalent modifications.*

13. Describe the phosphorylation of phosphorylase by *phosphorylase kinase.*

14. Explain the relationships between *phosphorylase a* and *phosphorylase b,* their *T (tense)* and *R (relaxed)* forms, and the allosteric effectors that mediate their interconversions in skeletal muscle. Outline the molecular bases for the relative inactivities of the T states.

15. Contrast the regulation of liver phosphorylase and muscle phosphorylase.

16. Contrast the important structural features of phosphorylase *a* and phosphorylase *b*. Note the variety of binding sites, their functional roles, and the critical location of the phosphorylation and AMP-binding sites near the subunit interface.

17. Describe the major compositional features of phosphorylase kinase and its activation by *protein kinase A (PKA)*. Explain the effects of *calmodulin* and Ca^{2+} on glycogen metabolism in muscle and liver.

Epinephrine and Glucagon Signal the Need for Glycogen Breakdown
(Text Section 21.3)

18. Compare the effects of *glucagon* and *epinephrine* on glycogen metabolism in liver and in muscle.

19. List the sequence of events from the binding of hormones by their receptors to the phosphorylation of glycogen synthase and phosphorylase. Explain the roles of *G proteins, cAMP,* and PKA in these processes.

20. Explain the role of *protein phosphatase 1 (PP1)* in glycogen metabolism.

21. Describe the progression of regulatory mechanisms during the evolution of the glycogen phosphorylases.

Glycogen is Synthesized and Degraded by Different Pathways (Text Section 21.4)

22. Explain the roles of *UDP-glucose* and *inorganic pyrophosphatase* in the synthesis of glycogen.

23. Outline the steps in the synthesis of glycogen, name the pertinent enzymes, and note the requirement for a *primer*. Describe the actions of *glycogenin* and *glycogen synthase*.

24. Explain why glycogen lacks *glucose residues* that can be *reduced*.

25. Explain the functional importance of *branching* in the glycogen molecule.

26. Describe the regulation of glycogen synthase by *reversible covalent modification*.

27. Discuss the efficiency of glycogen as a storage form of glucose.

Glycogen Breakdown and Synthesis are Reciprocally Regulated
(Text Section 21.5)

28. Contrast the effects of phosphorylation on glycogen synthase and glycogen phosphorylase. Appreciate the *reciprocal regulation strategies* employed and the consequences of *amplification cascades*.

29. Explain the role of *protein phosphatase 1* (PP1) in the control of the activities of glycogen phosphorylase and glycogen synthase.

30. Outline the effects of *insulin* on glycogen metabolism. Rationalize the existence of distinct pathways for the biosynthesis and degradation of glycogen.

31. Describe the events that lead to the inactivation of phosphorylase and the activation of glycogen synthase by glucose in the liver. Note the role of phosphorylase *a* as the glucose sensor in liver cells and the participation of phosphorylase *a* and PP1 in glucose sensing.

32. Provide examples of glycogen storage diseases, and relate the biochemical defects with the clinical observations. Use the disease discovered by von Gierke to show how a deficiency in one of several different enzymes can cause the same disease.

SELF-TEST

Introduction

1. Answer the questions about the glycogen fragment in Figure 21.1.

   FIGURE 21.1   Fragment of glycogen. (*R* represents the rest of the glycogen molecule.)

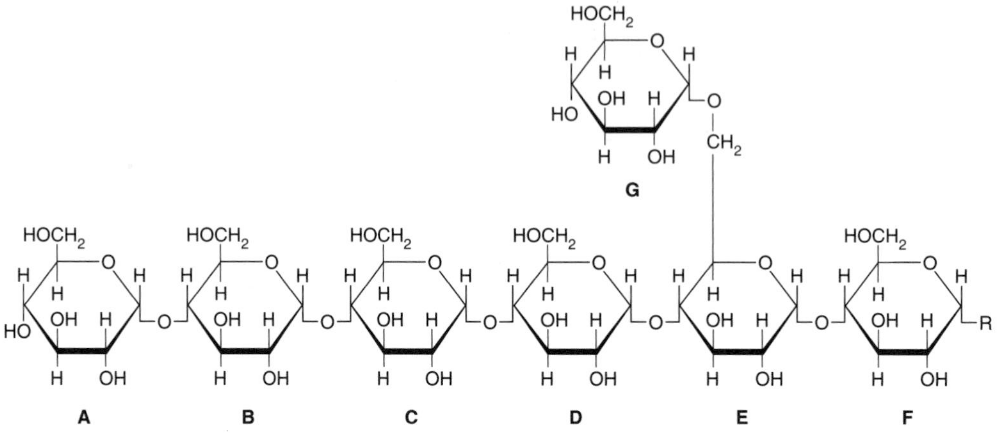

   (a) Which residues are at nonreducing ends? *G, A*
   (b) An α-1,6 glycosidic linkage occurs between which residues? *G, E*
   (c) An α-1,4 glycosidic linkage occurs between which residues? *A BCDEF*
   (d) Is the glycogen fragment a substrate for phosphorylase *a*? Explain. *No – needs to be over 4 units from*
   (e) Is the glycogen fragment a substrate for the debranching enzyme? Explain. *yes.*
   (f) Is the glycogen fragment a substrate for the branching enzyme? Explain. *No – needs to transfer 7 residues from a chain of 11*

2. Which of the following statements about glycogen storage are NOT correct?

   (a) Glycogen is stored in muscles and liver.
   (b) Glycogen is a major source of stored energy in brain.
   (c) Glycogen reserves are less rapidly depleted than fat reserves during starvation.
   (d) Glycogen nearly fills the nucleus of cells that specialize in glycogen storage.
   (e) Glycogen storage occurs in the form of dense granules in the cytoplasm of cells.

3. Is the largest total mass of glycogen found in the liver or the muscle? *muscle*

Glycogen Breakdown Requires the Interplay of Several Enzymes

4. Explain why the phosphorolytic cleavage of glycogen is more energetically advantageous than its hydrolytic cleavage.

5. Which of the following statements about the role of pyridoxal phosphate in the mechanism of action of phosphorylase are correct?

   (a) It interacts with orthophosphate.
   (b) It acts as a general acid–base catalyst.
   (c) It orients the glycogen substrate in the active site.
   (d) It donates a proton directly to the O-4 of the departing glycogen chain.
   (e) It binds water at the active site.

don't have to use ATP w/ phosphorolytic cleavage. (already phosphorylated)

6. Match the enzymes that degrade glycogen in the left column with the appropriate properties in the right column.

(a) Phosphorylase
(b) α-1,6-Glucosidase
(c) Transferase

(1) is part of a single polypeptide chain with two activities.
(2) cleaves α-1,4 glucosidic bonds.
(3) releases glucose.
(4) releases glucose 1-phosphate.
(5) moves three sugar residues from one chain to another.
(6) requires ATP.

7. The phosphoglucomutase reaction is similar to the phosphoglyceromutase reaction of the glycolytic pathway. Which of the following properties are common to both enzymes?

(a) Both have a phosphoenzyme intermediate.
(b) Both use a glucose 1,6-bisphosphate intermediate.
(c) Both contain pyridoxal phosphate, which donates its phosphate group to the substrate.
(d) Both transfer the phosphate group from one position to another on the same molecule.

8. The activity of which of the following enzymes is NOT required for the release of large amounts of glucose from liver glycogen?

(a) glucose 6-phosphatase
(b) fructose 1,6-bisphosphatase
(c) α-1,6-glucosidase
(d) phosphoglucomutase
(e) glycogen phosphorylase

9. Answer the following questions about the enzymatic degradation of amylose, a linear α-1,4 polymer of glucose that is a storage form of glucose in plants.

(a) Would phosphorylase act on amylose? Explain.
(b) Would the rates of glucose 1-phosphate release from an amylose molecule by phosphorylase relative to that from a glycogen molecule having an equivalent number of glucose monomers be equal? Explain.
(c) If the amylose were first treated with an endosaccharidase that cleaved some of its internal glycosidic bonds, how might the rate of production of glucose 1-phosphate be affected?

10. Starting from a glucose residue in glycogen, how many net ATP molecules will be formed in the glycolysis of the residue to pyruvate?

(a) 1
(b) 2
(c) 3
(d) 4
(e) 5

Phosphorylase Is Regulated by Allosteric Interactions and Reversible Phosphorylation

11. Consider the diagram of the different conformational states of muscle glycogen phosphorylase in Figure 21.2. Then answer the questions.

FIGURE 21.2 Conformational states of phosphorylase in muscle.

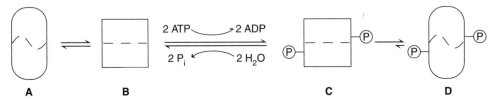

(a) Which are the active forms of phosphorylase?
(b) Which form requires high levels of AMP to become activated?
(c) Which conversion is antagonized by ATP and glucose 6-phosphate?
(d) What enzyme catalyzes the conversion of C to B?

12. How does the regulation of phosphorylase in the liver differ from the scheme for phosphorylase regulation in muscle shown in Figure 21.2?

13. Indicate which of the following substances have binding sites on phosphorylase. For those that do, give their major roles or effects.
    (a) calmodulin
    (b) glycogen
    (c) pyridoxal phosphate
    (d) Ca^{2+}
    (e) AMP
    (f) P_i
    (g) ATP
    (h) glucose

14. Explain the roles of protein kinase A and calmodulin in the control of phosphorylase kinase in the muscle.

Epinephrine and Glucagon Signal the Need for Glycogen Breakdown

15. Place the following steps of the reaction cascade of glycogen metabolism in the proper sequence.
    (a) phosphorylation of protein kinase A
    (b) formation of cyclic AMP by adenylate cyclase
    (c) phosphorylation of phosphorylase b
    (d) hormone binding to target cell receptors
    (e) phosphorylation of phosphorylase kinase

16. Why are enzymatic cascades, such as those that control glycogen metabolism and the clotting of blood, of particular importance in metabolism?

17. Compare and contrast the action of epinephrine and glucagon in glycogen metabolism.

Glycogen Is Synthesized and Degraded by Different Pathways

18. Which of the following features are common to both glycogen synthesis and glycogen breakdown?

    (a) Both require UDP-glucose.
    (b) Both involve glucose 1-phosphate.
    (c) Both are driven in part by the hydrolysis of pyrophosphate.
    (d) Both occur on cytoplasmic glycogen granules.
    (e) Both use the same enzyme for branching and debranching.

19. If glycogen synthase can add a glucose residue to a growing glycogen molecule only if the glucose chain is at least four units long, how does a new glycogen molecule start?

20. Why is the existence of distinct biosynthetic and catabolic pathways for glycogen important for the metabolism of liver and muscle cells?

21. Is it true or false that branching in the structure of glycogen increases the rates of its synthesis and degradation? Explain.

22. Which of the following statements about glycogen synthase are correct?

    (a) It is activated when it is dephosphorylated.
    (b) It is activated when it is phosphorylated.
    (c) It is activated when it is phosphorylated and in the presence of high levels of glucose 6-phosphate.
    (d) It is activated when it is phosphorylated and in the presence of high levels of AMP.

Glycogen Breakdown and Synthesis are Reciprocally Regulated

23. Which of the following statements about the hormonal regulation of glycogen synthesis and degradation are correct?

    (a) Insulin increases the capacity of the liver to synthesize glycogen.
    (b) Insulin is secreted in response to low levels of blood glucose.
    (c) Glucagon and epinephrine have opposing effects on glycogen metabolism.
    (d) Glucagon stimulates the breakdown of glycogen, particularly in the liver.
    (e) The effects of all three of the regulating hormones are mediated by cyclic AMP.

24. Which of the following are effects of glucose on the metabolism of glycogen in the liver?

    (a) The binding of glucose to phosphorylase a converts this enzyme to the inactive T form.
    (b) The T form of phosphorylase a becomes susceptible to the action of phosphatase.
    (c) The R form of phosphorylase b becomes susceptible to the action of phosphorylase kinase.
    (d) When phosphorylase a is converted to phosphorylase b, the bound phosphatase is released.
    (e) The free phosphatase dephosphorylates and activates glycogen synthase.

25. What would increased epinephrine do to protein phosphatase 1 (PP1) in muscle, and how would muscle glycogen metabolism be affected?

26. Explain the effect of insulin on the activity of protein phosphatase 1 and the subsequent effects on glycogen metabolism.

27. Explain how a defect in phosphofructokinase in muscle can lead to increased amounts of glycogen having a normal structure. Patients with this defect are normal except for having a limited ability to perform strenuous exercise.

28. For the defect in Question 27, explain why there is not a massive accumulation of glycogen.

ANSWERS TO SELF-TEST

Introduction

1. (a) A and G
    (b) G and E
    (c) All the bonds are α-1,4 glycosidic linkages except for the one between residues G and E.
    (d) No. The two branches are too short for phosphorylase cleavage. Phosphorylase stops cleaving four residues away from a branch point.
    (e) Yes. Residue G can be hydrolyzed by α-1,6-glucosidase (the debranching enzyme).
    (f) No. The branching enzyme transfers a block of around 7 residues from a nonreducing end of a chain at least 11 residues long. Furthermore, the new α-1,6 glycosidic linkage must be at least four residues away from a preexisting branch point at a more internal site. The fragment of glycogen in Figure 21.1 does not fulfill these requirements.

2. b, c, d

3. Although the concentration of glycogen is higher in liver, the larger mass of muscle stores more glycogen *in toto*.

Glycogen Breakdown Requires the Interplay of Several Enzymes

4. The phosphorolytic cleavage of glycogen produces glucose 1-phosphate, which can enter into the glycolytic pathway after conversion to glucose 6-phosphate. These reactions do not require ATP. On the other hand, the hydrolysis of glycogen would produce glucose, which would have to be converted to glucose 6-phosphate by hexokinase, requiring the expenditure of an ATP. Therefore, harvesting the free energy stored in glycogen by phosphorolytic cleavage rather than a hydrolytic one is more efficient because it decreases the ATP investment.

5. a, b

6. (a) 2, 4   (b) 1, 3   (c) 1, 2, 5. None of these enzymes requires ATP.

7. a. Answer d is incorrect because the phosphate group at one position on the small substrate molecule is transferred to and remains for one cycle of reaction on phosphoglucomutase, and the phosphate at the other position on the small molecule product comes from a preexisting phosphate on enzyme. For a given glucose 1-phosphate substrate, the phosphate on the product, glucose 6-phosphate, is not the same one that was present on the substrate; it came from the enzyme.

8. b

9. (a) Yes; phosphorylase would act on amylose by removing one glucose residue at a time from the nonreducing end.

(b) No; the rate of degradation of amylose would be much slower than that of glycogen because amylose would have only a single nonreducing end available for reaction, whereas glycogen has many ends.

(c) The increased number of ends available to phosphorylase as a result of cleaving the chain into pieces with the endosaccharidase would allow a more rapid production of glucose 1-phosphate by phosphorylase.

10. c. A glucose molecule that is degraded in the glycolytic pathway to two pyruvate molecules yields two ATP; however, the formation of glucose-1-P from glycogen does not consume the ATP that would be required for the formation of glucose-6-P from glucose. Thus, the net yield of ATP for a glucose residue derived from glycogen is three ATP.

Phosphorylase Is Regulated by Allosteric Interactions and Reversible Phosphorylation

11. (a) A and D
    (b) B
    (c) B to A
    (d) protein phosphatase 1

    The phosphorylated form of glycogen phosphorylase is phosphorylase *a,* which is mostly present in the active conformation designated D in Figure 21.2. In the presence of high levels of glucose, phosphorylase *a* adopts a strained, inactive conformation, designated C in the figure. The dephosphorylated form of the enzyme is called phosphorylase *b.* Phosphorylase *b* is mostly present in an inactive conformation, labeled B in the figure. When AMP binds to the inactive phosphorylase *b,* the enzyme changes to an active conformation, designated A in the figure. The effects of AMP can be reversed by ATP or glucose 6-phosphate.

12. AMP doesn't activate liver phosphorylase (the B to A conversion shown in Figure 21.2), and glucose shifts the equilibrium between the activated phosphorylase *a* toward the inactivated form (the D to C conversion).

13. (b) Glycogen, as the substrate, binds to the active site; there is also a glycogen particle binding site that keeps the enzyme attached to the glycogen granule.

    (c) Pyridoxal phosphate is the prosthetic group that positions orthophosphate for phosphorolysis and acts as a general acid-base catalyst.

    (e) AMP binds to an allosteric site and activates phosphorylase *b* in muscle.

    (f) P_i binds to the pyridoxal phosphate at the active site and attacks the α-1,4 glycosidic bond. Another P_i is covalently bound to serine 14 by phosphorylase kinase. This phosphorylation converts phosphorylase *b* into active phosphorylase *a.*

    (g) ATP binds to the same site as AMP and blocks its effects in muscle; therefore, energy charge affects phosphorylase activity.

    (h) Glucose inhibits phosphorylase *a* in the liver by changing the conformation of the enzyme to the inactive T form.

    Answers (a) and (d) are incorrect because calmodulin and Ca^{2+} bind to phosphorylase kinase rather than to phosphorylase.

14. Protein kinase A, which is itself activated by cAMP, phosphorylates phosphorylase kinase to activate it. Phosphorylase kinase can also be activated by the binding of Ca^{2+} to its calmodulin subunit. On binding Ca^{2+}, calmodulin undergoes conformational changes that activate the phosphorylase kinase. The activated kinase in turn activates glycogen phosphorylase. These effects lead to glycogen degradation in active muscle.

Epinephrine and Glucagon Signal the Need for Glycogen Breakdown

15. d, b, a, e, c

16. Enzymatic cascades lead from a small signal, caused by a few molecules, to a large subsequent enzymatic response. Thus, small chemical signals can be amplified in a short time to yield large biological effects. In addition, their effects can be regulated at various levels of the cascade.

17. Both epinephrine and glucagon stimulate glycogen breakdown by interacting with 7TM receptors and activating a cAMP signal-transduction cascade that ultimately activates glycogen phosphorylase. Glucagon activates glycogen breakdown in the liver. Epinephrine primarily activates glycogen breakdown in the muscle. Epinephrine can also act on the liver by initiating both the cAMP cascade and a phosphoinositide cascade.

Glycogen is Synthesized and Degraded by Different Pathways

18. b, d

19. The primer required to start a new glycogen chain is formed by the enzyme glycogenin, which has a glucose residue covalently attached to one of its tyrosine residues. Glycogenin uses UDP-glucose to add approximately eight glucose residues to itself to generate a primer that glycogen synthase can extend.

20. The separate pathways for the synthesis and degradation of glycogen allow the synthesis of glycogen to proceed despite a high ratio of orthophosphate to glucose 1-phosphate, which energetically favors the degradation of glycogen. In addition, the separate pathways allow the coordinated reciprocal control of glycogen synthesis and degradation by hormonal and metabolic signals.

21. True. Since degradation and synthesis occur at the nonreducing ends of glycogen, the branched structure allows simultaneous reactions to occur at many nonreducing ends, thereby increasing the overall rates of degradation or biosynthesis.

22. a, c

Glycogen Breakdown and Synthesis Are Reciprocally Regulated

23. a, d

24. a, b, d, e

25. Increased epinephrine activates PKA, which phosphorylates a subunit of PP1 and thus reduces the ability of PP1 to act on its protein targets. Furthermore, inhibitor 1 is also phosphorylated by PKA so that it too decreases PP1 activity, albeit by a different mechanism. Inactivated PP1 leads to increased levels of activated (phosphorylated) phosphorylase and inactivated (phosphorylated) glycogen synthase. Glycogen breakdown would be stimulated under these conditions.

26. Insulin results in the activation of PP1. The hormone activates an insulin-sensitive protein kinase that phosphorylates a subunit of PP1, rendering the phosphatase more active. The activated phosphatase dephosphorylates phosphorylase, protein kinase, and glycogen synthase. These changes result in a decrease in glycogen degradation and the stimulation of glycogen synthesis.

27. Since a defect in phosphofructokinase does not impair the ability of muscle to synthesize and degrade glycogen normally, the structure of glycogen will be normal. However, the utilization of glucose 6-phosphate in the glycolytic pathway is impaired,

and it equilibrates with glucose 1-phosphate; therefore, some net accumulation of glycogen will occur. The inability to perform strenuous exercise is probably a result of the impaired glycolytic pathway in muscle and the diminished production of ATP.

28. Although the impaired use of glucose 6-phosphate in glycolysis will lead to the storage of extra glycogen, it will not become excessive because the increased concentration of glucose 6-phosphate will inhibit hexokinase and hence the sequestering of glucose in muscle.

PROBLEMS

1. A patient can perform nonstrenuous tasks but becomes fatigued with physical exertion. Assays from a muscle biopsy reveal that glycogen levels are slightly elevated relative to normal. Crude extracts from muscle are used to determine the activity of glycogen phosphorylase at various levels of calcium ion for the patient and for a normal person. The results of the assays are shown in Figure 21.3. Briefly explain the clinical and biochemical findings for the patient.

FIGURE 21.3 Response of glycogen phosphorylase to calcium ion in a patient and in a normal person.

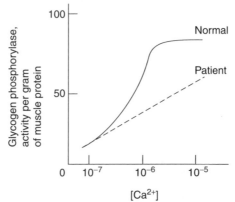

2. A strain of mutant mice is characterized by limited ability to engage in prolonged exercise. After a high-carbohydrate meal, one of these mice can exercise on a treadmill for only about 30% of the time a normal mouse can. At exhaustion, blood glucose levels in the mutant mouse are quite low, and they increase only marginally after rest. When liver glycogen in fed mutant mice is examined before exercise, the polymers have chains that are highly branched, with average branch lengths of about 10 glucose residues in either α-1,4 or α-1,6 linkage. Glycogen from exhausted normal mice has the same type of structure. Glycogen from exhausted mutant mice is still highly branched, but the polymer has an unusually large number of single glucose residues with α-1,6 linkages. Practically all the chains with α-1,4 linkages are still about 10 residues in length. Explain the metabolic and molecular observations for the mutant mice.

3. Your colleague discovers a fungal enzyme that can liberate glucose residues from cellulose. The enzyme is similar to glycogen phosphorylase in that it utilizes inorganic phosphate for the phosphorolytic cleavage of glucose residues from the nonreducing ends of cellulose. Why would you suspect that other types of cellulases may be important in the rapid degradation of cellulose?

4. Consider a patient with the following clinical findings: fasting blood glucose level is 25 mg per 100 ml (normal values are from 80 to 100 mg per 100 ml); feeding the patient glucose results in a rapid elevation of blood glucose level, followed by a normal return to fasting levels; feeding the patient galactose or fructose results in the elevation of blood glucose to normal levels; the administration of glucagon fails to generate hyperglycemia; biochemical examination of liver glycogen reveals a normal glycogen structure.

   (a) Which of the enzyme deficiencies described in Table 21.1 of the text (page 634) could account for these clinical findings?
   (b) What additional experiments would you conduct to provide a specific diagnosis for the patient?

5. Vigorously contracting muscle often becomes anaerobic when the demand for oxygen exceeds the amount supplied through the circulation. Under such conditions, lactate may accumulate in muscle. Under anaerobic conditions a certain percentage of lactate can be converted to glycogen in muscle. One line of evidence for this synthesis involves the demonstration of activity for malic enzyme, which can use CO_2 to convert pyruvate to malate, using NADPH as an electron donor.

   (a) Why is lactate produced in muscle when the supply of oxygen is insufficient?
   (b) In muscle, pyruvate carboxylase activity is very low. How could malic enzyme activity facilitate the synthesis of glycogen from lactate?
   (c) Why would you expect the conversion of lactate to glycogen to occur only after vigorous muscle contraction ceases?
   (d) Is there an energetic advantage to converting lactate to glycogen in muscle rather than using the Cori cycle for sending the lactate to the liver, where it can be reconverted to glucose and then returned to muscle for glycogen synthesis?

6. Cyclic nucleotide phosphatases are inhibited by caffeine. What effect would drinking a strong cup of coffee have on glycogen metabolism when epinephrine levels are dropping in the blood?

7. During the degradation of branched chains of glycogen, a transferase shifts a chain of three glycosyl residues from one branch to another, exposing a single remaining glycosyl residue to α-1,6-glucosidase activity. Free glucose is released, and the now unbranched chain can be further degraded by glycogen phosphorylase.

   (a) Estimate the free-energy change of the transfer of glycosyl residues from one branch to another.
   (b) About 10 percent of the glycosyl residues of normal glycogen are released as glucose, whereas the remainder are released as glucose 1-phosphate. Give two reasons why it is desirable for cells to convert most of the glycosyl residues in glycogen to glucose 1-phosphate.
   (c) Patients who lack liver glycosyl transferase have been studied. Why would you expect liver extracts from such people to perhaps lack α-1,6-glucosidase activity?

8. You are studying a patient with McArdle's disease, which is described on page 635 of the text. Explain what you would expect to find when you carry out each of the following analyses.

   (a) fasting level of blood glucose
   (b) structure and amount of liver glycogen
   (c) structure and amount of muscle glycogen
   (d) change in blood glucose levels upon feeding the patient galactose
   (e) change in blood lactate levels after vigorous exercise
   (f) change in blood glucose levels after administration of glucagon
   (g) change in blood glucose levels after administration of epinephrine

9. An investigator has a sample of purified muscle phosphorylase b that she knows is relatively inactive.

    (a) Suggest two *in vitro* methods that could be used to generate active phosphorylase from the inactive phosphorylase b.

    (b) After the phosphorylase is activated, the investigator incubates the enzyme with a sample of unbranched glycogen in a buffered solution. She finds that no glycosyl residues are cleaved. What else is needed for the cleavage of glycosyl residues by active phosphorylase?

10. Arsenate can substitute in many reactions for which phosphate is the normal substrate. However, arsenate esters are far less stable than phosphate esters, and they decompose spontaneously to arsenate and an alcohol:

$$R\text{—}OAsO_3^{2-} + H_2O \rightarrow R\text{—}OH + AsO_4^{2-}$$

    (a) In which of the steps of glycogen metabolism might arsenate be used as a substrate?

    (b) What are the energetic consequences of utilizing arsenate as a substrate in glycogen degradation?

11. As described on page 634 of the text, the ratio of glycogen phosphorylase to protein phosphatase 1 is approximately 10 to 1. Suppose that in some liver cells the overproduction of the phosphatase results in a ratio of one to one. How will such a ratio affect the cell's response to an infusion of glucose?

12. A young woman cannot exercise vigorously on the treadmill without leg pains and stiffness. During exercise, lactate levels do not increase in her serum, in contrast to results of exercise in normal subjects. As is the case with normal subjects, no significant hypoglycemia is observed when the patient exercises or fasts. Analyses of muscle biopsy samples show that glycogen content is about 10 times greater than normal in the young woman, but the level of muscle phosphorylase activity is normal. Other experiments with biopsy samples show that rapid incorporation of ^{14}C from radioactive glucose into fructose 6-phosphate and glycogen is observed, but very little incorporation of radioisotope into lactate is seen. When ^{14}C-pyruvate is incubated with another sample of the homogenate, the radioisotope is readily incorporated into glycogen.

    What specific deficiency in a metabolic pathway could contribute to these observations? Propose two additional studies that could confirm your conclusion.

13. In 1952, Dr. D. H. Andersen described a seriously ill infant with an enlarged liver as well as cirrhosis. When epinephrine was administered, a relatively low elevation in the patient's blood glucose levels was noted. Several days later when the infant was fed galactose, normal elevation of glucose was observed in the circulation. The infant died at the age of 17 months, and at autopsy Dr. Andersen found that glycogen from the liver, while present in unusually high concentration, was relatively insoluble, making it difficult to extract. She sent a sample of the liver glycogen to Dr. Gerty Cori. In an experiment designed to characterize the glycogen, Dr. Cori incubated a sample with orthophosphate (P_i) and two normal liver enzymes, active glycogen phosphorylase and debranching enzyme. She found that the ratio of glucose 1-phosphate to glucose released from the glycogen sample was 100:1, while the ratio from normal glycogen is 10:1.

    (a) What enzyme of glycogen metabolism is most likely to be deficient in the liver tissue of the infant? Write a concise explanation for your answer, and relate it to the relative insolubility of the glycogen in the autopsy sample.

    (b) Dr. Andersen, aware that a number of enzyme deficiencies might cause a glycogen-storage disease, sought to rule out a deficiency of a particular enzyme in the infant by studying the elevation of glucose levels after feeding galactose. What is that enzyme, and how does normal elevation of blood glucose after galactose feeding rule out a deficiency of that enzyme in the infant?

14. While muscle cells in tissue culture can be stimulated to break down glycogen only minimally when incubated in a solution containing cyclic AMP, they are more readily stimulated by compounds like dibutyryl cyclic AMP (Figure 21.4). Explain the difference in response of cells to these two substances.

FIGURE 21.4 Dibutyryl-cyclic AMP.

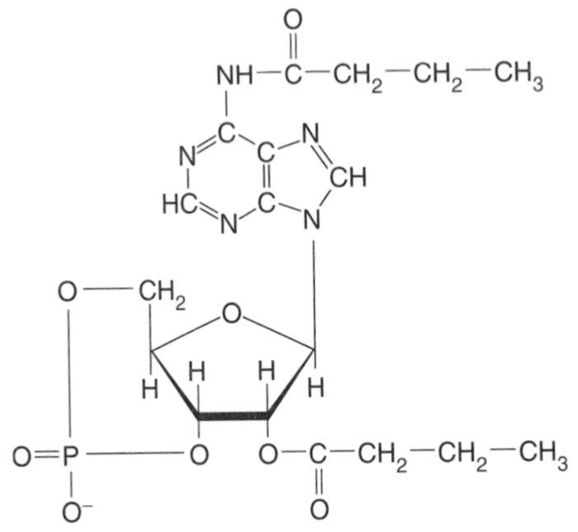

15. One method for the analysis of glycogen involves incubating a sample with methyl iodide, which methylates all free hydroxyl groups. Acid hydrolysis of exhaustively methylated glycogen yields a mixture of methyl glucosides, which can be separated and analyzed. Considering the various types of glycogen-storage diseases listed in Table 21.1 of the text, which of them could be diagnosed using exhaustive methylation and acid hydrolysis of glycogen?

16. Patients with Cori's disease lack debranching enzyme, and therefore the structure of liver and muscle glycogen is unusual, with short outer branches. Design an assay that would enable you to demonstrate the presence of short branches in glycogen from one of these patients. Also explain how you would demonstrate that the debranching enzyme is deficient in these patients.

17. Table 21.1 in the text lists eight diseases of glycogen metabolism, all of which affect the level of glycogen in muscle and liver or the structure of the polysaccharide in one or both of those tissues. Another rare disease of glycogen metabolism is caused by a deficiency in liver glycogen synthase. After fasting, affected subjects have low blood glucose. Hyperglycemia and high blood lactate are observed after a meal.

   (a) Briefly explain how these symptoms could be caused by glycogen synthase deficiency.
   (b) Under normal nutritional conditions, glycogen comprises about 4 percent of the wet weight of liver tissue in normal subjects. What proportion of glycogen in liver would you expect in a patient who lacks liver glycogen synthase?

18. Phosphoglucomutase converts the product of glycogen phosphorylase, glucose 1-phosphate, to the glycolytic pathway component glucose 6-phosphate. The reaction catalyzed by phosphoglucomutase proceeds by way of a glucose 1,6-bisphosphate intermediate.

   (a) What would happen to phophoglucomutase activity if the glucose 1,6-bisphosphate intermediate were to dissociate from the enzyme before completion of the reaction?

  (b) Would glucose 1,6-bisphosphate dissociation be equivalent to the hydrolysis of the serine phosphate on the enzyme? Explain why.

  (c) Suppose that a phosphoglucomutase in the dephosphoenzyme form arose. How might the enzyme be reactivated?

19. Why is it important for water to be excluded from the active site of glycogen phosphorylase?

20. Amylose is the linear polymer of glucose residues joined by α-1,4 glycosidic linkages found as a form of starch in plants. Why do you think it would serve less well as an energy source in muscle and liver than does glycogen?

21. Recall that Cu (II) can be reduced to Cu (I) by the aldehyde group of the open configuration of glucose to form Cu_2O (see page 325 in text). This reaction is the defining property of reducing sugars. Predict what would happen if you treated glycogen recently synthesized *de novo* with a Cu(II)-containing solution. Would you expect any reducing sugar reactions?

ANSWERS TO PROBLEMS

1. Calcium ion normally activates muscle phosphorylase kinase, which in turn phosphorylates muscle phosphorylase. In the patient, glycogen phosphorylase activity is less responsive to Ca^{2+} than it is in the normal subject. It is likely that Ca^{2+} cannot activate phosphorylase kinase in the patient, perhaps because the δ subunit (calmodulin) of the enzyme is altered in some way. As a result, there are too few molecules of enzymatically active glycogen phosphorylase to provide the rate of glycogen breakdown that is needed to sustain vigorous muscle contraction. Elevated levels of muscle glycogen should be expected when glycogen phosphorylase activity is lower than normal.

2. The longer chains of glucose residues in α-1,4 linkage and the unusually high number of single glucose residues in α-1,6 linkage suggest that although transferase activity is present, α-1,6-glucosidase activity is deficient in the mutant strain. For such chains, far fewer ends with glucose residues are available as substrates for glycogen phosphorylase. (Recall that glycogen phosphorylase cannot cleave α-1,6 linkages.) This limited ability to mobilize glucose residues means that less energy is available for prolonged exercise.

3. Cellulose is an unbranched polymer of glucose residues with β-1,4 linkages. Therefore, each chain has only one nonreducing end that is available for phosphorolysis by the fungal enzyme. Compared with the rate of breakdown of molecules of glycogen, whose branched chains provide more sites for the action of glycogen phosphorylase, the generation of glucose phosphate molecules from cellulose by means of the fungal enzyme alone could be quite slow. Therefore, you would expect to find endocellulases that generate additional nonreducing ends in cellulose chains.

4. (a) A low-fasting blood-glucose level indicates a failure either to mobilize glucose production from glycogen or to release glucose from the liver. However, the elevations in blood-glucose levels after feeding the patient glucose, galactose, or fructose indicate that the liver can release glucose derived from the diet or formed from other monosaccharides. The lack of response to glucagon indicates that the enzymatic cascade for glycogen breakdown is defective. Therefore, you would suspect a deficiency of liver glycogen phosphorylase or phosphorylase kinase. Of the diseases described in Table 21.1 of the text, both type VI and type VIII

could account for the findings, which include increased amounts of glycogen with normal structure.

(b) The direct assay of the activities of glycogen phosphorylase and phosphorylase kinase would enable you to make a specific diagnosis. For these purposes, a liver biopsy would be necessary.

5. (a) Muscle cells produce lactate from pyruvate under anaerobic conditions to generate NAD^+, which is required to sustain the activity of glyceraldehyde 3-phosphate dehydrogenase in the glycolytic pathway.

(b) The low activity of muscle pyruvate carboxylase means that other pathways for the synthesis of oxaloacetate must be available. The formation of malate, which is then converted to oxaloacetate (see page 509 of the text), enables the muscle cell to carry out the synthesis of glucose 6-phosphate via gluconeogenesis. Glycogen can then be synthesized through the conversion of glucose 6-phosphate to glucose 1-phosphate, the formation of UDP-glucose, and the transfer of the glucose residue to a glycogen primer chain.

(c) Yes. Energy for vigorous muscle contraction under anaerobic conditions is derived primarily from the conversion of glycogen and glucose to lactate. The simultaneous conversion of lactate to glycogen would simply result in the unnecessary hydrolysis of ATP.

(d) In liver, the conversion of two lactate molecules to a glucose residue in glycogen through gluconeogenesis requires seven high-energy phosphate bonds; six are required for the formation of glucose 6-phosphate from two molecules of lactate, and one is needed for the synthesis of UDP-glucose from glucose 1-phosphate. The conversion of lactate to glycogen in muscle requires two fewer high-energy bonds because the formation of oxaloacetate through the action of malic enzyme does not require ATP. Recall that pyruvate carboxylase requires ATP for the synthesis of oxaloacetate from pyruvate.

6. When epinephrine levels in the blood decrease, the synthesis of cyclic AMP decreases. Existing cyclic AMP is degraded by cyclic nucleotide phosphatases. The inhibition of these enzymes by caffeine prolongs the degradation of glycogen because the remaining cyclic AMP continues to activate protein kinase, which in turn activates phosphorylase kinase. Glycogen phosphorylase is, in turn, activated by phosphorylase kinase. Sustained activation of phosphorylase results in continued mobilization of glucose residues from glycogen stores in liver.

7. (a) Because the bonds broken and formed during the transferase action are both α-1,4 glycosidic bonds, the free-energy change is likely to be close to zero.

(b) The generation of glucose 1-phosphate rather than glucose means that one less ATP equivalent is required for the conversion of a glucose residue to two molecules of pyruvate. The glucose residue released does not have to be phosphorylated for subsequent metabolism when phophorolysis produces it as glucose 1-phosphate directly. In addition, the phosphorylation of glucose ensures that the molecule cannot diffuse across the cell membrane before it is utilized in the glycolytic pathway.

(c) The glucosidase and the transferase activities are both found on the same 160-kd polypeptide chain. A significant alteration in the structure of the domain for glucosidase in the bifunctional enzyme could also impair the functioning of the transferase domain.

8. (a) In McArdle's disease, muscle phosphorylase is deficient but liver phosphorylase is normal. Therefore, you would expect glucose and glycogen metabolism in the liver to be normal and the control of blood glucose by the liver also to be normal.

(b) Normal glyogen metabolism in the liver means that both the amount of liver glycogen and its structure would be the same as in unaffected people.

(c) Defective muscle glycogen phosphorylase means that glycogen breakdown is impaired. Moderately increased concentrations of muscle glycogen could be expected, although the structure of the glycogen should be similar to that in unaffected people.

(d) Galactose can be converted to glucose 6-phosphate in the liver, which can then export glucose to the blood. Because defective muscle phosphorylase has no effect on galactose metabolism, you would expect similar elevations in blood glucose after the ingestion of galactose in normal and affected people.

(e) During vigorous exercise, blood lactate levels normally rise as muscle tissue exports the lactate generated through glycogen breakdown. The defect in muscle phosphorylase limits the extent to which glycogen is degraded in the muscle. This in turn reduces the amount of lactate exported during exercise, so the rise in blood lactate levels would not be as great in the affected person.

(f) Glucagon exerts its effects primarily on liver, not muscle. In patients with McArdle's disease, blood glucose levels increase normally in response to glucagon.

(g) A slight increase in blood glucose concentration may occur after epinephrine administration, because liver is somewhat responsive to this hormone. Epinephrine does have a greater glycogenolytic effect on muscle, but you would not expect to see any change in blood glucose concentration when it is administered. The reason is that, even if glycogen breakdown is accelerated (which is unlikely to occur in patients with McArdle's disease), the glucose 6-phosphate produced cannot be converted to glucose for export into the circulation, because muscle lacks the enzyme glucose 6-phosphatase.

9. (a) The investigator could activate the phosphorylase by adding AMP to the sample or by using active phosphorylase kinase and ATP to phosphorylate the enzyme.

 (b) Inorganic phosphate is also required for the conversion of glycosyl residues in glycogen to glucose 1-phosphate molecules.

10. (a) Arsenate can substitute for inorganic phosphate in the glycogen phosphorylase reaction, generating glucose arsenate esters.

 (b) When P_i is used as a substrate for glycogen phosphorylase, glucose 1-phosphate is generated. The glucose 1-arsenate esters that are generated when arsenate is used as a substrate spontaneously hydrolyze to yield glucose and arsenate. The conversion of glucose to pyruvate requires one more ATP equivalent than does the conversion of glucose 1-phosphate to pyruvate.

11. The normal 10-to-1 ratio means that glycogen synthase molecules are activated only after most of the phosphorylase a molecules are converted to the inactive b form, which ensures that the simultaneous degradation and synthesis of glycogen does not occur. A phosphorylase to phosphatase ratio of one to one means that, as soon as a few phosphorylase molecules are inactivated, phosphatase molecules that are no longer bound to phosphorylase begin to convert glycogen synthase molecules to the active form. Glycogen degradation and synthesis then occur simultaneously, resulting in the wasteful hydrolysis of ATP.

12. From the clinical observations, it appears that the pathway from pyruvate to glucose 6-phosphate and on to glycogen is functional and that gluconeogenesis is working normally in liver (there is no hypoglycemia during fasting or exercise, when demands for glucose increase). Although muscle glycogen content is higher, normal phosphorylase activity indicates that glycogen could be phosphorylized normally. You should then consider whether there is a deficiency in the glycolytic pathway, because lactate

does not accumulate during exercise and it is not labeled when ^{14}C-glucose is administered. Labeled fructose 6-phosphate can be made from radioactive glucose in the biopsy sample, but knowledge about subsequent glycolytic reactions is not available. There could be a significant block at the level of phosphofructokinase or beyond. Such a deficiency would mean that while normal demands for glucose can be taken care of, a high rate of glycolytic activity during vigorous exercise cannot be accommodated. You should consider analyzing for additional radioactive glycolytic intermediates when glucose is administered, then testing for deficiency of one or more glycolytic enzymes using biopsy tissues. The description of the disorder corresponds most closely to a known condition for a deficiency in muscle phosphofructokinase (Type VII glycogen-storage disease). One might also argue that lactate dehydrogenase could be absent, explaining why no lactate is generated during exercise. However, in cases in which muscle lactate dehydrogenase is defective, affected subjects cannot exercise vigorously, but they have no accumulation of glycogen in their muscle tissue.

13. (a) The branching enzyme was deficient in the infant. This enzyme removes blocks of glucosyl residues from a chain of α-1 ⟶ 4-linked residues and transfers them internally to form a branch with an α-1 ⟶ 6 link to a polymer chain. The most important clue to the deficiency is found in the ratio of glucose 1-phosphate to glucose, which is 10 times higher in glycogen from the affected infant than from a normal polymer sample. Recall that glucose 1-phosphate is produced through the action of phosphorylase, which phosphorylizes α-1 ⟶ 4 linkages, while glucose is produced when the glycogen debranching enzyme hydrolyzes a glucose in α-1 ⟶ 6 linkage at a branch point. Normal glycogen has a branch at every 10 or so glycosyl residues, so treatment with a mixture of normal phosphorylase and debranching enzyme will yield a 10:1 ratio of glucose 1-phosphate to glucose. The autopsy sample yielded a ratio of 100:1, suggesting that there are far fewer branches in the sample. This conclusion is consistent with the relative insolubility of the infant's glycogen, which, with fewer branches, is more like amylopectin, a linear glucosyl polymer which has limited solubility in water.

    (b) The pathway for galactose metabolism includes its conversion, through steps that include epimerization, to glucose 6-phosphate. Thus feeding galactose should result in an increased concentration of glucose 6-phosphate in the liver cell. If glucose 6-phosphatase were deficient, glucose 6-phosphate would not be converted to glucose, so that the levels of blood glucose would not be elevated after galactose feeding. Dr. Andersen considered a glucose 6-phosphatase deficiency because of the limited increase in blood glucose levels after administration of epinephrine, so she used galactose feeding to increase glucose 6-phosphate levels in liver cells. When glucose levels rose in the blood, she concluded that glucose 6-phosphatase levels were normal. She subsequently considered other deficiencies that would result in storage of abnormal amounts of liver glycogen.

14. Like other nucleotides, cyclic AMP is polar and negatively charged at neutral pH. It therefore crosses plasma membranes at a relatively low rate. The presence of two hydrophobic acyl chains on the molecule make it much more hydrophobic, so that it can more easily dissolve in the bilayer and more readily enter the cytosol.

15. Type IV glycogen-storage disease, in which glycogen with a much lower number of α-1,6 glycosidic linkages is produced, could be analyzed using methylation and hydrolysis. Any glucose residue derived from a branch point will have methyl groups at C-2 and C-3, while all other residues (with one exception) will emerge from hydrolysis as 2,3,6-O-trimethyl glucose molecules. The glucose at the reducing end of the glycogen molecule will be converted to a tetramethlyglucoside. In normal subjects, the ratio of

trimethylglucose to dimethyl glucose should be about 10 to 1, while glycogen from a person with a deficiency in the branching enzyme will have a much higher ratio.

16. A short outer branch in a glycogen molecule has only a small number of α-1,4-glucosyl residues on the nonreducing side of a branch or an α-1,6 link. Incubating such a glycogen molecule with active phosphorylase and P_i will liberate only limited amounts of glucose 1-phosphate, compared with the number liberated from normal glycogen. Recall that phosphorylase cannot free glucose molecules that are within four residues of a branch point in glycogen. To demonstrate that phosphorylase action is limited by short outer branches, you can incubate another sample with purified debranching enzyme and phosphorylase, and you would expect to see an increase in production of glucose 1-phosphate. To demonstrate a debranching enzyme deficiency in a patient, you could treat normal glycogen with active muscle phosphorylase and muscle extracts from a patient with Cori's disease. If debranching enzyme activity is low, only limited amounts of glucose 1-phosphate will be produced. Larger numbers of glucose 1-phosphate molecules will be released from a normal glycogen sample treated with active phosphorylase and muscle cell extracts from a normal person.

17. (a) Lack of glycogen synthase implies that the ability of the liver to store glucose as glycogen is impaired. After fasting, when blood glucose concentrations are low, liver glycogen is normally converted to glucose 6-phosphate, which is converted to glucose and exported to the blood. Low glycogen levels in liver tissue would make it impossible for liver to maintain proper glucose levels in the blood. After a meal containing carbohydrates, the liver would be unable to convert glucose to glycogen. Even though glucokinase may convert glucose to glucose 6-phosphate, the high concentration of that substrate may cause accelerated conversion back to glucose through the action of glucose 6-phosphatase. Glucose levels would then increase in the circulation. The elevation of lactate levels in blood suggests that any glucose metabolized in the liver is preferentially converted to lactate rather than to glycogen.

    (b) As discussed, liver cells deficient in glycogen synthase would be unable to synthesize large amounts of glycogen. You would therefore expect the percentage of glycogen in affected people to be lower. In the few patients with the disorder, glycogen makes up less than 1 percent of liver tissue.

18. (a) If the glucose 1,6-bisphosphate were to dissociate from the enzyme, the enzyme would not have a phosphate on the serine hydroxyl that is necessary for activity. The dephosphorylated enzyme would lack the phosphate need for transfer to the incoming glucose 1-phosphate to form the bisphosphate intermediate and could not catalyze the mutase reaction.

    (b) Yes, both bisphosphate dissociation or phophoenzyme hydrolysis would lead to an inactive, unphosphorylated enzyme.

    (c) Since a phophoglucomutase carrying a phosphate group on a specific serine is required for activity, some means of producing the phosphosenzyme is required. A protein kinase could replace the covalently bound enzyme phosphate or a phosphglucokinase enzyme that produced glucose 1,6-bisphospate, which would bind to and phosphorylate phosphglucomutase, could also form the phosphorylated enzyme. The latter is a known mechanism.

19. Were water to enter the active site of glycogen phosphorylase, the enzyme would hydrolyze rather than phosphorolyze the glycogen. The result would be the production of a free glucose rather than a glucose 1-phosphate. For the free glucose to be metabolized, even to be catabolized for energy, it would have to be phosphorylated, which would consume an ATP molecule.

20. Because amylose is linear, it has but one nonreducing end per molecule that could be removed by glycogen phosphorylase. Glycogen, which is branched, with α-1,6 linkages every approximately 10 residues, has many more ends on which glycogen phosphorylase can act. Thus, more glucose 1-phosphate molecules can be released quickly per glycogen molecule compared with amylose. In addition, the branches make the glycogen more soluble so that it is more easily accessible to glycogen phosphorylase.

21. Glycogen is a treelike molecule in which glucose residues are linked in α-1,4 glycosidic linkages with branches occurring as α-1,6 glycosidic linkages at approximately every tenth residue. Recall also that the free aldehyde at the C-1 position of glucose is the agent that reduces the Cu(II). Since the C-1 of all the sugars in glycogen are in a glycosidic linkage they are unable to react with the divalent copper. Thus, glycogen itself would not act as a reducing sugar. You may ask, What about the first glucose residue that is at the root of the tree—the one to which the first α-1,4 glycosidic-linkage of the chain is formed? Recall that glycogen synthase needs a primer to start a new glycogen chain and that the dimeric protein glycogenin creates that primer by adding the glycosyl moiety of UDP-glucose to itself. The C-1 atom of the first glucose residue of the resultant primer is covalently attached to the protein through a phenolic hydroxyl on a tyrosine residue of the protein. Thus, glycogen that has been synthesized *de novo* has no free aldehyde groups and would itself give no reaction with Cu(II). See Smythe, C., and Cohen, P., *Eur. J. Biochem.* (1991) **200**:625– 631.

Fatty Acid Metabolism

In the discussion of the generation and storage of metabolic energy, the text has thus far focused on the carbohydrates (Chapters 16, 17, 20, and 21). In Chapter 22, the authors turn to the fatty acids as metabolic fuels. After reminding you of the structure of a tri-acylglycerol, they explain why such fats are the most concentrated biological energy stores. The transport of fatty acids from the intestine and the pathway of the oxidation of fatty acids, which liberates the energy of fatty acids and makes it available to the cell, are then presented. The oxidation of unsaturated fatty acids is described, and the formation and role of the ketone bodies as acetyl transport molecules in the circulation are discussed. The text then describes how both saturated and unsaturated fatty acids are synthesized. The energetics of the oxidation and synthesis of fatty acids are given, and an outline of the control of these processes is provided. The chapter concludes with an introduction to the eicosanoid hormones, which are derived from fatty acids. A review of Chapter 12 will remind you of the structures and nomenclature systems of the fatty acids, the role of lipids in membranes, and the effect of the fatty acids in determining membrane fluidity.

LEARNING OBJECTIVES

When you have mastered this chapter, you should be able to accomplish the following objectives.

Introduction

1. Define *fatty acid* (Chapter 12) and list the four major physiologic functions they serve.

2. Derive the structure of a *saturated* or *unsaturated fatty acid* from its systematic name. Specify the α, β, and ω carbon atoms and designate the position of a double bond in a fatty acid when given either its Δ or its ω number (Chapter 12).

3. Recognize the structures of *palmitate, stearate, palmitoleate, oleate, linoleate,* and *linolenate* (Chapter 12).

4. Describe the structure of a *triacylglycerol.*

5. Provide an overview of the synthesis and catabolism of fatty acids by listing the types of chemical reactions used.

Triacylglycerols Are Highly Concentrated Energy Stores (Text Section 22.1)

6. Explain why *triacylglycerols* are highly concentrated forms of *stored metabolic energy.* Appreciate that the *adipocyte* (fat cell) is specialized to store them and that *chylomicrons* carry fatty acids from the intestine to other tissues. Describe the role of *bile salts* in *micelle* formation.

The Utilization of Fatty Acids as Fuel Requires Three Stages of Processing (Text Section 22.2)

7. Describe the *lipolysis* of triacylglycerols by *lipases.* Explain the role of *cyclic AMP* in the regulation of lipase in *adipose cells.* Appreciate that serum albumin carries the fatty acids from the adipocyte to other tissues. Outline the conversion of glycerol to *glycerol 3-phosphate* and *dihydroxyacetone phosphate* and appreciate the physiological importance of these reactions.

8. Describe the reaction that links *coenzyme A (CoA)* to a fatty acid. Explain the roles of the *acyl adenylate* and *pyrophosphatase* in the reaction.

9. Explain the involvement of *carnitine* in the transport of fatty acids from the cytoplasm into mitochondria.

10. List the four reactions of the *β-oxidation pathway* of fatty acid catabolism and identify their substrates and products. Explain the function of NAD^+ and *FAD* in these reactions.

11. Calculate the *energy yield* in ATP molecules for the β oxidation of a given fatty acid.

Unsaturated and Odd-Chain Fatty Acids Require Additional Steps for Degradation (Text Section 22.3)

12. Indicate the two reactions, in addition to those of the β-oxidation pathway, used to oxidize naturally occurring unsaturated fatty acids. Explain how they allow the continuation of the β-oxidation pathway.

13. Outline the oxidation reactions of an *odd-numbered fatty acid*.

14. Explain the role of *vitamin B$_{12}$ (cobalamin)* in the pathway by which *propionyl CoA* is converted to succinyl CoA. List the three types of reactions carried out by cobalamin enzymes.

15. Identify the reactions of fatty acid degradation that occur in the *mitochondria* and in the *peroxisomes*.

16. Explain the consequences of limiting *oxaloacetate* concentrations on the oxidation of fatty acids. Name and identify the structures of the *ketone bodies*.

17. Describe the synthesis and normal catabolism of the ketone bodies. Explain why only the liver exports *acetoacetate* and *3-hydroxybutyrate* and appreciate the role of ketone bodies in normal human metabolism.

18. Describe the effect of high levels of *acetoacetate* on fat metabolism in adipose tissue.

19. Provide the biochemical basis for the inability of animals to convert fatty acids into glucose.

Fatty Acids Are Synthesized by Fatty Acid Synthase (Text Section 22.4)

20. Contrast fatty acid oxidation and *fatty acid synthesis*.

21. List the substrates and products of the committed step in fatty acid synthesis and describe its catalytic mechanism. Appreciate the role of *biotin* in the *acetyl CoA carboxylase* reaction.

22. Name the common component of *acyl carrier protein (ACP)* and CoA, give its functions, and describe the overall functions of ACP and CoA in fatty acid metabolism.

23. Describe the four reactions of the elongation cycle of fatty acid synthesis. Explain how *malonyl CoA* provides the driving force for the condensation of acetyl units with the growing acyl chain.

24. Contrast the enzymatic machinery for fatty acid biosynthesis in bacteria with that in eukaryotes. Outline the movements of the elongating acyl chain on the mammalian *fatty acid synthetase* dimer during fatty acid biosynthesis.

25. Calculate the energy cost of the synthesis of a given fatty acid.

26. Describe the transport of acetyl groups across the inner mitochondrial membrane in the form of *citrate* and explain its purpose. Account for the synthesis of *NADPH* during the conversion of oxaloacetate into pyruvate in the cytosol.

27. List the sources of the NADPH used in fatty acid synthesis.

The Elongation and Unsaturation of Fatty Acids Are Accomplished by Accessory Enzyme Systems (Text Section 22.5)

28. Describe the elongation and desaturation reactions that can occur on preformed fatty acids. Explain why linoleate and linolenate are essential in the diet.

29. List the different kinds of *eicosanoid hormones*. Outline their metabolic relationships and biological functions.

30. Describe the effects of *acetylsalicylate (aspirin)* on the synthesis of eicosanoids.

Acetyl CoA Carboxylase Plays a Key Role in Controlling Fatty Acid Metabolism
(Text Section 22.6)

28. Discuss the different modes of regulation of *acetyl CoA carboxylase*. Explain the reciprocal control of fatty acid synthesis and degradation through local (intracellular), hormonal, and adaptive regulation.

SELF-TEST

1. Which of the following are major physiologic functions of free fatty acids?
   (a) They stabilize the structure of membranes.
   (b) They serve as precursors of phospholipids and glycolipids.
   (c) They serve as fuel molecules.
   (d) They are precursors of triacylglycerols.
   (e) They are precursors of certain hormones and intracellular messengers.

2. For each of the following four naturally occurring fatty acids, give the systematic name, the common name, and the abbreviations. Use the Δ convention for the name and abbreviations of the unsaturated compounds. Also indicate the position of the double bond closest to the methyl end of the chain using the ω convention.

   (a) $CH_3(CH_2)_{14}CO_2H$

   (b)
$$CH_3(CH_2)_7\overset{\overset{\displaystyle H}{|}}{C}=\overset{\overset{\displaystyle H}{|}}{C}(CH_2)_7CO_2H$$

   (c)
$$CH_3(CH_2)_4\overset{\overset{\displaystyle H}{|}}{C}=\overset{\overset{\displaystyle H}{|}}{C}CH_2\overset{\overset{\displaystyle H}{|}}{C}=\overset{\overset{\displaystyle H}{|}}{C}(CH_2)_7CO_2H$$

   (d)
$$CH_3CH_2\overset{\overset{\displaystyle H}{|}}{C}=\overset{\overset{\displaystyle H}{|}}{C}CH_2\overset{\overset{\displaystyle H}{|}}{C}=\overset{\overset{\displaystyle H}{|}}{C}CH_2\overset{\overset{\displaystyle H}{|}}{C}=\overset{\overset{\displaystyle H}{|}}{C}(CH_2)_7CO_2H$$

3. Arrange the following fatty acids in the order of increasing melting point.
   (a) oleate
   (b) stearate
   (c) linoleate
   (d) palmitate
   (e) linolenate

4. (a) Do condensation, reduction, dehydration, and reduction occur during fatty acid degradation or synthesis?
   (b) How many carbon atoms are added to or removed from a fatty acid during its synthesis or degradation, respectively?

5. What two properties make triacylglycerols more efficient than glycogen for the storage of metabolic energy?

6. Which of the following statements about the triacylglycerols stored in adipose tissue are correct?
   (a) They are hydrolyzed to form fatty acids and dihydroxyacetone.
   (b) They are hydrolyzed by a lipase that is activated by covalent modification.
   (c) They release fatty acids that can be oxidized to CO_2 and H_2O to provide energy to the cell.
   (d) They can yield a precursor of glucose.
   (e) They are mobilized by epinephrine or glucagon.

7. (a) Draw a thioester bond and explain its role in the β oxidation of fatty acids.

   (b) How does the $\Delta G^{\circ\prime}$ value for the hydrolysis of acetyl coenzyme A compare with that for the hydrolysis of ATP? What is the significance of this value with respect to fatty acid metabolism?

   (c) Describe the mechanism for the formation of acyl CoA.

   (d) How is pyrophosphatase involved in the activation of fatty acids for β oxidation?

8. Place the following incomplete list of reactions or locations during the β oxidation of fatty acids in the proper order.

   (a) reaction with carnitine

   (b) fatty acid in cytosol

   (c) activation of fatty acid by joining to CoA

   (d) hydration

   (e) NAD^+-linked oxidation

   (f) thiolysis

   (g) acyl CoA in mitochondrion

   (h) FAD-linked oxidation

9. Explain the involvement of carnitine in the β oxidation of fatty acids.

10. Calculate the approximate yield in ATP molecules of the complete oxidation of hexanoic acid (C6:0).

11. Indicate whether the following statement is true or false and explain your answer: After a meal rich in carbohydrates, acetyl CoA levels rise and ketone body synthesis increases.

12. Which of the following statements about acetoacetate and 3-hydroxybutyrate are correct?

    (a) They are normal fuels for heart muscle and the renal cortex.

    (b) They are synthesized in the liver.

    (c) They can give rise to acetone.

    (d) They contain four carbon atoms and require three acetyl CoA molecules for their synthesis.

    (e) They can be regarded as water-soluble, transportable forms of citrate in the blood.

13. Methylmalonyl CoA mutase

    (a) converts D-methylmalonyl CoA to L-methylmalonyl CoA.

    (b) contains biotin.

    (c) involves a homolytic bond cleavage.

    (d) contains a derivative of vitamin B_{12}.

    (e) transforms a *cis*-Δ^3 double bond into a *trans*-Δ^2 double bond.

14. Match the reactant or characteristic in the right column with the appropriate pathway in the left column.

    (a) fatty acid oxidation

    (b) fatty acid synthesis

    (1) acyl CoA

    (2) occurs in the cytosol

    (3) uses NAD^+

    (4) D-3-hydroxyacyl derivative involved

    (5) pantetheine involved

    (6) malonyl CoA

    (7) single polypeptide with multiple activities involved

    (8) uses FAD

15. Explain the requirement for bicarbonate in fatty acid biosynthesis.

16. Which of the following statements about citrate are correct?
    (a) It transports reducing power from the mitochondria into the cytosol.
    (b) It inhibits gluconeogenesis.
    (c) It activates the first enzyme of fatty acid biosynthesis.
    (d) It transports acetyl groups from the mitochondria into the cytosol.
    (e) It supplies the CO_2 required for formation of malonyl CoA.

17. The fatty acid synthase of mammals is a dimer consisting of identical subunits, each of which contains all the activities necessary to synthesize fatty acids from malonyl CoA and acetyl CoA. Why is a single subunit unable to carry out the reactions?

18. Possible advantages of multifunctional polypeptide chains, that is, polypeptide chains having more than one active site, include which of the following?
    (a) enhanced stability beyond that expected for a noncovalent complex of the same activities on separate polypeptide chains
    (b) fixed stoichiometric relationships among the different enzymatic activities because of their coordinate synthesis
    (c) enhanced specificity and decreased side reactions because the product of each active site is in the immediate vicinity of the active site carrying out the next reaction in the sequence
    (d) enhanced versatility because the product of any one active site could be used by any other active site in its immediate vicinity to generate a variety of products
    (e) accelerated overall reaction rate because of the proximity of the active sites

19. Which of the following answers completes the sentence correctly? The major product of the fatty acid synthase complex in mammals is
    (a) oleate.
    (b) stearate.
    (c) stearoyl CoA.
    (d) linoleate.
    (e) palmitate.
    (f) palmitoyl CoA.

20. Calculate the ATP and NADPH requirements for the synthesis of lauric acid (C12:0) from acetyl CoA.

21. Which of the following statements about acetyl CoA carboxylase are correct?
    (a) It is active in the phosphorylated form.
    (b) It is partially active in the phosphorylated form in the presence of citrate.
    (c) It is phosphorylated by a cAMP-dependent protein kinase.
    (d) It is stimulated by a high-energy charge.
    (e) It is converted from an inactive form to an active form by protein phosphatase 2A.

22. Which of the following statements about desaturases in humans are correct?
    (a) They cannot introduce double bonds into a fatty acid that already contains a double bond.
    (b) They cannot introduce double bonds between the Δ^9 position and the ω end of the chain.
    (c) They convert the essential fatty acid linoleate into arachidonate.
    (d) They use an isozyme of the FAD-linked dehydrogenase of the β-oxidation cycle to form double bonds.

23. Which of the following statements about the eicosanoid hormones are NOT correct?

    (a) The major classes of eicosanoid hormones include prostaglandins, leukotrienes, thromboxanes, and prostacyclins.

    (b) The eicosanoid hormones are derived from arachidonic acid.

    (c) The eicosanoid hormones are very potent and exert global effects because they are widely distributed by the circulatory system.

    (d) The prostaglandins have a variety of physiologic effects.

    (e) Prostaglandins are derived directly from phospholipids.

24. Explain why aspirin is a potent anti-inflammatory agent.

ANSWERS TO SELF-TEST

1. b, c, d, e.

2. (a) hexadecanoic acid; palmitic acid; C16:0

    (b) cis-Δ^9-octadecenoic acid; oleic acid; C18:1 cis-Δ^9; ω-9

    (c) cis,cis-Δ^9, Δ^{12}-octadecadienoic acid; linoleic acid; C18:2 cis,cis-$\Delta^{9,12}$; ω-6

    (d) cis,cis,cis- (or all cis-) $\Delta^{9,12,15}$-octadecatrienoic acid; linolenic acid; C18:3 cis,cis,cis-$\Delta^{9,12,15}$; ω-3

    Note: Question 2 shows the undissociated form of the fatty acids; they would be dissociated at physiologic pH values.

3. e, c, a, d, b. The order of e, c, a, b is determined by the content of double bonds of the C_{18} fatty acids. Double bonds decrease intermolecular interactions and packing in the solid state and thus depress the melting points of fatty acids. The order of (d) relative to (b) is determined by the chain length C_{16} versus C_{18} of the saturated fatty acids.

4. (a) Synthesis. These four steps result in the addition of a 2-carbon alkane unit to a growing fatty acid.

    (b) Two.

5. Triacylglycerols contain a high proportion by weight of fatty acids. Fatty acids are highly reduced and consequently have a higher energy content (37.7 kjole or 9 kcal/g) than does glycogen (16.8 kjoule or 4 kcal/g), which is composed of carbohydrate residues containing numerous oxygen atoms. In addition, fats are anhydrous whereas glycogen is hydrated, so on the basis of actual storage weight, triacylglycerols contain six times more calories per gram than glycogen does.

6. b, c, d, e. The hydrolysis of triacylglycerols yields glycerol, not dihydroxyacetone, and glycerol can be converted into glyceraldehyde 3-phosphate, which can ultimately give rise to glucose. Several hormones affect the hormone-sensitive lipase of adipose tissue via a cyclic AMP-modulated phosphorylation that activates the enzyme.

7. (a) A thioester bond follows. The thioester linkage joins the fatty acid to CoA, which acts as a tag or handle by which the enzymes of the β-oxidation path can recognize, bind, and act on the saturated alkane chains of the fatty acids. Furthermore, since the thioester linkage is a "high-energy" bond, it can transfer the acyl group to carnitine—a reaction that is necessary to deliver the acyl group from the cytosol to the mitochondrial matrix for oxidation.

Thioester bond

(b) Since the $\Delta G^{\circ\prime}$ is approximately -31.4 kJ (-7.5 kcal/mol), it is comparable to that of the hydrolysis of ATP. The relatively large and negative value for the free energy of hydrolysis for acetyl CoA indicates that energy must be supplied to synthesize it and that, conversely, it can serve as an "activated" donor of acetyl groups.

(c) The carboxyl group of the fatty acid is first activated by reaction with ATP to form an acyl adenylate, which contains a mixed anhydride linkage between the carboxylate and the 5′-phosphate of AMP, with the release of PP_i. In a second step, also catalyzed by acyl CoA synthetase, the acyl group is transferred to the sulfhydryl group of CoA to form the thioester bond and release AMP.

(d) The hydrolysis of PP_i couples the cleavage of a second high-energy bond to the formation of the thioester bond to make its formation exergonic. In effect, two ATPs are used to make one acyl CoA.

8. b, c, a, g, h, d, e, f

9. Acyl CoA is formed in the cytosol, and the enzymes of the β-oxidation pathway are in the matrix of the mitochondrion. The mitochondrial inner membrane is impermeable to CoA and its acyl derivatives. However, a translocase protein can shuttle carnitine and its acyl derivatives across the inner mitochondrial membrane. The acyl group is transferred to carnitine on the cytosol side of the inner membrane and back to CoA on the matrix side. Thus, carnitine acts as a transmembrane carrier of acyl groups.

10. After activation of hexanoic acid to hexanoyl CoA, two rounds of β oxidation are required to produce 3 acetyl CoA molecules. The two cycles also produce 2 $FADH_2$ and 2 NADH molecules. Each acetyl CoA yields 10 ATP molecules on complete oxidation, each $FADH_2$ produces 1.5 ATP, and each NADH makes 2.5 ATP—a total of 38 ATP. Activation uses 2 ATP molecules because AMP is one of the products of the reaction, so the net yield is approximately 36 ATP.

11. False. When carbohydrates are abundant, oxaloacetate levels are high and condensation of acetyl CoA and oxaloacetate produces citrate. Citrate is used for energy production as well as for fatty acid biosynthesis. Ketone bodies are produced when acetyl CoA is abundant but oxaloacetate is depleted.

12. a, b, c, d. Choice (c) is correct because 3-hydroxybutyrate is in equilibrium with acetoacetate, the actual source of acetone. Choice (e) is incorrect because ketone bodies are the transportable form of acetyl units in blood.

13. a, c, d

14. (a) 1, 3, 5, 8   (b) 1, 2, 4, 5, 6, 7. Acyl CoA is involved in both the synthesis and the oxidation of fatty acids.

15. The irreversible and committed step of fatty acid biosynthesis is the formation of malonyl CoA from acetyl CoA and HCO_3^- by acetyl CoA carboxylase. HCO_3^- is fixed to form a dicarboxylic acid at the expense of an ATP cleavage. This facilitates the subsequent condensation reactions with activated acyl groups to form an acetoacyl-ACP by releasing CO_2 to help drive the reaction.

16. c, d

17. The reactions of elongation require the interactions of domains from different subunits of the dimer to form active sites at the interfaces of the subunits. One monomer holds the growing acyl chain while the other is linked to the incoming activated acetyl unit.

18. a, b, c, e. For a discussion relevant to the correct answers, see pages 659–661 in the text.

19. e

20. For a C12:0 fatty acid, 6 acetyl CoA molecules are required. One serves as a primer forming the ω end of the chain and five undergo condensation reactions as their malonyl CoA derivatives. Formation of each malonyl CoA requires 1 ATP molecule, and each cycle of elongation uses 2 NADPH molecules. Thus, 5 ATP and 10 NADPH are required.

21. b, d, e. Acetyl CoA carboxylase is inactivated by phosphorylation. This effect is partially abolished by citrate, which acts as an allosteric activator. Phosphorylation–dephosphorylation of this enzyme is under the control of hormones, whose action is mediated by a protein kinase that is dependent on AMP, not cAMP. High-energy charge stimulates acetyl CoA carboxylase, while low-energy charge, that is, high AMP levels, inhibits it.

22. b, c. For (c) additional elongations as well as desaturations are required.

23. c. Answer (c) is incorrect because the eicosanoid hormones have very short half-lives and therefore exert local rather than global effects. Answer (e) is correct because phospholipids supply the arachidonate for prostaglandin synthesis.

24. Aspirin acetylates a specific Ser residue in the cyclooxygenase component of prostaglandin synthase. Thus aspirin inhibits the synthesis of prostaglandins, thromboxanes, and prostacyclins, which mediate the inflammatory response.

PROBLEMS

1. Many plants have enzyme systems that catalyze the formation of a cis double bond in oleic acid at one or more positions between C-9 and the terminal methyl group. The fact that these enzyme systems exist in plants is of great significance to animals. Why?

2. (a) Stumpf and his colleagues have described an α-oxidation system in plant leaves and seeds in which fatty acid oxidation occurs at the α carbon. Molecular oxygen is used in the α-oxidative decarboxylation of a free fatty acid, which yields a fatty aldehyde that is one carbon shorter than the original fatty acid. The fatty aldehyde is in turn oxidized to the corresponding fatty acid, with NAD^+ serving as an electron acceptor. These steps are repeated, resulting in the complete oxidation of the fatty acid. Suppose that the NADH generated through the α oxidation of palmitate is reoxidized in the mitochondrial electron transport chain. Compare the yield of ATP generated by the α oxidation of palmitate with that generated by β oxidation of the same fatty acid. Assume that the products of the final round of oxidation are carbon dioxide and acetic acid.

   (b) If fatty acid oxidation occurs via the α-oxidation route, will odd-numbered fatty acids be glucogenic, that is, capable of forming glucose? Why?

3. Although most components of the diet contain fatty acids with unbranched chains, some plant tissues contain fatty acids with methyl groups at odd-numbered carbons in the acyl chain. These fatty acids cannot be broken down through β oxidation.

   (a) Which step in β oxidation is likely to be blocked when branched-chain fatty acids are substrates?

   (b) Some tissues, including brain tissue, can carry out the limited α oxidation of a fatty acid with one or more methyl groups at odd-numbered carbons. Using the pathway discussed in problem 2, show how one round of α oxidation enables a cell to bypass the block to β oxidation. Use as your substrate a molecule of palmitate with a methyl branch at C-3.

4. The oxidation by microbes of long-chain alkanes, which are found in crude oil, is a subject of study because of concern about oil spills. In many bacteria, alkane oxidation occurs within the outer membrane. A monooxygenase enzyme uses molecular oxygen and an oxidizable substrate, such as NADH, to convert an alkane to a primary alcohol. Studies show that three additional reactions are required for the primary alcohol to undergo β oxidation. Propose a pathway for the conversion of a long-chain primary alcohol to a substrate that can undergo β oxidation. Include cofactors and electron acceptors that might be required.

5. Malonyl CoA, labeled with ^{14}C in the methylene carbon, is used in excess as a substrate in a system in vivo for the synthesis of palmitoyl CoA, which is catalyzed by a yeast fatty acid synthase complex. Acetyl CoA and other substrates are also present in the system, but acetyl CoA carboxylase is not. Which carbons in palmitoyl CoA will be labeled?

6. A deficiency of carnitine acyltransferase I in human muscle causes cellular damage and recurrent muscle weakness, especially during fasting or exercise. A deficiency of the enzyme in the liver causes an enlarged and fatty liver, hypoglycemia, and a reduction in the levels of ketone bodies in blood. Explain the likely causes of these symptoms.

7. One intermediate in the conversion of propionyl CoA to succinyl CoA is methylmalonyl CoA, the structure of which follows. This compound is an analog of malonyl CoA. In people who are unable to convert propionyl CoA to succinyl CoA, high levels of methylmalonyl CoA are observed. What effect could such levels of methylmalonyl CoA have on fatty acid metabolism?

$$\begin{array}{c} COO^- \\ | \\ H_3C-C-H \\ | \\ C-S-CoA \\ | \\ O \end{array}$$

L-Methylmalonyl CoA

8. Animals cannot synthesize glucose from even-numbered fatty acids, which make up the bulk of the fatty acids in their diet.
   (a) How can odd-numbered fatty acids be used for the net synthesis of glucose in animals?
   (b) Triacylglycerols can be used as precursors of glucose. Give two reasons why this is possible.
   (c) Why are most of the fatty acids found in animal tissues composed of an even number of carbon atoms?
   (d) Some bacteria synthesize odd-numbered fatty acids. What CoA derivative is required, in addition to acetyl CoA and malonyl CoA, for the synthesis of an odd-numbered fatty acid?

9. (a) Describe how malonyl CoA affects the balance between the rates of synthesis and β oxidation of fatty acids in a liver cell.
   (b) Show that failure to regulate these two processes reciprocally could result in the wasteful hydrolysis of ATP.

10. Plant seeds contain triacylglycerols in organelles called *spherosomes*. During germination, lipases located in the spherosome membrane convert triacylglycerol to monoacylglycerols, free fatty acids, and glycerol. Both free fatty acids and monoacylglycerols enter the glyoxysome, whereas most of the glycerol is metabolized in the plant cell cytosol. A membrane-bound lipase in the glyoxysome converts monoacylglycerols to free fatty acids and glycerol.
    (a) Describe two possible metabolic fates of glycerol in the cytosol.
    (b) What is the fate of fatty acids in the glyoxysome?

(c) When a germinating plant begins to carry out photosynthesis, the number of gly-oxysomes in the germinating plant decreases rapidly. Why?

(d) Plant tissues with high numbers of mitochondria also have high concentrations of carnitine, but there is little correlation between numbers of glyoxysomes and carnitine concentrations in germinating tissue. What does this observation suggest about the role of carnitine in fatty acid metabolism in these two organelles?

(e) Another difference between plant glyoxysomes and plant mitochondria is that glyoxysomes cannot oxidize acetyl CoA, whereas mitochondria can. How is this observation related to the metabolism of fatty acids in these two organelles?

11. People concerned about their weight must pay attention not only to triacylglyceride intake but also to the consumption of starch, glucose, and other carbohydrates. Although carbohydrates can be converted to glycogen in liver, muscle, and other tissues, only about 5 percent of the energy stored in the body is present as glycogen. What happens to most carbohydrates that are consumed in excess of caloric need?

12. Wakil's pioneering studies on fatty acid synthesis included the crucial observation that bicarbonate is required for the synthesis of palmitoyl CoA. He was surprised to find that very low levels of bicarbonate could sustain palmitate synthesis; that is, there was no correlation between the amount of bicarbonate required and the amount of palmitate produced. Later he also found that ^{14}C-labeled bicarbonate is not incorporated into palmitate. Explain these observations.

13. Many of the enzymes of the β-oxidation pathway have relatively broad specificities for fatty acyl chain lengths. Why is this important for the economy of the cell?

14. Liver tissue carries out the synthesis of ketone bodies from fatty acids. Suppose a liver cell converts palmitic acid to acetoacetate and then exports it to the circulation. How many molecules of ATP per molecule of palmitate converted to acetoacetate are available to the liver cell?

15. In tissue culture, cells that are deficient in NADP$^+$-linked malate enzyme can be isolated. They exhibit a slightly lower rate of fatty acid synthesis when compared with normal cells. However, cells lacking citrate lyase are very difficult to isolate. Why?

16. An unusual sphingolipid contains a 22-carbon, polyunsaturated fatty acid called *clupanodonic acid*, or *7,10,13,16,19-docosapentaenoic acid*. In mammals, both the mitochondrial and endoplasmic reticular acyl chain elongation and desaturation systems can synthesize clupanodonate from linolenate.

(a) What steps are required to synthesize clupanodonate from linolenate?

(b) Why are mammals unable to synthesize clupanodonate from linoleate?

17. Hydrogenating oils to saturate the double bonds in their fatty acids (see problem 23) in order to increase their melting temperatures causes some of the cis double bonds to convert into the trans conformation. Predict what would happen if a monoenoic fatty acid with a trans-Δ^{10} bond were produced, ingested, and degraded by the β-oxidation pathway. If another of the ingested fatty acids contained a cis-Δ^{11} double bond, what would be the outcome of these processes? What effect would the presence of the double bond have on the yield of ATP obtained by the β oxidation of these fatty acids?

18. Certain desert mammals can survive long periods of drought by consuming plants and seeds and then generating water by metabolizing the fuels they provide.

(a) Briefly describe how water is generated through intermediary metabolism. Include the sources of oxygen and hydrogen and describe reactions that lead to the formation of water.

(b) While parts of mature plants are a reliable source of carbohydrates and proteins, plant seeds contain high quantities of triacylglycerols and free fatty acids. Would plants or seeds be better for generating metabolic water? Why?

(c) Suppose that a desert rat metabolizes 30 g of palmitoyl CoA from seeds. How many milliliters of water can be generated from the process?

19. Explain why the metabolism of a C_{15} fatty acid can lead to the net synthesis of glucose, but the metabolism of a C_{16} fatty acid cannot.

20. Compare the effects of high levels of intracellular citrate on pathways of fatty acid and carbohydrate metabolism. Explain how its transport from the mitochondrion to the cytosol is essential for the action of citrate on both sets of pathways.

21. You are examining mitochondria from muscle cells of an infant who has a deficiency in one of the enzymes in the fatty acid oxidative pathway. The mitochondria consume oxygen normally when incubated with pyruvate and malate, with succinate, or with palmitoyl CoA (in the presence of carnitine), but the rate of oxygen utilization is decreased when the mitochondria are incubated with linoleoyl CoA in the presence of carnitine. Blood levels of carnitine in the patient are low, while the levels of an unusual acylcarnitine derivative are present in blood and urine. Analysis of this acylcarnitine species using mass spectroscopy reveals that it is *trans*-Δ^2, *cis*-Δ^4 decadienoyl (C10:2)-acylcarnitine. The infant suffers from hypotonia (lack of muscle tone) and slow weight gain.

What enzyme is deficient in the cells of the infant? Explain the observed symptoms on the basis of such a deficiency and how you might treat such a disorder.

22. In mammals, acetyl CoA from fatty acid oxidation cannot be used for the net synthesis of pyruvate or oxaloacetate, which in turn means that net glucose synthesis from acetyl CoA is impossible. However, glucose can be radioactively labeled when ^{14}C-labeled acetate is introduced into human tissue culture cells and converted to acetyl CoA by acetyl CoA synthetase. Radioactive fatty acids can also be used to label glucose. Why? If the methyl carbon of acetate is labeled, where will glucose be labeled?

23. To improve the shelf life (resistance to oxidation) and "crunchiness" of baked products, partially hydrogenated fatty acids are used in recipes. Vegetable oils containing unsaturated fatty acids are treated at high temperature (typically 260°C) and high pressure with hydrogen to reduce the double bonds to single bonds. If the hydrogenation process went to completion, no double bonds would remain and the fatty acids would be fully saturated. However, incomplete reduction leads to isomerization of some of the remaining double bonds from their normal, natural cis configuration to the trans configuration, which confers the desired properties on the oil. Assuming that the original oil had palmitoleate (cis-Δ^9- hexadecanoate) as its only unsaturated fatty acid and that after partial hydrogenation some of the remaining cis-Δ^9 double bonds were now trans-Δ^9 double bonds, what would you predict concerning the β oxidation of the mixture of fully saturated palmitate (16:0) and the remaining 16:1 fatty acid isomers? What do you think trans fatty acids might do to the physical properties of the oil?

ANSWERS TO PROBLEMS

1. Because animals lack an enzyme that can introduce double bonds beyond the C-9 position in a fatty acid, they cannot synthesize linoleate and linolenate *de novo*. These unsaturated fatty acids are precursors for a number of other necessary fatty acids, as well as the eicosanoid hormones. Animals therefore rely on their diet as the source of linoleate and linolenate, which are synthesized only in plants.

2. (a) The net yield from the β oxidation of palmitate is 106 molecules of ATP, as discussed on pages 647 and 648 of the text. If 1 molecule of NADH is generated for 15 of the 16 carbons of palmitate, then the yield of ATP is 2.5×15, or 37.5. For α oxidation, activation of the acetate molecule requires 2 ATP. Subsequent oxidation of acetyl CoA generates 10 ATP molecules. Thus 45.5 molecules of ATP are generated by the α oxidation of a molecule of palmitate.

(b) In β oxidation of odd-numbered fatty acids, the products include propionyl CoA, which can be converted to succinyl CoA, a glucogenic substrate. However, α oxidation of an odd-numbered fatty acid would yield carbon dioxide as well as a single molecule of acetate or acetyl CoA, neither of which is glucogenic.

3. (a) As shown in Figure 22.1, the oxidation of a fatty acid with a methyl group at C-3 proceeds to the formation of the L-hydroxymethylacyl CoA derivative. Subsequent oxidation of the β carbon to the ketoacyl derivative is blocked by the methyl group. Compare this pathway with the one shown in Figure 22.9 of the text (page 647).

FIGURE 22.1 Formation of L-hydroxymethylacyl CoA through β oxidation of branched-chain acyl CoA.

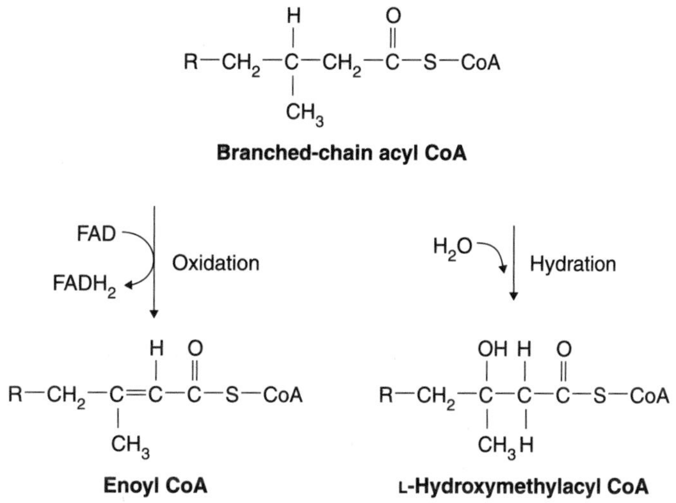

(b) The oxidation of the α carbon in the palmitate derivative followed by decarboxylation of the molecule yields a fatty acid that has a methyl group at an even-numbered carbon. Activation of the fatty acid to form an acyl CoA derivative followed by oxidation at the β carbon results in the generation of propionyl CoA and a shortened acyl derivative, lauroyl CoA (C:12).

FIGURE 22.2 The α oxidation and oxidative decarboxylation of a branched-chain fatty acid allow generation of intermediates that can enter normal oxidative pathways.

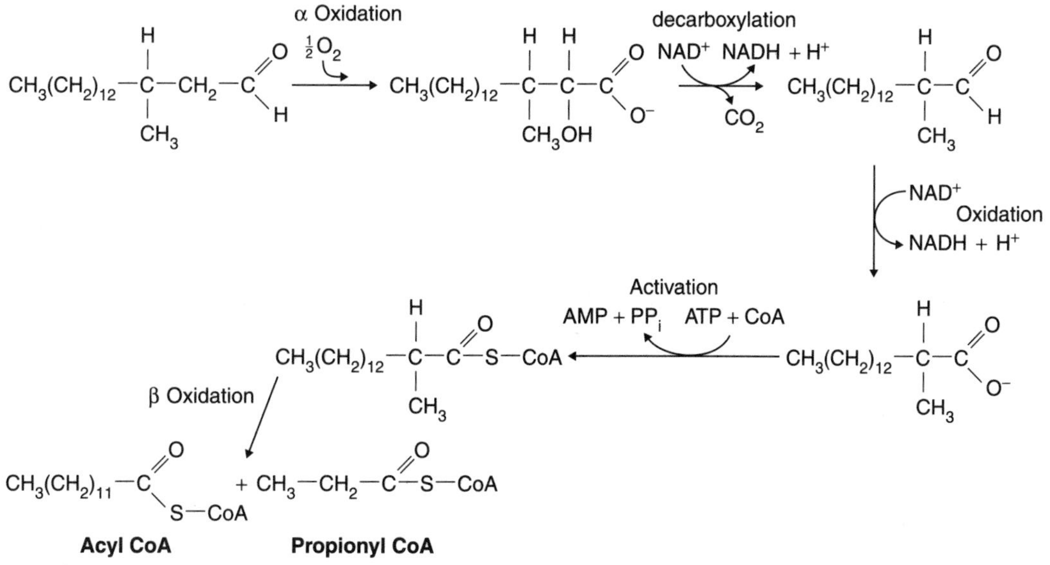

4. The normal route for β oxidation in bacteria utilizes acyl CoA derivatives, which are formed from free fatty acids. To convert a primary alcohol to a free fatty acid, two oxidative steps are needed, each requiring an electron acceptor. In *Corynebacterium*, NAD^+-dependent dehydrogenases catalyze the sequential conversion of a primary alcohol to a fatty acid, with the corresponding aldehyde as an intermediate. Conversion of the free fatty acid to an acyl CoA derivative requires two equivalents of ATP (because ATP is converted to AMP and PP_i), as well as coenzyme A. The reaction is catalyzed by acyl CoA synthase.

5. As shown in Figure 22.26 of the text, acetyl-ACP and malonyl-ACP condense to form acetoacetyl-ACP. Carbons 4 and 3 of acetoacetyl-ACP are not labeled because they are derived from acetyl CoA. These two carbons will become carbons 15 and 16 of palmitate. Only C-2 of acetoacetyl-ACP will be labeled because it is derived from the methylene carbon of malonyl-ACP. When the second round of synthesis begins, butyryl-ACP condenses with a second molecule of methylene-labeled malonyl-ACP, which contributes C-1 and C-2 of the newly formed six-carbon ACP derivative. In this compound, C-2 and C-4 will be labeled. Chain elongation continues until palmitoyl-ACP is formed. Each even-numbered carbon atom, except for carbon 16 (at the ω end), will be labeled.

6. Carnitine acyltransferase I facilitates the transfer of long-chain fatty acids into the mitochondrion by catalyzing the formation of fatty acyl carnitine molecules. The failure to form such molecules means that long-chain fatty acids are not available for cellular oxidation. In muscle, exercise or fasting increases dependence on fatty acids as a source of energy, so the inability to metabolize them interferes with cellular functions, causing cramps, weakness, and muscle damage. Liver cells also require formation of fatty acyl carnitine molecules to oxidize fats in mitochondria. If fatty acids cannot be utilized, they will remain in the cytosol, where high concentrations of them cause cell enlargement and interfere with other functions. Liver cells must then use glucose as a source of energy instead of exporting it to other cells. Because liver cells use acetyl CoA, which is derived primarily from fatty acid oxidation, as a precursor of ketone bodies, the failure to oxidize fatty acids will result in a reduction in the rate of ketone body synthesis. This in turn will exacerbate the symptoms of hypoglycemia because tissues that normally use ketone bodies as a source of energy, such as cardiac muscle and renal cortex, will have to rely more heavily on glucose as a source of energy.

7. Because malonyl CoA is a substrate for fatty acid synthase, competition from methylmalonyl CoA could cause a decrease in the rate of palmitoyl CoA synthesis in the cytosol, which could in turn lead to an increase in the concentration of acetyl CoA because palmitoyl CoA inhibits acetyl CoA carboxylase. In addition, high levels of methylmalonyl CoA could interfere with transport of long-chain fatty acyl chains into mitochondria by inhibiting carnitine acyltransferase, as does malonyl CoA. Thus, both the synthesis and the oxidation of fatty acids could be inhibited by methylmalonyl CoA.

8. (a) The oxidation of an odd-numbered fatty acid yields acetyl CoA molecules as well as one molecule of propionyl CoA, which can be converted to succinyl CoA, a component of the citric acid cycle. Although two-carbon compounds like acetyl CoA cannot be used for the net synthesis of glucose, succinyl CoA can contribute net carbons to the citric acid cycle, enabling oxaloacetate and ultimately glucose to be formed through gluconeogenesis.

   (b) Triacylglycerols are converted to glycerol and three free fatty acids through the action of lipases. Glycerol can be converted to glucose by way of dihydroxyacetone phosphate. Odd-numbered fatty acids found in triacylglycerols can also be used for net synthesis of glucose, whereas even-numbered fatty acids cannot.

(c) During fatty acid synthesis, most organisms use acetyl CoA as a source of the ω carbon and its adjacent carbon in the acyl chain. Two of the three carbons of malonyl CoA are incorporated during each cycle of acyl chain elongation. Thus the resulting fatty acid will contain an even number of carbon atoms.

(d) To produce an odd-numbered fatty acid, at least one odd-numbered CoA intermediate must be incorporated in its entirety during fatty acid synthesis. Propionyl CoA can be used by certain bacteria for the initial condensation step with malonyl CoA in fatty acid synthesis. The resulting five-carbon acyl intermediate is then extended in two-carbon units to yield an odd-numbered fatty acid.

9. (a) Malonyl CoA is a key substrate for the synthesis of fatty acids; when it is abundant, synthesis is stimulated. In addition, high levels of this intermediate inhibit carnitine acyltransferase I, thereby limiting the entry of fatty acyl chains into the mitochondrion, where they are oxidized. A decrease in the concentration of malonyl CoA leads to a decrease in the rate of fatty acid synthesis and an increase in the rate of fatty acid oxidation in the mitochondrion.

(b) The overall equation for the synthesis of palmitoyl CoA is

$$8 \text{ acetyl CoA} + 7 \text{ ATP} + 14 \text{ NADPH} \rightarrow$$
$$\text{palmitoyl CoA} + 14 \text{ NADP}^+ + 7 \text{ CoA} + 7 \text{ H}_2\text{O} + 7 \text{ ADP} + 7 \text{ P}_i$$

The overall equation for the oxidation of palmitoyl CoA is

$$\text{palmitoyl CoA} + 7 \text{ FAD} + 7 \text{ NAD}^+ + 7 \text{ CoA} + 7 \text{ H}_2\text{O} \rightarrow$$
$$8 \text{ acetyl CoA} + 7 \text{ FADH}_2 + 7 \text{ NADH} + 7 \text{ H}^+$$

Assuming that NADPH is equivalent in reducing power to NADH, that a molecule of $FADH_2$ yields 1.5 ATP during electron transport and oxidative phosphorylation, and that a molecule of NADH yields 2.5 ATP, then 42 ATP molecules are required to synthesize a molecule of palmitoyl CoA, whereas 28 ATP are generated by the conversion of palmitoyl CoA to 8 molecules of acetyl CoA. There is a net loss of 14 ATP molecules if the two processes occur simultaneously.

10. (a) Glycerol is converted to dihydroxyacetone phosphate, which in turn can serve as a source of glucose or can be converted to acetyl CoA.

(b) Fatty acids serve as a source of acetyl CoA, which is used in the glyoxylate cycle and gluconeogenesis.

(c) The primary function of glyoxysomes is to utilize fatty acids from triacylglycerols for the synthesis of glucose, which is used as a source of other molecules by the developing plant. Once leaf development enables the plant to generate glucose by photosynthesis, glyoxysomes are no longer needed.

(d) Carnitine functions in the transport of long-chain fatty acids from the cytosol to the interior of the mitochondrion. The observation suggests that, although carnitine may be important in mitochondrial transport of fatty acyl chains, the compound is not involved in the movement of fatty acyl chains into the glyoxysome. It is also possible that glyoxysomes metabolize fatty acids with shorter acyl chains, for which transport facilitated by carnitine is not necessary.

(e) The fate of fatty acids is different in glyoxysomes and in mitochondria. Both organelles carry out β oxidation of fatty acids to acetyl CoA; however, in glyoxysomes, acetyl CoA is a precursor of glucose, whereas mitochondria oxidize acetyl CoA to CO_2 and H_2O to generate ATP.

11. Carbohydrates consumed in excess of caloric need are converted to acetyl CoA, which in turn serves as a source of fatty acids. The concurrent synthesis of glycerol from carbohydrates such as glucose and fructose provides the second precursor needed for the synthesis of triacylglycerols, which are the primary storage form of energy in humans. Excess carbohydrate is converted to fat.

12. Bicarbonate is a source of carbon dioxide for the reaction catalyzed by acetyl CoA carboxylase, in which malonyl CoA is formed. Malonyl CoA is then used as a source of two-carbon units for fatty acyl chain elongation, and the carbon atom derived originally from bicarbonate is released as CO_2. Carbon dioxide is then rapidly converted to bicarbonate, which is used again for the synthesis of another molecule of malonyl CoA. Thus, the carbon atom derived from bicarbonate can be used many times for the production of malonyl CoA, but it is never incorporated into the growing acyl chain, so it does not appear in palmitate.

13. If each enzyme could operate only on fatty acyl CoA derivatives of a particular chain length, then as many as eight sets of enzymes would be required to carry out the β oxidation of palmitate. The fact that most enzymes of the β-oxidation pathway can use acyl CoA molecules of different chain lengths as substrates means that the cell needs to synthesize fewer different enzymes to carry out fatty acid oxidation.

14. To synthesize acetoacetate from palmitate, liver cells must carry out β oxidation of the 16-carbon acyl chain, generating 8 molecules of acetyl CoA, which will in turn generate 4 molecules of acetoacetate. A total of 7 NADH and 7 $FADH_2$ molecules are generated per molecule of palmitate converted to acetyl CoA. The 14 reduced cofactors are equivalent to 28 ATP molecules. Because 2 molecules of ATP are needed to activate palmitate, the net yield of ATP per palmitate is 26.

15. Both malate enzyme and citrate lyase are part of the shuttle system that transports two-carbon units from the mitochondrion to the cytosol. Malate enzyme also generates reducing power in the form of NADPH, which is used for fatty acid synthesis; however, the pentose phosphate pathway (Section 20.3 of the text) also serves as a source of NADPH, so fatty acid synthesis can continue even if malate enzyme is deficient. Recall from pages 551–552 of the text that malate can cross the mitochondrial membrane. Citrate lyase is more critical to fatty acid synthesis because it is required to generate acetyl CoA from citrate in the cytosol. Without cytosolic acetyl CoA, fatty acid synthesis cannot take place, and the cells cannot grow and divide.

16. (a) To synthesize clupanodonate from linolenate, the acyl chain must be elongated from 18 to 22 carbons, and two new double bonds must be introduced into the chain. Although the details of the various mammalian desaturation systems are not completely understood, it appears that a double bond at C-6 can be introduced when a double bond at C-9 is available, and a double bond at C-5 can be introduced when one at C-8 is available. Thus, the probable sequence of reactions includes the introduction of a double bond at C-6 of linolenate (yielding a 18:4 cis-Δ^6, Δ^9, Δ^{12}, Δ^{15}-acyl chain) followed by chain elongation to a 20-carbon derivative. The introduction of a double bond at C-5 then gives an acyl chain denoted as 20:5 cis-Δ^5, Δ^8, Δ^{11}, Δ^{14}, Δ^{17}. The final reaction required to yield clupanodonate is chain elongation to the 22-carbon fatty acyl chain.

(b) Linoleate has cis double bonds at C-9 and C-12. Elongation to a 22-carbon chain would yield an acyl chain with double bonds at C-13 and C-16. To form clupanodonate, a double bond at C-19 is needed, but mammals lack the enzymes required to introduce double bonds beyond C-9. Thus, linoleate cannot be used for the synthesis of clupanodonate.

17. Four rounds of β oxidation of a fatty acid with a trans-Δ^{10} double bond would yield a *trans-Δ^2*-enoyl CoA derivative. This compound is the natural intermediate formed by an acyl CoA dehydrogenase. It would be hydrated by enoyl CoA hydratase to form the L-3-hydroxyacyl CoA derivative. For the fatty acid with a cis-Δ^{11} double bond, four rounds of β oxidation would produce a cis-Δ^3 double bond, which would not serve as a substrate for enoyl CoA hydratase. An isomerase would convert this bond into the trans-$\Delta 2$ configuration to allow subsequent metabolism. Since the double bond already exists in the fatty acids and does not arise from β oxidations, one less $FADH_2$ would be formed. Consequently, approximately 1.5 fewer ATP molecules would be produced for each preexisting double bond.

18. (a) The source of oxygen for formation of water during respiration is atmospheric oxygen, whereas the sources of hydrogen include oxidizable foodstuffs such as carbohydrates and fats. These substances are oxidized to generate "energy-rich" electrons, which are in turn used to reduce oxygen to generate water. The principal terminal reaction in the process occurs in the mitochondrion, where electrons are transferred from cytochrome *c* to oxygen to generate oxidized cytochrome *c* and water. Also important in water generation is the formation of ATP from ADP and inorganic phosphate, where a molecule of water is generated during the formation of each ATP molecule.

   (b) The more reduced the carbons of a substrate, the larger the number of electrons available during metabolism and the more water generated. Most carbon atoms of fatty acids are saturated and therefore highly reduced, so that they are a better source of available electrons. Carbohydrate molecules like glucose, whose carbons are at the alcohol level of oxidation or, in the case of the C-1 atom, at the aldehyde level, provide fewer electrons during terminal oxidation. Seeds, which contain a high percentage of fats, are therefore a better source than mature plants for the generation of water.

   (c) First determine the number of moles of palmitate that are converted to CO_2 and water. The molecular weight of the molecule ($C_{16}H_{31}O_2$) is 255 g mol^{-1}.

   $$30 \text{ g}/255 \text{ g mol}^{-1} = 0.12 \text{ mole palmitate oxidized}$$

   Then, calculate the number of moles of water produced by the complete oxidation of palmitate.

   For palmitoyl CoA, the text shows on pages 625–626 that oxidation of the molecule gives 7 $FADH_2$, 7 NADH, and 8 acetyl CoA molecules, utilizing 7 molecules of water.

   On page 510, the text shows that oxidation of acetyl CoA in the citric acid cycle yields 3 NADH, 1 $FADH_2$, and 1 GTP, equivalent to 1 ATP, utilizing 2 water molecules. Thus, the 8 acetyl CoA molecules produced from palmitoyl CoA give 24 NADH, 8 $FADH_2$, and 8 ATP equivalents, utilizing 16 water molecules.

   The total number of reduced electron carriers from the oxidative process is 31 NADH and 15 $FADH_2$. Recall that 2.5 ATP molecules are produced when NADH is oxidized in the electron transport chain, and 1.5 ATP are generated from $FADH_2$ oxidation.

   One water molecule is gained per ATP molecule formed plus 1 water molecule per pair of e$^-$ molecules oxidized.

   The overall equation for the production of NAD$^+$, ATP, and water from palmitoyl CoA is

$$31 \text{ NADH} + 15.5 \text{ O}_2 + 77.5 \text{ ADP} + 77.5 \text{ P}_i + 108.5 \text{ H}^+ \longrightarrow$$
$$31 \text{ NAD}^+ + 77.5 \text{ ATP} + 108.5 \text{ H}_2\text{O}$$

and for $FADH_2$ it is

$$15 \text{ FADH}_2 + 7.5 \text{ O}_2 + 22.5 \text{ ADP} + 22.5 \text{ Pi} + 37.5 \text{ H}^+ \longrightarrow$$
$$15 \text{ FAD} + 22.5 \text{ ATP} + 37.5 \text{ H}_2\text{O}$$

The total number of water molecules produced is 146, and the net water produced is $(146 - 23) = 123$ molecules of water per palmitate oxidized or 123 moles of water per mole of palmitate. The molecular weight of water is 18.0 g mol^{-1}.

Thirty grams of palmitoyl CoA is equivalent to 0.12 mole of palmitate, which generates $0.12 \times 123 = 14.8$ moles of water when oxidized. At 18 g mol^{-1}, 14.8 moles of water equals 266 g, or 266 ml, of water.

19. The oxidation of a C_{16} fatty acid (palmitate) leads to the formation of eight molecules of acetyl CoA. Acetyl CoA, which contains two carbon atoms, is oxidized to two CO_2 in the citric acid cycle, so the net number of carbons entering and leaving the cycle is zero. Thus, no net carbons are available to enter the gluconeogenic pathway. On the other hand, oxidation of a C_{15} fatty acid generates seven acetyl CoA molecules plus one molecule of propionyl CoA. This compound is converted by carboxylation, epimerization, and conversion to succinyl CoA, a *four*-carbon compound that is an intermediate in the citric acid cycle. Succinyl CoA contributes two extra carbons to the gluconeogenic pathway, leading to the net synthesis of glucose.

20. High levels of citrate signal that glucose utilization is no longer necessary and that adequate carbon atoms are available for synthesis of palmitoyl CoA. Citrate inhibits phosphofructokinase 1 activity, decelerating the rate of glycolysis. On the other hand, citrate stimulates the activity of acetyl CoA carboxylase, so increased production of malonyl CoA leads to stimulation of fatty acid synthesis. The transport of citrate from the mitochondrial matrix to the cytosol is important because both phosphofructokinase 1 and acetyl CoA carboxylase are located in the cytosol.

21. The most likely deficiency is a lack of 2,4-dienoyl CoA reductase, an enzyme that is essential for the degradation of unsaturated fatty acids with double bonds at even-numbered carbons. Such fatty acids include linoleate (9-*cis*,12-*cis* 18:2). Four rounds of oxidation of linoleoyl CoA generate a 10-carbon acyl CoA that contains a *trans*-Δ^2 and a *cis*-Δ^4 double bond. This intermediate is a substrate for the reductase, which converts the 2,4-dienoyl CoA to *cis*-Δ^3-enoyl CoA. A deficiency of 2,4-dienoyl reductase leads to an accumulation of *trans*-Δ^2,*cis*-Δ^4-decadienoyl CoA molecules in the mitochondrion. The observation that carnitine derivatives of the 2,4-dienoyl CoA are found in blood and urine provides evidence that these molecules accumulate in the mitochondrion and are then attached to carnitine. Formation of carnitine decadienoate allows the acyl molecules to be transported across the inner mitochondrial membrane into the cytosol and then into the circulation.

Mitochondria from the patient function normally, taking up oxygen as they carry out oxidation of various substrates including palmitate, a saturated fatty acid. However, incubation of those mitochondria with linoleate results in reduced oxygen uptake, because the absence of the reductase molecule allows only a limited number of rounds of β oxidation to occur before the 2,4-dienoyl molecule is formed. Lack of muscle tone could mean that there are difficulties in oxidizing fuel molecules needed to provide energy for muscle contraction. If carnitine levels in cells are lower because many of them are esterified to decadienoate molecules, the result is a virtual deficiency of carnitine. The ability of the cell to transport other long-chain fatty acids across the inner mitochondrial membrane is limited under these conditions. Impairment of fatty acid oxidation means that fewer ATP molecules are available for muscular activity.

One immediate strategy for dealing with this disorder is to limit linoleate in the diet. However, linoleate is a starting point for other unsaturated fatty acids including arachidonate, a precursor of eicosanoid hormones. Limiting linoleate in the diet of a person with the reductase deficiency would have to be carried out carefully, to avoid a deficiency of an essential fatty acid.

22. Radioactive acetyl CoA can be generated by direct synthesis from ^{14}C-acetate or from β oxidation of radioactive fatty acids, such as uniformly labeled palmitate. Examination of the reactions of the citric acid cycle reveals that neither of the two carbons that enter citrate from acetate is removed as carbon dioxide during the first pass through the cycle. Labeled carbon from ^{14}C-methyl-labeled acetate appears in C-2 and C-3 of oxaloacetate because succinate is symmetrical, with either methylene carbon in that molecule labeling C-2 or C-3 of oxaloacetate. The conversion of oxaloacetate to phosphoenolpyruvate yields PEP labeled at C-2 or C-3 as well. Formation of glyceraldehyde 3-phosphate and its isomer dihydroxyacetone phosphate gives molecules both labeled at carbons 2 and 3. Condensation by aldolase gives fructose 1,6-bisphosphate radioactively labeled at carbons 1, 2, 5, and 6. The corresponding four carbons will then be labeled in glucose 6-phosphate or glucose. No net synthesis of glucose will have occurred, but the label will have been incorporated.

23. The β-oxidation pathway would convert the palmitate to eight acetyl CoA molecules. The natural palmitoleate (cis-Δ^9- hexadecanoate) would have three two-carbon units removed until it became cis-Δ^3- enoyl CoA)—a 10-carbon compound. A cis-Δ^3 double bond cannot be further metabolized as such, so the enzyme cis-Δ^3- CoA isomerase converts it into trans-Δ^2-enoyl CoA, which is a normal constituent in the β oxidation pathway. The trans-Δ^2-enoyl CoA can be further oxidized to completion to yield five more CoA molecules. The situation is different for the trans-Δ^9-hexadecanoate. It would be oxidized by β oxidation to yield trans-Δ^3-enoyl CoA. The cis-Δ^3- CoA isomerase converts the trans-Δ^3-enoyl CoA to trans-Δ^2-enoyl CoA that would then be subject to further, normal catabolism. Thus with the partial hydrogenation of this particular oil, the presence of a trans double bond causes no problem for complete catabolism because enzymes exist to isomerize them to metabolizable forms. Some trans fatty acids arising from partial hydrogenation of other oils are not recognized as normal physiological components and are deposited in blood vessels.

Trans fatty acids are more extended than those with cis double bonds giving them properties more like fully saturated fatty acids. They would increase the melting temperature of the oil (hardening it) and, if incorporated into the phospholipids of membranes, would decrease the fluidity of the lipid membrane.

Protein Turnover and Amino Acid Catabolism

Organisms derive energy from both stored and exogenous fuels. The catabolism of carbohydrates (Chapter 16) and fats (Chapter 22) have been discussed in previous chapters. In Chapter 23, the authors explain the role of proteins in energy metabolism. Although proteins are not stored as fuels per se as carbohydrates and fats are, supplies of proteins in excess of those needed to provide biosynthetic precursors are degraded for energy or are converted into fats or carbohydrates. Most of the amino groups of excess amino acids are converted into urea through the urea cycle, whereas their carbon skeletons are transformed into acetyl CoA, pyruvate, or one of the citric acid cycle intermediates. After a discussion of how dietary proteins are degraded and the rates at which proteins are degraded, the authors discuss the process of ubiquination, by which proteins are targeted for degradation. The N-terminal amino acid of a protein strongly determines its half-life. When tagged by ubiquitin, a large protease complex called the proteosome carries out the degradation of the protein. The proteosome then cleaves off the ubiquitin intact, so it can be recycled.

Once a protein is cleaved into individual amino acids, the amino acids are either incorporated into newly synthesized proteins or degraded to specific compounds for entry into an energy transduction pathway. If entry into an energy transduction pathway is its fate, the nitrogen(s) must first be removed. The α-amino groups of most amino acids are transferred to α-ketoglutarate to form glutamate by transamination (catalyzed by aminotransferases), and the α-amino group of glutamate is then converted to ammonia by an oxidative deamination. The authors describe the reaction mechanism of aminotransferases and the role the coenzyme pyridoxal phosphate (PLP) plays in this enzyme and others. The urea cycle is introduced next, which carries out the condensation of ammonia, the α-amino group of aspartate, and CO_2 to form urea—a nontoxic excretory product of nitrogen in higher animals. The urea cycle is linked to the citric acid cycle (Chapter 17) due to its production of the citric acid cycle intermediate fumarate. The physiological effects of defects in the urea cycle are also explored.

Because there are 20 amino acids, the catabolic pathways of their carbon skeletons are numerous and of varied types. The authors describe how the carbon atoms of each amino acid are funneled into one or more of seven primary products. Two of these, acetyl CoA and acetoacetyl CoA, can be converted to ketone bodies (Chapter 22), and the remaining five can be converted into glucose (Chapter 16), all of which can be oxidized in energy-generating pathways. The two groups of products lead to the glycogenic-ketogenic classification of the amino acids. The chapter ends emphasizing the importance of carrying out amino acid catabolism by examining the pathological consequences of defects or deficiencies in some of the enzymes involved in catabolism of amino acids and the synthesis of urea.

LEARNING OBJECTIVES

When you have mastered this chapter, you should be able to accomplish the following objectives.

INTRODUCTION

1. State the fate of exogenously supplied amino acids that are not used for biosynthesis.

Proteins Are Degraded to Amino Acids (Text Section 23.1)

2. List the *essential amino acids* in human beings.
3. Discuss dietary protein digestion and degredation.
4. Compare the *half-lives* of proteins in mammals.

Protein Turnover Is Tightly Regulated (Text Section 23.2)

5. Identify the features of *ubiquitin*.
6. Explain the functions of the three enzymes (E1, E2, and E3) that participate in the attachment of ubiquitin to proteins targeted for degradation.
7. List the advantages of *poly-ubiquination* of proteins targeted for degradation.
8. Contrast the effect of different N-terminal amino acids on the half-lives of proteins in yeast.
9. Give three examples of defects in normal E3 function that lead to diseases in humans.
10. Describe the subunit structure and function of the *26S proteosome*.
11. Explain how protein degradation plays a role in regulation of *NF-κB*.
12. Discuss how differences between human and bacterial proteosomes is leading to the development of new antibiotics.

The First Step in Amino Acid Degradation Is the Removal of Nitrogen (Text Section 23.3)

13. Name the major organ responsible for *amino acid degradation* in mammals.
14. Describe the reactions catalyzed by the *aminotransferases (transaminases)* and state the major function of these reactions.
15. Write the equations for the transamination reactions catalyzed by *aspartate aminotransferase* and *alanine aminotransferase*.

16. Describe the reaction catalyzed by *glutamate dehydrogenase,* outline its regulation, and state its major function. Note that the participation of NAD^+ in the reaction links nitrogen metabolism and energy generation.

17. Recognize the structures of *pyridoxal phosphate (PLP)* and *pyridoxamine phosphate (PMP),* indicate the reactive functional groups on each, and name the dietary precursor of PLP.

18. Describe the aminotransferase reaction mechanism and explain the involvement of a *Schiff base* and PLP. List the other kinds of reactions catalyzed by PLP and describe the common features of PLP catalysis. Define *stereoelectronic control.*

19. Explain how the α-amino groups of serine and threonine can be directly converted into NH_4^+.

20. Describe the transport of nitrogen from muscle to liver.

Ammonium Ion Is Converted into Urea in Most Terrestrial Vertebrates
(Text Section 23.4)

21. Define the terms *ureotelic, uricotelic,* and *ammonotelic.*

22. Name the molecule that brings nitrogen and carbon into the *urea cycle.*

23. Describe the caramoyl phosphate synthetase mechanism. Account for the ATP requirement of the urea cycle.

24. Name the enzymes of the urea cycle, note their intracellular locations, and indicate the molecular connection between this cycle and the citric acid cycle. Recognize the reaction catalyzed by each enzyme.

25. Distinguish the structural and functional properties of the two isozymes of *carbamoyl phosphate synthetase* in mammals.

26. Explain how deficiencies in several different enzymes of the urea cycle give rise to *hyperammonemia* and outline strategies for coping with deficiencies in urea synthesis.

Carbon Atoms of Degraded Amino Acids Emerge as Major Metabolic Intermediates (Text Section 23.5)

27. State the strategy used by humans for catabolizing the carbon atom skeletons of amino acids and name the seven major metabolic products formed.

28. Describe the basis for the *glycogenic-ketogenic* designation of the amino acids and classify each amino acid accordingly. Appreciate the limitations of this classification.

29. List the amino acids that give rise to *pyruvate, oxaloacetate, fumarate, succinyl CoA, acetyl CoA,* and *acetoacetyl CoA* respectively.

30. List the amino acids that give rise to α-*ketoglutarate.*

31. Describe the role of *tetrahydrobioprotein* and *S-adenosylmethionine* in amino acid degredation.

Inborn Errors in Metabolism Can Disrupt Amino Acid Degradation
(Text Section 23.6)

32. Describe *alcaptonuria.* Relate Garrod's hypothesis concerning this disorder.

33. Explain the biochemical bases of *phenylketonuria* and *maple syrup urine disease.* Describe some of the consequences of these diseases.

SELF-TEST

Introduction

1. Which of the following answers complete the sentence correctly? Surplus dietary amino acids may be converted into

   (a) proteins.
   (b) fats.
   (c) ketone bodies.
   (d) glucose.
   (e) a variety of biomolecules for which they are precursors.

Proteins Are Degraded to Amino Acids

2. Which of the following statements are true about essential amino acids in humans?

   (a) Humans can't synthesize amino acids containing rings and therefore histidine and the aromatic amino acids are essential.
   (b) Alanine and Proline are the only aliphatic amino acids that are not essential in humans.
   (c) All essential amino acids must come from the diet; amino acids released through cellular protein breakdown are excreted.
   (d) Humans can't synthesize sulfur-containing amino acids.

3. Order the following events in the correct sequence in the digestion of dietary proteins.

   (a) proteolysis by peptidases
   (b) proteolysis by pepsin in stomach
   (c) release of free amino acids into bloodstream
   (d) proteolysis in lumen of intestine by aminopeptidases
   (e) acidic denaturation of proteins in stomach.
   (f) transport of free amino acids and di- and tripeptides into intestinal cells

4. Which of the following is not a feature of ubiquitin?

   (a) It is a small protein whose structure is a mixture of alpha helices, beta strands, and turns.
   (b) It is found in all eukaryotic cells.
   (c) Targeted proteins are attached to ubiquitin via lysine 48.
   (d) It is highly conserved among eukaryotes.

Protein Turnover Is Tightly Regulated

5. Which of the following features characterize the 26S proteosome?

   (a) consists of a 20S catalytic subunit and a 19S regulatory subunit
   (b) releases only free amino acids as products
   (c) digests ubiquitin along with the protein to which it is attached
   (d) contains six ATPases

6. Match each disease at left with its link to amino acid metabolism at right.

(a)	Early-onset Parkinson disease	(1)	Treated with a proteosome inhibitor
(b)	Angelman syndrome	(2)	Deficiency in phenylalanine hydroxylase
(c)	Cervical carcinomas	(3)	Deficiency in branched-chain
(d)	Multiple myeloma		dehydrogenase

(e) Maple syrup urine disease
(f) Phenylketonuria

(4) Inappropriate activation of E3
(5) Defective E3

The First Step in Amino Acid Degradation Is the Removal of Nitrogen

7. Which of the following compounds serves as an acceptor for the amino groups of many amino acids during catabolism?

(a) glutamine
(b) asparagine
(c) α-ketoglutarate
(d) oxalate

8. Which of the following answers completes the sentence correctly? The removal of α-amino groups from amino acids for conversion to urea in animals may occur by

(a) transamination.
(b) reductive deamination.
(c) oxidative deamination.
(d) transamidation.

9. From which of the following amino acids are α-amino groups removed by dehydratases?

(a) histidine
(b) tryptophan
(c) serine
(d) glutamine
(e) threonine

10. Which of the following answers completes the sentence correctly? The products of an aminotransferase-catalyzed reaction between pyruvate and glutamate would be

(a) aspartate and oxaloacetate.
(b) aspartate and α-ketoglutarate.
(c) alanine and oxaloacetate.
(d) alanine and α-ketoglutarate.

11. Explain the role of pyridoxal phosphate in aminotransferase reactions. Be sure to describe the Schiff base and the ketimine that are involved in the mechanism.

acts as an amino carrier

Ammonium Ion Is Converted into Urea in Most Terrestrial Vertebrates

12. Considering all forms of life, which of the following are major excretory forms of the α-amino groups of amino acids?

(a) urea
(b) uracil
(c) ammonia
(d) uric acid

13. How many moles of ATP are required to condense two moles of nitrogen and one mole of CO_2 into one mole of urea via the urea cycle? How many high-energy bonds are used in this process? Do both atoms of nitrogen enter the cycle as NH_4^+?

14. What would be the net effect of linking the following two transamination reactions, and why would such coupling be useful when excess proteins were being catabolized for energy generation?

Alanine + α-ketoglutarate $\rightarrow$ pyruvate + glutamate
Oxaloacetate + glutamate $\rightarrow$ aspartate + α-ketoglutarate

15. Describe the role of ornithine in the urea cycle.

16. Explain how hyperammonemia, the increased concentration of NH_4^+ in the serum, can arise from defects in more than one enzyme of the pathway that forms urea.

Carbon Atoms of Degraded Amino Acids Emerge as Major Metabolic Intermediates

17. Match the catabolic products in the right column with the amino acids in the left column from which they can be derived.

(a) alanine 5
(b) aspartate 4 7
(c) glutamine 3
(d) phenylalanine 2 7
(e) leucine 6 2 6
(f) valine 5 1

(1) succinyl CoA
(2) acetoacetate
(3) α-ketoglutarate
(4) oxaloacetate
(5) pyruvate
(6) acetyl CoA
(7) fumarate

18. Classify the following amino acids as glycogenic (G), ketogenic (K), or both (GK).

(a) leucine K
(b) alanine G
(c) tyrosine KG
(d) serine G

(e) histidine G
(f) isoleucine KG
(g) aspartate G
(h) phenylalanine KG

19. What common feature is shared by the catabolism of fatty acids having an odd number of carbon atoms and the catabolism of the amino acids isoleucine, methionine, and valine?

20. Catabolism of which of the following amino acids requires the direct involvement of O_2?

(a) histidine
(b) phenylalanine
(c) tyrosine
(d) isoleucine
(e) glutamine

21. For each of the following types of chemical reactions, list one enzyme in an amino acid degragative pathway that catalyzes that type of reaction.

(a) dehydration
(b) oxidative deamination
(c) transamination
(d) hydroxylation
(e) hydrolysis
(f) oxidative decarboxylation

Inborn Errors in Metabolism Can Disrupt Amino Acid Degradation

22. Which of the following statements is true of the metabolic disease phenylketonuria?

(a) The disease is caused by an inability to synthesize phenylalanine.
(b) The disease can be caused by a deficiency in phenylalanine hydroxylase.
(c) The disease can be caused by a deficiency in tetrahydrobiopterin.
(d) The disease is treated with a high phenylalanine diet.
(e) The disease leads to a buildup of phenylalanine in the body.

ANSWERS TO SELF-TEST

1. All are correct. The amino acids that are needed for protein synthesis and as precursors for other biomolecules are used directly for those purposes. The carbon skeletons of any in excess can be converted into acetyl CoA or glucose, depending on the particular amino acid, and thus into products that are derivable from these two basic molecules.

2. b. Answer (a) is incorrect for a couple of reasons. First of all, proline contains a ring and yet is not essential, so clearly humans can synthesize an amino acid containing a ring. Of the three aromatic amino acids, tyrosine is non-essential, but only because humans can synthesize it from phenylalanine, an essential amino acid. This process will be discueed in Chapter 24. Answer (d) is incorect because the sulfur-containing cysteine is non-essential.

3. The correct order is e, b, d, f, a, c.

4. The correct answers are (a) and (d). (b) is incorrect because the proteosome also releases small peptides and (c) is incorrect because ubiquitin is spared degradation so it can be recycled.

5. c. Targeted proteins are attached to ubiquitin via several lysine residues on the targeted proteins. The carboxy-terminal glycine on ubiquitin is the site of the attachment.

6. (a) 5, (b) 5, (c) 4, (d) 1, (e) 3, (f) 2

7. (c.) The transamination of several different amino acids with α-ketoglutarate forms glutamate and α-keto acids that can subsequently be catabolized.

8. (a,) c

9. (c,e.) Serine and threonine are deaminated by dehydratases that take advantage of the β-hydroxyl of these amino acids to carry out a dehydration followed by a rehydration to release NH_4^+.

10. (d.) A five-carbon amino acid will yield a five-carbon ϵ-keto acid as a result of a transamination; in this case, glutamate yields ϵ-ketoglutarate. The other partner in the reaction, pyruvate, will yield alanine as a product.

11. Pyridoxal phosphate (PLP) acts as an amino carrier in transamination reactions. It is covalently bound to the α-amino group of a lysine residue in the enzyme by a Schiff base or aldimine bond, that is, by a carbon-nitrogen double bond between the α-amino group of the lysine and a carbon of PLP (see page 682 of the text). The enzyme catalyzes a displacement of the ϵ-amino group of the lysine of the enzyme and forms an analogous aldimine bond with the α-amino group of an amino acid. Through a deprotonation and a reprotonation, the aldimine is isomerized to a ketimine in which the carbon-nitrogen bond is between the α-amino nitrogen and the α-carbon of the amino acid. The addition of H_2O to the ketimine releases the carbon skeleton of the amino acid as an α-keto acid and leaves the coenzyme in the enzyme-bound pyridoxamine form, which now contains the α-amino group of the amino acid. After dissociation of the α-keto acid, a different α-keto acid binds to the enzyme to form a new ketimine, and the overall process reverses, resulting in the transfer of the enzyme-bound amino group to the keto acid to form a new amino acid and regenerate the pyridoxal phosphate.

12. a, c, d. Uracil is a pyrimidine component of RNA and is not a nitrogen excretory product. Uric acid is a purine derivative excreted by uricotelic organisms.

13. Three moles of ATP are directly involved in the synthesis of urea. Two are converted to ADP and P_i by carbamoyl phosphate synthetase, and one is converted to AMP and PP_i by argininosuccinate synthetase. Two ATP would be required to convert AMP back into ATP, so a total of four high-energy bonds are used. Only one molecule of nitrogen enters as NH_4^+; the other enters in the α-amino group of aspartate, which can be formed by a transamination between oxaloacetate and glutamate.

14. The sum of the two reactions would be alanine + oxaloacetate $\longrightarrow$ aspartate + pyruvate, and the net effect would be that the α-amino group of alanine would appear as the α-amino group of aspartate, one of the two direct donors of nitrogen into the urea cycle—the other is carbamoyl phosphate. The α-amino groups of many other amino acids can be similarly collected on aspartate and fed into the cycle.

15. Ornithine serves as the carrier on which the urea molecule is constructed. The α-amino group of ornithine has a carbamoyl group added to it by ornithine transcarbamoylase to form citrulline. Subsequent steps of the cycle add aspartate to bring in the second nitrogen atom and also regenerate ornithine when arginase cleaves arginine to form urea.

16. A defect in carbamoyl phosphate synthesis would cause increased NH_4^+ concentrations, as would a defect in any of the four reactions that condense carbamoyl phosphate with ornithine and regenerate the ornithine. Essentially, blocking a biochemical pathway at any of its steps may lead to increased concentrations of any of the precursors or members of the pathway.

17. (a) 5 (b) 4, 7. Aspartate can be converted to fumarate via the urea cycle. (c) 3 (d) 2, 7 (e) 2, 6 (f) 1

18. (a) K   (b) G   (c) GK   (d) G   (e) G   (f) GK   (g) G   (h) GK

19. Both give rise to methylmalonyl CoA, which can, in turn, be converted into succinyl CoA and ultimately into glucose.

20. b, c. Monoxygenases and dioxygenases are involved in the conversion of phenylalanine to tyrosine and the subsequent opening of the aromatic ring during tyrosine catabolism.

21. (a) serine and threonine dehydratase
    (b) glutamate dehydrogenase
    (c) aspartate and alanine aminotransferases
    (d) phenylalanine hydroxylase
    (e) asparaginase
    (f) branched-chain α-ketoacid dehydrogenase complex

22. Answers (b), (c), and (e) are correct. Phenylketonuria is a metabolic disorder arising from an absence or deficiency in the enzyme phenylalanine hydroxylase or (more rarely) its cofactor tetrahydrobiopterin. It results in the buildup of phenylalanine in the body and is treated with a diet low in phenylalanine.

PROBLEMS

1. Three enzymes are needed for the attachment of ubiquitin to proteins targeted for degradation: E1, E2, and E3. For each enzyme state its name and give a brief description of its role in ubiquitin attachment.

2. Birds require arginine in their diet. Would you expect to find the production of urea in these animals? Explain.

3. Which would you expect to have a greater effect on the rate of urea biosynthesis, a defect in fumarase activity or a defect in alanine aminotransferase?

4. Pyridoxal phosphate or related metabolites are required growth factors for *Lactobacillus* species (Morishita et al., *J. Bacteriol.* 148[1981]:64–71). For example, when amino acids such as alanine or glutamate are used as the sole source of nutrition, these bacilli do not grow nor do they generate metabolic energy unless pyridoxal phosphate or its metabolites are supplied. Explain these observations.

5. Why is glutamate dehydrogenase a logical point for the control of ammonia production in cells?

6. During the process of glomerular filtration in the kidney, amino acids, as well as other metabolites, enter the lumen of the kidney tubule. Normally, a large portion

of these amino acids are reabsorbed into the blood through the action of membrane-bound carrier systems that are specific for different classes of amino acids. Cystinuria is a disorder whose symptoms include urinary excretion with unusually high concentrations of cystine as well as excess amounts of ornithine, lysine, and arginine. Cystine is a dibasic amino acid composed of two cysteine molecules joined by a disulfide linkage. Patients with this disorder often have urinary tract stones, which are caused by the limited solubility of cystine. A related disorder found in other people is characterized by the appearance of ornithine, lysine, and arginine in the urine, although the levels of urinary cystine are normal.

(a) What is the most likely source of cystine in cells?

(b) What common structural feature of the four amino acids—cystine, ornithine, lysine, and arginine—is recognized by the carrier in the kidney tubule membrane?

(c) How many carrier systems may exist for these molecules?

(d) Other amino acidurias are due to a deficiency in one or more of the enzymes in the catabolic pathway for an amino acid. This deficiency leads to higher concentrations of the amino acid in the blood and a corresponding increase in the concentrations in the glomerular filtrate. In this case, the capacity of the reabsorption system is surpassed, causing some amino acid to be lost in the urine. How could you distinguish between a defect in amino acid metabolism and a defect in a renal transport system?

7. Why is the catabolism of isoleucine said to be both glucogenic and ketogenic?

8. Brain cells take up tryptophan, which is then converted to 5-hydroxytryptophan by tryptophan hydroxylase, an enzyme whose activity is similar to that of phenylalanine hydroxylase. Aromatic amino acid decarboxylase then catalyzes the formation of the potent neurotransmitter 5-hydroxytryptamine, also called *serotonin*. In the blood, tryptophan is bound to serum albumin, with an affinity such that about 10% of the tryptophan is freely diffusable. The rate of tryptophan uptake by brain cells depends on the concentration of free tryptophan. In these cells, tryptophan concentration is normally well below that of the K_M for tryptophan hydroxylase. Aspirin and other drugs displace tryptophan from albumin, thereby increasing the concentration of free tryptophan.

(a) What cofactor is required for the activity of tryptophan hydroxylase?

(b) Dietary deficiencies in pyridoxin and related metabolites can induce a number of symptoms, including those that appear to be related to derangements in serotonin metabolism. What enzyme could be affected by a deficiency of vitamin B6?

(c) What effect does aspirin have on tryptophan metabolism in brain cells?

9. In many microorganisms, glutamate dehydrogenase (GDH) participates in the catabolism of glutamate by generating ammonia and α-ketoglutarate, which undergoes oxidation in the citric acid cycle. However, when *E. coli* is grown with glutamate as the sole source of carbon, the synthesis of GDH protein is strongly repressed. Under these conditions, aspartase, an enzyme that catalyzes the removal of ammonia from aspartate to form fumarate, is required for the cell to grow in glutamate. Propose a cyclic pathway for the catabolism of glutamate that includes aspartate.

10. When *E. coli* is grown in glucose and ammonia, GDH synthesis is accelerated and the enzyme is active. Under these conditions, what role does GDH play in bacterial metabolism?

11. After an overnight fast, muscle tissue proteolysis generates free amino acids, many of which pass into the blood. Among the amino acids that are found in the blood are alanine, glutamate, and glutamine, all of which are rapidly taken up by the liver. What happens to these amino acids when they enter hepatic cells?

12. Propionyl CoA and methylmalonyl CoA both inhibit N-acetylglutamate synthase activity in slices of liver tissue. What clinical symptom would you expect to see in patients suffering from methylmalonic aciduria as a result of this inhibition?

13. Dialysis of purified glutamate aminotransferase can be used to remove pyridoxal phosphate from the enzyme, but the dissociation of the cofactor from the enzyme is very slow. Why would the addition of glutamate to the enzyme solution increase the rate of dissociation of the cofactor from the enzyme?

14. Early work by Esmond Snell on the enzymes that employ pyridoxal phosphate included experiments in which free pyridoxal was heated with amino acids like glutamate. Snell found that the α-amino group of glutamate was transferred to pyridoxal, generating pyridoxamine. Why was this an important clue to the function of pyridoxal as a cofactor?

15. The enzyme serine dehydratase uses pyridoxal phosphate (PLP) in the direct deamination of serine. In this reaction, the α-carbon of serine undergoes a two-electron oxidation through α-elimination. Show how PLP participates in the process by writing a mechanism for serine dehydration and deamination.

16. A small number of infants who have phenylketonuria have normal levels of phenylalanine hydroxylase activity, but on normal diets they continue to accumulate phenylalanine as well as other metabolites, including phenylpyruvate, phenyllactate, and phenylacetate. They also have high levels of quinonoid dihydrobiopterin.

    (a) What is the probable enzyme deficiency in these infants? Rationalize the deficiency with the observed clinical symptoms.
    (b) Write brief pathways for the formation of the phenylalanine metabolites found in these infants.

Phenylpyruvate   Phenyllactate   Phenylacetate

17. The mechanism of proteolysis by the β subunits of the proteosome is through a threonine-dependent nucleophilic attack. In addition to the catalytic threonine, conformational changes in a loop outside the active site are critical to catalysis. Using chymotrypsin and HIV protease as a model (sections 8.5 and 9.1 in the text), speculate on ways in which an inhibitor could be designed that would inactivate the proteosome. How could the *Mycobacterium tuberculosis* proteoseome be targeted without also inhibiting the human proteosome?

ANSWERS TO PROBLEMS

1. E1, ubiquitin-activating enzyme: adenylates ubiquitin on its terminal carboxylate with release of PP_i. Transfers ubiquitin to a sulfhydryl on the enzyme with release of AMP, forming a thioester bond. E2, ubiquitin-conjugating enzyme: receives ubiquitin from E1. Ubiquitin is also bound to a sulfhydryl of E2 via a thioester bond. E3, ubiquitin-protein ligase: transfers ubiquitin from E2 to the ε-amino group of a target protein.

2. In the urea cycle, arginine is cleaved to yield urea and ornithine. The fact that birds require arginine in their diet indicates that they are unable to synthesize it for utilization in protein synthesis. As a result, they are also unable to synthesize urea to dispose of ammonia; instead, they synthesize uric acid. Birds do have carbamoyl phosphate synthetase activity; however, it is located in the cytosol, and it catalyzes the formation of carbamoyl phosphate, which is then utilized for pyrimidine synthesis.

3. Fumarase activity has an effect on the urea cycle because it is needed, along with malate dehydrogenase, for the regeneration of oxaloacetate, which in turn undergoes transamination to form aspartate. The amino group of aspartate contains one of the two nitrogen atoms that are used to synthesize urea. Alanine aminotransferase is one of a number of aminotransferases that can transfer amino groups from amino acids to α-ketoglutarate to generate glutamate. Subsequent deamination of glutamate provides ammonia for the urea cycle. If all the other aminotransferases in the cell are active, then alanine aminotransferase would not be particularly essential. Thus, a defect in fumarase activity would have the greater effect on the rate of urea biosynthesis.

4. To utilize amino acids as sources of oxidative energy or to generate glucose through gluconeogenesis, the bacilli must carry out transamination reactions to dispose of ammonia (or to use it for the biosynthesis of other nitrogen-containing compounds) as well as to generate α-keto acids that can be used in the citric acid cycle or other pathways. Pyridoxal phosphate is a required cofactor for the aminotransferase enzymes, and in bacteria in which it cannot be synthesized, it must be derived from the growth medium. Otherwise, amino acids cannot be metabolized. In this case, where an amino acid is the only source of carbon and of nitrogen, the bacilli will not be able to introduce the amino acid into any catabolic pathway. Pyridoxal phosphate also functions as a cofactor for a large number of other enzymes, including decarboxylases, racemases, aldolases, and deaminases. Therefore, deficiencies in pyridoxal phosphate would also adversely affect a large number of other pathways.

5. The deamination of amino acids occurs through the action of transaminases as well as through the action of glutamate dehydrogenase (GDH). GDH is the only enzyme that catalyzes the oxidative deamination of an L amino acid. It deaminates glutamate, whose precursor, α-ketoglutarate, is the ultimate acceptor of amino groups from almost all the amino acids. In addition, while the reactions catalyzed by the transaminases are freely reversible, the GDH reaction is far from equilibrium; it is therefore a logical activity to control because large changes in its velocity can be achieved with small changes in the concentrations of allosteric effectors like ATP or NADH. Thus, GDH is the enzyme of choice for the control of ammonia synthesis.

6. (a) Disulfide linkages exist in many proteins, and when they are hydrolyzed by proteases to yield free amino acids, cystine is often one of the products.
   (b) Like cystine, the other three amino acids—arginine, ornithine, and lysine—have two basic groups. The carrier systems probably recognize and bind these groups for transport.
   (c) From the disorders described, it is likely that there are two transport systems. One carries all four dibasic acids; when it is defective, reabsorption of all four species fails, allowing all four to spill over into the urine. Another system transports ornithine, arginine, and lysine but not cystine; its failure to function accounts for the appearance of these three amino acids but not cystine in the urine.
   (d) A defect in the catabolic pathway for a particular amino acid causes elevation in the concentration of that single amino acid in blood and in urine as well, unless other amino acids carried by the same kidney transport system are lost to the urine. A defect in renal reabsorption means that all those amino acids that share

the affected carrier system will be lost to the urine. Their concentration in blood will be lower than normal, while their concentration in urine will be higher.

7. The catabolic pathway for isoleucine leads to the formation of acetyl CoA and propionyl CoA. Acetyl CoA can be utilized for the net synthesis of fatty acids or ketone bodies, but it cannot be used for the net synthesis of glucose; thus, it is said to be ketogenic. In contrast to acetyl CoA, propionyl CoA is converted to succinyl CoA, which can be utilized through part of the citric acid cycle and the gluconeogenic pathway to give the net formation of glucose. The distinction between the two types of substrates is somewhat arbitrary, however. For example, succinyl CoA can also be converted via pyruvate and acetyl CoA to citrate, which, when transported to the cytosol, serves as the source of carbons for the synthesis of fatty acids. Thus, glucogenic substrates can, under certain conditions, be ketogenic; however, ketogenic substrates cannot be glucogenic, unless a cell has a functional glyoxylate pathway.

8. (a) Tetrahydrobiopterin is utilized as a reductant by many hydroxylase enzymes, including phenylalanine hydroxylase and tryptophan hydroxylase.

   (b) Pyridoxal phosphate, which is derived from dietary pyridoxine (vitamin B_6), is a cofactor for a number of enzymatic reactions that occur at the α carbon of an amino acid including decarboxylations (see page 683 of the text). In this case, the cofactor participates in the decarboxylation of 5-hydroxytryptophan by the enzyme aromatic amino acid decarboxylase to form serotonin. A deficiency of vitamin B6 would lead to a reduction in the rate of synthesis of serotonin.

   (c) The higher the concentration of free tryptophan, the greater the rate of uptake of the amino acid by the brain cells. Because the normal concentration of tryptophan in these cells is below that of the K_M for tryptophan hydroxylase, an influx of more tryptophan into the cells provides more substrate for the enzyme. You would therefore expect an increase in the level of 5-hydroxytryptophan production.

9. A possible catabolic pathway for glutamate that includes aspartate follows:

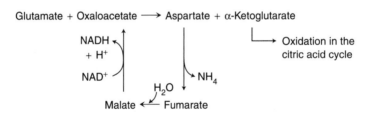

   Glutamate undergoes transamination with oxaloacetate to generate α-ketoglutarate and aspartate. Oxidation of α-ketoglutarate is carried out in the citric acid cycle, whereas aspartate is cleaved to yield ammonia and fumarate. Fumarate is converted to malate, which is then oxidized to oxaloacetate in the citric acid cycle so that it can be regenerated to serve as an acceptor of the amino group from glutamate. It should be noted that glutamate will also be used as a source of amino groups by the aminotransferases and other enzymes involved in biosynthesis.

10. In cells grown in glucose and ammonia, GDH catalyzes the assimilation of ammonia by incorporating it into α-ketoglutarate to yield glutamate, which serves as a source of amino groups for other biosynthetic reactions.

11. In the liver, alanine, glutamate, and glutamine are utilized as sources of carbon for gluconeogenesis. Glutamine is deaminated to yield glutamate and ammonia. Glutamate undergoes oxidative deamination to form ammonia and α-ketoglutarate, a substrate for gluconeogenesis. Pyruvate is generated from the transamination of ala-

nine, and carboxylation of the α-keto acid yields oxaloacetate, another source of carbon for gluconeogenesis. The ammonia generated by the conversion of the amino acids to their corresponding α-keto acids is used for the synthesis of urea. Glucose synthesized by liver can be returned through the blood to the muscle, where it serves as a source of energy. Thus, muscle uses the amino acids as a means of contributing to the generation of glucose in liver as well as a means of transporting ammonia to the liver, where the synthesis of urea can be carried out.

12. The activity of mitochondrial carbamoyl phosphate synthetase (CPS) depends on the availability of N-acetylglutamate, which is generated from glutamate and acetyl CoA. A reduction in the availability of the activating molecule will lead to a decrease in the activity of CPS, which utilizes ammonia and bicarbonate for the synthesis of carbamoyl phosphate. This, in turn, leads to an increase in the level of ammonia in blood and urine. Over two-thirds of patients with methylmalonic aciduria are hyperammonemic.

13. In the native enzyme, the dissociation of the cofactor from the enzyme during dialysis is very slow because PLP is covalently bound to a lysine residue. The addition of glutamate, a substrate for the transamination reaction, leads to the formation of a Schiff base between glutamate and PLP, which means that the cofactor is no longer covalently attached to the enzyme molecule. Although PLP is bound to the enzyme by noncovalent forces, these forces are not as strong as a covalent bond, so during dialysis, the rate of dissociation of the cofactor from the enzyme is increased.

14. Snell's observations of the formation of pyridoxamine by heating with α-amino acids suggested that pyridoxal is involved in transamination reactions. As an enzyme cofactor, it could transfer an α-amino group from an amino acid to the α-keto group of an α-keto acid. It is now established that the action of pyridoxal phosphate in aminotransferase enzymes includes formation of pyridoxamine phosphate during the catalytic cycle (see page 682 of the text).

15. In the accompanying figure, PLP in a Schiff base linkage with a lysine residue in the enzyme forms a new Schiff base link with serine. A hydrogen atom is removed from the α-carbon, and the hydroxyl group is removed by elimination from the β-carbanion, generating aminoacrylate attached to pyridoxal phosphate. Hydrolysis of the Schiff base to give aminoacrylate and PLP is followed by tautomerization to the imino form. This compound hydrolyzes spontaneously to form pyruvate and ammonia. PLP is once again linked covalently to the enzyme.

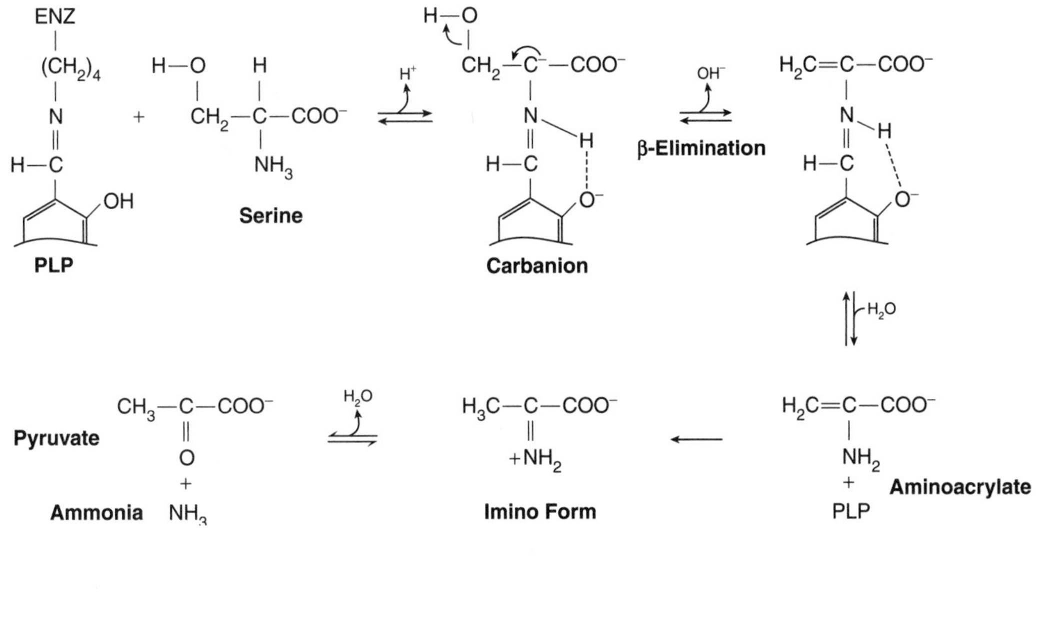

16. (a) The enzyme that is deficient in these infants is dihydropteridine reductase, which converts quinonoid dihydrobiopterin to tetrahydrobiopterin, using NADH as a substrate. In the phenylalanine hydroxylase reaction, tetrahydrobiopterin, the reductant in the conversion of phenylalanine to tyrosine, is oxidized to quinonoid dihydrobiopterin. The reductase enzyme regenerates tetrahydrobiopterin so that it can be used for further use in tyrosine formation. Cells that are deficient in the reductase cannot carry out efficient conversion of phenylalanine to tyrosine because they cannot regenerate tetrahydrobiopterin.

(b) Phenylpyruvate can be generated from phenylalanine by transamination, and reduction of phenylpyruvate by NADH or NADPH generates phenyllactate, in a reaction similar to that catalyzed by lactate dehydrogenase. Phenylacetate can be generated by oxidative decarboxylation of phenylpyruvate, in a reaction reminiscent of the conversion of pyruvate to acetyl CoA, catalyzed by pyruvate dehydrogenase.

17. An irreversible, mechanism-based inhibitor such as diisopropylfluorophosphate (DIPF) or a sulfonyl fluoride could be used to covalently label a reactive threonine in much the same way as they modify the serine in serine proteases. The rest of the molecule would have to be designed to bind specifically to the proteosome active site. Since both eukaryotic and archaeal proteosomes contain the reactive threonine, some other feature might need to be used to target it specifically. Some inhibitors of the HIV protease use interactions with flexible regions of the protein to add specificity. This method could be used to differentiate the eukaryotic and archaeal proteosomes by way of the conformational changes in the loop. This approach has indeed been followed in research such as that published by Lin et al. in *Nature* 461, 621–626 (1 October 2009).

The Biosynthesis of Amino Acids

In this chapter, the biosynthetic origins of the amino acids are explained, beginning with the need for a source of nitrogen for the amino acids. This need is met by nitrogen fixation, which is the process of converting atmospheric nitrogen in the form of N_2 to NH_4^+. The enzyme that carries out this difficult task, nitrogenase, is discussed in detail, including the role of an unusual molybdenum-iron cofactor. The authors then explain how NH_4^+ is incorporated into the amino acids glutamate and glutamine via the enzymes glutamate dehydrogenase and glutamine synthetase. These two amino acids are major nitrogen donors in a range of biosynthetic pathways, including those of the remaining amino acids whose synthesis is discussed next. While the pathways for the synthesis of the amino acids are diverse, they have in common the fact that their carbon skeletons come from intermediates in glycolysis, the pentose phosphate pathway or the citric acid cycle. This leads to a grouping of the amino acids into one of six biosynthetic families based on their starting material: oxaloacetate, pyruvate, ribose-5-phosphate, α-ketoglutarate, 3-phosphoglycerate, and phosphoenolpyruvate/erythrose 4-phosphate. The synthesis of the members of each family is examined in detail, including the role of three important cofactors involved in some of the syntheses: pyridoxal phosphate, tetrahydrofolate, and S-adenosylmethionine. The latter two are carriers of single carbon atoms in metabolism. The authors also explain that the lack of some biosynthetic pathways in humans has led to the dietary requirement for nine amino acids. The examination of amino acid synthesis concludes with a general discussion of how metabolic pathways are controlled via feedback inhibition, using examples from amino acid metabolism to illustrate the relevant principles.

The chapter concludes with a look at the role of the amino acids as precursors of many important biomolecules. The synthesis of glutathione, a sulfhydryl buffer and detoxifying agent, and nitric oxide, a short-lived signal molecule, are examined. The final topic is the biosynthesis and degradation of the porphyrin heme. The multistep synthetic pathway beginning with glycine and succinyl CoA is examined, as is the mechanism for degradation of excess heme. The physiological consequences of disorders in heme biosynthesis (collectively known as porphyrias) are discussed.

LEARNING OBJECTIVES

When you have mastered this chapter, you should be able to accomplish the following objectives.

Introduction

1. Recognize that nitrogen in the form of *ammonia* is the source of nitrogen for all amino acids, and amino acids, in turn, are the nitrogen source for many other biomolecules.

2. Appreciate the need for amino acids to be synthesized in the correct enationmeric form, and that the stereochemistry at the α-carbon is established by a *transamination* reaction that includes *pyridoxal phosphate (PLP)*.

Nitrogen Fixation: Microorganisms Use ATP and a Powerful Reductant to Reduce Atmospheric Nitrogen to Ammonia (Text Section 24.1)

3. Define *nitrogen fixation* and name the groups of organisms that can carry out this conversion. Compare the reduction of N_2 in industrial reactions with that carried out in biology.

4. Describe the *nitrogenase complex* and explain the roles of its *reductase* and *nitrogenase* components. Note the function of the *FeMo-cofactor*.

5. Explain the energy requirement for nitrogen fixation and write the equation giving the *stoichiometry* of the overall reaction.

6. Outline the key roles of *glutamate* and *glutamine* in the assimilation of NH_4^+ into amino acids and describe the reactions of *glutamate dehydrogenase, glutamine synthetase,* and *glutamate synthase*. Recognize the functions of *ATP* and *NADPH* in these processes.

Amino Acids Are Made from Intermediates of the Citric Acid Cycle and Other Major Pathways (Text Section 24.2)

7. Classify the amino acids into six *biosynthetic families* and identify their *seven precursors*. Name the metabolic pathways from which these precursors originate.

8. Identify the *essential* amino acids for humans and explain why they are essential.

9. Describe the single-step biosyntheses of alanine, aspartate, and glutamate.

10. Compare the formation of asparagine from aspartate to that of glutamine from glutamate.

11. Outline the syntheses of glutamine, proline, and arginine from *α-ketoglutarate*.

12. Outline the syntheses of serine, glycine, and cysteine from *3-phosphoglycerate*.

13. Explain the roles of pyridoxal phosphate, *tetrahydrofolate,* and S-adenosylmethionine in amino acid biosyntheses. Explain how PLP generates amino acids with an L configuration at the Cα center.

14. Identify the structure of tetrahydrofolate and indicate the reactive part of the molecule. Draw the structures of the single-carbon groups that can be carried on tetrahydrofolate and provide examples of reactions that generate and use them. Describe the sources of this cofactor in humans.

15. Draw the structure of *S*-adenosylmethionine and describe its synthesis. Indicate the reactive part of the molecule and describe the basis of its *high methyl group–transfer potential*.

16. Outline the *activated methyl cycle* and describe the roles of *methylcobalamin* and *ATP* in the cycle. Give examples of important derivatives from *S*-adenosylmethionine.

17. Describe the synthesis of cysteine from *homocysteine* and serine. Explain the most common cause and consequences of elevated levels of homocysteine in humans.

18. Outline the biosyntheses of phenylalanine, tyrosine, and tryptophan in *E. coli*. Describe the roles of *phosphoenolpyruvate, erythrose 4-phosphate,* and *phosphoribosylpyrophosphate* in these reactions.

19. Point out the locations of shikimate and chorismate in the shikimate pathway.

20. Describe the structure of *tryptophan synthetase* and the role of *substrate channeling* in its catalytic reaction.

Feedback Inhibition Regulates Amino Acid Biosynthesis (Text Section 24.3)

21. Define the *committed step* of a metabolic pathway and recognize that it is often the target of *feedback regulation*.

22. Note the main features of control of branched pathways by *feedback inhibition and activation, enzyme multiplicity,* and *cumulative feedback*.

23. Describe the cumulative feedback control of *glutamine synthetase* from *E. coli*. Explain the mechanisms and functions of the *reversible covalent modifications* and describe the advantage of employing an *enzymatic cascade* in regulating this reaction.

Amino Acids Are Precursors of Many Biomolecules (Text Section 24.4)

24. Give examples of important biomolecules that are derived from amino acids.

25. Draw the structure of *glutathione* and describe its cycle of oxidation/reduction. Indicate the functions of glutathione and describe the involvement of *selenium* in the *glutathione peroxidase* reaction.

26. Outline the synthesis of *nitric oxide (NO)* and explain its function.

27. Name the two molecular precursors of the *porphyrins* in mammals and outline the biosynthesis and degradation of *heme*.

28. Explain the molecular defects in *congenital erythropoietic porphyria* and *acute intermittent porphyria*.

SELF-TEST

Nitrogen Fixation: Microorganisms Use ATP and a Powerful Reductant to Reduce Atmospheric Nitrogen to Ammonia

1. Define *nitrogen fixation* and explain why it is crucial to the maintenance of life on earth.

2. Place the following components, reactants, and products of the nitrogenase complex

reaction in their correct sequence during the electron transfers of nitrogen fixation:

(a) oxidized ferredoxin

(e) N_2

(b) reductase component

(f) reduced ferredoxin

(c) nitrogenase component

(g) electron source

(d) NH_3

3. Write the net equation for nitrogen fixation and describe the sources of the electrons and ATP.

4. Match the structural components or features in the right column with the appropriate component of the nitrogenase reaction.

(a) reductase component

(b) nitrogenase component

(1) MoFe-cofactor

(2) [4Fe-4S] cluster

(3) ATP-ADP binding site

(4) N_2-binding site

(5) $\alpha_2\beta_2$ tetramer

(6) dimer of identical subunits

5. Match the enzyme with the reaction it catalyzes.

(a) glutamine synthetase

(b) glutamate dehydrogenase

(c) glutamate synthase

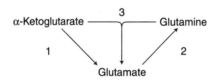

6. Which of the reactions shown in question 5 require the following?

(a) NH_4^+

(b) ATP

(c) NADH

(d) NADPH

7. All organisms can incorporate NH_4^+ into glutamate and glutamine using glutamate dehydrogenase and glutamine synthetase. Why do prokaryotes have an additional enzyme, glutamate synthase, to perform this function?

Amino Acids Are Made from Intermediates of the Citric Acid Cycle and Other Major Pathways

8. Which of the following amino acids are essential dietary components for an adult human?

(a) alanine

(b) aspartate

(c) histidine

(d) tryptophan

(e) leucine

(f) phenylalanine

(g) glutamine

(h) asparagine

(i) glutamate

(j) threonine

(k) methionine

9. Which of the following amino acids are derived from pyruvate?
   (a) phenylalanine
   (b) alanine
   (c) tyrosine
   (d) histidine
   (e) valine
   (f) leucine
   (g) cysteine
   (h) glycine

10. Which of the following amino acids are derived from α-ketoglutarate?
    (a) glutamate
    (b) proline
    (c) cysteine
    (d) aspartate
    (e) glutamine
    (f) arginine
    (g) ornithine
    (h) serine

11. Which of the following compounds provide the carbon skeletons of the six biosynthetic families of amino acids? Name the metabolic pathways from which each of the precursor compounds originates.
    (a) pyruvate
    (b) oxaloacetate
    (c) α-ketoglutarate
    (d) succinate
    (e) 2-deoxyribose
    (f) 3-phosphoglycerate
    (g) ribose 5-phosphate
    (h) glucose 6-phosphate
    (i) phosphoenolpyruvate
    (j) erythrose 4-phosphate
    (k) α-ketobutyrate

12. Explain the role of the conserved lysine and arginine residues in the active site of transaminases. Be sure to include how they aid in the stereochemical specificity of the reaction.

13. Three coenzymes are involved in carrying activated one-carbon units. Match the activated group in the right column with the appropriate coenzyme in the left column.
    (a) tetrahydrofolate
    (b) *S*-adenosylmethionine
    (c) biotin

    (1) $-CH_3$
    (2) $-CH_2-$
    (3) $-CHO$
    (4) $-CHNH$
    (5) $-CH=$
    (6) $-CO_2^-$

14. Which of the following answers completes the sentence correctly? The major source of one-carbon units for the formation of the tetrahydrofolate derivative N^5,N^{10}-methylenetetrahydrofolate is the conversion of
    (a) methionine to homocysteine.
    (b) deoxyuridine 5′-phosphate to deoxythymidine 5′-phosphate.
    (c) 3-phosphoglycerate to serine.
    (d) serine to glycine.

15. Why does *S*-adenosylmethionine have a higher methyl group–transfer potential than N^5-methyltetrahydrofolate?

16. *S*-adenosylmethionine is involved directly in which of the following reactions?
    (a) methyl transfer to phosphatidyl ethanolamine
    (b) synthesis of glycine from serine
    (c) DNA methylation
    (d) conversion of homocysteine into methionine
    (e) synthesis of ethylene in plants

17. How many high-energy bonds are expended during the synthesis of *S*-adenosylmethionine from ATP and methionine?

18. The conversion of homocysteine into methionine involves which of the following co-factors?
    (a) N^5-methyltetrahydrofolate
    (b) N^5, N^{10}-methylenetetrahydrofolate
    (c) methylcobalamin
    (d) pyridoxal phosphate

19. Which of the following statements about vitamin treatment for patients suffering from high homocysteine levels is correct?
    (a) They help reduce oxidative damage caused by an excess of homocysteine.
    (b) Folic acid helps activate cystathionine β-synthase, decreasing the concentration of homocysteine.
    (c) They decrease the activity of two major metabolic pathways involving homo-cysteine.
    (d) Vitamins B6 and B12 support conversion of homocysteine into other metabolites.

20. Which of the following are intermediates in the pathway for the biosynthesis of both phenylalanine and tryptophan?
    (a) anthranilate
    (b) chorismate
    (c) shikimate
    (d) prephenate

21. The two binding sites for indole on tryptophan synthetase subunits α and β are about 25 Å apart. Explain how indole is transferred between these sites.

Amino Acid Biosynthesis Is Regulated by Feedback Inhibition

22. In the following biosynthetic pathway A → B → C → D → E → F → G, which is likely to be the committed step? Which compound is likely to inhibit the committed step?

23. Threonine deaminase
    (a) contains two identical regulatory domains.
    (b) contains two differentiated regulatory domains.
    (c) is controlled by separate regulatory subunits as in phosphoglycerate dehydrogenase.
    (d) is regulated by serine as well as isoleucine and valine.

24. Since glutamine is an important source of nitrogen in biosynthetic reactions, the enzyme that synthesizes it is carefully regulated. Which of the following compounds act as inhibitors of glutamine synthetase in *E. coli*?
    (a) tryptophan
    (b) histidine
    (c) carbamoyl phosphate
    (d) glucosamine 6-phosphate

(e) AMP
(f) CTP
(g) alanine
(h) glycine

25. How does covalent modification contribute to the regulation of glutamine synthetase in *E. coli*?

Amino Acids Are Precursors of Many Biomolecules

26. Glutathione is composed of which of the following amino acids?

(a) glutamine
(b) glutamate
(c) methionine
(d) cysteine
(d) glycine

27. Which of the following answers complete the sentence correctly? Glutathione

(a) cycles between oxidized and reduced forms in the cell.
(b) is involved in the detoxification of H_2O_2 and organic peroxides.
(c) donates amide groups from its γ-glutamyl residue during biosynthetic reactions.
(d) contains an Se atom.

28. Which of the following statements about nitric oxide (NO) is correct?

(a) It is used as an inhalation anesthetic.
(b) It is a long-lived signal molecule.
(c) It is produced from asparagine.
(d) Its synthesis requires NADPH and O_2.
(e) Its synthesis requires ATP and NH_4^+.

29. Which of the following are intermediates or precursors in the synthesis of heme?

(a) α-aminolevulinic acid
(b) bilirubin
(c) porphobilinogen
(d) biliverdin
(e) glycine
(f) succinyl CoA
(g) Fe^{2+}

ANSWERS TO SELF-TEST

1. Nitrogen fixation is the process by which nitrogen present in the atmosphere as N_2 is enzymatically converted to NH_3 by some bacteria and blue-green algae. This process is crucial to all other organisms because they can use only NH_4^+, and not N_2, as the source of nitrogen for biosynthesis.

2. g, f, a, b, c, e, d

3. The net equation for nitrogen fixation is

$$N_2 + 8\,e^- + 16\,ATP + 16\,H_2O \rightarrow 2\,NH_3 + 16\,ADP + 16\,P_i + 8\,H^+ + H_2$$

The eight electrons needed to reduce N_2 are supplied by oxidative processes in non-photosynthetic nitrogen-fixing organisms and by light energy from the sun in photosynthetic nitrogen-fixing organisms. The ATP requirement is met by the usual oxidative or photosynthetic mechanisms of the cells.

4. (a) 2, 3, 6 (b) 1, 2, 4, 5

5. (a) 2 (b) 1 (c) 3

6. (a) 1, 2 (b) 2 (c) None (d) 1, 3. Glutamate dehydrogenase uses NADPH when catalyzing reductive aminations and NAD^+ when carrying out oxidative deaminations.

7. Glutamate synthase catalyzes the reductive amination of α-ketoglutarate in a reaction with glutamine to form two glutamates. Glutamate can also be made from NH_4^+ and α-ketoglutarate using glutamate dehydrogenase. However, this route requires high concentrations of NH_4^+ because of the high K_M of the enzyme for NH_4^+. Prokaryotes can use glutamine synthetase, which has a low K_M for NH_4^+, to form glutamine when NH_4^+ concentrations are low. Thus, by using an additional enzyme, they can form glutamate from glutamine and α-ketoglutarate when NH_4^+ is scarce.

8. c, d, e, f, j, k

9. b, e, f

10. a, b, e, f, g. Recall that ornithine is a precursor of arginine in the urea cycle and is derived from α-ketoglutarate.

11. The biosynthetic precursors are a, b, c, f, g, i, and j. The pathways they originate from are as follows:

Glycolysis: pyruvate (a)
  3-phosphoglycerate (f)
  phosphoenolpyruvate (i)

Citric acid cycle: oxaloacetate (b)
  α-ketoglutarate (c)

Pentose phosphate pathway: ribose 5-phosphate (g)
  erythrose 4-phosphate (j)

12. The lysine forms a Schiff base with the PLP cofactor and the arginine interacts with the α-carboxylate group of the ketoacid. The interaction between the arginine and the α-carboxylate helps orient the substrate so that the lysine residue transfers a proton to the bottom face of the quinonoid intermediate, generating an aldimine with an L configuration at the Cα center.

13. (a) 1, 2, 3, 4, 5   (b) 1   (c) 6

14. d. Furthermore, since 3-phosphoglycerate can give rise to serine, you can see how carbohydrates can provide activated one-carbon units via a glucose → 3-phosphoglycerate → serine → glycine pathway.

15. The positive charge on the sulfur atom of *S*-adenosylmethionine activates the methyl sulfonium bond and makes methyl group transfer from *S*-adenosylmethionine energetically more favorable than from N^5-methyltetrahydrofolate.

16. a, c, e. The cofactor for reactions (b) and (d) is tetrahydrofolate.

17. Three high-energy bonds are expended. The adenosyl group of ATP is condensed with methionine to form a carbon-to-sulfur bond with the release of P_i and PP_i, which is hydrolyzed to 2 P_i.

18. a, c. Homocysteine transmethylase uses a vitamin B_{12}–derived cofactor.

19. d

20. b, c

21. Tryptophan synthetase contains a 25-Å-long channel between the active sites of adjacent α and β subunits. This channel allows the diffusion of the intermediate, indole, through the protein from one binding site to the other without diffusing away from the enzyme. This alleviates the potential problem of the hydrophobic indole molecule diffusing across the plasma membrane and out of the cell were it allowed to leave the enzyme.

22. A ⟶ B. Control of the first step conserves the first compound, A, in the sequence and also saves metabolic energy by preventing subsequent reactions in the pathway. Compound G would likely inhibit the committed step. The end product of a biosynthetic pathway often controls the committed step.

23. b. The regulatory domain of threonine deaminase is homologous to that of phosphoglycerate dehydrogenase, but while phosphoglycerate dehydrogenase contains a dimer of two subunits, threonine deaminase has a single-chain regulatory domain. Unlike that of phosphoglycerate dehydrogenase, the regulatory domain of threonine deaminase contains two differentiated regulatory domains, one that binds isoleucine and one that binds valine. It is phosphoglycerate dehydrogenase that is regulated by serine.

24. All the choices are correct. When all eight compounds are bound to the enzyme, it is almost completely inactive. The control of this enzyme is an excellent example of cumulative feedback inhibition.

25. Glutamine synthetase can be covalently modified by the attachment of an AMP to each of its 12 subunits. The more adenylylated the enzyme becomes, the more susceptible it is to feedback inhibition by the compounds listed in Question 21. Thus, covalent modification modulates the sensitivity of the enzyme to its effectors. An added level of control exists in this system; adenylyltransferase, the enzyme that adenylylates glutamine synthetase, is itself covalently modified.

26. b, d, e

27. a, b. Answer (d) is incorrect because the enzyme glutathione peroxidase, rather than glutathione, contains an Se analog of cysteine.

28. d. Note that (c) is incorrect because arginine rather than asparagine is the precursor of NO.

29. a, c, e, f, g

PROBLEMS

1. The *glyA⁻* mutation in Chinese hamster ovary cells in tissue culture makes these cells partially dependent on glycine. The mutation affects the mitochondrial form of serine transhydroxymethylase, which catalyzes the conversion of serine to glycine, with tetrahydrofolate serving as an acceptor of the hydroxymethyl group. Would you expect heme synthesis to be adversely affected in *glyA⁻* mutants? Why?

2. The essential amino acids are those that cannot be synthesized de novo in humans. Given an abundance of other amino acids in the diet, the α-keto acid analogs that correspond to the essential amino acids can substitute for these compounds in the diet.

   (a) What do these observations tell you about the steps in the synthesis of essential amino acids that may be missing in humans?

   (b) If ^{15}N-labeled alanine is supplied in the diet, many other amino acids in the body will contain at least a small amount of the label within 48 hours. What enzymes are primarily responsible for this observation?

3. In muscle, glutamine synthetase is very active, catalyzing the formation of glutamine from glutamate and ammonia at the expense of a molecule of ATP. In the liver, the rate of formation of glutamine is very low, but a high level of glutaminase activity, which generates ammonia and glutamate, is observed. How would you explain the difference in the levels of enzyme activity in these two organs?

4. Consider three forms of bacterial glutamine synthetase: GS, the deadenylylated form; GS–(AMP)$_1$, a form with one AMP unit per 12 subunits; and GS–(AMP)$_{12}$, the fully adenylylated form.

   (a) Which of these forms is most sensitive to feedback inhibition by several of the final products of glutamine metabolism, such as tryptophan or histidine? Why is it important that the activity of the most sensitive form not be *completely* inhibited by tryptophan?

   (b) Which form has the lowest K_M for ammonia?

   (c) Why is it important that adenylyl transferase not carry out adenylylation and deadenylylation of glutamine synthetase at the same time?

   (d) Glutamine synthetase in mammals is not subject to the same type of complex regulation that is seen in bacteria. Why?

5. Most of the proteins synthesized in mammals contain all 20 common amino acids. More protein is degraded than is synthesized when even one essential amino acid is missing from the diet.

   (a) Under such conditions, how could an increase in the rate of protein degradation provide the missing amino acid?

   (b) How does an increase in the rate of protein degradation contribute to increased levels of nitrogen excretion?

6. The diagram below outlines the biosynthesis of a compound that is required for the oxidation of fatty acids in the mitochondrion.

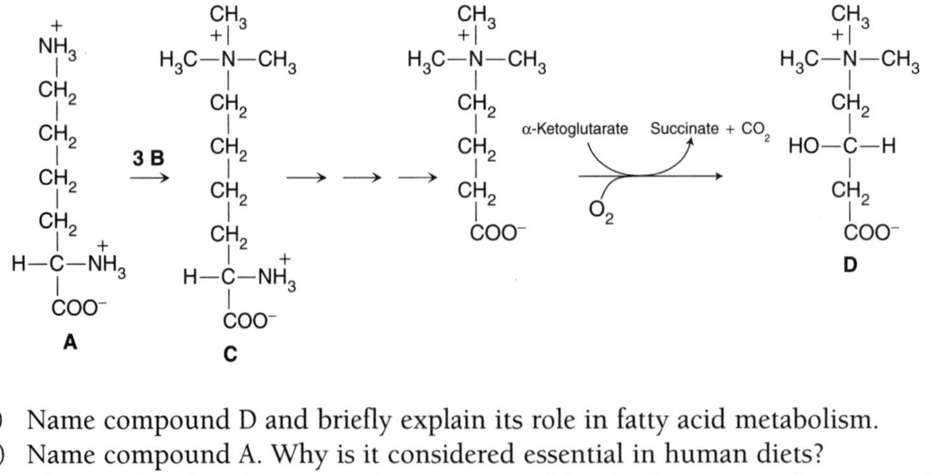

   (a) Name compound D and briefly explain its role in fatty acid metabolism.

   (b) Name compound A. Why is it considered essential in human diets?

   (c) Three molecules of compound B are required for the formation of compound C. Its synthesis depends on the availability of an essential amino acid. Name that amino acid and then name compound B and compound C.

7. A pathway for the synthesis of ornithine from glutamate is shown in Figure 24.1.

FIGURE 24.1   Biosynthesis of ornithine from glutamate.

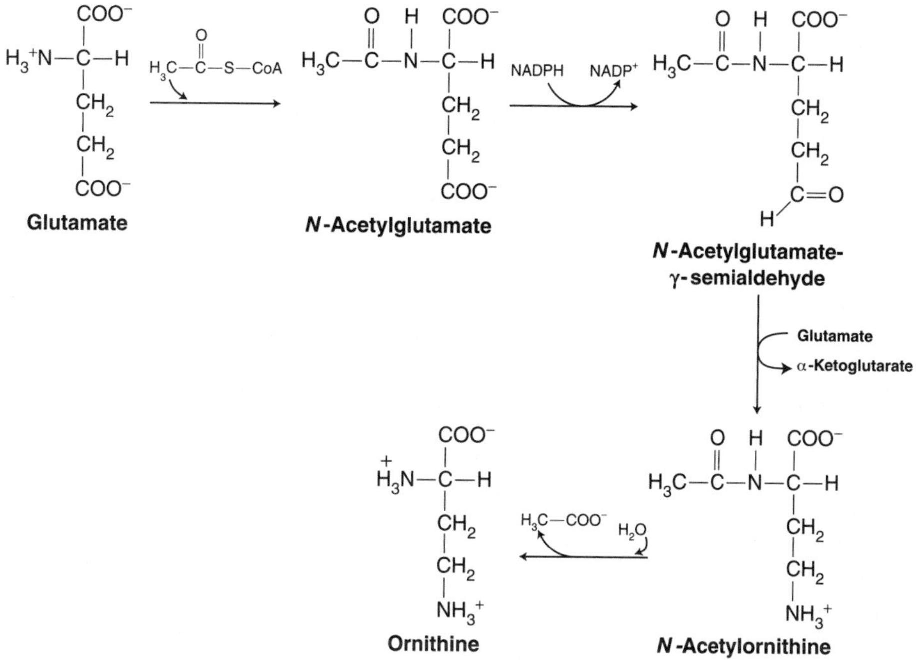

(a) Why can this pathway also be considered to be part of the de novo pathway for the synthesis of arginine?

(b) Inspect the pathway for proline biosynthesis given on page 714 of the text, and then explain why the N-acetylation of glutamate is needed for the synthesis of ornithine.

8. Elevated levels of ammonia in blood can result from deficiencies in one or another of the enzymes of the urea cycle. Measures taken to relieve hyperammonemia have included limiting intake of dietary proteins, administering α-keto analogs of several of the naturally occurring L-amino acids, or administering other compounds designed to exploit pathways of nitrogen metabolism and excretion.

(a) In trying to determine why a patient has hyperammonemia, which organ should you check first for normal function? Why?

(b) Why would limiting protein intake assist in relieving chronic hyperammonemia? Why would eliminating dietary proteins *altogether* (without any other supplement) probably increase the level of hyperammonemia?

(c) Write a brief rationale for using α-keto acid analogs in treating hyperammonemia, mentioning a particular group of enzymes essential to your explanation. Would it be better to use α-keto analogs of essential or nonessential amino acids? Why?

9. Plants synthesize all 20 common amino acids de novo. Glyphosate, a weed killer sold under the trade name Roundup, is an analog of phosphoenolpyruvate that specifically inhibits 3-enolpyruvylshikimate 5-phosphate synthase, a key enzyme of the pathway for chorismate biosynthesis. This compound is a very effective plant herbicide, but has virtually no effect on mammals. Why?

10. In *B. subtilis*, the pathway from chorismate to tryptophan is feedback-inhibited by tryptophan, which suppresses anthranilate synthase activity. Mutant *B. subtilis* that lacks tryptophan synthetase can grow on minimal medium only when supplemented with exogenous tryptophan. Under these conditions, none of the intermediates in the tryptophan biosynthetic pathway from anthranilate to indole 3-glycerol phosphate are produced. However, when the bacteria have depleted the medium of tryptophan, the levels of those intermediates increase, even though there is no net production of tryptophan. Why?

11. In the biosynthesis of branched amino acids, threonine deaminase (threonine dehydratase) is inhibited by isoleucine and activated by valine. This controls the relative amounts of the two amino acids by controlling the amount of α-ketoglutarate available for isoleucine synthesis. Use information from Chapter 23 to describe an analogous hypothetical control mechanism that would act by controlling the levels of pyruvate instead of α-ketoglutarate. Using your knowledge of energy metabolism from Chapters 16 and 17, speculate on why this mechanism is not used.

ANSWERS TO PROBLEMS

1. Glycine is an obligatory precursor of heme; in the reaction catalyzed by δ-aminolevulinate synthase, glycine condenses with succinyl CoA to form δ-aminolevulinate. The reduction in the concentration of glycine in the cell caused by the *glyA⁻* mutation will cause a decrease in the rate of heme synthesis.

2. (a) The fact that α-keto acid analogs can substitute for essential amino acids means that the carbon skeletons of the essential amino acids are not synthesized in humans. Many studies have shown that one or more of the enzymes needed for the synthesis of these structures are missing.

   (b) The enzymes that are primarily responsible for the distribution of the ^{15}N label among the other amino acids are the aminotransaminases, which catalyze the interconversions of amino acids and their corresponding α-keto acids. The redistribution of the label begins with the transamination of alanine, with α-ketoglutarate serving as the amino acceptor to yield pyruvate and glutamate. Glutamate then serves as an amino donor for other α-keto acids. In order for an essential amino acid to be labeled, you must postulate the transamination of that amino acid to yield the corresponding α-keto acid analog, followed by the donation of a labeled amino group from glutamate or another donor of amino groups.

3. Ammonia, which is generated as part of the process of amino acid catabolism in muscle, is toxic and must be removed from the cells. This could be done through the synthesis of urea, but that process occurs only in the liver. In muscle cells, therefore, glutamine synthetase catalyzes the formation of glutamine, which is an efficient and nontoxic carrier of ammonia. This accounts for the high activity of that enzyme in muscle. The glutamine is transported by the blood to the liver, where glutaminase and aspartate aminotransferase work together to generate aspartate and two molecules of ammonia from glutamine, hence the high activity of glutaminase in the liver. Aspartate and ammonia are both used by the liver for the synthesis of urea, a nontoxic and disposable form of ammonia.

4. (a) The fully adenylylated form of glutamine synthetase is the most sensitive to molecules like tryptophan and histidine. Because glutamine is utilized for the synthesis of a variety of compounds, complete inhibition of the enzyme by only one of those products, such as tryptophan, would inappropriately inhibit the synthesis of all the others. Thus, the enzyme is cumulatively inhibited by at least eight different nitrogen-containing compounds.

(b) The deadenylylated form, which is not subject to cumulative feedback inhibition, is generated in response to increases in the cellular concentrations of α-ketoglutarate (the precursor of glutamate) and ATP. These molecules signal the need for glutamine synthesis, even when other nitrogen-containing compounds are present. Under these conditions, the deadenylylated form of the enzyme binds ammonia even when ammonia concentrations are relatively low; that is, it has a relatively low K_M for ammonia.

(c) The simultaneous adenylylation and deadenylylation of glutamine synthetase would result in a loss of feedback control of the enzyme because the adenylylated form is subject to cumulative inhibition whereas the deadenylylated form is not. In addition, it would lead to the wasteful hydrolysis of ATP, since every round of adenylylation and deadenylylation generates AMP and inorganic pyrophosphate from ATP.

(d) Mammals acquire many nitrogen-containing compounds, such as tryptophan and histidine, in their diet rather than through de novo biosynthesis, so glutamine synthetase does not play so prominent a role in the nitrogen metabolism of mammals as it does in that of bacteria. Complex regulation of the enzyme is therefore not needed in mammals.

5. (a) Many experiments have shown that under normal conditions cells continuously synthesize and degrade proteins. Although both essential and nonessential amino acids are continuously recycled during these processes, reutilization is not completely efficient; thus, additional amino acids are needed. In mammals, there are no reservoirs of free amino acids; the only sources of essential amino acids are dietary proteins or the proteins of the body tissues. If an essential amino acid is not available from the diet, cells appear to accelerate the hydrolysis of their own proteins in order to generate the missing essential amino acid. How the rate of cellular proteolysis is accelerated in response to a deficiency of an essential amino acid is not understood.

(b) An increased rate of protein degradation generates a higher concentration of free amino acids. During the oxidation of those amino acids not used for synthesis of other proteins, ammonia will be produced. An elevation in ammonia concentration in the body stimulates the formation of urea, causing the level of nitrogen excretion to increase.

6. (a) Compound D is carnitine, which, when esterified to the acyl group of a long-chain fatty acid, shuttles it from the cytosol to the matrix of the mitochondrion, where fatty acid oxidation takes place.

(b) Compound A is the essential amino acid lysine; it is termed *essential* because it cannot be synthesized de novo in humans. Lysine and other essential amino acids must be obtained from the diet.

(c) Compound C is trimethyllysine, and the methyl groups that are attached to lysine are likely to be derived from compound B, *S*-adenosylmethionine, the major donor of methyl groups in biosynthetic reactions. The methyl group of *S*-adenosylmethionine is derived from methionine, an essential amino acid.

7. (a) Ornithine is a precursor of arginine, as part of the pathway for the synthesis of urea. Thus the pathway for the synthesis of ornithine from glutamine, along with part of the urea cycle pathway, can together be considered as a de novo pathway for the synthesis of arginine. Arginine can in turn be used for the synthesis of urea, or it can serve instead as one of the amino acids used for polypeptide synthesis.

   (b) In the pathway for proline biosynthesis, glutamic-γ-semialdehyde cyclizes with the loss of water to form Δ1-pyrroline-5-carboxylate. However, in the pathway for the formation of ornithine shown in Figure 24.1, an *N*-acetylated derivative of the semialdehyde molecule is formed. The *N*-acetyl group blocks the condensation of the amino group with the aldehyde group, thereby preventing the formation of the pyrroline ring. This allows the pathway to proceed toward the synthesis of ornithine.

8. (a) You should assess liver function, because enzymes of the urea cycle are found primarily in this organ. In addition, liver takes up amino acids such as alanine, glutamate, and glutamine, which are in effect nontoxic forms of ammonia generated by muscle and other tissues. Amino groups of these compounds, along with carbon dioxide from carbamoyl phosphate, are precursors of urea in the liver.

   (b) During digestion, dietary proteins are hydrolyzed to their component amino acids. Those amino acids that are not needed immediately for protein synthesis or for the biosynthesis of other nitrogen-containing compounds are degraded. One of the products of amino acid degradation is ammonia. Usually ammonia is metabolized through conversion to nitrogen carriers such as alanine, glutamate, and glutamine, and it is ultimately utilized for the synthesis of urea when the urea cycle is operating. Thus limiting protein intake in a patient with a deficiency in urea synthesis would be expected to reduce ammonia production in liver and other tissues.

   Protein turnover and amino acid degradation constantly take place in the tissues, and essential amino acids (those that cannot be synthesized de novo in human tissues) must be generated either from dietary sources or from additional breakdown of body proteins. A complete restriction of dietary protein would accelerate body protein breakdown and would exacerbate the condition of hyperammonemia.

   (c) The α-keto acid analogs can serve as acceptors for amino groups from glutamate and other amino acids in reactions catalyzed by transaminases or aminotransferases. Nonessential amino acids formed as the result of this process may be themselves eliminated, degraded (often generating more ammonia), or else used for biosynthesis of other nitrogen-containing compounds. Employing analogs of essential amino acids might be preferable, because tissues are more likely to require them for protein synthesis or other biosyntheses.

9. Chorismate is an intermediate in the biosynthesis of the aromatic amino acids tryptophan, phenylalanine, and tyrosine. Mammals do not synthesize these amino acids from chorismate. Instead, they obtain the essential aromatic amino acids tryptophan and phenylalanine from the diet, and they can synthesize tyrosine from phenylalanine. Glyphosate is an effective herbicide because it prevents synthesis of aromatic amino acids in plants. But the compound has no effect on mammals because they have no active pathway for de novo aromatic amino acid synthesis.

10. Since anthranilate synthase is inhibited by tryptophan, exogenous tryptophan from the medium inhibits the production of downstream intermediates in the biosynthetic pathway by halting the first step in the pathway. When exogenous tryptophan is depleted, intracellular levels of tryptophan decrease and anthranilate synthase activity is no longer inhibited. An increase in anthranilate production leads to an increase in the production of other intermediates in the pathway until equilibria are established among those compounds. The block at the step catalyzed by tryptophan synthetase prevents large accumulations of the intermediates, but under these conditions their concentrations are higher than in the presence of exogenous tryptophan.

11. Section 23.5 describes the one-step conversion of serine into pyruvate by serine dehydratase. This step is analagous to the conversion of threonine into α-ketoglutarate by threonine dehydratase. Pyruvate can then react with hydroxyethyl-TPP ultimately form valine. This enzyme could be regulated through inhibition by valine and activation by isoleucine. Pyruvate is a less likely control step for amino acid biosynthesis however, because it is a key control point in energy metabolism.

Nucleotide Biosynthesis

In this chapter, the authors complete their treatment of the biosyntheses of the major classes of macromolecular precursors by describing the synthesis of the purine and pyrimidine nucleotides. Besides being the precursors of RNA and DNA, these compounds serve a number of other important roles that are reviewed in the opening paragraph of the chapter. Nucleotide nomenclature is reviewed in the introduction to the chapter, as is an outline for the synthesis of nucleotides through de novo and salvage pathways.

The chapter begins with the synthesis of the pyrimidine nucleotides. The pyrimidine ring is synthesized de novo from bicarbonate, aspartate, and ammonia (usually from glutamine) prior to attachment to a ribose sugar. The authors go through the synthesis step by step, paying particular attention to the enzyme carbamoyl phosphate synthetase (CPS), which synthesizes carbamoyl phosphate from bicarbonate and ammonia and catalyzes the committed step in eukaryotic pyrimidine synthesis. The next step in the synthesis is catalyzed by aspartate transcarbamoylase (ATCase), an enzyme that was discussed in Chapter 10. This reaction, the formation of carbamoylasparate from carbamoyl phosphate and aspartate, is the committed step in prokaryotic pyrimidine synthesis. A condensation and an oxidation reaction complete the formation of orotate, which is then coupled to a phosphoribose by reaction with 5-ribosyl-1-pyrophosphate (PRPP) to form the pyrimidine nucleotide orotidylate. Decarboxylation of orotidylate gives uridine monophosphate (UMP), which can be phosphorylated by nucleoside mono- and diphosphate kinases to form UDP and UTP, respectively. Amination of UTP forms cytidine triphosphate (CTP) and completes the synthesis of the pyrimidine ribonucleotides.

Next the authors turn to synthesis of the purine nucleotides. Unlike synthesis of pyrimidines, synthesis of purine nucleotides builds upon the ribose ring. As in the pyrimidine ring system, the ribose sugar is donated by the activated form of ribose 5-phosphate, 5-phosphoribosyl-1-pyrophosphate (PRPP). The two purine nucleotides AMP and GMP have a common precursor, inosine 5'-monophosphate (IMP). The authors discuss the synthesis of this initial purine product and then formation of AMP and GMP. The reactions that allow cells to salvage free purines and the control of purine biosynthesis are presented. The regulation of nucleotide biosynthesis through feedback inhibition is discussed later in the chapter.

Two reactions that are required to form the precursors of DNA are described in detail: ribonucleotide reductase converts ribonucleotides to deoxyribonucleotides, and thymidylate synthase methylates dUMP to form dTMP. The authors present the mechanisms and cofactors of these enzymes and explain how some anticancer drugs and antibiotics function by inhibition of dTMP synthesis and thus the growth of cells. Nucleotides also serve important roles as constituents of NAD^+, $NADP^+$, FAD, and coenzyme A (CoA), so the syntheses of these cofactors are described briefly. The chapter concludes with an explanation of how the purines are catabolized and some of the pathological conditions that arise from defects in the catabolic pathway of the purines.

LEARNING OBJECTIVES

When you have mastered this chapter, you should be able to accomplish the following objectives.

Introduction

1. List the major biochemical roles of the *nucleotides*.

2. Distinguish among the *purine* and *pyrimidine nucleosides* and *nucleotides*.

3. Define *de novo* and *salvage pathways* for biosynthesis of nucleotides.

The Pyrimidine Ring Is Assembled de Novo or Recovered by Salvage Pathways
(Text Section 25.1)

3. Draw the structure of a pyrimidine ring and identify the precursors that provide each carbon and nitrogen atom of the ring. Note the numbering of the ring atoms.

4. Discuss the structure and mechanism of *carbamoyl phosphate synthetase (CPS)*. Explain the role of *carbamoyl phosphate* in pyrimidine biosynthesis.

5. Discuss the role of substrate channeling in the mechanism of CPS.

6. Write the *aspartate transcarbamoylase* reaction and outline the remaining reactions that form *orotate*. Outline the conversion of orotate to *uridylate (UMP)*.

7. Explain how nucleoside *mono-* and *diphosphate kinases* interconvert the nucleoside mono-, di-, and triphosphates.

8. Describe the reaction that converts *UTP* to *CTP*.

9. Use the example of *thymine* to discuss the use of salvage pathways to recover pyrimidine bases from breakdown products of DNA and RNA. Discuss why *thymidine kinase* is a good therapeutic target for viral therapy.

Purine Bases Can Be Synthesized de Novo or Recycled by Salvage Pathways
(Text Section 25.2)

10. Draw the structure of a purine ring and identify the precursors that provide each carbon and nitrogen atom of the ring. Note the numbering of the ring atoms.

11. Describe the committed step in de novo purine biosynthesis and the enzyme that catalyzes it.

12. Explain the mechanism for replacing a carbonyl oxygen with an amino group, and list the potential sources of the nitrogen atom in these reactions.

13. Outline the synthesis of *inosine 5′-monophosphate (IMP),* noting the sources of the atoms and the cofactors involved. List the steps that require ATP.

14. Describe the synthesis of *adenylate (AMP)* and *guanylate (GMP)* from IMP. List the cofactors and intermediates of the reactions.

15. Discuss the evidence that enzymes involved in many metabolic pathways are physically associated with each other. Explain how assembly of enzymes in purine biosynthesis is thought to be controlled.

16. Outline the synthesis of purine nucleotides by the *salvage reactions* and explain why these reactions are energetically advantageous.

Deoxyribonucleotides Are Synthesized by the Reduction of Ribonucleotides Through a Radical Mechanism (Text Section 25.3)

17. Describe the subunit structure of *ribonucleotide reductase.* Outline its reaction, mechanism, and regulation. Give examples of radicals used by other ribonucleotide reductases.

18. Explain the roles of *NADH, thioredoxin,* and *thioredoxin reductase* in the ribonucleotide reductase mechanism. Include the role of disulfide bonds in the reduction of the *thiyl radical.*

19. Describe the *thymidylate synthase* reaction. Account for the source of the methyl group and describe the change in the oxidation state of the transferred carbon atom that occurs during the reaction.

20. Explain the role of *dihydrofolate reductase* in the synthesis of deoxythymidylate.

21. Account for the ability of *fluorouracil* to act as a *suicide inhibitor.* Describe the inhibitory mechanism of *methotrexate* and *aminopterin.* Explain how these three compounds interfere with the growth of cancer cells. List the mechanisms by which a cell could become resistant to methotrexate. Explain the antibiotic activity of *trimethoprim.*

Key Steps in Nucleotide Biosynthesis Are Regulated by Feedback Inhibition
(Text Section 25.4)

22. Outline the regulation of the biosynthesis of the purine and pyrimidine nucleotides and name the committed steps in the pathways.

23. Describe the control of deoxyribonucleotide synthesis by ribonuclease reductase regulation.

Disruptions in Nucleotide Metabolism Can Cause Pathological Conditions
(Text Section 25.5)

24. Describe the reactions of the *nucleotidases* and *nucleoside phosphorylases*.

25. Describe the link between *adenosine deaminase* and *severe combined immunodeficiency (SCID)*.

26. Outline the conversions of AMP and *guanine* to *uric acid*. Describe the role of *xanthine oxidase* in these processes.

27. Describe the major clinical findings in patients with *gout* and explain the rationale for the use of *allopurinol* to alleviate the symptoms of the disease.

28. Describe the *antioxidant* role of *urate*.

29. Name the biochemical lesion that leads to the *Lesch-Nyhan syndrome* and describe the symptoms of the disease.

SELF-TEST

Introduction

1. Describe the physiological roles of the nucleotides.

2. Which of the following answers completes the sentence correctly? Cytosine is a

   (a) purine base.
   (b) pyrimidine base.
   (c) purine nucleoside.
   (d) pyrimidine nucleoside.

3. Which of the following are nucleotides?

   (a) deoxyadenosine
   (b) cytidine
   (c) deoxyguanylate
   (d) uridylate

The Pyrimidine Ring Is Assembled de Novo or Recovered by Salvage Pathways

4. Which of the following statements about the carbamoyl phosphate synthetase of mammals, which is used for pyrimidine biosynthesis, are true?

   (a) It is located in the mitochondria.
   (b) It is located in the cytosol.
   (c) It uses NH_4^+ as a nitrogen source.
   (d) It uses glutamine as a nitrogen source.
   (e) It requires *N*-acetylglutamate as a positive effector.

5. Which of the following statements about 5-phosphoribosyl-1-pyrophosphate (PRPP) are true?

   (a) It is an activated form of ribose 5-phosphate.
   (b) It is formed from ribose 1-phosphate and ATP.
   (c) It has a pyrophosphate group attached to the C-1 atom of ribose in the α configuration.
   (d) It is formed in a reaction in which PP_i is released.

6. How is orotate, a free pyrimidine, converted into a nucleotide? Is this reaction considered to be a salvage reaction or a biosynthetic one?

7. Which of the following enzymes are involved in converting the nucleoside 5'-monophosphate (NMP) products of the purine or pyrimidine biosynthetic pathways into their 5'-triphosphate (NTP) derivatives?
   (a) purine nucleotidase
   (b) nucleoside diphosphate kinase
   (c) nucleoside monophosphate kinases
   (d) nucleoside phosphorylase

8. How is the exocyclic amino group on the N-4 position of cytosine formed?

9. Which of the following is not true about the thymine salvage pathway?
   (a) Thymine is converted into the nucleoside thymidine by thymidine phosphorylase in the first step of the pathway.
   (b) The activity of thymidine phosphorylase fluctuates with the cell cycle.
   (c) Thymidine kinase phosphorylates thymidine to form thymine monophosphate.
   (d) The mammalian and viral versions of thymidine kinase are highly homologous.

Purines Can Be Synthesized de Novo or Recycled by Salvage Pathways

10. Which of the following compounds directly provide atoms to form the purine ring?
    (a) aspartate
    (b) carbamoyl phosphate
    (c) glutamine
    (d) glycine
    (e) CO_2
    (f) N^5,N^{10}-methylenetetrahydrofolate
    (g) N^{10}-formyltetrahydrofolate
    (h) NH_4^+

11. Which of the following answers completes the sentence correctly? The first product of purine nucleotide biosynthesis that contains a complete purine ring (hypoxanthine) is
    (a) AMP.
    (b) GMP.
    (c) IMP.
    (d) xanthylate (XMP).

12. The conversion of IMP to AMP requires which of the following?
    (a) ATP
    (b) GTP
    (c) aspartate
    (d) glutamine
    (e) NAD^+

13. The conversion of IMP to GMP requires which of the following?
    (a) ATP
    (b) GTP
    (c) aspartate
    (d) glutamine
    (e) NAD^+

14. Describe the general mechanism used by cells to replace a carbonyl group with an amino group in nucleotide biosynthesis.

15. Which of the following reactants and products are involved in the salvage reactions of purine biosynthesis?
    (a) IMP $\rightarrow$ AMP
    (b) IMP $\rightarrow$ GMP
    (c) adenine $\rightarrow$ AMP
    (d) guanine $\rightarrow$ GMP

16. During a purine salvage reaction, what is the source of the energy required to form the C–N glycosidic bond between the base and ribose?

17. Show which of the nucleotides in the right column regulate each of the conversions in the left column.

    (a) ribose 5-phosphate $\rightarrow$ PRPP   (1) AMP
    (b) PRPP $\rightarrow$ 5-phosphoribosylamine   (2) GMP
    (c) phosphoribosylamine $\rightarrow$ IMP   (3) IMP
    (d) IMP $\rightarrow$ adenylosuccinate
    (e) IMP $\rightarrow$ xanthylate (XMP)

18. Explain how GFP has been used to study enzyme association in purine biosynthesis.

Deoxyribonucleotides Are Synthesized by the Reduction of Ribonucleotides Through a Radical Mechanism

19. Which of the following statements about ribonucleotide reductase are true?

    (a) It converts ribonucleoside diphosphates into 2′-deoxyribonucleoside diphosphates in humans.
    (b) It catalyzes the homolytic cleavage of a bond.
    (c) It accepts electrons directly from $FADH_2$.
    (d) It receives electrons directly from thioredoxin.
    (e) It contains two kinds of allosteric regulatory sites—one for control of overall activity and another for control of substrate specificity.

20. Select from the following those compounds that are precursors of 2′-deoxythymidine-5-triphosphate (dTTP) in mammals and place them in their correct biosynthetic order.

    (a) OMP   (f) dUDP
    (b) UMP   (g) dUTP
    (c) UDP   (h) dTMP
    (d) UTP   (i) dTDP
    (e) dUMP   (j) dTTP

21. Define *suicide inhibitor* and give an example from pyrimidine biosynthesis.

22. Methotrexate and trimethoprim are both inhibitors of dihyrofolate reductase. Why is trimethoprim the drug of choice in treating a human microbial infection?

Key Steps in Nucleotide Biosynthesis Are Regulated by Feedback Inhibition

23. What is the committed step in purine biosynthesis and which of the following compounds are involved in the control of the purine biosynthetic pathway?

    (a) IMP   (d) GMP
    (b) AMP   (e) PRPP
    (c) OMP

24. For each of the following, indicate the effect it would have on ribonucleotide reductase activity

    (a) dATP   (c) TTP
    (b) ATP   (d) dGTP

Disruptions in Nucleotide Metabolism Can Cause Pathological Conditions

25. What is the biochemical deficiency that leads to severe combined immunodeficiency disease?

    (a) nucleoside phosphorylase
    (b) xanthine oxidase
    (c) adenosine deaminase
    (d) all of the above.

26. Which of the following compounds would give rise to urate if they were catabolized completely in humans?

    (a) ADP-glucose   (e) CoA
    (b) GDP-mannose   (f) FAD
    (c) CDP-choline   (g) UMP
    (d) UDP-galactose

27. What is the benefit of high serum levels of urate in humans, given that too much urate leads to gout?

ANSWERS TO SELF-TEST

1. The nucleotides (1) are the activated precursors of DNA and RNA; (2) are the source of derivatives that are activated intermediates in many biosyntheses; (3) include ATP, the universal currency of energy in biological systems, and GTP, which powers many movements of macromolecules; (4) include the adenine nucleotides, which are components of the major coenzymes NAD^+, FAD, and CoA; and (5) serve as metabolic regulators.

2. b

3. c, d. Nucleotides are nucleosides that contain one or more phosphate substituents on their ribose or deoxyribose moieties.

4. b, d. The mitochondrial carbamoyl phosphate synthetase used for urea synthesis is activated by N-acetylglutamate and uses N_4^+ as the nitrogen source.

5. a, c

6. Orotate condenses with PRPP in a reaction catalyzed by orotate phosphoribosyl transferase to form the nucleotide orotidylate (OMP). Orotidylate decarboxylase converts OMP to the more abundant nucleotide UMP. The reaction occurs during de novo pyrimidine biosynthesis and is therefore not a salvage reaction.

7. b, c. Several different specific nucleoside monophosphate kinases phosphorylate dNMPs and NMPs, using ATP as the phosphoryl donor. A single enzyme, nucleoside diphosphate kinase, uses the phosphorylation potential of ATP to convert the dNDPs and NDPs to dNTPs and NTPs. The ubiquitous adenylate nucleotides are interconverted by adenylate kinase (myokinase).

8. CTP is formed by amination of UTP. The carbonyl oxygen at C-4 of UTP is replaced with an amino group via the formation of an enol phosphate ester intermediate. In *E.*

coli, NH_4^+ serves as the source of the nitrogen atom that displaces the phosphate group, whereas the amide group of glutamine serves this purpose in mammals.

9. d. Answer d is incorrect because the mammalian and viral versions of TK are quite different, a fact that has been utilized in the development of anti-viral drugs.

10. a, c, d, e, g

11. c

12. b, c

13. a, d, e

14. A carbonyl oxygen is converted into a phosphoryl ester or a substituted phosphoryl ester through a reaction with a high-energy phosphate (ATP or GTP) to form a mono- or diphosphate ester. The phosphate or pyrophosphate group can then be readily displaced by the nucleophilic attack of the nitrogen atom from NH_3, the side-chain amide group of glutamine, or the α-amino group of aspartate. The resulting adduct is an amino group or can be converted into one.

15. c, d

16. The activated form of ribose 5-phosphate (R-5-P), PRPP, reacts with the purine base to form the nucleotide and release PP_i. The displacement and subsequent hydrolysis of PP_i drives the formation of the *N*-glycosyl bond. ATP ultimately provides the energy through its reaction with R-5-P to form PRPP.

17. (a) 1, 2, 3   (b) 1, 2, 3   (c) none   (d) 1   (e) 2

18. Researchers have fused different enzymes of the purine biosynthetic pathway with the fluorescent protein GFP. They then observed the GFP in cells that had been switched to growth media lacking in purines to see how the cellular localization would change upon initiation of purine synthesis. They observed that the enzymes were spread diffusely throughout the cell in the absence of purine synthesis, but were localized to cytoplasmic granules called purinosomes upon intiation of purine synthesis. Additional evidence suggests a role for phosphorylation in the assembly.

19. a, b, d, e. NADPH provides electrons via thioredoxin.

20. a, b, c, f, g, e, h, i, j. In mammals, NDPs are converted to dNDPs by ribonucleotide reductase. Thus, UDP is converted to dUDP, which is converted to dUTP by nucleoside diphosphate kinase. A specific pyrophosphatase hydrolyzes dUTP to dUMP, which is then converted to dTMP by thymidylate synthase. The dTMP is converted to dTTP. A priori, you might have expected the dUDP product of the ribonucleotide reductase reaction to be converted directly to dUMP. However, in fact, cells contain a dUTP pyrophosphatase to prevent dUTP from serving as a DNA precursor, and it is this enzyme that functions in the dTTP biosynthetic pathway.

21. A suicide inhibitor is a substrate that is converted by an enzyme into a substance that is capable of reacting with and inactivating the enzyme. In pyrimidine biosynthesis, thymidylate synthase converts fluorouracil into a derivative that becomes covalently attached to the enzyme and thereby inactivates it.

22. Since trimethoprim binds to the mammalian dihydrofolate reductase much less tightly than to the enzyme of susceptible microorganisms, it causes fewer deleterious effects to humans than does methotrexate.

23. The conversion of PRPP into phosphoribosylamine by *glutamine phosphoryl amido-transferase* is the committed step in purine biosynthesis. Compounds a, b, and d are involved in the regulation.

24. All of the molecules bind to the specificity site, but only ATP and dATP bind to the activity site.
    (a) Lowers activity when bound to activity site; enhances reduction of UDP and CDP when bound to specificity site.
    (b) Reverses effect of dATP on activity site; same effect as dATP on specificity site.
    (c) Enhances reduction of GDP; inhibits pyrimidine reduction.
    (d) Enhances reduction of ATP.

25. a, b, e, f. Each of these compounds contains a heterocyclic purine base.

26. Urate levels in humans are often close to the solubility limit, which leads to gout when salts of urate crystallize (resulting in damage to joints and kidneys). There is a significant benefit to high concentrations of urate, however, as urate is a highly effective scavenger of reactive oxygen species (ROS). ROS can cause damage in cells contributing to cancer and the effects of aging. Urate is about as effective as ascorbate (vitamin C) as an antioxidant.

27. (c)

PROBLEMS

1. Why might covalently linked (multifunctional) enzymes, such as those of the pyrimidine biosynthetic pathway of mammals, be advantageous to an organism?

2. Mammalian lymphocytes that lack adenosine deaminase neither grow nor divide. The level of dATP in these cells is 100 times higher than that in normal lymphocytes, and the synthesis of DNA in the cells is impaired.
   (a) How is adenosine converted to dATP? Assume that the first step is catalyzed by a specific nucleoside kinase.
   (b) How does the elevation in dATP concentration in the abnormal lymphocytes affect the synthesis of DNA?

3. Clinicians who use F-dUMP and methotrexate *together* in cancer treatment find that the combined effects on cancer cells are not synergistic. Suggest how the administration of methotrexate could interfere with the action of F-dUMP.

4. Elevated levels of ammonia in the blood can be caused by a deficiency of mitochondrial carbamoyl phosphate synthetase or a deficiency of any of the urea cycle enzymes. These two types of disorders can be distinguished by the presence of orotic acid or related metabolites in the urine.
   (a) Why is it possible to determine the basis of hyperammonemia in this way?
   (b) Why would a deficiency of cytoplasmic carbamoyl phosphate synthetase not cause hyperammonemia? What problems would such an enzyme deficiency cause? How would you treat a patient who has a deficiency in cytoplasmic carbamoyl phosphate synthetase?

5. The degradation of thymine yields β-aminoisobutyrate, as shown in the figure below, which can be converted to succinyl CoA and then degraded in the citric acid cycle. What cofactors are needed to convert β-aminoisobutyrate to succinyl CoA?

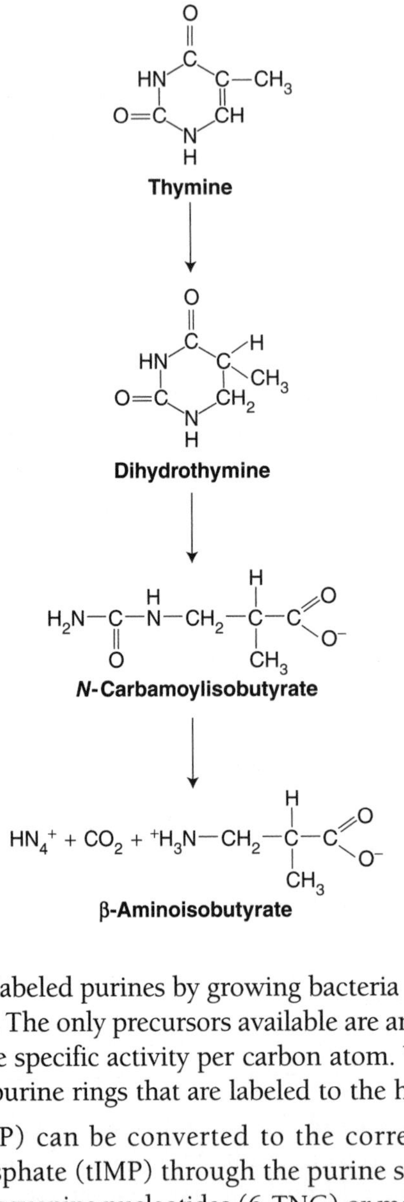

Thymine

Dihydrothymine

N-Carbamoylisobutyrate

β-Aminoisobutyrate

6. You wish to prepare 14C-labeled purines by growing bacteria in a medium containing a suitably labeled precursor. The only precursors available are amino acids that are all uniformly labeled to the same specific activity per carbon atom. Which of the amino acids would you use to obtain purine rings that are labeled to the highest specific activity?

7. 6-Mercaptopurine (6-MP) can be converted to the corresponding nucleotide 6-thioinosine-5′-monophosphate (tIMP) through the purine salvage pathway. tIMP can then be converted to 6-thioguanine nucleotides (6-TNG) or methylated to form Me-tIMP. Methotrexate (MTX) increases incorporation of 6-TNGs through the salvage pathway (Bokkerink et al., *Hematol. Blood Transf.* 33(1990):110–117). Both 6-MP and MTX are clinically useful anticancer agents and have been used together for many years in the treatment of childhood leukemia in part because of the synergistic effect they have on each other.

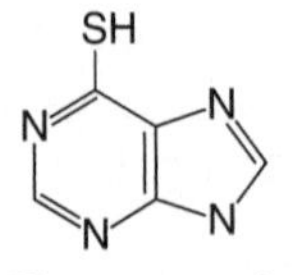

6-Mercaptopurine

(a) Briefly describe the salvage reactions required to convert 6-MP to the corresponding nucleotide.

(b) What step in the de novo biosynthesis of purines is likely to be inhibited by tIMP?

(c) How could the presence of MTX increase the incorporation of 6-TNG into DNA and RNA?

8. Hydroxyurea, a potent chelator of ferric ions, has been shown to interfere with DNA synthesis, and it is used as an antitumor agent. What is a likely target enzyme for hydroxyurea?

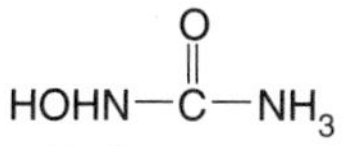

$$HOHN-\overset{\overset{\displaystyle O}{\|}}{C}-NH_3$$
Hydroxyurea

9. Methotrexate, a folate antagonist, interferes with nucleic acid biosynthesis. Would you expect it to inhibit purine or pyrimidine biosynthesis or both processes? Explain.

10. Nucleoside phosphorylases catalyze the interconversion of bases and nucleosides through the following reactions:

$$\text{Ribose 1-phosphate } + \text{ base } \longrightarrow \text{ ribonucleoside } + \text{ P}_i$$

or

$$\text{Deoxyribose 1-phosphate } + \text{ base } \longrightarrow \text{ deoxyribonucleoside } + \text{ P}_i$$

The equilibrium constant for each of these reactions is close to 1.

(a) The pathway for the incorporation of radioactive thymine into bacterial DNA includes a step catalyzed by nucleoside phosphorylase. It has often been observed that the incorporation of thymine into DNA is enhanced when deoxyadenosine or deoxyguanosine is added to the medium. Can you explain this observation? Why might deoxyguanosine be preferable to deoxyadenosine?

(b) In cells that cannot carry out de novo synthesis of IMP, inosine can be utilized to produce IMP but only through an indirect salvage route because of the absence of inosine kinase. Suggest an alternative pathway for the formation of IMP from inosine. Among the enzymes you will need are nucleoside phosphorylase and phosphoribomutase, which isomerizes ribose 1-phosphate to ribose 5-phosphate.

11. Hypoxanthine-guanine phosphoribosyl transferase (HGPRT), a salvage enzyme of nucleotide metabolism, uses 5′-phosphoribosylpyrophosphate (PRPP) to convert hypoxanthine to IMP and guanine to GMP. A deficiency of this enzyme can lead to an increased level of purine synthesis, excess formation of uric acid, and hyperuricemia, or gout.

(a) How might a deficiency in HGPRT stimulate purine synthesis?

(b) Under what conditions might one expect a deficiency of hypoxanthine-guanine phosphoribosyl transferase to affect the rate of *pyrimidine* nucleotide synthesis? How could you estimate the rate of pyrimidine nucleotide synthesis in humans?

12. In mammals, the committed step for pyrimidine synthesis is catalyzed by carbamoyl phosphate synthetase, while in bacteria, the committed step is the formation of N-carbamoyl-aspartate, catalyzed by aspartate transcarbamoylase.

(a) Account for these differences in mammals and bacteria.

(b) Bacterial carbamoyl phosphate synthetase is only partially inhibited by UMP. Why?

13. The synthesis of deoxythymidylate can proceed not only from dUMP but also from dCMP. The route from dCMP begins with the formation of dCDP from CDP, catalyzed

by ribonucleotide reductase, followed by dephosphorylation to dCMP, and the deamination of dCMP to form dUMP, catalyzed by dCMP deaminase.

$$dCMP \ + \ H_2O \ + \ H^+ \longrightarrow dUMP \ + \ NH_3^+$$

dCMP deaminase is an allosteric enzyme that is stimulated by dCTP and inhibited by dTTP. Account for these effects and relate them to the regulation of ribonucleotide reductase by deoxynucleoside triphosphates.

ANSWERS TO PROBLEMS

1. The clustering of two or more enzymes (active sites) in a single polypeptide chain ensures that their synthesis is coordinated and helps assure that they will assemble into a coherent complex. Also, the proximity of the active sites means that side reactions are minimized as substrates are channeled from one active site to another. Finally, a multifunctional complex with covalently linked active sites is likely to be more stable than a complex formed by noncovalent interactions.

2. (a) Adenosine is phosphorylated to AMP, with ATP serving as the phosphoryl donor, in a reaction carried out by a specific nucleoside kinase. The conversion of AMP to ADP through the action of a specific nucleoside monophosphate kinase is accomplished, with ATP again utilized as a phosphate donor. Ribonucleotide reductase catalyzes the reduction of ADP to dADP, which is then converted to dATP by nucleotide diphosphokinase.

   (b) High concentrations of dATP displace ATP from the overall activity site on ribonucleotide reductase, which lowers the rate of synthesis of all four deoxyribonucleoside diphosphates. This, in turn, leads to a depletion of deoxyribonucleoside triphosphates, which are the substrates for DNA synthesis.

3. Methotrexate blocks the regeneration of tetrahydrofolate from dihydrofolate, which is produced during the synthesis of thymidylate. The failure to regenerate tetrahydrofolate means that those biochemical reactions in the cell that depend on one-carbon metabolism cannot be carried out. One of the products of tetrahydrofolate is methylenetetrahydrofolate, which is used as a substrate by thymidylate synthetase and is required for inhibition of the enzyme by F-dUMP. A deficiency of methylenetetrahydrofolate means that F-dUMP cannot irreversibly inactivate thymidylate synthetase. Conversely, F-dUMP prevents the formation of dihydrofolate, thereby abolishing the adverse effects caused by the depletion of tetrahydrofolate in the cell.

4. (a) The presence of orotic acid, a precursor of pyrimidines, in the urine suggests that carbamoyl phosphate synthesized in mitochondria is not utilized there. Instead, carbamoyl phosphate enters the cytosol, where it stimulates an increase in the rate of synthesis of precursors of pyrimidines, including orotic acid. An excess of carbamoyl phosphate arises in mitochondria whenever any of the urea cycle enzymes are deficient. Such a condition will lead to hyperammonemia, as well as to the accumulation of carbamoyl phosphate. Although a deficiency in mitochondrial carbamoyl phosphate synthetase leads to hyperammonemia, it cannot lead to an accumulation of mitochondrial carbamoyl phosphate and therefore does not stimulate pyrimidine synthesis in the cytosol.

   (b) Cytoplasmic carbamoyl phosphate synthetase is involved primarily in the pathway for pyrimidine synthesis, not for the assimilation of ammonia. Recall that, in the cytosol, the substrate for the formation of carbamoyl phosphate is glutamine, not ammonia. A deficiency of carbamoyl phosphate synthesis in the cytosol would cause a depletion of pyrimidines. Such a deficiency is treated by administration of uracil or uridine, which are precursors of UMP and CMP.

5. The transamination of β-aminoisobutyrate to form methylmalonate semialdehyde requires pyridoxal phosphate as a cofactor. This reaction is similar to the conversion of ornithine to glutamate γ-semialdehyde. Then NAD+ serves as an electron acceptor for the oxidation of methylmalonate semialdehyde to methylmalonate. The conversion of methylmalonate to methylmalonyl CoA requires coenzyme A. The final reaction, in which methylmalonyl CoA is converted to succinyl CoA, is catalyzed by methylmalonyl CoA mutase, an enzyme that contains a derivative of vitamin B_{12} as its coenzyme.

6. Examination of the pathway for purine synthesis shows that only glycine is incorporated intact into the purine ring at the C-4 and C-5 positions. Therefore, glycine is a good choice as the radiolabeled precursor. Serine can also be considered because it is a precursor of glycine and the ultimate donor of C-1 groups to tetrahydrofolate, and activated tetrahydrofolate derivatives participate in two reactions in the formation of purines. Whether serine is a better choice than glycine depends on the relative amounts of the two unlabeled amino acids in the cell.

7. (a) 6-Mercaptopurine is converted to the mononucleotide through the action of hypoxanthine-guanine phosphoribosyl transferase (HGPRT), which uses PRPP to add 5′-phosphoribose to the purine ring. The resulting compound is 6-thioinosine-5′-monophosphate, an analog of IMP.

    (b) IMP (and presumably its analog 6-MP) inhibits Gln-PRPP aminotransferase, which catalyzes the committed step of purine biosynthesis. The 6-MP metabolite Me-tIMP may be involved in the inhibition of Gln-PRPP aminotransferase as well (Stet et al., *Biochem. Journal* 304(1994):163–168).

    (c) MTX inhibits folate-dependent enzymes in the de novo purine biosynthetic pathway leading to an accumulation of PRPP. This increase in substrate availability for the salvage pathway leads to a greater conversion of 6-MP into 6-TNG and therefore a higher level of incorporation of 6-TNGs into DNA. Also, since 6-MP is a substrate for HGPRT—see part (a)—it can compete with endogenous purine bases for HGPRT, leading to a decrease in the synthesis of AMP and GMP.

8. Hydroxyurea inhibits ribonucleotide reductase. By sequestering ferric ions, hydroxyurea destabilizes the organic free radical in the R2 subunit of the enzyme. The inhibition of enzyme activity leads to a depletion of deoxyribonucleoside diphosphates, which are normally converted to deoxyribonucleoside triphosphates, the substrates for DNA synthesis.

9. Methotrexate and aminopterin, a similar compound, are analogs of dihydrofolate (DHF) and inhibitors of dihydrofolate reductase, an enzyme that converts DHF to tetrahydrofolate (THF). The thymidylate synthase reaction converts N^5,N^{10}-methylenetetrahydrofolate to DHF in the process of methylating dUMP to form dTMP. In the presence of one of the inhibitors, this reaction functions as a sink that reduces the THF level of the cell by converting THF to DHF. Since THF derivatives are substrates in two reactions of purine metabolism and one of pyrimidine metabolism, both pathways are affected by the inhibitor.

10. (a) Deoxyribonucleosides such as deoxyadenosine can be converted to the free base and deoxyribose 1-phosphate by nucleoside phosphorylase. Increased levels of deoxyribose 1-phosphate are then available for the formation of deoxythymidine from thymine in the reverse reaction catalyzed by nucleoside

phosphorylase. Deoxyguanosine might be preferable to deoxyadenosine because the conversion of elevated levels of deoxyadenosine to dAMP and then to dATP could lead to the inactivation of ribonucleotide reductase, which is sensitive to the concentration of dATP.

(b) Inosine is cleaved to produce hypoxanthine and ribose 1-phosphate through the action of nucleoside phosphorylase; note that inorganic phosphate is required for this reaction. Hypoxanthine-guanine phosphoribosyl transferase converts free hypoxanthine to IMP by condensation with PRPP. PRPP can be derived from ribose 1-phosphate in two steps: (1) the conversion of ribose 1-phosphate to ribose 5-phosphate, which is catalyzed by phosphoribomutase; and (2) the formation of PRPP from ribose 5-phosphate and ATP, which is catalyzed by PRPP synthetase.

11. (a) When active, HGPRT consumes PRPP as it catalyzes the synthesis of GMP and IMP. Decreased flux through this reaction raises the steady-state level of PRPP, thereby increasing the activity of PRPP amidotransferase, which catalyzes the initial step in purine synthesis. Increased activity may make the amidotransferase resistant to feedback inhibition by AMP and GMP, the end products of the purine biosynthetic pathway.

(b) As noted above, decreased activity of HGPRT increases the concentration of PRPP. This increases the rate of pyrimidine synthesis at the orotate phosphoribosyl transferase reaction, if PRPP levels are normally subsaturating for that enzyme. Orotate incorporation into nucleotide pools or into nucleic acids would give a reasonable estimate of de novo pyrimidine nucleotide synthesis.

12. (a) Bacteria use a single form of carbamoyl phosphate synthetase not only for the synthesis of pyrimidines but also for the synthesis of arginine. Arginine biosynthesis begins with glutamate and includes formation of citrulline from ornithine and carbamoyl phosphate, a pathway that resembles urea formation in mammals. Two forms of carbamoyl phosphate synthetase, a cytoplasmic form for pyrimidine synthesis and a mitochondrial form for arginine and urea synthesis, are employed in mammals. Because the two pathways are compartmentalized, the formation of carbamoyl phosphate in the cytosol is regarded as the committed step for pyrimidine synthesis.

(b) UMP does not completely inhibit bacterial carbamoyl phosphate synthetase because that inhibition would interfere with arginine production.

13. An increase in dCTP levels signals that the cell has ample deoxynucleotides for DNA synthesis and that there is a need for thymidylate synthesis. An increase in dTTP levels signals that the activity of thymidylate synthase can be decreased, and the inhibition of dCMP deaminase by dTTP reduces the input of dUMP into the pathway. While ribonucleotide reductase is subject to regulation by other deoxynucleotides, it is not subject to allosteric regulation by dCTP. Instead it appears that regulation of dCMP deaminase provides a second control point for the generation of deoxynucleotides in the cell.

Chapter 26

The Biosynthesis of Membrane Lipids and Steroids

This chapter describes the biosynthesis of membrane lipids, steroids, and other important lipid molecules, such as bile salts, and vitamin D. As background material for this chapter, you should review the earlier chapters on cell membranes (Chapter 12) and fatty acid metabolism (Chapter 22), paying particular attention to the structure and properties of lipids and the central role of acetyl CoA in the metabolism of lipids. Chapter 26 begins with a discussion of the formation of triacylglycerols, phosphoglycerides, and sphingolipids from the simple precursors glycerol 3-phosphate, fatty acyl CoAs, and polar alcohols, for example, choline, inositol, and sugars. The text then describes the synthesis of cholesterol from acetyl CoA via the important intermediate isopentenyl pyrophosphate. The regulation of a key enzyme in the biosynthetic pathway as well as other modes of regulation of cholesterol metabolism are also outlined. Cholesterol is the precursor of bile salts as well as steroid hormones. The cholesterol and triacylglycerols synthesized in the liver and intestines are transported by lipoproteins to peripheral tissues. Cholesterol from dietary sources is also moved from the intestine to the liver by lipoproteins. Therefore, the classification, the properties, and the mechanisms by which the lipoproteins deliver lipids to cells are discussed next. A biochemically based strategy for controlling one form of abnormal cholesterol metabolism is also described. Finally, the authors describe the synthesis of steroid hormones and vitamin D from cholesterol.

LEARNING OBJECTIVES

When you have mastered this chapter, you should be able to accomplish the following objectives:

Introduction

1. Name the three major lipid-based components of biological membranes (Chapter 12).

2. Explain the biological significance of *cholesterol*.

3. List the primary biological functions of *triacylglycerols* and *phospholipids* (Chapter 22).

Phosphatidate is a Common Intermediate in the Synthesis of Phospholipids and Triacylglycerols (Text Section 26.1)

4. Describe the roles of *phosphatidate, glycerol 3-phosphate, lysophosphatidate,* and *diacylglycerol (DAG)* in the synthesis of triacylglycerols and phospholipids and identify their sources.

5. Note the significance of *CDP-diacylglycerol* and *CDP-ethanolamine,* the activated precursors in the biosyntheses triacylglycerols and phospholipids.

6. Describe the biosyntheses of *phosphatidyl serine, phosphatidyl ethanolamine, phosphatidyl choline,* and *phosphatidyl inositol.*

7. Restate the physiologic roles of phosphatidyl inositol and its degradation products, *inositol 1,4,5-trisphosphate* and *diacylglycerol* (Chapter 14).

8. Summarize the steps in the biosynthesis of *sphingosine* from *palmitoyl CoA* and *serine.*

9. Outline the synthesis of *sphingomyelin, cerebrosides,* and *gangliosides* from sphingosine. Note the use of activated sugars and acidic sugars.

10. Provide examples of how sphingolipids confer diversity on lipid structure and function.

11. Discuss the general degradation pathway of gangliosides and the biochemical basis of *Tay-Sachs disease* and *respiratory distress syndrome.*

12. Discuss the proposed role of *phosphatidic acid phosphatase (PAP)* and *diacylglycerol kinase* in regulation of lipid synthesis, including the consequences of an abundance or lack of PAP function in mammals.

Cholesterol is Synthesized from Acetyl Coenzyme A in Three Stages (Text Section 26.2)

13. Describe the physiologic roles of *cholesterol.*

14. List the major stages in cholesterol biosynthesis and give the key intermediates. Recognize the structure of cholesterol.

15. Compare the synthetic paths leading from acetyl CoA to *mevalonate* and to the *ketone bodies* (Chapter 22). Note the role of *3-hydroxy-3-methylglutaryl CoA reductase (HMG CoA reductase)* as the major regulatory enzyme in cholesterol biosynthesis.

16. Describe the conversion of mevalonate into *isopentenyl pyrophosphate.*

17. Outline the condensation reactions leading from isopentenyl pyrophosphate to *squalene.* Describe the mechanisms of these condensation reactions.

18. Discuss the *cyclization* of squalene and the formation of cholesterol from *lanosterol.* Note the role of O_2 in the formation of cholesterol.

The Complex Regulation of Cholesterol Biosynthesis Takes Place at Several Levels (Text Section 26.3)

19. List the sources of cholesterol and discuss the four mechanisms of regulation of cholesterol biosynthesis through HMG CoA reductase..

20. List the various classes of *lipoproteins* together with their lipid and protein components. Describe their lipid transport functions.

21. Discuss the diagnostic value of measuring serum levels of LDL and HDL. Explain the mechanism by which HDL protects against *arteriosclerosis*.

22. Summarize the steps in the delivery of cholesterol to cells *via the low-density-lipoprotein (LDL) receptor*. Discuss the regulation of cellular functions by the *LDL pathway*.

23. Describe the proposed domain structure of the LDL receptor derived from the primary sequence of this protein.

24. Discuss the biochemical defects of the LDL receptor that result in *familial hypercholesterolemia*.

25. Summarize approaches used to reduce *serum cholesterol*.

Important Derivatives of Cholesterol Include Bile Salts and Steroid Hormones (Text Section 26.4)

26. Describe the physiologic roles and the general structures of the *bile salts*.

27. List the five major classes of *steroid hormones*, their physiologic functions, and their sites of synthesis. Outline their biosynthetic relationships.

28. Give the numbering scheme for the carbon atoms of cholesterol and its derivatives, and distinguish between the α- and β-oriented groups and cis or trans *ring fusions*.

29. Describe the *hydroxylation reactions* involving *cytochrome P450*. Indicate the role of these *monooxygenase reactions* in steroid biosynthesis, the *detoxification* of *xenobiotic compounds*, and the generation of *carcinogens*.

30. Describe the synthesis of *pregnenolone* from cholesterol, the conversion of pregnenolone into *progesterone*, and the subsequent reactions leading to *cortisol* and *aldosterone*.

31. Outline the synthesis of *androgens* and *estrogens* from progesterone.

32. Discuss the synthesis and the physiologic role of *vitamin D*. Explain the consequences of a deficiency in vitamin D.

SELF-TEST

Introduction

1. Which of the following are components of biological membranes?
   (a) free fatty acids
   (b) sphingolipids
   (c) triacylglycerols
   (d) phospholipids
   (e) cholesterol
   (f) proteins

Phosphatidate is a Common Intermediate in the Synthesis of Phospholipids and Triacylglycerols

2. Which of the following reactions are significant sources of glycerol 3-phosphate that is used in lipid synthesis?

   (a) reduction of dihydroxyacetone phosphate
   (b) oxidation of glyceraldehyde 3-phosphate
   (c) phosphorylation of glycerol
   (d) dephosphorylation of 1,3-bisphosphoglycerate
   (e) reductive phosphorylation of pyruvate

3. Match the lipids in the left column with the major synthetic precursors or intermediates in the right column.

   (a) triacylglycerol
   (b) phosphatidyl ethanolamine
       (mammals)

   (1) phosphatidate
   (2) diacylglycerol
   (3) acyl CoA
   (4) glycerol 3-phosphate
   (5) CDP-diacylglycerol
   (6) CDP-ethanolamine

4. Explain the role of the CDP derivatives in the synthesis of phosphoglycerides.

5. Calculate the number of "high-energy" phosphate bonds that are expended in the formation of phosphatidyl choline from diacylglycerol and choline in mammals.

6. Which of the following is a common reaction used for the formation of phosphatidyl ethanolamine?

   (a) decarboxylation of phosphatidyl serine
   (b) reaction of CDP-ethanolamine with a diacylglycerol
   (c) demethylation of phosphatidyl choline
   (d) reaction of ethanolamine with CDP-diacylglycerol
   (e) reaction of CDP-ethanolamine with CDP-diacylglycerol

7. Which of the following is a lipid with a signal-transducing activity?

   (a) phosphatidyl choline
   (b) phosphatidyl serine
   (c) plasminogen activator
   (d) phosphatidyl inositol 4,5-bisphosphate
   (e) phospholipase A_2

8. Which of the following structural components is found in glyceryl ester phospholipids?

   (a) two long hydrocarbon chains
   (b) acetyl group
   (c) phosphate group
   (d) ether linkage
   (e) α, β-double bond
   (f) glycerol group
   (g) long fatty acyl chain

9. Which of the following phospholipases would you expect to cleave the R_1-containing chain from the phospholipid shown in Figure 26.1?

   (a) phospholipase A_1
   (b) phospholipase A_2
   (c) phospholipase C
   (d) phospholipase D
   (e) none of the above

FIGURE 26.1 A phospholipid.

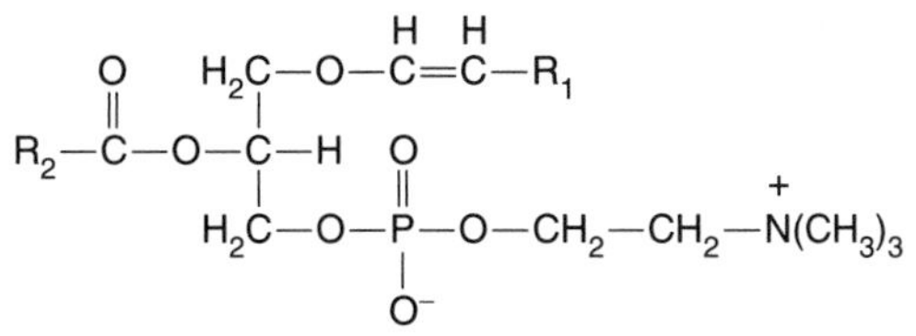

$$
\begin{array}{c}
\text{O} \qquad \text{H}_2\text{C}-\text{O}-\overset{\text{H}}{\underset{}{\text{C}}}=\overset{\text{H}}{\underset{}{\text{C}}}-\text{R}_1 \\
\text{R}_2-\text{C}-\text{O}-\text{C}-\text{H} \quad \text{O} \\
\qquad\qquad \text{H}_2\text{C}-\text{O}-\overset{}{\underset{\text{O}^-}{\text{P}}}-\text{O}-\text{CH}_2-\text{CH}_2-\overset{+}{\text{N}}(\text{CH}_3)_3
\end{array}
$$

10. Which of the following is NOT a precursor or intermediate in the synthesis of sphingomyelin?

    (a) palmitoyl CoA
    (b) lysophosphatidate
    (c) CDP-choline
    (d) acyl CoA
    (e) serine

11. Match the lipids in the left column with the appropriate activated precursors in the right column.

    (a) sphingomyelin
    (b) ganglioside
    (c) phosphatidyl serine

    (1) acyl CoA
    (2) CDP-choline
    (3) CDP-diacylglycerol
    (4) CMP-*N*-acetylneuraminate
    (5) UDP-sugar

12. In which compartment of the cell does ganglioside G_{M2} accumulate in Tay-Sachs patients? What is the biochemical defect?

13. When phosphatidic acid phosphatase (PAP) activity is high,

    (a) phosphatidate is dephosphorylated.
    (b) phosphatidylethanolamine, phosphatidylserine, and phosphatidylcholine are produced.
    (c) phosphatidylinositol and cardiolin are produced.
    (d) expression of genes in phospholipid synthesis is increased.
    (e) obesity can result.

Cholesterol Is Synthesized from Acetyl Coenzyme A in Three Stages

14. From the following compounds, identify the intermediates in the synthesis of cholesterol and list them in their proper sequence.

    (a) geranyl pyrophosphate
    (b) squalene
    (c) isopentenyl pyrophosphate
    (d) mevalonate
    (e) cholyl CoA
    (f) farnesyl pyrophosphate
    (g) lanosterol

15. Which of the following are common features of the syntheses of mevalonate and ketone bodies?

    (a) Both involve 3-hydroxy-3-methylglutaryl CoA (HMG CoA).
    (b) Both require NADPH.
    (c) Both require the HMG CoA cleavage enzyme.

(d) Both occur in the mitochondria.

(e) Both occur in liver cells.

16. Select the appropriate characteristics in the right column for the three stages in the synthesis of cholesterol in the left column.

(a) mevalonate to isopentenyl pyrophosphate   (1) releases PP_i
(b) isopentenyl pyrophosphate to squalene   (2) requires NADPH
(c) squalene to cholesterol   (3) requires O_2
         (4) releases CO_2
         (5) requires ATP

17. Yeast cells growing aerobically synthesize sterols and incorporate them into membranes. However, under anaerobic conditions, yeast cells do not survive unless they are provided with an exogenous source of sterols. Explain the metabolic basis for this nutritional requirement.

The Complex Regulation of Cholesterol Biosynthesis Takes Place at Several Levels

18. The key step in cholesterol biosynthesis is the conversion of 3-hydroxy-3-methylglutaryl CoA to mevalonate. Which of the following are ways in which this reaction can be modulated?

(a) covalent modification of HMG CoA reductase through phosphorylation
(b) controlling the rate of translation of the mRNA encoding HMG CoA reductase
(c) controlling the rate of transcription of the gene encoding HMG CoA reductase
(d) proteolytic degradation of HMG CoA reductase
(e) deletion and duplication of the gene encoding HMG CoA reductase

19. Which of the following is correct about SCAP and SREBP regulation of cholesterol biosynthesis?

(a) SCAP and SREBP form a complex when cholesterol levels are high.
(b) Binding of cholesterol to SREBP leads to nuclear translocation, where SREBP binds to an SRE on the 5' side of the HMG CoA reductase gene.
(c) Proteases are used to release SREBP to the nucleus.
(d) The cytoplasmic domain of SREBP is a nuclear transcription factor for HMG CoA reductase.

20. Match the appropriate components or properties in the right column with the lipoproteins in the left column.

(a) chylomicron   (1) contains apoprotein B-100
(b) VLDL   (2) contains apoprotein B-48
(c) LDL   (3) contains apoprotein A
(d) HDL   (4) transports endogenous cholesterol esters
         (5) transports dietary triacylglycerols
         (6) transports endogenous triacylglycerols
         (7) is degraded by lipoprotein lipase
         (8) is taken up by cells via receptor-mediated mechanisms
         (9) is a precursor of LDL
        (10) may remove cholesterol from cells

21. Which of the following events occur in the LDL pathway in fibroblasts? Place them in their proper sequential order.

(a) breakdown of LDL in lysosomes
(b) endocytosis of LDL along with LDL receptors

(c) degradation of LDL receptors in lysosomes
(d) binding of LDL to LDL receptors
(e) return of LDL receptors to the plasma membrane

22. Exons in the gene for the LDL receptor give rise to structurally diverse domains. What is the likely function of the cysteine-rich amino-terminal domain?

(a) carbohydrate binding   (d) growth-factor binding
(b) LDL Binding   (e) clathrin binding
(c) Ca^+ binding   (f) structure stabilization

23. Explain how LDL regulates the cholesterol content in fibroblasts.

24. Assume that LDL is produced normally in a patient but that the apoprotein B-100 domain that recognizes the receptor is functionally defective, which prevents the binding of LDL to its receptor. What outcome would this defect have on cholesterol metabolism in peripheral cells?

Important Derivatives of Cholesterol Include Bile Salts and Steroid Hormones

25. The physiologic roles of bile salts include which of the following?

(a) They aid in the digestion of lipids.
(b) They aid in the digestion of proteins.
(c) They facilitate the absorption of sugars.
(d) They facilitate the absorption of lipids.
(e) They provide a means for excreting cholesterol.

26. Which of the following are common features in the structures of cholesterol and glycocholate?

(a) Both have three hydroxyl groups.   (d) Both contain a carboxylate group.
(b) Both contain four fused rings.   (e) Both contain double bonds.
(c) Both have a hydrocarbon side chain. (f) Both contain a sulfur atom.

27. Explain the structural characteristics of bile salts that make them effective biological detergents.

28. For the sterol structure in Figure 26.2, answer the questions that follow the figure.

FIGURE 26.2   A sterol.

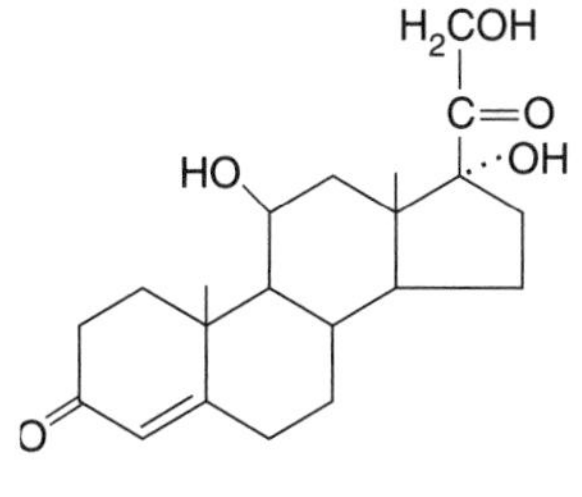

(a) Name the sterol.
(b) It is synthesized from what *via* three hydroxylation reactions?
(c) It has how many fewer carbon atoms than cholesterol?
(d) Its concentration will be diminished if there is a deficiency of 21-hydroxylase-true or false? Explain why.

29. Hydroxylation reactions involving cytochrome P450 have which of the following characteristics?

(a) They require a proton gradient.
(b) They involve electron transport from NADPH to O_2.

(c) They activate O_2 by binding it to adrenodoxin.

(d) They transfer one oxygen atom from O_2 to the substrate and form water from the other oxygen atom.

(e) They occur in adrenal mitochondria and liver microsomes.

30. Explain how foreign aromatic compounds are detoxified and excreted by mammals. Describe a possible deleterious effect of this process.

31. Match the steroid hormones in the left column with the characteristics in the right column that distinguish them from one another.

(a) aldosterone
(b) estrogen
(c) testosterone

(1) has 18 carbon atoms
(2) has 19 carbon atoms
(3) has 21 carbon atoms
(4) contains an aromatic ring
(5) contains an aldehyde group at C-18

32. Name the principle catabolic form of excreted cholesterol.

33. Which of the following statements about active vitamin D are NOT correct?

(a) It has the same fused ring system as cholesterol.
(b) It requires hydroxylation reactions for its synthesis from cholecalciferol.
(c) It is important in the control of calcium and phosphorus metabolism.
(d) It can be synthesized from cholesterol in the presence of UV light.
(e) It can be derived from the diet.

34. List the physiological consequences of a deficiency in Vitamin D.

ANSWERS TO SELF-TEST

1. b, d, e, f. Answers (a) and (c) are incorrect because neither neutral fats nor free fatty acids appear in membranes.

2. a, c. Reduction of dihydroxyacetone phosphate is the primary route.

3. (a) 1, 2, 3, 4   (b) 1, 2, 3, 4, 6. Phosphatidyl ethanolamine can also be formed through an exchange reaction of ethanolamine with phosphatidyl serine.

4. CDP-diacylglycerol and the CDP-alcohols are activated intermediates that allow the formation of phosphate ester bonds in phosphoglycerides, a process that is otherwise highly exergonic. ATP supplies the energy to form these compounds. Recall that UDP-sugars are used in a similar manner in the synthesis of carbohydrates (see pages 628–629 of the text).

5. Summing the individual reactions:

Choline + ATP $\rightarrow$ phosphorylcholine + ADP

Phosphorylcholine + CTP $\rightarrow$ CDP–choline + PP_i

CDP–choline + diacylglycerol $\rightarrow$ CMP + phosphatidyl choline

Choline + ATP + CTP + diacylglycerol $\rightarrow$ ADP + PP_i + CMP + phosphatidyl choline

Two high-energy bonds (from ATP and CTP) are directly consumed in these reactions. In addition, pyrophosphate is hydrolyzed by pyrophosphatase, driving the net reaction farther to the right. A total of three high-energy bonds would be consumed to regenerate ATP and CTP from ADP and CMP.

6. b

7. d

8. a, c, f, g

9. e. Phospholipases are specific for *ester* bonds; therefore, none will cleave the *ether* bond on the C-1 carbon of this plasmalogen phospholipid. Phospholipases A_2, C, and D cleave specific ester linkages: A_2 cleaves the R_2-containing chain to release the fatty acid, C cleaves the phophodiester bond to produce the choline (R_3) -phosphate, and D cleaves the phosphodiester bond to produce the R_3-alcohol. If the phospholipid had been a glycerol ester phospholipid, *e.g.*, phosphatidyl ethanolamine, phospholipase A_1 would have cleaved the ester bond to yield the R_1-containing fatty acid. Phospholipase C hydrolyzes phosphatidyl inositol 4, 5-bisphosphate to produce inositol 1, 4, 5,-trisphosphate and diacylglycerol, which are intracellular second messengers (Chapter 14).

10. b

11. (a) 1, 2   (b) 1, 4, 5   (c) 1, 3

12. The degradative enzymes for gangliosides are located in lysosomes; therefore, ganglioside G_{M2} will accumulate in the lysosomes of Tay-Sachs patients. The enzyme, a specific β-N-acetylhexosaminidase that removes the terminal sugar GalNAc from the ganglioside is deficient in these people.

13. a, b, and e.

14. All the compounds given are intermediates in the biosynthesis of cholesterol except for (e) cholyl CoA, which is a catabolic derivative of cholesterol and a precursor of bile salts. The proper sequence is d, c, a, f, b, g.

15. a, e

16. (a) 4, 5   (b) 1, 2   (c) 2, 3

17. A key intermediate in the biosynthesis of cholesterol and related sterols is squalene, an open-chain isoprenoid hydrocarbon. It is converted to squalene 2,3-epoxide, which in turn is converted to lanosterol. The conversion of squalene to the 2,3-epoxide is catalyzed by a monooxygenase, and molecular oxygen is a required component for this reaction. Under anaerobic conditions, yeast cells cannot synthesize sterols because they lack oxygen, a substrate for the monooxygenase reaction.

18. a, b, c, d

19. c, d

20. (a) 2, 5, 7   (b) 1, 6, 7, 9   (c) 1, 4, 8   (d) 3, 4, 10

21. All the events except (c) occur in the LDL pathway. The proper sequence is d, b, a, e.

22. c

23. The main source of cholesterol for cells outside the liver and intestine is from circulating LDL. Cholesterol released during the degradation of LDL suppresses the formation of new LDL receptors, thereby decreasing the uptake of exogenous cholesterol by the cell.

24. A defect in apoprotein B-100 that prevents the binding of LDL to the cell-surface receptor would result in the stimulation of the synthesis of endogenous cholesterol and LDL receptors and a decrease in the synthesis of cholesterol esters via the ACAT reaction. Indeed, the cellular and physiologic consequences of such a mutation may be similar to those seen in familial hypercholesterolemia.

25. a, d, e

26. b, e

27. Bile salts are effective detergents because they contain both polar and nonpolar regions. They have several hydroxyl groups, all on one side of the ring system, and a polar side chain that allow interactions with water. The ring system itself is nonpolar and can interact with lipids or other nonpolar substances. Bile salts are planar, amphipathic molecules, in contrast with such detergents as sodium dodecyl sulfate (see page 72 of the text), which are linear.

28. (a) cortisol
    (b) progesterone
    (c) six
    (d) true. A deficiency of 21-hydroxylase will impair hydroxylation at C-21 of progesterone, which will prevent the normal synthesis of cortisol and mineralocorticoids from progesterone.

29. b, d, e

30. In mammals, foreign aromatic molecules are hydroxylated by the cytochrome P450-dependent monooxygenases that are present in the endoplasmic reticulum of the liver cells. The hydroxylated derivatives are more water-soluble and have functional groups for the attachment of very polar substances, such as glucuronate, that allow them to be excreted in urine. The action of the cytochrome P450 system sometimes converts potential carcinogenic compounds into highly carcinogenic derivatives.

31. (a) 3, 5 (b) 1, 4 (c) 2

32. The water-soluble bile salt glycocholate is a major cholesterol breakdown product.

33. a

34. The most well-known consequence of a lack of vitamin D in childhood is the disease rickets, which results in inadequate calcification of cartilage and bone. In adults a deficiency can lead to bone softening and weaking called osteomalacia. New research suggests that a lack of vitamin D can impead muscle performance, play a role in cardiovascular disease, and increase risk of developing cancer and some autoimmune diseases.

PROBLEMS

1. Why is synthesis of cholesterol de novo dependent on the activity of ATP-citrate lyase?

2. An infant has an enlarged liver and spleen, cataracts, and anemia and exhibits general retardation of development. Mevalonate is found in the urine. Investigation reveals a deficiency of mevalonate kinase, which catalyzes the formation of 5-phosphomevalonate from mevalonate.
   (a) Why is urinary excretion of mevalonate consistent with a deficiency of mevalonate kinase?
   (b) How would a deficiency of mevalonate kinase affect cholesterol synthesis in this infant?
   (c) What level of HMG CoA reductase activity, relative to normal, would you expect to find in cells isolated from the infant? Briefly explain your answer.

3. Normally, most of the bile acids that are secreted into the intestine undergo reabsorption and are returned to the liver. Cholestyramine is a positively charged resin that binds bile acids in the intestinal lumen and prevents their reabsorption.
   (a) To examine the effects of cholestyramine on LDL metabolism, two fractions of LDL were prepared: one was covalently labeled on tyrosine residues with ^{125}I; the

other was labeled with ^{131}I and treated with cyclohexanedione, which interferes with LDL binding to the LDL receptor. When rabbits were given cholestyramine, hepatic uptake of ^{125}I-labeled LDL was enhanced relative to normal, whereas the uptake of ^{131}I-labeled LDL was unchanged relative to that in rabbits that had not been given cholestyramine. Briefly explain the relationship between the action of cholestyramine and LDL uptake in the liver.

(b) The administration of cholestyramine usually results in a 15 to 20% reduction in levels of circulating LDL, whereas the administration of a combination of cholestyramine and mevacor (lovastatin) can often yield a 30 to 40% reduction. Why?

4. The presence of apoprotein E in lipoproteins enables them to be taken up by hepatic cells. Provide a brief explanation for each of the following observations made of a person with a deficiency in apoprotein E synthesis.

(a) elevated levels of plasma triacylglycerols and cholesterol, coupled with the presence of chylomicron remnants and IDL. The latter particles persist in the bloodstream much longer than in normal people

(b) abnormally low levels of LDL in the blood

(c) abnormally high levels of LDL receptors in liver cells

(d) a marked reduction in levels of circulating chylomicron remnants and IDL when the diet is low in cholesterol and fat

5. Pregnant women often have increased rates of triacylglycerol breakdown and, as a result, have elevated levels of ketone bodies in their blood. Why do they also often exhibit an increase in plasma lipoprotein levels?

6. Hopanoids are pentacyclic molecules that are found in bacteria and in some plants. As an example, a typical bacterial hopinoid is shown below. Organisms that make hopanoids use a pathway similar to that for cholesterol synthesis. The biosynthetic pathway for hopane includes the formation of squalene, followed by more steps to form the final product itself, a C30 compound. Hopane is similar to lanosterol (page 770 of the text), but it lacks the hydroxyl group.

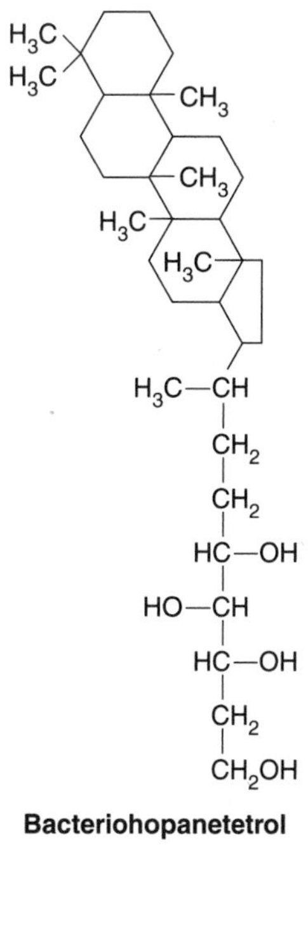

Bacteriohopanetetrol

(a) How many molecules of mevalonic acid are required for the synthesis of hopane?

(b) Squalene can undergo concerted cyclization to form hopane in a reaction that is catalyzed by a unique type of squalene cyclase. The reaction is initiated by a proton and does not require oxygen. Compare this step with the formation of lanosterol from squalene. Why could it be argued that the synthesis of hopanoids preceded the synthesis of sterols in evolution?

7. Your colleague has discovered a compound that is a very powerful inhibitor of HMG CoA reductase, and she has evidence that the drug will completely block the synthesis of mevalonate in liver. Why is this compound unlikely to be useful as a drug?

8. The liver is the site of the synthesis of plasma phospholipids and lipoproteins. Rats maintained on a diet deficient in choline often develop fat deposits in liver tissue. How could choline deficiency be related to this aberration in lipid metabolism?

9. Among the sugar residues found in a blood group ganglioside is fucose. Experiments utilizing isolated Golgi membranes and ribonucleoside triphosphates show that fucose can be incorporated into the ganglioside only when GTP is available. What is the role of GTP in fucose incorporation?

10. Suppose a cell is deficient in phosphatidate phosphatase, which catalyzes the formation of diacylglycerol from phosphatidate. What effects on lipid metabolism would you expect?

11. Desmolase is involved in the synthesis of pregnenol (see page 783 of the text).

(a) Would you expect virilization among patients who have desmolase deficiency?

(b) Why is enlargement of the adrenal glands common among such patients?

12. In the adult form of Gaucher's disease, glucosylcerebrosides accumulate in liver, spleen, and bone-marrow cells. Although the common galactosylceramides and their derivatives are found in the tissues of affected patients, accumulations of galactosylcerebrosides or their metabolites are not found, nor do ceramides accumulate. What enzyme activity is probably deficient in patients with Gaucher's disease?

13. Cells of the adrenal cortex have very high concentrations of LDL receptors. Why?

14. At low concentrations of phospholipid substrates in water, the reaction catalyzed by a phospholipase occurs at a rather low rate. The reaction rate accelerates when the concentrations of the phospholipid substrates increase to the point that micelles are formed. How is this property of phospholipases related to their activity in the cell?

15. Glucagon has been shown to reduce the activity of HMG CoA reductase. Why is this observation consistent with the overall effect of glucagon on cellular metabolism?

16. People who have elevated levels of low-density lipoproteins (LDL) in their serum can be treated in a number of ways. These include restriction of dietary intake of cholesterol, ingestion of positively charged resin polymers that inhibit intestinal reabsorption of bile salts, and administration of lovastatin, a competitive inhibitor of 3-hydroxy-3-methylglutaryl CoA reductase.

(a) Briefly explain how each of these treatments reduces serum LDL levels.

(b) Why should none of these treatments be used for patients who are homozygous for a defect in LDL receptors?

17. Currently there are two established methods for dealing with hypercholesterolemia in patients homozygous for LDL receptor deficiency.

(a) The first is plasma apheresis, whereby the plasma and blood cells of a patient are separated in a continuous flow device, and the plasma is passed over a column that removes lipoproteins containing apoprotein B-100. Which lipoproteins are removed by this procedure, and how could their removal lower plasma cholesterol levels?

(b) A more extreme method of dealing with extreme hypercholesterolemia in FH homozygotes is liver transplantation. The rationale for this procedure is based on the observation that over 70% of total body LDL receptors are in the liver. In the small group of patients who have undergone liver transplants, LDL cholesterol levels are substantially reduced. In one particular case, the rate of LDL turnover increased about threefold, and one patient became responsive to lovastatin after the transplant. Why did increased LDL turnover and lovastatin response indicate that the transplantation procedure was successful?

18. Provide a physiologic rationale for each of the following responses to an increase in the rate of transport of unesterified cholesterol into a mammalian cell:

(a) stimulation of the synthesis of cholesteryl oleate esters
(b) suppression of the activity of HMG CoA reductase
(c) suppression of the synthesis of LDL receptors

19. Glycerol phosphate acyltransferase can convert 3,4-dihydroxybutyl-1-phosphonate, an analog of glycerol 3-phosphate, to diacylbutyl-1-phosphonate, which is an analog of phosphatidate. Would this analog be more likely to inhibit triglyceride synthesis or CDP-diacylglycerol synthesis? Briefly explain your answer.

$$H_2C-OH$$
$$HO-C-H \quad O$$
$$CH_2-CH_2-P-O^-$$
$$O^-$$

3, 4-Dihydroxybutyl-1-phosphonate

20. During the uptake of low-density lipoprotein (LDL) by a liver cell, LDL-receptor protein complexes are internalized by endocytosis. The endosomes then fuse with lysosomes, where protein components of LDL are hydrolyzed to free amino acids, while cholesterol esters are hydrolyzed by a lysosomal acid lipase. The LDL receptor itself is not affected by lysosomal enzymes.

(a) Briefly describe what would happen to cholesterol metabolism in a cell deficient in lysosomal acid lipase.
(b) Why is it important that LDL receptors are not degraded by lysosomal enzymes?

21. Glycerol kinase catalyzes the conversion of free glycerol into glycerol 3-phosphate, using ATP as a phosphoryl donor. Although liver tissue has high levels of the enzyme, the activity of glycerol kinase in adipose tissue is low. How do these differences contribute to the balance between carbohydrate and triglyceride metabolism in mammals?

22. Niemann-Pick disease is an inherited disorder of sphingomyelin breakdown due to a deficiency of sphingomyelinase. This enzyme, found in all tissues and easily assayed in white blood cells collected from a patient, converts sphingomyelin to ceramide and phosphocholine. Phosphocholine is highly soluble in water, while sphingomyelin is more soluble in chloroform. Assuming that you can obtain sphingomyelin labeled with 14C in any desired carbon atoms, design an assay that would allow you to confirm a diagnosis of Niemann-Pick disease.

23. Chronic alcoholics derive a significant fraction of their calories from the metabolism of ethanol in the liver, the first step of which is its reaction with NAD^+ to form acetaldehyde and $NADH + H^+$—a reaction catalyzed by liver alcohol dehydrogenase. The acetaldehyde is itself oxidized to acetate in a reaction by an NAD^+-dependent acetaldehyde

dehydrogenase that produces more NADH. Thus, excessive metabolism of ethanol leads to a marked increase in the NADH/ NAD$^+$ ratio. Relate this fact to the development of "fatty" liver (droplets of triacylglycerol are deposited in the liver) in alcoholics.

ANSWERS TO PROBLEMS

1. ATP-Citrate lyase catalyzes the formation of acetyl CoA in the cytosol (see page 709 of the text). Acetyl CoA is used for the synthesis of HMG CoA, which gives rise to mevalonate for the synthesis of cholesterol in the cytosol.

2. (a) A deficiency of mevalonate kinase activity means that mevalonate cannot be used as a precursor of 5-phosphomevalonate. If no other pathways can use mevalonate, its concentration in the liver will increase until it spills into the blood and then, in turn, into the urine. Furthermore, because the activity of HMG CoA reductase is increased in this infant (see answer c), the rate of mevalonate synthesis will be stimulated.

   (b) You would expect the rate of cholesterol synthesis to be depressed because the pathway is blocked at the step in which 5-phosphomevalonate is formed.

   (c) You would expect to find a higher-than-normal level of HMG CoA reductase activity. A depressed rate of cholesterol synthesis lowers the amount of cholesterol in the cell. HMG CoA reductase activity increases because cholesterol reduces both its synthesis and its activity.

3. (a) The experiments show that rabbits given cholestyramine have higher rates of removal of LDL from the blood and that hepatic uptake of LDL depends on the ability of the lipoprotein to bind to the LDL receptor. One explanation for this is that cholestyramine interferes with the return of bile acids to the liver, stimulating the synthesis of more bile acids from cholesterol. An increased demand for cholesterol stimulates the synthesis of LDL receptors, which take up more cholesterol-containing LDL particles from the blood.

   (b) Cholestyramine stimulates the hepatic uptake of cholesterol-containing LDL, but it has no direct effect on cholesterol synthesis de novo in the liver. Mevacor inhibits HMG CoA reductase, thereby depressing the rate of cholesterol biosynthesis. The subsequent requirement for cholesterol leads to a further increase in the number of LDL receptors, which in turn can take up more LDL from the circulation.

4. (a) Chylomicrons, chylomicron remnants, and IDL normally contain apoprotein E. A deficiency of the apoprotein means that hepatic uptake of chylomicron remnants and IDL is impaired, so these particles persist in the circulation. Because both these types of particles contain triacylglycerols and cholesterol, circulating levels of these compounds are also elevated.

   (b) Both chylomicron remnants and IDL particles serve as precursors of VLDL in the liver. When the uptake of the VLDL precursors by hepatic tissue is impaired by an apoprotein E deficiency, the rate of synthesis and export of VLDL particles is reduced. Since VLDL are LDL precursors in circulation, LDL are reduced.

   (c) As discussed in answer (b), VLDL synthesis in the liver is impaired. Additional LDL receptors are synthesized because their synthesis is no longer repressed by VLDL-derived cholesterol.

   (d) A diet low in cholesterol and fat will reduce the rate of formation of chylomicrons, which are precursors of chylomicron remnants.

5. Increased levels of ketone bodies, such as acetoacetate, imply that the levels of acetyl CoA and HMG CoA, both precursors of cholesterol, are also elevated. Cholesterol synthesis is stimulated by an increase in the availability of these substrates. The subsequent decreased demand for dietary cholesterol results in an elevation in cholesterol-containing lipoproteins.

6. (a) Mevalonic acid, a six-carbon compound, is a precursor of isopentenyl pyrophosphate (IPP), which contains five carbon atoms. IPP serves as the basic unit for the formation of squalene, a 30-carbon compound. Six molecules of IPP are needed for the synthesis of a molecule of squalene, which is in turn the precursor of hopane. Thus, six molecules of mevalonic acid are required.

   (b) Aerobic processes, such as the synthesis of sterols, probably evolved later than anaerobic processes and only after free oxygen became available. Thus, the synthesis of hopane from squalene, an anaerobic process, probably preceded the synthesis of sterols, such as lanosterol, from squalene.

7. The synthesis of mevalonate is required not only for the synthesis of cholesterol but also for the synthesis of a number of other important compounds derived from isopentenyl pyrophosphate, including ubiquinone (CoQ), an important component of the electron transport chain. Therefore, the complete blockage of mevalonate synthesis, even if adequate cholesterol is available in the diet, would be ill-advised.

8. Choline, which is ordinarily supplied by the diet and is synthesized only to a limited extent in mammals, is a constituent of phosphatidyl choline, an important component of membranes and lipoproteins. A deficiency in dietary choline, which could not be completely replaced by the endogenous methylation of phospatidyl ethanolamine, could interfere with the synthesis and export of lipoproteins like VLDL, which is a carrier of triacylglycerols to peripheral tissues. Failure to export fats such as triacylglycerols leads to their accumulation in the liver.

9. Nucleotide sugars, such as UDP-glucose, serve as donors during the incorporation of sugar residues into gangliosides. In this case, it appears that the donor of fucose residues is GDP-fucose, which is synthesized from fucose and GTP.

10. You would expect to see reduced rates of synthesis of triacylglycerols, which use diacylglycerols as acceptors of activated acyl groups. In addition, phosphatidyl choline synthesis is dependent on the availability of diacylglycerols as acceptors of choline phosphate from CDP-choline.

11. (a) You would not expect desmolase deficiency to lead to virilization. An increase in androgen production, which causes virilization in both males and females, is due to elevated levels of 17 α-hydroxyprogesterone. The pathway from cholesterol to 17 α-hydroxyprogesterone includes a step catalyzed by desmolase, which cleaves the bond between C-20 and C-22 in 20α, 22β dihydroxycholesterol to form pregnenolone (see page 783 of the text). A deficiency of desmolase would therefore decrease the rate of androgen synthesis.

    (b) Pregnenolone is a precursor of the glucocorticoids, which exert feedback control on the activity of the adrenal cortex. Desmolase deficiency leads to diminished production of glucocorticoids. Failure of the normal feedback mechanism leads to increased ACTH production and to enlargement of the adrenal glands.

12. Because glucosylcerebrosides accumulate but galactosylcerebrosides do not, you would suspect that the defect involves ganglioside breakdown rather than ganglioside synthesis. The defect involves the step that removes glucose from the cerebroside to yield free ceramide, or N-acyl sphingosine. The enzyme that carries out this step is a glycosyl hydrolase; it is also called β-glucosidase.

13. Cells of the adrenal cortex utilize cholesterol for the synthesis of a number of steroid hormones, including cortisol. Although these cells can themselves synthesize cholesterol, it is often also necessary for additional cholesterol to be obtained from plasma lipoproteins. A high concentration of LDL receptors enables cortical cells to take up LDL, which contains cholesterol, rapidly.

14. In the cell, a phospholipase would most often encounter substrates that are part of an aggregate, such as those phospholipids found in membranes. Thus, the enzyme should be expected to function at a higher rate with aggregates or assemblies of lipid molecules because their local concentrations would be higher than if they were individually free in solution.

15. The presence of glucagon is a signal that carbohydrate and triacylglycerol catabolism is needed to generate energy in the organism. Under such conditions, one would expect biosynthetic reactions to be suppressed because energy charge is low. Low energy charge means high AMP levels that would activate an AMP-dependent protein kinase leading to the phosphorylation of HMG CoA reductase.

16. (a) Restricting the level of dietary cholesterol lowers the input of exogenous cholesterol into lipoproteins, so fewer LDL molecules are present in serum. Compounds that inhibit bile salt reabsorption from the gut stimulate additional synthesis of bile acids from cholesterol in the liver, decreasing concentrations of the sterol in liver cells and, by causing an increase in the number of receptors on the cell surface, stimulating LDL uptake from the circulation. Finally, mevacor reduces the rate of cholesterol biosynthesis de novo, which can also stimulate uptake of LDL from the bloodstream. Any of these treatments could help in reducing circulating levels of cholesterol as LDL.

    (b) Homozygotes have virtually no functional LDL receptors. Such people are unable to internalize significant amounts of LDL, which means that circulating levels of that lipoprotein are elevated in the blood. In addition, an absence of LDL receptors means that endogenous cholesterol fails to enter the liver cell to suppress de novo synthesis. Dietary restriction could reduce exogenous LDL levels somewhat but would not prevent formation of LDL and other lipoproteins arising from cell turnover of cholesterol. Sequestrants of bile salts could cause some depletion of liver cell cholesterol but again would not stimulate LDL uptake from the circulation. Mevacor will suppress cholesterol synthesis de novo, but this suppression would again not be compensated for by uptake from the circulation. Homozygotes with two nonfunctional receptor genes are therefore resistant to compounds that inhibit LDL synthesis and stimulate uptake. Thus, none of these measures has a great effect on reducing circulating LDL levels in these patients.

17. (a) Protein B-100 is found in VLDL, IDL, and LDL, and all three lipoproteins are removed from the plasma. Each contains cholesterol, and, in addition, VLDL and IDL are regarded as precursors of LDL, so that total cholesterol concentration in the blood would be lowered by apheresis. This method can reduce LDL cholesterol levels by 70%. Treatment must be repeated about every two weeks, and the effects of long-term apheresis are problematic.

    (b) Increased LDL turnover indicates that the transplanted liver is producing normal LDL receptors that can take up LDL from the circulation and can facilitate its conversion to other lipoproteins. A response to lovastatin also indicates that functional LDL receptors are available to accelerate uptake in response to diminished de novo cholesterol synthesis, which is inhibited by lovastatin. It should be noted that liver transplant operations are hazardous, especially in FH homozygotes who usually have advanced atherosclerosis. The capability of normal liver cells to contribute functional LDL receptors has stimulated interest in gene therapy designed to target a normal LDL gene to liver cells of FH homozygotes.

18. (a) Formation of cholesteryl esters provides the cell with a means of storing cholesterol until it is needed for membrane biosynthesis or other purposes.

(b) Suppression of the activity of HMG CoA reductase leads to a decrease in cholesterol synthesis de novo, which occurs when the cell has sufficient endogenous cholesterol to not need to synthesize the steroid on its own.

(c) Suppression of LDL receptor synthesis leads to a gradual reduction in the number of LDL receptors in the cell because the receptors undergo a relatively constant rate of degradation. Reduction in LDL receptor levels means that fewer LDL particles will be able to enter the cell, resulting in a decrease in the entry of endogenous cholesterol.

19. The conversion of phosphatidate to a triacylglycerol is initiated by hydrolysis of the phosphate group and the formation of diacylglycerol. The C—P bond in the phosphonate cannot be cleaved by the phosphatase (also known as *phosphatidate phosphohydrolase*), so it is likely that triglyceride synthesis could be impaired by the analog. On the other hand, formation of CDP-diacylglycerol involves the formation of an anhydride bond between the phosphates of phosphatidic acid and cytidylic acid (see page 761 of the text). Since the phosphonate is not cleaved in this reaction, formation of the phosphonyl analog of CDP-diacylglycerol seems possible even if at a much slower rate, or at least the synthesis of normal substrates would not be impaired. Phosphonolipids are known in trace amounts in mammals but are found more extensively in some invertebrates, including the protozoan *Tetrahymena*, where they may represent up to 25% of total phospholipid (see D. E. Vance and J. Vance [eds.], *Biochemistry of Lipids, Lipoproteins and Membranes*. Elsevier, 1991, p. 205).

20. (a) Both the LDL receptor gene and the gene for HMG CoA reductase contain sterol regulatory elements that are responsive to free cholesterol. A reduction in free cholesterol release from the lysosome leads to an increase in LDL receptor production and to increased HMG CoA reductase activity. Both these consequences lead to an increase in cholesterol concentrations in the cell through an increased rate of LDL entry and accelerated cholesterol synthesis. Lysosomal accumulation of cholesteryl esters and triglycerides can eventually destroy the cell. One form of this disorder, termed *Wolman disease*, is characterized by liver enlargement, digestive difficulties, and enlargement and deterioration of the adrenal glands. The disease is usually fatal within a year after birth.

(b) LDL receptors, after their release from lysosomes, return to the cell surface, where they take up other LDL particles and bring them back to lysosomes. A round trip takes about 10 minutes. Destruction of LDL receptors by lysosomal enzymes would make it necessary to synthesize the 115-kd glycoprotein at a much faster and energetically wasteful rate.

21. Free glycerol in mammals is produced in adipocytes when triglycerides are converted to free fatty acids and glycerol by hormone-sensitive lipases. Hormone-responsive lipolysis occurs when glucose levels are low and fatty acids are needed as fuels. Under these conditions, the low activity of glycerol kinase in adipocytes prevents unnecessary resynthesis of triacylglycerols from free fatty acids and glycerol 3-phosphate via phosphatidic acid. Triacylglycerol synthesis is more likely to occur when glucose levels are high. Adipocytes can then synthesize glycerol 3-phosphate from dihydroxyacetone

phosphate produced from glucose during glycolysis. Thus, adipocytes are unable to synthesize triglycerides unless glucose or another suitable carbon source is available.

22. Extracts of white cells from the blood of a person suspected of having a sphingo-myelinase deficiency are incubated in a buffered solution with radioactive sphingomyelin labeled with ^{14}C in the methyl groups of the phosphocholine moiety. Then the incubation mixture is extracted with chloroform. Any radioactive phosphocholine liberated by the enzyme will remain in the upper aqueous layer while intact radioactive sphingomyelin will be extracted into the lower chloroform layer. Using white cells from patients with Neimann-Pick disease, incubation with cell extracts followed by chloroform extraction results in little or no radioactivity in the aqueous phase, confirming the deficiency of sphingomyelinase activity.

23. As stated, the metabolism of ethanol leads to a more reduced state in the hepatocyte; that is, the ratio of NADH to NAD^+ is increased. Increased NADH levels lead to inhibition of fatty acid oxidation (NADH is a product of fatty acid oxidation) and signal the capacity for biosynthesis requiring reducing equivalents. Decreased fatty acid oxidation leads to the accumulation of triacylglycerols. In addition, the increased levels of acetate and NADH promote the synthesis of fatty acids and subsequently triacylglycerol to exacerbate the problem. The excess triacylglycerols coalesce into droplets in the fatty liver.

The Integration of Metabolism

Chapter 27

This chapter, which concludes the two major sections of the text devoted to metabolism, provides an integrated view of mammalian metabolism and a review of the principal themes of metabolism. The chapter starts with an examination of caloric homeostasis, or the ability to maintain adequate but not excessive energy stores. The authors discuss the current obesity epidemic in the developed world, which is a consequence of failure to maintain caloric homeostasis, and the pathological consequences of obesity. The authors then turn to the role that the brain plays in caloric homeostasis. They introduce the leptin and reintroduce insulin, two signal molecules that act on the brain to control appetite. This section ends with a discussion of defects in leptin signaling that may lead to obesity. The authors next turn to the most common metabolic disease in the world, diabetes. They distinguish type 1 and type 2 diabetes, then discuss the mechanisms by which each develop. They focus on type 2 diabetes, for which obesity is a significant predisposing factor. The authors then discuss the biochemical benefits of exercise. The fuel choices that the body makes during exercise and how those choices differ between aerobic and anaerobic activity complete this section. The ways in which the body responds to a series of physiological conditions, such as the well-fed state, the early fasting state, and the refed state are the next topic in the chapter. The metabolic consequences of prolonged starvation are discussed, with attention to the priorities of metabolism in starvation, such as maintaining a blood-glucose level above 2.2 mM, and preserving protein by shifting the fuel being used from glucose to fatty acids and ketone bodies. The chapter concludes with the ways in which ethanol alters energy metabolism in the liver and the adverse consequences of excess consumption.

LEARNING OBJECTIVES

Introduction

1. Define *caloric homeostasis.*

2. Outline how *leptin, glucose, insulin,* and food intake modulate metabolic pathways in the following organs: brain, intestine, muscle, pancreas, adipose tissue, and liver.

3. Recall the function of insulin and *glucagon* in energy regulation.

Caloric Homeostasis Is a Means of Regulating Body Weight (Text Section 27.1)

4. Relate the Second Law of Thermodynamics to energy consumption and expenditure. Explain the physiological and health consequences of the Second Law of Thermodynamics.

5. Give two explanations for the current obesity epidemic.

6. Calculate the calories (and amount of food ingested) required to maintain a particular body weight over time.

The Brain Plays a Key Role in Caloric Homeostasis (Text Section 27.2)

7. Explain how *cholecystokinin* and *glucagon-like peptide 1* induce feelings of satiety in the brain. Include coordination between the intestine, pancreas, and the brain.

8. Explain how *ghrelin* is thought to act as an appetite-enhancing peptide.

9. Distinguish between short-term and long-term regulation of caloric homeostasis.

10. State the organ from which leptin and insulin are secreted and then contrast their functions in long-term caloric homeostasis.

11. Explain the role of the signaling molecule leptin in maintaining caloric homeostasis and appetite control. Recall the regulatory enzyme *AMPK* (Chapter 22), through which leptin and *adiponectin* exert their effects. Describe the experiment that led to the discovery of leptin.

12. Describe how leptin resistance may be a contributing factor to obesity.

Diabetes Is a Common Metabolic Disease Often Resulting from Obesity
(Text Section 27.3)

13. Explain the characteristic of *diabetes mellitus* and distinguish between type 1 and type 2 diabetes.

14. Outline the insulin signaling pathway in normal cells. List the three steps involved in turning the pathway off.

15. List the cluster of pathologies that characterize *metabolic syndrome.*

16. Describe the sequence of cellular events in peripheral tissues that lead to diet-induced insulin resistance.

17. Outline the pathway that leads to the release of insulin in normal pancreatic cells. Include the role of ATP in regulating the pathway. Explain how over-nutrition leads to death of pancreatic cells and therefore type 2 diabetes.

18. Describe the metabolic derangements in diabetes mellitus resulting from relative insulin insufficiency and glucagon excess.

Exercise Beneficially Alters the Biochemistry of Cells (Text Section 27.4)

19. Explain the molecular mechanisms of how exercise increases mitochondrial biogenesis and fat metabolism.

20. Discuss the different patterns of fuel use in short- and long-distance running.

21. Explain the use of fuels by resting and active muscle and list the approximate fuel reserves of muscles in terms of mmol of ~P available.

Food Intake and Starvation Induce Metabolic Changes (Text Section 27.5)

22. Describe how the blood-glucose level is controlled by the liver in response to glucagon and insulin in *well-fed*, *fasting*, and *refed* states. Discuss the contributions of muscle and adipose tissue to the regulation of the blood-glucose level.

23. Compare the approximate fuel reserves of blood, liver, brain, muscle, and adipose tissue in kilojoules and kilocalories.

24. Describe the metabolic changes that occur after one and three days of *starvation*. Discuss the metabolic adaptations that occur after prolonged starvation; note especially the shift in brain fuels and the decreased rate of protein degradation.

Ethanol Alters Energy Metabolism in the Liver (Text Section 27.6)

25. Describe the two pathways by which ethanol is metabolized and note the deleterious effects of large levels of ethanol metabolism.

26. Discuss how excess ethanol consumption disrupts vitamin metabolism. Compare the cause and symptoms of *Wernicke-Korsakoff syndrome* with that of *beriberi*.

27. Explain the biochemical basis for scurvy and recognize why it can occur in alcoholics.

SELF-TEST

Introduction

1. The metabolism of which of the following tissues is directly regulated by insulin?
    (a) Adipose tissue
    (b) Brain
    (c) Intestine
    (d) Liver
    (e) Muscle

Caloric Homeostasis Is a Means of Regulating Body Weight

2. List some of the health consequences of being overweight or obese.

3. Approximately how many additional calories would a woman who requires 2000 kcal per day to maintain her weight have to eat per day to gain 10 pounds over a period of 5 years?
    (a) 11
    (b) 4050
    (c) 111
    (d) 41

The Brain Plays a Key Role in Caloric Homeostasis

4. For each of the following metabolic processes, indicate the hormone and organ that controls it, and whether the process is increased or decreased by binding the hormone.
   (a) Insulin secretion
   (b) Food intake
   (c) Insulin biosynthesis
   (d) Satiety
   (e) Body weight
   (f) β-cell proliferation
   (g) β-cell survival

5. Which of the following is correct regarding the effects of leptin in the brain?
   (a) Fasting results in AgRP inhibition of MSH activity, which activates appetite-suppressing neurons.
   (b) Fasting results in a decrease in leptin levels, which facilitates release of NPY and AgRP.
   (c) Leptin prevents release of NPY and AgRP, supressing the desire to eat.
   (d) Leptin proteolyses POMC into MSH, a hormone that activates appetite-suppressing neurons.

6. Which of the following is incorrect about the regulation of fatty acid metabolism by AMP-dependent protein kinase (AMPK)?
   (a) AMPK phosphorylates acetyl CoA carboxylase to inactivate it and stop fatty acid synthesis.
   (b) Insulin stimulates fatty acid synthesis by activating acetyl CoA carboxylase.
   (c) Leptin leads to the activation of AMPK and acetyl CoA carboxylase, thereby increasing fatty acid oxidation.
   (d) Leptin and adiponectin act through AMPK, whereas insulin and glucagon act directly on acetyl CoA carboxylase.

7. Explain how suppressors of cytokine signaling (SOCS) modulate the action of insulin and leptin.

Diabetes Is a Common Metabolic Disease Often Resulting from Obesity

8. Use an "S" to indicate the following metabolic processes that are stimulated by and an "I" to indicate those that are inhibited by the action of insulin.
   (a) gluconeogenesis in liver
   (b) entry of glucose into muscle and adipose cells
   (c) glycolysis in the liver
   (d) intracellular protein degradation
   (e) glycogen synthesis in liver and muscle
   (f) uptake of branched-chain amino acids by muscle
   (g) synthesis of triacylglycerols in adipose tissue

9. For each of the following, indicate whether they are part of the activation or deactivation of the insulin pathway.
   (a) Akt is phosphorylated by PDK.
   (b) Insulin receptor is autophosphorylated.
   (c) IRS-1 is phosphorylated on serine residues.
   (d) IRS-1 is phosphorylated on tyrosine residues.
   (e) PIP2 is phosphorylated by PI3K.

(f) PIP3 is dephosphorylated by PTEN.

(g) Phosphatases dephosphorylate the insulin receptor.

10. Which of the following is a reason for the development of insulin insensitivity?

(a) Inability of muscle cells to process large amount of fats leads to increased export of fats into the bloodstream.

(b) Increase in DAG leads to dephosphorylation of IRS-1 and decreased insulin signaling.

(c) Increase in ceramide activates stress-induced pathways that interfere with insulin signaling.

(d) Increase in DAG and ceramide activate stress-induced kinases, which open the GLUT4 glucose transporter.

11. Which of the following occurs in the unfolded protein response (UPR) that leads directly to development of type 2 diabetes?

(a) General protein synthesis is activated to produce more proinsulin.

(b) Chaperone synthesis is stimulated.

(c) Misfolded proteins are removed from the ER and are subsequently delivered to the proteasome for destruction.

(d) Apoptosis is triggered, leading to cell death.

12. Why is diabetic ketosis common in type 1 diabetes, but not type 2?

Exercise Beneficially Alters the Biochemistry of Cells

13. Place the following in order of which they occur during exercise.

(a) Calcium is released from the sarcoplasmic reticulum.

(b) Efficiency of fatty acid metabolism increases.

(c) Metabolic reprogramming takes place, which leads to mitochondrial biogenesis.

(d) There's an increase in synthesis of proteins required for fatty acid metabolism.

(e) Transcription factors are activated by calcium-dependent kinases and AMPK.

14. List the following metabolic pathways or sources in the order of decreasing ATP production rate during strenuous exercise.

(a) muscle glycogen to CO_2

(b) liver glycogen to CO_2

(c) muscle glycogen to lactate

(d) adipose tissue fatty acids to CO_2

(e) muscle creatine phosphate

15. From the energy sources given in Question 14, select the ones that provide most of the energy in:

(1) a 100-meter race.

(2) a 1000-meter race.

(3) a marathon race.

Food Intake and Starvation Induce Metabolic Changes

16. Which of the following statements about metabolism in the brain are INCORRECT?

(a) It uses fatty acids as fuel in the fasting state.

(b) It uses only glucose as a fuel source in the resting state.

(c) It lacks fuel reserves.

(d) It can use acetoacetate and 3-hydroxybutyrate under starvation conditions.

(e) It releases lactate during periods of intense activity.

17. When fuels are abundant, the liver does not degrade fatty acids; rather, it converts them into triacylglycerols for export as very low-density lipoproteins (VLDL). Explain how β-oxidation of fatty acids and the formation of ketone bodies from fatty acids are prevented under these conditions.

18. Explain the allosteric effects of glucose on glycogen metabolism.

19. The blood-glucose level of a normal person, measured after an overnight fast, is approximately 80 mg/100 ml. After a meal rich in carbohydrate, it rises to about 120 mg/100 ml and then declines to the fasting level. The approximate time course of these changes and the inflection points is shown in Figure 27.1. After examining the figure, complete the following sentences:

   (a) The increase in the glucose level from A to B is due to
   (b) The decrease in the glucose level from B to C is due to
   (c) The leveling off of the glucose level from C to D is due to
   (d) The slight overshoot that is sometimes observed at C can be explained by

FIGURE 27.1   Blood-glucose levels after a meal rich in carbohydrate.

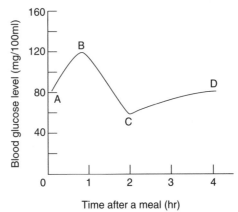

20. Match the fuel storage forms in the left column with the most appropriate characteristics from the right column.

   (a) glycogen
   (b) triacylglycerols
   (c) protein

   (1) largest storage form of calories
   (2) most readily available fuel during muscular activity
   (3) major source of precursors for glucose synthesis during starvation
   (4) depleted most rapidly during starvation
   (5) not normally used as a storage form of fuel

21. Relative to the well-fed state, fuel utilization after three days of starvation shifts in which of the following ways?
   (a) More glucose is consumed by the brain.
   (b) Adipose tissue triacylglycerols are degraded to provide fatty acids to most tissues.
   (c) The brain begins to use ketone bodies as fuels.
   (d) Proteins are degraded in order to provide three-carbon precursors of glucose.
   (e) Glycogen is stored as a reserve fuel.

22. Metabolic adaptations to prolonged starvation include which of the following changes relative to the metabolic picture after three days of starvation?

    (a) The rate of lipolysis (mobilization of triacylglycerols) in the adipose tissue increases.
    (b) The glucose output by the liver decreases.
    (c) The ketone body output by the liver decreases.
    (d) The utilization of glucose by the brain decreases as the utilization of ketone bodies increases.
    (e) The rate of degradation of muscle protein decreases.

23. Show the changes in blood-glucose levels you would expect to see for an insulin-dependent diabetic patient after a meal rich in carbohydrate by plotting their time course on Figure 27.1 (see Question 19). Explain your answer.

24. Which of the following occur in people with untreated diabetes?

    (a) Fatty acids become the main fuel for most tissues.
    (b) Glycolysis is stimulated and gluconeogenesis is inhibited in the liver.
    (c) Ketone body formation is stimulated.
    (d) Excess glucose is stored as glycogen.
    (e) Triacylglycerol breakdown is stimulated.

25. Is it true or false that in diabetes the brain shifts to ketone bodies as its major fuel? Explain.

26. Select the statements from the right column that best describe the metabolism of each organ, tissue, or cell in the left column.

    (a) brain

    (b) muscle

    (c) adipose tissue

    (d) liver

    (1) releases glycerol and fatty acids into the blood during fasting periods
    (2) in a normal nutritional state, utilizes glucose as the exclusive fuel
    (3) synthesizes ketone bodies when the supply of acetyl CoA is high
    (4) can release lactate into the blood
    (5) utilizes α-keto acids from amino acid degradation as an important fuel
    (6) can store glycogen but cannot release glucose into the blood
    (7) can synthesize fatty acids, triacylglycerols, and VLDL when fuels are abundant

Ethanol Alters Energy Metabolism in the Liver

27. Which of the following are consequences of ethanol consumption?

    (a) accumulation of NADH
    (b) accumulation of NADPH
    (c) generation of acetaldehyde
    (d) generation of lactate
    (e) metabolism of triacylglycerols in the liver
    (f) regeneration of glutathione

28. Which of the following is an enzyme involved in ethanol metabolism?
    (a) alcohol dehydrogenase
    (b) aldehyde dehydrogenase
    (c) α-ketoglutarate dehydrogenase
    (d) cytochrome P450
    (e) isocitrate dehydrogenase

ANSWERS TO SELF-TEST

1. a, d, e

2. Coronary heart disease; type 2 diabetes; cancers of the endometrial tissues, breast and colon; hypertension; dyslipidemia; stroke; liver and gallbladder disease; sleep apnea; respiratory problems; osteroarthritis; gynecological problems.

3. (a) First we convert pounds to kilograms: 10 pounds = 4.5 kg. If we assume that all of the weight was gained through the consumption of fatty acids derived from lipids as in the text example, then the amount of calories needed to gain the 10 pounds = (4500 g weight) * (9 kcal g1^{-1}) = 40,500 kcal. To calculate the amount needed per day we divide by 3650 days, which equals 40,500 kcal/(3650 days) = 11.1 kcal.

4. (a) GLP-1, pancreas, increased
   (b) CCK and GLP-1, brain, decreased
   (c) GLP-1, pancreas, increased
   (d) CCK and GLP-1, brain, increased
   (e) CCK and GLP-1, brain, decreased
   (f) GLP-1, pancreas, increased
   (g) GLP-1, pancreas, increased

5. b,c. Answer (a) is incorrect because although AgRP does inhibit MSH activity under fasting conditions, that keeps appetite-suppressing neurons from being activated. Answer (d) is incorrect because leptin is a hormone, not a protease.

6. Only answer c is incorrect. Leptin leads to activation of AMPK, which inactivates acetyl CoA carboxylase through phosphorylation.

7. SOCS binds to members of signal-transduction pathways, inactivating the pathway. SOCS prevents IRS-1 from transmitting the signal to PI3K from the insulin receptor. In the case of leptin resistance, SOCS activity appears to increase, which prevents activation of signaling components downstream of leptin. This leads to overeating as the appetite is not suppressed as it should be.

8. (a) I   (b) S   (c) S   (d) I   (e) S   (f) S   (g) S

9. a, b, d, e lead to activation of the insulin pathway. Answers c, f, and g lead to deactivation.

10. Answer a is incorrect because the large amount of fats accumulate in the cytoplasm, not the bloodstream. Answer b is incorrect because DAG activates PKC, which leads to phosphorylation of IRS-1. Answer d is incorrect because increases in DAG and ceramide lead to closure of the GLUT4 transporter.

11. Answer b is correct, but does not lead directly to type 2 diabetes.

12. Diabetic ketosis occurs when the body shifts from carbohydrates to fats as a fuel source. A lack of insulin leads to the uncontrolled breakdown of lipids and proteins, resulting in the ketogenic state. Ketones build up and overwhelm the body's ability to maintain acid-base balance. In type 2 diabetes, insulin is active enough to prevent lipolysis in liver and adipose tissue.

13. a, e, d, c, b

14. e, c, a, d, b. Liver glycogen and adipose tissue fatty acids as fuels for active muscle are the slowest and are about equivalent in terms of the maximal rate of ATP production (6.2–6.7 mmol/s). This rate is probably limited by the slow transport of the fuels from the storage sites to the muscle.

15. (1) c, e   (2) a, c, e   (3) a, b, d

16. a, e

17. The selection of the pathway depends on whether the fatty acids enter the mitochondrial matrix, the compartment of β-oxidation and ketone body formation. When citrate and ATP concentrations are high, as in the fed state, the activity of acetyl CoA carboxylase is stimulated. The resulting malonyl CoA, which is a precursor for fatty acid synthesis, inhibits carnitine acyltransferase I, which translocates fatty acids from the cytosol into the mitochondria for oxidation.

18. Phosphorylase a binds glucose, which makes this enzyme susceptible to the action of phosphatase. The resulting phosphorylase b is inactive; therefore, glycogen degradation is decreased. Since phosphorylase b does not bind phosphatase, phosphatase is released and activates glycogen synthase by dephosphorylating it, leading to the production of glycogen.

19. (a) The increase in the glucose level from A to B is due to the absorption of dietary glucose.
    (b) The decrease in the glucose level from B to C is due to the effects of insulin, which is secreted in response to increased blood glucose. Glucose is removed from the blood by the liver, which synthesizes glycogen, and by muscle and adipose tissue, which store glycogen and triacylglycerols.
    (c) The leveling off of the glucose level from C to D is due to the increased secretion of glucagon and the diminished concentration of insulin. Glucagon maintains blood-glucose levels by promoting gluconeogenesis and glycogen degradation in the liver and by promoting the release of fatty acids, which partially replace glucose as the fuel for many organs.
    (d) The slight overshoot that is sometimes observed at C can be explained by the continued effects of insulin, which are not yet balanced by the metabolic effects of glucagon.

20. (a) 2, 4   (b) 1   (c) 3, 5

21. b, c, d

22. b, d, e. Answer (a) is incorrect because lipolysis remains essentially constant.

23. See Figure 27.2. In diabetic patients, the level of insulin is too low and that of glucagon is too high, so after a meal the glucose levels will reach higher values than in a normal person. Also the removal of glucose from blood will be slower, and the fasting glucose levels in the blood may remain higher.

FIGURE 27.2 Blood-glucose levels for a diabetic patient and a normal person after a meal rich in carbohydrate.

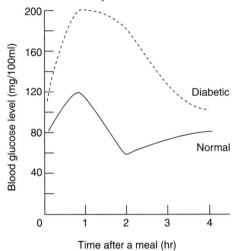

24. a, c, e

25. False. Although ketone body concentrations in blood may become high in diabetes, glucose is even more plentiful. Therefore, the brain continues to use glucose as its major fuel.

26. (a) 2   (b) 4, 6   (c) 1   (d) 3, 5, 7

27. a, c, and d. Ethanol consumption leads to production of NADH through the oxidation of ethanol to acetaldehyde and acetaldehyde to acetate. The increase in NADH inhibits gluconeogenesis and leads to accumulation of lactate. NADPH is used by liver cytochrome P450-dependent pathways and inhibits the regeneration of glutathione.

28. a, b, and d. While α-ketoglutarate dehydrogenase and isocitrate dehydrogenase are affected by ethanol, it is an indirect effect.

PROBLEMS

1. Cardiac muscle exhibits a high demand for oxygen, and its functioning is severely impaired when coronary circulation is blocked.

   (a) Considering the energy-generating substrates available to and used by the heart under normal circumstances, why is oxygen required by heart muscle?

   (b) Suppose that in an intact animal you can measure the concentrations of biochemical metabolites in the arteries leading to the cardiac muscle and in the veins carrying blood away from heart tissue. If the supply of oxygen to heart tissue is reduced, what differences in arterial and venous glucose concentrations will you observe? What metabolite will be elevated in heart muscle that has an insufficient supply of oxygen?

2. An infant suffering from a particular type of organic acidemia has frequent attacks of vomiting and lethargy, which are exacerbated by infections, fasting, and the consumption of protein or fat. During these episodes, the patient suffers from hypoglycemia, which can be alleviated by injections of D-3-hydroxybutyrate. In addition, concentrations of ketone bodies in the blood are extremely low. The patient also has elevated concentrations of a number of organic acids in both blood and urine. Among these acids are 3-hydroxy-3-methylglutarate, β-methylglutaconate, and isovalerate.

From the evidence of the buildup of these compounds, the enzyme that converts 3-hydroxy-3-methylglutaryl CoA (HMG CoA) to acetoacetate and acetyl CoA (HMG CoA cleavage enzyme) is probably deficient. (See Chapter 26).

(a) How could this enzyme deficiency lead to a reduction in the concentration of ketone bodies?

(b) How does fasting exacerbate the symptoms of the disorder?

(c) The consumption of fat causes a noticeable increase in the concentration of 3-hydroxy-3-methylglutarate. Why?

(d) How can the administration of D-3-hydroxybutyrate relieve hypoglycemia?

3. Patients who remain unconscious after a serious surgical operation are given 100 to 150 g of glucose daily through the intravenous administration of a 5% solution. This amount of glucose falls far short of the daily caloric needs of the patient. What is the benefit of the administration of glucose?

4. A biochemist in the Antarctic is cut off from his normal food supplies and is forced to subsist on a diet that consists almost entirely of animal fats. He decides to measure his own levels of urinary ketone bodies, beginning on the day he starts the high-fat diet. What changes in urinary ketone body levels will he find?

5. In liver tissue, insulin stimulates the synthesis of glucokinase. What implications does this have for a person who has an insulin deficiency?

6. Within a few days after a fast begins, nitrogen excretion accelerates to a relatively high level. After several weeks, the rate of nitrogen excretion falls to a lower level. The excretion of nitrogen then continues at a relatively constant rate until the body is depleted of triacylglycerol stores; then the rate of urea and ammonia excretion again rises to a very high level.

(a) What events trigger the initial surge of nitrogen excretion?

(b) Why does the nitrogen excretion rate decrease after several weeks of starvation?

(c) Explain the increase in nitrogen excretion that occurs when lipid stores are exhausted.

7. Among the difficulties caused by prolonged fasting are metabolic disorders caused by vitamin deficiencies. What vitamins are needed during starvation to ensure that cells can continue to carry out the metabolic adaptations discussed in Section 27.3 of the text?

8. Young men who are championship marathon runners have levels of body fat as low as 4%, whereas most casual runners have levels ranging from 12% to 15%. Why would marathoners be at greater risk during prolonged fasting?

9. Assume that a typical 70-kg man expends about 2000 kcal of energy per day. If the energy for his activities were all derived from ATP, how many grams of ATP would have to be generated on a daily basis? How many grams of glucose would be required to drive the formation of the needed amount of ATP? The molecular weight of ATP is 500, and that of glucose is 180. One mole of glucose generates 686 kcal of energy when completely oxidized, and 7.3 kcal are required to drive the synthesis of one mole of ATP. Assume that 40% of the energy from the oxidation of glucose can be used for ATP synthesis.

10. In chronically malnourished people or in healthy people who have missed one or two meals, hypoglycemia develops rapidly with the ingestion of moderate amounts of ethanol. A reduced rate of hepatic glucose synthesis is observed, along with increases in intracellular ratios of lactate to pyruvate, of glycerol 3-phosphate to dihydroxyacetone phosphate, of glutamate to α-ketoglutarate, and of D-3-hydroxybutyrate to

acetoacetate. In a well-fed person whose liver contains normal amounts of glycogen, ethanol infusion is less likely to induce hypoglycemia. The rate of hepatic glucose production is relatively normal. Increases in intracellular ratios of the pairs of compounds named above do occur, however.

(a) How do elevated intracellular ratios of NADH to NAD$^+$ in response to ethanol infusion lead to an increase in ratios of the pairs of compounds named above?

(b) Briefly describe how an increase in the ratios of any of the pairs of compounds could impair glucose synthesis in the liver of a malnourished person.

(c) Why is ethanol infusion less likely to impair hepatic glucose production when liver glycogen content is normal?

11. During starvation, the rate of liver-cell lipolysis accelerates, and the concentration of acetyl CoA increases. These changes are accompanied by an increase in the activity of pyruvate carboxylase, which converts pyruvate to oxaloacetate. Give two reasons why it is desirable to increase oxaloacetate concentrations in starved cells. Also name a source of pyruvate during starvation, when glucose availability in the liver is low.

12. For several hours after birth, premature infants are particularly susceptible to hypoglycemia and are also unable to rapidly generate ketone bodies. Describe how each of the characteristics below would contribute to hypoglycemia, low circulating levels of ketone bodies, or both.

(a) a large brain-to-body-weight ratio
(b) a small store of liver glycogen
(c) Low specific activity of cytosolic carnitine long-chain acyl-CoA transferase in liver
(d) Very low levels of liver phosphoenolpyruvate carboxykinase

13. The Cori cycle (see Section 16.4 of the text) is especially important during early phases of starvation, in which lactate molecules generated in the peripheral tissues are sent to the liver for use in gluconeogenesis.

(a) Why is it important for fatty acid oxidation to occur in the liver while the Cori cycle is operating?

(b) Suppose that lactate from muscle were completely oxidized to carbon dioxide and water in the liver. How would this make it more difficult for that organ to maintain glucose homeostasis during the early phases of starvation?

(c) Give two advantages of the alanine cycle in muscle and liver, in comparison with the Cori cycle.

ANSWERS TO PROBLEMS

1. (a) Under normal conditions, heart muscle consumes acetoacetate and D-3-hydroxybutyrate, both of which are converted to two molecules of acetyl CoA. Oxygen is required for terminal oxidation of acetyl CoA in the citric acid cycle.

(b) Cardiac cells deprived of oxygen are unable to generate metabolic energy by oxidizing acetyl CoA. An alternative source of energy is glucose, which can be converted to lactate under anaerobic conditions, with the subsequent generation of ATP. Thus, under anaerobic conditions, glucose uptake by heart muscle increases; the concentration of glucose in cardiac veins will decrease relative to glucose levels in coronary arteries. The concentration of lactate in the coronary veins is also elevated, compared with the levels in the coronary arteries.

2. (a) The production of the ketone bodies acetoacetate and D-3-hydroxybutyrate is dependent on the activity of HMG CoA cleavage enzyme, which is deficient in the infant.

(b) Fasting causes an increase in acetyl CoA production through the increased rates of lipolysis that occur in the attempt to generate sources of metabolic energy. Any increase in acetyl CoA concentration stimulates HMG CoA synthesis, and the inability to convert HMG CoA to acetoacetate leads to an increase in concentrations of 3-hydroxy-3-methylglutarate, which is excreted by the cells into the plasma.

(c) Consumption of fats such as triacylglycerols leads to the increased production of acetyl CoA, from the β-oxidation of fatty acids. As noted in the answer to (b), elevation in the level of acetyl CoA stimulates the production of HMG CoA, leading to an increase in the concentration of 3-hydroxy-3-methylglutarate.

(d) Because the liver is unable to generate normal levels of ketone bodies, those tissues that normally utilize them are required to use other substrates as metabolic fuels. Since glucose is the substrate of choice in such situations, increased demand for blood glucose results in hypoglycemia. D-3-Hydroxybutyrate can be oxidized into the ketone body acetoacetate, and both can be used as fuel sources for the brain and heart. The administration of D-3-hydroxybutyrate therefore provides an alternative source of metabolic energy, conserving glucose and reducing hypoglycemia

3. The limited amount of glucose is given to prevent the hydrolysis of muscle protein during fasting. It serves as a source of energy for the brain and blood cells; otherwise, during fasting, body proteins are hydrolyzed to provide carbon atoms for the generation of glucose by gluconeogenesis in the liver and kidney.

4. A high-fat diet will stimulate ketone body formation because the oxidation of fatty acids causes an increase in acetyl CoA concentration, which in turn stimulates the production of ketone bodies. The overall profile of ketone body production may resemble that found during starvation because a source of carbon for glucose production will be lacking. However, dietary fats will serve as a source of energy instead of the triacylglycerols stored in body tissue.

5. Under normal conditions, glucokinase acts to phosphorylate glucose when concentrations of the hexose are high. The failure to synthesize sufficient quantities of glucokinase means that the liver cannot control the levels of glucose in blood, compounding the other difficulties caused by insulin deficiency described in the text.

6. (a) During the first few days of starvation, the brain continues to utilize glucose. Glycogen stores are exhausted, so the primary source of carbon atoms for gluconeogenesis is amino acids. Because the concentration of free amino acids in the tissues is limited, body proteins are broken down to provide the amino acids to support gluconeogenesis. Nitrogen excretion increases because the amino groups of those amino acids are eliminated as urea.

(b) After several weeks of fasting, the brain adapts to the utilization of ketone bodies as a source of energy, so less glucose is required. The resulting reduction in gluconeogenesis means a reduction in the rate of oxidation of amino acids and in the production of ammonia and urea.

(c) When triacylglycerol stores are depleted, the body relies on body proteins not only as a source of glucose for the brain but also as a source of energy for all other tissues. These requirements cause a great increase in the rate of body protein catabolism, with corresponding increases in amino acid oxidation and nitrogen excretion. Often more than a kilogram per day in weight is lost, indeed causing a threat to life.

7. Most of the vitamins and cofactors discussed in previous chapters of the text would be needed during starvation because many of the essential metabolic pathways must continue to operate. Among the most obvious vitamins needed for those pathways are pyridoxal phosphate (for the transamination of amino acids), niacin and riboflavin (for electron transport), thiamin (for the oxidative decarboxylation of pyruvate, α-ketoglutarate, and the branched-chain amino acids), biotin (for the carboxylation of pyruvate), and cobalamin (for the conversion of methylmalonyl CoA to succinyl CoA).

8. The greater the percentage of body fat, the larger the reserves of triacylglycerols, which are the primary source of metabolic energy during fasting. Once these reserves are depleted, the body accelerates the breakdown of muscle protein as an energy source. Extensive hydrolysis of muscle tissue threatens many vital body functions and can lead to death.

9. Since 7.3 kcal are required to drive the synthesis of 1 mole of ATP, the caloric value of the energy in each mole of ATP is 7.3 kcal. Therefore, the amount of ATP needed per day is

$$\frac{2000 \text{ kcal/day}}{7.3 \text{ kcal/mol ATP}} = 274 \text{ mol ATP per day}$$

$$274 \text{ mol ATP/day} \times 500 \text{ g/mol} = 137 \text{ kg ATP per day}$$

If the oxidation of 1 mole of glucose yields 686 kcal of energy, and if 40% can be used to drive the synthesis of ATP, glucose yields (0.4)(686 kcal/mol) = 274 kcal/mol of usable energy. Therefore, the amount of glucose needed per day is

$$\frac{2000 \text{ kcal/day}}{274 \text{ kcal/mol glucose}} = 7.4 \text{ mol glucose per day}$$

$$7.3 \text{ mol glucose/day} \times 180 \text{ g/mol} = 1.31 \text{ kg glucose per day}$$

Normal fuel stores available in the blood to the typical 70-kg man include about 250 g of glycogen and approximately 60 g of glucose. These figures make it evident that humans can rely on stores of carbohydrate for only a short time.

10. (a) Each pair of compounds referred to in the problem is interconvertible through oxidation–reduction reactions linked to the NADH/NAD$^+$ redox pair. An increased NADH/NAD$^+$ ratio will therefore suppress net oxidation reactions because the NAD$^+$ needed to serve as an electron acceptor (e.g., for conversion of lactate to pyruvate) is limited in concentration. Thus an ethanol-induced increase in NADH will suppress conversion of lactate to pyruvate, of glycerol 3-phosphate to dihydroxyacetone phosphate, of glutamate to α-ketoglutarate, and of D-3-hydroxybutyrate to acetoacetate.

(b) During starvation the liver carries out gluconeogenesis using lactate, amino acids, and α-glycerol phosphate (from triacylglycerols) as initial substrates. These compounds are converted via NAD$^+$-linked oxidation to compounds like pyruvate, α-ketoglutarate, and dihydroxyacetone phosphate, all directly in the pathway to glucose synthesis. The inability of the cell to produce these and other compounds (like oxaloacetate) because of the lack of NAD$^+$ would result in a low level of glucose production. The acetoacetate/D-3-hydroxybutyrate pair is not involved in gluconeogenesis; these are ketone bodies. However, the inability to convert D-3-hydroxybutyrate to acetoacetate would interfere with the terminal oxidation of these compounds as well.

(c) In well-nourished people, ethanol infusion will elevate intracellular NADH levels, and gluconeogenesis through the liver's utilization of amino acids and other compounds will be somewhat impaired. However, stored hepatic glycogen serves as a source of glucose in response to any drop in blood sugar levels. Because liver glycogen levels are virtually depleted after 24 to 36 hours of starvation, ethanol-induced hypoglycemia can develop rapidly in malnourished people.

11. Increased concentrations of oxaloacetate are needed to provide more acceptors of acetyl groups from acetyl CoA to form citrate, ensuring that the citric acid cycle can operate at higher capacity. This makes it able to oxidize the increasing amounts of acetyl CoA present in the cell. Elevated levels of oxaloacetate are also required to provide more molecules that can serve as precursors for gluconeogenesis. During starvation, liver cells increase their rate of glucose formation through glycogenolysis and gluconeogenesis, in order to provide more glucose to peripheral tissues. Gluconeogenesis depends primarily on the availability of oxaloacetate molecules, which are converted first to phosphoenolpyruvate and then ultimately to glucose. One source of pyruvate during starvation is alanine, produced by the degradation of muscle tissue proteins. Alanine, along with glutamine, serves as a carrier of carbon atoms and nitrogen from muscle to liver. Alanine is converted to pyruvate by aminotransferase enzymes, which use another α-keto acid as an acceptor of the amino group from alanine.

12. (a) Because brain tissue preferentially uses glucose as a fuel, the demand for glucose in neonates is disproportionately high compared with that of an older person with a lower brain-to-body weight ratio.

   (b) Low reserves of liver glycogen mean that the ability of the liver to export glucose synthesized from glycogen is limited.

   (c) Neonates are unable to generate sufficient levels of ketone bodies as an alternative fuel during hypoglycemia because the low specific activity of carnitine acyl transferase limits transport of long-chain fatty acids across the mitochondrial membrane. The depletion of fatty acids in the mitochondrion means that only limited amounts of acetyl CoA from fatty chain oxidation are available for synthesis of acetoacetyl CoA and β-hydroxybutyrate.

   (d) The liver cannot effectively carry out gluconeogenesis because the activity of phosphoenolpyruvate carboxykinase (which carries out synthesis of PEP from oxaloacetate), a key enzyme in glucose synthesis, is present only at a very low level.

13. (a) Because most of the ATP used for gluconeogenesis in the liver is generated by β-oxidation of fatty acids.

   (b) Compounds of the Cori cycle, such as lactate and pyruvate, provide a readily available source of carbon for gluconeogenesis in the liver. If those molecules were unavailable, additional carbon atoms from proteolysis would be required to sustain the level of glucose required through production by gluconeogenesis.

   (c) Formation of alanine in muscle allows transport of an atom of nitrogen in nontoxic form to the liver, where it can be disposed of as urea. Conversion of pyruvate to alanine means that the electrons in NADH normally consumed in the conversion of pyruvate to lactate can now be sent to the mitochondrion to be used to drive ATP synthesis.

DNA Replication, Repair, and Recombination

The text returns to the topic of the flow of genetic information and considers the detailed biochemical mechanisms underlying this complex process. Chapter 4 introduced DNA and RNA and outlined the storage, duplication, and expression of genetic information. In Chapter 5, the enzymes and techniques used to analyze, construct, and clone DNA were presented. You should review these chapters to prepare for studying Chapter 28, which describes how DNA is replicated. Pay particular attention to DNA structure, the supercoiling of DNA, and DNA polymerase I (discussed in Chapter 4) and to DNA ligase (discussed in Chapter 5).

Chapter 28 covers the biochemistry of the transmission of genetic information from parent to progeny, describing DNA and the enzyme systems that replicate, recombine, and maintain its sequences. The chapter opens by pointing out the need for faithful copying and the obstacles to fidelity that a cell must overcome to duplicate its duplex DNA. The general description in Chapter 4 of the biochemistry of DNA polymerases is expanded to explain the roles the template, primer, and metal ions play in their activities. The chemical basis for the fidelity of DNA chain extension by the polymerases is also presented, and the helicases that unwind DNA are described. The text then provides a detailed description of the topology of covalently closed circular DNAs and the topoisomerases that modulate their linking numbers. The authors then discuss the replication fork; replication initiation; and RNA-primed, semidiscontinuous DNA elongation. The mechanisms and role of DNA ligases and DNA helicases are also given. The structure and important roles of DNA polymerases I and III in replication are described in detail. The special problems of replication arising from the sheer amount of DNA and number of linear DNA molecules in a eukaryotic cell are introduced, and the nature and functions of telomeres and telomerase are described. The precise organization and timing of the origination of DNA replication in eukaryotes and in prokaryotes are explained and contrasted. Examples of pathological deficiencies of DNA repair in humans, the relationship of mutations and DNA repair impairments to carcinogenesis, and a test system for detecting potential carcinogens through their mutagenic action on bacteria is outlined. The causes of damage to DNA and the mechanisms of its repair are also presented.

To provide the new sequences of nucleotides in DNA upon which evolution can act, not only mutation, but also recombination between different DNA molecules occurs. The breakage and joining of fragments of DNA with similar sequences also rearrange gene orders within the chromosome and are mechanisms for repairing damaged DNA and regulating gene expression. The activity of the RecA protein and a description of a key intermediate in recombination, the Holliday junction, and the recombinases that form and resolve it are presented. (Review the material on plasmids and bacteriophages in Chapter 5 for a better understanding of this section.)

LEARNING OBJECTIVES

When you have mastered this chapter, you should be able to accomplish the following objectives:

Introduction

1. Reprise the Watson-Crick hypothesis for DNA replication and outline some problems facing a cell in creating an exact duplicate copy of its double-strand DNA genome. Appreciate that enzymes and DNA-binding proteins play essential roles in solving the challenges of DNA replication.

DNA Replication Proceeds by the Polymerization of Deoxyribonucleoside Triphosphates Along a Template (Text Section 28.1)

2. Outline the key features of the reactions catalyzed by *DNA polymerases*. Define *template* and *primer* as they relate to DNA polymerases.

3. Relate the *3′ → 5′ nuclease* activity of DNA polymerases to the fidelity of DNA replication.

4. Appreciate the common structures and the evolutionary relationships among DNA polymerases.

5. Explain the roles of Mg^{2+} in the reaction catalyzed by DNA polymerases.

6. Account for the fidelity with which a DNA polymerase selects the correct incoming *deoxyribonucleotide triphosphate (dNTP)* substrate in terms of *shape complementarity* and *hydrogen-bond donors* and *acceptors* in the minor groove.

7. Explain the roles of *RNA* in DNA replication and describe the enzymes that form and remove *RNA primers* from the genome.

8. Define *continuous replication* and *discontinuous replication* and relate these processes to the *leading* and *lagging strands* of replicating DNA. Describe an *Okazaki fragment*.

9. Draw a *replication fork* and describe the reactions and the movements of the DNA strands that occur during replication.

10. List the substrates and outline the reaction mechanisms of the *DNA ligases*

11. Describe how *helicases* separate the strands of duplex DNA.

12. Understand that AMP-PNP is used as a non-hydrolyzable analog of ATP.

DNA Unwinding and Supercoiling Are Controlled by Topoisomerases
(Text Section 28.2)

13. Define the *linking number (Lk)* of a circular DNA molecule and relate *supercoiling* to the *electrophoretic* and *centrifugal mobility* of the molecule.

14. Write the equation relating the linking number to the *twisting number (Tw)* and *writhing number (Wr)* of a DNA *topoisomer*. Explain how a negative *superhelix density* can facilitate the unwinding of the helix. Describe the partitioning of the free energy of a negatively supercoiled molecule into *Tw* and *Wr*.

15. Describe the three steps catalyzed by *topoisomerases*; distinguish between *type I* and *type II* topoisomerases; and describe the substrates, products, and mechanisms of *topoisomerase I and II*.

DNA Replication Is Highly Coordinated (Text Section 28.3)

16. Define *processivity* as it relates to DNA polymerases and describe the role of the *sliding-clamp β_2* subunit of DNA polymerase III in retarding dissociation of the enzyme from the template.

17. List the distinctive features of the *DNA polymerase III holoenzyme,* and describe how an asymmetric dimer of the enzyme, along with other proteins, coordinates the synthesis of the leading and lagging strands of the daughter duplexes. Appreciate the structural complexity of the replication machinery and explain the essential role of *DNA polymerase I* and DNA ligase in this process.

18. Describe the function and features of the nucleotide sequence of *oriC* and note that it is the unique site of *bidirectional replication initiation* in E. coli. List the proteins that interact with the DNA in this region of the chromosome, and give the reactions they catalyze and functions they serve.

19. Summarize the reactions and identify the proteins at the replication fork that carry out DNA replication.

20. Describe the special problems arising from DNA length and the *cell cycle* in eukaryotes. Define *replication* and explain the roles of *multiple replication origins* and the *telomeres* in eukaryotic DNA replication.

21. Describe how *telomerase* makes DNA of defined sequence in the absence of a DNA template.

22. Explain why telomerase is a potential target for anti-cancer therapy.

Many Types of DNA Damage Can Be Repaired (Text Section 28.4)

23. Outline the ways in which the DNA of an organism can be damaged or the DNA of its progeny be an unfaithful copy.

24. Describe the structure of the *pyrimidine dimer* formed by *ultraviolet light*. Distinguish among *substitution, insertion,* and *deletion mutations.*

25. Explain why the formation of *mismatched base pairs* during replication can cause a mutation. List other types of errors that occur during replication of DNA. Describe the involvement of *error-prone* DNA polymerases in repairing replication errors and appreciate their limitations.

26. Describe *trinucleotide-repeat expansions* during DNA replication and relate them to *Huntington disease.*

27. List some agents that damage DNA.

28. Outline the general strategies for repairing damaged DNA and relate them to the *repair* of particular damage, for instance, *mismatches*, *pyrimidine dimers*, alkylated bases. Distinguish between *base-excision* repair and *nucleotide-excision* repair.

29. Explain why *thymine* rather than *uracil* is used in DNA.

30. Relate defective DNA repair to *cancer* and relate *tumor-suppressor genes* to DNA repair. List two defining features of cancer cells and relate them to DNA damage.

31. Describe *xeroderma pigmentosum* and *hereditary nonpolyposis colorectal cancer*, their causes, pathological consequences, and relationships to DNA repair.

32. Relate *mutagens* and *carcinogens* and outline the *Ames Salmonella mutagen assay*.

DNA Recombination Plays Important Roles in Replication, Repair, and Other Processes (Text Section 28.5)

33. Explain when *recombination* plays a role in DNA repair, normal cellular processes, and DNA technology.

34. Describe the *Holliday model* for *homologous recombination*. Explain how resolution of the *Holliday junction intermediate* can form different recombinant DNA products.

35. Outline the reactions catalyzed by *recombinases* and note their mechanistic similarity to the topoisomerases.

SELF-TEST

DNA Replication Proceeds by the Polymerization of Deoxyribonucleoside Triphosphates Along a Template

1. Which of the following statements about DNA polymerases are correct?
   (a) They add deoxyribonucleotide units to the 3'-hydroxyl of a primer.
   (b) They use the template strand to help determine which deoxyribonucleotide unit to add to the growing DNA chain.
   (c) They contain a $3' \rightarrow 5'$ nuclease that cleaves phosphodiester bonds of misincorporated deoxyribonucleotides.
   (d) They check the size of an incoming deoxyribonucleotide triphosphate (dNTP) to help ensure that the correct, complementary choice is made.
   (e) They bind one complementary dNTP and add a second complementary dNTP to initiate a new DNA chain.

2. Mg^{2+} serves the following function(s) in the reaction catalyzed by DNA polymerases:
   (a) It stabilizes the pentacoordinate transition state of the phosphdiester bond formation.
   (b) It precipitates inorganic phosphate (P_i) arising from the reaction.
   (c) It interacts with the 3'-hydroxyl of the incoming dNTP.
   (d) It forms a bridge between the 3'-hyroxyl of the primer and a phosphate in the dNTP.
   (e) It stabilizes the negative charge on the departing pyrophosphate, which is derived from the incoming dNTP.

3. Why are helicases required during DNA replicaton? Is ATP required for their action?

DNA Unwinding and Supercoiling Are Controlled by Topoisomerases

4. The topological features of circular DNA may affect which of the following?
   (a) the electrophoretic mobility of the DNA
   (b) the sedimentation properties of the DNA
   (c) its affinities toward proteins that bind to the DNA
   (d) the susceptibility of the strands of the DNA to unwinding
   (e) the susceptibility of the DNA to the action of DNA ligase

5. Which of the following statements about DNA molecules that are topoisomers are correct?
   (a) They are bound to topoisomerases.
   (b) They differ from one another topologically only in that they have different linking numbers.
   (c) They may be separated from one another by electrophoresis.
   (d) They have identical molecular weights.
   (e) They are topological or spatial isomers.

6. Which of the following statements about topoisomerases are correct?
   (a) They alter the linking numbers of topoisomers.
   (b) They break and reseal phosphodiester bonds.
   (c) They require NAD^+ as a cofactor to supply the energy to drive the conversion of a supercoiled molecule to its relaxed form.
   (d) They form covalent intermediates with their DNA substrates.
   (e) They can, in the case of a particular type of topoisomerase, use ATP to form negatively supercoiled DNA from relaxed DNA in E. coli.

7. In the equation $Lk = Tw + Wr$, the convention is that right-handed twists are positive numbers, but right-handed "writhes" or supercoils are negative numbers. Why not have both right-handed structures be positive?

8. Why are the antibiotics novobiocin, ciprofloxin, and nalidixic acid, which inhibit DNA gyrase, useful in treating bacterial infections in humans?

DNA Replication Is Highly Coordinated

9 Match the properties or functions in the right column with a DNA polymerase in the left column.

   (a) DNA polymerase I   (1) involved in replication
   (b) DNA polymerase III   (2) requires a primer and a template
                                 (3) involved in DNA repair
                                 (4) makes most of the DNA phosphodiester bonds during replication
                                 (5) removes the primer and fills in gaps during replication

10. Which of the following statements about DNA replication in E. coli are correct?
    (a) It occurs at a replication fork.
    (b) It starts at a unique locus on the chromosome.
    (c) It proceeds with one replication fork per replicating molecule.
    (d) It is bidirectional.
    (e) It involves discontinuous synthesis on the leading strand.
    (f) It uses RNA transiently as a template.

11. Which of the following statements about DNA polymerase III holoenzyme from *E. coli* are correct?

    (a) It elongates a growing DNA chain hundreds of times faster than does DNA polymerase I.
    (b) It associates with the parental template, adds a few nucleotides to the growing chain, and then dissociates before initiating another synthesis cycle.
    (c) It maintains a high fidelity of replication, in part, by acting in conjunction with a subunit containing a $3' \rightarrow 5'$ exonuclease activity.
    (d) When replicating DNA, it is a molecular assembly composed of at least 10 different kinds of subunits.

12. Explain how the β_2 subunit of DNA polymerase III holoenzyme contributes to the processivity of the DNA synthesis machinery.

13. Which of the following statements about DNA ligase are correct?

    (a) It forms a phosphodiester bond between a 5'-hydroxyl and a 3'-phosphate in duplex DNA.
    (b) It requires a cofactor, either NAD^+ or ATP, depending on the source of the enzyme, to provide the energy to form the phosphodiester bond.
    (c) It catalyzes its reaction by a mechanism that involves the formation of a covalently linked enzyme adenylate.
    (d) It catalyzes its reaction by a mechanism that involves the activation of a DNA phosphate through the formation of a phosphoanhydride bond with AMP.
    (e) It is involved in DNA replication, repair, and recombination.

14. Why is RNA synthesis essential to DNA synthesis in *E. coli*?

15. Match the functions or features related to DNA replication in *E. coli* listed in the right column with the molecules or structures in the left column.

    (a) replication fork
    (b) *oriC*
    (c) lagging strand
    (d) leading strand
    (e) Okazaki fragment
    (f) DnaB helicase
    (g) single-strand binding protein (ssb)
    (h) DNA gyrase
    (i) primase
    (j) DNA polymerase III holoenzyme
    (k) ε subunit of DNA polymerase III
    (l) DNA polymerase I
    (m) DNA ligase

    (1) synthesis direction is opposite that of replication fork movement
    (2) unwinds strands at the origin of replication in association with dnaA and dnaC proteins
    (3) is synthesized continuously
    (4) synthesizes most of DNA
    (5) is synthesized discontinuously
    (6) relieves positive supercoiling
    (7) is the locus of DNA unwinding
    (8) hydrolyzes ATP to reduce the linking number of DNA
    (9) binds dnaA, dnaB, and dnaC proteins
    (10) fills in gaps where RNA existed
    (11) is the point of initiation of synthesis
    (12) joins lagging strand pieces to each other
    (13) contains a $5' \rightarrow 3'$ exonuclease that removes RNA primers
    (14) is an RNA polymerase
    (15) performs "proofreading" on most of the DNA synthesized
    (16) stabilizes unwound DNA
    (17) uses NAD+ to form phosphodiester bonds

16. The duplication of the ends of linear, duplex DNA in humans presents a problem to the replicative machinery of a cell. Describe this problem, its cause, and the way the cell overcomes it.

17. Give two reasons why eukaryotes must use multiple origins of replication (in contrast to prokaryotes, which usually use one).

Many Types of DNA Damage Can Be Repaired

18. How could the tautomerization of a keto group on a guanine residue in DNA to the enol form lead to a mutation?

19. Explain why most nucleotides that have been misincorporated during DNA synthesis in *E. coli* do not lead to mutant progeny.

20. Match the usual type of mutation or physiologic consequence in the right column with the appropriate mutagen or mutagenic process in the left column.

    (a) alkylating agents (activated aflatoxin B_1)
    (b) deaminating agents
    (c) ultraviolet light
    (d) psoralens
    (e) x-rays
    (f) oxidizing agents (hydroxyl radicals)

    (1) interstrand cross-links
    (2) intrastrand cross-links
    (3) single-strand and double-strand breaks
    (4) mispaired base pairs
    (5) replication and transcription blockage

21. What general property of DNA allows the repair of some residues damaged through the action of mutagens?

22. Which of the following enzymes or processes can be involved in repairing DNA in *E. coli* damaged by UV light–induced formation of a thymine dimer?

    (a) DNA ligase seals the newly synthesized strand to undamaged DNA to form the intact molecule.
    (b) The UvrABC enzyme (excinuclease) hydrolyzes phosphodiester bonds on both sides of the thymine dimer.
    (c) DNA polymerase I fills in the gap created by the removal of the oligonucleotide bearing the thymine dimer.
    (d) The UvrABC enzyme recognizes a distortion in the DNA helix caused by the thymine dimer.
    (e) A photoreactivating enzyme absorbs light and cleaves the thymine dimer to re-form two adjacent thymine residues.

23. What role does N^5, N^{10}-methenyltetrahydrofolate play in the action of direct repair of DNA by DNA photolyase?

    A. It methylates uracil to produce thymine
    B. It adds a methylene bridge to the thymine dimmer
    C. It absorbs a photon to provide the energy for the reaction
    D. It has no actual role in the reaction.

24. Given that the base T requires more energy to synthesize than U, and A pairs equally well with U or T, what is the probable reason that DNA contains A·T base pairs instead of A·U base pairs?

25. Explain how mutations in genes encoding proteins likely to be involved in DNA repair, such as those defective in xeroderma pigmentosum and hereditary nonpolyposis colorectal cancer, may contribute to the onset of cancer.

26. How might a trinucleotide expansion in a gene affect the primary structure of the protein normally encoded by that gene?

27. Explain how some strains of *Salmonella* are used to detect carcinogens. How is an extract from human liver involved in this test?

DNA Recombination Plays Important Roles in Replication, Repair, and Other Processes

28. Which of the following statements about genetic recombination are correct?
    (a) It generates new combinations of genes.
    (b) It can move a segment of DNA from one chromosome to another (for example, from a virus to a host cell).
    (c) It is mediated by the breakage of DNA and the rejoining of the resulting fragments.
    (d) It generates genome sequence variability upon which natural selection can act.

29. How many strands of DNA are present at the junction of a Holliday junction?

30. Recombinases
    (a) make and break phophodiester bonds in a reaction requiring ATP.
    (b) pair homologous DNA molecules prior to strand breakage to form a recombination synapse.
    (c) form covalent complexes with their DNA substrates.
    (d) have a reaction intermediate reminiscent of those of the DNA ligases.
    (e) are related to type I topoisomerases by divergent evolution.

ANSWERS TO SELF-TEST

1. a, b, c, d

2. a, d, e

3. Duplex B-DNA is a stable molecule at physiological temperature and helicases are required to unwind the two strands of the helix so that each can serve as a template for DNA polymerases. Because DNA is so stable, energy is required in the form of ATP hydrolysis to drive helicase action.

4. a, b, c, d. In regard to answer (e), supercoiling is ordinarily a property of covalently closed circular DNA—that is, of DNA in which there are no discontinuities in either strand of the helix. Hence, these molecules are not substrates for DNA ligase, because they lack ends.

5. b, c, d, e. Answer (a) is not correct because the DNA topoisomer need not necessarily be bound by a topoisomerase.

6. a, b, d, e. Although all topoisomerases break and reseal phosphodiester bonds, an external energy source is not always required. Relieving the torsional stress in a negatively supercoiled DNA molecule by relaxing it with topoisomerase I is exergonic and requires no energy input, whereas introducing negative supercoils with DNA gyrase is endergonic and must be coupled to ATP hydrolysis. The particular catalytic mechanisms of given topoisomerases determine whether they are coupled to ATP hydrolysis.

7. This is not just a convention. Right-handed "twists" are positive mainly because that is the predominant structure in DNA; normal "B-form" DNA is right-handed. But defining that automatically means that right-handed "writhes" must be negative because

topologically they cancel each other out; a right-handed writhe will diminish the linking number whereas a left-handed writhe will increase it. It is easier to understand this by using models because it is rather difficult to visualize.

8. These compounds interfere with the essential helix-destabilizing function of DNA gyrase in bacterial DNA replication. By inhibiting its action, they prevent the gyrase from relieving the positive supercoils that build up ahead of the moving replication fork. Human cells lack an enzyme similar to gyrase, and thus they are relatively unharmed by these antibiotics.

9. (a) 1, 2, 3, 5 (b) 1, 2, 4

10. a, b, d. Answer (c) is incorrect because, although not explicitly stated in the text, the replicating *E. coli* chromosome has two replication forks that synthesize the DNA bidirectionally from the unique *oriC* origin. Answer (e) is incorrect because only the lagging strand is synthesized discontinuously. Answer (f) is incorrect because RNA serves as a primer and not as a template.

11. a, c, d. Answer (b) is incorrect because DNA polymerase III holoenzyme is a highly processive enzyme that synthesizes extensively before dissociating from its template.

12. The β_2 subunit forms a torus, with the duplex DNA in its aperture. The β_2 ring acts as a sliding clamp that holds the replication machinery on the DNA.

13. b, c, d, e. Answer (a) is incorrect because the enzyme joins a 3′-hydroxyl to a 5′-phosphate. Answer (d) is correct because, although not completely described in the text, DNA ligase activity is required to seal the discontinuities in DNA arising during DNA replication, repair, and recombination.

14. Because DNA polymerases are unable to initiate DNA chains de novo and because they require a primer with a 3′-hydroxyl group, short RNA chains are used as primers to start DNA replication on the leading strand at the origin of replication and to initiate the Okazaki fragments of the lagging strand. RNA polymerases can start RNA chains by adding a nucleotide to an initiating NTP, but they do so with relatively low accuracy. RNA-initiated DNA chains facilitate high-fidelity replication at the beginning sequences of new chains because they allow DNA polymerase I to replace the RNA with DNA, using the information in the complementary strand and both of its exonucleases in a nick translation reaction (page 823 of the text).

15. (a) 7 (b) 7, 9, 11 (c) 1, 5 (d) 3 (e) 1, 3 (f) 2 (g) 16 (h) 6, 8 (i) 14 (j) 4 (k) 15 (l) 1, 10, 13 (m) 12, 17. Answer (5) is not a correct match with (e) because each Okazaki fragment is synthesized continuously.

16. Because DNA polymerases extend their growing polynucleotide chains only in the 5′ $\rightarrow$ 3′ direction and because the two strands of the parental DNA duplex are antiparallel, removal of the RNA primer that is paired with the 3′ end of the parental template DNA would leave an overhanging 3′ DNA strand with no means of having its complement synthesized. Ordinary DNA polymerases are unable to initiate DNA chains de novo. Each round of replication would consequently shorten the DNA because a portion could not be copied. To circumvent this problem, human DNA chromosomes have a segment of repeating G-rich DNA (telomeres) at their ends. In addition, a special enzyme, telomerase, which is an RNA-dependent DNA polymerase (reverse transcriptase) that carries its own RNA template, can extend the uncompleted end at each round of replication. The RNA template renews the repeating telomere sequence so that the DNA is not shortened.

17. The great amount of genomic DNA in eukaryotes (6×10^9 bp in a human) requires that polymerases start at multiple sites in order to complete the DNA synthesis in a biologically relevant period of time. In addition, the genome of eukaryotes comprises

multiple separate DNA molecules—23 pairs of chromosomes in a human, so sites must be provided on each molecule. The first requirement accounts for the vast majority of the ~30,000 origins in a human cell.

18. The rare enol tautomer of G could base-pair with a T in the template to allow its incorporation into a growing DNA strand during replication. If the proofreading process missed this erroneous incorporation, the resulting daughter DNA duplex would contain a G·T base pair. During the next round of replication, the T would direct the incorporation of an A into its complementary daughter strand. The final result would be the substitution of an A·T base pair for the original G·C base pair.

19. The proofreading $3' \rightarrow 5'$ nuclease of the ε subunit of DNA polymerase III holoenzyme removes most of the misincorporated nucleotides that do not form a base pair with the template. The polymerase activity of the enzyme then has a second chance to incorporate the correct nucleotide. Additionally, DNA repair systems exist that can detect and repair a mismatched base pair resulting from a misincorporation during synthesis.

20. (a) 4 (b) 4 (c) 2, 5 (d) 1, 5 (e) 3, 5 (x-rays also generate hydroxyl radicals, and thus can lead to mispairing of oxidized bases; in addition, double-stranded breaks can prevent replication and transcription) (f) 4

21. Since DNA is double-strand, damage to one strand of the DNA can often be repaired by using the undamaged complementary strand as a template to direct new incorportion of correct deoxynucleotides in place of the removed incorrect ones.

22. a, b, c, d, e. Although not mentioned in the text, the uvrD protein is a helicase that removes the 12-nucleotide-long oligonucleotide bearing the thymine dimer.

23. C. Tetrahydrofolate is generally a cofactor for one-carbon transfers, and in fact one derivative (methylene or CH_2) is used in the synthesis of thymine. But DNA photolyase uses this cofactor in an unusual way, as a collector of photons, as described in the text.

24. C spontaneously deaminates to form U in DNA. This change would lead to a mutation during the next round of replication of the DNA, since U would pair with A, changing what was a C·G base pair into an A·U base pair. The repair machinery of a cell that used U normally in its DNA would be unable to distinguish the U in an A·U base pair arising from a C deamination from one formed during "normal" replication. The methyl group on T, which is almost universally used in DNA, distinguishes it from the uracils formed by deamination so that the uracils can be repaired.

25. The inability to effectively repair mutagenic lesions in DNA may lead to their accumulation. As a consequence, genes regulating cellular proliferation may malfunction and thereby cause cancer.

26. The increased number of triplet sequences would give rise to a protein with a corresponding repeat of the amino acid encoded by the trinucleotide. In the case of Huntington disease, the sequence CAG is increased in number within the coding region of the gene. This leads to the gene product protein huntingtin, which has an inserted stretch of glutamine residues that were encoded by the increased number of CAG codons. These extra amino acids alter the function of the protein.

27. Special strains of *Salmonella* have been developed to detect substitution, insertion, and deletion mutations in their DNA as a result of exposure to exogenously supplied chemicals. Mutagens can alter the DNA in these strains and thus convert them from auxotrophs, which are unable to grow in the absence of histidine, to prototrophs. The revertants can grow on media lacking histidine and are detected with high sensitiv-

ity. Since there is a correlation between mutagenicity and carcinogenicity, these strains are used as an inexpensive initial test of the carcinogenic potential of a compound. Because animals sometimes metabolize innocuous compounds and convert them to carcinogens, incubation of a suspect chemical with a human liver extract before using the bacterial test can sometimes mimic what would happen to the chemical in vivo. This adjunct to the test expands its capacity to detect potential human carcinogens.

28. a, b, c, d

29. Four. The Holliday junction is formed from the four strands of two interacting duplex DNA molecules that, as a result of the initial reactions of recombination, become joined to form one molecule. The Holliday junction is resolved when recombination is completed and two separate duplexes are reformed. Although not mentioned in the text, the products can sometimes have regions of duplex where one strand of DNA is from one parent and the other strand from the other parent, that is, a heteroduplex is formed.

30. b, c, e. Recombinases have reaction mechanisms similar to those of the topoisomerases in that a covalent enzyme–DNA phosphodiester bond is formed. The bond linking the enzyme to the DNA has a high free energy of hydrolysis and is therefore capable of resynthesizing the phosphodiester bond broken during its formation. Consequently, recombinases, unlike DNA ligases and DNA polymerases, do not require an external source of energy to form a phosphodiester bond.

PROBLEMS

1. The T7 helicase described in the text (Figs. 28.11, 28.12, and 28.13) has the same "P-loop NTPase" structure as the proteins described in Chapter 9 (Fig. 9.51). What purpose does this structure serve in the action of a helicase?

2. Mismatch repair is a DNA repair system in which non-Watson Crick base pairs are detected, and one of the nucleotides is replaced to fix the mismatch. But with this process happening independently of replication, how can the enzymes determine which base is likely to be correct and which is an error?

3. Suppose that two polynucleotide chains are joined by DNA ligase in a reaction mixture to which ATP labeled with ^{32}P in the α-phosphoryl (the innermost) group has been added as an energy source. What products of the reaction would be expected to carry the radioactive label? Explain.

4. Suppose that negatively supercoiled DNA with $Lk = 23$, $Tw = 25$, and $Wr = -2$ is acted on by topoisomerase I. After one catalytic cycle, what would be the approximate values of Lk, Tw, and Wr?

5. Suppose that negatively supercoiled DNA with $Lk = 23$, $Tw = 25$, and $Wr = -2$ is acted on by DNA gyrase and ATP. After one catalytic cycle, what would be the approximate values of Lk, Tw, and Wr?

6. What property of DNA polymerase I leads to the observation that $polA1$ mutants of *E. coli* are more sensitive to ultraviolet light than are the wild-type cells? [Hint: The $5' \rightarrow 3'$ exonuclease activity of DNA polymerase I can remove damaged nucleotides from DNA as well as destroying the RNA primers on Okazaki fragments.]

7. Suppose that a single-strand circular DNA with the base composition 30% A, 20% T, 15% C, and 35% G serves as the template for the synthesis of a complementary strand by DNA polymerase.

(a) Give the base composition of the complementary strand.

(b) Give the overall base composition of the resulting double-helical DNA.

8. An early and initially attractive mechanism proposed for genetic recombination was the copy-choice model. It suggested that recombination between two parental DNA duplexes occurs during DNA replication when DNA polymerase switches or jumps from one parental duplex to the other, producing a recombinant daughter DNA duplex that contains sequences derived from the templates of two different DNA duplexes. The copy-choice model is now known to act infrequently. One experimental finding inconsistent with the copy-choice model is the observation that when *E. coli* bacteria are infected by T4 bacteriophage of two different genotypes whose DNAs are distinctly marked, one by ^{32}P and the other by bromouracil, recombinant DNAs containing both markers are found under conditions in which DNA synthesis is blocked. How are these findings inconsistent with the copy-choice model but consistent with the breakage–reunion model for genetic recombination?

9. Relate genetic recombination to exon shuffling, that is, the rearrangement of exons to form new proteins.

10. Suppose that a plasmid with a single origin of replication on its circular chromosome and containing only genes A, B, C, and D begins to replicate rapidly at time $t = 0$. At $t = 1$, there are twice as many copies of genes B and C as there are copies of genes D and A. Is it possible to establish the order of the four genes on the plasmid? Explain.

11. Suppose that a bacterial mutant is found to replicate its DNA at a very low rate. Upon analysis, it is found to have normal levels of activity of DNA polymerases I and III, DNA gyrase, and DNA ligase. It also makes normal amounts of the wild-types of dnaA, dnaB, dnaC, and SSB proteins. The sequence of the *oriC* region of its chromosome is found to be wild type. What defect might account for the abnormally low rate of DNA replication in this mutant? Explain.

12. Which of the following mutations in a polypeptide chain could have been induced by a single hit of the mutagen 5-bromouracil, which rarely forms a base pair with guanine, on the coding strand DNA? (See the genetic code on page 133 of the text.)

    (a) Phe → Glu   (c) Phe → Leu
    (b) Asp → Ala   (d) Met → Lys

13. Hydroxylamine reacts readily with cytosine to yield a product that base pairs with adenine. Which of the following mutations could result from the action of a single-hit reaction of hydroxylamine with DNA? (See the genetic code on page 133 of the text.)

    (a) Gln → Asn   (c) His → Tyr
    (b) Glu → Lys   (d) Gly → Asp

14. The drug fluorouracil is used as an anticancer agent. It irreversibly inactivates the enzyme thymidylate synthase. Explain how this treatment retards the growth of tumor tissue. Will the growth of normal cells be affected as well?

15. Mammalian cells of two differing genotypes can be fused, usually in the presence of Sendai virus, to form multinucleate cells (heterokaryons) containing nuclei of both genotypes. When fibroblasts from two patients suffering from xeroderma pigmentosum were fused, the resulting heterokaryons showed no deficiency in DNA repair. What conclusions can be drawn from this observation? Explain.

16. Physical studies on the interaction of the β_2 subunit of DNA polymerase III holoenzyme show that the β_2 subunit binds much more tightly to circular than to linear DNA molecules.

    (a) Propose an explanation for this observation.

(b) What do you think would happen if the circular DNA were treated with a double-strand hydrolyzing endonuclease?

17. Eukaryotic DNA can be highly methylated at the C-5 position of cytosine. The degree of methylation is inversely correlated with gene expression (page 945 of the text). Although the exact role of C-5 methylation in gene expression has not been determined, it is known that these C-5–methylated cytosines can cause mutations. How?

18. Acyclovir is an antiviral agent used to reduce the pain and promote the healing of skin lesions resulting from adult chicken pox. Its structure is shown in Figure 28.2.

FIGURE 28.2   Structure of acyclovir.

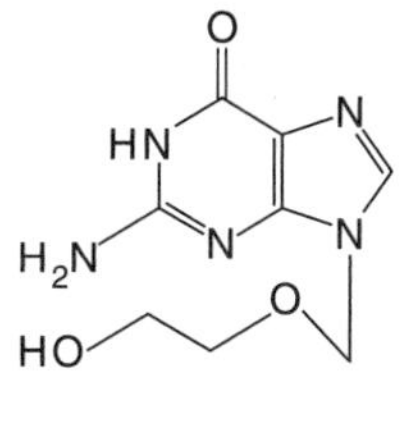

(a) What nucleoside does this drug resemble?
(b) Why must the drug be administered in dephosphorylated form?
(c) Acyclovir has very few side effects because it inhibits DNA replication only in herpes-infected cells. This is because all herpes viruses encode a thymidine kinase gene that is able to activate the drug. What is this activating reaction?
(d) Once acyclovir is activated, how does it inhibit DNA replication?
(e) Why are cells uninfected by virus relatively unaffected by acyclovir?
(f) The herpes virus can become insensitive to acyclovir therapy by mutations in either of two genes. What might they be?

19. A covalently closed circular DNA plasmid DNA molecule with 3000 base-pairs is completely relaxed in the solution in which it exists, that is, its $Lk = 0$. Would introducing two negative supercoils ($Lk = -2$) into the molecule require energy? Explain how you arrived at your answer. Name an enzyme that can put negative supercoils into the plasmid.

ANSWERS TO PROBLEMS

1. At the end of Chapter 9 there is a discussion of the "P-loop NTPase" superfamily, which describes a wide variety of proteins and enzymes that use this structure to propel "springs, motors, and clocks." Suffice it to say that the P-loop binds to the phosphates in ATP or other nucleoside triphosphates, and that the binding can lead to large conformational changes. So the task of a helicase, which is like unraveling two twisted threads, is accomplished by binding single strands and pulling them apart. And this is accomplished by binding and hydrolyzing molecules of ATP at the P-loops. The wide variety of other proteins and enzymes in this family all make use of NTP binding and concomitant conformational changes.

2. In the text there is a description of the system in which MutS and MutL detect a mismatch and then recruit MutH to cut the strand with the incorrect nucleotide (text Fig. 28.36). While it is not spelled out in the textbook, many forms of both DNA and RNA are extensively methylated. At least in gram negative bacteria, "hemimethylated" DNA provides a clue. The older strand has been around long enough to accumulate methyl groups, whereas the newly made strand has not. Therefore the

mismatch repair system (in E. coli and other bacteria) "believes" the older strand and the MutH nicks and initiates repair in the newer, unmethylated strand.

3. In the overall reaction, ATP is hydrolyzed to AMP and pyrophosphate. Only AMP would be labeled. The phosphate involved in the formation of the phosphodiester bond is furnished by the polynucleotide chain and does not arise from ATP.

4. $Lk = 24$, $Tw = 25$, and $Wr = -1$. Topoisomerase I increases the linking number of DNA by 1 each catalytic cycle (pages 826–828 of the text). This increase comes about at the expense of unwinding the negative supercoil.

5. $Lk = 21$, $Tw = 25$, and $Wr = -4$. DNA gyrase catalyzes a reaction in which both DNA strands are broken, the linking number is decreased by 2, and the number of negative supercoils is correspondingly increased by 2. The answers given for this and the preceding question are approximate because solution conditions affect the distribution between twists and supercoils.

6. The *polA1* mutants are extraordinarily sensitive to ultraviolet irradiation because they are deficient in the $5' \rightarrow 3'$ exonuclease activity of DNA polymerase I and are therefore impaired in DNA repair. They have only 1% of the activity of their wild-type counterpart and cannot efficiently remove thymine dimers formed by UV light. They can, however, replicate their DNA at normal rates because DNA polymerase III is the enzyme that is primarily responsible for DNA replication. An enzyme, RNaseH, which hydrolyzes RNA only when it is base paired to DNA, likely replaces the $5' \rightarrow 3'$ exonuclease in processing the Okazaki fragments.

7. (a) 30% T, 20% A, 15% G, 35% C
   (b) 25% A, 25% T, 25% C, 25% G. The base composition of the double strand is the average of that of the two single strands.

8. First, if copy-choice were a correct model, no recombinant phage should be produced in the absence of new DNA synthesis. Second, according to that model, no recombinant DNA duplexes should contain both bromouracil and ^{32}P. However, the Holliday model for homologous recombination (pages 845–846 of the text) accounts for how different labels from different DNA molecules could occur in the same progeny molecule.

9. Exons often encode protein domains. Genetic recombination can lead to rearrangements in the order of exons in a gene. Upon expression, such rearranged genes could give rise to proteins with new domain orders and possibly new capabilities (pages 136–137 of the text).

10. The order cannot be unambiguously established from the information given. Two possibilities are shown in Figure 28.3.

FIGURE 28-3 Two possible gene arrangements for problem 11.

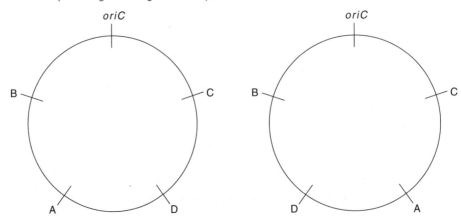

11. A decrease in the activity of primase would account for the low rate of DNA replication. Synthesis of DNA itself requires the prior synthesis of RNA primers. Also, decreased rates of dNTP synthesis could slow replication.

12. The mutagen 5-bromouracil changes A·T pairs to G·C pairs and G·C pairs to A·T pairs. The mutation in (c) could be induced by 5-bromouracil. For example, the DNA sequence AAA, which codes for phenylalanine, could be changed to the sequence AAG, which codes for leucine. The other mutations could not arise from treatment with 5-bromouracil. Remember that the genetic code presented in the text is expressed in terms of RNA. The sequence UUU on RNA corresponds to the sequence AAA on the informational strand of DNA. Leucine is encoded by the sequence CUU on RNA, which corresponds to the sequence AAG on the informational strand of DNA. Remember also that, unless otherwise specified, nucleotide sequences are written in the $5' \longrightarrow 3'$ direction.

13. Hydroxylamine causes the unidirectional change of C·G pairs to T·A pairs. The mutation in (a) cannot result from the action of hydroxylamine. Those in (b), (c), and (d) might. In (b), TTC (Glu) could change to TTT (Lys). In (c), ATG (His) could be converted to ATA (Tyr). In (d), ACC (Gly) could change to ATC (Asp), or GCC (Gly) could change to GTC (Asp).

14. Because thymidylate synthase is inactivated, the supply of dTTP is insufficient to support the synthesis of DNA at normal rates. If DNA synthesis is suppressed, so too will be the rate of division of the tumor cells. This type of treatment takes advantage of the fact that tumor cells divide more rapidly than do normal cells. The dosage of the drug is adjusted so that it will primarily affect more rapidly dividing cells. However, the division of some normally rapidly dividing cells, for example, those lining the intestinal tract and blood forming cells, may be retarded as well.

15. The fibroblasts from the two patients show complementation (the defect in each is remedied by the other), so it is likely that the two patients suffer from different genetic variants of xeroderma pigmentosum. The action of several genes is likely responsible for the excision and subsequent repair of damaged DNA. One patient, for example, could have produced a normal nuclease that excises damaged DNA but have been deficient in a ligase. The other patient could have produced normal ligase but have been deficient in nuclease activity. There are at least nine different complementation groups among xeroderma patients.

16. (a) A possible explanation is that the β_2 subunit falls off the end of a linear molecule whereas it is trapped on the circular molecule because it forms a torus around the DNA.
    (b) Treatment of circular DNA with an endonuclease cleaving both strands would convert it to a linear DNA, and the β_2 protein could dissociate from the free end, thereby decreasing its apparent affinity.

17. C-5 methyl cytosine can spontaneously deaminate just as cytosine can. When C-5 methyl cytosine deaminates, it forms thymidine, not uracil. Therefore uracil N-glycosylase, a DNA repair enzyme, will not recognize this product of deamination as an inappropriate base and will not remove it from the DNA, causing a transition mutation.

18. (a) guanosine
    (b) The phosphorylated form could not cross cell membranes.
    (c) The activation involves adding phosphates onto the distal —OH group at the expense of ATP hydrolysis by kinases to produce a compound resembling 5′-GTP. (Although one of the enzymes responsible for the activation is called *thymidine kinase,* it can activate other nucleosides and some of their analogs as well.)

(d) The triphosphate of acyclovir will serve as a substrate for DNA polymerase, and acyclovir nucleotide residues will be incorporated into growing polynucleotide chains in the place of guanosine resiues. Acyclovir will, however, cause premature termination of nascent polynucleotide chains because it lacks a free $-OH$ group onto which further nucleotides can be linked by phosphodiester bonds.

(e) Thymidine kinase is encoded by a viral, not a host, gene. Therefore, uninfected cells will lack the susceptible enzyme.

(f) Mutations that impair the ability of either the viral thymidine kinase or DNA polymerase to use the analog would render infected cells insensitive to the agent.

19. Yes, energy would be required because introducing a negative supercoil is the energetic equivalent of unwinding a turn of the helix. Duplex DNA is stable because the complementary strands are held together by stacking forces and hydrogen bonds. Separating the strands requires energy. Heat can supply the energy in vitro or helicases that use ATP can provide it within the cell. The prokaryotic enzyme DNA gyrase could be used to introduce the negative supercoils in the plasmid DNA. ATP would be hydrolyzed in the process. Recall that the supercoiling would be partitioned between supercoils and partial unwinding of the helix (page 826 of the text).

20. Z-DNA is rare in the cell because it is not a stable form of DNA. Recall that high salt concentrations are required to stabilize Z-DNA when it is formed by short oligo-deoxyribonucleotides. Also recall that sequences of alternating pyrimidines and purines are required to form Z-DNA. Negative supercoiling can also drive the formation of stretches of Z-DNA within B-DNA if the appropriate sequences are present. To investigate whether Z-DNA had a biological function, you could search genomes for sequences of alternating sequences that might form it. Their presence might suggest the possibility of Z-DNA existence in the cell. You could also look in vitro for proteins that would bind Z-DNA. Were you to find Z-DNA–specific binding proteins, that would suggest possible function for this form of DNA. Alexander Rich and coworkers have identified proteins that contain domains capable of binding specifically to Z-DNA. Their existence is taken as evidence that Z-DNA plays a role in DNA metabolism, although the details are not yet clear. See Rich, A., and Zhang, S. (2003) *Nature Reviews Genetics* 4:566–572.

RNA Synthesis and Processing

The conversion of DNA nucleotide sequences into RNA sequences is an early step in the expression of genetic information (see Chapter 4). Chapter 29 describes the DNA-dependent RNA polymerases that catalyze this reaction, and the various ways in which the product RNA transcript must sometimes be cleaved and modified before it becomes functional.

The chapter begins with an overview of the three stages of RNA synthesis: initiation, elongation, and termination. The subunit structure of RNA polymerase from *E. coli* is described and a distinction between the core and holoenzymes is made. The nature of prokaryotic promoters is given, and the function of the σ subunit of RNA polymerase in specific transcript initiation is explained. The important role that supercoiling plays in transcription initiation is emphasized. The structure of the ternary elongating complex, consisting of the template DNA, RNA polymerase, and product RNA, is described. Two mechanisms of transcription termination, one of which requires the ρ protein, are detailed. Two antibiotics, rifampicin and actinomycin D that inhibit transcription by different mechanisms are described. The cleavage and modification reactions involved in the maturation of ribosomal and transfer RNAs are outlined.

The authors next turn to the process of transcription in eukaryotes, which is more complex than in prokaryotes. The impossibility of coupling transcription and translation in eukaryotes as occurs in prokaryotes because of eukaryotic subcellular separation in the nucleus and cytoplasm, is pointed out, as is the more complex nature of eukaryotic transcription regulation. The three eukaryotic RNA polymerases that carry out transcription are described and related to the kinds of RNA they synthesize. The roles of the eukaryotic TATA box and the TATA-box-binding protein in forming the Transcription Factor IIFD (TFIID) complex that is active in basal transcription are explained, as are some other eukaryotic promoters and enhancers and the proteins that bind them. The reactions that modify the 5′ and 3′ ends of typical eukaryotic mRNA transcripts to cap and add a poly(A) tail are described, as are ways in which the nucleotide sequence of certain mRNAs can be modified by base alterations and insertions (a process called RNA editing). Splicing—the molecular machinery and reactions that remove introns from eukaryotic mRNA—is described along with the consequences of alternative splicing reactions. The chapter concludes with the discovery of catalytic RNA and the mechanism of RNA-catalyzed self-splicing in *Tetrahymena*.

LEARNING OBJECTIVES

When you have mastered this chapter, you should be able to accomplish the following objectives.

Introduction

1. Define *transcription*.
2. Name the three *stages* of *RNA synthesis* and list the functions of RNA polymerase in these processes.

RNA Polymerases Catalyze Transcription (Text Section 29.1)

3. Describe the *subunit structure* of *RNA polymerase* from *E. coli* and assign functions to the individual subunits.
4. Recognize the convention for numbering the nucleotides in the DNA template with regard to the *transcription start site*. Distinguish between the *template* (or *antisense*) *strand* and the *coding* (or *sense*) *strand* of the duplex DNA template.
5. Note the *consensus sequences* around the -35 and -10 positions and the *upstream element* of *E. coli* promoters. Contrast the rates of transcript initiation on *strong* and *weak promoters* in *E. coli*.
6. Know that RNA polymerases backtrack and correct errors.
7. Explain how the σ *factor* enables RNA polymerase to *recognize promoters*. Distinguish between the σ^{70} and σ^{32} subunits, contrast the sequences of *standard promoters* and *heat-shock promoters,* and provide examples of how σ factors can determine which genes are expressed.
8. Distinguish between *closed* and *open promoter complexes*.
9. Detail the de novo initiation of chain growth by RNA polymerase, name the usual initiating ribonucleoside, and describe the chemical nature of the 5′ end of the RNA.
10. Describe the model for the *transcription bubble*. State the *number of base pairs in the RNA–DNA hybrid*. Appreciate the *rate of RNA chain elongation* in terms of both the nucleotides added and the distance on the template traversed by RNA polymerase.
11. Define the term *riboswitch,* and explain how mRNA can sense levels of metabolites or signaling molecules.
12. Contrast ρ-*dependent* and ρ-*independent* transcription termination. Outline the mechanisms of the ρ protein and explain the role of *ATP hydrolysis* in its function.
13. Describe the mechanisms of *inhibition of transcription* by *rifampicin* and *actinomycin D*.
14. Outline the processing and modification of the precursors of rRNA and tRNA in prokaryotes.

Transcription In Eukaryotes Is Highly Regulated (Text Section 29.2)

15. Note the spatial and temporal differences in transcription and translation between prokaryotes and eukaryotes. Consider the regulatory implications of these differences.

16. Contrast *mRNA processing* in prokaryotes and eukaryotes. Outline the major reactions that process the *primary RNA transcripts* of eukaryotes.

17. Describe the *RNA polymerases* of eukaryotes, locate them within the cell, and list the kinds of RNA they synthesize. Account for the toxic effects of α -amanitin, and describe how it can be used to differentiate the three eukaryotic RNA polymerases.

18. List the salient *sequence elements* of *eukaryotic promoters*. Contrast the nucleotide sequences and locations of the *TATA box* of eukaryotes and the -10 sequence of prokaryotes.

19. Explain the key role of *TATA-box-binding protein (TBP)* in assembling active *TFII transcription complexes*. Describe the role of phosphorylation of the *carboxy-terminal domain (CTD)* of RNA polymerase II.

20. Outline the combinatorial activity of *transcription factors* and other DNA-binding regulatory proteins in directing RNA polymerase to specific genes. Indicate the properties of *enhancer* sequences.

The Transcription Products of Eukaryotic Polymerases Are Processed
(Text Section 29.3)

21. Draw the structure of the *5′ end* of a typical *eukaryotic mRNA* and distinguish *caps 0, 1, and 2.* Outline the reactions required to cap the primary transcript.

22. Describe the events leading to the production of mRNA with a *poly(A) tail.*

23. Describe how *microRNAs* are produced by cleavage of larger transcripts.

24. Describe *RNA editing* and provide examples of its effects on gene expression.

25. Describe the *splicing of eukaryotic mRNA*, give the *consensus sequences* at the *splice site junctions*, and designate the other nucleotide *sequence elements* involved in the process. List the functions of this *posttranscriptional modification*, and explain the importance of *alternative splicing* in gene expression.

26. Describe the role of *transesterification reactions* in splicing, and compare the number of phosphodiester bonds broken and formed during mRNA splicing.

27. Describe the *spliceosome* and detail the structural and catalytic involvement of *small nuclear ribonucleoprotein particles* (snRNPs) in mRNA splicing.

The Discovery of Catalytic RNA Was Revealing in Regard to Both Mechanism and Evolution (Text Section 29.4)

28. Outline the reactions that occur during the *self-splicing* conversion of a ribosomal RNA (rRNA) precursor from *Tetrahymena* into the mature rRNA plus other linear and circular products. Explain the role of *guanosine* or a *guanylyl nucleotide* in the self-splicing reaction.

29. Contrast the *group I* and *group II self-splicing* introns. Compare *spliceosome-catalyzed* splicing with self-splicing.

SELF-TEST

RNA Polymerases Catalyze Transcription

1. Give the subunit composition of the RNA polymerase of E. coli for both the holoenzyme and the core enzyme.

2. Match the subunit of the RNA polymerase of *E. coli* in the left column with its putative function during catalysis from the right column.

   (a) α
   (b) β
   (c) β′
   (d) σ⁷⁰

   (1) binds the DNA template
   (2) binds regulatory proteins and sequences
   (3) binds NTPs and catalyzes bond formation
   (4) recognizes the promoter and initiates synthesis

3. Which of the following statements about *E. coli* promoters are correct?
   (a) They may exhibit different transcription efficiencies.
   (b) For most genes they include variants of consensus sequences.
   (c) They specify the start sites for transcription on the DNA template.
   (d) They have identical and defining sequences.
   (e) They are activated when C or G residues are substituted into their −10 regions by mutation.
   (f) Those that have sequences that correspond closely to the consensus sequences and are separated by 17 base pairs are very efficient.

4. The sequence of a duplex DNA segment in a longer DNA molecule is

   5′-ATCGCTTGTTCGGA-3′

   3′-TAGCGAACAAGCCT-5′

   When this segment serves as a template for E. coli RNA polymerase, it gives rise to a segment of RNA with the sequence 5′-UCCGAACAAGCGAU-3′. Which of the following statements about the DNA segment are correct?

   (a) The top strand is the coding strand.
   (b) The bottom strand is the sense strand.
   (c) The top strand is the template strand.
   (d) The bottom strand is the antisense strand.

5. The text states that RNA polymerases backtrack for proofreading. Why would this be energetically unfavorable? What is the error rate of RNA synthesis?

6. RNA utilizes uracil instead of thymine. Why?

7. Which of the following statements about the σ subunit of RNA polymerase are correct?
   (a) It enables the enzyme to transcribe asymmetrically.
   (b) It confers on the core enzyme the ability to initiate transcription at promoters.
   (c) It decreases the affinity of RNA polymerase for regions of DNA that lack promoter sequences.
   (d) It facilitates the termination of transcription by recognizing hairpins in the transcript.

8. When growing *E. coli* are subjected to a rapid increase in temperature, a new and characteristic set of genes is expressed. Explain how this alteration in gene expression occurs.

9. Match the regions of a ρ-independent transcription termination signal in a DNA template in the left column with the structures or the functions performed by the encoded transcript segments in the right column.

(a) GC-rich palindromic region
(b) AT-rich region

(1) oligo(U) stretch in RNA
(2) hairpin in RNA
(3) promotes the dissociation of RNA–DNA hybrid helix
(4) causes the enzyme to pause

10. Which of the following statements about the ρ protein of *E. coli* are correct?

(a) It is an ATPase that is activated by binding to single-strand DNA.
(b) It recognizes specific sequences in single-strand RNA.
(c) It recognizes sequences in the DNA template strand.
(d) It causes RNA polymerase to terminate transcription at template sites that are different from those that lead to ρ-independent termination.
(e) It acts as a RNA–DNA helicase.

11. Match the functions in the right column with the appropriate antibiotic inhibitor of *E. coli* transcription in the left column.

(a) rifampicin
(b) actinomycin D

(1) interacts with the template
(2) interacts with nascent mRNA polymerase
(3) prevents initiation
(4) prevents elongation
(5) intercalates into mRNA hairpins

12. Explain how a mutation might give rise to an *E. coli* that is resistant to the antibiotic rifampicin.

Transcription in Eukaryotes Is Highly Regulated

13. Which of the following statements about eukaryotic mRNAs are correct?

(a) They are derived from larger RNA precursors.
(b) They result from extensive processing of their primary transcripts before serving as translation components.
(c) They usually have poly(A) tails at their 5′ ends.
(d) They have a cap at their 3′ ends.
(e) They are often encoded by noncontiguous segments of template DNA.

14. Match the descriptions in the right column with the appropriate eukaryotic DNA-dependent RNA polymerase(s) in the left column.

(a) RNA polymerase I
(b) RNA polymerase II
(c) RNA polymerase III

(1) is located in the nucleolus
(2) is located in the nucleoplasm
(3) makes mRNA precursors
(4) makes tRNA precursors
(5) makes 5S rRNA
(6) makes 18S, 5.8S, and 28S rRNA precursors
(7) is strongly inhibited by α-amanitin
(8) synthesizes RNA in the 5′ → 3′ direction
(9) is composed of several subunits
(10) has a subunit with repeated amino acid sequences subject to phosphorylation

15. List the major sequence features of promoters for eukaryotic mRNA genes.

16. Describe the role of TATA-box-binding protein (TBP, TFIID) in forming the basal transcription apparatus. What properties does TBP confer on this apparatus?

17. Which of the following statements about enhancers are correct?

    (a) They function as promoters.
    (b) They function in either orientation in the DNA.
    (c) They function on either side of the activated promoter.
    (d) They function even when located many base pairs away from the promoter.
    (e) They function only in specific types of cells.

The Transcription Products of Eukaryotic Polymerases Are Processed

18. Which of the following statements about the poly(A) tails that are found on most eukaryotic mRNAs are correct?

    (a) They are added as preformed polyriboadenylate segments to the 3′ ends of mRNA precursors by an RNA ligase activity.
    (b) They are encoded by stretches of polydeoxythymidylate in the template strand of the gene.
    (c) They are added by RNA polymerase II in a template-independent reaction using ATP as the sole nucleotide substrate.
    (d) They are added by poly(A) polymerase using dATP as the sole nucleotide substrate.
    (e) They are cleaved from eukaryotic mRNAs by a sequence-specific endoribonuclease that recognizes the RNA sequence AAUAAA.

19. What functions are the caps and tails of mRNAs thought to perform?

20. Which of the following statements about apolipoprotein B (apo B) are correct?

    (a) The apo B-48 form is formed by the proteolytic cleavage of the primary (apo B-100) translation product.
    (b) Apo B-48 and apo B-100 are formed in different tissues.
    (c) Apo B-48 arises from the expression of a form of the gene for apo B-100 that has been shortened by nonhomologous recombination.
    (d) The transcript of the apo B-100 gene is spliced to remove a segment and form apo B-48.
    (e) A specific nucleotide in the apo B-100 transcript is altered, thereby creating a stop codon in the mRNA.

21. Which of the following are important sequence elements in the splicing reactions that produce eukaryotic mRNAs?

    (a) exon sequences located between 20 and 50 nucleotides from the 5′ splice site
    (b) exon sequences located between 20 and 50 nucleotides from the 3′ splice site
    (c) intron sequences located between 20 and 50 nucleotides from the 5′ splice site
    (d) intron sequences located between 20 and 50 nucleotides from the 3′ splice site
    (e) intron sequences at the 5′ splice site
    (f) intron sequences at the 3′ splice site

22. Eukaryotic mRNA splicing involves which of the following?

    (a) the formation of 2′ → 5′ phosphodiester bonds
    (b) a sequence-specific endoribonuclease that hydrolyzes the phosphodiester bond at the junctions of the intron with the exon

(c) the spliceosome

(d) the coupling of phosphodiester bond formation to ATP hydrolysis

(e) the formation of lariat intermediates

The Discovery of Catalytic RNA Was Revealing in Regard to Both Mechanism and Evolution

23. Place the following events in the order in which they occur during the formation of mature rRNA in *Tetrahymena*.

    (a) The 3′-hydroxyl of a guanine nucleoside attacks the phosphodiester bond at the 5′ splice site leaving the 5′ (upstream) exon with a free 3′-hydroxyl and attaching the guanine nucleoside to the 5′ end of the intron.

    (b) The transcript from the rRNA gene folds into a specific structure.

    (c) The rRNA transcript specifically binds a guanine nucleoside or nucleotide.

    (d) Self-splicing occurs within the intron to form L19 RNA.

    (e) The 3′-hydroxyl of the 5′ exon attacks the bonds at the 3′ splice junction to form the spliced rRNA and eliminate the intron.

24. Match the statements in the right column with the appropriate type(s) of splicing in the left column.

    (a) group I splicing

    (b) group II splicing

    (c) nuclear mRNA splicing

    (1) A $2' \rightarrow 5'$ phosphodiester bond is involved.

    (2) Nuclease and ligase activities are required.

    (3) Transesterification reactions break and form bonds.

    (4) A guanine nucleoside or nucleotide is required.

    (5) Spliceosomes are required.

    (6) Lariat intermediates are involved.

    (7) snRNPs are involved.

ANSWERS TO SELF-TEST

1. The holoenzyme has the subunit composition $\alpha_2\beta\beta'\omega\sigma$. The core enzyme lacks the σ subunit.

2. (a) 2   (b) 3   (c) 1   (d) 4

3. a, b, c, f. The promoters of most *E. coli* genes include variants of defining consensus sequences that are centered at about the -35 and -10 positions. The nearer the sequences of a promoter are to the consensus sequence and the nearer the separation between them is to the optimal 17-bp spacing, the more efficient the promoter. The -10 consensus sequence is TATAAT. The substitution of a C or G into the sequence would likely lower the efficiency of a promoter.

4. b, c. The sense (bottom) strand of the template DNA has the same sequence as the mRNA.

5. Backtracking involves breaking hydrogen bonds between an RNA-DNA base pair, but if the base pair is incorrect there will often be fewer hydrogen bonds to break. The

text states that the final error rate is between one mistake in 10^4 and one in 10^5 nucleotides added. Compare this to the DNA error rate (after various corrective processes have been applied, including mismatch repair) of one in 10^9–10^{10}.

6. DNA is passed onto progeny, and is an archive of genes for the cell. In comparison, RNA is "single use" and much like a protein, it is made to do a job in the cell. If DNA is the encyclopedia, then perhaps RNA is a blueprint used at a construction site, and if the blueprint gets torn or dirty in use, then the harm is minimal. The same point is made in the previous question. Uracil can be produced by oxidative degradation of cytosine, and in RNA the new uracil can't be told from original uracil. In DNA thymine provides a methyl marker that allows the incorrect uracils to be removed.

7. a, b, c. The σ subunit recognizes promoter sites, decreases the affinity of the enzyme for regions of DNA lacking promoter sequences, and facilitates the specific, oriented initiation of transcription. Orienting the binding of the enzyme to the DNA results in only one of the two DNA strands functioning as a template for RNA transcription; that is, it gives rise to asymmetric transcription.

8. The temperature increase induces the synthesis of a new σ factor, σ^{32}, which directs RNA polymerase to promoters that have -10 and -35 sequences different from those recognized by σ^{70}. Transcription from these promoters gives rise to characteristic heat-shock proteins.

9. (a) 2, 4   (b) 1, 3

10. d, e. The ρ protein recognizes and binds stretches of RNA that are devoid of hairpins and are at least 72 nucleotides long. It acts to hydrolyze ATP and to unwind the RNA–DNA hybrid in the transcription bubble.

11. (a) 2, 3   (b) 1, 4. Actinomycin D intercalates only into duplex DNA.

12. Rifampicin must bind to the β subunit of RNA polymerase to inhibit the enzyme. A mutation in the gene encoding this subunit that would interfere with the binding of the antibiotic but not with polymerization would produce a rifampicin-resistant cell.

13. a, b, e. Answers (c) and (d) are incorrect because the poly(A) tail is found at the 3′ end and a cap is found at the 5′ end of typical eukaryotic mRNA. Segments of the primary transcript are discarded by RNA splicing to bring together RNA sequences encoded by noncontiguous regions of the template; that is, introns are removed to form a product smaller than the primary transcript.

14. (a) 1, 6, 8, 9   (b) 2, 3, 7, 8, 9, 10   (c) 2, 4, 5, 8, 9. Both RNA polymerases I and III are involved in rRNA synthesis, with polymerase I synthesizing the 18S, 5.8S, and 28S precursor transcripts and polymerase III synthesizing the 5S rRNA.

15. A TATA box centered at about -25 from the start site of transcription and consisting of a variant of the consensus heptanucleotide sequence *TATAAAA* is usually essential for promoter activity. Nearby, an initiator element (Inr) defines the transcriptional start site. In promoters lacking a TATA box, a downstream promoter element is sometimes paired with the Inr site on its 3′ side. In addition, many genes have a CAAT box, GC boxes, or both elements located between -40 and -150. These sequences can function in either orientation, that is, be in either DNA strand. Genes that are expressed constitutively, as distinct from those whose expression is regulated, tend to have GC boxes. Activating sequences farther upstream of these promoter elements are necessary for the functioning of most promoters.

16. TBP serves a critical role as an enucleating center for the assembly of the minimal molecular apparatus (the basal transcription assembly) required for transcription by RNA polymerase II. After TBP binds the TATA box, other TFII proteins and RNA poly-

merase II join the supramolecular complex to render it transcriptionally competent. In addition to enabling the assembly of the apparatus through its ability to recognize and bind the TATA box, TBP binds this asymmetric sequence in one orientation that "points" the RNA polymerase in the right direction, thereby defining the strand of DNA that will serve as the template.

17. b, c, d, e. Answer (a) is incorrect because enhancer sequences do not serve as promoters per se. They will not by themselves enable RNA polymerase II to initiate a transcript and will function only when the basal transcription apparatus exists.

18. None of the statements is correct. Poly(A) polymerase uses ATP, not dATP, to add a stretch of A residues to the 3′-hydroxyl formed by the cleavage of an mRNA precursor by a specific ribonuclease that recognizes the upstream sequence AAUAAA within particular mRNA sequence contexts.

19. The 5′ caps are thought to contribute to the stability and efficiency of translation of the mRNA. The poly(A) tails probably perform the same functions, albeit by different, unknown mechanisms.

20. b, e. The apo B system illustrates one kind of RNA editing. The pre-mRNA transcript has its sequence altered—in this case, a specific cytidine is deaminated to uracil to create the stop codon. Other editing mechanisms involve the addition and removal of U residues, an A-to-G change, and an adenosine-to-inosine change.

21. d, e, f

22. a, c, e. Answers (b) and (d) are incorrect because hydrolysis of phosphodiester bonds is not involved during mRNA splicing up to the point of intron removal and lariat formation; only transesterification reactions occur. Consequently, ATP is not required for the synthesis of phosphodiester bonds.

23. b, c, a, e, d

24. (a) 3, 4   (b) 1, 3, 6   (c) 1, 3, 5, 6, 7. The splicing of eukaryotic tRNAs requires protein nucleases to cut the phosphodiester bonds at the intron–exon junctions and protein RNA ligases to form the bonds joining the exons.

PROBLEMS

1. The rate constant for the binding of RNA polymerase holoenzyme to a promoter on a long DNA molecule is greater than that for the collision of two small molecules in solution. Since small molecules diffuse through solutions more rapidly than large ones, how can this be true?

2. Explain how FMN can interact directly with mRNA to inhibit FMN production via a kind of RNA dependent "feedback inhibition." Refer to Figure 29.15. This interactive RNA loop is called a *riboswitch*. Are there other examples of RNA processes directly controlling the use of mRNA to produce proteins?

3. On page 853 of the text, it states that RNA polymerases "detect termination signals that specify where a transcript ends." Consider the ways in which transcription terminates in bacteria, and suggest an argument that the above statement might be considered incorrect.

4. The rho protein (ρ) that is involved in termination of bacterial transcription has six subunits and rotates around the emerging RNA strand as it "climbs" it, all propelled by conformational changes due to ATP hydrolysis. What sort of proteins come to mind as possible homologues?

5. One would expect an analog of 5'-ATP that lacks an oxygen at the 3' position of its ribose (3'-deoxy-5'-ATP; see Figure 29.1) to interrupt RNA formation because it cannot form phosphodiester bonds at its 3' position. Could such a compound be used to ascertain the direction of chain growth in RNA synthesis? Explain.

FIGURE 29.1 Structure of 3'-deoxy-5'-ATP.

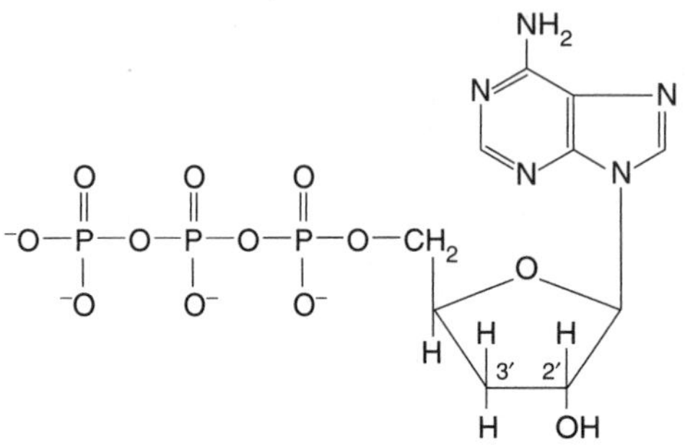

6. The ciliated protozoan *Tetrahymena* contains an enzyme that can synthesize 5'-pseudouridine monophosphate from a mixture of PRPP and uracil. For a time it was thought that this enzyme was instrumental in the synthesis of transfer RNAs in *Tetrahymena*. Explain why this is not the case.

7. When mammalian genes are cloned, a strategy that is frequently followed involves the isolation of mRNA rather than DNA from a cell and the preparation of a complementary DNA (cDNA) by the enzyme reverse transcriptase. Suppose that mRNA isolated from a cell specialized for the production of protein X is used as a template for the production of cDNA. What major difference or differences would you expect to find between the structure of that cDNA and genomic DNA for protein X?

8. Rifampicin specifically inhibits the initiation of transcription in prokaryotes and may therefore be used in humans as a therapeutic antibacterial agent. Would you expect actinomycin D to be useful in antibacterial therapy? Why or why not?

9. The mRNAs produced by mammalian viruses undergo modification at the 5' and 3' ends in a fashion similar to that of eukaryotic mRNA. Why do you think this is the case?

10. Sketch the most stable secondary structure that could be assumed by the oligonucleotide AAGGCCCUACGGGGCCG.

11. Suppose that human DNA is cleaved into fragments approximately the size of a given mature human messenger RNA and that mRNA–DNA hybrids are then prepared. The corresponding procedure is then carried out for *E. coli*. When the mRNA–DNA hybrids from each species are examined with an electron microscope, which will show the greater degree of hybridization? Explain.

12. What are snRNPs, and how are they involved in the eukaryotic mRNA splicing reaction?

13. Figure 29.40 in the text gives an example of how a base-change mutation (A → G) within an intron in the β-globin gene can produce a 5' splice site downstream from the normal 5' splice site. The result is that extra amino acids not normally present in the protein are inserted into the chain before synthesis is terminated by a stop codon.

   (a) Corresponding mutants in exons are also known in which base change mutations introduce 5' splice sites upstream from the normal 5' sites (see Figure 29.2A; X marks the location of the new 5' splice site). Sketch the resulting processed mRNA. What changes in amino acid sequence would you expect to result?

FIGURE 29.2A Creation of a new 5′ splice site in exon 2 of β-globin.

(b) In some forms of thalassemia, the creation of a new 5′ splice site within intron 2 activates a "cryptic" 3′ splice site upstream from the 5′ site (see Figure 29.2B; X marks the location of the new 5′ splice site and the cryptic 3′ side is labeled). Sketch the resulting processed mRNA. How would the resulting changes in amino acid sequence differ from those in (A)?

FIGURE 29.2B Activation of a cryptic 3′ splice site in intron 2 of β-globin.

14. In an attempt to determine whether a given RNA was catalytically active in the cleavage of a synthetic oligonucleotide, the following experimental results were obtained. When the RNA and the oligonucleotide were incubated together, cleavage of the oligonucleotide occurred. When either the RNA or the oligonucleotide was incubated alone, there was no cleavage. When the RNA was incubated with higher concentrations of the oligonucleotide, saturation kinetics of the Michaelis-Menten type were observed. Do these results demonstrate that the RNA has catalytic activity? Explain.

15. Negative supercoiling affects transcription at many promoters in *E. coli*. In addition to facilitating the unwinding of the DNA helix necessary to form a transcription bubble, how might negative supercoiling affect transcription either positively or negatively?

ANSWERS TO PROBLEMS

1. RNA polymerase holoenzyme has lower affinity for nonspecific DNA sequences than for promoter sequences. The nonspecific affinity, however, allows the enzyme to bind to "random-sequence" DNA and then "slide" along the molecule in a unidimensional random walk until it encounters a promoter sequence for which its binding affinity is higher. Diffusion in one dimension is much faster than diffusion in three dimensions, thereby explaining the observed rapid rate constant for the binding of RNA polymerase holoenzyme to promoter sequences. If one measured the encounter of the polymerase with the nonspecific regions of the DNA rather than with promoter sequences, the value of the rate constant would be much lower and would fit our expectations for a three-dimensional, diffusion-limited reaction between macromolecules.

2. The caption to Figure 29.15 provides all of the necessary information about how this riboswitch works. In the presence of FMN a "hairpin-poly-U" terminator forms so that the remainder of the mRNA is not even transcribed, so the protein is never built. This is very similar to a process called "attenuation" (see Chapter 31, Figs. 31.33 and 31.34) except that a ribosome is involved in the optional formation of the "hairpin

poly-U" terminator. It is also reminiscent of the iron-response element (IRE) (Fig. 31.37), except that in that case proteins are also involved. A riboswitch is unusual in that nothing is included except the mRNA and a metabolite molecule.

3. One method of termination is "hairpin-poly-U" (see text Fig. 29.14). When the hairpin forms, the formation stresses the weak A-U base pairs and the RNA is "yanked" from the active site of the enzyme. If formation of the hairpin is prevented (for example in the riboswitch, Fig. 29.15) then the sequence itself does not cause termination. So just what is the enzyme doing to cause termination? Arguably it is the RNA itself that is removing itself from the enzyme and causing termination. Similarly, ρ the rho protein (Fig. 29.17) may pull the RNA from the active site when it recognizes a target sequence.

4. The text points out that the rho protein is related to helicases, which also can have a circular hexameric structure, and can "pull the string" on a nucleic acid (Fig. 28.12). The discussion of helicases states that they are related to a large family of P-loop NTPase proteins, which include several interesting molecules. One of the most interesting connections is the relationship between the rho protein and the F1 ATPase in mitochondria, which is also a circular arrangement of six subunits and a rotary mechanism. This relationship has long been known [*J. Bacteriol.* (1994) 176:5033]. Of course F1 is forced to run "backwards," which causes ATP to be synthesized rather than broken down.

5. A 3′-deoxy analog of ATP could be used to establish the direction of chain growth. In 5′ → 3′ growth the analog would donate a nucleotide containing 3′-deoxyadenosine, and the polynucleotide chain would be terminated as a result. No additional nucleotides could be added because of the lack of a 3′-OH group on the terminal 3′-deoxyadenosine. In 5′ → 3′ growth the nucleotide could not be added to the growing polynucleotide chain because of the lack of the 3′-OH group.

6. Nascent polynucleotides formed by RNA polymerases contain only the four usual bases. Subsequently, some of the bases are chemically modified. Were unusual nucleotides to be incorporated into a growing RNA chain, this would in turn require the presence of unusual bases on DNA. The pseudouridine found in transfer RNAs is formed by breaking the nitrogen–carbon bond linking uracil to ribose and forming a carbon–carbon bond instead. Only certain uracils are modified in this manner, owing to their position in the three-dimensional structure of the RNA and to the specificity of the enzymes that carry out the modification.

7. The cDNA prepared from mRNA would have a long poly(T) tail, unlike genomic DNA. Remember that the poly(A) tail is added to the 3′ end of mammalian mRNA and that there is no counterpart on DNA. A second striking difference would be that the cDNA would contain no intervening sequences (introns) and would therefore be much shorter than the corresponding sections of genomic DNA. (Remember that most mammalian genes are mosaics of introns and exons.) A third difference would be found if any RNA editing were involved. An edited mRNA could generate a cDNA with nucleotides that did not correspond to those in genomic DNA.

8. In order to be useful as a therapeutic antibacterial agent, a compound must selectively inhibit processes in prokaryotes but leave the corresponding processes in eukaryotes (including those in mitochondria) largely unaffected. Because rifampicin selectively inhibits the initiation of transcription in prokaryotes but not in eukaryotes, it is useful as an antibacterial agent. Actinomycin D is an intercalating agent that binds to DNA duplexes and inhibits both DNA replication and transcription, although it has

a greater inhibiting effect on transcription than on replication. It cannot discriminate between the duplex DNA of bacteria and that of humans, however, and will therefore bind to both. Because it disrupts eukaryotic as well as prokaryotic processes, it is not very useful as an antibacterial agent. It is sometimes used as an anticancer agent, however, because of its ability to slow the replication rate of human DNA.

9. Viruses use the host's enzyme system to replicate their DNA and to synthesize their proteins. Since eukaryotic translation systems must synthesize viral protein, the structure of viral mRNAs must mimic that of the host mRNA.

10. The structure is a stem-and-loop ("lollipop") hairpin structure, as shown in Figure 29.14.

FIGURE 29.14 Stable secondary structure for the oligonucleotide in problem 8.

11. The mRNA–DNA hybrid of *E. coli* will show greater hybridization because it is produced continuously from a DNA template without processing. Because of the presence of intervening sequences in human DNA, there will be regions in the human RNA–DNA hybrids where no base pairing occurs.

12. The snRNPs are small ribonucleoprotein particles that occur in the nucleus. Each is composed of a small RNA molecule and several characteristic proteins, some of which are common to different snRNPs. Distinct snRNPs recognize and bind to splice junctions and the branch site and are involved in assembling the spliceosome in an ATP-dependent manner. They are requisite components of the splicing apparatus, and the RNA components of some of them are probably catalytically active. The RNAs of some snRNPs form hydrogen bonds with sequences within introns and exons to help to juxtapose properly the reacting splice junctions.

13. (a) The spliced gene arising from the mutation would be shorter than the correctly spliced version (see below) and some amino acids (those between the X in the figure and the right-hand edge of the box labeled "Exon 2") would be omitted from the region of the protein encoded by nucleotides between the newly introduced and normal 5′ splicing sites. Since the splicing machinery might also recognize the unchanged, normal 5′ splice site, some normal mRNA could also be formed.) Also, the introduction of a new splicing site could lead to changes in the reading frame, and therefore to gross changes in amino acid sequence, polypeptide chain length, or both.

Exon 2	Exon 3

(b) The resulting processed mRNA would be longer than the normal processed β-globin gene, introducing the possibility of a misfolded or malfunctioning protein or a truncated protein if a stop codon were introduced from the intron sequence. As in (a), a change in reading frame could also result. Again, some normal mRNA might be formed by recognition of the unchanged, original 5′ splice site.

14. These results alone do not establish that the RNA has catalytic activity, because a catalyst must be regenerated. It is entirely possible that the results observed could be accounted for by a stoichiometric, as opposed to a catalytic, interaction between RNA and the oligonucleotide, in which the RNA may "commit suicide" as the oligonucleotide is cleaved. In such an interaction, a portion of the RNA would participate chemically in the cleavage of the oligonucleotide, but it would also be cleaved itself as a part of the reaction. Four reaction products would accumulate, two resulting from the cleavage of RNA and two from the cleavage of the oligonucleotide. To show that this particular RNA was catalytic, it would be necessary to demonstrate that it turns over and is regenerated in the course of the reaction. You could perform experiments in which the putative catalytic RNA was present in much lower molar concentrations than those of the substrate oligonucleotide. If more substrate oligonucleotide molecules were hydrolyzed than the total number of the presumed catalytic RNA, you would have evidence that true turnover had occurred.

15. The interaction of the σ subunit of RNA polymerase with a promoter requires the precise positioning of the atomic determinants on the surface of the DNA at the −10 and −35 regions to which it binds. Particular hydrogen bond donors and acceptors in the grooves of the DNA in these regions must be positioned so that the amino acids on the protein can interact optimally with them. The degree of supercoiling of a DNA molecule affects the precise positioning of these groups because the base pairs are moved with respect to one another by twisting the DNA. Thus, supercoiling could either promote or hinder the interaction of the polymerase with the DNA by bringing the groups in the −10 region and in −35 regions of the DNA of particular promoters into a configuration that was better or worse for interaction. (If you are interested in this topic, see Hatfield, W.G., and Benham, C.J. (2002) *Annu. Rev. Genet.* 36:175–203.

Protein Synthesis

I n Chapter 30 the mechanism of protein synthesis, a process called translation, is examined in detail. Translation is a complicated process in which the four-letter alphabet of nucleic acids is translated into the 20-letter alphabet of proteins. The chapter begins with an introduction to the major components of translation—mRNA, tRNA, ribosomes, and aminoacyl-tRNA synthetases. The detailed structures and conformations of tRNAs, the adaptor molecules that recognize both the codons on the mRNAs and the enzymes that attach the corresponding amino acids, are discussed first. The wobble hypothesis is then presented to explain the lack of strict one-to-one Watson-Crick base-pairing interactions among the three nucleotides of the tRNA anticodons and the mRNA codons. Next the authors explain how amino acids are activated for the subsequent formation of peptide bonds through their attachment to tRNAs by the two classes of aminoacyl-tRNA synthetases. The exquisite specificity of these reactions is explored, in terms of correct binding of amino acids and tRNAs to a given synthetase. Threonyl-tRNA synthetase is used as an example of specificity at the level of amino acid selection. This enzyme discriminates between threonine and the isosteric valine and the isoelectronic serine, using a combination of selective binding at the active site and proofreading after aminoacylation. Several aminoacyl-tRNA synthetases are used as examples of ways in which the correct tRNA is chosen, ranging from those which require multiple contact points (glutaminyl-tRNA synthetase) to alanyl-tRNA synthetase, which will recognize a "microhelix" containing only the acceptor stem and a hairpin loop.

The authors next turn to the structure and composition of the ribosome, a molecular machine that coordinates charged tRNAs, mRNA, and proteins, leading to protein synthesis. The fact that the ribosome is now recognized to be a ribozyme, with the RNA components playing the major role in catalysis, is introduced. The experiments that showed the polarities of polypeptide formation and the translation of mRNA are presented next. Then initiation is described, and the roles of a specialized initiator tRNA, the mRNA start codon, and 16S rRNA sequences are outlined. The spatial and functional relationships of the sites on the ribosome that bind aminoacyl-tRNAs and peptidyl-tRNAs, the peptide-bond–forming reaction, the role of GTP, and the mechanism of the translocation of the peptidyl-tRNA from site to site on the ribosome are presented in the description of the elongation stage of protein synthesis.

The critical role that protein factors play in translation is discussed next, including initiation, elongation, and release factors. The termination of translation is outlined, and the role of release factors that recognize translation stop codons is described. The chapter closes with a brief overview of translation in eukaryotes, emphasizing the major contrasting features with respect to translation in prokaryotes. Differences in the initiator tRNA, the selection mechanism of the initiator codon, the ribosomes, and the overall complexity of the process are highlighted. Next, the mechanisms of several potent inhibitors of translation and the mechanism of the bacterial toxin that causes diphtheria is presented. The chapter ends with a discussion of how proteins are targeted to different cellular destinations. The role of signal sequences and signal-recognition particle are discussed as are transport vesicles.

LEARNING OBJECTIVES

When you have mastered this chapter, you should be able to accomplish the following objectives.

Introduction

1. List the key participants in protein synthesis.

Protein Synthesis Requires the Translation of Nucleotide Sequences into Amino Acid Sequences (Text Section 30.1)

2. Explain the effect of error frequency on synthesis of different-size proteins.

3. Draw the *cloverleaf structure* of a *tRNA* and identify the regions containing the *anticodon* and the *amino acid attachment site*.

4. List the features common to all tRNAs.

5. Relate the two-dimensional cloverleaf representation of the tRNA structure to its three-dimensional configuration.

6. Explain how some *codons are recognized by more than one anticodon,* that is, how they interact with more than one species of *aminoacyl-tRNA*. List the base-pairing interactions allowed according to the *wobble hypothesis.*

Aminoacyl-Transfer RNA Synthetases Read the Genetic Code (Text Section 30.2)

7. Write the two-step reaction sequence of the *aminoacyl-tRNA synthetases.* Enumerate the high-energy phosphate bonds that are consumed in the overall reaction.

8. Describe the mechanisms of amino acid selection and *proofreading* that contribute to the accuracy of the attachment of the appropriate amino acid to the correct tRNA.

9. Describe the different modes of recognition of the correct tRNA molecule by aminoacyl-tRNA synthetases.

10. Outline the distinguishing properties of *class I* and *class II* aminoacyl-tRNA synthetases.

The Ribosome is the Site of Protein Synthesis (Text Section 30.3)

11. List the kinds and numbers of macromolecular components of the prokaryotic ribosome. Give the *mass, sedimentation coefficient,* and *dimensions* of the *ribosome* of E. coli.

12. Outline the three-dimensional structure of a ribosome.

13. List the evidence that suggests that the RNA components of ribosomes have active roles in protein synthesis.

14. Recount the experiments that established the two ways in which the start of protein synthesis is determined.

15. Name and describe the functions of the three tRNA binding sites on the *70S ribosome.*

16. Name the major *initiator codon* and the amino acid it encodes. Explain the roles of the nucleotide sequences in *16S rRNA, mRNA,* and *tRNA* in *selecting the initiation codon* rather than the identical codon that encodes an internal amino acid.

17. Distinguish between the *initiator tRNA (tRNA$_i$)* and *tRNA$_m$* and outline the conversion of methionine into *formylmethionyl-tRNA$_f$*.

18. List the components of the *70S initiation complex* and indicate the roles of the *initiation factors (IF)* and GTP in its formation.

19. Account for the fact that internal AUG codons are not read by the initiator tRNA.

20 Outline the elongation stage of protein synthesis and describe the roles of the *elongation factors (EFs)* and GTP in the process. Locate the *aminoacyl-tRNAs* and *peptidyl-tRNAs* in the *A or P sites of the ribosome* during one cycle of elongation.

21. Describe how the GTP–GDP cycle of *EF-Tu* controls its affinity for its reaction partners. Relate the GTP-GDP cycle of EF-Tu with that of G proteins (Chapter 14).

22. Explain the role of EF-Tu in determining the accuracy and timing of protein synthesis.

23. Describe the mechanism of peptide bond formation in the *peptidyl transferase center.*

24. Outline the *translocation* steps that occur after the formation of a peptide bond and describe the roles of *EF-G* and GTP.

25. Name the *translation stop codons,* describe the termination of translation, and explain the roles of the *release factors (RFs)* in the process.

Eukaryotic Protein Synthesis Differs from Prokaryotic Protein Synthesis Primarily in Translation Initiation (Text Section 30.4)

26. Contrast eukaryotic and prokaryotic ribosomes with respect to composition and size.

27. Contrast the mechanisms of translation initiation in prokaryotes and eukaryotes. Note the different initiator tRNAs, AUG codon selection mechanisms, and numbers of IFs and RFs.

28. Discuss the structure of eukaryotic mRNA including a possible advantage in such a structure

29. Compare prokaryotic and eukaryotic elongation and termination.

30. Describe the physiological consequences of defects in *elF4-G* and *elF2.*

A Variety of Antibiotics and Toxins Can Inhibit Protein Synthesis (Text Section 30.5)

31. Provide examples of *antibiotics that inhibit translation,* and describe their mechanisms of action.

32. Describe the mechanisms by which *diphtheria toxin* and ricin inhibit protein synthesis in eukaryotes.

Ribosomes Bound to the Endoplasmic Reticulum Manufacture Secretory and Membrane Proteins. (Text Section 30.6)

33. Discuss sorting of proteins into different cellular locations.

34. Describe the translocation of proteins across the ER, highlighting the four components involved.

35. Explain the mechanism of protein transport by transport vesicles including the roles of cargo receptors, coat proteins, and γ-SNARE.

SELF-TEST

Protein Synthesis Requires the Translation of Nucleotide Sequences into Amino Acid Sequences

1. Which of the following statements about functional tRNAs are correct?
   (a) They contain many modified nucleosides.
   (b) About half their nucleosides are in base-paired helical regions.
   (c) They contain fewer than 100 ribonucleosides.
   (d) Their anticodons and amino acid accepting regions are within 5 Å of each other.
   (e) They consist of two helical stems that are joined by loops to form a U-shaped structure.
   (f) They have a terminal AAC sequence at their amino acid accepting end.

2. Explain why tRNA molecules must have both unique and common structural features.

3. How many errors would be expected to occur in a 1000-amino acid protein given an error frequency of 10^{-3}?

4. Which of the following answers completes the sentence correctly? The wobble hypothesis
   (a) accounts for the conformational looseness of the amino acid acceptor stem of tRNAs that allows sufficient flexibility for the peptidyl-tRNA and aminoacyl-tRNA to be brought together for peptide-bond formation.
   (b) accounts for the ability of some anticodons to recognize more than one codon.
   (c) explains the occasional errors made by the aminoacyl-tRNA synthetases.
   (d) explains the oscillation of the peptidyl-tRNAs between the A and P sites on the ribosome.
   (e) assumes steric freedom in the pairing of the first (5′) nucleotide of the codon and the third (3′) nucleotide of the anticodon.

Aminoacyl-Transfer RNA Synthetases Read the Genetic Code

5. Which of the following statements about the aminoacyl-tRNA synthetase reaction are correct?
   (a) ATP is a cofactor.
   (b) GTP is a cofactor.

(c) The amino acid is attached to the 2′- or 3′-hydroxyl of the nucleotide cofactor (ATP).

(d) The amino group of the amino acid is activated.

(e) A mixed anhydride bond is formed.

(f) An acyl ester bond is formed.

(g) An acyl thioester bond is formed.

(h) A phosphoamide (P–N) bond is formed.

6. The $\Delta G°′$ of the reaction catalyzed by the *aminoacyl-tRNA synthetases* is

(a) ~0 kcal/mol.

(b) <0 kcal/mol.

(c) >0 kcal/mol.

7. Considering the correct answer to question 6, explain how aminoacyl-tRNAs can be produced in the cell.

8. Match the *class I* and *class II* aminoacyl-tRNA synthetases with the appropriate items in the right column.

(a) class I

(b) class II

(1) are generally dimeric

(2) are generally monomeric

(3) acylate the 2′-hydroxyl of the tRNA

(4) generally acylate the 3′ hydroxyl of the tRNA

(5) contain β strands at the activation domain

9. Indicate the possible ways in which different aminoacyl-tRNA synthetases may recognize their corresponding tRNAs.

(a) by recognizing the anticodon

(b) by recognizing specific base pairs in the acceptor stem

(c) by recognizing the 3′ CCA sequence of the tRNA

(d) by recognizing both the anticodon and acceptor stem region

(e) by recognizing extended regions of the L-shaped molecules

10. In an experiment, it was found that Cys-tRNA^{Cys} can be converted to Ala-tRNA^{Cys} and used in an in vitro system that is capable of synthesizing proteins.

(a) If the Ala-tRNA^{Cys} were labeled with ^{14}C in the amino acid, would the labeled Ala be incorporated in the protein in the places where Ala residues are expected to occur? Explain.

(b) What does the experiment indicate about the importance of the accuracy of the aminoacyl-tRNA synthetase reaction to the overall accuracy of protein synthesis?

The Ribosome is the Site of Protein Synthesis

11. Assuming that each nucleoside in the left column is in the first position of an anticodon, with which nucleoside or nucleosides in the right column could it pair during a codon–anticodon interaction if each of the nucleosides on the right is in the third position (3′ position) of a codon?

(a) adenosine

(b) cytidine

(c) guanosine

(d) inosine

(e) uridine

(1) adenosine

(2) cytidine

(3) guanosine

(4) uridine

12. Which of the following statements about an *E. coli* ribosome are correct?
    (a) It is composed of two spherically symmetrical subunits.
    (b) It has a large subunit comprising 34 kinds of proteins and two different rRNA molecules.
    (c) It has a sedimentation coefficient of 70S.
    (d) It has two small subunits, one housing the A site and the other the P site.
    (e) It has an average diameter of approximately 200 Å.
    (f) It has a mass of approximately 270 kd, one-third of which is RNA.

13. What is the significance of the reconstitution of a functional ribosome from its separated components?

14. Which of the following statements about translation are correct?
    (a) Amino acids are added to the amino terminus of the growing polypeptide chain.
    (b) Amino acids are activated by attachment to tRNA molecules.
    (c) A specific initiator tRNA along with specific sequences of the mRNA ensures that translation begins at the correct codon.
    (d) Peptide bonds form between an aminoacyl-tRNA and a peptidyl-tRNA positioned in the A and P sites, respectively, of the ribosome.
    (e) Termination involves the binding of a terminator tRNA to a stop codon on the mRNA.

15. An experiment is carried out in which labeled amino acids are added to an in vitro translation system under the direction of a single mRNA species. Samples are withdrawn at different times, and the labeling patterns below are observed in the *completed* polypeptide chains. The dashes (-) represent unlabeled amino acids, X represents labeled amino acids, and A and B represent the ends of the intact protein.

    Time 1 (early)   A - - - - - - - - - - - - - - XXB
    Time 2   A - - - - - - - - - - - XXXXB
    Time 3   A - - - - - - - XXXXXXB
    Time 4 (late)   A - - - - XXXXXXXXB

    Which of the following statements about these proteins are correct?

    (a) The labeled amino acids are added in the B-to-A direction.
    (b) The labeled amino acids are added in the A-to-B direction.
    (c) A is the amino terminus of the protein.
    (d) B is the amino terminus of the protein.

16. Given an in vitro system that allows protein synthesis to start and stop at the ends of any RNA sequence, answer the following questions:
    (a) What peptide would be produced by the polyribonucleotide 5′-UUUGUUUUUGUU-3′? (See the table with the genetic code on the inside back cover of the textbook.)
    (b) For this peptide, which is the N-terminal amino acid and which is the C-terminal amino acid?

17. What experiment showed that each initiator region displays an AUG (or GUG or UUG) codon?

18. What sequence on the 16s rRNA binds to the Shine-Delgarno sequence on mRNA?
    (a) CCUCC
    (b) CAGGU
    (c) GGAGG
    (d) There is no specific sequence, it is merely a purine-rich region of the 16s rRNA.

19. What is the role of the vitamin folate in prokaryotic translation?

20. Match the functions or characteristics of prokaryotic translation in the right column with the appropriate translation components in the left column.

(a) IF1
(b) IF2
(c) IF3
(d) EF-Tu
(e) EF-Ts
(f) EF-G
(g) peptidyl transferase
(h) RF1
(i) RF2

(1) moves the peptidyl-tRNA from the A to the P site
(2) delivers aminoacyl-tRNA to the A site
(3) binds to the 30S ribosomal subunit
(4) recognizes stop codons
(5) forms the peptide bond
(6) delivers fMet-tRNA$_f^{Met}$ to the P site
(7) cycles on and off the ribosome
(8) binds GTP
(9) prevents the combination of the 50S and 30S subunits
(10) is involved in the hydrolysis of GTP to GDP
(11) associates with EF-Tu to release a bound nucleoside diphosphate
(12) hydrolyzes peptidyl-tRNA
(13) modifies the peptidyl transferase reaction

21. Which of the following statements about occurrences during translation are correct?
(a) The carboxyl group of the growing polypeptide chain is transferred to the amino group of an aminoacyl-tRNA.
(b) The carboxyl group of the amino acid on the aminoacyl-tRNA is transferred to the amino group of a peptidyl-tRNA.
(c) Peptidyl-tRNA may reside in either the A or the P site.
(d) Aminoacyl-tRNAs are shuttled from the A to the P site by EF-G.

22. About 5% of the total bacterial protein is EF-Tu. Explain why this protein is so abundant.

23. For each of the following steps of translation, give the nucleotide cofactor involved and the number of high-energy phosphate bonds consumed.
(a) amino acid activation
(b) formation of the 70S initiation complex
(c) delivery of aminoacyl-tRNA to the ribosome
(d) formation of a peptide bond
(e) translocation

24. Which of the following statements about release factors are correct?
(a) They recognize terminator tRNAs.
(b) They recognize translation stop codons.
(c) They cause peptidyl transferase to use H_2O as a substrate.
(d) They are two proteins in E. coli, each of which recognizes two mRNA triplet sequences.

Eukaryotic Protein Synthesis Differs from Prokaryotic Protein Synthesis Primarily in Translation Initiation

25. Which of the following statements about eukaryotic translation are correct?
(a) A formylmethionyl-tRNA initiates each protein chain.
(b) It occurs on ribosomes containing one copy each of the 5S, 5.8S, 18S, and 28S rRNA molecules.
(c) The correct AUG codon for initiation is selected by the base-pairing of a region

on the rRNA of the small ribosomal subunit with an mRNA sequence upstream from the translation start site.

(d) It is terminated by a release factor that recognizes stop codons and hydrolyzes GTP.

(e) It involves proteins that bind to the 5′ ends of mRNAs.

(f) It can be regulated by protein kinases.

26. Which of the following is a correct statement regarding the disease vanishing white matter (VWM)?

(a) It is caused by a mutation in eIF4-G.

(b) It is caused by a mutation in eIF2.

(c) The mutation is found in all cells in the body but the effects are limited to brain tissue.

(d) The mutation occurs only in cells in the brain.

A Variety of Antibiotics and Toxins Can Inhibit Protein Synthesis

27. Many antibiotics act by inhibiting protein synthesis. How can some of these be used in humans to counteract microbial infections without causing toxic side effects due to the inhibition of eukaryotic protein synthesis?

28. Increasing the concentration of which of the following would most effectively antagonize the inhibition of protein synthesis by puromycin?

(a) ATP

(b) GTP

(c) aminoacyl-tRNAs

(d) peptidyl-tRNAs

(e) eIF3

29. Which of the following statements about the diphtheria toxin are correct?

(a) It is cleaved on the surface of susceptible eukaryotic cells into two fragments, one of which enters the cytosol.

(b) It binds to peptidyl transferase and inhibits protein synthesis.

(c) It reacts with ATP to phosphorylate eIF2 and prevent the insertion of the Met-tRNA$_i$ into the P site.

(d) It reacts with NAD$^+$ to add ADP-ribose to eEF2 and prevent movement of the peptidyl-tRNA from the A to the P site.

(e) One toxin molecule is required for each translation factor inactivated, that is, it acts stoichiometrically.

Ribosomes Bound to the Endoplasmic Reticulum Manufacture Secretory and Membrane Proteins

30. Which of the following proteins are usually synthesized by free ribosomes?

(a) lysosomal proteins

(b) cytoplasmic proteins

(c) secretory proteins

(d) integral membrane proteins

31. Place the following in the order in which a newly synthesized protein is transported out of the lumen of the ER.

(a) Cis-Golgi

(b) Secretory granules

(c) Trans-Golgi

(d) Transport vesicles

ANSWERS TO SELF-TEST

1. a, b, c. The molecules consist of two helical stems, each of which is made of two stacked helical segments. However, the molecules are L-shaped, and the anticodon and amino acid accepting regions are some 80 Å from each other. Functional tRNAs have a CCA sequence, not an AAC sequence at their 3′ termini.

2. Transfer RNAs need common features for their interactions with ribosomes and elongation factors but unique features for their interactions with the activating enzymes.

3. One error would be expected. An ε value of 10^{-3} means 1 error per 1000 amino acids.

4. b

5. a, e, f. The carboxyl group of the amino acid is activated in a two-step reaction via the formation of an intermediate containing a mixed anhydride linkage to AMP. The amino acid is ultimately linked by an ester bond to the 2′- or 3′-hydroxyl of the tRNA.

6. a. Since the standard free energy of the hydrolysis of an aminoacyl-tRNA is nearly equal to that of the hydrolysis of ATP, the reaction has a $\Delta G^{\circ\prime} \sim 0$; that is, it has an equilibrium constant near 1.

7. In the cell, the hydrolysis of PP_i by pyrophosphatase shifts the equilibrium toward the formation of aminoacyl-tRNA.

8. (a) 2, 3, 5 (b) 1, 4, 5. Both classes of enzymes contain β sheet domains at their activation domains.

9. a, b, d, e. There are many different ways in which aminoacyl-tRNA synthetases recognize their specific tRNAs. Answer (c) is incorrect because the 3′ CCA sequence is common to all tRNAs and cannot be used to distinguish among them.

10. (a) No. The tRNA, acting as an adaptor between the amino acid and mRNA, would associate with Cys codons in the mRNA through base-pairing between codon and anticodon. The labeled alanine would incorporate only at sites encoded by the Cys codons and not at those encoded by Ala codons.

    The experiment demonstrates that the tRNA and not the amino acid reads the mRNA. Thus, if the activating enzyme mistakenly attaches an incorrect amino acid to a tRNA, that amino acid will be incorporated erroneously into the protein.

    (b) Answer (e) is incorrect because the ambiguities in base-pairing occur between the third nucleotide of the codon and the first nucleotide of the anticodon.

11. (a) 4 (b) 3 (c) 2, 4 (d) 1, 2, 4 (e) 1, 3. (See Table 30.2.)

12. b, c, e. Answer (f) is incorrect because two-thirds of the 2700-kd mass of a ribosome is rRNA.

13. It shows that the components themselves contain all the information necessary to form the structure and that neither a template nor any other factors are involved. Thus, the ribosome serves as model from which we might learn the general principles involved in self-assembly. Reassembly allows systematic study of the roles of the individual components through the determination of the effects of substitutions of mutant or altered individual proteins or rRNAs.

14. b, c, d. Answer (a) is incorrect because the incoming activated aminoacyl-tRNA, in the A site of the ribosome, adds its free amino group to the activated carboxyl of the growing polypeptide on a peptidyl-tRNA in the P site. Answer (e) is incorrect because termination does not involve tRNAs that recognize translation stop codons but rather protein release factors that recognize these and cause peptidyl transferase to donate the growing polypeptide chain to H_2O rather than to another aminoacyl-tRNA.

15. b, c. Although longer incubation times result in completed proteins that have labeled polypeptides closer to their amino terminus, the chains actually grow in the amino-to-carboxyl direction. When the labeled amino acid is introduced into the system, it begins adding to the carboxyl ends of the growing chains that are already present in all stages of completion. The completed chains in samples withdrawn after a short time will have labeled polypeptides only near their carboxyl end. As time passes, more and more polypeptides that began adding labeled polypeptides near their amino terminals will become complete chains.

16. (a) Phenylalanylvalylphenylalanylvaline
    (b) Phe is the N-terminal amino acid and Val is the C-terminal amino acid.

17. The initiator regions from a number of mRNAs were isolated using pancreatic ribonuclease to digest mRNA–ribosome complexes (formed under conditions in which protein synthesis could begin but elongation could not take place).

18. a. Answers b and c are actual Shine-Delgarno sequences or parts of one (Figure 30.18).

19. After folic acid is converted to N^{10}-formyltetrahydrofolate, it acts as a carrier of formyl groups and is a substrate for a transformylase reaction that converts Met-tRNA$_f$ to fMet-tRNA$_f$—the initiator tRNA.

20. (a) 3, 7, 9   (b) 3, 6, 7, 8, 10   (c) 3, 7, 9   (d) 2, 7, 8, 10   (e) 11   (f) 1, 7, 8, 10   (g) 5, 12   (h) 4, 7, 13   (i) 4, 7, 13

21. a, c. The aminoacyl-tRNA in the A site becomes a peptidyl-tRNA when it receives the carboxyl group of the growing polypeptide chain from the peptidyl-tRNA in the P site. After the free tRNA leaves, the extended polypeptide on its new tRNA is then moved to the P site by EF-G. Answer (d) is incorrect because transfer RNAs bearing aminoacyl derivatives with free amino groups are never found in the P site.

22. The large amounts of EF-Tu in the cell bind essentially all of the aminoacyl-tRNAs and protect these activated complexes from hydrolysis.

23. (a) ATP, 2   (b) GTP, 1   (c) GTP, 1   (d) none   (e) GTP, 1. With regard to the answer for (d), the formation of a peptide bond per se does not require a cofactor. The energy for the exergonic reaction is supplied by the activated aminoacyl-tRNA.

24. b, c, d. Each of the two release factors of E. coli recognizes two of the three translation stop codons and interacts with the synthesis machinery such that peptidyl transferase donates the polypeptide chain to H_2O and thus terminates synthesis by hydrolyzing the ester linkage of the protein to the tRNA.

25. b, d, e, f. Eukaryotic ribosomes usually scan the mRNA from the 5′ end for the first AUG codon, which then serves to initiate the synthesis. Answer (e) is correct because proteins that bind to the cap of the mRNA are involved in the association of the ribosome with the mRNA.

26. b, c. Answer a is incorrect because defective eIF4-G has been implicated in fragile-X syndrome not VWM.

27. The inhibition of the prokaryotic translation and not that of the eukaryote can result from differences between their respective ribosomes. Some antibiotics interact with the RNA components that are unique to bacterial ribosomes and, consequently, can inhibit bacterial growth without affecting the human cells.

28. c. Puromycin is an analog of aminoacyl-tRNA. It inhibits protein synthesis by binding to the A site of the ribosome and accepting the growing polypeptide chain from the peptidyl-tRNA in the P site and thus terminating polymer growth. Because

aminoacyl-tRNAs compete with the puromycin for the A site, increasing their concentration would lessen the extent of inhibition.

29. d. Answer (e) is incorrect because the toxin acts catalytically and is thus extremely deadly; one toxin molecule can inactivate many translocase molecules by modifying them covalently.

30. b. Free ribosomes synthesize proteins that remain in the cell, either in the cytoplasm or in organelles such as the nucleus or mitochondria.

31. d, a, c, b

PROBLEMS

1. (a) The template strand of DNA known to encode the N-terminal region of an *E. coli* protein has the following nucleotide sequence: GTAGCGTTCCATCAGATTT. Give the sequence for the first four amino acids of the protein.

   (b) Suppose that the sense strand of the DNA known to encode the amino acid sequence of the N-terminal region of a mammalian protein has the following nucleotide sequence: CCTGTGGATGCTCATGTTT. Give the amino acid sequence that would result.

2. The nucleotide sequence on the sense strand of the DNA that is known to encode the carboxy terminus of a long protein of *E. coli* has the following nucleotide sequence: CCATGCAAAGTAATAGGT. Give the resulting amino acid sequence.

3. Suppose that a particular aminoacyl-tRNA synthetase has a 10% error rate in the formation of aminoacyl-adenylates and a 99% success rate in the hydrolysis of incorrect aminoacyl-adenylates. What percentage of the tRNAs produced by this aminoacyl-tRNA synthetase will be faulty?

4. Students of biochemistry are frequently distressed by "Svedberg arithmetic," that is, for instance, by the fact that the 30S and 50S ribosomal subunits form a 70S particle rather than an 80S particle. Why don't the numbers add up to 80?

5. The possible codons for valine are GUU, GUC, GUA, and GUG.

   (a) For each of these codons write down all the possible anticodons with which it might pair (use the wobble rules in Table 30.2 in the text).

   (b) How many codons could pair with anticodons having I as the first base? How many could pair with anticodons having U or G as the first base? How many could pair with anticodons beginning with A or C?

6. What amino acid will be specified by a tRNA whose anticodon sequence is IGG?

7. According to the wobble principle, what is the *minimum* number of tRNAs required to decode the six leucine codons—UUA, UUG, CUU, CUC, CUA, and CUG? Explain.

8. Coordination of the threonine hydroxyl by an active site Zn in the threonyl-tRNA synthetase allows discrimination between threonine and the isosteric valine (Sankaranarayanan et al., *Nat. Struct. Biol.* 7[2000]:461–465). Given the similarity of serine and threonine (Ser lacks only the methyl group of Thr), if this is the only mechanism for amino acid discrimination available, threonyl-tRNA synthetase mistakenly couples Ser to threonyl-tRNA at a rate several-fold higher than it does threonine. Since this would lead to unacceptably high error rates in translation, how it is it avoided?

9. Mutations from codons specifying amino acid incorporation to one of the chain-terminating codons, UAA, UAG, or UGA, so-called *nonsense* mutations, result in the synthesis of shorter, usually nonfunctional, polypeptide chains. It was discovered that some strains of bacteria can protect themselves against such mutations by having mutant tRNAs that can recognize a chain-terminating codon and insert an amino acid instead. The result would be a protein of normal length that may be functional, even though it may contain an altered amino acid residue. How can bacteria with such mutant tRNA molecules ever manage to terminate their polypeptide chains successfully?

10. Change of one base pair to another in a sense codon frequently results in an amino acid substitution. Change of a C–G to a G–U base pair at the 3:70 position of tRNA^{Cys} causes that tRNA to be recognized by alanyl-tRNA synthetase.

    (a) What amino acid substitution or substitutions would result with the mutated tRNA^{Cys} present?

    (b) How does the pattern differ from that resulting from base substitutions within a codon?

11. The methionine codon AUG functions both to initiate a polypeptide chain and to direct methionine incorporation into internal positions in a protein. By what mechanisms are the AUG start codons selected in prokaryotes?

12. Laboratory studies of protein synthesis usually involve the addition of a radioactively labeled amino acid and either natural or synthetic mRNAs to systems containing the other components. To observe the formation of protein, advantage is taken of the fact that proteins, but not amino acids, can be precipitated by solutions of trichloroacetic acid. Thus, one can observe the extent to which radioactivity has been incorporated into "acid-precipitable material" as a function of time to estimate the rate of formation of protein. In one such experiment, poly(U) is used as a synthetic mRNA in an in vitro system derived from wheat germ (a eukaryote).

    (a) What labeled amino acid would you add to the reaction mixture?

    (b) What product will be formed?

    For each of the following procedures, explain the results observed. Assume that in a complete system 3000 cpm (counts per minute) are found in acid-precipitable material at the end of 30 minutes and that values below 150 cpm are not significantly above the background level.

    (c) 85 cpm is recovered when RNase A is added to the complete system.

    (d) 2900 cpm is recovered when chloramphenicol is added to the complete system.

    (e) 300 cpm is recovered when cyclohexamide is added to the complete system.

    (f) 640 cpm is recovered when puromycin is added to the complete system.

    (g) 1518 cpm is recovered when puromycin and extra wheat germ tRNA are added to the complete system.

    (h) 120 cpm is recovered when poly(A) is used instead of poly(U).

ANSWERS TO PROBLEMS

1. (a) The sequence of the first four amino acids of the protein is (formyl)Met-Glu-Arg-Tyr. As the name implies, the template (antisense) strand of DNA serves as the template for the synthesis of a complementary mRNA molecule. (Remember that by conven-

tion nucleotide sequences are always written in the 5′ to 3′ direction unless otherwise specified.) The template strand of DNA and the mRNA synthesized are as follows:

DNA template strand: 5′-GTAGCGTTCCATCAGATTT-3′

mRNA: 3′-CAUCGCAAGGUAGUCUAAA-5′

Remember that the codons of an mRNA molecule are read in the 5′-to-3′ direction. Because this particular nucleotide sequence specifies the N-terminal region of an *E. coli* protein, the first amino acid must be (formyl)-methionine, which may be encoded by either AUG or GUG. Because there is no GUG and only a single AUG in the mRNA sequence, the location of the initiation codon can be established unambiguously. The portion of the mRNA sequence encoding protein and the first four amino acids it encodes are

mRNA: 5′-AUG-GAA-CGC-UAC-3′

Amino acid sequence: (Formyl)Met-Glu-Arg-Tyr

(b) The expected amino acid sequence is Met-Leu-Met-Phe. The nucleotide sequences on DNA and mRNA are

Sense strand of DNA: 5′-CCTGTGGATGCTCATGTTT-3′

mRNA: 5′-CCUGUGGAUGCUCAUGUUU-3′

In eukaryotes the first triplet specifying an amino acid is almost always the AUG that is closest to the 5′ end of the mRNA molecule. In this example there are two AUGs, so there will be two Met residues in the polypeptide that is produced. The reading frame and the resulting amino acids are as follows:

mRNA: 5′-CCUGUGG-AUG-CUC-AUG-UUU-3′

Amino acid sequence: Met-Leu-Met-Phe

2. The sequence is His-Ala-Lys. The DNA and mRNA sequences are

Sense strand of DNA: 5′-CCATGCAAAGTAATAGGT-3′

mRNA: 5′-CCAUGCAAAGUAAUAGGU-3′

Since this sequence specifies the carboxyl end of the peptide chain, it must contain one or more of the chain-termination codons: UAA, UAG, or UGA. UAA and UAG occur in tandem in the sequence, so we can infer the reading frame. The mapping of the amino acid residues to the mRNA is as follows:

mRNA: 5′-C-CAU-GCA-AAG-UAA-UAG-GU-3′

Amino acid sequence: His-Ala-Lys

3. The percentage of tRNAs that will be faulty is 0.11%. For every 1000 aminoacyl-adenylates that are produced, 100 are faulty and 900 are correct. The 900 correct intermediates will be converted to correct aminoacyl tRNAs because the intermediates are tightly bound to the active site of the aminoacyl-tRNA synthetase. Of the 100 incorrect aminoacyl-adenylates, 99 will be hydrolyzed and will therefore not form aminoacyl tRNAs. Only one will survive to become an incorrect aminoacyl tRNA. The fraction of incorrect aminoacyl tRNAs is therefore 1/901, or 0.11%.

4. The Svedberg unit (S) is a sedimentation coefficient, which is a measure of the velocity with which a particle moves in a centrifugal field. It represents a hydrodynamic property of a particle, a property that depends on, among other factors, the

size and shape of the particle. When two particles come together, the sedimentation coefficient of the resulting particle should be less than the sum of the individual coefficients because there is no frictional resistance between the contact surfaces of the particles and the centrifugal medium.

5. (a) The possible anticodons with which the codons might pair are as follows:

Codon	Possible anticodon
GUU	AAC, GAC, IAC
GUC	GAC, IAC
GUA	UAC, IAC
GUG	CAC, UAC

   (b) Three codons could pair with the anticodon beginning with I; two codons could pair with an anticodon beginning with U or G; only one codon could pair with an anticodon beginning with A or C.

6. Proline. The three codons that will pair with IGG—CCU, CCC, and CCA—all specify proline.

7. A minimum of three tRNAs would be required. One tRNA having the anticodon UAA could decode both UUA and UUG. For the other four codons, which have C in the first position and U in the second, there are two different combinations of two tRNAs each that could decode them. The first combination would be two tRNAs that have anticodons with A in the second position and G in the third, one with I in the first position to decode CUU, CUC, and CUA, and the other with U or C in the first position to decode CUG. The second combination would be two tRNAs that have anticodons with A in the second position and G in the third, one with G in the first position to decode CUU and CUC, and the other with U in the first position to decode CUA and CUG.

8. Threonyl-tRNA synthetase has a proofreading mechanism. Any Ser-tRNA^{Thr} that is mistakenly formed is hydrolyzed by an editing site 20 Å from the activation site. The "decision" to hydrolyze the aminoacyl-tRNA appears to depend on the size of the amino acid substituent. If it is smaller than the correct amino acid, the amino acid fits into the hydrolytic site and is cleaved. If it is the same size as the correct amino acid, it does not fit and is not destroyed. Discrimination between amino acids that are larger than the correct one or are not isoelectronic with it occurs at the aminoacylation step.

9. If two different legitimate stop codons are present in tandem, it would be extremely improbable that mutant tRNAs would exist for both and would simultaneously bind to each of them and thereby prevent proper chain termination.

10. (a) The tRNA^{Cys} will become loaded with Ala rather than Cys, and as a result will insert Ala rather than Cys into polypeptide chains.
    (b) In the case of a base change within a codon, only a single amino acid of a single polypeptide is changed. In the case of a tRNA recognition mutation, amino acid substitutions at many positions of many polypeptides would occur.

11. A purine-rich mRNA sequence, three to nine nucleotides long (called the Shine-Dalgarno sequence), which is centered about 10 nucleotides upstream of (to the 5′ side of) the start codon, base-pairs with a sequence of complementary nucleotides near the 3′ end of the 16S rRNA of the 30S ribosomal subunit. This interaction plus the association of fMet-tRNA$_f$ with the AUG in the P site of the ribosome sets the mRNA reading frame.

12. (a) Poly(U) codes for the incorporation of phenylalanine. Therefore, labeled phenylalanine must be added to the reaction mixture.

(b) Polyphenylalanine will be formed.

(c) RNase A will digest poly(U) almost completely to 3'-UMP, thus destroying the template for polyphenylalanine synthesis. Also the tRNA will be digested and the ribosomes damaged. No protein synthesis will take place.

(d) Chloramphenicol inhibits the peptidyl transferase activity of the 50S ribosomal subunit in prokaryotes but has no effect on eukaryotes so synthesis in a eukaryote is unaffected.

(e) Cyclohexamide inhibits the peptidyl transferase activity of the 60S ribosomal subunit in eukaryotes so synthesis is largely blocked.

(f) Puromycin mimics an aminoacyl-tRNA and causes premature polypeptide chain termination leading to a low level of protein synthesis.

(g) The addition of extra wheat germ tRNA reduces the inhibiting effect of puromycin, since they both compete for the A site on ribosomes. Therefore synthesis is increased over that in experiment (f).

(h) Poly(A) directs the synthesis of polylysine; since there is no lysine (either labeled or unlabeled) in the system, no product can be detected.

The Control of Gene Expression in Prokaryotes

Chapter 31

In this chapter, the authors describe the biochemistry underlying several mechanisms that control gene expression in prokaryotes and eukaryotes. They point out that control of transcription is the primary mechanism of gene regulation, and that the specific binding of proteins to particular regulatory DNA sequences is the basis of the control. The lactose (lac) operon in bacteria is an example of a mechanism in which the initiation of transcription is regulated. Negative control is exerted through the binding of a repressor protein to DNA carrying the lac operator. Positive control of the lac operon is accomplished by a complex of CAP protein and cyclic AMP (cAMP). The complex binds near the lac promoter in the absence of the repressor and through protein-protein interactions stimulates the activity of RNA polymerase at the lac promoter.

Two examples of regulation of transcription by switching patterns of gene expression are discussed. The first is that of bacteriophage λ, which switches between its lytic and lysogenic phases by coordinated expression of the λ repressor and the Cro protein. The second example is that of quorum sensing in bacteria. This process involves the release of chemicals called autoinducers into the medium surrounding the cell. These autoinducers are taken up by surrounding cells where they activate transcription of genes in the neighboring cells, allowing them to change their gene-expression patterns in response to the number of other cells in their environments. The authors finish this section with the introduction of biofilms, complex communities of prokaryotes that are promoted by quorum-sensing mechanisms.

In the final section of the chapter, the authors describe the phenomenon of attenuation of translation. The trp operon in bacteria is provided as an example of posttranscriptional gene regulation. This mechanism, which is used by several amino acid biosynthetic operons, relies on alternative RNA secondary structures and on the coupling of transcription and translation in prokaryotes. In preparation for studying this chapter, you should review the sections of Chapters 4 and 5 on gene expression and on introns and exons. Material in Chapter 28 of the text describes the functional groups on DNA that can serve as determinants for specific interactions with proteins. These examples provide some of the principles by which the proteins described in this chapter can interact specifically with sequences of DNA.

LEARNING OBJECTIVES

When you have mastered this chapter, you should be able to accomplish the following objectives:

Introduction

1. Define *gene expression* and indicate the primary level of its regulation during expression of the genetic information.

2. Distinguish between *constitutive* and *regulated* gene expression.

Many DNA-Binding Proteins Recognize Specific DNA Sequences (Text Section 31.1)

3. Differentiate protein-encoding and *regulatory DNA sequences.*

4. Describe the role of *symmetry matching* in *protein-DNA interactions.*

5. Describe the *helix-turn-helix* motif and relate it to DNA binding.

Prokaryotic DNA-Binding Proteins Bind Specifically to Regulatory Sites in Operons (Text Section 31.2)

6. Outline the metabolism of lactose in *E. coli.* Draw the structure of *lactose*, describe its entry into the cell, and write the equations for the reactions catalyzed by *β-galactosidase,* providing both the substrates and the products.

7. Recount the observations by Jacob and Monod that led to the concept of the *lac operon* and its regulation. Define "operon."

8. Draw the *genetic map* of the lac operon and outline the functions of the *promoter, repressor, operator,* and *inducer* in controlling the production of *polycistronic lac mRNA.* Distinguish between regulatory and structural genes.

9. Describe the role of the *lac* repressor in determining whether a bacterium is *inducible* or constitutive for *β-galactosidase.*

10. Describe the *subunit structure* of the *lac* repressor and relate it to the *symmetrical sequence* of the *lac* operator. Compare the affinities of the *lac* repressor for *lac* operator DNA and nonspecific DNA. Describe the effect of the binding of *allolactose* or *isopropylthiogalactoside* by the repressor on its affinity for the operator.

11. Appreciate that many other gene-regulatory networks in prokaryotes functions like the *lac* operon.

12. Explain the functions of *cyclic AMP (cAMP)* and the *catabolite activator protein (CAP)* in modulating the expression of the *lac* operon.

13. Diagram the relative positions of cAMP–CAP complex, RNA polymerase, and *lac* repressor on the DNA template. Explain their effects on one another and on the DNA structure.

Regulatory Circuits Can Result in Switching Between Patterns of Gene Expression (Text Section 31.3)

14. Compare the features of gene expression in the *lytic* and *lysogenic* pathways of *bateriophage λ*.

15. Describe the structure and function of the *λ repressor*.

16. Describe binding of the λ repressor to the *λ right operator*. Describe the relative affinities of the three binding sites in the operator and explain the effect cooperative binding to the three sites has on binding the repressor. Compare conditions in which the λ repressor is present in the cell at moderate and high concentrations.

17. Explain the causes of a switch between the lytic and lysogenic pathways. Relate the process to a genetic circuit, with *Cro* and the λ repressor controlling the switches between the two states.

18. Define the term *quorum sensing* and use the bacteria *Vibrio fisheri* as an example of it. Explain the role of *autoinducers* in the process and how they allow a bacterium to determine the density of the *V. fisheri* population in its environment.

19. Describe *biofilms* and relate the role quorum sensing plays in their formation.

Gene Expression Can Be Controlled at Posttranscriptional Levels (Text Section 31.4)

20. Use the *tryptophan operon* to explain the mode of transcriptional regulation called *attenuation*.

21. Compare the *leader peptide* sequences in the threonine, phenylalanine, tryptophan, and histidine operons.

SELF-TEST

Many DNA-Binding Proteins Recognize Specific DNA Sequences

1. What features of a DNA regulatory element would endow it with properties enabling a homodimeric regulatory protein (one composed of two identical subunits) to bind it specifically?

2. The helix-turn-helix motif
   (a) is a protein-folding pattern.
   (b) is observed in a variety of prokaryotic DNA-binding proteins.
   (c) contains a recognition helix that inserts itself into the minor groove of DNA.
   (d) is often observed in proteins that bind DNA as dimers.

Prokaryotic DNA-Binding Proteins Bind Specifically to Regulatory Sites in Operons

3. Which of the following are common mechanisms used by bacteria to regulate their metabolic pathways?

   (a) control of the expression of genes
   (b) control of enzyme activities through allosteric activators and inhibitors
   (c) formation of altered enzymes by the alternative splicing of mRNAs
   (d) deletion and elimination of genes that specify enzymes
   (e) control of enzyme activities through covalent modifications

4. Which of the following statements about β-galactosidase in E. coli are correct?

   (a) It is present in varying concentrations depending on the carbon source used for growth.
   (b) It is a product of a unit of gene expression called an *operon*.
   (c) It hydrolyzes the β-1,4-linked disaccharide lactose to produce galactose and glucose.
   (d) It forms the β-1,6-linked disaccharide allolactose.
   (e) It is activated allosterically by the nonmetabolizable compound isopropylthio-galactoside (IPTG).
   (f) Its levels rise coordinately with those of galactoside permease and thiogalacto-side transacetylase.

5. Match each feature or function in the right column with the appropriate DNA sequence element of the *lac* operon in the left column.

   (a) *i*
   (b) *p*
   (c) *o*
   (d) *z*
   (e) *y* and *a*
   (f) CAP binding site

   (1) contains a specific binding sequence for the *lac* repressor
   (2) encodes a galactoside permease
   (3) contains a binding sequence for the cAMP–CAP complex
   (4) encodes a protein that interferes with the activation of RNA polymerase
   (5) encodes a protein that binds allolactose
   (6) contains a specific binding sequence for RNA polymerase
   (7) encodes β-galactosidase
   (8) is a regulatory gene
   (9) is the *lac* promoter
   (10) encodes thiogalactoside transacetylase
   (11) is the *lac* operator
   (12) encodes the *lac* repressor

6. Explain the stoichiometric relationship between the concentrations of β-galactosidase, galactoside permease, and thiogalactoside transacetylase in E. coli.

7. When E. coli is added to a culture containing both lactose and glucose, which of the sugars is metabolized preferentially? What is the mechanism underlying this selectivity?

8. What happens after the first-used sugar is depleted during the experiment described in question 7?

9. Which of the following statements about the cAMP–CAP complex are correct?

   (a) It protects the −87 to −49 sequence of the *lac* operon from nuclease digestion.

(b) It protects the -48 to $+5$ sequence of the *lac* operon from nuclease digestion.

(c) It protects the -3 to $+21$ sequence of the *lac* operon from nuclease digestion.

(d) It affects RNA polymerase activity in a number of operons.

(e) Upon binding to the *lac* operon, it contacts RNA polymerase.

Regulatory Circuits Can Result in Switching Between Patterns of Gene Expression

10. For each of the following, indicate whether it is a feature of the lysogenic or lytic pathway of bacteriophage λ.

    (a) Most of the genes in the viral genome are transcribed.
    (b) The viral genome is incorporated into the bacterial DNA.
    (c) Most of the genes in the viral genome are unexpressed.
    (d) A large number of viral particles are produced.

11. Which of the following events occurs in the presence of high levels of λ repressor?

    (a) λ repressor binds to O_R1 and O_R2, but not O_R3.
    (b) λ repressor binds to all three right operator sites.
    (c) The λ repressor bound to O_R1 blocks access to the promoter on the right side of the operator sites, repressing transcription of Cro.
    (d) The λ repressor bound to O_R2 stimulates transcription of the λ repressor promoter.
    (e) λ repressor transcription is blocked.

12. What is the switch between the two stable states of bacteriophage λ?

13. Place the following steps in the order they occur in quorum sensing by *V. Fisheri*.

    (a) Luciferase is produced.
    (b) LuxR binds AHL.
    (c) Transcription of LuxA, LuxB, and LuxI is increased.
    (d) *V. Fisheri* release an autoinducer called AHL.

Gene Expression Can Be Controlled at Posttranscriptional Levels

14. Regulation of the *trp* operon involves which of the following?

    (a) controlling the amount of polycistronic mRNA formed at the level of transcription initiation
    (b) controlling the amount of polycistronic mRNA at the level of transcription termination
    (c) the sequential and coordinate production of five enzymes of tryptophan metabolism from a single mRNA
    (d) the sequential and coordinate production of five enzymes of tryptophan metabolism from five different mRNAs produced in equal concentrations
    (e) the production of transcripts of different sizes, depending on the level of tryptophan in the cell

15. Which of the following statements concerning the *trp* operon leader RNA, which has 162 nucleotides preceding the initiation codon of the first structural gene of the operon, are correct?

    (a) A deletion mutation in the DNA encoding the 3′ region of the leader RNA gives rise to increased levels of the biosynthetic enzymes forming Trp.

(b) A short open reading frame, containing Trp codons among others, exists within the leader RNA.

(c) The leader RNA encodes a "test" peptide whose ability to be synthesized monitors the level of Trp-tRNA in the cell.

(d) The leader RNA may form two alternative and mutually exclusive secondary structures.

(e) The structure of the leader RNA in vivo depends on the position of the ribosomes translating it.

16. Which of the following is correct about leader sequences in amino acid operons?

(a) An abundance of one amino acid residue in the leader peptide sequence leads to attenuation.

(b) Operons that synthesize amino acids all contain leader sequences.

(c) Leader peptides contain at least 10 of the same amino acids in a row.

(d) Leader sequences are found on the 5′ end of the operon.

17. Describe the mechanism of attenuation in an amino acid operon.

ANSWERS TO SELF-TEST

1. The regulatory element would likely have a nearly perfect inverted repeat sequence giving it a twofold axis of symmetry that could be matched to the twofold symmetry of the homodimeric binding protein. In addition, the sequence itself would comprise an array of hydrogen-bond donors and acceptors plus hydrophobic methyl groups that were recognizable by the regulatory protein to provide the sequence specificity of binding.

2. a, b, d. Answer (c) is incorrect because the recognition helix inserts into the wider major groove rather than the narrower minor groove.

3. a, b, e. Answer (c) is incorrect because the splicing of mRNA is rare in bacteria.

4. a, b, c, d, f. Although IPTG is an inducer for the synthesis of β-galactosidase, it is neither a substrate nor an allosteric activator of the enzyme, so answer (e) is incorrect.

5. (a) 4, 5, 8, 12   (b) 6, 9   (c) 1, 11   (d) 5, 7   (e) 2, 10   (f) 3. For (d), 5 is correct because allolactose is the product of a reaction catalyzed by β-galactosidase and, as a product, it binds to the enzyme.

6. The genes encoding these three enzymes are transcribed as a polycistronic mRNA, so there are approximately equal numbers of copies of the mRNA sequences specifying each of the enzymes.

7. Glucose is metabolized preferentially because it results in a decrease in the synthesis of cAMP by adenylate cyclase. The lack of cAMP prevents the formation of the cAMP–CAP complex, which is necessary for the efficient transcription of the *lac* operon and other catabolite-repressible operons.

8. When glucose is depleted, the concentration of cAMP rises. The cAMP–CAP complex forms and binds to the CAP binding site just upstream of the RNA polymerase binding site in the *lac* promoter. At the same time, some lactose has entered the cell, has been converted to allolactose by β-galactosidase, and is bound by the *lac* repressor so

that it no longer binds to the *lac* operator. RNA polymerase now binds to the *lac* promoter even more effectively because of protein–protein interactions with the cAMP–CAP complex. The enzymes and permease of the *lac* operon are expressed fully; consequently, lactose readily enters the cell and is efficiently metabolized.

9. a, d, e. Answers (b) and (c) are incorrect because they correspond to the binding sequences for RNA polymerase and *lac* repressor, respectively.

10. Answers a and d are features of the lytic pathway and b and c are features of the lysogenic pathway.

11. b and e. Answers a, c, and e occur when λ repressor is present in moderate levels.

12. When the repressor is high and Cro is low, the phage switches to the lytic state. The reverse causes a switch to the lysogenic state.

13. The steps are d, b, c, and finally a.

14. a, b, c, e. Answer (a) is correct; although not mentioned in the text, the *trp* operon contains an operator, and interaction with a repressor, in addition to attenuation, is involved in transcription regulation. Attenuation provides a rapid, sensitive fine-tuning mechanism on top of the control exerted by the repressor-operator interaction. When the RNA is not terminated at the attenuator, a single polycistronic mRNA, which encodes five enzymes, is produced, so answer (d) is incorrect.

15. a, b, c, d, e. The discovery of a deletion mutation in front of the first structural gene of the operon, and not in the operator, was the first clue that a control mechanism operating at the level of transcription termination was involved in regulating the expression of the trp biosynthetic enzymes. This deletion changed the potential mRNA structures so that the rho-independent transcription-termination structure could no longer form.

16. a and d. Answer b is incorrect because only some of the operons coding for amino acid biosynthetic genes are regulated by attenuator sites. Answer c is incorrect because some leader peptides (tryptophan and histidine, for example) are less than 10 amino acids.

17. In *E. coli* translation begins shortly after transcription of mRNA. A short leader peptide on the N-terminus of the nascent polypeptide chain contains multiple copies of the amino acid whose biosynthetic genes are being transcribed. If the amino acid is plentiful, translation continues at a relatively fast rate compared to transcription and an RNA structure forms that terminates transcription. If levels of the amino acid are low, the ribosome stalls at the leader peptide, and an alternative RNA structure forms that does not terminate transcription.

PROBLEMS

1. What property of enzymes makes them more suitable than, say, structural proteins for studies of the genetic regulation of protein synthesis? Explain.

2. When lactose is used as an inducer a lag occurs before the enzymes of the lactose operon are synthesized. With IPTG synthesis starts without a lag. Explain this observation. (See Figure 31.1.)

FIGURE 31.1 Kinetics of β-galactosidase induction by lactose and IPTG.

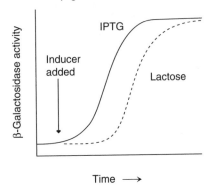

3. Some of the known constitutive mutations of the lactose operon occur in the operator sequence rather than the regulator gene.
   (a) Would you expect such an o^c mutant to be dominant or recessive to its wild-type o^+ allele? Explain.
   (b) Is a constitutive mutation in an operator *cis*-acting or *trans*-acting in its effects? Explain.
   (c) Design an experiment involving the genes i^+, o^c, o^+, and z^+ that would confirm your answer to part (b). Assume that it is possible to detect whether enzymes are produced in diploid (++) or haploid (+) amounts.

4. Since the permease required for the entry of lactose into *E. coli* cells is itself a product of the lactose operon, how might the first lactose molecules enter uninduced cells? Explain.

5. Design an experiment to show that lactose stimulates the synthesis of new enzyme molecules in *E. coli* rather than fostering the activation of preexisting enzyme molecules, for example, by zymogen activation.

6. The three enzymes of the lactose operon in *E. coli* are not produced in precisely equimolar amounts following induction. Rather, more galactosidase than permease is produced, and more permease than transacetylase is produced. Propose a mechanism to account for this that is consistent with known facts about the lactose operon.

7. Assume that the following allelic possibilities exist for the *i* genes and *o* sequence of the lactose operon of *E. coli*:

$$i^+ = \text{wild-type regulator gene}$$
$$i^c = \text{regulator constitutive mutation, makes inactive repressor}$$
$$i^s = \text{repressor substance insensitive to inducer}$$
$$o^+ = \text{wild-type operator}$$
$$o^c = \text{operator consitutive mutation}$$

In addition, assume that the mutations z^-, y^-, and a^- lead to nonfunctional enzymes Z, Y, and A, respectively.

For each of the following, predict whether active enzymes Z, Y, and A will or will not be produced. For partially diploid cells, assume semidominance; that is, the enzyme activity in a diploid cell will be twice that found in a haploid cell. Use the following answer code:

$$0 \quad = \text{active enzyme absent}$$
$$+ \quad = \text{active enzyme present in haploid amounts}$$
$$+ + = \text{active enzyme present in diploid amounts}$$

		Without IPTG			With IPTG	
		Z	Y	A	Z	Y	A
(a)	$i^+o^+z^+y^+a^+$	____	____	____	____	____	____
(b)	$i^co^+z^+y^+a^+$	____	____	____	____	____	____
(c)	$i^so^+z^+y^+a^+$	____	____	____	____	____	____
(d)	$i^+o^cz^+y^+a^+$	____	____	____	____	____	____
(e)	$\dfrac{i^+o^+z^+y^+a^+}{i^co^+z^+y^+a^+}$	____	____	____	____	____	____
(f)	$\dfrac{i^+o^+z^+y^+a^+}{i^co^+z^-y^+a^+}$	____	____	____	____	____	____
(g)	$\dfrac{i^+o^+z^+y^+a^+}{i^+o^cz^+y^+a^+}$	____	____	____	____	____	____
(h)	$\dfrac{i^so^+z^+y^+a^+}{i^+o^+z^+y^-a^-}$	____	____	____	____	____	____
(i)	$\dfrac{i^so^+z^+y^+a^+}{i^+o^cz^-y^+a^+}$	____	____	____	____	____	____
(j)	$\dfrac{i^+o^+z^+y^+a^+}{i^co^cz^+y^+a^-}$	____	____	____	____	____	____

8. Assume that the structural genes x and y, which code for the repressible enzymes X and Y, are subject to negative control by a regulator gene r, which produces a substance S whose ability to bind to the operator gene o is modified by cosubstance T. Assume that the following allelic possibilities exist for genes r and o:

$$o^+ = \text{wild-type operator gene}$$
$$o^o = \text{operator gene unable to bind S (with or without T)}$$
$$r^+ = \text{wild-type regulator gene}$$
$$r^o = \text{inactive regulator gene product (with or without T)}$$
$$r^s = \text{regulator gene product always active (with or without T)}$$

In addition, assume that the mutations x^- and y^- in the two structural genes lead, respectively, to nonfunctional enzymes X and Y.

(a) Are enzymes X and Y likely to be biosynthetic or degradative? Justify your answer.
(b) Is substance S active or inactive in the presence of T? Explain.

(c) For each of the following, predict whether active enzymes X and Y will or will not be produced under the specified conditions. For partially diploid cells, assume semidominance; that is, the enzyme activity in a diploid cell will be twice that found in a haploid cell. Use the following answer code:

$$0 \quad = \text{active enzyme absent}$$
$$+ \quad = \text{active enzyme present in haploid amounts}$$
$$++ \quad = \text{active enzyme present in diploid amounts}$$

	Without T		With T	
	X	Y	X	Y
(1) $r^+o^+x^+y^+$	___	___	___	___
(2) $r^+o^+x^-y^+$	___	___	___	___
(3) $r^so^+x^+y^+$	___	___	___	___
(4) $r^so^0x^+y^+$	___	___	___	___
(5) $r^+o^0x^+y^+$	___	___	___	___
(6) $r^oo^+x^+y^+$	___	___	___	___
(7) $\dfrac{r^+o^+x^+y^+}{r^+o^+x^-y^+}$	___	___	___	___
(8) $\dfrac{r^+o^+x^+y^+}{r^oo^+x^+y^+}$	___	___	___	___
(9) $\dfrac{r^so^+x^-y^-}{r^oo^+x^+y^+}$	___	___	___	___
(10) $\dfrac{r^+o^0x^+y^-}{r^oo^+x^+y^+}$	___	___	___	___

9. The kinetics of induction of enzyme X are shown in Figure 31.2. What percentage of total cellular protein is due to enzyme X in induced cells when 60 μg of total bacterial protein has been synthesized?

FIGURE 31.2   Kinetics of induction of enzyme X.

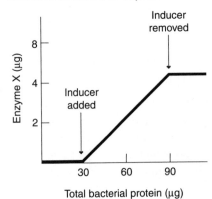

10. Assume that the dissociation constant K for the repressor-operator complex is 10^{-13} M and that the rate constant for association of operator and repressor is 10^{10} M^{-1} s^{-1}.

Calculate the rate constant k_{diss} for the dissociation of the repressor–operator complex. What is the $t_{1/2}$ (half-time of dissociation, or half-life) of the repressor–operator complex?

11. In systems of genetic regulation involving positive control, a regulatory gene produces a substance that enhances rather than inhibits transcription. Are there elements of positive control in the lactose operon of *E. coli*? Explain.

12. An operon for the biosynthesis of amino acid X in a certain bacterium is known to be regulated by a mechanism involving attenuation. What can one confidently predict about the amino acid sequence in the leader peptide for that operon? Explain.

13. In order to prove that regulation by attenuation occurs in vivo, Charles Yanofsky and others studied tryptophan synthesis regulation in a series of *E. coli* mutants. For each mutant described below predict the expression of tryptophan synthesis genes in the presence or absence of tryptophan:

 Mutant A: Mutations with decreased, but detectable, tryptophan-tRNA synthetase activity

 Mutant B: Mutations in AUG or Shine-Dalgarno sequence of leader peptide sequence

 Mutant C: Same as mutant B but with the leader peptide fully expressed on a plasmid

 Mutant D: Mutations replacing the two Trp codons with leu codons

 Mutant E: Mutant E with a mutation in the Leu-tRNA synthetase gene

 Mutant F: Mutant E that constitutively synthesizes leucine

14. Suppose that a system regulating the expression of a single copy DNA leads to the synthesis of an enzyme having a turnover number (k_{cat}) of $10^4 \, s^{-1}$. Each DNA copy is transcribed into 10^3 molecules of mRNA and that each of the mRNA molecules is translated into 10^5 molecules of enzyme protein. How many molecules of substrate are converted into product per second for each wave of transcription that sweeps over the DNA?

15. Proteins that interact with a specific sequence of DNA usually remain bound under conditions of low ionic strength during electrophoresis in gels formed with low percentages of acrylamide. (Low percentages of acrylamide form gels with pores sufficiently large to allow entry of large macromolecular complexes.) Under these conditions, protein–DNA complexes usually migrate more slowly than does the free DNA. Suppose you have a sample of a DNA fragment radioactively labeled with ^{32}P that contains an entire promoter sequence. Describe how you could use the sample and gel electrophoresis to isolate transcription factors from extracts of protein from eukaryotic cells. How might you compensate for the possibly competing binding of histones to the DNA? How could you use unlabeled samples of your promoter fragment to demonstrate the specificity of the interaction between the fragment and a putative transcriptional factor? Might you also detect proteins that do not themselves bind directly to the DNA?

16. All proteins that bind DNA at specific sites also show affinity for DNA at sequences other than the ones that comprise their targets. In general, what effect do you think that nonspecific DNA could have on the ability of a repressor to regulate an operon? Consider the *lac* operator-repressor system in *E. coli* in which the operator is 35 bp within 4.6×10^6 bp total DNA. What would the nonspecific DNA do to the concentration of the free repressor in the cell?

17. Antibiotic resistance occurs when bacteria that are not resistant to the antibiotic are killed, leaving behind bacteria that have developed resistance to the antibiotic. Eventually only the antibiotic-resistant bacteria are left. Speculate on how antibiotics could be developed that target quorum sensing, especially in bacteria that form biofilms.

ANSWERS TO PROBLEMS

1. Enzymes are catalysts, and thus small amounts can be readily detected. An enzyme having a turnover number of $300,000\ s^{-1}$ will provide an assay that is 300,000 times as sensitive as that for a structural protein, which must be assayed stoichiometrically, that is as a single molecule. Many cellular proteins are produced in amounts that are too small to be detected by direct chemical methods.

2. The actual inducer of the lactose operon in vivo is 1,6-allolactose. (See page 926 of the text.) The lag represents the time it takes for lactose to be converted into 1,6-allolactose by residual β-galactosidase. IPTG itself directly acts as an inducer. Therefore, no lag is observed.

3. (a) Imagine a partial diploid that has one $o+$ and one oc sequence. The o^+ gene will bind a repressor, so the structural genes on its chromosome will not be expressed. The oc sequence will not bind a repressor, so the structural genes on its chromosome will always be expressed. Thus, an oc mutant would be dominant to its wild-type o^+ allele.

   (b) Repressors do not bind oc sequences. Only the structural genes on the same chromosome as the oc mutant will be affected, a *cis*-acting effect.

   (c) One could prepare a partial diploid with the following genotype

$$\frac{i^+o^+z^+}{i^+o^cz^+}$$

   If the effect of the mutation is cis, the haploid amount of enzyme Z will be produced in the absence of inducer. (Its synthesis will be specified by the chromosome containing $i^+o^cz^+$.) If the effect is trans, the diploid amount of Z will be produced in the absence of inducer.

4. Very low levels of lactose operon enzymes are synthesized even in the absence of an inducer (see page 926 of the text, which indicates that few enzymes are produced in the absence of inducer).

5. Add IPTG to *E. coli* cells growing in a medium containing a carbon source other than lactose in both the presence and the absence of an inhibitor of prokaryotic protein synthesis, like chloramphenicol. If zymogen activation is involved, chloramphenicol will not inhibit induction. If the synthesis of new protein is involved (as it is), induction will not be observed in the presence of chloramphenicol.

6. Differential expression of the three structural genes in the lactose operon must be at the level of translation and not transcription since a single, polycistronic mRNA molecule is formed. Following induction, mRNA transcripts containing genetic information for all three genes are produced. Some ribosomes might drop off the messenger at the end of the structural genes, with a smaller number reading through the more distal genes.

7. (a) 0   0   0   +   +   +
   (b) +   +   +   +   +   +
   (c) 0   0   0   0   0   0
   (d) +   +   +   +   +   +
   (e) 0   0   0   ++   ++   ++
   (f) 0   0   0   +   ++   ++
   (g) +   +   +   ++   ++   ++
   (h) 0   0   0   0   0   0
   (i) 0   +   +   0   +   +
   (j) +   +   0   ++   ++   +

8. (a) The enzymes will be biosynthetic. The clue to this is provided by the statement that the enzymes are repressible. (Degradative enzymes are often inducible, whereas biosynthetic enzymes are often repressible.)

   (b) Because this operon is under negative control, substance S must act as a repressor. Substance S is active as a repressor in the presence of T, so T functions as a corepressor.

   (c) (1) + + 0 0
   (2) 0 + 0 0
   (3) 0 0 0 0
   (4) + + + +
   (5) + + + +
   (6) + + + +
   (7) + ++ 0 0
   (8) + ++ 0 0
   (9) 0 0 0 0
   (10) ++ + + 0

9. From the graph in Figure 31-2, we see that 2 μg of enzyme X is present when the total bacterial protein present is equal to 60 μg. Thus, the percentage of enzyme X is 100% × 2/60 = 3.3%.

10. Remembering that the dissociation constant K is equal to the ratio of the off rate to the on rate for a reaction,

$$[RO] \rightleftharpoons [R] + [O]$$

$$\frac{[R][O]}{[RO]} = K = 10^{-13} \text{ M} \frac{k_{diss}}{10^{10} \text{ M}^{-1}\text{s}^{-1}}$$

$$k_{diss} = 10^{-3}\text{s}^{-1}$$

$$t_{\frac{1}{2}} = \frac{0.693}{k_{diss}} = \frac{0.693}{10^{-3}\text{s}^{-1}} = 693 \text{ s} \cong 11.6 \text{ min}$$

11. Catabolite activator protein (CAP) is a positive control element. When the level of glucose in cells is low, the level of cAMP is high, leading to the formation of cAMP−CAP complex. The cAMP−CAP complex binds to DNA in the promoter region, creating an entry site for RNA polymerase. The result is the transcription of the lactose operon (providing that no repressor is present). (See Figure 31.13 on page 928 of the text.)

12. The leader sequence should contain codons for amino acid X. If sufficient X is present in the cell, there will be sufficient X-tRNA^X for the synthesis of the leader peptide (as well as for the synthesis of other X-containing proteins in the cell). Therefore, there will be no need to biosynthesize the enzymes needed to produce X.

13. Mutant A: There will be increased transcription of the Trp synthesis genes in both the presence and absence of tryptophan. The low levels of Trp-tRNA synthetase will slow down the rate of translation regardless of the levels of tryptophan. It is the stalling or slowing of the ribosome in segment 1 that is important for regulation, not just the inability to synthesize the leader peptide.

Mutant B: Since the ribosome will never start translation, transcription will always terminate regardless of the levels of tryptophan.

Mutant C: The results will be the same as in mutant B. Attentuation relies on the coupling of transcription and translation. Providing the leader peptide in *trans* would have no effect on the level of attenuation.

Mutant D: Removing the Trp codons would lose all regulation by tryptophan, and Trp synthesis would be regulated by the levels of leucine. In this mutation, the Trp synthesis genes would not be transcribed even in the absence of tryptophan, and this strain, like mutants B2D, would always require tryptophan to grow, unless the strain was grown in the absence of leucine.

Mutant E: There would be constitutive expression of the Trp synthesis genes, even in the presence of tryptophan or leucine.

Mutant F: This mutant would behave like mutants B–E and never express the genes for Trp synthesis.

14. For each transcription, the number of molecules of substrate that are converted into product per second is given by $10^4 \text{ s}^{-1} \times 10^3 \times 10^5 = 10^{12} \text{ s}^{-1}$

15 Assays that exploit the differential electrophoretic mobility of protein-DNA complexes and free DNA are called *gel-shift* or *electrophoretic-mobility-shift assays*. In these experiments, specific DNA sequences are allowed to associate with putative DNA-binding proteins from cell extracts or from fractionated samples from the extracts. They are then subjected to gel electrophoresis along with control samples of labeled fragment by itself. The radioactivity on the gels is visualized by autoradiography on film or in a phosphoimaging machine. Labeled samples that are retarded on the gel are candidates for sequence-specific protein–DNA complexes. To minimize retardation caused by nonspecific binding of proteins, such as histones, to the labeled fragment, large amounts of unlabeled, random sequences of DNA can be added to the sample before electrophoresis. This DNA binds the nonspecific proteins, but is not observed on the gel because it is not labeled. Because sequence-specific proteins usually have a much higher affinity for a specific sequence than those proteins binding nonspecifically, the addition of the extra DNA does not usually interfere with the gel-shift assay. Experiments which assess specificity of the DNA–protein interaction can be conducted by including an excess amount of unlabeled DNA which contains the same specific sequence in an analyzed sample. If the interaction is truly sequence-specific, the unlabeled DNA should bind the protein and by specific competition abolish the shift of the labeled sequence on the gel. Gel-shift analysis is a widely employed technique. Although it is not a true equilibrium technique and therefore cannot easily provide true thermodynamic equilibrium binding constants, it reliably indicates the relative affinities of proteins for specific DNA sequences. You might also detect proteins that are associated tightly through protein-protein interactions with the DNA–binding proteins, for example, you might detect coactivators or corepressors.

16. Depending on the equilibrium dissociation constants of the repressor for its target and for nonspecific sequences, the number of target sequences present, and the ratio of specific to nonspecific DNA present, much of the repressor could be sequestered through binding to the nonspecific DNA. Such nonspecific binding would decrease the concentration of free repressor that could interact with the operator. Since the DNA binding reaction depends on the concentrations of the interacting repressor and DNA, lowering the repressor's concentration would decrease the fraction of operator bound and thereby repressed. The *lac* repressor-operator system provides a good example of this situation. See Kao-Huang, Y., Revzin, A., Butler, A.P., O'Connor, P, Noble,

D.W., and von Hippel, P. (1977) *Proc. Natl. Acad. Sci. USA* 74:4228-4232. These authors found that nonspecific DNA competed effectively with the specific sequence for the repressor. When one copy of *lac* operator is present in *E. coli*, ~4.6 × 10⁶ different, nonspecific binding sites will be present. Think of this by considering that the repressor binds to a 35-bp-long operator site. By sliding this 35-bp-long window along the DNA, one bp at a time, there are ~4.6 × 10⁶ nonspecific sites that can compete with the one correct site. The ability of the operon to be transcribed depends on the average fraction of free operator present, and nonspecific DNA significantly affected that fraction. Thus, nonspecific DNA can play an important role in the functioning of a repressor–operator interaction in the cell. See the article if you wish to see a detailed, quantitative analysis of the phenomenon.

17. Bacteria that form biofilms are of particular medical importance because organisms that form them are often resistant to the immune response as well as to antibiotics. Quorum sensing appears to play a major role in the formation of biofilms in that cells are able to sense other cells in their environments and to promote the formation of communities with particular compositions. The opportunistic bacteria *Pseudomonas aeruginosa*, for example, uses quorum sensing to coordinate the formation of biofilms. If a molecule could be designed that binds to the autoinducer receptor, but blocks dimerization, and therefore blocks activation of transcription, quorum sensing would also be blocked. This should prevent the bacteria from forming biofilms, and give the body's immune system time to remove the infection through normal immune pathways. This process is called quorum quenching. An advantage to this approach compared to traditional antibiotic design is that there would be few evolutionary forces that select for resistance. Non-resistant strains could still multiply, and resistant strains would have to compete with them. There would be no strong survival advantage to resistant mutations.

The Control of Gene Expression in Eurkaryotes

I n this chapter, the authors describe the biochemistry underlying gene expression in eukaryotes, using Chapter's 31 discussion of prokaryotic gene expression for comparison.
They discuss how the abundance of DNA, chromosome structure, and the existence of
the cellular differentiation and posttranscriptional gene regulatory mechanisms complicate the control of gene expression. After a brief discussion that provides an overview of
the major differences between gene regulation and gene-regulatory proteins in prokaryotes and eukaryotes, the authors describe the structure of chromatin, which consists of
DNA tightly bound to basic proteins called histones. A nucleosome consists of DNA
wrapped around an octameric protein core composed of one of four core histones.
Individual nucleosomes are linked together by intervening stretches of DNA. When genes
are not being expressed, the nucleosomes are packed, along with other proteins, to form
a highly condensed chromosome. The text describes the modifications of chromatin structure that occur during transcription.

The authors next move to a class of proteins called transcription factors. Most eukaryotic genes are not expressed unless they are activated by the binding of a transcription factor to a specific site on the DNA. The authors provide a brief introduction to
homedomains [spelling?], leucine-zippers, and zinc finger transcription factors and point
out that each uses an α helix to make specific contacts with DNA. Activation domains and
enhancers are discussed as mechanisms for regulating transcription. The text describes the
steroid-hormone estrogen receptor that uses a zinc-finger motif to bind specifically estrogen response elements in DNA and thereby directly activate transcription. This process involves remodeling of chromatin structure so that the DNA is more open near the
transcription start sites of actively transcribed DNA. The authors next turn to the role of
coactivators in catalyzing the acetylation of histones and thereby modulating gene

expression. The authors end the chapter by using the examples of iron transport and storage and microRNAs to discuss posttranscriptional control of gene expression.

In preparation for studying this chapter, you should review the sections of Chapters 4 and 5 on gene expression and on introns and exons. Material in Chapter 28 of the text describes the functional groups on DNA that can serve as determinants for specific interactions with proteins. Chapter 31 should be reviewed as it discusses in detail the much simpler prokaryotic control of gene expression. These examples provide some of the principles by which the proteins described in this chapter can interact specifically with sequences of DNA.

LEARNING OBJECTIVES

When you have mastered this chapter, you should be able to accomplish the following objectives:

Introduction

1. State the relative amounts of genomic DNA in the haploid genomes of *E.coli*, *S. cerevisiae* (yeast), and *H. sapiens* (human) and give the estimates for the numbers of proteins they encode.

2. Recognize that presence of many different cell types in most eukaryotes is a manifestation of the complexity of *chromatin* structure.

3. Define *epigenome*.

Eukaryotic DNA is Organized into Chromatin (Text Section 32.1)

4. Describe the composition of chromatin. List the types of *histones* and describe their general characteristics. Note the evolutionary stability of the sequences of the H2A, H2B, H3, and H4 histones and assign the types of histones to their locations within the *nucleosome*.

5. Describe the composition and structure of the nucleosome and relate the nucleosome to the proposed structure of the chromatin fiber.

6. Distinguish between the DNA associated with the nucleosome core and that in the internucleosome linker. Discuss how treatment with DNase helped clarify these regions.

Transcription Factors Bind DNA and Regulate Transcription Initiation (Text Section 32.2)

7. Contrast the roles of eukaryotic transcription factors with those in prokaryotes.

8. Describe the modular nature of transcription factors. Describe the function of each type of domain.

9. Describe the structures of DNA-binding motifs such as the *homeodomain*, *leucine zipper*, and *zinc-finger* structures and explain how they interact with DNA.

10. Describe the features of *activation domains*, and recognize *mediators* as an important target.

11. Explain the molecular basis for combinatorial control of transcription.

12. Describe *enhancers* and outline their mechanism of action.

13. Discuss the importance of the development of *induced pluripotent stem (iPS) cells*.

The Control of Gene Expression Can Require Chromatin Remodeling
(Text Section 32.3)

14. Relate hypersensitivity to DNase I to chromatin structure.

15. Discuss the implications of the chromatin *immunoprecipitation* experiments with yeast GAL4 concerning chromatin structure.

16. Relate the covalent modification of DNA by *methylation* to gene expression. Describe CpG islands.

17. Contrast the mechanism of the *steroid hormones* with hormones initiating their actions through interactions with a transmembrane receptor.

18. Describe the domain structure of *nuclear hormone receptors*.

19. Describe the zinc-finger structure of nuclear hormone receptors and describe how it is involved in the interaction of the estrogen–estrogen receptor complex interaction with *estrogen response elements (ERE)*.

20. Outline the role of *coactivators* in transcription by nuclear hormone receptors.

21. Distinguish between *agonists* and *antagonists*. Characterize *anabolic steroids* as agonists of the androgen receptor. Contrast mechanisms of positive and negative gene regulation in eukaryotes.

22. Explain how *tamoxifen* and *raloxifene* serve as an anticancer agents. Recognize the origin of their characterization as *estrogen receptor modulators (SERMs)*.

23. Outline the effects of the *acetylation* and *deacetylation* of *histone tails* on chromatin structure and gene transcription.

24. Recognize the variety of histone modifications and their role in the *histone code*.

Eukaryotic Gene Expression Can Be Controlled at Posttranscriptional Levels
(Text Section 32.4)

25. Outline the role of RNA secondary structure in the regulation of iron metabolism in animals. Describe the roles of *transferrin, transferrin receptor, ferritin,* the *iron-response element (IRE)*, and *IRE-binding protein*.

26. Relate the IRE-binding protein to *aconitase* and *iron sensing*.

27. Define *microRNAs (miRNA)* and describe the discovery of the first microRNA.

28. Relate the mechanism of microRNAs to that of *RNA interference (RNAi)*. Distinguish between the RNA sources in the two types of posttranscriptional regulation.

SELF-TEST

Introduction

1. Match the organism listed in the left column with the amount of DNA in its haploid genome from the right column.

   (a) *E. coli* (bacterium)   (1) 4600 kb
   (b) *S. cerevisiae* (yeast)   (2) 3000 Mb
   (c) *H. sapiens* (human)   (3) 12 Mb

2. Which of the following is not a characteristic of the epigenome?

    (a) Differences in chromatin structure.
    (b) Covalent modifications of the DNA sequence.
    (c) Distribution of genes coding for steps in a given pathway.
    (d) Differential expression of genes based on cell type.

Eukaryotic DNA is Organized into Chromatin

3. What is a primary consequence of the folding of chromatin on gene regulation?

4. Which of the following statements about histones are correct?

    (a) They are highly basic because they contain many positively charged amino acid side chains.
    (b) They are extensively modified after their translation.
    (c) In combination with DNA, they are the primary constituents of chromatin.
    (d) They account for approximately one-fifth of the mass of a chromosome.

5. Which of the following statements about nucleosomes are correct?

    (a) They constitute the repeating units of a chromatin fiber.
    (b) Each contains a core of eight histones.
    (c) They contain DNA that is surrounded by a coating of histones.
    (d) They occur in chromatin in association with approximately 200 base pairs of DNA, on average.

6. Describe the structure of the nucleosome.

7. Does the formation of nucleosomes account for the observed packing ratio of human metaphase chromosomes? Explain.

Transcription Factors Bind DNA and Regulate Transcription Initiation

8. Specific combinatorial control of transcription

    (a) is enabled by specific interactions between transcription factors and specific DNA sequences.
    (b) allows a given regulatory protein to have different effects depending upon the neighboring proteins with which it is associated.
    (c) is effected by transcription factors some of which do not themselves interact with DNA.
    (d) depends upon the assembly of multicomponent nucleoprotein complexes.
    (e) results from the ability of one protein to recruit another to a complex.

9. Match each transcription factor in the left column with its characteristics in the column on the right.

    (a) Basic-leucine-zipper
    (b) Homeodomain
    (c) C_2-His$_2$ zinc finger

    (1) α-helix inserts into major groove
    (2) Dimers stabilized by hydrophobic interactions
    (3) Helix-turn-helix structure
    (4) Forms heterodimers
    (5) Forms homodimers
    (6) Occurs in sets
    (7) Single, long α-helix
    (8) Structure is a tandem set of a small α-helix and β-sheet domain

10. For each of the following, indicate whether it is a feature of an activator (A), an enhancer (E), or neither (N).
    (a) They recruit other proteins to promote transcription.
    (b) They are often redundant; part of them can be deleted with little loss of function.
    (c) They are modular.
    (d) They can act synergistically.
    (e) They perturb local chromatin structure.

The Control of Gene Expression Can Require Chromatin Remodeling

11. Which of the following statements about eukaryotic genes that are actively being transcribed are correct?
    (a) They are cell-type specific.
    (b) They are highly condensed.
    (c) They are more susceptible to hydrolysis by DNAase I than are silent genes.
    (d) They are developmentally regulated.
    (e) They can be detected by chromatin immunoprecipitation.

12. The DNA methylation involved in gene regulation
    (a) requires S-adenosylmethionine as a source of methyl groups.
    (b) occurs at 5′-CpG-3′ sequences.
    (c) uses N^5,N^{10}-methylenetetrahydrofolate to form the 5-methyl group of thymine.
    (d) converts cytosine in DNA to 5-methylcytosine.
    (e) is less frequent at sites adjacent to actively transcribed genes.

13. Which of the following statements about steroid hormones are correct?
    (a) They bind to a seven-helix transmembrane receptor to initiate a series of phosphorylations that culminate in gene transcription.
    (b) Upon binding to their specific receptor proteins, they enable the receptors to bind specific DNA sequences.
    (c) They activate specific protein kinases and protein phosphatases.
    (d) They are recognized by members of the nuclear receptor superfamily of proteins.
    (e) They require plasma membrane transporters to go from the blood to the cytosol.

14. Nuclear hormone receptors
    (a) are dimers.
    (b) bind to response elements, which are specific DNA sequences at or near the genes the hormones control.
    (c) undergo conformational changes when they bind their ligand.
    (d) contain zinc-finger domains.
    (e) interact with coactivators and corepressors in the presence of their ligands.

15. The tails of histones
    (a) when acetylated have lower affinity for DNA.
    (b) are involved in recruiting chromatin-remodeling engines that move nucleosomes.
    (c) when acetylated, serve as substrates for histone deacetylases.
    (d) have their positive charges reduced by acetylation.
    (e) when acetylated interact with the bromodomain of many eukaryotic transcription factors when that domain is brominated.

Eukaryotic Gene Expression Can Be Controlled at Posttranscriptional Levels

16. What are the biochemical similarities and differences between an iron-response element (IRE) and an estrogen-response element (ERE)?

17. Describe the structure and mechanism of the IRP and compare with that of the mitochondrial aconitase.

18. Which of the following is involved in the mechanism of microRNA posttranslational regulation?
    (a) miRNAs bind to complementary mRNA sequences and prevent them from being translated.
    (b) Double-stranded RNAs are cleaved into 21-nucleotide fragments, the single-stranded components of which form RISC complexes that cleave complementary mRNAs.
    (c) The Argonaute family of proteins are RNases that cleave sequences complementary to the microRNA.
    (d) The single-stranded miRNAs are generated from larger genetically encoded precursors.

ANSWERS TO SELF-TEST

1. (a) 1 (b) 3, and (c) 2. Note that the length of the bacterial chromosome is expressed here in kilobases, not megabases.

2. c.

3. The tight folding of chromatin renders many of the sites on DNA inaccessible to the proteins that must be assembled to form an active transcription complex. Chromatin structure decreases the amount of DNA available to nonchromatin proteins. Remodeling of chromatin makes some of these sites accessible.

4. a, b, c. Answer (d) is incorrect because histones make up nearly half the mass of a chromosome.

5. a, b, d. A nucleosome core consists of ~145 base pairs of DNA wrapped around a histone octamer. The nucleosome cores are connected by linker DNA, which contains from fewer than 20 to more than 100 base pairs (bp), the exact length depending on the organism and the tissue. The average length is ~200 bp.

6. The nucleosome core has a disk shape and is composed of eight histone molecules. The octameric core of histones has ~145 base pairs of DNA wound about it in approximately 1 3/4 turns of a left-handed torroidal supercoil. Two copies each of histones H2A, H2B, H3, and H4 are on the inside of the toroidal coil, whereas histone H1 is associated with the DNA where it emerges from the core. Each core histone has a basic tail that protrudes from the core structure. In total, ~200 bp of DNA is present per nucleosome.

7. No. As mentioned, each nucleosome is associated with approximately 200 base pairs of DNA. If this DNA were coiled into a sphere with a diameter of approximately 100 Å, it would be condensed from 200 base pairs × 3.4 Å per base pair = 680 Å of linear DNA to 100 Å, which is a packing ratio of about 7. The chromatin fiber, which is composed of a helical array of nucleosomes, must be formed, and the resulting 360 Å coils must themselves be looped and folded. Scaffolding proteins, topoisomerases, and small basic molecules, such as the polyamines, also contribute to the ultimate compaction of 104 that is observed in metaphase chromosomes.

8. a, b, c, d, e. Although a critical feature of combinatorial control is mediated by protein-protein interactions, some components of the transcription complex must interact specifically with DNA in order to locate the transcriptional assemblage to the proper region on the DNA. For instance, some transcription factors bind to enhancers far from the site of transcription initiation.

9. (a) 1, 2, 5, 7; (b) 1, 3, 4; (c) 1, 6, 8.

10. (a) A, E; (b) A; (c) A, E; (d) E; (e) N. Answer e is not a feature of either because DNA-binding proteins that influence transcription initiation through the control of enhancers are the molecules that cause the DNA perturbation, not the enhancers themselves.

11. a, c, d, e. Answer (b) is incorrect because actively transcribed genes are less compact than those that are transcriptionally silent.

12. a, d, e. Answer (b) is wrong because the methylation takes place at 5'-CpG-3' sequences. Answer (c) is wrong because AdoMet is the methyl donor for postsynthetic DNA methylation.

13. b, d. Answers (a) and (c) are wrong because they are properties of a more numerous [right word?] class of hormones that act by initiating phosphorylation cascades within cells after binding outside the cell to a transmembrane receptor.

14. a, b, c, d, e.

15. a, b, c, d. Answer (e) is incorrect because, although named "bromodomain," bromination has nothing to do with the action of this acetyllysine-binding protein structure. The name derives from the *brahma* gene in *Drosophila*, where the archetype bromodomain was found.

16. The IRE is a sequence in the 5'-untranslated region of the mRNA that encodes the ferritin molecule. The IRE-binding protein (IRE-BP) binds to the IRE and blocks translation. The ERE is a DNA sequence to which the estrogen receptor-estrogen complex binds to facilitate transcription.

17. The IRP is 30% identical in amino acid sequence with the mitochrondrial aconitase and is an active aconitase, whose function was not previously understood. The iron-sulfur cluster at the center of the IRP is unstable, and under conditions of low iron dissociates from the protein, allowing binding of the IRE. The mitochondrial aconitase in contrast has a stable iron-sulfur center and catalyzes the isomerization of citrate to isocitrate (Chapter 17).

18. c,d. Answer b is incorrect. This is part of the mechanism of RNAi.

PROBLEMS

1. Would you expect the interaction between protamines and DNA to be enhanced or diminished in solutions that have highly ionic strength, that is, high salt concentrations? Protamines, which are found in very high concentrations in sperm where they participate in condensation of the DNA, are low molecular weight compounds rich in groups with high pKa values, that is, they are basic compounds. Explain the basis for your answer. Does the action of histone acetyltransferases (HATs) use a similar principle? Explain.

2. One measure of the evolutionary divergence between two proteins is the number of amino acid differences between them. It can be argued that a better measure would

be the minimum number of mutational events that must have occurred to result in those differences.

(a) The differences in the amino acid sequences of histone H4 between calf thymus and pea seedlings are as follows:

AA position	Pea seedlings	Calf thymus
60	Ile	Val
77	Arg	Lys

What minimum number of mutational events accounts for these differences? Give the changes that must occur in both mRNA and DNA. (Refer to the genetic code inside the back cover of the text.)

(b) Comment on the nature of the changes, based on your knowledge of amino acid chemistry. What conclusion follows about the function of H4?

3. Chromatin that is transcriptionally active (*euchromatin*) is disperse in structure, whereas chromatin that is transcriptionally inactive (*heterochromatin*) is compact. When the nuclei of chicken globin-producing cells were treated briefly with pancreatic DNase, the adult globin genes were selectively destroyed, but the genes for embryonic globins and ovalbumin remained intact. In contrast, when the nuclei of oviduct cells were treated with DNase, the ovalbumin genes were destroyed. Explain these results.

4 The chromatin of globin-producing cells can be treated with micrococcal nuclease under conditions that cleave DNA almost exclusively in the linker regions between intact nucleosomes. When the resulting nucleosomes are isolated and their DNA is examined, it is found to contain DNA sequences for the synthesis of globin. Are these results consistent or inconsistent with the explanation for the results in problem 16? Explain.

5. Would you expect a zinc deficiency in eukaryotes to be associated with any sort of developmental abnormality? Explain.

6. You have isolated the *M* gene, which is involved in muscle cell differentiation and, after introducing it into cultured mammalian cells, you wish to study its expression. In sequencing the region upstream from the open reading frame you notice that there are a number of CG-rich regions, from 100 to 1000 bp upstream.

(a) When you grow the cells in presence of 5-azacytidine, the *M* gene is expressed, whereas cells grown in the absence of the analog do not express it. Provide an explanation for these observations, relating them to the CG-rich regions and how they would be affected by 5-azacytidine. (Hint: How might changing the 5 position in the heterocyclic ring of cytidine from carbon to nitrogen affect its chemistry?)

(b) You decide to carry out some experiments to verify your explanation of the results with 5-azacytidine. Isolating the *M* gene upstream sequence that contains the CG-rich regions, you place it into a bacterial plasmid immediately next to the gene for the enzyme chloramphenicol acetyl transferase (CAT), which is not found in mammalian cells and can be easily assayed. You then amplify this plasmid construct in, and isolate it from, bacteria. When the isolated plasmid molecules are used to infect transiently the muscle cells in culture medium lacking 5-azacytosine, CAT is formed. You then isolate the upstream DNA sequence from cell-derived chromosomes as well as from the plasmid, and digest samples from both isolates with the restriction endonucleases HpaII and MspI. The DNA from the digests is separated using gel electrophoresis. What data would you expect to obtain from these experiments? You will need to look up the cutting specificities of these restriction endonucleases at http://rebase.neb.com/rebase/rebase.html

7. Proteins that interact with a specific sequence of DNA usually remain bound under conditions of low ionic strength during electrophoresis in gels formed with low percentages of acrylamide. (Low percentages of acrylamide form gels with pores sufficiently large to allow entry of large macromolecular complexes.) Under these conditions, protein–DNA complexes usually migrate more slowly than does the free DNA. Suppose you have a sample of a DNA fragment radioactively labeled with ^{32}P that contains an entire promoter sequence. Describe how you could use the sample and gel electrophoresis to isolate transcription factors from extracts of protein from eukaryotic cells. How might you compensate for the possibly competing binding of histones to the DNA? How could you use unlabeled samples of your promoter fragment to demonstrate the specificity of the interaction between the fragment and a putative transcriptional factor? Might you also detect proteins that do not themselves bind directly to the DNA?

8. Would you be surprised if your analysis of the gene regulatory machinery of an eukaryotic cell indicated that DNA sequences far removed (1 or more kbp) from the site of transcription initiation were involved? Explain.

9. In sex determination in humans, female is the default state. To become male, genes must be activated that lead to the development of the testes and external male genitalia and suppression of the development of what would ultimately become the female sex organs. Many of these genes are on the Y chromosome, which is absent in genotypic (XX) females. The steroid hormone testosterone, an androgen, is involved in this process. What would you predict would happen to a genetic male (XY) fetus whose testosterone receptor had a mutation in its C-terminal domain that rendered that domain resistant to binding the androgen?

10. A group of molecular biologists showed that the presence of three eukaryotic DNA sequences below can activate transcription of a reporter gene, such as the gene encoding chloramphenicol transacetylase.

    Sequence X: 5′-TAATTGCGCAATTA-3′

    3′-ATTAACGCGTTAAT-5′

    Sequence Y: 5′-TAATTGCTACGGTA-3′

    3′-ATTAACGATGCCAT-5′

    Sequence Z: 5′-TACCGTATACGGTA-3′

    3′-ATGGCATATGCCAT-5′

    Gel-shift assays, described in problem 21, were used to purify *three* dimeric protein activators, each of which bound to one of the three DNA sequences shown above. Further investigation revealed that although three dimeric proteins could be purified, only *two* polypeptide monomers could be isolated. How is this possible?

11. Unlike bacterial RNA polymerases, eukaryotic polymerases have relatively low affinity for their promoters and therefore often depend on several activator proteins for initiation of transcription. Thus, while many bacterial genes are subject to negative regulation by repressor proteins, eukaryotic genes are more likely to be under positive regulatory control. The reasons for this difference in the mode of regulation may be related to the great difference in genome sizes. For example, the human haploid genome is ~650 times larger than that of *E. coli* and may contain over 25,000 genes. What are the advantages of positive regulation in the control of gene expression in eukaryotes?

12. A human cell has approximately 1300 times more DNA than an *E. coli* cell, assuming one set of chromosomes in the human and one chromosome in the bacterium (~6 $\times 10^9$ bp vs. 4.6×10^6 bp). What biological process may render the eukaryotic genome functionally equivalent to the prokaryotic one with respect to the location of targeted nucleic acid sequences by specific-binding proteins? Explain.

13. Describe the experiment carried out by Shinya Yamanaka that demonstrated that four genes in embryonic stem cells could induce pluripotency (iPS). Comment on the potential therapeutic use of iPS cells.

ANSWERS TO PROBLEMS

1. Just as histones can be dissociated from DNA with salt, the interaction between protamines and DNA is diminished in highly ionic solutions because the salt in solution disrupts the ionic interactions between the ligands and the DNA. Protamines are arginine-rich proteins whose positively charged guanidinium groups can associate with the negatively charged phophodiester bonds in a DNA helix. These electrostatic interactions bind the protamines or histones tightly to the polynucleotide. The positively and negatively charged ions that result from the addition of a salt to an aqueous solution compete with DNA-ligand interactions and, hence, weaken them. By acetylating the primary amino groups on the side chains of lysine, HATs remove their positive charges and thereby weaken the interaction between the histone and the DNA. (See page 949 of the text.) Thus, the physicochemical principle, reduction of charge–charge interactions is the same whether accomplished through the action of HATs or the addition of salts.

2. (a) There would be a minimum of two mutations involved. The possible codons for Ile and Val are as follows:

Ile	Val
AUU	GUU
AUC	GUC
AUA	GUA
	GUG

A single change from A to G in the first position of the codon would give a substitution of Val for Ile. This corresponds to a change from an $A \cdot T$ to a $G \cdot C$ base pair on DNA.

The possible codons for Arg and Lys are as follows:

Arg	Lys
CGU	AAA
CGA	AAG
CGG	
CGC	
AGA	
AGG	

Again a single change, from G to A in the second position of the codon, would account for the amino acid difference. This corresponds to a change from a $G \cdot C$ to an $A \cdot T$ base pair on DNA.

(b) These are conservative changes. Both Lys and Arg have positively charged side chains, and both Ile and Val are hydrophobic. Therefore, we would expect virtually no structural or functional difference in the H4 of calf thymus and pea seedlings. However, the two organisms are clearly different in other respects.

3. In cells that are actively synthesizing adult globins, chromatin is dispersed so that the globin genes may be transcribed into mRNA. Accordingly, this region is sensitive to DNase. In an adult cell specialized for the production of globins, neither embryonic globins nor ovalbumin is produced to a significant extent. Accordingly, the regions of DNA carrying the information for these genes are compact and are therefore not sensitive to DNase. Conversely, in those oviduct cells making ovalbumin, the ovalbumin gene is destroyed by DNase but not the globin genes.

4. The results are consistent. The genes for globin synthesis are contained within the nucleosomes, many of which are required to cover the globin genes. When these genes become transcriptionally active, the chromatin becomes dispersed and the linker regions between the nucleosomes becomes susceptible to cleavage by micrococcal nuclease.

5. The transcription of many eukaryotic genes is activated by proteins containing from one or more zinc fingers, each of which is a ~30-residue-long amino acid sequence containing (usually) two cysteines and two histidines coordinated to a zinc ion. For example, zinc is involved in the structure of DNA-binding domains of the nuclear hormone receptors. (See page 944 of the text.) It is also involved in a large number of other enzyme-catalyzed reactions, including the conversion of acetaldehyde to ethanol, the formation of bicarbonate ion, and the cleavage of peptides by chymotrypsin. Because of its essential roles in gene expression and in cellular metabolism, it is likely that zinc deficiencies could lead to significant developmental abnormalities.

6. (a) The presence of CG-rich regions (islands) in the upstream region of the M gene indicates that these are sequences that might be subject to methylation and can thereby control transcription by influencing promoter activity. In many cells, about three-fourths of CG sequences are methylated, whereas those sequences in transcriptionally active regions are less methylated. In your experiments, cells grown in the presence of 5-azacytidine incorporate the analog into DNA, and the azacytosine residue cannot be methylated at N-5 by specific DNA methyltransferases because the atom at the 5 position is N, not the normally present C. The undermethylated region is therefore more susceptible to transcriptional activation. Transcription of the M gene leads to expression of the gene in those cells grown in 5-azacytosine. Cells grown in the absence of the analog are more likely to have methylated CG-rich sequences in the promoter region and are less likely to express the gene you are studying.

(b) If your explanation in (a) is correct, you would expect the CG sequences in the upstream island to be methylated in samples taken from cells, but not methylated in samples taken from the plasmid grown in bacteria (methylation of C residues occurs frequently in vertebrates but is rare in prokaryotes). Thus, expression of the gene from the CAT gene must be due to the activity of its promoter. You would expect that both restriction enzymes would cleave your upstream sequence isolated from the bacterial plasmid, while only MspI would cleave the sequence isolated from chromosomes. The fragment would be resistant to cleavage by HpaII, which does not cleave CmCGG sequences, that is, sequences methylated at the interior C of the 5 position.

7. Assays that exploit the differential electrophoretic mobility of protein-DNA complexes and free DNA are called *gel-shift* or *electrophoretic-mobility-shift assays*. In these experiments, specific DNA sequences are allowed to associate with putative DNA-binding proteins from cell extracts or from fractionated samples from the extracts. They are then subjected to gel electrophoresis along with control samples of labeled frag-

ment by itself. The radioactivity on the gels is visualized by autoradiography on film or in a phosphoimaging machine. Labeled samples that are retarded on the gel are candidates for sequence-specific protein–DNA complexes. To minimize retardation caused by nonspecific binding of proteins, such as histones, to the labeled fragment, large amounts of unlabeled, random sequences of DNA can be added to the sample before electrophoresis. This DNA binds the nonspecific proteins, but is not observed on the gel because it is not labeled. Because sequence-specific proteins usually have a much higher affinity for a specific sequence than those proteins binding nonspecifically, the addition of the extra DNA does not usually interfere with the gel-shift assay. Experiments which assess specificity of the DNA–protein interaction can be conducted by including an excess amount of unlabeled DNA which contains the same specific sequence in an analyzed sample. If the interaction is truly sequence-specific, the unlabeled DNA should bind the protein and by specific competition abolish the shift of the labeled sequence on the gel. Gel-shift analysis is a widely employed technique. Although it is not a true equilibrium technique and therefore cannot easily provide true thermodynamic equilibrium binding constants, it reliably indicates the relative affinities of proteins for specific DNA sequences. You might also detect proteins that are associated tightly through protein-protein interactions with the DNA–binding proteins, for example, you might detect coactivators or corepressors.

8. You would not be surprised because the distant sequences might well be enhancers that bind transcription factors that themselves associate with the core transcription machinery by looping the DNA to achieve proximity to the transcription start site.

9. The testosterone–nuclear hormone receptor complex could not form and the genes necessary to promote virilization and suppress feminization would not function properly. The outcome would be a genetic male who developed into a phenotypic female. Recall (see pages 944–945 of the text) that the ligand binding domain of nuclear hormone receptors is near their C termini. The inability to form the testosterone–receptor complex results in a male genotype expressing a phenotype similar to one arising from a missing Y chromosome, that is, being a female—in neither case is a functional androgen receptor formed. The disorder arising from the situation described is called testicular feminization, and many other biochemical and developmental factors beyond those mentioned here are involved.

10. The most likely possibility is that each of the two monomers can form homodimers, each of which could bind to one of the sequences above. In addition to forming homodimers, the two monomers associate to form a heterodimer, which binds to a third sequence. You would expect an activator that is a homodimer to bind to a DNA sequence that has dyad symmetry (is an inverted repeat), whereas a heterodimer would bind to an asymmetrical sequence that contains sequences common to each of the individual symmetric sequences. Inspection of the three sequences shows that sequence X and Z have dyad symmetry, while sequence Y is asymmetrical, containing sequences that represent half of X and half of Z. Thus one homodimer binds to and activates sequence X, while another homodimer associates with the symmetric sequence Z. A heterodimer, composed of each of the two monomers, binds to and activates the asymmetric sequence Y, which contains half of each sequence X and Z. This problem illustrates the important principle that heterodimer formation can allow recognition of DNA sequences that do not have dyad symmetry, thereby increasing the potential for regulation of expression. The cyclic AMP–response element binding protein CREB is an example of a homodimer DNA binding protein. The oncogenes *fos* and *jun* form transcription-regulatory proteins that can associate with themselves to form homodimers or with each other to form a heterodimer-the situation described in this problem.

11. In bacteria, negative regulation requires synthesis of a specific repressor that blocks transcription of a gene or an operon. To carry out negative regulation of genes in a human genome, ~25,000 repressor proteins would need to be synthesized, which would be an inefficient means of controlling transcription. Because most eukaryotic genes are not in operons and are normally inactive with regard to transcription, selective activation through synthesis of a small group of activator proteins is used to promote transcription of a particular array of genes needed by the cell at a certain time. Another reason for positive control may be related to the fact that a larger genome presents the possibility that a relatively short DNA sequence for a regulatory protein would be present in multiple and possibly wrong locations, bringing about inappropriate or unneeded gene activation. This can be avoided by requiring that several positive regulatory proteins form a complex that can specifically activate a gene and promote its transcription, thereby reducing the possibility of incorrect gene activation.

12. Chromatin formation. The structure of chromatin in eukaryotes renders much of their DNA inaccessible to DNA-binding proteins, leaving only the remainder free enough to be searched and bound by these proteins. The net effect is to decrease markedly the effective size of the searchable eukaryotic genome by having only a portion of it accessible for scanning at a given time. Of course, remodeling the chromatin by the mechanisms described in the text can release this condensed DNA so that it becomes available for gene regulation at the appropriate times during the cell cycle or during development.

13. The experiment involved introducing DNA encoding four transcription factors into fibroblasts (skin cells). Fibroblasts are cells that have differentiated from pluripotent cells into cells with fixed characteristics and function. Pluripotent cells have the ability to differentiate into many different cell types. In a differentiated cell, most of the genes in the cells are no longer actively transcribed. Until recently, this process of silencing genes during differentiation was thought to be irreversible. The four transcription factors were identified, along with dozens of others, as contributing to pluripotency of embryonic stem cells. When the four genes were introduced into the fibroblast cells, they de-differentiated into cells that appeared to have characteristics very nearly identical with those of embryonic stem cells. This result appears to have opened the way for development of a new class of therapeutic agents. In theory, a patient's fibroblast cells could be converted into iPS cells, then treated to differentiate them into the desired cell type. These new cells could be transplanted back into the patient. The book gives the example of using nerve cells to treat a neurodegenerative disease.

Sensory Systems

Chapter 33

This chapter describes the functioning of the five major senses—smell, taste, vision, hearing, and touch—on a molecular level. All are shown to rely on mechanisms involved in transduction of other sorts of signals (hormones, neurotransmitters, etc.). Olfaction, taste, and vision utilize G-protein-linked 7TM receptors. Hearing and touch have different receptors but appear to share ankyrin repeats as part of their structures.

The human genome contains sequences for hundreds of different odorant receptors (OR), each of which is a 7TM receptor. When an odorant arrives with the proper shape to bind, a G protein binds to GTP and triggers adenylate cyclase. The stimuli send signals to areas of the brain, and the perception appears to be decoded by a combinatorial mechanism. In other words a familiar scent may be the result of a dozen ORs firing at once. The bitter receptors on the tongue form a very similar family of 7TM receptors, but it appears that all of the signals converge on a single area of the brain, so that many different molecules can trigger the same bitter signal. Glucose and other sugars trigger a similar process for sweetness and glutamate binds to its own 7TM receptor for the savory umami flavor. Ion channels account for salty (Na^+) and sour (H^+) tastes.

Vision is mediated by rhodopsin, a 7TM receptor with retinal bound at the center. The change in shape when light isomerizes retinal is exactly the same as when other 7TM receptors bind to their appropriate ligands. And the result inside the cell is parallel, but in this case the G protein (transducin) triggers the cleavage of cyclic GMP. Color is perceived by cone photoreceptors very similar to rhodopsin, but modified to absorb maximally in the red, green, and blue regions. Color blindness is due to homologous recombination of the photoreceptor genes.

Hearing and touch use related systems to sense mechanical stimuli. Displacement of hair cells in the cochlea causes direct stimulation of nerves by opening ion channels. A similar receptor in *Drosophila,* known as NompC, appears to be an ion channel with 29 ankyrin repeats at the amino terminus, inside the cell. Touch appears to utilize similar receptors for touch, plus receptors for temperature and pain. Capsaicin is the active compound in hot peppers, and the capsaicin receptor (VR1) has been studied. It also appears to be an ion channel with ankyrin repeats, and is involved in sensing pain. The authors finish with a discussion of the magnetic sense, pheromones, and circadian rhythms.

LEARNING OBJECTIVES

When you have mastered this chapter, you should be able to accomplish the following objectives.

A Wide Variety of Organic Compounds Are Detected by Olfaction
(Text Section 33.1)

1. Describe the properties of a typical *odorant,* and explain the relationship between shape and smell.

2. Define the term *specific anosmia.*

3. Explain how scientists grew to suspect that *G proteins* are involved with olfaction. Furthermore, explain how we know there are hundreds of different *odorant receptors* (OR) in humans.

4. Describe what happens when an OR binds to an odorant. Include details about $G_{(olf)}$, cAMP, and calcium.

5. Describe what is meant by a *combinatorial mechanism,* and how it applies to decoding of olfactory signals.

Taste Is a Combination of Senses That Function by Different Mechanisms
(Text Section 33.2)

6. List the five *primary tastes*. Describe the different receptor types for each taste, and name compounds that stimulate each receptor.

7. Identify *gustducin, PROP,* and *T2R1.*

8. Explain why a large family of *bitter receptors* does not produce a broad spectrum of different bitter flavors.

9. Present the evidence that suggests that the family of *sweet receptors* is closely related and parallel to the family of bitter receptors.

10. Describe the *glutamate receptor,* which is responsible for sensing the *umami* taste.

11. Describe the structure of an *amiloride sensitive sodium channel*. Compare this sodium channel to the potassium channel shown in figures on page 384 of the text. Explain how ion channels aid in detecting salty tastes.

12. Explain how protons interact with various ion channels to send the *sour* sensation to the brain.

Photoreceptor Molecules in the Eye Detect Visible Light (Text Section 33.3)

13. Distinguish the function and number of *cone* and *rod photoreceptor cells.*

14. Describe the structure of a rod cell.

15. Define the roles of *rhodopsin, opsin,* and *11-cis-retinal* in visual signal tranduction.

16. Explain how light absorption affects the structure of retinal.

17. Identify *transducin,* and describe the reaction that is stimulated in response to light in rod cells.

18. Briefly describe how rod cells recover after being stimulated by light.

10. Explain how the red, blue, and green cone receptors resemble one another, and how they differ.

20. Describe the relationship between DNA recombination and color blindness.

Hearing Depends on the Speedy Detection of Mechanical Stimuli (Text Section 33.4)

21. Compare the speeds at which visual and sound detection occur. Point out the implications for the mechanism of sound detection.

22. Define *cochlea, hair cell, stereocilia,* and *tip link* and describe their function in hearing.

23. Relate the evidence that systems found in *Drosophila* may be homologous to auditory sensors in vertebrates. Explain the role *ankyrin* appears to play.

24. Compare the roles of ion channels in mechanical responses in prokaryotes.

Touch Includes the Sensing of Pressure, Temperature, and Other Factors (Text Section 33.5)

25. Identify *nociceptor, capsaicin,* and *VR1.*
26. Define *circadian rhythm.*
27. Explain how the capsaicin receptor was isolated. Describe its response to pH and temperature.

SELF-TEST

A Wide Variety of Organic Compounds Are Detected by Olfaction

1. If we have 850 odorant receptor (OR) genes in the human genome, but only 40% of them are functional, then:
   (a) How many functional genes would there be?
   (b) What percentage of the complete human genome is taken up by OR genes?

2. At the start of Section 33.1 the structures of R- and S-carvone are presented. One of these enantiomers smells like spearmint and the other like caraway seeds. How can two "identical" compounds have such different sensory qualities?

3. Which would NOT be an indicator that a certain gene codes for an OR?
   (a) codes for 7TM protein
   (b) genes found in cells of nasal epithelium
   (c) part of a large and diverse family of similar genes

4. Is it likely that there is a single OR responsible for a smell like orange peel or chocolate?

5. Functional magnetic resonance focuses on what in the brain?
   (a) electron flow in nerves
   (b) temperature of active brain regions
   (c) speed of blood flow
   (d) hemoglobin versus oxyhemoglobin in the brain
   (e) turnover of neurotransmitters in the brain

Taste Is a Combination of Senses That Function by Different Mechanisms

6. Which of the following is INCORRECT?
   (a) Like OR, bitter receptors form a large family of 7TM proteins.
   (b) There are many subtle shades of bitter linked with the different receptors.
   (c) Bitter and sweet taste receptors are closely related.
   (d) Bitter receptors are found mostly on the back of the tongue.

7. Compare the structures of tetrodotoxin (see Section 13.4 of the text) and amiloride. (See section 33.2 of the text.) What structural similarity do you see? Are there similarities in the actions of the two compounds?

8. How does the umami receptor differ the sweet receptor?

Photoreceptor Molecules in the Eye Detect Visible Light

9. Match the two main regions of the rod cell in the left column with the appropriate functions or properties from the right column.
   (a) outer segment
   (b) inner segment

   (1) contains discs
   (2) carries out normal cellular processes
   (3) contains the photoreceptors that absorb light

10. Which of the following statements about retinal, the chromophore of rhodopsin, is INCORRECT?
    (a) The unprotonated Schiff base absorbs maximally at 440 nm and higher.
    (b) In the dark, it is covalently bound to opsin through a Schiff base linkage.
    (c) In the dark, it is present as the 11-*cis*-retinal isomer.
    (d) When bound to rhodopsin, it absorbs light maximally at 500 nm.
    (e) It becomes the all-cis isomer after absorbing light.

11. Place the following events in the excitation of rhodopsin by light in their correct sequence.
    (a) metarhodopsin II
    (b) conversion of all-*trans*-retinal to 11-*cis*-retinal
    (c) triggering enzyme cascade
    (d) conversion of 11-*cis*-retinal to all-*trans*-retinal
    (e) activation of transducin G protein
    (f) bathorhodopsin

12. Which of the following statements about $G_{(olf)}$, gustducin, and transducin are true?

    (a) All are G proteins.

    (b) All are associated with 7TM sensory receptors.

    (c) All bind calcium ions.

    (d) All bind GTP, which is hydrolyzed to GDP.

13. Explain briefly the major roles of the following participants in the enzymatic cascade that is triggered by the photoexcitation of rhodopsin.

    (a) transducin
                         (c) activated phosphodiesterase

    (b) cyclic GMP (cGMP)

14. With its seven transmembrane helices, rhodopsin has a structure similar to that of which of the following integral membrane proteins?

    (a) nicotinic acetylcholine receptor channel

    (b) Na^+-K^+ pump

    (c) sarcoplasmic Ca^{2+}-ATPase

    (d) hexokinase

    (e) bacteriorhodopsin light-driven pump

    (f) photoreceptors of retinal cones

15. How does the photoexcited system return to the dark state?

16. Which of the following statements about human color vision are correct?

    (a) It is mediated by three different chromophores.

    (b) It is mediated by three different photoreceptors.

    (c) It involves only the cone cells.

    (d) It involves only the rod cells.

    (e) It involves seven-transmembrane-helix proteins.

17. Which colors do not correspond to human visual pigments? What would be different if this question were about color vision in chickens?

    (a) red
                            (d) blue

    (b) yellow
                      (e) violet

    (c) green

18. Why is the proportion of color-blind males so much higher than that of color-blind females?

Hearing Depends on the Speedy Detection of Mechanical Stimuli

19. The text states (see Section 33.4 of the text) that a likely ortholog of the transduction channel used in human hearing has been found in fruit flies. What is an ortholog? How similar is its use in flies and people?

20. How does the speed of sound detection give clues as to its mechanism?

Touch Includes the Sensing of Pressure, Temperature, and Other Factors

21. Birds appear to make use of the Earth's magnetic field while migrating. How could this be proven?

22. Why are birds relatively insensitive to capsaicin?

23. Why does ingestion of foods containing capsaicin create a feeling of having eaten something hot?

ANSWERS TO SELF-TEST

1. There would be some 260 functional genes. If the human genome has 40,000 genes, then one could either calculate the percentage of functional genes (260/40,000 = 0.65%) or calculate the percentage of "apparent" genes—which is justifiable because the figure of 40,000 includes all "apparent" genes. This yields 750/40,000 = 1.875 %. Either way, this is a surprisingly large chunk of the genome, and it shows how important the sense of smell must be in higher animals.

2. Even though they look very much the same on paper, in three dimensions they would bind to an active site very differently. Thus they interact with different OR and smell completely different.

3. (b) is wrong because all cells in an individual have the same germ-line DNA. Thus OR receptors are found everywhere. If cDNA can be made from cells in the nasal epithelium, that means that mRNA is being expressed there, and thus that would be an indication that the gene might be an OR.

4. No. Even when a scent is known to be triggered by a single chemical (like vanilla) there will be an array of receptors that perceive it, and the sensation will be decoded by a combinatorial process.

5. d

6. b

7. Both contain the guanidinium moiety. This cation would be attracted to sodium channels of all types, and then the rest of the molecule will block the channel. So both compounds function as the "cork" in the "bottle."

8. The umami receptor consists of two subunits, T1R1 and T1R3. The heterodimeric sweet receptor consists of T1R2 and T1R3 subunits. Studies have shown that if the T1R1 receptor is disrupted in mice they still respond to sweet tastants, but not longer respond to aspartate.

9. (a) 1, 3   (b) 2

10. e

11. d, f, a, e, c, b

12. c.

13. (a) Photoexcited rhodopsin binds inactive transducin (T–GDP), catalyzes the exchange of GTP for GDP, and releases T_α–GTP. This form of transducin then activates cGMP phosphodiesterase. Hydrolysis of GTP bound to T_α deactivates phosphodiesterase and allows the binding of $T_{\beta\gamma}$, regenerating T–GDP.
    (b) In the dark, cGMP keeps the cation-specific channels open. Activated phosphodiesterase hydrolyzes cGMP to 5'-GMP, leading to the closing of the channels.
    (c) When phosphodiesterase is activated by transducin, it hydrolyzes cGMP, which leads to the closing of the cation-specific channels and to hyperpolarization of the plasma membrane.

14. a, f

15. The return to the dark state requires the deactivation of both cGMP phosphodiesterase and photoexcited rhodopsin and the formation of cGMP. Phosphodiesterase is deactivated by the hydrolysis of GTP bound to T_α to return transducin to the inactive state. Photoexcited rhodopsin is deactivated by rhodopsin kinase, which catalyzes the phosphorylation of photoexcited rhodopsin at multiple sites. The phosphorylated rhodopsin

binds arrestin, which blocks the binding of transducin. Guanylate cyclase catalyzes the synthesis of cGMP from GTP. Guanylate cyclase is inhibited by high Ca^{2+} levels, and therefore light-induced lowering of Ca^{2+} reactivates cGMP formation.

16. b, c, e

17. b, e. Chickens have a violet pigment so the answer would be (b) only.

18. The red and green visual pigment genes are both on the X chromosome. Women get two chances to "get it right" because they have two copies of X. Men get only one chance because they are XY. Homologous recombination ensures that many X chromosomes have a missing visual pigment gene.

19. Review Section 6.1 of text. Orthologs arise by gene duplication and generally diverge in function. Paralogs arise by a speciation event and generally retain similar function in different species. The use of the protein in flies and humans appears remarkably similar; both are sensing vibrations or disturbances in the air.

20. Sound is detected in less than a millisecond, which means it can't use second messengers in its mechanism. Several milliseconds are needed for such mechanisms.

21. One way would be to make magnetic "hats" and see if the birds can be made to fly perpendicularly to their normal routes. This may not work, because birds appear to pay attention to the stars, and use the magnetic field as a kind of calibration. One recent article suggests that some species use the magnetic field to tell them when to fatten up for a long flight with little available food (*Nature* 414[2001]:35).

22. Capsaicin synthesis is a defense mechanism in plants. Since plants need birds to spread seeds, they've chosen a defense that won't affect birds.

23. The VR1 receptor opens at temperatures above 40°C and is responsible for our feeling that something is hot. Since capsaicin acts through the same receptor, the sensory sensation is that of heat. Capsaicin and temperature work synergistically, and capsaicin lowers the temperature at which the VR1 receptor opens. Therefore we will feel the heat sensation of capsaicin more acutely when the temperature of the food is higher. Unlike temperature, capsaicin causes no actual tissue damage.

PROBLEMS

1. A very common specific anosmia is the inability to smell musk. When exposed to the pure compound, some 10% of the population smell nothing, and another 20% find the smell unpleasant. What is happening on the molecular level to explain this?

2. Wine drinkers sometimes find that a bottle is "corked." This means that a defective cork has leached trichloro-anisole, or TCA, into the wine. This compound ruins the wine, making it smell like wet cardboard. Some wine collectors report that as much as 20% of the bottles they open are corked, while others hardly ever encounter the problem. How would you account for the individual differences? How could the problem of corked wines be solved?

3. Why are specific anosmias so common, but the lack of some particular taste (would we call it specific "dis-gustia"?) is quite rare? Consider the bitter receptor in your answer.

4. The fruit of the African Miracle Berry bush (*Synsepalum dulcificum*) has no flavor. But after it has been chewed, acidic foods taste sweet and not sour. A lemon tastes delicious, and a tomato is sweeter than an apple. The active substance appears to be a protein, and the mechanism has not been discovered. What are some possible mechanisms?

5. One close relative of the umami receptor is the NMDA receptor found in the brain. The NMDA receptor is one of several glutamate receptors in the central nervous system. NMDA stands for N-methyl-D-aspartate, a compound that does not exist in nature. Why do you think a glutamate receptor would be named for a compound not found in the brain?

6. Late at night, police investigated a car pulled over to the side of the highway. When they interviewed the driver, he stated that he had seen streams of blue light coming out of his dashboard. At first the officers thought that the man had taken hallucinogenic drugs, but then he told them he had taken sildenafil (Viagra) earlier in the evening. Why does Viagra sometimes have visual side effects? Why are they most likely to occur at night? Sildenafil works by inhibiting a phosphodiesterase known as PDE5, an enzyme that breaks down cGMP.

7. Which of the compounds in Figure 33.1 would be expected to have the smallest absorbance per mole in the visible light range? Give the reason for your choice.

FIGURE 33.1 Four light-absorbing compounds.

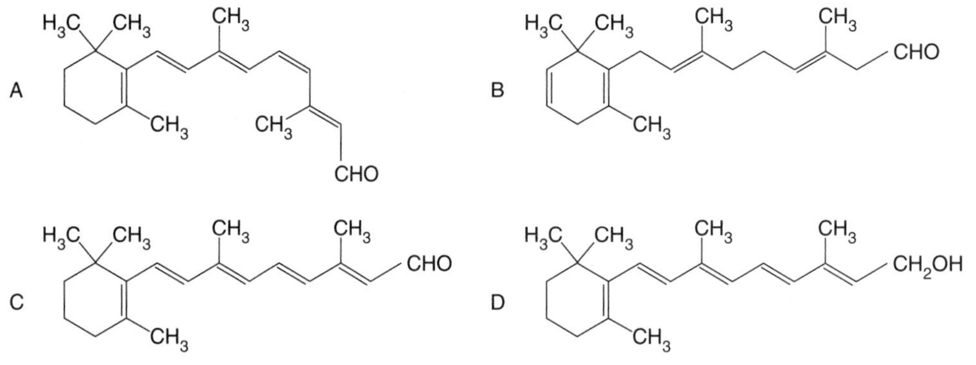

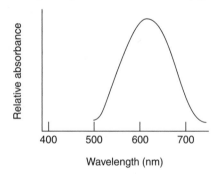

8. Figure 33.2 shows the absorption spectrum of a light-absorbing pigment that is involved in color vision.
   (a) What color light is absorbed by the pigment?
   (b) When the pigment is extracted into an organic solvent, what is the color of the resulting solution?

FIGURE 33.2 Absorption spectrum of a color vision pigment.

9. 11-*cis*-Retinal is covalently linked to a lysine side chain of opsin by the formation of a Schiff base:

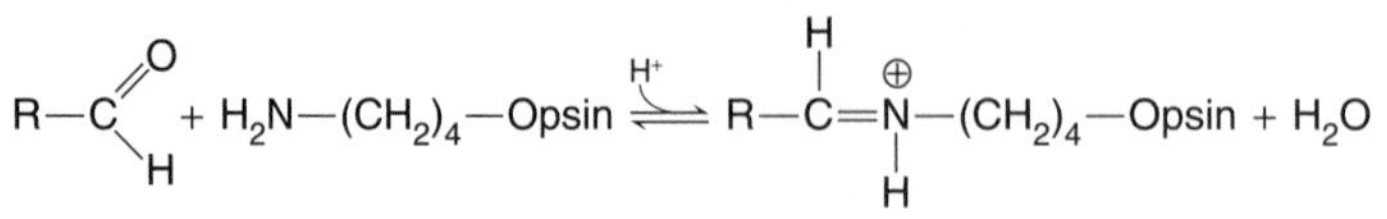

$$R-C{\overset{O}{\underset{H}{}}} + H_2N-(CH_2)_4-Opsin \overset{H^+}{\rightleftharpoons} R-C=\overset{\overset{H}{|}}{\underset{H}{\overset{\oplus}{N}}}-(CH_2)_4-Opsin + H_2O$$

Such linkages can typically be stabilized in the laboratory by reduction with borohydride:

$$R-\overset{\overset{H}{|}}{\underset{\underset{H}{|}}{C}}=\overset{\oplus}{N}-(CH_2)_4-Opsin \xrightarrow{BH_4^-} R-CH_2-\overset{}{\underset{\underset{H}{|}}{N}}-(CH_2)_4-Opsin$$

Would you expect a rod cell preparation that has been treated with borohydride to be active in the cycle of visual excitation? Why or why not?

10. A color vision pigment that absorbs red light is chemically cleaved to separate the retinal from the protein. The same is done for a pigment that absorbs blue light. Then a new pigment is constituted using the retinal from the red-absorbing pigment and the opsin from the blue-absorbing pigment. What color of light will be absorbed by the new pigment?

11. Many people who are hard of hearing or completely deaf have been helped by cochlear implants. How is it possible to mimic the natural hearing process?

12. Zostrix is one of several skin creams that contain capsaicin. Rubbing the cream on the less sensitive parts of the body (knee, elbow, neck) normally does not produce a burning sensation, and can provide relief from neuralgia or arthritis pain. One must avoid contacting eyes, nose, or mouth because capsaicin is the hot principle of the chili pepper and can cause painful burning of mucus membranes. Why would capsaicin work as an analgesic, when it directly stimulates pain receptors?

ANSWERS TO PROBLEMS

1. The 10% who are anosmic for musk either lack the receptor or have a mutation that changes its specificity. People who can smell musk normally describe the scent as sweet and floral. But for the 20% who dislike it, it smells like old wet newspapers or something stored in the attic too long. It seems that a different receptor must have mutated and developed the ability to bind to musk. The structure of natural musk is interesting; it is the cyclic lactone of ω-hydroxypalmitate. If one appreciates the sensory differences these genetic variations cause in humans, it brings home the truth of the old saying "de gustibus non disputandem est" (to each his own taste).

2. Just as with musk, there are many people with specific anosmia to TCA. This can either lead to an inability to smell it, or a diminished ability to smell it. The unlucky people who smell it very intensely often find that they can't drink their expensive wine because of a cheap little cork. The obvious solution is to switch to plastic corks, or screw-on caps. Some people argue that the gentle oxidation that occurs through a cork allows the wine to age better. Others say that the disadvantages of real corks greatly outweigh the advantages.

3. Loss of an OR leads to the inability to smell that odor, or the diminished ability to smell it. But because of the way bitter receptors are "wired," loss of one kind of bitter receptor has essentially no effect on the perception of bitter tastes. All of the many receptors send their signals to the same area of the brain, so nothing is really lost when one or two bitter receptors mutate and lose their functionality.

4. Some people think that the protein binds directly to sweet receptors and stimulates them when exposed to acid. It also might be possible that it is a glycoprotein, and some

sugar hydrolyzes off in the presence of acid. The compound has been proposed as an aid to dieters, but artificial sweeteners solve the problem more directly. Artichokes contain a substance that produces similar changes in the taste of other foods.

5. NMDA isn't the preferred transmitter for the NMDA receptor. But that receptor is distinguished by its ability to bind and respond to the artificial compound, NMDA. Joe Z. Tsien, a researcher at Princeton, attracted much attention in the popular press when he found a way to enhance the functioning of NMDA receptors and produced "smart" mice, which could solve problems better and faster than normal mice (*Nature* 401[1999]:63).

6. The mechanism of action of Viagra is based on its ability to inhibit PDE-5. In other words, it prevents cGMP from being cleaved to yield GMP. If levels of the drug are high enough, then cGMP levels in the retina are propped up at a high level because the local enzyme, PDE-6, becomes inhibited, so that normal cycling cannot occur. In a small percentage of the population this yields blue-tinted or blurry vision, sometimes with more extreme manifestations. Scientists are working on second-generation Viagra(tm)-like drugs which will be more specific for PDE-5, and not stimulate the retina as much.

7. Compounds that absorb visible light significantly have long sequences of alternating single and double bonds, that is, they are conjugated. Compound B is unconjugated, so it would have negligible absorbance in the visible range of the spectrum. Compound A is 11-*cis*-retinal, C is all-*trans*-retinal, and D is all-*trans*-retinol. All of these would have significant absorbance in the visible range.

8. (a) Red light is absorbed by the pigment.
   (b) A solution of the pigment would be blue or blue-green. The pigment absorbs red light, but it transmits light at the blue end of the spectrum.

9. The photoexcitation of rhodopsin leads initially to the isomerization of 11-*cis*-retinal to all-*trans*-retinal and ultimately to the cleavage of all-*trans*-retinal from the protein. In a rod cell preparation that has been treated with borohydride, the Schiff base would be stabilized by reduction, so the removal of the all-*trans*-retinal would be impaired. As a result, the cycle of visual excitation could not occur. (See Figures 33.21 and 33.22 and Section 33.3 of the text for a discussion of these events.)

10. The new pigment will absorb blue light. The retinal is the same in all color vision pigments. The protein component, however, varies among the pigments, giving absorption maxima at different wavelengths.

11. Physiologists learned enough about the sort of nerve signals generated by the cochlea that it was possible to simulate them. As long as enough nerves are present in the inner ear to convey the signals, an external device will perceive auditory vibrations and transduce them into electrical signals just like the hair cells in the cochlea. So the electrical impulses received by the brain are the same as would be experienced by a hearing person.

12. Even though the pain is not perceived when capsaicin is rubbed on a knee or elbow, the receptors are being stimulated. And enough capsaicin applied regularly appears to flatten the response of the pain receptors on the surface of the skin. This means that there is less response to real pain, which occurs under the skin.

The Immune System

Chapter 34

Chapter 34 deals with the immune system. The cells and proteins of this system cooperate to detect and inactivate foreign (nonself) molecules, microorganisms, and viruses. The *innate immune system* is an ancient system found in all multi-cellular plants and animals. The *"Toll-like receptors"* (TLR) in this system recognize *"pathogen-associated molecular patterns"* (PAMP), including cell-wall glycolipids from gram-negative bacteria and double-stranded RNA from some viruses. The evolutionarily more recent *adaptive immune system* is found mainly in vertebrates and has a two-pronged response. The *humoral immune response* acts through soluble antibodies, secreted into the circulatory system, that bind antigens with high specificity and affinity. The *cellular immune response* acts through receptor proteins similar to antibodies on the surface of specialized cells (T lymphocytes) that bind to peptides presented on another kind of cell-surface protein, the *major-histocompatibility-complex (MHC)* proteins. The immune system affords a good example of a system governed by the basic principles of evolution.

After defining essential immunological terms, the authors discuss the structure of immunoglobulin G, including the molecular details of antibody–antigen recognition. Next they relate the variable and constant regions of immunoglobulins to the organization of the genes that encode them. They explain how somatic recombination of a large variety of V-, D-, and J-segment genes with a few C genes, plus imprecise joining, can generate the enormous diversity of antibody molecules. The five classes of immunoglobulins with their characteristic polypeptide compositions and functions are introduced. These five classes are generated by gene rearrangements, in a process called class switching. Alternative mRNA splicing leads to the formation of membrane-bound or soluble immunoglobulins. Finally the authors turn to the cellular immune response. They describe the functions of cytotoxic T cells and helper T cells, and discuss the diversity, polypeptide composition, and structures of MHC proteins and T-cell receptors. They conclude the chapter with a discussion of the *human immunodeficiency virus (HIV)* infection of helper T cells, and a discussion of autoimmune disease and how the system normally avoids it.

Preceding chapters in the text that deal with protein structure (Chapters 2 and 3), evolution (Chapter 6), molecular recognition (Chapters 9 and 10), and flow of genetic information (Chapters 4 and 5) constitute important background for this chapter. Note that some basic information about the immune system, and a discussion of the production of monoclonal antibodies, are covered in Section 3.3.

LEARNING OBJECTIVES

When you have mastered this chapter, you should be able to accomplish the following objectives.

1. Discriminate between the innate and the adaptive immune systems.

2. Name three different PAMPs recognized by TLRs in the innate immune system.

3. Contrast the usual targets of the *humoral immune response* and the *cellular immune response*.

4. Distinguish between cytoxic T cells ("*killer T cells*") and *helper T cells* and describe their functions.

5. Describe the basic evolutionary principles that govern the immune system.

Antibodies Possess Distinct Antigen-Binding and Effector Units
(Text Section 34.1)

6. Relate the intact structure of an *immunoglobulin G (IgG)* molecule to the F_{ab} and F_c fragments produced by proteolysis. Describe the functions performed by the different regions of IgG.

7. Sketch the polypeptide chains of an IgG molecule, and relate the *heavy chain (H)–light chain (L) subunit composition* (H_2L_2) to the F_{ab} and F_c fragments of the molecule. Understand the function of the *hinge region* of IgG.

8. Describe the features of the *constant (C), variable (V),* and *hypervariable* amino acid sequences of the L and H chains of IgG molecules.

9. List the different classes of immunoglobulins and give their functions. Note the common occurrence of (κ) and (λ) L chains in all classes and the (α, μ, δ, γ, or ε) H chains that provide the structural bases for the function of each class.

Antibodies Bind Specific Molecules Through Hypervariable Loops
(Text Section 34.2)

10. Discuss the biological distribution of proteins that contain the immunoglobulin-fold structural motif.

11. Relate the hypervariable sites of the H and L chains of IgG to the *complementarity-determining regions (CDRs)* of the immunoglobulin. Describe the function of the constant regions of the H and L chains of IgG.

12. Describe the *domain structure* of IgG as revealed by crystallography, and note the presence of the *immunoglobulin fold* as a common structural feature. Locate the *antigen-binding sites*.

13. Summarize the types of bonds that form complexes between immunoglobulins and antigens. Note the similarities of *combining sites* of immunoglobulins with the active sites of enzymes.

Diversity Is Generated by Gene Rearrangements (Text Section 34.3)

14. Describe the contributions of the number of immunoglobulin genes (the *germ-line repertoire*), *somatic recombination*, and *somatic mutation* in the generation of antibody diversity. Describe the roles of the *V, C, D,* and *J segment genes* in these processes.

15. Calculate the diversity of immunoglobulin structures that arise from the *combinatorial association* of different genes and from somatic mutation.

16. Describe the process of *affinity maturation* and how it can lead to a 1000-fold increase in binding affinity during the course of a disease.

17. Define *ITAM* and explain its significance in linking *oligomerization* of B-lymphocyte surface antibodies to secretion of soluble antibodies.

18. Discuss the mechanism and use of the drug *cyclosporin*.

19. Define *hapten, epitope, antigen,* and *immunogen*.

20. Describe the phenomenon of *class switching*, and note its significance in maintaining constant recognition specificity among the immunoglobulin classes.

Major-Histocompatibility-Complex Proteins Present Peptide Antigens on Cell Surfaces for Recognition by T-Cell Receptors (Text Section 34.4)

21. Describe the formation and recognition of foreign peptides displayed on cell surfaces in complexes with MHC proteins.

22. Describe the interaction of a peptide with the peptide-binding site of HLA-A2. Mention the typical length of the peptide and the *anchor residues* for this site.

23. Compare the structures and features of *T-cell receptors* with those of the immunoglobulins. Note the sizes and conformations of the epitopes that are recognized by each kind of protein.

24. Explain the origins of the diversity of T-cell receptors and why these receptors can be even more diverse than immunoglobins.

25. Outline a model that accounts for the recognition of a *combined epitope* by a T-cell receptor.

26. Explain the role of *CD8 coreceptors* in the activation of *cytotoxic T cells*, and describe the functions of *perforin* and *granzymes* in leading to the cell's death from apoptosis.

27. Explain the role of *CD4 coreceptors* in the activation of *helper T cells*, and describe the functions of *cytokines* such as *interleukin-2* and *interferon-γ*.

28. Describe the subunit structures of the class I and class II MHC proteins. Locate the peptide-binding sites.

29. Describe the three classes of proteins (*class I, class II, and class III*) encoded by the *MHC genes*. Note the diversity of the class I and class II MHC proteins and their significance in *transplantation rejection*.

30. Describe HIV and the disease it causes—*AIDS*.

31. Outline the mechanism of infection and lysis of helper T cells by HIV.

The Immune System Contributes to the Prevention and the Development of Human Diseases (Text Section 34.5)

32. Explain the selection process, both positive and negative, that is applied to T cells in the thymus.

33. List three *autoimmune diseases*. Explain the consequences when the immune system fails to distinguish between self and nonself.

34. Describe the role that the immune system plays in cancer prevention.

SELF-TEST

Antibodies Possess Distinct Antigen-Binding and Effector Units

1. Match the structure or feature listed in the right column with the appropriate IgG fragment on the left.

   (a) F_{ab}
   (b) F_c

   (1) contains an antigen combining site
   (2) one is formed per IgG molecule
   (3) contains an H-chain fragment
   (4) contains an intact L chain
   (5) mediates effector functions such as complement fixation in the intact IgG
   (6) two are formed from an IgG molecule
   (7) forms a precipitate upon binding an antigen

2. Which of the following statements about the L and H chains of IgG are correct?

   (a) The H chains of IgG molecules have variable and constant regions of amino acid sequences.
   (b) The H chains are responsible for segmental flexibility.
   (c) The constant region of L chains exists in two forms (κ and λ).
   (d) The variable region of the L chain has a counterpart of the same length and amino acid sequence in the variable region of the H chain.

3. Match the immunoglobulin class listed in the left column with its property or function from the right column.

   (a) IgA
   (b) IgD
   (c) IgE
   (d) IgG
   (e) IgM

   (1) most prevalent soluble antibodies
   (2) unknown function
   (3) first soluble antibodies to appear in serum after immunization
   (4) protect against parasites
   (5) major antibodies in tears, saliva, and mucus

Antibodies Bind Specific Molecules Through Hypervariable Loops

4. Fill in the blanks: A molecule of IgG contains _____ immunoglobulin domains. Each heavy chain has _____ of these sandwiches, and each light chain has _____.

5. The immunoglobulin fold is made up of

   (a) seven alpha-helical segments.
   (b) a beta-barrel.

(c) a sandwich of two parallel beta sheets.

(d) a sandwich of two antiparallel beta sheets.

(e) a beta saddle domain.

6. Which of the following statements about IgG structure are correct?

(a) Each of the two antigen-combining sites on an IgG molecule can bind to a structurally distinct epitope.

(b) Both interchain and intrachain disulfide bonds stabilize IgG structure.

(c) Both the L and the H chains of IgG contain domains with similar structures.

(d) The hypervariable regions of the L chain are the sole determinants for the binding of the IgG to the specific antigen.

7. Which of the following are properties of antigen-binding sites of IgG?

(a) They are located between the two sheets of β strands of the V domains.

(b) They are made up of loops formed by the complementarity-determining regions of both the V_L and the V_H domains.

(c) They may undergo conformational changes upon binding of the hapten or antigen.

(d) They contain a specific amino acid residue that covalently binds to the antigen.

(e) They form numerous weak electrostatic, hydrogen-bond, van der Waals, and hydrophobic interactions with the antigen surface.

8. X-ray analysis has revealed that when small antigens bind to antibodies

(a) they usually fit into a cleft.

(b) they contact all six CDRs.

(c) they can bind to the F_c end of the antibody.

(d) they are attracted by the same noncovalent bonds found in enzyme/substrate interactions.

9. Explain why the antibodies produced in an animal in response to a given antigen display a range of binding constants for the antigen eliciting them.

Diversity Is Generated by Gene Rearrangements

10. If the mRNA encoding the L chain of an IgG molecule were isolated, radiolabeled, and hybridized to genomic DNA that has been isolated from either the plasma cell producing the antibody or from germ-line cells from the same organism, what would you observe with respect to the relative locations of the L-chain gene sequences on the two DNAs?

11. Some antibody diversity arises from the combination of one V gene and one C gene from pools containing numerous different copies of each. Why doesn't this mechanism account completely for the observed diversity?

12. Match the following gene segments with their approximate number in the germ line.

(a) V_κ	(1) 5
(b) V_H	(2) 6
(c) J_κ	(3) 27
(d) J_H	(4) 40
(e) D	(5) 51

13. Small foreign molecules do not usually elicit the formation of soluble antibodies, and the cellular immune system also ordinarily responds only to macromolecules. Explain how an antibody can sometimes be directed against a small foreign molecule.

14. Match the immunological term in the left column with its description or definition in the right column.

    (a) antigen
    (b) antibody
    (c) epitope
    (d) hapten

    (1) particular site on an immunogen to which an antibody binds
    (2) protein synthesized in response to an immunogen
    (3) macromolecule that elicits antibody formation
    (4) small foreign molecule that elicits antibody formation

15. Which of the following statements about class switching are correct?

    (a) RNA splicing joins the sequences that encode $V_H DJ_H$ regions to sequences that encode different class C_H regions.
    (b) Plasma cells that initially synthesize IgM switch to form IgG with the same antigen specificity.
    (c) Class switching doesn't affect the variable region of the H chains.
    (d) Class switching allows a given recognition specificity of an antibody to be coupled with different effector functions.

Major-Histocompatibility-Complex Proteins Present Peptide Antigens on Cell Surfaces for Recognition by T-Cell Receptors

16. Which are properties of the peptide-binding site of HLA-A2, a MHC class I protein?

    (a) consists of a deep groove with a β sheet floor and α-helical walls
    (b) can bind all peptides of 7 to 11 amino acid residues with equal affinity
    (c) interacts specifically with two anchor residues of the peptide and nonspecifically with the rest of the amino acid residues
    (d) binds peptides that retain their α-helical conformation
    (e) has a very high affinity for peptides

17. Which of the following statements about T-cell receptors are correct?

    (a) T-cell receptors recognize soluble foreign molecules in the extracellular fluid.
    (b) T-cell receptors recognize T cells.
    (c) For a T-cell receptor to recognize a foreign molecule, the molecule must be bound by proteins encoded by the genes of the major histocompatibility complex.
    (d) The T-cell receptor is structurally similar to an IgG immunoglobulin in that it has two H and two L chains.
    (e) The T-cell receptor is encoded by genes that arise through the recombination of a repertoire of V, J, D, and C DNA sequences.
    (f) T-cell receptors primarily recognize fragments derived from foreign macromolecules.

18. Match the receptor proteins in the left column with appropriate structural features in the right column.

    (a) class I MHC proteins
    (b) class II MHC proteins
    (c) T-cell receptors

    (1) contain an immunoglobulin fold motif
    (2) contain domains homologous to the V and C domains of immunoglobulins
    (3) contain a β-microglobulin chain
    (4) contain two transmembrane polypeptide chains
    (5) contain one transmembrane chain

(6) are encoded by six different genes that are highly polymorphic

(7) are encoded by V, D, and J gene segments

19. Match the type of T cell with the corresponding feature listed in the right column.

(a) helper T cell
(b) cytotoxic T cell

(1) detects foreign peptides presented on cell surfaces
(2) recognizes class I MHC protein plus peptide
(3) recognizes class II MHC protein plus peptide
(4) expresses CD8
(5) expresses CD4
(6) stimulates B lymphocytes and other cells
(7) lyses infected cells

20. Which of the following statements about the MHC proteins are correct?

(a) MHC proteins play a role in the rejection of transplanted tissues.
(b) One class of MHC proteins is present on the surfaces of nearly all cells, binds fragments of antigens, and presents them to T-cell receptors.
(c) MHC proteins are encoded by multiple genes.
(d) One class of MHC proteins provides components of the complement system.
(e) The genes encoding MHC proteins produce three classes of soluble proteins.

21. Which one of the following is NOT a property of the human immunodeficiency virus (HIV)?

(a) It is a retrovirus, that is, it has an RNA genome that produces viral DNA in the host cell.
(b) It contains a bilayer membrane with two kinds of glycoproteins.
(c) It interacts with the CD4 coreceptor of helper cells through the gp120 glycoprotein.
(d) It injects its RNA into the cell and releases the coat into the medium surrounding the cell.
(e) It impairs and destroys the host cell by increasing its permeability.

The Immune System Contributes to the Prevention and the Development of Human Diseases

22. T cells are subject to both positive and negative selection during fetal development in vertebrates. Why are both needed?

23. Human cancer cells are very much like normal human cells. How then can the immune system play a role in cancer prevention?

ANSWERS TO SELF-TEST

1. (a) 1, 3, 4, 6 (b) 2, 3, 5. Answer (7) is inappropriate for either fragment because the F_c fragment lacks an antigen-binding site and the F_{ab} fragment contains only one. The insoluble lattice of antigen–antibody molecules forms because intact IgG molecules each have two antigen-binding sites and can therefore link several antigens together.

2. a, b, c. Answer (d) is incorrect because the variable sequences of the L and H chains in a given IgG are different from one another.

3. (a) 5 (b) 2 (c) 4 (d) 1 (e) 3

4. 12, 4, 2

5. (d), but (b) isn't entirely wrong. Some people describe the beta-sandwich as a "collapsed barrel."

6. b, c. Answer (a) is incorrect because the two antigen-combining sites on an IgG molecule are directed toward the same epitope. Thus, an IgG can bind only one kind of antigen. Answer (d) is incorrect because the hypervariable regions of both H and L chains form the antigen-combining site.

7. b, c, e

8. a, d

9. The antigen stimulates several B lymphocytes that bear different surface antibodies that recognize it to differentiate into plasma cells and secrete antibodies. Each plasma cell secretes a different kind of antibody that binds the antigen through a unique array of noncovalent interactions between the hypervariable regions of the antibody and the epitope bound.

10. The mRNA probe would hybridize to one region on the DNA from the plasma cell, but to widely separated regions on the germ-line DNA. The gene on the plasma cell DNA encodes the intact gene sequence for the V, J, and L regions of the L chain, including introns. The genes encoding the V, J, and L regions of the L chain are in distant locations in the germ-line DNA because the DNA has not yet rearranged to bring these regions into proximity.

11. There aren't enough unique V and C genes to provide a sufficient number of different sequences when they are recombined in all the possible combinations. Joining (J) and diversity (D) genes increase the number of possible combinations.

12. (a) 4 (b) 5 (c) 1 (d) 2 (e) 3. (See Section 34.4 of the text.)

13. If the small molecule (hapten) becomes attached to a macromolecule (carrier), it can act as an immunogen and serve as an epitope (haptenic determinant) to which an antibody can be selected to bind. (See Section 34.4 of the text.)

14. (a) 3 (b) 2 (c) 1 (d) 4

15. b, c, d. Answer (a) is incorrect because the sequence rearrangements of class switching take place through DNA recombination.

16. a, c, e. Peptides are bound in "extended" conformation.

17. c, e, f. Answer (a) is incorrect because T-cell receptors recognize fragments of foreign macromolecules only when the antigen is bound on the surface of a cell. Answer (d) is incorrect because the T-cell receptor is composed of one α and one β chain, each having sequences that are homologous to the V regions of the chains of immunoglobulins.

18. (a) 1, 3, 5, 6 (b) 1, 4, 6 (c) 1, 2, 4, 7

19. (a) 1, 3, 5, 6 (b) 1, 2, 4, 7

20. a, b, c, d. Answer (e) is incorrect because MHC proteins are all bound to the cell surface and are not soluble. Answer (d) is correct because class III MHC proteins contribute to the complement cascade; they are mentioned in Section 34.5 of the text.

21. d. This answer is incorrect because the HIV RNA and other core components enter the helper T cells as the viral membrane fuses with the cell membrane.

22. The positive selection winnows out the T-lymphocytes that do not bind to any of the available MHC-peptide complexes. This step explains the Nobel-prize–winning observations, by Peter C. Doherty and Rolf M. Zinkernagel (1996), who observed that cytotoxic T-cells, which would kill virus infected cells from the mouse they came from, would not kill mouse cells infected with the same virus, but from an unrelated mouse (*Nature* 251[1974]:547). Most developing T-cells are discarded during the "positive selection" phase including cells that would respond to MHC proteins from other individuals, but not those present in "self" cells. The negative selection process then removes T-cells that bind too tightly to "self" peptides complexed with MHC proteins.

23. As the text points out, cancer cells will sometimes produce proteins that are inappropriate for the developmental stage of the individual, like the CEA associated with colorectal cancer. And sometimes unique abnormal proteins can be produced. A widely used test for prostate cancer is the blood test for *prostate specific antigen (PSA)*. The fact that human cancer cells are basically human cells poses a serious challenge for cancer treatment, and explains the fact that chemotherapy is generally rather difficult and uncomfortable. Antibiotics, directed against prokaryotic organisms, can be quite safe. But chemotherapeutic drugs generally have to inhibit processes that occur both in cancer cells and normal cells, so the therapeutic ratio is less favorable.

PROBLEMS

1. The human genome contains only some 25,000 genes, but millions of antibodies are produced by gene rearrangements as described in the text. This diversity-generating system is confined to jawed vertebrates, but all higher animals (down to the sponge) have some kind of self versus nonself recognition system. What happens when a crab or an octopus or a jellyfish is infected? How would its immune system cope?

2. Assuming that antigen–antibody precipitates have lattice-like structures (see Figure 34.7 in the text), draw simple sketches showing possible arrangements of antigen and antibody molecules in a precipitate in which the ratio of antibodies to antigens is (a) 1.14 and (b) 2.83.

3. When polyacrylamide gel electrophoresis of a monoclonal antibody preparation is conducted, a single sharp band appears. When the antibody preparation is treated with β-mercaptoethanol, two bands appear. Why is this the case?

4. Pepsin cleaves IgG molecules on the carboxyl-terminal side of the interchain disulfide bonds between heavy chains. How many physical pieces would result from the cleavage of IgG by pepsin? How many of the pieces derive from the F_c region of IgG?

5. Suppose that antibody is prepared against the extracellular portion of a particular hormone receptor known to have intracellular tyrosine kinase activity, and known to carry out autophosphorylation of its tyrosine kinase. Suppose further that addition of this antibody to target cells elicits intracellular responses similar to those obtained by addition of hormone itself. Propose a mechanism by which addition of antibody may mimic the effects of adding hormone. (Hint: F_{ab} fragments do not elicit the hormone-like response.)

6. DNP or 2,4-dinitrophenyl is used as a hapten (artificial epitope) in many experiments on the immune system. The addition of a bifunctional DNP affinity-labeling reagent (one with two affinity-labeling groups) to myeloma protein (making the protein an

artificial antigen) produces light and heavy chains that are cross-linked through Tyr 34 and Lys 54, respectively. What conclusion is suggested by this observation?

7. Most antigens are polyvalent, that is, they have more than one antibody-binding site. In the case of macromolecules that contain regular, repeating sequences, like polysaccharides, it is easy to understand how a molecule might have multiple binding sites. In the case of proteins with nonrepeating sequences, it is more difficult to envision how polyvalence might be accounted for. Yet proteins with single polypeptide chains are polyvalent as antigens. What feature of antibody production accounts for this behavior?

8. Quantitative measures of the interactions between antigens and antibodies are frequently given as association constants, the reciprocal of dissociation constants. (See Problem 34.6 in the text.) The association constant for the binding of a given hapten to an antibody is 10^9 M^{-1}, and the second-order rate constant for its binding is 10^8 $M^{-1} s^{-1}$. Calculate the rate constant for the dissociation of the hapten from the antibody.

9. Propose a method using the technique of affinity chromatography that would allow one to select lymphocytes producing antibody to one particular antigen from a heterogeneous population of immature lymphocytes. Explain the rationale behind your proposal.

10. Suppose that dinitrophenol is attached to a protein with many potential DNP binding sites and that the resulting antigen is used to stimulate antibody production in rabbits. When serum is harvested and the immunoglobulin fraction is purified and mixed together with antigen, no precipitate forms, yet fluorescence measurements reveal the presence of antigen–antibody complexes. Explain this paradox.

11. The system that supplies MHC proteins with peptides is called "cut and display." (See Section 34.5 of the text.) Distinguish between *what* is cut and *where* it is cut in cells with MHC class I and MHC class II proteins.

12. In a 1995 experiment, researchers had male college students sleep in plain white T-shirts for two nights. They then had female volunteers sniff the shirts and rate them as to how attractive they smelled. Because the researchers had tissue-typed all participants, they were able to determine that the women preferred the scent of men whose MHC proteins were the most different from their own. Why would this be logical, on an evolutionary basis? In a 2002 experiment, along parallel lines, other researchers found that women favored men who smelled like their fathers. Can this also be considered reasonable?

13. The text states that the mutation rate of HIV is more than 65 times higher than that of the influenza virus. (See Section 34.5 of the text.) What are the medical implications of this? What about the evolutionary implications?

14. Although the immune system is designed so that antibodies are not ordinarily directed toward one's own tissue components, sometimes that process goes awry, leading to so-called *autoimmune* diseases. One human disease thought to involve such an autoimmune mechanism is myasthenia gravis, a relatively common disorder in which antibodies directed toward acetylcholine receptors lead to a decrease in receptor numbers. Those suffering from myasthenia gravis show weakness and fatigability of skeletal muscle. Eventually death results from loss of function of breathing muscles. Medical therapy for people suffering from myasthenia gravis may include two types of drugs, immunosuppressive agents and inhibitors of acetylcholinesterase. Give a brief rationale for the use of each.

ANSWERS TO PROBLEMS

1. Even without a diversity-generating system, higher invertebrates will have a rather large number of variable sequences in the germ-line DNA. These will allow for an immune response, although without the power and flexibility of the vertebrate system. Some invertebrates live for many years, but most have a fairly short life-span, which probably somewphat diminishes the importance of the immune system. For all the species mentioned, including the jellyfish, remember that the innate immune system is a universal feature of higher animals. Thus there will be a system of Toll-like receptors (TLRs) that will detect telltale molecules (PAMPs) from invading viruses, bacteria, or fungi.

2. The sketches for the two precipitates are shown in Figure 34.1. In (A) the Ab/Ag ratio is 8/7 = 1.14. In (B) the ratio is 17/6 = 2.83.

3. Polyacrylamide gel electrophoresis separates proteins on the basis of size. β-mercaptoethanol reduces the disulfide bonds that link the light and heavy antibody chains. Thus, the light and heavy chains are separated from one another on the gel.

4. Three pieces result from the cleavage of IgG by pepsin, as shown in Figure 34.2. One piece contains both F_{ab} units. The other two pieces derive from the bisection of the F_c region.

FIGURE 34.1   Possible lattice structures for Ab/Ag ratios of (A) 1.14 and (B) 2.83.

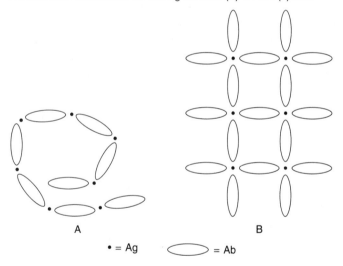

A   B

• = Ag   ⬭ = Ab

FIGURE 34.2   Cleavage of IgG by pepsin.

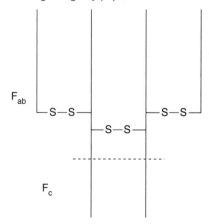

5. Addition of hormone to receptors in some instances causes the hormone-receptor complex to dimerize. The tyrosine kinase domains of two such approximated receptor monomers will then become capable of cross-phosphorylation, with a consequent increase in tyrosine kinase activity. The addition of bivalent antibodies would also have the effect of drawing monomeric receptor units together in such a way that they could cross-phosphorylate and thus trigger intracellular effects.

6. The observation suggests that both Tyr 34 on the light chain and Lys 54 on the heavy chain are involved in binding of the DNP hapten.

7. A given antigenic protein will stimulate the production of a mixed population of antibodies, with each type of antibody being specific for a different region in the tertiary structure of the antigenic protein.

8. Since $K_a = k_{on}/k_{off}$, substituting the given values yields

$$10^9 \text{ M}^{-1} = \frac{10^8 \text{ M}^{-1}\text{s}^{-1}}{k_{off}}$$

$$k_{off} = 10^{-1}\text{s}^{-1}$$

9. One could attach antigen to insoluble beads, and then pass the lymphocyte preparation over a column containing the beads. Lymphocytes capable of producing antibody toward the particular antigen would have a receptor complementary to that antigen on their cell surface and hence would be bound to the beads. Other lymphocytes would pass through the column without binding. Bound lymphocytes could then be released from the column by the addition of free antigen, which would compete with bound antigen for combination with the cell-surface receptors. The fact that such a purification method works is a consequence of the fact that many types of lymphocytes are produced by the immune system, each having a specific receptor on the cell surface capable of reacting with some (previously unencountered) antigen. Combination of antigen with that cell-surface receptor causes the lymphocyte to divide and stimulates antibody production.

10. The results would occur if the haptens were clustered so densely on the antigen that bivalent antibodies combined preferentially with two neighboring haptens on a given antigen molecule. Such behavior is actually found in some systems and is termed *monogamous bivalency.*

11. Nearly all cells have MHC class I proteins. Protein degradation by proteasomes occurs constantly in the cytoplasm. Both self and nonself proteins will be degraded, and the peptides will be translocated across the ER membrane where they will encounter nascent MHC class I proteins. The combination of peptide and MHC will eventually be displayed on the cell's exterior, and nonself peptides will attract cytotoxic T-cells.

    In contrast, MHC class II proteins are only found in B-cells from the immune system. The peptides do not originate in the cytoplasm but from foreign proteins bound by antibodies outside the cell, and brought into the cell in endosomes where acidic hydrolysis occurs. The nonself peptides attract helper T-cells, which stimulate the B-cell to reproduce, thus bolstering the immune response.

12. It is surprising that the pattern of MHC proteins is detectable in the scent of an individual, but not surprising that females would prefer males with maximally different MHCs. This would ensure that the offspring would have widely varying MHCs and thus perhaps improved survival chances in an epidemic of some virulent disease. The data in the

study are not overwhelmingly convincing, and there was some variation in the results according to whether the women were taking birth control pills. The work was done by Claus Wedekind (*Proc. Roy. Soc. Lond.* B 260 [1995]:245). The more recent paper by Suma Jacob (*Nature Gen.* 30[2002]:175) contradicted Wedekind's conclusions, but corresponded to the part of the study on women taking birth control pills. It seems possible that "dating" women would be interested in adventure (maximum difference) while "settled" or pregnant women (birth control pills simulate pregnancy) would prefer the familiar scents of home.

13. Each year, there is a different flu shot because we know that the influenza virus has changed, and the previous year's shot wouldn't work any more. The HIV virus changes so rapidly that after an individual has been infected for several years, there are numerous strains within the body. Obviously it is difficult to find a way to design a vaccine when the target is constantly moving. AIDS is used as an example by some writers on evolution, because it seems to adapt to the life-style of the infected population. It certainly fits the model of reproduction with variation and subsequent selection. This would allow it to change in whatever ways it can to keep spreading the infection from person to person. These facts, plus the fact that it is a retrovirus that actually becomes part of the genes of infected cells, makes it terribly hard to eradicate or even cure.

14. Use of immunosuppressive agents will retard the synthesis of antibodies against acetylcholine receptors (as well, of course, as inhibiting the synthesis of many other useful and protective antibodies). Administration of acetylcholinesterase inhibitors will lead to an increase in acetylcholine concentrations at motor end plates. This increase will lead in turn to an increased number of acetylcholine–acetylcholine receptor complexes by mass action, and will therefore compensate somewhat for the decrease in receptor concentration.

Molecular Motors

This chapter describes the transduction of chemical energy into mechanical energy—for example, the use of ATP hydrolysis to drive the contraction of muscles or to move cells, or the exploitation of transmembrane proton-motive force to rotate bacterial flagella. The authors describe how nanometer motions of proteins can be converted into the coordinated movements of cellular organelles, bacteria, and even animals themselves.

The authors begin with the fact that most molecular-motor proteins are based on the P-loop NTPase structure. Myosin, kinesin, and dynein are described briefly and compared. Then the subunit structures of all three are shown in detail, and the mechanism of myosin's dramatic flexing in response to ATP hydrolysis is explained. The authors then describe the structure of vertebrate skeletal muscle by showing how thick and thin protein filaments give rise to the striated appearance revealed in micrographs. Muscle contraction occurs by the oriented sliding of these filaments past one another. The thick filaments are primarily myosin and the thin filaments are principally an actin polymer with associated troponin and tropomyosin molecules. They next show the structure of the F-actin polymer, which is composed of a linear coiled array of G-actin monomers. The opposite polarities of the thick and thin filaments within a sarcomere indicate how coordinated molecular movement can result in the shortening of myofibrils. Reconstituted, moving assemblies of myosin-coated beads traversing along actin filaments reveal that myosin is the motor driving movement along the track. The repeated association and dissociation of the S1 heads of myosin with actin and the conformational changes in myosin that are effected by the binding of ATP, its hydrolysis to ADP and P_i, and the release of the hydrolysis products suggest how the power stroke occurs.

Microtubules are found in nearly every cell and serve multiple structural and functional roles. They are composed primarily of α-tubulin and β-tubulin monomers that form tubular structures which have large diameters (300 Å). Microtubules also contribute the basic macromolecular assembly of the axoneme, which is the fundamental structural component of the cilia and flagella of eukaryotic cells. Dynein and kinesin interact with microtubules to bend cilia and flagella and to move vesicles along microtubules, respectively. The rapid association of GTP-tubulin with microtubules and the rapid dissociation of GDP-tubulin from the ends of tubulin polymers in conjunction with the GTPase activity of the tubulins explains the dynamic instability of microtubules. Vesicle transport in neurons is explained as their ATP-dependent, kinesin-driven movement along microtubule tracks, which is analogous to the myosin–actin interactions that slide muscle filaments past one another.

The chapter ends with a description of the flagellar motor of bacteria and a discussion of how the interplay of two kinds of protein components forming the motor give rise to directional rotation as a consequence of protons moving from outside the membrane into the cytosol. The system is related to bacterial chemotaxis, which can reverse the rotation of the flagella by phosphorylation of CheY. The examples given in this chapter demonstrate how energy in one form (chemical) can be converted into another form (kinetic) by the regulated activities of proteins.

LEARNING OBJECTIVES

When you have mastered this chapter, you should be able to accomplish the following objectives.

Most Molecular-Motor Proteins Are Members of the P-Loop NTPase Superfamily (Text Section 35.1)

1. Name the two sources of energy that power coordinated molecular movement.

2. Give a general description of *P-loop NTPases*, and (using the text's index) list several examples.

3. Sketch the general structure of the polypeptide backbone of a *myosin* molecule. Describe the subunit composition of myosin, and note its *α-helical coiled-coil structure* and its *dual globular head*.

4. Describe the fragmentation of myosin into *light meromyosin* (LMM) and *heavy meromyosin* (HMM) by proteolysis and the further fragmentation of HMM into its *S1* and *S2 subfragments*. Sketch the polypeptide structures of the fragments, and associate them with the *ATPase activity*, the *actin-binding sites*, and the *thick-filament-forming domains* of the intact molecule.

5. Compare *kinesin* to *dynein* and myosin.

6. Describe the large conformational difference between myosin-ADP and myosin-ATP (same as myosin-ADP-vanadate). Locate *switch I*, *switch II*, and the *relay helix* in relation to the P-loop, and explain how they cause the protein to flex.

7. Describe the differences in protein affinity between myosin, kinesin, and the α subunit of heterotrimeric G-protein.

Myosins Move Along Actin Filaments (Text Section 35.2)

8. Define the terms *sarcomere* and *myofibril*.

9. Identify the *A band, H zone, M line, I band,* and *Z line* of a sarcomere in an electron micrograph of a myofibril.

10. Relate the locations of the *thick filaments* and *thin filaments* to the I band, the A band, and the H zone of a sarcomere.

11. Relate the structures of the thick and thin filaments to their compositions of *actin, myosin, tropomyosin,* and the *troponin* complex. Note the *cross-bridges* between the thick and thin filaments.

12. Describe muscle contraction in terms of the *sliding-filament model,* and relate contraction to changes in the sizes of the A band, I band, and H zone.

13. Distinguish between G-actin and *F-actin.* Define *critical concentration* as it relates to the polymerization of actin. Contrast the roles of the ATPase activities of actin and myosin.

14. Recount the experiment that displayed the unidirectional movement of myosin molecules along an actin cable. Describe the experiment indicating that the force of muscle contraction is generated by the S1 head of myosin.

15. Provide an overview of a model for the mechanism of the *power stroke* during muscle contraction. Explain the roles of actin and the ATPase activity of myosin in the model.

Kinesin and Dynein Move Along Microtubules (Text Section 35.3)

16. Sketch the general structure of a *microtubule,* noting its diameter and showing its alternating subunits of *α-tubulin* and *β-tubulin.*

17. Describe the structure of the *axoneme,* and sketch the *9 + 2 array* of the microtubule doublets and singlets that form its basic ring motif.

18. Note the role of *dynamic instability* in growing microtubules. Describe the effects of *taxol* on the polymerization and depolymerization of microtubules.

19. Describe the *kinesin*-dependent movement of *vesicles* and *organelles* along microtubules. Distinguish between the plus and minus ends of a microtubule.

A Rotary Motor Drives Bacterial Motion (Text Section 35.4)

20. Outline the structure and function of *bacterial flagella.* Distinguish between the mechanism by which eukaryotic and prokaryotic flagella generate motile force.

21. Explain the role of *proton-motive force* in flagellar rotation. Present a model that explains the production of rotary motion from the effects of a proton gradient on the transmembrane flagellar motor.

22. Define *chemotaxis* and outline the sequence of events constituting it.

23. Describe the *flagella* and the *motors* of the chemotactic apparatus of *E. coli.* Relate their actions to the smooth swimming and tumbling motions of a bacterium in response to a *temporal gradient* (change over time) of *attractant* or *repellent* substances.

24. Describe the effect of *phosphorylation* of the *che* gene product *CheY* on flagellar rotation.

SELF-TEST

Most Molecular-Motor Proteins Are Members of the P-Loop NTPase Superfamily

1. Match the proteins in the left column with their descriptions in the right column.
   (a) myosin   (1) enormous protein
   (b) kinesin   (2) helps separate chromosomes
   (c) dynein   (3) important in muscle contraction

2. Which of the following statements about HMM and LMM are true?
   (a) HMM and LMM are formed from myosin by tryptic cleavage.
   (b) HMM can be cleaved by papain to yield two globular proteins that polymerize to form the thin filaments.
   (c) LMM is an α-helical coiled coil composed of two polypeptide chains that can form filaments in vitro.
   (d) HMM contains the two globular heads of the myosin molecule, hydrolyzes ATP, and binds actin in vitro.

3. Energy is required to drive the contraction of striated muscle, the beating of flagella or cilia, and the intracellular transport of vesicles along microtubules. Which of the following statements about energy transduction in these systems are correct?
   (a) The proton-motive force across the plasma membrane surrounding the cell provides the energy for these movements.
   (b) The binding of ATP to proteins, which then undergo a conformational change, provides the energy for these processes.
   (c) The hydrolysis of ATP is coupled to the phosphorylation of tyrosine residues on the proteins of these systems to drive them.
   (d) The hydrolysis of protein-bound ATP and the release of ADP $+$ P_i lead to conformational transitions that complete a movement cycle.
   (e) The binding of GTP to oriented proteins at the interface of the moving assemblies drives these processes.

Myosins Move Along Actin Filaments

4. Which of the following answers complete the sentence correctly? Actin
   (a) is formed from a 42-kd monomer (G-actin).
   (b) monomers can exist with either ATP or ADP bound to themselves.
   (c) exhibits an ATPase activity that helps power muscle contraction.
   (d) in its F form binds myosin in an oriented manner.
   (e) in its F form with myosin has a barbed and a pointed end.

5. Figure 35.1 is a schematic diagram of a longitudinal segment of a skeletal muscle myofibril. Label the structures indicated in the figure by matching them with the listed choices.
   (1) I band
   (2) A band
   (3) thin filaments
   (4) M line
   (5) H zone
   (6) thick filaments
   (7) Z line

FIGURE 35.1

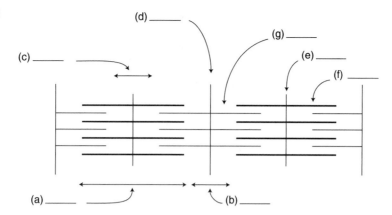

6. Which of the following statements about myosin are true?
   (a) Myosin binds to polymerized actin.
   (b) In vitro, myosin assembles spontaneously into the thin filaments.
   (c) Myosin is an ATPase.
   (d) Myosin has domains that interact with one another to effect its physiological functions.
   (e) Myosin is composed of two polypeptide chains, one of which forms an α-helical coiled coil and the other a globular head.

7. Assign the proteins in the right column to the appropriate myofibrillar component in the left column.
   (a) thin filament
   (b) thick filament
       (1) tropomyosin
       (2) myosin
       (3) actin
       (4) troponin complex

8. Which proteins are homologs of actin?
   (a) hexokinase
   (b) MreB
   (c) Hsp-70
   (d) myosin

9. Which of the following statements concerning events related to the power stroke of muscle contraction are correct?
   (a) The hydrolysis of ATP to ADP and P_i by myosin is fast relative to the release of the ADP and P_i from the protein.
   (b) The binding of actin to myosin stimulates the ATPase activity of myosin by facilitating the release of ADP and P_i.
   (c) Actin and myosin are joined by cross-bridges that are stabilized by the binding of ATP to the myosin head domains.
   (d) Repeated cycles of ATP binding, ATP hydrolysis, and the resulting association and dissociation of cross-bridges and conformational changes in myosin contribute to the contractile process.
   (e) In the region of overlapping thick and thin filaments of a sarcomere, the cross-bridges will either all be formed or all be dissociated, depending on the phase of the power stroke.

10. Considering only the power stroke of skeletal muscle contraction and the events that precede and follow it, place the following states or processes that occur in going from the resting state to the contracted state and back again in their proper order.
    (a) The thick filament moves with respect to the thin filament.
    (b) The S1 heads of myosin interact with actin, P_i is released, and myosin changes its conformation.
    (c) ATP binds to myosin.
    (d) The S1 heads of myosin are dissociated from the thin filament and contain bound ADP and P_i.
    (e) ATP is hydrolyzed to ADP and P_i, and myosin undergoes a conformational change.
    (f) ADP dissociates from the myosin to complete the power stroke.

11. Which of the domains of myosin is primarily responsible for generating the force of skeletal muscle contraction?
    (a) the hinge between the S1 and S2 domains
    (b) the hinge between the S2 and the LMM domains
    (c) the LMM domain
    (d) the S1 globular head
    (e) the S2 domain that connects the S1 domain to the α-helical coiled coil
    (f) the α-helical coiled coil

Kinesin and Dynein Move Along Microtubules

12. Which of the following statements concerning microtubules are correct?
    (a) Microtubules are filaments composed of α-helical coiled-coil polypeptide chains.
    (b) Microtubules contain α-tubulin and β-tubulin protomers that are disposed in a helical array around a hollow core to form a cylindrical structure.
    (c) The cilia and flagella of eukaryotic cells contain nine microtubule doublets that surround a pair of microtubule singlets.
    (d) The outer microtubules in an axoneme are linked together and to an ATPase called *dynein*.
    (e) The powered movements of dynein in an axoneme shorten the structure.

13. A neuron can move a vesicle approximately a meter from the central cell body to the end of an axon in a day. How are microtubules involved in this process, what provides the energy for the movement, and what protein is directly involved in the movement?

A Rotary Motor Drives Bacterial Motion

14. Contrast the protein compositions and molecular movements of the flagella of bacteria and of eukaryotic cells.

15. How would the movements of a starved bacterium (one whose energy stores had been depleted) be affected by having the pH value of the media in which it resides lowered below that of its cytoplasm?

16. Match the major components of the chemotaxis system of *E. coli* in the left column with the appropriate descriptions from the right column.

    (a) chemoreceptor
    (b) processing system
    (c) flagellar motor
    (d) flagellum

    (1) contains several rings, a hook, and a rod
    (2) has binding sites for attractants or repellents
    (3) includes cytosolic peripheral membrane proteins
    (4) contains flagellin and adopts a helical configuration

17. For a bacterium moving toward an increasing concentration of an attractant, which of the following statements are correct?

   (a) Tumbling will be less frequent.
   (b) Tumbling will be more frequent.
   (c) The counterclockwise rotation of flagella will occur more frequently.
   (d) The clockwise rotation of flagella will occur more frequently.

18. The proposed model for the transduction of chemotactic signals via CheY, the tumble regulator, includes which of the following?

   (a) Phosphorylated CheY promotes the counterclockwise rotation of the motor.
   (b) The activation of CheY requires ATP.
   (c) Attractants block the CheY pathway, and smooth swimming results.

ANSWERS TO SELF-TEST

1. (a) 3   (b) 2   (c) 1

2. a, c, d. Answer (b) is incorrect because the two globular heads arising from papain digestion cannot polymerize, as they lack α-helical coiled-coil structures.

3. b, d

4. a, b, d, e. Answer (c) is incorrect because the ATPase activity of actin is involved in the formation and disassembly of F-actin.

5. (a) 2   (b) 1   (c) 5   (d) 7   (e) 4   (f) 6   (g) 3. The M lines lie halfway between the Z lines and correspond to the middle of the bare zone separating regions where oppositely oriented myosin molecules point toward each other. (See Figure 35.17 in the text.)

6. a, c, d. Answer (b) is incorrect because myosin assembles to form the thick filaments. Answer (e) is incorrect because myosin is composed of six polypeptide chains. Two heavy chains intertwine their C-terminal portions to form the α-helical coiled coil, with their N-terminal segments forming two globular heads. Four more polypeptide chains of two kinds associate with the globular heads.

7. (a) 1, 3, 4   (b) 2

8. a, b, and c. *Heat shock protein* (Hsp-70) is not mentioned in the current chapter, but its homology with actin is discussed in Chapter 6 of the text.

9. a, b, d. Answer (c) is incorrect because the binding of ATP to actomyosin dissociates the cross-bridges. Answer (e) is incorrect because, at any instant, numerous cross-bridges will be in all stages of forming and breaking because the process is dynamic and asynchronous.

10. d, b, a, f, c, e, d

11. d. Although the whole molecule is required for muscle contraction, the activities of the S1 head domains provide the biochemical activity for the power generation. The cyclic changes in the conformation of the head domains and in their affinities for actin, ATP, ADP, and P_i are the basis of the energy transduction.

12. b, c, d. Answer (a) is incorrect because microtubules are assembled from relatively globular tubulin subunits and lack the α-helical coiled-coil structure of myosin and the intermediate filaments. Answer (e) is incorrect because dynein movements lead to the bending, not the contraction, of the axoneme.

13. The microtubules, which form a network of fibers that traverse the cell, provide tracts along which vesicles and organelles can move. ATP hydrolysis by the protein kinesin

acts as a molecular engine, in a manner analogous to the way myosin acts, to power the movements.

14. Bacterial flagella are formed principally from flagellin subunits, whereas eukaryotic flagella contain microtubules, among other components. Eukaryotic flagella bend owing to an internally generated force. Bacterial flagella are rotated by a motor attached to one of their ends.

15. The bacterium would commence swimming because protons would move from the higher concentration outside the cell through the intramembrane motors of the flagella, causing their counterclockwise rotation and thereby coordinated movement.

16. (a) 2   (b) 3   (c) 1   (d) 4. The flagellar motor is shown in Figure 35.28 in the text.

17. a, c

18. b, c. Answer (a) is incorrect because phosphorylated CheY induces clockwise rotation of the flagella, leading to tumbling.

PROBLEMS

1. In the text, compare Figure 35.9 to Figure 9.51. We can assume that there is some relationship between adenylate kinase and myosin. Which of these proteins is likely to be older? Why?

2. Apply your knowledge of the α helix and protein folding to answer this question: What features of the tails of myosin molecules contribute to their ability to interact with one another to form thick filaments?

3. Design an experiment involving ATP labeled in the γ phosphoryl group with ^{32}P that would show that actin stimulates the hydrolysis of ATP by myosin.

4. Decide whether each of the following will remain unchanged or will decrease upon muscle contraction. Assume that the sliding-filament model applies. Refer to Figure 35.17 in the text.
   (a) the distance between adjacent Z lines
   (b) the length of the A band
   (c) the length of the I band
   (d) the length of the H zone

5. The symmetry of thick and thin filaments in a sarcomere is such that six thin filaments ordinarily surround each thick filament in a hexagonal array. (See Figure 35.17 in the text.) In electron micrographs of cross-sections of fully contracted muscle, the ratio between thin and thick filaments has been found to be double that of resting muscle.
   (a) Propose an explanation for this observation based on the sliding-filament model.
   (b) How might the explanation you have given in part (a) also account for the long-appreciated fact that a fully contracted muscle is, paradoxically, "weaker" than a resting muscle?

6. Some people have defective dynein, which causes an "immotile-cilia syndrome." This leads to chronic respiratory disorders and also infertility in males. Explain.

7. Colchicine, and the mold products vincristine and vinblastine, interfere with the polymerization of microtubules. Vincristine and vinblastin are widely used in the treatment of rapidly growing cancers. Explain the basis for their effects, remembering that the mitotic spindles of dividing cells are composed of microtubules.

8. The processive nature of kinesin motion is shown in Figure 35.25 in the text. There are two possible interpretations of how the two "feet" of kinesin walk down the "tightrope" of the microtubule. One is called the "inchworm" model, and the other the "hand-over-hand" model. The first one can be visualized as walking with one foot, say the left foot, always in front, and the other as normal walking, left foot and then right foot in front. How do you think scientists were able to show which model was correct?

9. A single kinesin molecule can move a single vesicle from the nucleus of a nerve cell to the end of its axon. Each "step" in the kinesin cycle is 8 nm, and each cycle uses one ATP.

   (a) How many cycles of kinesin binding and ATP hydrolysis will it take for a vesicle to reach the little toe from the upper cervical region of a tall human? Assume that the length of the nerve cell is 6 ft.

   (b) Given that the rate of vesicle transport is 400 mm per day, how long will it take the vesicle to reach the little toe?

   (c) How long does each step take and what is the rate of ATP hydrolysis?

10. Myosin, kinesin, and dynein all contain a globular head which harbors ATPase activity and a tail region which can be thought of as a protein-binding region. The specificity of these protein-binding domains changes the role of each of these three proteins in tissues. What is the specificity of each of these three protein-binding domains, and how does it relate to the roles of myosin, kinesin, and dynein in tissues?

11. Figure 35.2 below depicts the response of bacteria to two chemotactic agents, X and Y. Which agent is likely to be an attractant, and which is likely to be a repellent? Give the reasons for your choices.

FIGURE 35.2   Bacterial response to two chemotactic agents.

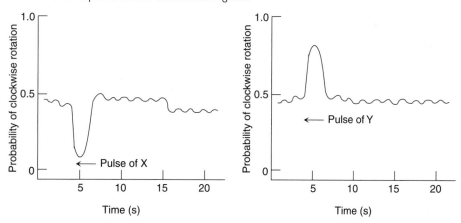

12. Section 35.3 of the text points out that kinesins carry cellular cargo from the center of the cell outward along microtubules (toward the "plus" end of the microtubule, anterograde transport) and dyneins carry cargo in the opposite direction along microtubules (retrograde transport). Because neurons are the longest and thinnest cells, this "freight transport" system is extremely important, and there are some neuropathic diseases which are known to be caused by mutations in important proteins that make up the system. Recent research seems to link Alzheimer disease to transport processes, but Chapter 2 (Section 2.4) describes the accumulation of amyloid plaques of polypeptide Aβ in Alzheimer disease. How can both theories be correct?

ANSWERS TO PROBLEMS

1. Both proteins flex dramatically when ATP binds to the P-loop area. The change in myosin is amplified by nearby structures like the relay helix, and by the length of the lever arm, but the change is generally parallel. Adenylate kinase maintains an equilibrium between ATP and ADP and must be extremely ancient. Myosin is confined to eukaryotes, and hence should be somewhat younger.

2. The absence of proline, which would interfere with α-helix formation, and the abundance of regularly spaced leucine, alanine, and glutamate residues in seven, long-repeating motifs that form long α helices with hydrophobic pockets and knobs on one face and charged residues on the opposite face allow two such helices in two myosin molecules to form a long coiled-coil rod. Because one turn of the α helix occupies 3.6 residues, seven residues represents approximately two turns, with hydrophobic R-groups on one side and the more polar glutamate on the other.

3. Incubate the labeled ATP with myosin and measure the amount of labeled inorganic phosphate liberated as a function of time without actin. Then add actin. The result should be a burst in the amount of labeled P_i released. (See Figure 35.16 in the text for the ATP/ADP cycle.)

4. (a) The distance between adjacent Z lines will decrease.
   (b) The length of the A band will remain unchanged.
   (c) The length of the I band will decrease.
   (d) The length of the H zone will decrease. See Figure 35.3 below.

FIGURE 35.3   Schematic illustration of (A) uncontracted and (B) partially contracted sarcomeres.

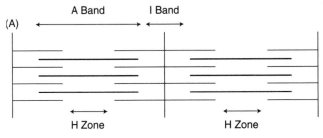

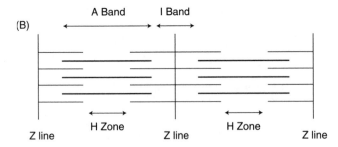

5. (a) The number of thin filaments per thick filament doubles because the thin filaments override one another when the sarcomere is fully contracted. See Figure 35.4.

FIGURE 35.4   Schematic illustration of a fully contracted sarcomere.

(b) Thin filaments and thick filaments must have the same polarity to interact with one another to form cross-bridges, but the polarity of the thick filaments reverses halfway between the Z lines. Thus, in the region where the thin filaments override one another in a fully contracted sarcomere, the ends of the thin filaments have the wrong polarity to interact with an adjacent thick filament. Hence, a slightly smaller number of cross-bridges are formed, so the fully contracted muscle develops less tension, that is, it is "weaker."

6. Defective dynein molecules not only immobilize the cilia of the respiratory tract, preventing inhaled particles from being swept out of the lungs, but also render sperm immotile.

7. Cancer cells have a greater than normal rate of cell division. In the presence of vincristine or vinblastine, mitotic spindle fibers fail to form, and cell division is retarded. Since cancer cells grow more rapidly, they are more sensitive to these drugs.

8. Normal human walking makes use of hip and knee joints to allow the body to remain aimed straight ahead. Scientists realized that the "hand-over-hand" model would require kinesin to rotate about its own axis, turning 180° with each step, whereas the "inchworm" model would allow kinesin to move with very little rotation. So scientists at Brandeis University attached a length of microtubule to the top of kinesin and took photographs that showed that as the microtubule was carried along, it stayed aimed in more or less the same direction. Thus it appears that the "inchworm" model is the correct one. (*Science* 295[2002]:844). Look at Figure 35.25 in the text—which version is shown there? Early editions showed the "hand-over-hand" version, which is incompatible with the results of the Brandeis experiment.

9. (a) $$\frac{6 \text{ ft} \times 12 \text{ in/ft} \times 2.54 \text{ cm/in} \times 0.01 \text{ m/cm}}{8 \times 10^{-9} \text{ m/step}} = 2.3 \times 10^8 \text{ steps}$$

   and 2.3×10^8 molecules of ATP hydrolyzed.

   (b) $$\frac{6 \text{ ft} \times 12 \text{ in/ft} \times 2.54 \text{ cm/in} \times 0.01 \text{ m/cm}}{0.4 \text{ m/day}} = 4.6 \text{ days}$$

   (c) $$\frac{4.6 \text{ days} \times 24 \text{ h/day} \times 60 \text{ min/h} \times 60 \text{ s/min}}{2.3 \times 10^8 \text{ steps}} = 1.7 \times 10^{-3} \text{ s/steps}$$

   and the inverse of 1.7×10^{-3} seconds/step gives 588 ATPs hydrolyzed per second.

10. The tail of myosin binds other myosin molecules to form the thick filaments of striated muscle. Without this self-aggregating property of myosin, striated muscle would not form. The tail regions of kinesin or dynein bind the surface of a vesicle that is

being transported. Changing their tail regions would change the vesicles or molecules that are being transported. Finally, cellular and ciliary dynein must have different tail regions so that the former will transport vesicles and the latter will polymerize with tubulin in cilia.

11. X is likely to be an attractant and Y is likely to be a repellent. The addition of X causes a decrease in clockwise rotation (an increase in counterclockwise rotation), which leads to smooth swimming. The addition of Y causes an increase in clockwise rotation (a decrease in counterclockwise rotation), which leads to tumbling.

12. It helps to think of the neuronal transport system as a railroad, with tubulin as the track, and kinesin as the outward bound train. The cargo is "containerized" in vesicles, and the vesicles are attached to kinesin by a protein known as AβPP (amyloid beta protein precursor). Normally the cargo is taken to the far (distal) part of the axon and unloaded. But if something disrupts the system, then there is a pile-up in the near (proximal) part of the axon and the neuron swells up with undelivered material. Accumulation of AβPP in the cell stimulates production of the enzyme that cleaves it—and one of the cleavage products of AβPP is Aβ. So it appears that difficulties with kinesin transport can lead to the amyloid plaques observed in Alzheimer disease (AD). Recent studies have shown a correlation between AD and polymorphisms in KLC-1 (kinesin light chain 1). The transport hypothesis is also supported by a recent paper by Stokin et al., in which axonal swelling and transport problems were observed more than a year before the appearance of plaques in mice. (G. B. Stokin et al., Axonopathy and transport deficits early in the pathogenesis of Alzheimer's disease, *Science* **307**[2005] 1282–1288, nicely explained and discussed in the short paper by X. Zhu et al., Alzheimer's disease: an intracellular movement disorder? *Trends Mol. Med.* **11**[2005] 391–393.)

Drug Development

Biochemistry is the study of chemical processes in living cells. Drugs are chemicals that interact with these processes, enhancing or inhibiting them. So drug design is a field that links biochemistry with medicine and is a very appropriate topic for a biochemistry text. The authors describe the history of drug development. Throughout the nineteenth and most of the twentieth century, scientists simply didn't know enough to design drugs to match cellular targets. Drug discoveries tended to be serendipitous: scientists would randomly encounter drugs that worked. As the nature of drug action was understood better, screening methods came into play. Only in the last decade or so has it been routinely possible to select molecular targets and design drugs to interact with them.

The authors discuss drug–ligand binding, the EC_{50}, and the *dissociation constant K_d*. Drugs that are active in vitro often fail in vivo because of *ADME*, or absorption, distribution, metabolism, and excretion factors. One way to predict the ADME characteristics of a compound is to apply *Lipinski's rules*. Compounds that obey these rules should have good *oral bioavailability*. Drugs and other *xenobiotic compounds* must be excreted from the body. Phase I processes (including cytochrome P450 oxidases) oxidize the compounds, and the Phase II processes produce conjugates with *glutathione, glucuronic acid*, or *sulfate*. These processes make it more likely that metabolites will be picked up by the kidneys and excreted. Various types of toxicity, due either to drugs or their metabolites, are discussed.

Serendipity in drug design is illustrated using penicillin and chlorpromazine. Viagra was also the result of observation by chance of an unintended side effect of a drug that had been rationally designed. *Drug screening* is illustrated by the cholesterol-lowering statin drugs. Large numbers of similar compounds with varied side chains can be produced by *combinatorial chemistry* methods including *split-pool synthesis*. Finally, *structure-based drug design* is described using HIV protease inhibitors and COX2 inhibitors as examples. Anti-HIV drugs have greatly reduced the death toll from AIDS.

Knowledge of the human genome and the developing field of proteomics have led to great new possibilities in the field of drug design. We now know that there are more than 500 *protein kinases* and 450 non-olfactory *7TM receptors* encoded in the human genome. Many would provide suitable targets for new drugs. Other genomes also provide fertile ground for new drug ideas. The mouse genome is quite similar to the human genome, and *knockout mice* lacking any given gene can be produced and studied. Such studies have already led to discoveries of important new drugs. Furthermore, the genomes of pathogens can be studied for targets to inhibit. It is also true that understanding genetic differences between individuals can lead to use of drugs that are more efficacious or less toxic for each patient. Then *clinical trials* are described. There are three phases to a clinical trial. The first involves a few healthy individuals, the second a few appropriate patients, and the third thousands of patients. It is important to make use of *double-blind studies* and *placebos* if possible. Satisfying all the current FDA requirements typically costs hundreds of millions of dollars per drug. Finally, drug resistance is addressed. Antibiotics and HIV protease inhibitors provide a powerful selective force that produces evolutionary change in the targeted microorganisms. We can see changes in the structure of HIV-protease enzymes, sharing of enzymes like β-lactamase (which inactivates penicillin) by bacteria, and use of the multiple-drug resistance transporter in human cancer cells. Fortunately, new concepts in drug design may keep us a step ahead of the evolution of drug resistance.

LEARNING OBJECTIVES

When you have mastered this chapter, you should be able to accomplish the following objectives.

The Development of Drugs Presents Huge Challenges (Text Section 36.1)

1. Compare and contrast the two main paths to drug discovery.

2. Describe how drug–target binding relates to the Michaelis Menten equation.

3. Define K_d, K_d^{app}, EC_{50}, and EC_{90}.

4. Explain what a drug's *ADME* properties are and why they are important considerations in drug design.

5. State *Lipinski's rules*, and define the terms *partition coefficient* and *bioavailability*.

6. Describe the basic facts of drug distribution in the body, including the role of *serum albumin* and the *blood–brain barrier*.

7. Distinguish between the two main pathways of *drug metabolism: oxidation* and *conjugation*. Define *phase I* and *phase II transformations*.

8. Define the LD_{50} and describe four distinct ways that a compound can have toxic effects.

Drug Candidates Can Be Discovered by Serendipity, Screening, or Design (Text Section 36.2)

9. Cite specific examples of drugs discovered by *serendipity*, by *screening*, and by *design*.

10. Define *high-throughput screening*, *combinatorial chemistry*, and *split-pool synthesis*.

11. Give two examples of *structure-based drug design*, and describe *Fischer's lock and key model* for drug–protein interactions.

The Analyses of Genomes Hold Great Promise for Drug Discovery
(Text Section 36.3)

12. Explain how knowledge of the human genome can provide drug developers with many new targets, especially among *protein kinases* and *7TM* proteins.

13. Describe how the ability to produce *knock-out mice* can lead to specific goals in drug design.

14. Give examples of how drugs can be designed to combat specific pathogens.

15. Explain how individual genetic differences in the *β1-adrenergic receptor* can affect a patient's response to *metoprolol*.

16. Give an example of how ADME can vary in individuals, resulting in different responses to a given drug.

The Development of Drugs Proceeds Through Several Stages (Text Section 36.4)

17. Outline the three phases of a clinical trial. Discuss the use of *double-blinded studies* and *placebos*.

18. Describe various ways in which drug targets can evolve drug resistance.

SELF-TEST

The Development of Drugs Presents Huge Challenges

1. In Figure 36.4 in the text, graphs show the normal substrate of an enzyme acting as an inhibitor to drug binding. What sort of inhibition would this correspond to?

2. Lipinski's rules say that to be well absorbed and distributed, a drug should have a partition coefficient less than 5. What does a partition coefficient of 5 mean?

3. Dopamine is an important neurotransmitter in the brain. Parkinson's disease results from a deficiency of dopamine. Yet giving oral doses of dopamine to patients with Parkinson's has no effect. Why?

4. Drugs can be solubilized and marked for excretion by conjugation with which compounds?
   (a) phosphate
   (b) sulfate
   (c) glucuronic acid
   (d) ribose
   (e) glutathione

5. A given drug has an $EC_{50} = 20$ µM and an $LD_{50} = 4$ mM. What is the therapeutic ratio for this drug?
   (a) 5
   (b) 2
   (c) 20
   (d) 200
   (e) 2000

6. You have a very specific enzyme inhibitor. You know it binds tightly to the target enzyme. If you want to use it as a drug, why do you still have to do whole-animal studies?

Drug Candidates Can Be Discovered by Serendipity, Screening, or Design

7. The first anti-inflammatory drug isolated from willow bark was salicylic acid. How does aspirin differ? Why is it a much better drug?

8. Screening possible compounds as drugs can lead to the discovery of highly effective and useful drugs. But you need to screen hundreds or thousands of compounds to succeed. Where would all these compounds come from?

The Analyses of Genomes Hold Great Promise for Drug Discovery

9. The human genome encodes about 23,000 proteins. The text singles out protein kinase enzymes and 7TM receptors as promising targets for drug design. Why aren't all 23,000 proteins similarly promising targets?

10. Why are knockout mice such useful indicators for suggesting new drugs? Why not use a simpler organism like *Caenorhabditis elegans*, whose genes can be easily silenced one at a time using RNAi (antisense RNA)?

11. The text describes individual variation in response to thiopurine drugs. These differences are due to genetic variations in drug

    (a) absorption.
    (b) distribution.
    (c) metabolism.
    (d) excretion.

The Development of Drugs Proceeds Through Several Stages

12. Which phase of clinical drug trials involves healthy volunteers, not patients?

    (a) phase 1
    (b) phase 2
    (c) phase 3

13. The full cost of developing a drug is said to be about

    (a) $50,000
    (b) $100,000
    (c) $500,000
    (d) $1,000,000
    (e) $500,000,000

14. Cancer patients often take a variety of drugs. Sometimes the cancer cells become simultaneously resistant to many or all of them. How?

ANSWERS TO SELF-TEST

1. Competitive inhibition. One way to tell is that the K_d increases with increasing "inhibitor." Also, although you can't really extrapolate a curve, it appears that all the curves are tending toward the same V_{max}. Finally, the caption to the graph says, "Inhibitors compete with substrates for binding sites." This is the definition of competitive inhibition.

2. A partition coefficient of 5 would mean that when the drug is shaken with a mixture of octanol and water, the octanol would contain 100,000 times as much of the drug as the aqueous phase. Clearly this would be a very hydrophobic compound.

3. Although there could be various plausible answers to this question, it has long been known that dopamine cannot cross the blood–brain barrier. Sometimes small modifications can solve this sort of problem, and in fact L-DOPA does cross the blood–brain barrier and then is decarboxylated to yield dopamine. Unfortunately this is a treatment of last resort for Parkinson because administering L-DOPA "burns out" the dopaminergic neurons.

Dopamine   L-DOPA

4. b, c, e. Phosphate transfers are relatively common in biochemistry, but phosphates tend to be used in the cell rather than excreted. Ribose is not acidic and would not cause compounds to be picked up by the kidney and excreted.

5. d. 4000 μM / 20 μM = 200

6. The drug may have problems, because of ADME properties, which don't show up in vitro. But it also may have toxic side effects despite being an inhibitor of a specific enzyme. For example, in the text, the authors mention that many compounds designed for other uses turn out to bind to ion channels and thus have serious cardiac side effects. Only whole-animal studies can reveal such effects.

7. Aspirin is acetylated. That makes it less acidic, so it causes less damage to the stomach lining (although damage still may occur). As described in the text, instead of just acting as an inhibitor of prostaglandin H2 synthase, it covalently modifies the active site by acetylating it. This makes it a much more potent drug than salicylic acid.

8. Historically, researchers have screened a large array of natural products. Selman Waksman had the idea that soil organisms should produce bactericidal molecules, so he screened thousands of soil cultures before discovering streptomycin. Recently drug manufacturers have made use of methods of combinatorial chemistry, including split-pool synthesis. This allows a rather small number of reactions to produce a large number of related molecules, perfect for "high throughput screening."

9. The two types of proteins singled out in the text, 7TM receptors and protein kinase enzymes, are both involved in signal transduction. This means that they have the power to change the whole profile of metabolism in a cell or organ. Many protein kinases are controlled by G-protein–linked 7TM receptors. A protein kinase will turn several enzymes off or on. Targeting the individual enzymes wouldn't produce as great a change. Furthermore, there are many structural proteins like α-keratin (in hair) and storage proteins like myoglobin that would be unlikely targets for drugs.

10. As the text points out, because mice are vertebrates and mammals, they are quite close to humans, so the ability to "knock out" individual genes to learn what the gene does in the body is very valuable. Considering that *C. elegans* is smaller than an eyelash and obviously an invertebrate, there is a surprising amount of useful information in the genome. Many genes from *C. elegans* have relatives in the human genome, but mice are much closer even though a little more difficult to work with.

11. c

12. a

13. e

14. As described in the text, the cells can start to express multidrug resistance proteins (MDR), also known as ABC transporters (see Chapter 13 for details). These transporters randomly pump small molecules out of the cell, so the resistance is not specific but broad spectrum.

PROBLEMS

1. The hyperbolic graph shown in the chapter (Figure 36.3 in the text) for drug-target binding resembles the curve for Michaelis-Menten enzymes. What other biochemical processes yield similar curves?

2. A drug is known to work well on its target in vitro and also in vivo when injected into mice. Yet when it is administered to humans in pill form, there is no effect whatever. What are some possible causes to investigate?

3. The text describes drugs with long or short half-lives in the body. Many drugs are eliminated with "first-order" kinetics, meaning that they have a given half-life in the blood. Other drugs, including ethanol, have a fixed rate of disappearance. In such cases the curve in Figure 36.12 would be a straight line. What would cause a drug to follow "zero-order"kinetics and disappear at a rate rather than a half-life?

4. Acetaminophen toxicity results from glutathione depletion, which does not occur under standard conditions. Normally, about 90% of acetaminophen is conjugated (on the phenolic hydroxyl) with either glucuronic acid or sulfate. Only about 4% reacts first with cytochrome P450 to produce the benzoquinone imine, which is then conjugated with glutathione. With higher doses, much larger percentages of acetaminophen are forced to react with cytochrome P450, potentially depleting the liver's glutathione stores. Heavy alcohol consumers,who may have impaired liver function, are especially endangered. For them, the toxic of acetaminophen can be as low as 4 or 5 grams per day. The label information on a package of 500 mg of a common brand of acetaminophen says, "Take 2 caplets every 4 to 6 hours, not more than 8 caplets in any 24-hour period." What would you estimate that the therapeutic ratio is for acetaminophen in drinkers?

5. Designing drugs to match specific targets is very much a product of the age of genomics and proteomics. Most of the "classic" drugs were discovered by either serendipity or screening. The text tells the story of penicillin. Using the Internet and other sources, what can you learn about the discovery of sulfanilamide and streptomycin?

6. The development of COX2 inhibitors was a triumph for modern methods of structure-based drug design. Why hasn't this been possible until recently? Why will it become more and more important?

7. Because bacteria and viruses have small genomes, the genomes of many pathogens have already been sequenced. Why is this extremely promising for drug design?

8. The text discusses the rather large individual variations in $\beta 1$-adrenergic receptors and the resulting variability in response to antihypertensive drugs such as metoprolol. In fact, it appears that African Americans are less responsive to cardiac and blood pressure drugs. Using the Internet or other resources, can you find out how the FDA responded to this situation?

9. In the early investigation of the HIV virus, researchers in Paris, France, and in Maryland sequenced viral RNA genomes. The two groups of researchers found the identical sequence. Many people considered this fact suspicious. Why?

10. Bacterial resistance to common antibiotics is becoming more and more common. Often the resistance involves the appearance of enzymes like β-lactamase (for penicillin resistance) or a folate transporter (for sulfanilamide resistance). How do bacteria develop resistance in such complicated ways? Do they evolve the capability to make these new enzymes and proteins?

ANSWERS TO PROBLEMS

1. Any time a single ligand is binding to a single site on a single protein, we expect to see a hyperbolic curve. Thus when myoglobin binds to oxygen, the curve is hyperbolic, and when a channel allows a specific ligand to cross a membrane, the curve should look the same. For example the GLUT glucose transporters mentioned in Chapter 16 are saturable and specific (like most agents of facilitated diffusion) and would have a hyperbolic curve when the rate of diffusion across the membrane is plotted against the concentration of glucose.

2. The most obvious problem would be inactivation by stomach acid. But there are various other possible problems. There have been instances where tablets were compressed to such solidity that they could not dissolve and passed through the digestive tract intact. The problem does not state whether mice could absorb the drug when administered orally; perhaps it is not taken up through the intestinal epithelium. Researchers should try administering the drug as a powder and should also do in vitro experiments to see what effects acid would have.

3. Ethanol is metabolized by the enzyme alcohol dehydrogenase, which is known to have the very low turnover number (or k_{cat}) of about 2 per second. This means that the enzyme is saturated and operating at V_{max} at essentially all detectable levels of ethanol concentration. When the rate doesn't vary with substrate concentration, you can't have a half-life, and the disappearance will be "zero order."

4. 500 mg times 8 caplets (the maximum daily dose) equals 4 grams. In individuals with impaired liver function, this amount might be toxic, that is LD_{50} would be about equal to EC_{50}, making the therapeutic index approximately 1. Acetaminophen is described as having a "narrow" therapeutic ratio, "narrow" usually meaning less than 2.0. Of course, the actual value varies with the individual patient.

5. Sulfa drugs: In 1934, Gerhard Domagk discovered that a dye used for cloth cured streptococcal infections in mice. When his daughter became ill with a streptococcal infection, he injected her with the dye and she immediately got better. The active substance turned out to be sulfanilamide. This is a classic story of serendipity.

   Streptomycin: Selman Waksman screened more than 10,000 soil cultures for antibiotic activity. He finally isolated streptomycin, which became the drug of choice for treating tuberculosis. This is an excellent example of the use of screening.

6. Before genome sequencing became common, relatively few protein sequences were known, and only a small percentage of those had been solved for their three-dimensional structures. Genome sequencing means that *all* protein sequences in an organism are known, and modern methods of analysis are producing 3D structures at an increasing rate. Once you can model the binding site for the normal ligand or the active

site of an enzyme, you should be able to synthesize structures to block that site. For example the COX2 site had an "elbow" bend that is matched by "V-shaped" molecules like celecoxib and rofecoxib (shown in the text).

7. Antibiotics exploit differences in bacterial and human metabolism. One clear example is sulfanilamide, which blocks the synthesis of tetrahydrofolate (THF). Humans don't synthesize THF but require it as a vitamin. Many well-known antibiotics inhibit bacterial ribosomes (streptomycin, tetracycline, erythromycin, and so on), which are different from eukaryotic ribosomes. Knowing more about the structures of bacterial and viral proteins will allow scientists to pinpoint the differences and develop new drugs. Another possibility is that individual proteins from dangerous pathogens can be synthesized and then used to raise antibodies for a vaccine. The story of the SARS virus in the text is an excellent example of how quickly scientists can respond to novel pathogens using modern techniques of molecular biology.

8. The FDA approved a new heart failure drug specifically for patients who self-identify as "black." The drug is called BiDil, and it consists of a combination of hydralazine, an antihypertensive, and isosorbide dinitrate, an antianginal drug. This development probably foreshadows a new era of "individualized" medicine.

9. In contrast to genomes composed of DNA, retroviral RNA genomes change very rapidly. The text points out that within a single infected person each position of the genome mutates 1000 times a day. Thus it would seem relatively unlikely that two viral samples from a given population would have identical sequences, much less viral samples from two continents. The great variability of RNA genomes naturally leads to altered proteins and active sites in the face of selective forces such as protease inhibitors. As for the early researchers (Robert Gallo and Luc Montagnier), today many people think that the viral samples were somehow cross contaminated and that both sequenced viruses came from the same patient.

10. Bacteria "swap" genetic information routinely. Parts of the genome can be transmitted during conjugation, and small plasmids can even be shared between bacteria of different species. In earlier chapters the authors discussed specific plasmids like pBR 322. Plasmids used in molecular biology generally contain resistance genes; and so do other plasmids found in bacteria as extrachromosomal DNA. So by somehow ingesting a plasmid, a bacterium can suddenly become resistant to several different drugs. American medical practice prescribes high doses of antibiotics taken for several days. This makes survival of partially resistant bacteria unlikely. In other countries, the more relaxed practice allows these bacteria to survive and reproduce, which leads inevitably to resistant bacteria.

Expanded Solutions To Text Problems

CHAPTER 1

1. The hydrogen bond donors are the NH and NH_2 groups.
   The hydrogen bond acceptors are the carbonyl oxygens and those ring nitrogens that are not bonded to hydrogen or to (deoxy)ribose.

2. Interchange the positions of the single and double bonds in the six-membered ring.

3. (a) electrostatic interactions.
   (b) van der Waals interactions.

4. $\Delta G = \Delta H - T\Delta S$. $T = 298$ K. A process can occur without violating the Second Law if ΔG is less than zero.

part	ΔH kJ mol^{-1}	ΔS kJ mol^{-1} K^{-1}	$T\Delta S$ kJ mol^{-1}	ΔG kJ mol^{-1}	ΔG kcal mol^{-1}	answer
a	−84.0	0.125	37.25	−121.25	−28.98	yes
b	−84.0	−0.125	−37.25	−46.75	−11.17	yes
c	84.0	0.125	37.25	46.75	11.17	no
d	84.0	−0.125	−37.25	121.25	28.98	no

5. $\Delta G = \Delta H - T\Delta S$. $T = 298$ K.

   $$T\Delta S = \Delta H - \Delta G = [-251 - (-54)] \text{ kJ mol}^{-1} = -197 \text{ kJ mol}^{-1}.$$

   $$\Delta S_{\text{system}} = (-197{,}000 \text{ J mol}^{-1})/(298 \text{ K}) = -661 \text{ J mol}^{-1} \text{ K}^{-1} \ (-158 \text{ cal mol}^{-1} \text{ K}^{-1}).$$

   $$\Delta S_{\text{surroundings}} = -\Delta H/T = (251{,}000 \text{ J mol}^{-1})/(298 \text{ K}) = +842 \text{ J mol}^{-1} \text{ K}^{-1}.$$
   $$(+201 \text{ cal mol}^{-1} \text{ K}^{-1})$$

   (Note also that $\Delta S_{\text{universe}} = \Delta S_{\text{surroundings}} + \Delta S_{\text{system}}$
   $$= (842 - 661) \text{ J mol}^{-1} \text{ K}^{-1} = 181 \text{ J mol}^{-1} \text{ K}^{-1} \ (43 \text{ cal mol}^{-1} \text{ K}^{-1}).)$$

6. $pH = -\log_{10}[H^+]$.

   (a) $[H^+] = 0.1$ M. $pH = 1.0$.   (*Note:* $[OH^-] = 10^{-13}$ M.)
   (b) $[H^+] = 10^{-13}$ M. $pH = 13.0$.   (*Note:* $[OH^-] = 0.1$ M.)
   (c) $[H^+] = 0.05$ M. $pH = 1.3$.   (*Note:* $[OH^-] = 2 * 10^{-13}$ M.)
   (d) $[H^+] = 2 * 10^{-13}$ M. $pH = 12.7$.   (*Note:* $[OH^-] = 0.05$ M.)

7. $CH_3COOH \rightarrow H^+ + CH_3COO^-$. $K_a = 10^{-4.75}$.

   Let $x = [H^+]$.

   Then $[CH_3COO^-]$ is also equal to x, and $[CH_3COOH] = (0.1$ M $- x)$.

   $K_a = (x^2)/(0.1$ M $- x) = 10^{-4.75}$.

   $x^2 = (10^{-4.75})(0.1$ M $- x)$.

   $x^2 + (10^{-4.75})x - (10^{-5.75}) = 0$.

   Using the quadratic formula stated in the problem leads to $x = 1.33 * 10^{-3}$ M. **pH = <u>2.88</u>**.

   [*Note:* In this problem, it also is possible to avoid the quadratic formula by assuming that the extent of dissociation $x << 0.1$ M. Then $K_a \simeq (x^2)/(0.1$ M$) \simeq 10^{-4.75}$; and

   $x^2 \simeq 10^{-5.75}$. This assumption leads to the same answer:

   $x = 1.33 \times 10^{-3}$ M; and pH = **2.88**. When making such an assumption, it is important to check the assumption after completing the calculation.

   In this case, indeed, 1.33×10^{-3} M $<< 0.1$ M.]

8. The quadratic formula must be used in this case (because the assumption noted in problem 7 will not hold true). Following the approach in problem 7 leads to:

   $x^2 + (10^{-2.86})x - (10^{-3.86}) = 0$.

   Using the quadratic formula stated in problem 7 leads to $x = 1.11 * 10^{-2}$ M. **pH = <u>1.96</u>**.

9. Consider the equilibrium for the conjugate base:

   $$CH_3CH_2NH_2 + H_2O \rightarrow OH^- + CH_3CH_2NH_3^+.$$

   Given that K_a is $10^{-10.7}$, then K_b is $10^{-14}/K_a = 10^{-3.3}$.

   Let us solve for $[OH^-]$. The $[H^+]$ will be $10^{-14}/[OH^-]$.

   Let $x = [OH^-]$. Then $[CH_3CH_2NH_3^+]$ is also equal to x, and $[CH_3CH_2NH_2] = (0.1$ M $- x)$.

   (The concentration of water will not change and does not enter the calculations.) Then by analogy with the procedure in problems 7–8,

   $K_b = (x^2)/(0.1$ M $- x) = 10^{-3.3}$.

   $x^2 = (10^{-3.3})(0.1$ M $- x)$.

   $x^2 + (10^{-3.3})x - (10^{-4.3}) = 0$.

   Using the quadratic formula stated in the problem leads to $x = 7.33 * 10^{-3}$ M. so the $[H^+]$ is $10^{-14}/x = 1.46 * 10^{-12}$, and the **pH = <u>11.83</u>**.

10. $pH - pK_a = \log_{10}([A^-]/[HA])$.

    $(7.4 - 4.75) = \log_{10}([acetate^-]/[acetic\ acid]) = 2.65$

    so $[acetate^-]/[acetic\ acid]$ is $10^{2.65}$ which is = **<u>447</u>**.

    $(7.4 - 10.7) = \log_{10}([CH_3CH_2NH_2]/[CH_3CH_2NH_3^+]) = -3.3$

    so $[CH_3CH_2NH_2]/[CH_3CH_2NH_3^+]$ is $10^{-3.3}$ which is = **$5.0 * 10^{-4}$**.

11. $CH_3COOH \rightarrow H^+ + CH_3COO^-$. $K_a = 10^{-4.75}$.

When the pH is 4.0, $[H^+] = [CH_3COO^-] = 10^{-4.0}$ M.

$([CH_3COO^-]/[CH_3COOH]) = 10^{pH - pKa} = 10^{0.75}$.

$[CH_3COOH] = [CH_3COO^-]/10^{0.75} = 10^{-3.25} = 5.62 * 10^{-4}$ M.

The total added acetic acid is $[CH_3COO^-] + [CH_3COOH] = \underline{\mathbf{6.6 * 10^{-4} \text{ M}}}$.

12. The initial $[H^+]$ is 10^{-5} M in a volume of 0.1 L.

The number of moles of H^+ is 10^{-5} (moles/L) $* 0.1$ L $= 10^{-6}$ moles.

Final volume is 1.0 L. Final $[H^+]$ is 10^{-6} M. The pH is $-\log_{10}(10^{-6}) = \underline{\mathbf{6.0}}$.

13. The initial concentrations at pH 5.0 are: $[H^+] = 10^{-5}$ M, $[CH_3COO^-] = 10^{-5}$ M, and

$[CH_3COOH] = [CH_3COO^-]/10^{pH - pKa} = (10^{-5.0})/(10^{0.25}) = (10^{-5.25}) = 5.62 * 10^{-6}$ M.

Immediately after dilution, the concentrations (temporarily) become:

$[H^+] = 10^{-6}$ M, $[CH_3COO^-] = 10^{-6}$ M, and $[CH_3COOH] = 5.62 * 10^{-7}$ M.

The weak acid then adjusts to fit the relation $[CH_3COO^-]/[CH_3COOH] = K_a/[H^+]$.

Let x be the extent of further dissociation of CH_3COOH. Then: $[H^+] = (10^{-6} \text{ M} + x)$,

$[CH_3COO^-] = (10^{-6} \text{ M} + x)$, and $[CH_3COOH] = (5.62 * 10^{-7} \text{ M} - x)$.

Note that the value of K_a is $10^{-4.75} = 1.79 * 10^{-5}$. Entering values into the equation $[CH_3COO^-] [H^+] = K_a [CH_3COOH]$ gives:

$(10^{-6} + x) (10^{-6} + x) = (1.79 * 10^{-5}) (5.62 * 10^{-7} - x)$.

Rearranging: $x^2 + x(2.0 * 10^{-6} + 1.80 * 10^{-5}) + 10^{-12} - 1.01 * 10^{-11} = 0$.

(Collect terms): $x^2 + x(2.00 * 10^{-5}) - 1.11 * 10 11 = 0$.

Using the quadratic formula stated in problem 7 leads to: $x = 5.40 * 10^{-7}$ M.

The final $[H^+]$ is $(10^{-6} \text{ M} + 5.4 * 10^{-7} \text{ M}) = 1.54 * 10^{-6}$ M. $\underline{\mathbf{pH = 5.81}}$.

14. $pH = pK_a + \log_{10} ([A^-]/[HA])$.

$pK_a = pH - \log_{10} ([A^-]/[HA]) = 6.0 - \log_{10} (0.025/0.075) = 6.0 + 0.48$.

$pK_a = \underline{\mathbf{6.48}}$.

15. $pH = pK_a + \log_{10} ([A^-]/[HA])$.

$[A^-] = 0.001 \text{ M} - 0.0002 \text{ M} = 0.0008$ M.

$pH = 7.2 + \log_{10} (0.0008/0.0002) = 7.2 + 0.6 = \underline{\mathbf{7.8}}$.

16. $pH = pK_a + \log_{10} ([A^-]/[HA])$.

$([A^-]/[HA]) = 10^{(pH - pKa)}$.

$([HA]/[A^-]) = 10^{(pKa - pH)} = 10^{(8 - 6)} = 10^2 = \underline{\mathbf{100}}$.

17. $pH = pK_a + \log_{10} ([HPO_4^{-2}]/[H_2PO_4^-])$.

Following the logic of problem 10, $([H_2PO_4^-]/[HPO_4^{-2}]) = 10^{(pKa - pH)}$.
(The relevant pK_a is 7.21. See page 16 of the text.)

	pK_a	pH	$pK_a - pH$	$([H_2PO_4^-]/[HPO_4^{-2}])$
a.	7.21	7.00	0.21	1.6
b.	7.21	7.50	−0.29	0.5
c.	7.21	8.00	−0.79	0.16

18. As long as the added [HCl] is less than the initial [acetate⁻], one assumes that all of the added H^+ combines with acetate⁻ to form a stoichiometric amount of HA.

Initial acetate (M)	initial HCl (M)	[A-]	[HA]	pK_a	pH
0.1	0.0025	0.0975	0.0025	4.75	6.3
0.1	0.005	0.095	0.005	4.75	6.0
0.1	0.01	0.09	0.01	4.75	5.7
0.1	0.05	0.05	0.05	4.75	4.8
0.01	0.0025	0.0075	0.0025	4.75	5.2
0.01	0.005	0.005	0.005	4.75	4.8
0.01	0.01	0.00041*	0.010	4.75	3.4
0.01	0.05	0.00#	0.01		1.4

*When 0.01 M HCl is mixed with 0.01 M sodium acetate, consider:

$$HA \rightarrow H^+ + A^-. \quad K_a = 10^{-4.75}.$$

Let = $[HA] = (0.01\ M - y)$. Then $[A^-]$ and $[H^+]$ will equal y.

$$K_a = (y^2)/(0.01\ M - y) = 10^{-4.75}.$$

$$y^2 = (10^{-4.75})(0.01\ M - y) \cong (10^{-4.75})(0.01) \cong 10^{-6.75} \text{ when } y \text{ is} << 0.01\ M.$$

$$y \cong 10^{-6.75/2}.$$

$$y = [H+] \cong 4.2 * 10^{-4}\ M. \quad \underline{\textbf{pH} = \textbf{3.4}}.$$

#When 0.05 M HCl is mixed with 0.01 M sodium acetate, assume that 0.01 M H^+ combines stoichiometrically with acetate, leaving 0.04 M excess $[H^+]$. Find the pH when $[H^+]$ = 0.04 M. pH = $-\log_{10} (0.04)$ = **1.4**.

19. Use the relation: pH − pK_a = $\log_{10} ([CH_3COO^-]/[CH_3COOH])$.

Let $x = [CH_3COO^-]$. Then $[CH_3COOH] = (0.25 - x)$. The units of x are molar.

pH − pK_a = 0.25 = $\log_{10} (x/(0.25 - x))$.

$x/(0.25 - x) = 10^{0.25} = 1.78$. $x = 1.78(0.25 - x)$. $2.78x = 0.445$.

$x = 0.445/2.78 = 0.16$. $[CH_3COO^-] = \underline{0.16\ M}$.

$[CH_3COOH] = (0.25 - 0.16)M = \underline{0.09\ M}$.

For two liters, one needs **0.32 moles** CH_3COO^- and **0.18 moles** CH_3COOH.

In grams, for sodium acetate, $(0.32\ mol)(82.03\ g\ mol^{-1}) = \underline{\textbf{26.25 g}}$.

In grams, for acetic acid, $(0.18\ mol)(60.05\ g\ mol^{-1}) = \underline{\textbf{10.81 g}}$.

20. To make the full buffer concentration, all of the acetate must come from acetic acid, by means of titration with NaOH. The desired volume and buffer concentration are given in problem 19. Therefore, one adds the numbers of moles from problem 19 to determine:

$(0.32 + 0.18) = \underline{\textbf{0.50 moles acetic acid}}$. To get the required acetate concentration, a portion of the acetic acid must be titrated with **0.32 moles NaOH**. (See problem 19.)

In grams, for acetic acid, $(0.50\ mol)(60.05\ g\ mol^{-1}) = \underline{\textbf{30.03 g}}$ acetic acid.

In grams, for sodium hydroxide, $(0.32\ mol)(40.00\ g\ mol^{-1}) = \underline{\textbf{12.80 g}}$ NaOH.

21. The moles of sodium acetate used is:

$(41.02\ g)/(82.03\ g\ mol^{-1}) = \underline{\textbf{0.5 mol}}$ total (acetate + acetic acid).

Yes, the pH will be 5.0 because the 0.18 mol H^+ added will titrate 0.18 mol of the acetate to give the ratio of $[CH_3COO^-/CH_3COOH]$ specified in problem 19.

This methods results in 0.18 moles of extra NaCl being included in the 2 L solution (giving 0.09 M NaCl), so the buffer is not identical with the original.

22. The total number of moles of phosphate needed is: $(100 \text{ L})(0.001 \text{ mol L}^{-1}) = 0.1$ mol.

Recall the equilibrium: $H_2PO_4^- \rightarrow H^+ + HPO_4^{2-}$: $pK_a = 7.21$.

$pH - pK_a = 0.19 = \log_{10} ([HPO_4^{2-}]/[H_2PO_4^-])$.

Let x = moles HPO_4^{2-}. Then $(0.1 - x)$ = moles $H_2PO_4^-$.

$(x)/(0.1 - x) = 10^{0.19} = 1.55$.

Solving, $x = 0.155 - 1.55x$, such that $2.55x = 0.155$.

Then $x = 0.155/2.55 = 0.061$ mol, and $(0.1 - x) = 0.039$ mol.

In grams, for Na_2HPO_4, $(0.061 \text{ mol})(141.96 \text{ g mol}^{-1}) = \underline{\textbf{8.63 g}} \ Na_2HPO_4$.

In grams, for NaH_2PO_4, $(0.039 \text{ mol})(119.98 \text{ g mol}^{-1}) = \underline{\textbf{4.71 g}} \ NaH_2PO_4$.

23. The pH is = $pK_a + \log_{10} ([CH_3COO^-]/[CH_3COOH])$.

Moles acetic acid = $(0.060 \text{ g})/(60.05 \text{ g mol}^{-1}) = 10^{-3}$ mol CH_3COOH.

Moles acetate = $(14.59 \text{ g})/(82.03 \text{ g mol}^{-1}) = 0.178$ mol CH_3COO^-.

So the pH = $4.75 + \log_{10} (0.178/10^{-3}) = 4.75 + \log_{10} (178) = 4.75 + 2.25$.

Therefore, the pH indeed is $\underline{7.0}$. However, the buffering capacity is small because the pH value is far from the pK_a value (and this makes the ratio of $[CH_3COO^-]/[CH_3COOH]$ too large for the buffer to be useful).

24. Use the equation $E = (kq_1q_2)/(Dr)$. See text for the units.

When q_1 and q_2 are elementary charges, r is in Å, D is a dielectric constant, then k of 1389 gives the energy E in kJ/mol (while k of 332 gives E in kcal/mol). Substituting:

For D = 80 and k = 1389, E = $(1389)(-1)(-1)/(80*12) = 1.45$ kJ mol^{-1}.

For D = 80 and k = 332, E = $(332)(-1)(-1)/(2*12) = 0.35$ kcal mol^{-1}.

For D = 2 and k = 1389, E = $(1389)(-1)(-1)/(80*12) = 57.9$ kJ mol^{-1}.

For D = 2 and k = 332, E = $(332)(-1)(-1)/(2*12) = 13.8$ kcal mol^{-1}.

25. 0.1% (1/1000) of $3 * 10^9$ base pairs is $3 * 10^6$ base pairs, or three million base pair differences.

CHAPTER 2

1. (A) Proline, Pro, P; (B) tyrosine, Tyr, Y; (C) leucine, Leu, L; (D) lysine, Lys, K.

2. (a) All except the positively charged lysine (D) are hydrophobic. Leucine (C) is more hydrophobic than tyrosine or proline.

(b) Lysine.

(c) Tyrosine, in which the side-chain OH is the third ionizable group, and lysine, in which the side-chain NH_3^+ is the third ionizable group.

(d) The pK_a value is about 10 for both the tyrosine OH and the lysine NH_3^+.

(e) Tyrosine is *p*-hydroxy-phenylalanine.

3. (a) 6; (b) 2; (c) 3; (d) 1; (e) 4; (f) 5.

4. (a) Alanine has a smaller and less hydrophobic side chain than leucine and so is more soluble in water.

(b) The polar OH group on the tyrosine side chain makes Tyr more soluble in water than Phe.

(c) The polar OH group on the serine side chain makes Ser more soluble in water than Ala.

(d) The histidine aromatic ring contains two polar nitrogen atoms and is smaller than the ring of tryptophan. Furthermore, the histidine side chain is mildly basic and carries a significant fractional positive charge at neutral pH.

5. The polar OH groups of Ser, Tyr, and Thr; and the side-chain carboxylate of Glu have significant hydrogen-bonding potential. The purely hydrocarbon side chains of Ile, Phe, Ala, and Gly do not participate in hydrogen bonding.

6. (a) Alanine-glycine-serine; (b) Alanine; (c and d):

7.

At pH 5.5, the net charge is +1   At pH 7.5, the net charge is 0

The terminal amino group's +1 charge balances the terminal carboxyl group's charge of −1 at both pH 5.5 and pH 7.5. Since the pK_a of the histidine imidazole ring is about 6.5, the predominant state of the ring (approximately 90% of the population) will have a charge of +1 at pH 5.5 and a charge of 0 at pH 7.5. (See the Henderson-Hasselbalch equation in Chapter 1.)

8. There are 20 choices for each of the 50 amino acids, so the formula is $50^{20} = 5.0 \times 10^{21}$, a very large number!

9.

Aspartame at pH 7

10. The (nitrogen-α carbon-carbonyl carbon) repeating unit.

11. The side chain is the functional group attached to the α-carbon atom of an amino acid.

12. Amino acid composition refers simply to the amino acids that make up the protein. The order is not specified. Amino acid sequence is the same as the primary structure—the sequence of amino acids from the amino terminal to the carboxyl terminal of the protein. Many different proteins may have the same amino acid composition, but amino acid sequence identifies a unique protein.

13. (a) Since tropomyosin is double-stranded, each strand will have a mass of 35 kd. If the average residue has a mass of 110 d, there are 318 residues per strand (35,000/110), and the length is 477Å (1.5 Å/residue × 318).

    (b) Since 2 of the 40 residues form the end of the hairpin turn, 38 residues form the antiparallel β pleated sheet which is 19 residues long (38/2). In β pleated sheets, the axial distance between adjacent amino acids is 3.5 Å. Hence, the length of this segment is 66.5 Å (3.5 × 19).

14. Branching at the β carbon of the side chain (isoleucine), in contrast to branching at the γ carbon (leucine), sterically hinders the formation of a helix. This fact can be shown with molecular models.

15. Changing alanine to valine results in a bulkier side chain, which prevents the correct interior packing of the protein. Changing a nearby, bulky, isoleucine side chain to glycine apparently alleviates the space problem and allows the correct conformation to take place.

16. The amino acid sequence of insulin does not determine its three-dimensional structure. By catalyzing a disulfide-sulfhydryl exchange, this enzyme speeds up the activation of scrambled ribonuclease because the native form is the most thermodynamically stable. In contrast, the structure of active insulin is not the most thermodynamically stable form. The three-dimensional structure of insulin is determined by the folding of preproinsulin, which is later processed to mature insulin.

17. Appropriate hydrogen-bonding sites on the protease might induce formation of an intermolecular β pleated sheet with a portion of the target protein. This process would effectively fully extend α helices and other folded portions of the target molecule.

18. Being the smallest amino acid, glycine can fit into spaces too small to accommodate other amino acids. Thus, if sharp turns or limited spaces for amino acids occur in a functionally active conformation of a protein, glycine is required; no substitute will suffice. In view of this, it is not surprising that glycine is highly conserved.

19. To answer this question one needs to know some of the characteristics of the guanidinium group of the side chain of arginine and of the other functional groups in proteins. Most of the needed information is presented in Figure 2.9 in your textbook; note that the guanidinium group has a positive charge at pH 7 and contains several hydrogen bond donor groups. The positive charge can form salt bridges with the negatively charged groups of proteins (glutamate and asparatate side chains and the terminal carboxylate). As a hydrogen bond donor, the guanidinium group can react with the various hydrogen bond acceptors (glutamine, asparagine, aspartate, and the main chain carbonyl). It can also hydrogen bond with the hydroxyl group of serine and threonine. Hydroxyl groups accept hydrogen bonds much like water does.

20. The keratin of hair is essentially a bundle of long protein strands joined together by disulfide bonds. If these bonds are broken (reduced) by the addition of a thiol and the hair curled, the keratin chains slip past each other into a new configuration. When an oxidizing agent is added, new disulfide bonds are formed, thus stabilizing the new "curled" state.

21. There is a considerable energy cost for burying charged groups of non-hydrogen-bonded polar groups inside a hydrophobic membrane. Therefore, an α-helix with hydrophobic side chains is particularly suited to span a membrane. The backbone hydrogen-bonding requirements are all satisfied by intramolecular interactions within the α-helix. Good candidate amino acids with hydrophobic side chains would include Ala, Ile, Leu, Met, Phe, and Val. (Pro is also hydrophobic but will cause a bend in the helix.) Additionally, the aromatic (and amphipathic) amino acids Trp and Tyr are often

found toward the ends of membrane-spanning helices, near the phospholipid head groups in the membrane/water interface region.

22. The amino acids in such a membrane-spanning helix would be hydrophobic in nature. Particularly good choices would be those having large hydrophobic side chains, such as Leu, Ile, or Phe. An α helix is especially suited to crossing a membrane because all of the amide hydrogen atoms and carbonyl oxygen atoms of the peptide backbone take part in intrachain hydrogen bonds, thus stabilizing these polar atoms in a hydrophobic environment. When sequestered from water, these helix-stabilizing hydrogen bonds are stronger than when the helix contacts a water environment.

23. This example demonstrates that the pK_a values are affected by the environment. A given amino acid can have a variety of pK_a values, depending on the chemical environment inside the protein.

24. A possible explanation is that the severity of the symptoms corresponds to the degree of structural disruption. Hence, substitution of alanine for glycine might result in mild symptoms, but substitution of the much larger tryptophan might lead to little or no collagen triple-helix formation.

25. The protein is not at equilibrium, but is in a state where the peptide bond is "kinetically stable" against hydrolysis. This situation is due to the large activation energy for hydrolyzing a peptide bond.

26. One can effectively apply the Henderson-Hasselbalch equation successively to the amino group and to the carboxyl group and multiply the results to arrive at a ratio of 10^{-5}.

    So, considering first the amino group, $pH = pK + \log_{10} [NH_2]/[NH_3^+]$. With pH $= 7$ and $pK = 8$, one has $7 = 8 + \log_{10} [NH_2]/[NH_3^+]$ or $[NH_2]/[NH_3^+] = 10^{-1}$. Considering now the carboxyl group with pK of 3, one has $7 = 3 + \log [COO^-]/[COOH]$, or $[COO^-]/[COOH] = 10^{+4}$ or $[COOH]/[COO^-] = 10^{-4}$. Then to consider the two simultaneous ionizations that relate the zwitterionic form to the neutral form of an amino acid such as alanine, one needs to multiply the ratio of $[NH_2]/[NH_3^+]$ by the ratio of $[COOH]/[COO^-]$, that is, $(10^{-1}) \times (10^{-4}) = (10^{-5})$.

27. The presence of the larger sulfur atom (next to the β carbon of Cys) alters the relative priorities of the groups attached to the α carbon. The stereochemical arrangement of the β carbon with respect to the α hydrogen does not change, but the convention for assigning the R configuration changes when the Cβ-sulfur is present. (With methionine, the sulfur is too far removed for Cβ to influence the group priority.)

28. ELVIS IS LIVING IN LAS VEGAS!

29. No. Unlike the Pro nitrogen in X-Pro, the nitrogen of X in the peptide bond of Pro-X is *not* bonded between two tetrahedral carbon atoms. Therefore, the steric preference for the *trans* conformation will be similar to that of other (nonproline) peptide bonds.

30. Model A shows the reference structure for extended polypeptide chain with $\phi = 180°$ and $\psi = 180°$, so the answer is c. Models C and E have one torsion angle identical to model A and the other angle changed to 0°. In model C ϕ is changed to 0° (answer d), and in model E ψ is changed to 0° (answer b). Comparing model B with a reference for which $\phi = 0°$ (model C), we see a 60° *counter*clockwise rotation of ϕ, when viewed from Cα, so answer e is correct for model B. Finally, comparing model D with the $\phi = 0°$ reference in model C, we see a 120° clockwise rotation of ϕ, when viewed from Cα (answer a).

31. The reason is that the wrong disulfides formed pairs in urea. There are 105 different ways of pairing eight cysteine molecules to form four disulfides; only one of these combinations is enzymatically active. The 104 wrong pairings have been picturesquely termed "scrambled" ribonuclease.

CHAPTER 3

1. (a) The Edman method is best because it can be used repeatedly on the same peptide; hence, phenyl isothiocyanate.

   (b) Reversible denaturation is usually achieved with 8 m urea. However, if disulfide bonds are present, they must first be reduced with β-mercaptoethanol to obtain a *random coil* by urea treatment.

   The known cleavage specificities of chymotrypsin (c), CNBr (d), and trypsin (e) provide easy answers.

2. Whereas the hydrolysis of peptides yields amino acids, hydrazinolysis yields hydrazides

$$\overset{\displaystyle O}{\underset{\displaystyle }{(-\overset{\|}{C}-NHNH_2)}}$$

   of all amino acids *except* the carboxyl-terminal residue. The latter can be separated from the hydrazides by the use of anion exchange resin. (The hydrazides of aspartic and glutamic acids would also be picked up by the anion exchange resin; thus, a further purification step might be necessary.)

3. Ethyleneimine reacts with cysteine to form *S*-aminoethylcysteine, which has the following structure:

$$(^{+}H_3N-CH_2-CH_2-S-CH_2-\overset{\overset{\displaystyle NH_3^+}{\displaystyle |}}{\underset{\underset{\displaystyle H}{\displaystyle |}}{C}}-COO^-)$$

   Note that the cysteine side chain has increased in length and has added a plus charge. It closely resembles lysine in both size and charge, and therefore, its carboxyl peptide bonds are susceptible to hydrolysis by trypsin.

4. A 1-mg/ml solution of myoglobin (17.8 kd) is 5.62×10^{-5} M (1/17,800). The absorbance is 0.843 ($15,000 \times 1 \times 5.62 \times 10^{-5}$). Since this is the $\log_{10}$ of I_0/I, the ratio is 6.96 (the antilog of 0.843). Hence, 14.4% (1/6.96) of the incident light is transmitted. Note that when we say the ratio is 6.96, we are really saying 6.96/1. Inverting this ratio gives the 1/6.96.

   The above assumes a myoglobin mol. wt. of 17.8 kd. Most myoglobins have mol. wts. of about 16.8–17.8 kd.

5. The sample was diluted 1000-fold. The concentration after dialysis is thus 10^{-3} M or 0.001 M, or 1 mM. You could reduce the salt concentration still further by dialyzing your sample against a fresh volume of buffer solution, free of $(NH_4)_2SO_4$.

6. If the salt concentration becomes too high, the salt ions interact with the water molecules. Eventually, there will not be enough water molecules to interact with the protein, and the protein will precipitate. If there is lack of salt in a protein solution, the proteins may interact with one another—the positive charges on one protein with the negative charges on another or several others. Such an aggregate becomes too large to be solublized by water alone. If salt is added, the salt neutralizes the charges on the proteins, preventing protein-protein interactions.

7. Rod-shaped molecules have larger frictional coefficients than do spherical molecules. Because of this, the rod-shaped tropomyosin has a smaller (slower) sedimentation coefficient than does the spherical hemoglobin, even though it has a higher molecular weight. Imagine a metal pellet and a nail of equal weight and density sinking in a

syrup. The pellet will sink in an almost straight line, whereas the nail will twist and turn and sink more slowly.

8. From equation 2 on page 71 and the equation on page 76 of your text, we can derive the expression $s \propto m^{2/3}$, where s is the sedimentation coefficient and m is the mass, if we assume that the buoyancy $(1-v\rho)$ and viscosity (η) factors are constant. Then the mass (m) of a sphere is proportional to its volume (v). Since $v = 4/3\pi r^3$, m is $\propto r^3$ and $r \propto m^{1/3}$. Also, if $f = 6\pi\eta r$ (equation 2, page 71), $f \propto r \propto m^{1/3}$.

When the buoyancy factor is constant, s is proportional to m/f (equation on page 76) $\propto m/m^{1/3} \propto m^{2/3}$. Note that this says that the sedimentation coefficient is proportional to the two-thirds root of the mass.

Therefore, $s(80 \text{ kd})/s(40 \text{ kd}) = 80^{2/3}/40^{2/3} = 2^{2/3} = 1.59$.

9. The long hydrophobic tail on the SDS molecule (see p. 72) disrupts the hydrophobic interactions in the interior of the protein. The protein unfolds, with the hydrophobic R groups now interacting with SDS rather than with one another.

10. Electrophoretic mobilities are usually proportional to the $\log_{10}$ of the molecular weight (see Figure 3.10 on page 73 of the text). Note in Figure 3.10 that

$$\text{Slope} = \frac{D\log_{10} \text{ MW}}{D \text{ mobility}} = \frac{\log_{10} 92000 - \log_{10} 30000}{0.41 - 0.80} = \frac{\log_{10} x - \log_{10} 30000}{0.62 - 0.80}$$

where x is the molecular weight of the unknown. Solving,

$$\text{Log}_{10} x = \frac{(4.964 - 4.476)(-0.18)}{-0.39} + 4.476 = 4.701$$

and

$$x = 10^{4.701} = 50,000.$$

11. The protein may be glycosylated, which could have two effects. First, the molecular weight of the carbohydrate portion would not be considered in the amino acid sequence. Second, the binding of SDS to the carbohydrate portion and the relative migration of the carbohydrate portion often does not follow the same pattern as for unmodified polypeptides. Furthermore, membrane proteins also may behave differently from soluble proteins in terms of SDS binding and migration.

12. Assuming that cells with receptors that bind bacterial degradation products may also bind fluorescent derivatives of those products, you could synthesize a fluorescent-labeled derivative of a degradation product (perhaps some peptide of interest) and use this derivative to detect cells having receptors for this peptide.

13. (a) The digestion products will be AVGWR, VK, and S. The products have slightly different sizes (though all are quite small on a macromolecular scale) and somewhat different isoelectric points. These isoelectric points are approximately ½ (3.1 + 8.0) = 5.5 for serine (the pK values for ionizable groups from Table 3.1), approximately ½ (8.0 + 10.0) = 9.0 for VK, and approximately ½(8.0 + 12.0) = 10.0 for AVGWR. If high-pressure (high resolution) liquid chromatography is used, either an ion-exchange or a molecular-exclusion approach should work.

(b) The digestion products will be AV, GW, RV, and KS. Because all of the products are dipeptides (similar in size), an ion-exchange column should be used.

14. Antibody molecules bound to a solid support can be used for affinity purification of proteins for which a ligand molecule is not known or unavailable.

15. If the product of the enzyme-catalyzed reaction is highly antigenic, it may be possible to obtain antibodies to this particular molecule. These antibodies can be used to detect the presence of product by ELISA, providing an assay format suitable for the purification of this enzyme.

16. A probable explanation is that an inhibitor of the enzyme was removed during a particular purification step. When the inhibitor is absent, the apparent activity will increase. (Several other scenarios may be possible.)

17. Many proteins have similar masses but different sequences and different fragment lengths when digested with trypsin. The set of masses of tryptic peptides forms a detailed "fingerprint" of a protein that is very unlikely to occur at random in other proteins regardless of size. (A conceivable analogy is this: "As similarly sized fingers will give different individual fingerprints, so too will similarly sized proteins give different digestion patterns with trypsin!")

18. Isoleucine and leucine are isomers and, hence, have identical masses. Peptide sequencing by mass spectrometry as described in this chapter is incapable of distinguishing these residues. Further analytical techniques are required to differentiate these residues.

19. The specific activity is (total activity) divided by (total protein). The purification level is (specific activity) divided by the (initial specific activity). The yield is (total activity) divided by (initial total activity), multiplied by 100%. Answers are given in the table below.

Purification procedure	Total protein (mg)	Total activity (units)	Specific activity (units mg^{-1})	Purification level	Yield (%)
Crude extract	20,000	4,000,000	200	1.0	100
$(NH_4)_2SO_4$ precipitation	5,000	3,000,000	600	3.0	75
DEAE-cellulose chromatography	1,500	1,000,000	667	3.3	25
Size-exclusion chromatography	500	750,000	1,500	7.5	19
Affinity chromatography	45	675,000	15,000	75.0	17

20. Protein crystal formation requires the ordered arrangement of identically positioned molecules. Proteins with flexible linkers can introduce disorder into this arrangement and prevent the formation of suitable crystals. A ligand or binding partner may induce an ordered conformation to this linker and could be included in the solution to facilitate crystal growth. Alternatively, the individual domains separated by the linker may be expressed by recombinant methods and their crystal structures solved separately.

21. Two types of 15-kD subunits are present, one type beginning with N-terminal Ala and the other with N-terminal Leu. Pairs of these small 15-kD subunits are linked by covalent disulfide bonds that are broken only by the mercaptoethanol. The disulfide-linked subunits comprise the 30-kD species. (There is insufficient information to discern whether the 30-kD species consists of two different homodimers in which identical subunits are linked by disulfide bonds, or a unique heterodimer in which one Ala-initiated subunit is linked precisely to one Leu-initiated subunit.) Finally, two of the 30-kD species associate noncovalently to form the 60-kD particle that contains two copies each of two different 15-kD subunits. The 60-kD particle constitutes the native protein that is observed by molecular exclusion chromatography. (Urea disrupts the noncovalent subunit association, without breaking the disulfide bonds.)

The final quaternary structure may be described as either (A-A)(B-B), or (A-B)$_2$, where a hyphen (-) indicates a disulfide bond, "A" designates a subunit beginning with Ala, and "B" designates a subunit that begins with Leu.

22. The key question is whether the 30-kD units are two different homodimers or a unique heterodimer? (Either population would give a 50/50 mixture of dabsyl-Ala and dabsyl-Leu upon N-terminal analysis.) Therefore, one needs a method that will separate the (putative) A-B dimers from A-A and B-B dimers before the disulfide bonds are broken. While no method is absolutely (100%) certain to accomplish this, methods based on native-gel electrophoresis or high-resolution ion-exchange chromatography will provide opportunities for a favorable outcome. A good choice would be to use two-dimensional electrophoresis consisting of isoelectric focusing of 30-kD units (in the presence of 6 M urea), followed by mercaptoethanol/SDS-PAGE in the second direction. The possible outcomes are diagrammed below. The A-B unit would travel as a single entity in the first direction (isoelectric focusing), whereas the A-A and B-B units *may* be separable under high-resolution isoelectric focusing.

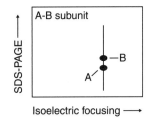
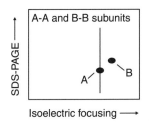

23. The difference between the predicted and the observed masses for this fragment is 28.0, exactly the mass shift that would be expected for a formyl group. Protein synthesis in *E. coli* begins with N-formyl methionine. Therefore this peptide MLNSFK, which begins with methionine, is likely formylated at its amino terminus and corresponds to residues 1–6 at the N-terminal of the intact protein.

24. Light was used to direct the synthesis of these peptides. Each amino acid added to the solid support contained a photolabile protecting group instead of a *t*-Boc protecting group at its α-amino group. Illumination of selected regions of the solid support led to the release of the protecting group, thus exposing the amino groups in these sites and making them reactive. The pattern of masks used in these illuminations and the sequence of reactants define the ultimate products and their locations. (See S. P. A. Fodor, J. L. Read, M. C. Pirrung, L. Stryer, A. T. Lu, & D. Solas. *Science* 251[1991]:767, for an account of light-activated, spatially addressable–parallel-chemical synthesis.)

25. Mass spectrometry is highly sensitive and capable of detecting the mass difference between a protein and its deuterated counterpart. Fragmentation techniques can be used to identify the amino acids that retained the isotope label. Alternatively, NMR spectroscopy can be used to detect the isotopically labeled atoms because the deuteron and the proton have very different nuclear-spin properties.

26. The peptide is AVRYSR.

Trypsin cleaves after R. The other R is at the C-terminal, and carboxypeptidase will not cleave the C-terminal R. Chymotrypsin cleaves after Y.

27. The full peptide is: S-Y-G-K-L-S-I-F-T-M-S-W-S-L.

The peptide will give these digestion patterns:

Carboxypeptidase cleaves the C-terminal L.

Cyanogen bromide: S-Y-G-K-L-S-I-F-T-M* and S-W-S-L (*cleavage after M, with conversion of the M to homoserine).

Chymotrypsin: S-Y, G-K-L, S-I-F, T-M, S-W, and S-L (cleavage after Y, L, F, M, and W).

Trypsin: S-Y-G-K and L-S-I-F-T-M-S-W-S-L (cleavage after K).

28. If the protein did not contain any disulfide bonds, then the electrophoretic mobility of the trypsin fragments would be the same before and after performic acid treatment: all the fragments would lie along the diagonal of the paper. If one disulfide bond were present, the disulfide-linked trypsin fragments would run as a single peak in the first direction, then would run as two separate peaks after performic acid treatment. The result would be two peaks appearing off the diagonal:

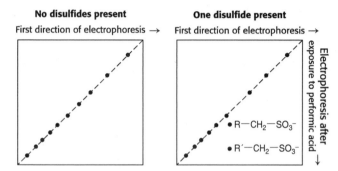

These fragments could then be isolated from the chromatography paper and analyzed by mass spectrometry to determine their amino acid composition and thus identify the cysteines participating in the disulfide bond.

CHAPTER 4

1. A nucleoside is a base attached to a ribose sugar. A nucleotide is a nucleoside with one or more phosphoryl groups attached to the ribose.

2. Hydrogen-bond pairing between the base A and the base T as well as hydrogen-bond pairing between the base G and the base C in DNA.

3. The [A] and the [T] are equal. Since the DNA is 20% T, it is also 20% A. The remaining 60% of bases then are divided equally between G and C. Therefore the DNA is 30% G and 30% C.

4. Nothing, because the base-pair rules do not apply to single-stranded nucleic acids.

5. By convention, when polynucleotide sequences are written, left to right means $5' \rightarrow 3'$. Since complementary strands are antiparallel, if one wishes to write the complementary sequence without specifically labeling the ends, the order of the bases must be reversed.
    (a) TTGATC
    (b) GTTCGA
    (c) ACGCGT
    (d) ATGGTA

6. (a) Since [A] + [G] account for 0.54 mole-fraction units, [T] + [C] must account for the remaining 0.46 (1 − 0.54). However, the individual mole fraction of [C] or [T] cannot be predicted.
    (b) Due to base pairing (A:T, G:C) in the complementary strand, [T] = 0.30, [C] = 0.24, and [A] + [G] = 0.46.

7. Two purines are too large to fit inside the double helix, and two pyrimidines are too small to form base pairs with each other.

8. The thermal energy causes the chains to wiggle about, which disrupts the hydrogen bonds between base pairs and the stacking forces between bases and thereby causes the strands to separate.

9. The probability that any sequence will appear is 4^n, where 4 is the number of nucleotides and n is the length of the sequence. Using the equation:

   $4^n = 3.0 * 10^9$, and solving for n yields:

   $n(\log_{10}4) = \log_{10}(3.0 * 10^9)$,   or $0.60\,n = 9.48$.

   $n = (9.48/0.60) = 15.74$, so on average, any particular DNA sequence of length between 15 and 16 bases is likely to appear only once in the human genome. (Many random 15-base sequences will appear slightly more than once (on average), while a random 16-base sequence is likely to appear either once or not at all.)

   (Note: For the ratio of logarithms, either the $\log_{10}$ terms or natural logrithms can be used, such that, equivalently: $n = (\ln(3.0 * 10^9))/(\ln 4) = (21.82/1.39) = 15.74$.)

10. One end of a nucleic acid polymer ends with a free 5´-hydroxyl group (or a phosphoryl group esterified to the hydroxyl group), and the other end has a free 3´-hydroxyl group. Thus, the ends are different. Two chains of DNA can form a double helix only if the chains are running in different directions—that is, have opposite polarity.

11. Although the individual bonds are weak, the population of thousands to millions of such bonds provides much stability. There is strength in numbers.

12. There would be too much charge repulsion from the negative charges on the phosphoryl groups. These charges must be countered by the addition of cations.

13. The three forms are the A-DNA, the B-DNA and the Z-DNA, with B-DNA being the most common. There are many differences (see Table 4.2). Some key differences are: A-DNA and B-DNA are right-handed, whereas Z-DNA is left-handed. A-DNA forms in less-hydrated conditions than does B-DNA. The A form is shorter and wider than the B form. For sections of RNA that are double-helical, the 2´-OH group of ribose interferes with the folding of the B-form helix, causing the A-form double helix to be favored for RNA.

14. To answer this question one must know that $2\mu m = 2 \times 10^{-6}$ m, that one Å $= 10^{-10}$ m, and that the distance between the base pairs is 3.4 Å. The length of a DNA segment (in this case 2×10^{-6} m) divided by the distance between the base pairs (3.4×10^{-10} m) gives the answer, 5.88×10^3 base pairs.

15. After 1.0 generation, one-half of the molecules would be ^{15}N-^{15}N, the other half ^{14}N-^{14}N. After 2.0 generations, one-quarter of the molecules would be ^{15}N-^{15}N, the other three-quarters ^{14}N-^{14}N. Hybrid ^{14}N-^{15}N molecules would not be observed.

16. (a) Thymine is the molecule of choice because it occurs in DNA, is not a component of RNA, and is not readily converted to cytosine or uracil. If they enter the cell, labeled deoxythymidine or dTTP would also be useful molecules. Its large negative charge prevents dTTP from entering most cells.

    (b) During DNA synthesis, the β- and γ-phosphorus atoms of the nucleoside triphosphates are lost as pyrophosphate. Since the α-phosphorous atom is incorporated into DNA, one should use dATP, dGTP, dTTP, and dCTP labeled with ^{32}P in the α position.

17. Only (c) would lead to DNA synthesis because (a) and (b) have no primer or open end to build on and (d) has no template extending beyond a free 3′-OH. Note: Single-stranded linear DNA can be used as a template for DNA synthesis because it can prime synthesis through hairpin formation at its 3′ end.

18. A retrovirus is a virus that has RNA as its genetic material. However, for the information to be expressed, it must first be converted into DNA, a reaction catalyzed by the enzyme reverse transcriptase. Thus, at least initially, information flow is opposite that of a normal cell: RNA $\rightarrow$ DNA rather than DNA $\rightarrow$ RNA.

19. A short polythymidylate chain would serve as a primer because T base pairs with A. Radioactive dTTP labeled in any position except the β- and γ-phosphates would be useful for following chain elongation.

20. After the synthesis of the complementary (–) DNA on the RNA template, the RNA must be disposed of by hydrolysis prior to the completion of the synthesis of the DNA duplex.

21. One should treat the infectious nucleic acid with either highly purified ribonuclease or deoxyribonuclease and then determine its infectivity. RNAse will destroy the infectivity if it is RNA; DNAse will destroy it if it is DNA.

22. Ultimately, this mutation results in half the daughter DNA duplexes being normal and half having a TA pair that had been CG. The first two rounds of replication at the mutant site will be as follows:

23. (a) From the 4 mononucleotides one can formulate 16 different dinucleotides. If you don't believe it, try it! From these dinucleotides you can make 64 different trinucleotides. Note that 64 is 4^3. There will be 4^4 (256) tetranucleotides. Proceeding in this manner we get to 4^8 (65,536) different octonucleotides (8-mers).

(b) A bit specifies two bases (say A and C), and a second bit specifies the other two (G and T). Hence, two bits are needed to specify a single nucleotide (or base pair) in DNA. An 8-mer stores 16 bits ($2^{16} = 65,535$), the *E. coli* genome (4×10^6 bp) stores 8×10^6 bits, and the human genome (2.9×10^9 bases) stores 5.8×10^9 bits of genetic information.

(c) A high-density diskette stores about 1.5 megabytes, which is equal to 1.2×10^7 bits. A large number of 8-mer sequences could be stored on such a diskette. The DNA sequence of *E. coli,* once known, could be written on a single diskette. Nearly 500 diskettes would be needed to record the human DNA sequence.

24. (a) Deoxyribonucleoside triphosphates versus ribonucleoside triphosphates.
(b) 5′ $\rightarrow$ 3′ for both.
(c) DNA serves as the template for both polymerases. During DNA replication by polymerase I each parent strand acts as a template for the formation of a new complimentary strand. Since each daughter molecule receives one strand from the parent DNA molecule, the template is said to be semiconserved. However, after guiding the synthesis of RNA by RNA polymerase, the DNA double helix remains intact. Hence the template is said to be conserved.
(d) DNA polymerase I requires a primer, whereas RNA polymerase does not.

25. Messenger RNA encodes the information that, on translation, yields a protein. Ribosomal RNA is the catalytic component of ribosomes, the molecular complexes that synthesize proteins. Transfer RNA is an adaptor molecule, capable of binding a specific amino acid and recognizing a corresponding codon. Transfer RNAs with attached amino acids are substrates for the ribosome.

26. (a) Because mRNA is synthesized antiparallel to the DNA template and A pairs with U and T pairs with A, the correct sequence is 5'-UAACGGUACGAU-3'.

    (b) Since the 5' end of an mRNA molecule codes for the amino terminus, appropriate use of the genetic code (see text, p. 109) leads to Leu-Pro-Ser-Asp-Trp-Met.

    (c) Since one has a repeating tetramer (UUAC) and a 3-base code, repetition will be observed at a 12-base interval (3 × UUAC). Comparison of this 12-base sequence with the genetic code leads to the conclusion that a polymer with a repeating tetrapeptide (Leu-Leu-Thr-Tyr) unit will be formed.

27. The instability of RNA in alkali is due to its 2'-OH group. In the presence of OH$^-$ the 2'-OH group of RNA is converted to an alkoxide ion (RO$^-$) by removal of a proton. Intramolecular attack by the 2'-alkoxide on the phosphodiester in RNA gives a 2',3'-cyclic nucleotide, cleaving the phosphodiester bond in the process. Further attack by OH$^-$ on the 2',3'-cyclic nucleotide produces a mixture of 2' and 3'-nucleotides. Note that the mechanism for ribonuclease action is quite similar (see Figure 9.17). Since DNA lacks a 2'-OH group, it is quite stable in alkali.

28.

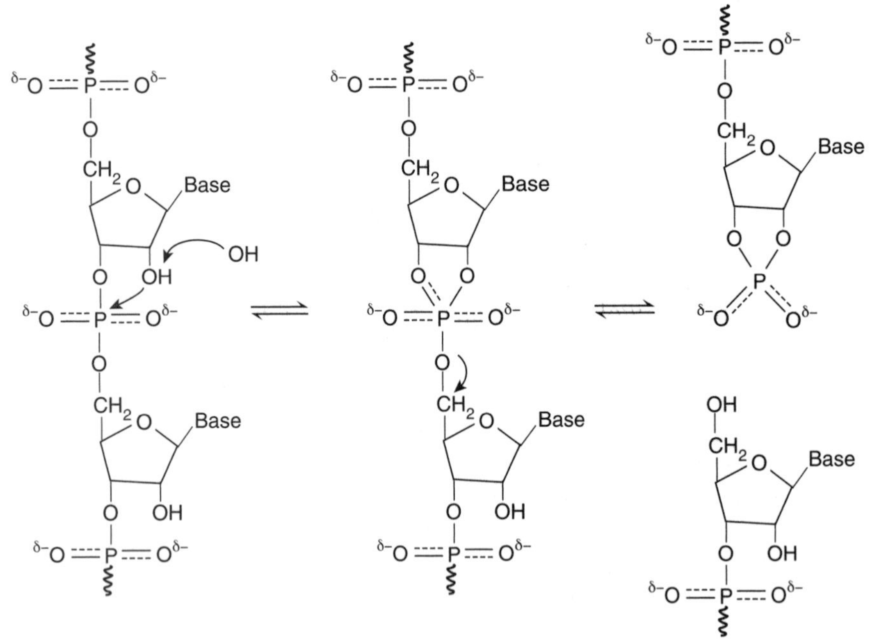

29. Gene expression is the process of converting the information encoded within a gene into a functional molecular form. For many genes, the functional information leads to production of a protein molecule. For protein production, therefore, gene expression will include both transcription and translation. (For a limited number of genes, the functional product is a molecule of RNA, and gene expression in these few cases includes only transcription.)

30. A consensus sequence can be regarded as the average of many similar sequences. The nucleotide bases in such a sequence represent the most common, but not necessarily the only, members found within the set of similar sequences.

31. Apparently cordycepin is converted to its 5′-triphosphate and incorporated into the growing RNA chain. This chain containing cordycepin now lacks a 3′-OH group; hence, RNA synthesis is terminated.

32. Only single-stranded mRNAs can serve as templates for protein synthesis. Since poly(G) forms a triple-stranded helix, it cannot serve as a template for protein synthesis.

33. Degeneracy of the code refers to the fact that most amino acids are encoded by more than one codon.

34. If only 20 of the 64 possible codons encoded amino acids, then a mutation that changed a codon would likely result in a nonsense codon, leading to termination of protein synthesis. With degeneracy, a nucleotide change might yield a synonym or a codon for an amino acid with similar chemical properties.

35. (a) Replication requires <u>DNA polymerase</u>, <u>dNTP</u> precursors, a <u>primer</u> (made from RNA by a special RNA polymerase called "primase") and a DNA template (not in the list).
    (b) Transcription requires <u>RNA polymerase</u>, <u>NTP</u> precursors, and a <u>promoter</u> region of a DNA template. (mRNA and tRNA are products of transcription.)
    (c) Translation requires <u>ribosomes</u> (that are assembled from proteins and <u>rRNA</u>), <u>tRNA</u> adapter molecules, and a template consisting of <u>mRNA</u>.

36. (a) <u>fMet</u> is the first "of many" amino acids in a protein synthesized by prokaryotic ribosomes.
    (b) A "Shine-Dalgarno" sequence helps to "locate the start" codon for translation in prokaryotes.
    (c) <u>Intron</u>(s) are "removed" from pre-mRNA, by splicing, to generate the mature mRNA.
    (d) <u>Exon</u>(s) are "joined" when introns are removed from pre-mRNA.
    (e) A <u>pre-mRNA</u> molecule contains exons and introns and is therefore a "discontinuous message."
    (f) A molecule of <u>mRNA</u> contains a "continuous message" that encodes the amino-acid sequence of a protein.
    (g) The <u>spliceosome</u> is a "uniter" of exons because it removes the introns from pre-mRNA.

37. Note that each complimentary strand is missing one of the four bases; d(TAC) lacks G and d(GTA) lacks C. Thus, incubation with RNA polymerase and only UTP, ATP, and CTP led to the synthesis of only poly(UAC), the RNA complement of d(GTA). When GTP was used in place of CTP, the complement of d(TAC), poly(GUA), was formed.

38. Since three different polypeptides are synthesized, the synthesis must start from three different reading frames. One of these will be in phase with the AAA in the sequence shown in the problem and will therefore have a terminal lysine, since UGA is a stop signal. The reading frame in phase with AAU will result in a polypeptide having an Asn-Glu sequence in it, and the reading frame in phase with AUG will have a Met-Arg sequence in it.

39. Highly abundant amino acid residues have the most codons (e.g., Leu and Ser each have six), whereas the least-abundant amino acids have the fewest (Met and Trp each have only one). Degeneracy (a) allows variation in base composition and (b) decreases

the likelihood that a substitution for a base will change the encoded amino acid. If the degeneracy were equally distributed, each of the 20 amino acids would have three codons. Both benefits (a and b) are maximized by the assignment of more codons to prevalent amino acids than to less frequently used ones.

40. Labeled ^{14}C-Ala will be incorporated in response to the UGU codon. From the beginning of the mRNA, using the genetic code, the codons specify:

UUU - Phe

UGC - Cys

CAU - His

GUU - Val

UGU - (Cys converted to Ala)

GCU - Ala

Hence, the resulting sequence of the radiolabeled peptide will be

Phe-Cys-His-Val-Ala*-Ala, with the Ala* carrying the ^{14}C label.

41. Exon shuffling is a molecular process that can lead to the generation of new proteins by the rearrangement of exons within genes. Because many exons encode functional protein domains, exon shuffling is a rapid and efficient means of generating new genes. If two such functional domains are brought together by shuffling, the result may in some cases yield a protein with new properties that could perform new biological functions or show improved performance for an existing function.

42. It shows that the genetic code and the biochemical means of interpreting the code are common to even very distantly related life forms. It also testifies to the unity of life; that all life arose from a common ancestor.

43. (a) Two point mutations would be required to change Lys to Asp (unlikely).
    (b) More likely would be a single point mutation changing lysine to asparagine (Asn). Because Asn is converted to Asp during acid hydrolysis, the protein chemist mistakenly assigned Asp instead of Asn.

44. The genetic code is degenerate. Of the 20 amino acids, 18 are specified by more than one codon. Hence, many nucleotide changes (especially in the third base of a codon) do not alter the nature of the encoded amino acid. Mutations leading to an altered amino acid are usually more deleterious than those that do not and hence are subject to more stringent selection.

45. GC base pairs are stabilized by three hydrogen bonds, whereas AT base pairs have only two hydrogen bonds. The graph indicates that DNA with 100% AT base pairs has T_m of approximately 70 °C, that DNA with 100% GC base pairs has T_m of approximately 110 °C, and furthermore that there is a linear relationship between the GC content and the T_m of DNA between about 70 °C and 110 °C. The higher content of GC base pairs in DNA leads to more hydrogen bonds and greater helix stability.

46. The C_0t value—representing initial concentration multiplied by time—corresponds to the complexity of the DNA sequence and therefore also to the information content of the DNA sequence. The sample of (poly U)/(poly A) contains a simple sequence with low information content, such that the complementary strands will renature rapidly to form the double helix. The more complex sequences require increasingly longer times (or higher concentrations, hence the "C_0t" parameter) to find their complementary strands and form the double helices.

CHAPTER 5

1. (a) The direction of movement on the gel is from top to bottom, with the smallest fragment, in this case G, moving most rapidly. Since the 5′ end carries the ^{32}P label, the 5′ → 3′ sequence is read from bottom to top, opposite the direction of movement. The sequence is 5′-GGCATAC-3′. It should be noted that the fastest-moving spot in the autoradiogram is the radioactive inorganic phosphate resulting from the destruction of the guanine at the 5′ terminus.

   (b) Note that in the Sanger dideoxy method (see Figure 5.4 on page 143 of the text), the new DNA strands that are subjected to electrophoresis are elongated 5′ → 3′. Since the larger molecules move more slowly, the results shown below are obtained. The DNA strand serving as template in the Sanger method is the complement of the strand shown here in the figure.

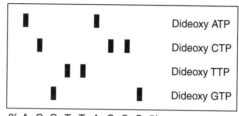

Direction of electrophoresis

Dideoxy ATP
Dideoxy CTP
Dideoxy TTP
Dideoxy GTP

3′ A C G T T A C C G 5′

2. Since *E. coli* lack the machinery to excise introns and splice exons, they would make a meaningless mRNA if presented with ovalbumin genomic DNA. Therefore, if you wish to express the ovalbumin gene in *E. coli,* you must use its cDNA, which contains the information in the eight exons, but no introns.

3. Consistent with its planar, aromatic structure, ethidium bromide is a DNA intercalator: it aligns itself between the paired bases in a DNA duplex.

4. The probability of finding a given specific DNA sequence is $^{1}/4^{n}$, where **n** is the number of nucleotides in one strand of the sequence that will be recognized by the restriction enzyme (because any of four nucleotides can be present at any given position in a random sequence). The average distance between cleavage sites in double-stranded DNA therefore is 4^{n} nucleotides along one strand. For *Alu*I, $4^{4} = 256$. For *Not*I, $4^{8} = 65,536$. (The sequence of the second strand of DNA is completely defined by complementary base pairing to the first strand; therefore we need only consider the probability of finding the correct recognition sequence on one strand.)

5. (a) No, because most human genes are much longer than 4 kb. One would obtain fragments containing only a small part of a complete gene.

   (b) No, chromosome walking depends on having *overlapping* fragments. Exhaustive digestion with a restriction enzyme produces nonoverlapping, short fragments.

6. Southern blotting of an *Mst*II digest would distinguish between the normal and mutant genes. The loss of a restriction site would lead to the replacement of two fragments on the Southern blot by a single longer fragment. Such a finding would not prove that GTG replaced GAG; other sequence changes at the restriction site could yield the same result.

7. Although the two enzymes cleave the same recognition site, they break different bonds within the 6-bp sequence. Cleavage by *Kpn*I yields a single-stranded overhang on the 3′ end of the + strand (upper strand in the diagram in the text), whereas cleavage by *Acc*65I produces an overhang on the 5′ end of the <u>same</u> strand. Neither enzyme leaves

an overhang on the opposing strand (− strand, or lower strand in the diagram in the text). Therefore the resulting sticky ends do not overlap.

8. A few years ago this would have been very difficult, if not impossible. However, the availability of automated solid-phase chemical methods for synthesizing DNA has made the impossible fairly easy. A simple strategy for generating many mutants is to synthesize a group of oligonucleotides that differ only in the sequence of bases in one triplet, or codon. For example, with the 30-mer described in this question, if a mixture of all four nucleotides is used in the *first and second rounds* of synthesis, the resulting oligonucleotides will begin with the sequence XYT (where X and Y denote A, C, G, or T). This will provide 16 different versions of the first triplet (codon) of the 30-mer, which will encode proteins containing either Phe, Leu, Ile, Val, Ser, Pro, Thr, Ala, Tyr, His, Asn, Asp, Cys, Arg, Ser, or Gly at the first position (see Table 4.5 in the text). In similar fashion, one can synthesize oligonucleotides in which two or more codons are simultaneously varied.

9. A number of questions could be asked about the nature of the original sample and the possibility of contamination before DNA amplification by PCR. Even if no contaminants were introduced during the handling of the sample, it is possible that the fossil might have contained remains from microorganisms or other species mixed with the dinosaur materials. Analysis of the DNA sequence and sequence complexity could be revealing, however, especially in relation to DNA sequences from modern reptiles and other known organisms. If sufficient length and number of DNA sequences would be available from several PCR experiments on the same (fossil) sample, then one would be able to narrow the phylogenetic classification of the type of organism from which the DNA originated with some confidence.

10. PCR amplification is greatly hindered by the presence of G–C-rich regions within the template. Owing to their high melting temperatures, these templates do not denature easily, preventing the initiation of an amplification cycle. In addition, rigid secondary structures prevent the progress of DNA polymerase along the template strand during elongation.

11. Higher hybridization temperatures require greater numbers of complementary base pairs between the primer and target DNAs. Conversely, lower hybridization temperatures are more permissive of sequence mismatch between the primer and the target. Let us suppose that particular yeast gene A indeed has a moderately diverged counterpart (with a moderately different sequence) in humans. If so, no PCR amplification will be observed when the hybridization temperature is too high, but a lower hybridization temperature would allow the target human DNA to be amplified in a PCR experiment that uses the yeast primer. (If the hybridization temperature (stringency) is too low, however, then spurious unrelated artifacts could also be amplified.)

12. For PCR to amplify a DNA duplex, the polymerase must be active (primed) in both directions if each strand is to be replicated. If the DNA is linear, known sequences are needed on both sides of the portion to be amplified. To explore DNA on both sides of a single known sequence, one can digest genomic DNA with a restriction enzyme and circularize the fragments. Then, by using a pair of primers that hybridize specifically with portions of the known sequence, one can use PCR to amplify only the fragments containing the known sequence. Note that in a circular DNA a single known sequence can be used to prime polymerase action in *both directions* away from the known sequence, resulting in complete replication of the duplex.

13. These results suggest that the DNA is composed of four repeating units. If these were transcribed into mRNA and the latter translated into protein, one would expect a protein molecule whose linear sequence was composed of four repeating peptides.

14. Use chemical synthesis or the polymerase chain reaction to prepare hybridization probes that are complementary to both ends of the known (previously isolated) DNA fragment. Challenge clones representing the library of DNA fragments with both of the hybridization probes. Select clones that hybridize to one of the probes but not to the other; such clones are likely to represent DNA fragments that contain only one end of the known fragment, along with the adjacent region of the particular chromosome.

15. The codon(s) for each amino acid can be used to determine the number of possible nucleotide sequences that encode each peptide sequence (see Table 4.5):

    Ala—Met—Ser—Leu—Pro—Trp:
    $4 * 1 * 6 * 6 * 4 * 1 = 576$ total sequences
    Gly—Trp—Asp—Met—His—Lys:
    $4 * 1 * 2 * 1 * 2 * 2 = 32$ total sequences
    Cys—Val—Trp—Asn—Lys—Ile:
    $2 * 4 * 1 * 2 * 2 * 3 = 96$ total sequences
    Arg—Ser—Met—Leu—Gln—Asn:
    $6 * 6 * 1 * 6 * 2 * 2 = 864$ total sequences

    The set of DNA sequences encoding the peptide Gly-Trp-Asp-Met-His-Lys would be most ideal for the probe.

16. Within a single species, individual dogs show enormous variation in body size and substantial diversity in other physical characteristics. Therefore, genomic analysis of individual dogs would provide valuable clues concerning the genes responsible for the diversity within the species.

17. On the basis of the comparative genome map shown in Figure 5.27, the region of greatest overlap with human chromosome 20 can be found on mouse chromosome 2.

18. In each PCR cycle, each new DNA strand is synthesized from a complementary strand that was made in the previous cycle. If the synthesis of one strand is inefficient, then the entire process will be inefficient, regardless of the rate of synthesis of the other strand. Because the experiment is performed in a single tube, the same hybridization temperature must be used for the priming of both strands. If the two primers would have very different values of T_m, then one strand could amplify much more readily than the other, and the efficient doubling of DNA at each cycle might not occur.

19. Careful comparison of the sequences reveals that there is a 7-bp region of complementarity at the 3′ ends of these two primers:

    5′-G GAT CGATGC TCGCGA-3′
               | | | | | | |
            3′-GAGCGCTGGGCTAGGA-5′

    In a PCR experiment, these primers would likely anneal to one another, preventing their interaction with the template DNA.

    During DNA synthesis by the polymerase, each primer would act as a template for the other primer, leading to the amplification of a 25-bp sequence corresponding to the overlapped primers.

20. Individual B is without symptoms because he has one gene that functions normally, even though his X mRNA is smaller than normal. Individual B appears to be heterozygous in the HindIII restriction experiment, perhaps having one functional and one nonfunctional gene X. Although B's mRNA for X is shorter than normal, his Y protein that is produced from this mRNA is of the normal size and apparently is functional because he has no symptoms.

Individuals C and D fail to express mRNA from gene X. Without the mRNA, they cannot make protein Y.

Individual E does express X mRNA but is unable to synthesize protein Y encoded by this mRNA (perhaps a regulatory mutation). (Alternatively, E could possibly make a defective protein Y that does not fold properly or is degraded rapidly and is not recognized by the antibody in the Western blot.)

Individual F makes X mRNA and Y protein of the proper size. Yet the Y protein apparently fails to function properly. This could be due to a point mutation that makes a single change in the amino acid sequence of Y at a location that is critical for function.

21. The gels should be read from the bottom to the top. The data indicate the sequence of the coding strand of the DNA. The normal sequence of codons from the first gel is GTG CTG TCT CCT GCC GAC AAG, which encodes Val-Leu-Ser-Pro-Ala-Asp-Lys.

Hemoglobin Chongqing differs in having CGG instead of CTG as the second codon. The corresponding amino changes from leucine to arginine.

Hemoglobin Karachi differs in having CCC instead of GCC as the fifth codon. The corresponding amino changes from alanine to proline.

Hemoglobin Swan River differs in having GGC instead of GAC as the sixth codon. The corresponding amino changes from aspartic acid to glycine.

22. This particular person is heterozygous for this particular mutation: one allele is wild type, whereas the other carries a point mutation at this position. Both alleles are PCR amplified in this experiment, yielding the "dual peak" appearance on the sequencing chromatogram.

CHAPTER 6

1. For the sequences shown, there are 27 identities and 2 gaps (over a total length of 98 residues in sequence 2). Based on a score of +10 for an identity and −25 for a gap, the score would be $270 - 50 = 220$. The percentage of identical residues can be calculated as $27/98 = 27.6\%$. One could also note that the % identity is higher over the first 75 residues ($26/75 = 34.7\%$). Based on the methods of sequence shuffling and statistical comparison (Figures 6.7 and 6.8 in the text), one can comfortably conclude that these scores *are* statistically significant.

2. Because tertiary structure is more highly conserved than is primary structure (Section 6.3.1 of the text), it is possible that these two proteins have retained a common three-dimensional structure, even while their sequences have *diverged* extensively from a common ancestor. Alternatively, the two proteins could represent an example of *convergent* evolution to a particular structure and function (Section 6.3.4 of the text) using sequences that are conspicuously different.

3. With identity-based scoring, the first sequence has one gap and no identities, hence a score of −25; whereas the second sequence has one gap and four identities, hence a score of $40 - 25 = +15$.

However, using the Blosum-62 substitution matrix, the scores would be +7 for the first sequence and −12 for the second. Here is the result of summing the pairwise values for alignments (a) and (b) using Figure 6.9 in the text (with −12 for a gap):

sequence 1	sequence 2	(a) score	sequence 1	sequence 2	(b) score
A	G	0	A	G	0
—	S	−12	S	S	6
S	N	1	N	N	8
N	D	2	L	D	−5
L	F	1	F	F	4
F	Y	4	D	Y	−5
D	E	2	I	E	−4
I	V	4	R	V	−4
R	K	3	L	K	−4
L	I	2	I	I	6
I	M	2	—	M	−12
G	D	−2	G	D	−2
	Sum	+7		Sum	−12

4. U sometimes pairs with G in these RNA sequences. For example, in the bacterial sequences, G7 pairs with U24, G11 with U20, and U12 with G19. Furthermore, in the eucaryotic sequences, U10 pairs with G21. Here is a possible structure for a U···G base pair:

5. 26,400 grams. This answer is obtained by considering the need for 4^{40} different molecules (because any of 4 bases can be at each of 40 positions). The number 4^{40} is equal to 1.2×10^{24}. One mole of the RNA will weigh (330×40) grams and will consist of Avogadro's number of molecules (6×10^{23}). Combining all data, one has:

$$(330 \text{ grams} \times 40 \times 1.2 \times 10^{24} \text{ molecules}) / (6 \times 10^{23} \text{ molecules}) = 26,400 \text{ grams}.$$

6. (See Section 6.3 of the text.) Biomolecules function at the level of three-dimensional structure, so from a functional point of view it is more important to conserve particular three-dimensional structures than one-dimensional sequences. Although mutations occur at the level of one-dimensional sequences, the effects of mutations are felt at the level of function. Therefore, many mutations will lead to sequence changes that are tolerable because they preserve a common three-dimensional structure.

7. **A**SNF**LD**KAGK
   ATDY**LE**KAGK   (Identities are underlined; score 60).

With six identities and no gaps, the score for the initial alignment would be 60. A very large number of shuffled versions of sequence 2 can be generated. Here are two examples, along with their scores for alignment with sequence 1:

ASNFL**D**KAG**K**
KYTAG**D**ELA**K**   (Identities are underlined; score 20).

ASNFL**D**KAGK
TKEAY**D**LKAG   (Identities are underlined; score 10).

8. (See Section 6.2.2 of the text.) Sequences that are longer than 100 amino acids *and* have greater than 25% identity are probably homologous. At the other end of the scale, sequences that are less than 15% identical are unlikely to have statistically significant sequence similarity (although they could nevertheless have similar three-dimensional structures—see Section 6.3 of the text). For pairs of sequences that show between 15% and 25% identity, further analysis is necessary to determine the significance of the alignment. Following these guidelines, these are the answers:

   a. (80% identity) Divergence from a common ancestor is probable.
   b. (50%) Divergence from a common ancestor is probable.
   c. (20%) Further analysis is needed.
   d. (10%) Divergence from a common ancestor is unlikely.

9. Replacement of cysteine, glycine, and proline never yields a positive score. Each of these residues exhibits unique features that cannot be fulfilled by any of its 19 counterparts. Cysteine is the only amino acid capable of forming disulfide bonds. Glycine is the only amino acid without a side chain and is highly flexible. Proline is the only secondary amino acid. In peptides and proteins the proline amine nitrogen becomes tertiary and highly constrains the protein backbone through the bonding of the nitrogen to its side chain, generating the five-membered pyrrolidine ring.

10. Yes. Three-dimensional structure is more highly conserved than is amino acid sequence. Sequence B is similar to both A and C. Therefore protein B is likely to have a three-dimensional structure similar to those of both A and C. If the A and C protein structures are both similar to B, then they are similar to each other.

11. To test for possible hairpin structures, try inverting the first half of the sequence and checking for possible Watson/Crick base-paired alignments between the inverted first half and the (non-inverted) second half of sequence. In the alignments below, the original sequences begin at the 5′ end and proceed around the hairpin to the 3′ end.

```
(1)   U A G A A U C U C C C -3′
      G     | | | | | | | |
          G   C U U A G A G G U U -5′

(2)   C A G A U U C C C C G -3′
      A     | | | | | | | |
          G   C U A A G G G C C G -5′

(3)   C A G G G A C U U A C -3′
      G     | | | | | | |
          G   C C C U G A A C C C -5′

(4)   U A G G C A G G U C A -3′
      A     | | | | | | | |
          G   C C G U C C A C U C -5′

(5)   U A G G G U G G U U C -3′
      G     | | | | | | |
          G C C C A C C A U A A -5′
```

12. To detect pairs of residues with correlated mutations, there must be variability in these sequences. If the alignment is overrepresented by closely related organisms, there may not be enough changes in their sequences to allow the identification of potential base-pairing patterns.

13. After RNA molecules have been selected and reverse transcribed, PCR is performed to introduce additional mutations into these strands. The use of this error-prone, thermostable polymerase in the amplification step would enhance the efficiency of this random mutagenesis.

14. The initial pool of RNA molecules used in a molecular-evolution experiment is typically much smaller than the total number of possible sequences. Hence, the best possible RNA sequences will likely not be represented in the initial set of oligonucleotides. Mutagenesis of the initial selected RNA molecules allows for the iterative improvement of these sequences for the desired property.

15. At the listed Web site (www.ncbi.nlm.nih.gov), search "protein" for "coli triose phosphate isomerase." Note the entry (or entries) that relate to *E. coli*. Here is the Genbank locus:

LOCUS P0A858 255 amino acids linear BCT

DEFINITION Triosephosphate isomerase (TIM) (Triose-phosphate isomerase).

ORGANISM Escherichia coli.

Sequence:

```
  1 mrhplvmgnw klngsrhmvh elvsnlrkel agvagcavai appemyidma kreaegshim
 61 lgaqnvdlnl sgaftgetsa amlkdigaqy iiighserrt yhkesdelia kkfavlkeqg
121 ltpvlciget eaeneagkte evcarqidav lktqgaaafe gaviayepvw aigtgksatp
181 aqaqavhkfi rdhiakvdan iaeqviiqyg gsvnasnaae lfaqpdidga lvggaslkad
241 afavivkaae aakqa //
```

Go to the site: http://www.ncbi.nlm.nih.gov/BLAST/. Select protein-protein Blast. On the new screen, paste the *E. coli* sequence (255 amino acids) into the search box. Choose database 'nr.' Click the button for 'Blast!' When searching the above sequence in a "protein-protein" Blast against the non-redundant database (nr), one needs to display 500 alignments in order to see the alignment between *E. coli* and *Homo sapiens*. Most of the entries for *Homo sapiens* show 107 identities with *E. coli*. One entry (Genbank locus AAH17917) shows 108 identities; the alignment of this entry with the *E. coli* sequence is shown below.

gi|17389815|gb|AAH17917.1| Triosephosphate isomerase 1 [Homo sapiens] Length = 249

Score = 186 bits (472), Expect = 7e-46

Identities = 108/233 (46%), Positives = 140/233 (60%), Gaps = 5/233 (2%).

```
Query   6   VMGNWKLNGSRHMVHELVSNLRKELAGVAGCAVAIAPPEMYIDMAKREAEGSHIMLGAQN   65
            V GNWK1NG + + EL+ L      A V APP  YID A+++ +  I + AQN
Sbjct   9   VGGNWKMNGRKQSLGELIGTLNAAKVP-ADTEVVCAPPTAYIDFARQKLDPK-IAVAAQN   66

Query  66   VDLNLSGAFTGETSAAMLKDIGAQYIIIGHSERRTYHKESDELIAKKFAVLKEQGLTPVL  125
            +GAFTGE S  M+KD GA ++++GHSERR    ESDELI +K A   +GL +
Sbjct  67   CYKVTNGAFTGEISPGMIKDCGATWVVLGHSERRHVFGESDELIGQKVAHALAEGLGVIA  126

Query 126   CIGETEAENEAGKTEEVCARQIDAVLKTQGAAAFEGAVIAYEPVWAIGTGKSATPAQAQA  185
            CIGE   E EAG TE+V   Q  +     A  +   V+AYEPVWAIGTGK+ATP QAQ
Sbjct 127   CIGEKLDEREAGITEKVVFEQTKVI--ADNAKDWSKVVLAYEPVWAIGTGKTATPQQAQE  184

Query 186   VHKFIRDHI-AKVDANIAEQVIIQYGGSVNASNAAELFAQPDIDGALVGGASL   237
            VH+ +R + + V +A+   I YGGSV +   EL +QPD+DG LVGGASL
Sbjct 185   VHEKLRGWLKSNVSDAVAQSTRIIYGGSVTGATCKELASQPDVDGFLVGGASL   237
```

In the above alignment, the query is the *E. coli* sequence. The subject is the human sequence. The center line shows identical residues (letters) and similar residues (+). Residues preceding #6 or following #237 in the query do not align when the two species are compared. Note that the human sequence exhibits two one-residue deletions, one two-residue deletion (following Ile-151), and one insertion (lysine 194).

CHAPTER 7

1. The whale swims long distances between breaths. A high concentration of myoglobin in the whale muscle maintains a ready supply of oxygen for the muscle between breathing episodes.

2. (a) Convert μm^3 to cm^3 because $1\ cm^3$ is $1\ mL$. $1\ \mu m = 10^{-4}$ cm. $1\ \mu m^3$ is $10^{-12}\ cm^3$.
   $87\ \mu m^3 = 87 * 10^{-12}\ cm^3$. $0.34\ g\ mL^{-1} * 8.7 * 10^{-11}\ mL = \underline{2.96 * 10^{-11}\ g}$.

   (b) Using molecular weight of about $65{,}000\ g\ mol^{-1}$ for hemoglobin, the number of moles of hemoglobin is about $(2.96 * 10^{-11}\ g)/(65{,}000\ g\ mol^{-1}) = 4.55 * 10^{-16}$ mol.

   Multiplying by Avogadro's number gives

   $(4.55 * 10^{-16}\ mol) * (6.02 * 10^{23}$ molecules $mol^{-1}) = \underline{2.74 * 10^8\ molecules}$.

   (c) 65 Å is $65 * 10^{-8}$ cm, or $6.5 * 10^{-7}$ cm.

   The volume of a cube that is $(65$ Å$)^3$ is $(6.5 * 10^{-7}\ cm)^3 = 2.75 * 10^{-19}\ cm^{-3}$.

   The number of these cubes that could fit into a cell volume of $8.7 * 10^{-11}\ cm^3$ (see part a, above) is about $(8.7 * 10^{-11}\ cm^3)/(2.75 * 10^{-19}\ cm^3) = 3.17 * 10^8$, which is only slightly greater than the number of molecules in the answer to part b, above. The hemoglobin concentration in red cells could <u>not</u> be much higher.

3. The total blood volume is about $(70\ kg)(70\ mL\ kg^{-1}) = 4900\ mL$.

   The total hemoglobin is about $(0.16\ g\ mL^{-1})(4900\ mL) = 784\ g$.

   One mole of hemoglobin ($\sim 65{,}000\ g\ mol^{-1}$) contains $(4\ mol)(55.8\ g\ mol^{-1})$ iron,

   or about $223\ g$ iron. The mass % iron in hemoglobin is therefore about

   $100\% * (223\ g)/(65000\ g) = 0.34\%$ iron.

   So the total iron in the 70-kg adult is about $(0.0034)(784\ g) = \underline{2.7\ g}$.

4. (a) The molecular weight of myoglobin is about $17{,}800\ g\ mol^{-1}$. Each myoglobin molecule binds one O_2 molecule. Therefore, the number of moles of myoglobin or O_2 per kg is:

   $(8\ g)/(17{,}800\ g\ mol^{-1}) = \underline{4.49 * 10^{-4}\ mol\ of\ O_2}$ per kg of human muscle. In grams, the answer is $(16\ g\ mol^{-1})(4.49 * 10^{-4}\ mol) = \underline{1.44 * 10^{-2}\ g\ O_2}$ per kg human muscle.

   The sperm whale muscle has 10-fold more oxygen, namely $\underline{4.49 * 10^{-3}}$ mol or $\underline{1.44 * 10^{-1}\ g\ O_2}$ per kg whale muscle.

   (b) The units of $(mol\ kg^{-1})$ are approximately equal to $(mol\ L^{-1})$ (or molar). Therefore, the ratio of bound to free oxygen is about $(4.49 * 10^{-3}\ mol\ O_2)/(3.5 * 10^{-5}\ mol\ O_2) = \underline{128}$.

5. (a) The pKa is lowered because the positive charge on the nearby lysine side chain will favor having a negative charge on the glutamate.
   (b) The pKa is raised because the negative charge on the nearby carboxyl will oppose the prospect of having a negative charge on the glutamate.
   (c) The pKa is raised because the nonpolar environment of the protein interior will favor the neutral state of the glutamic acid side chain.

6. Deoxy Hb A contains a complementary site (Phe 85 and Val 88 of the β chain), and so it can add on to a fiber of deoxy Hb S. The fiber cannot then grow further because the terminal deoxy Hb A molecule lacks the Val 6 sticky patch on the β chain.

7. $Y = pO_2{}^n / (pO_2{}^n + P_{50}{}^n)$. $P_{50} = 26$ torr. $n = 2.8$

    $Y_{lung} = $ when $pO_2 = 75$ torr is 95.1%.

    $Y_{tissue} = $ when $pO_2 = 20$ torr is 32.4%

    The difference: $Y_{lung} - Y_{tissue} = 62.7\%$ oxygen-carrying capacity.

8. A higher concentration of BPG will shift the oxygen-binding curve to the right, causing an increase in P_{50}. The larger value of P_{50} will promote dissociation of oxygen in the tissues, and will thereby increase the percentage of oxygen that is delivered to the tissues.

9. It appears that oxygen binding will cause the copper ions and their associated histidine ligands to move closer to each other, thereby also moving the helices to which the histidines are attached (in similar fashion to the conformational change in hemoglobin).

10. The modified hemoglobin should not show cooperativity. Although the imidazole in solution will bind to the heme iron (in place of histidine) and will facilitate oxygen binding, the imidazole lacks the crucial connection to the particular α-helix that must move in order to transmit the change in conformation.

11. Answer c. Inositol pentaphosphate provides a high concentration of negatively charged phosphate groups, similar chemistry as 2,3-BPG (with an even larger number of phosphate groups).

12. Hill method with $n = 1.8$ and $P_{50} = 10$ torr:

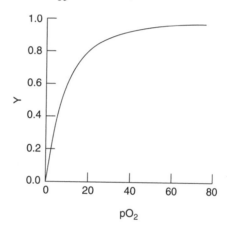

    Concerted model with $n = 2$, $L = 10{,}000$, $c = 0.01$, and $K_R = 10$ torr:

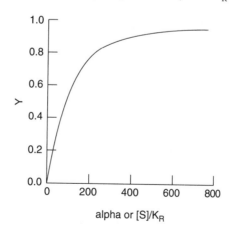

(*Note:* The resulting graphs are quite similar. The graph using the Hill method shows very slight evidence for cooperativity.)

13. Release of acid will lower the pH. At lower pH, oxygen dissociation in the tissues is promoted. The influence of low pH on oxygen dissociation is more pronounced than the rather small effect on oxygen binding in the lungs (see Figure 7.19 in the text), such that the efficiency of oxygen-carrying capacity of the red blood cells will increase. However, the enhanced release of oxygen in the tissues (lower Y when pO_2 is 20 torr in Figure 7.19 in the text) will increase the concentration of deoxy-Hb, thereby increasing the likelihood that the cells will sickle.

14. (a) $Y = 0.5$ when $pO_2 = 10$ torr. The graph below appears to indicate little or no cooperativity.

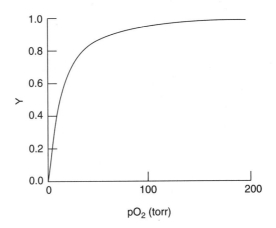

(b) The Hill plot (below) has an overall slope of about 1.17, with a slightly greater slope of about 1.3 in the central region. Therefore, $n \cong 1.3$. A value of n greater than one is evidence for cooperativity.

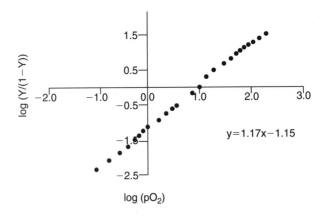

(c) Deoxy dimers of lamprey hemoglobin could have lower affinity for oxygen than do the monomers. If the binding of the first oxygen to a dimer causes dissociation of the dimer to give two monomers, then the process would be cooperative. In this mechanism, oxygen binding to each monomer would be easier than binding the first oxygen to a deoxy dimer.

15. (a) Curve 2, because increasing concentrations of carbon dioxide decrease the oxygen affinity, whereas decreasing concentrations of carbon dioxide increase the oxygen affinity (see the Bohr effect). (The O_2 binding remains somewhat cooperative, so curve 2 is favored over curve 1 for this answer.)

(b) Curve 4, because BPG decreases the oxygen affinity.

(c) Curve 2, increasing the pH will <u>decrease</u> the concentration of H^+ and increase the oxygen affinity. The effects of CO_2 and H^+ are similar (see the Bohr effect).

(d) Curve 1, because loss of quaternary structure will remove the cooperative behavior and cause the binding curve to resemble that of myoglobin.

16. The electrostatic interactions between BPG and hemoglobin would be weakened by competition with water molecules. The T state would not be stabilized. The oxygen affinity would increase and the binding would become less cooperative, making the release of oxygen to tissues more difficult.

CHAPTER 8

1. Enzymes enhance the rates of chemical reactions and are highly specific for their substrates.

2. A cofactor is needed. When the necessary cofactor binds to the apoenzyme, the holoenzyme is formed.

3. Coenzymes and metals are the main types of cofactors.

4. Vitamins are chemical precursors for coenzymes. Once ingested, vitamins are converted into coenzymes in the host organism.

5. Enzymes lower the energy requirement for forming the transition state during a reaction.

6. The intricate three-dimensional structure of an enzyme active site will recognize only specific substrates and will reject closely similar alternative molecules.

7. The activation energy that is required to reach a particular transition state is recovered when the transition state proceeds to the formation of products.

8. Even though energy is released when proteins are hydrolyzed, the proteins can nevertheless be quite stable if there is a large activation energy for the hydrolysis reaction. Protein synthesis must require energy to produce the high-energy protein products from the amino acid precursors.

9. Lysozyme helps to protect the fluid that surrounds eyes from bacterial infection.

10. Transition states are unstable. Consequently, molecules that resemble transition states are likely to be unstable and, hence, challenging to synthesize.

11. Use the equation: $\Delta G^{\circ\prime} = -RT \ln (K'_{eq})$.

R is $8.315 * 10^{-3}$ kJ mol^{-1} K^{-1}.   T is 298° K.

	K'_{eq}	$\ln K'_{eq}$	$G^{\circ\prime}$ (kJ mol^{-1})
a	1.0	0.00	0.0
b	10^{-5}	-11.51	28.5
c	10^4	9.21	-22.8
d	100	4.61	-11.4
e	0.1	-2.30	5.7

12. (a) The reaction will proceed to only a small extent.

Use $\Delta G^{\circ\prime} = -RT \ln K'_{eq}$, with R = 8.315 $*$ 10^{-3} kJ mol^{-1}°K^{-1} 1 and T = 298°K.

+7.5 kJ mol^{-1} = $-(8.315 \times 10^{-3}$ kJ^{-1} mol^{-1} K$^{-1})$(298 K)(ln [G1P]/[G6P]).

ln [G1P]/[G6P] = $-(7.5$ kJ mol$^{-1})/[(8.315 \times 10^{-3}$ kJ^{-1} mol^{-1} K$^{-1}) * (298$ K)]

= -3.03.

The ratio [G1P]/[G6P] = $e^{(-3.03)}$ = 0.0485.

Let x = [G1P]. Then [G6P] = (0.1 M $-$ x), and x / (0.1 M $-$ x) = 0.0485.

Solving, x = 0.0485(0.1 M $-$ x); and x = 0.00485 M $-$ 0.0485x.

1.0485x = 0.00485 M; and x = (0.00485 M)/1.0485 = 0.0046 M.

The final concentration of glucose-1-phosphate (x) is **0.0046 M**.

The final concentration of glucose-6-phosphate (0.1 $-$ x) is **0.0954 M**.

Consequently, the reaction proceeds only to a small extent as written.

(b) Supply G6P at a high rate and remove G1P at a high rate by other reactions. In other words, make sure that the [G6P]/[G1P] ratio is kept large.

13. K'_{eq} is (0.19/0.01) = **19**.

(T is 298 Kelvin. R is given in the answer to problem 11, above.)

$\Delta G^{\circ\prime} = -RT \ln (K'_{eq}) = -RT \ln(19) = $ **-7.30 kJ mol^{-1}** (or -1.74 kcal mol^{-1}).

14. The three-dimensional structure of an enzyme is stabilized by interactions with bound substrates, reaction intermediates, or products. This stabilization due to a bound ligand minimizes the thermal denaturation.

15. When the substrate concentration is close to the value of K_M, the enzyme displays significant catalytic activity yet also remains sensitive to changes in the substrate concentration.

16. Use the relation:

$$V_0 = \frac{V_{max}\ [S]}{K_M + [S]} \qquad \text{giving:} \qquad \frac{V_0}{V_{max}} = \frac{[S]}{K_M + [S]}$$

When [S] = K_M, (V_0 / V_{max}) = 0.50.

When [S] = 10 K_M, (V_0 / V_{max}) = 0.91.

When [S] = 20 K_M, (V_0 / V_{max}) = 0.95.

The ratio (V_0 / V_{max}) is approaching 1.0, but it never reaches 1.0.

17. The values for $\Delta G^{\circ\prime}$ are found by substituting the values for K'_{eq} into equation 3 on page 236 of the text.

(a) $\Delta G^{\circ\prime} = -RT \ln K'_{eq}$

= $-8.31 \times 298 \ln (1.5 \times 10^4)$

= -20.4 kJ/mol (-5.7kcal/mol)

(b) -1.00 kJ/mol (-0.24 kcal/mol)

(c) $+4.60$ kJ/mol ($+1.1$ kcal/mol)

(d) $+21.76$ kJ/mol ($+5.2$ kcal/mol)

18. For (a) and (b), proper graphing of the data given will provide the correct answers:

$K_M = 5.2 \times 10^{-6}$ M; $V_{max} = 6.84 \times 10^{-10}$ mol/min

(c) Turnover = mol S s^{-1}/mol E = $(6.84 \times 10^{-10})/[(60 \times 10^{-9})/29{,}600] = 337$ min^{-1}

19. Penicillinase, like glycopeptide transpeptidase, forms an acyl-enzyme intermediate with its substrate but transfers it to water rather than to the terminal glycine of the pentaglycine bridge.

20. For (a) and (b), proper graphing of the data given will provide the correct answers:

(a) V_{max} is 9.5 µmol/min. K_M is 1.1×10^{-5} M, the same as without inhibitor.

(b) Since K_M does not change, this is noncompetitive inhibition.

(c) To answer this question you need to obtain a value for V_{max} (47.6 µmol/min) from the graphs used in question 4(a) above. Because this is noncompetitive inhibition, use the equation on page 240 of the text as follows: 9.5 µmol/min = 47.6 µmol/min/ $(1 + 10^{-4}$ M/K_i). Solving for K_i one obtains the answer 2.5×10^{-5} M.

(d) Since an inhibitor does not affect K_M, the fraction of enzyme molecules binding substrate = [S]/(K_M + [S]), with or without inhibitor. For solution, see 4(e).

21. (a)

$$V = \frac{V_{max}[S]}{(K_M + [S])}$$

$$\frac{V(K_M + [S])}{[S]} = V_{max}$$

$$V + \frac{VK_M}{[S]} = V_{max}$$

$$V = V_{max} - \frac{VK_M}{[S]}$$

(b) The slope of a straight line is the constant that the x-coordinate is multiplied by in the equation for the straight line. Thus, in the Lineweaver-Burk plot, K_M/V_{max} is the slope; in the Eadie-Hofstee plot the slope is $-K_M$ because V/[S] is plotted on the x axis; see (a). By inspection, the y-intercept is V_{max}. The x-intercept is V_{max}/K_M because one is extrapolating to [S] = 0.

(c) Note that with a competitive inhibitor V_{max} (y-intercept) stays the same but K_M increases (the slope of 2 is greater than the slope of 1). In contrast, with a noncompetitive inhibitor, K_M does not change; 1 and 3 have the same slope (while V_{max} decreases).

22. The rates of utilization of A and B are given by equation 33 (page 234 in the text):

$$V_A = \left(\frac{k_2}{K_M}\right)_A [E][A]$$

and

$$V_B = \left(\frac{k_2}{K_M}\right)_B [E][B]$$

Hence, the ratio of these rates is

$$\frac{V_A}{V_B} = \frac{\left(\dfrac{k_2}{K_M}\right)_A [A]}{\left(\dfrac{k_2}{K_M}\right)_B [B]}$$

Thus, an enzyme discriminates between competing substrates on the basis of their values of k_2/K_M rather than of K_M alone. Note that the velocity is dependent on the constants (k_2/K_M) *and* the concentrations of enzyme and substrate.

23. A tenfold change in the equilibrium constant corresponds to a standard free-energy change ($\Delta G^{o'}$) of RT ln 10 = 5.69 kJ/mol (1.36 kcal/mol). If a mutant enzyme binds a substrate, S, 100-fold as tightly as does the native enzyme, more Gibbs free energy of activation ($\Delta G^{\ddagger}$) is needed to convert S to S‡ (transition state). In fact, the $\Delta G^{\ddagger}$ is increased by 11.38 kJ/mol (2.72 kcal/mol; RT ln 100) and the velocity of the reaction will be slowed down by a factor of 100.

24. By substituting $[S] = 0.1*K_M$ into the Michaelis-Menten equation,
    $v = (V_{max})([S]) / ([S] + K_M)$, we can show that:
    $v = (V_{max})(0.1)(K_M) / ((0.1 + 1.0)K_M)$, or $v = (1/11)V_{max}$.
    So with $v = 1.0$ µmol min^{-1}, $V_{max} = 11.0$ µmol min^{-1}.

25. (a) The information is necessary for determining the correct dose of succinylcholine to administer. The length of time for the paralysis to persist can be controlled by balancing the serum cholinesterase activity and the amount of succinylcholine administered.
    (b) The duration of the paralysis depends on the ability of the serum cholinesterase to clear the drug. If there were one-eighth the amount of enzyme activity, paralysis could last up to eight times longer.
    (c) K_M is the substrate concentration needed for the enzyme to perform catalysis at a rate that is $\frac{1}{2}V_{max}$. Consequently, for a given concentration of substrate, the reaction rate will be lower when K_M is higher. The patient having the mutant enzyme with the higher K_M will therefore clear the drug at a much slower rate, thereby causing the paralysis to last longer for a given dose of succinylcholine.

26. K_M will remain the same (center graph), and the apparent V_{max} will change with the different amounts of enzyme (y-intercept in center graph). Therefore, the correct answer is the center graph.

27. (a) The double-reciprocal plot will turn up to form a second line near the 1/v axis, giving an approximately "V-shaped" graph.
    (b) The decrease in reaction velocity at high substrate concentration could be due to an allosteric inhibition by substrate at a second binding site. The binding affinity of the second (allosteric) site for substrate could be lower than the affinity of the catalytic site for substrate.

28. The step catalyzed by E_A will be rate-limiting because the actual substrate concentration (10^{-4} M) is much *less* than that needed to achieve half-maximal reaction velocity (10^{-2} M) for this step.

29. The fluorescence spectroscopy reveals the existence of an enzyme-serine complex that has unique spectral properties (high quantum yield for PLP fluorescence). The existence of a second unique complex, enzyme-serine-indole, is also demonstrated by not only the reversal but also the further decrease in the intensity of the fluorescence emission.

30. The mechanism suggests that H^+ is behaving as a competitive inhibitor. Therefore, at sufficiently high substrate concentration, the substrate will overcome the inhibition, and the velocity, v_o, will equal V_{max}, independent of pH (part a). At a low (constant) substrate concentration, the observed v_o will follow a titration curve with a pK of 6.0 (parts b, c).

(a)

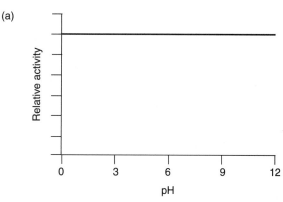

(b)

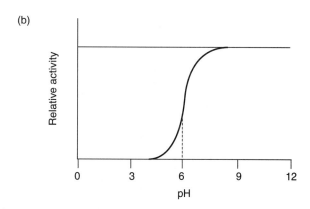

31. (a) The enzyme is unstable at 37°C. It unfolds or denatures as a function of time of storage at 37°C.

   (b) The PLP coenzyme partially protects the enzyme against the thermal unfolding. When PLP is bound to the enzyme, the rate of denaturation is slower.

CHAPTER 9

1. The answer concerns the different kinetic behavior of chymotrypsin toward amide and ester substrates. Substrate A is N-acetyl-L-phenylalanine *p*-nitrophenyl *amide*, rather than N-acetyl-L-phenylalanine *p*-nitrophenyl *ester* for which the initial "burst" activity was described in the text. The burst is observed if the first step of a reaction (in this case, acyl-chymotrypsin formation, together with release of *p*-nitrophenyl amine) is much faster than the second step (release of N-acetyl-phenylalanine and free chymotrypsin). With the amide substrate, however, the relative rates of the two steps are more nearly equal; therefore no burst is observed.

2. The Ala-64 subtilisin lacks the critical histidine in the catalytic triad of the active site and therefore cleaves most substrates much more slowly than does normal subtilisin. However, the histidine in substrate B can act as a general base; thereby the substrate itself partially compensates for the missing histidine on the mutant enzyme.

3. The statement is false (incorrect). Either mutation alone is such a serious impairment for the enzyme that the second mutation will be of little additional consequence.

4. A reasonable prediction is that the substrate specificity of the mutant protease would resemble that of trypsin. The mutant enzyme would be predicted to hydrolyze peptide bonds that follow either lysine or arginine in the sequence (i.e., peptide bonds whose carbonyl groups are from either lysine or arginine).

5. Small molecular buffers such as imidazole can diffuse into the active site of carbonic anhydrase and substitute for the proton shuttle function of His 64 near the zinc ion. Large molecular buffers cannot fit into the active site because of their steric bulk and therefore cannot compensate for the loss of the side chain of His 64.

6. The enzyme would *not* be useful. The probability of finding a particular unique restriction site of length 10 is $1/(4^{10}) = 1/1,048,576$, that is, about once per million base pairs of DNA. Therefore most viral genomes that contain only 50,000 base pairs have little statistical chance of having a site that would be recognized by this enzyme.

7. The increased rate would not be beneficial. Because only a small number of cuts (or even a single cut) of an invading foreign DNA molecule will be sufficient to inactivate the foreign DNA, the host cell would realize no practical benefit from a faster rate of hydrolysis. Specificity is more important than turnover number for restriction endonucleases.

8. In the absence of the gene for the corresponding methylase, there would be no benefit. The restriction endonuclease from the newly acquired gene would digest the host cell's own DNA.

9. (a) (Assuming that magnesium also is present), ATP and AMP will be generated from two molecules of ADP in a "reverse" adenylate kinase reaction. Enzymes catalyze both forward and reverse reactions.

   (b) The answer will require knowledge of an equilibrium constant for the reaction ATP + AMP $\rightleftharpoons$ 2 ADP. In this reaction, the two substrates together are approximately isoenergetic with the products. If one therefore assumes an equilibrium constant of one, then: $[ADP]^2/([ATP][AMP]) = 1$.

   Let $[ATP] = x$ at equilibrium. Then $[AMP] = x$, and $[ADP] = ((1\text{ mM}) - (2x))$.

   $(1 - 2x)^2/(x^2) = 1$.

   $(1 - 4x + 4x^2) = x^2$.

   $3x^2 - 4x + 1 = 0$.   Solve for x.

   Two answers emerge. Either $x = 1$, or $x = \frac{1}{3}$. However, $x = 1$ is physically unreasonable (impossible). Therefore $x = \frac{1}{3}$.

   The concentrations of [ATP], [ADP], and [AMP] therefore are all 0.3333 mM.

10. The chelator will remove the zinc from the active site of the enzyme. Without zinc, the carbonic anhydrase is inactive.

11. Aldehydes can react with one molecule of an alcohol to form a hemiacetal (see Chapter 11). Because the catalytic site of elastase contains an active serine hydroxyl group, it is reasonable that an aldehyde derivative of a peptide substrate of elastase would react with the serine -OH group to form a hemiacetal, which is a tetrahedral analogue of the transition state of the peptide hydrolysis reaction. (See also Robert C. Thompson and Carl A. Bauer. *Biochemistry* 18[1979]:1552–1558.)

12. The reaction is expected to be slower by a factor of 10 because the rate depends on the pK_a of the zinc-bound water. Therefore k_{cat} is estimated to be 60,000 s^{-1} at pH 6.0.

13. The EDTA binds the magnesium ions that are required for the reaction.

14. ATP hydrolysis is reversible within the active site. When ATP is hydrolyzed in the active site, ^{18}O from water is incorporated into ADP during the hydrolysis. Some ATP is then re-formed from ADP within the active site, and the labeled ATP is released back into solution.

15. If the aspartate were to be mutated, the protease would become inactive and the virus would not be viable. The mutations that confer the resistance must occur at nearby residues that contribute to the drug binding.

16. Apparently, the new water molecule can substitute for the hydroxyl group of serine 236 to mediate proton transfer from the attacking water molecule and to the γ-phosphoryl group of ATP.

17. (a)

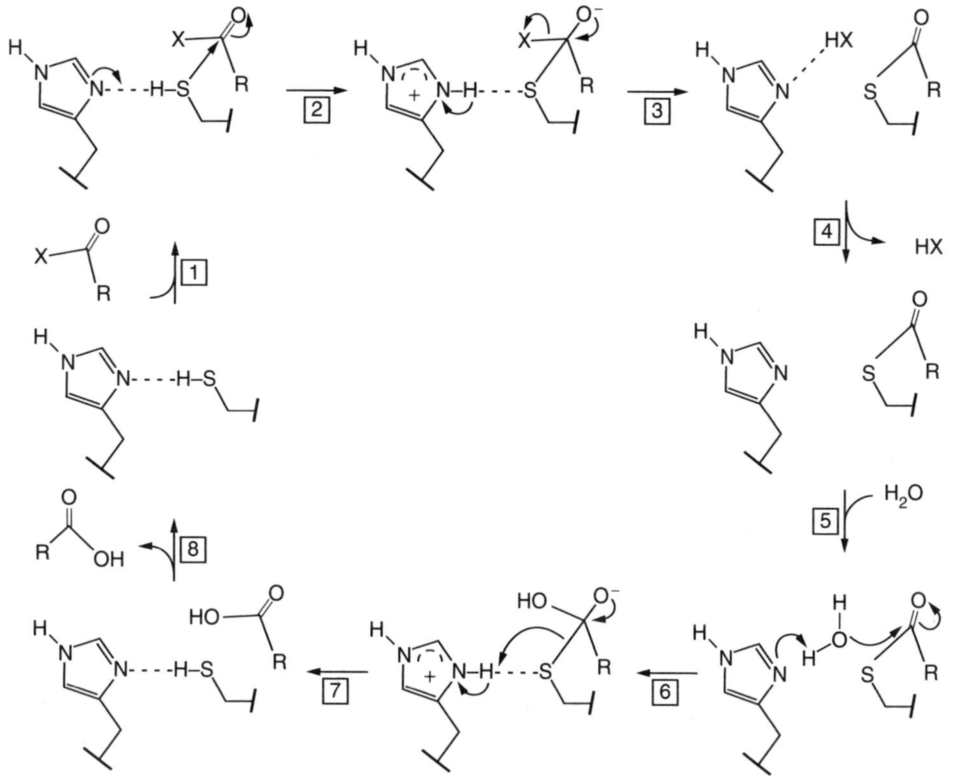

(b)

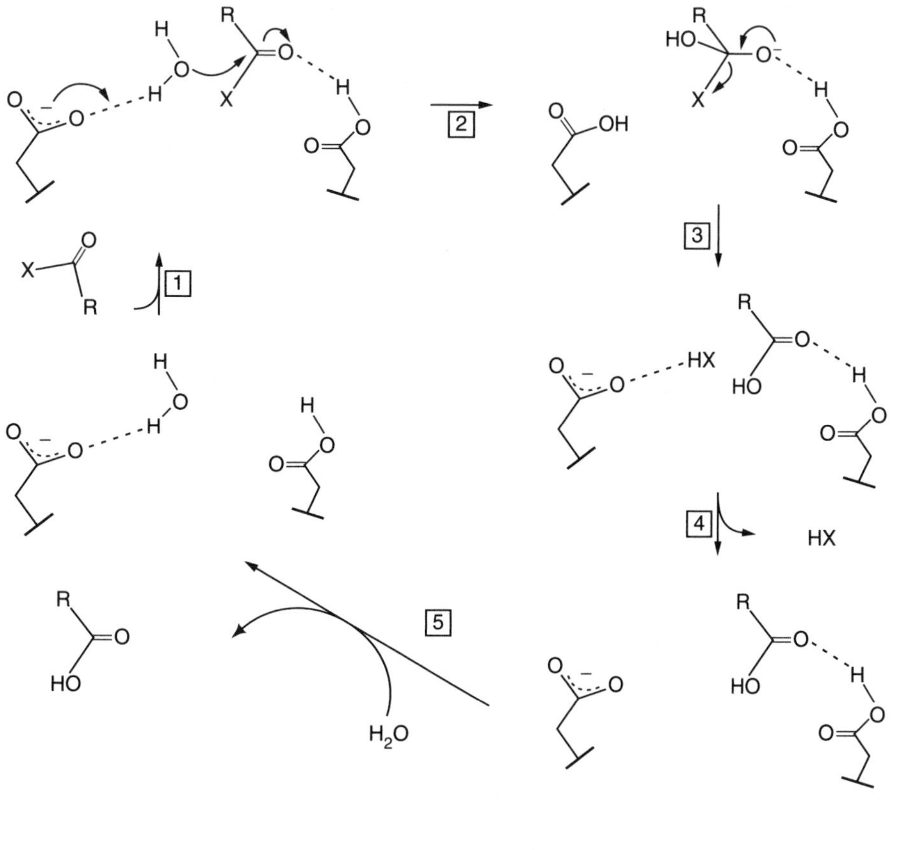

(c)

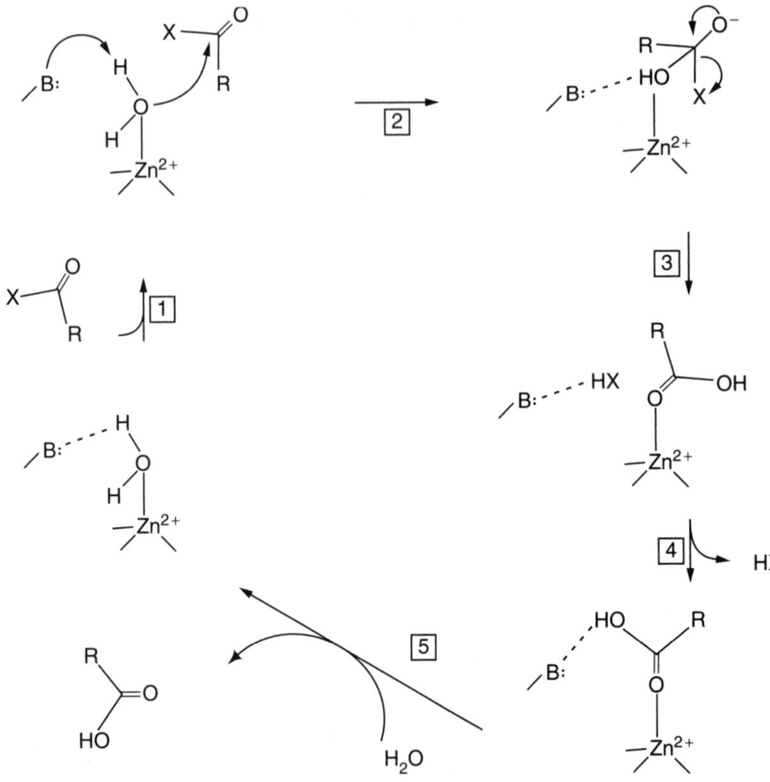

CHAPTER 10

1. The enzyme catalyzes the first step in the synthesis of pyrimidines. It facilitates the condensation of carbamoyl phosphate and aspartate to form *N*-carbamoylaspartate with release of inorganic phosphate.

2. Since histidine 134 is thought to stabilize the negative charge on the carbonyl oxygen in the transition state (Figure 10.7), the protonated imidazole ring (which carries a positive charge) must be the active species. That being the case, the enzyme velocity V should be half of V_{max} at a pH of 6.5 (the pK of an unperturbed histidine side chain in a protein). Raising the pH above 6.5 will remove protons from the imidazole ring, thus causing a decrease in V; lowering the pH below 6.5 will have a reverse effect.

3. Feedback inhibition is inhibition of an allosteric enzyme by the end product of the pathway controlled by the enzyme. This type of inhibition prevents the production of too much end product and the accompanying unnecessary consumption of substrates when the product is not required.

4. High concentrations of ATP might signal two overlapping situations. The high levels of ATP might suggest that some *nucleotides* are available for nucleic acid synthesis, and consequently, CTP should be synthesized. The high levels of ATP furthermore indicate that *energy* is available for nucleic acid synthesis, and this feature also indicates that CTP should be produced.

5. All of the enzyme would be in the R form all of the time. There would be no cooperativity. The mutant enzyme would be expected to obey Michaelis-Menten kinetics.

6. Similar to the case in problem 5, this mutant enzyme also would be essentially always in the R state. The mutant enzyme would show simple Michaelis-Menten kinetics.

7. CTP is formed by the addition of an amino group to UTP. It is plausible that UTP also would be capable of inhibiting ATCase.

8. Homotropic effectors are the substrates of allosteric enzymes, whereas heterotropic effectors are the regulators of allosteric enzymes. Therefore, homotropic effectors account for the sigmoidal nature of the velocity versus substrate concentration curve. Heterotropic effectors alter the midpoint of the curve (K_M value). Ultimately, both types of effectors exert their effects by altering the T/R ratio.

9. The reconstitution and self assembly show that the complex quaternary structure and the resulting catalytic and regulatory properties are ultimately encoded in the primary structures of the individual component subunits.

10. If substrates had been used, both the soluble and crystalline forms of the enzyme would catalyze the reaction. Intermediates would not accumulate on the enzyme or in the crystals. Substrates and products would diffuse away. Consequently, the crystalline enzyme would have been free of substrates or products.

11. (a) One can show that the change in [R]/[T] is the same as the ratio of the substrate affinities of the two forms. For example, the mathematical constant for the conversion of R to T_S is the same whether one proceeds R → T → T_S → or R → R_S →T_S. Let us assume that the constant for the conversion of R to T and R to R_S is 10^3. Since the affinity of R for S is 100 times that of T for S, it follows that the constant for the conversion of T to T_S is 10. The constant for the conversion of R_S to T_S is therefore equal to $10^3 \times 10/10^3$, or 10. Note that the binding of substrate with a hundred-fold tighter binding to R changes the R to T ratio from 1/1000 to 1/10.

    (b) Since the binding of *one* substrate molecule changes the [R]/[T] by a factor of 100, the binding of *four* substrate molecules will change the [R]/[T] by a factor of $100^4 = 10^8$. If [R]/[T] in the absence of substrate = 10^{-7}, the ratio in the fully liganded molecule will be $10^8 \times 10^{-7} = 10$.

12. Following the nomenclature in Section 10.1.5, L is [T]/[R], the ratio of T to R in the absence of ligand, and L = 10^5. With j ligands bound, a new L_j will equal $(K_R/K_T)^j*L$. Then we have $[R]L_j = [T]$. The fraction of molecules in the R state therefore is [R]/([R] + [T]), or [R]/([R] + [R]L_j), or $1/(1 + L_j)$. Now we can set up a table, using L = 10^5, $K_R = 5 \times 10^{-6}$ M, $K_T = 2 \times 10^{-3}$ M, and j from 0 to 4:

Ligands bound (j)	$L_j = (K_R/K_T)^j*L$	Fraction R = $1/(1 + L_j)$
0	$1.0 * 10^5$	10^{-5}
1	250	0.004
2	0.625	0.615
3	$1.6 * 10^{-3}$	0.998
4	$3.9 * 10^{-6}$	1.000

13. The concerted model, in contrast with the induced-fit (sequential) model, cannot account for negative cooperativity because, according to this model, the binding of substrate promotes a conformational transition of all subunits to the high-affinity R state. Hence, homotropic allosteric interactions must be cooperative if the concerted model holds. In the sequential (induced-fit) model, the binding of ligand changes the conformation of the subunit to which it is bound but not that of its neighbors. This conformational change in one subunit can *increase* or *decrease* the binding affinity of other subunits in the same molecule and can, therefore, account for negative cooperativity.

14. The binding of PALA switches ATCase from the T to the R state because it acts as a substrate analog. An enzyme molecule containing bound PALA has fewer free catalytic sites than does an unoccupied enzyme molecule. However, the PALA-containing enzyme will be in the R state and hence have higher affinity for the substrates. The dependence of the degree of activation on the concentration of PALA is a complex function of the allosteric constant L_0, and of the binding affinities of the R and T states for the analog and the substrates. For an account of this experiment, see J. Foote and H. K. Schachman, *J. Mol. Biol.* 186(1985):175.

15. The net outcome of the two reactions is the hydrolysis of ATP to ADP and P_i. ATP has a standard free energy of hydrolysis, $\Delta G^{o\prime}$, of -30.5 kJ mol^{-1} (-7.3 kcal mol^{-1}). Under typical cellular conditions, ΔG is approximately -50 kJ mol^{-1} (-12 kcal mol^{-1}). (See page 300 of the text.)

16. An isozyme is one member of a set of homologous enzymes within a single organism that catalyze the same reaction but differ slightly in structure and in their catalytic and regulatory properties.

17. Different isozymes can function at distinct locations or at different times to meet the specific physiological needs of particular tissues at particular times.

18. (a) 7; (b) 8; (c) 11; (d) 6; (e) 1; (f) 12; (g) 3; (h) 4; (i) 5; (j) 2; (k) 10; (l) 9.

19. When phosphorylation takes place at the expense of ATP, considerable energy is released. If captured, this energy could be used to alter the structure and hence the activity of a protein in dramatic fashion. Moreover, because ATP is the cellular energy currency, protein phosphorylation is linked to the energy status of the cell.

20. Covalent modifications are reversible, whereas proteolytic cleavage is irreversible.

21. The activation of zymogens involves the cleavage of one or more peptide bonds. In the case of pepsinogen, when the catalytic site is exposed by lowering the pH, it hydrolyzes the peptide bond between the percursor and pepsin moities. Note that this activation is *autocatalytic*. Therefore, the time required for activation of half the pepsinogen molecules is independent of the total number of the molecules present.

22. With failure to activate both trypsin and the subsequent chymotrypsin (which should be activated by trypsin), protein digestion to produce peptides and essential amino acids will be severely impaired. Protein (amino acid) deficiency and associated nutritional problems will be expected.

23. Because of its ring structure, the imino acid proline cannot be accommodated in the substrate binding sites of trypsin, chymotrypsin, or carboxypeptidase A. Therefore, these proteases fail to cleave peptide bonds involving proline.

24. The function of Factor X is to convert prothrombin to thrombin on phospholipid membranes derived from blood platelets. This proteolytic activation removes the amino-terminal fragment of prothrombin, which contains Ca^{+2}-binding sites, and releases thrombin to activate fibrinogen. Meanwhile, Factor X remains bound to the platelet membrane, where it can activate other prothrombin molecules, because during activation it retains the Ca^{+2}-binding γ-carboxyglutamate residues.

25. Apparently antithrombin III is a very poor substrate for thrombin. Remember, many enzyme inhibitors have high affinity for active sites. Thrombin, not prothrombin, can react with antithrombin III because it has available a fully formed active site.

26. One needs to remember α-helical coiled coils, introduced in Chapter 2 of your text-book (page 38). Examination of Figure 2.38 (page 43) suggests that near the axis of the superhelix some amino acid residues are located in the interior (hydrophobic) portion of the molecule. Since this is a long molecule made up of repeating units, one would expect to have hydrophobic side chains at regular intervals in the molecule.

27. Methionine 358 has a side chain that not only is essential for the binding of elastase by α₁-antitrypsin but also is most susceptible to oxidation by cigarette smoke. What is needed is a side chain resistant to oxidation yet having a strong binding affinity for elastase. A likely choice would be leucine, the side chain of which is much more stable than that of methionine but which has nearly the same volume and is very hydrophobic.

28. Inappropriate clot formation could block arteries in the brain, causing a stroke, or the heart, causing a heart attack.

29. Tissue-type plasminogen activator, or TPA, is a serine protease that leads to the dissolution of blood clots. TPA activates plasminogen, a zymogen that is bound to a fibrin clot, converting it into the active enzyme plasmin. Plasmin itself is also a serine protease, which then hydrolyzes the fibrin of the clot.

30. A mature clot is stabilized by amide linkages between the side chains of lysine and glutamine that are absent in a soft clot. The side-chain amide linkages are sometimes called "isopeptide" bonds and are formed by the enzyme transglutaminase.

31. The concerted model (in which all subunits change conformation in response to the first instance of substrate binding) predicts that the change in f_R should *precede* the change in Y. By contrast, the sequential model predicts that the fraction of subunits in the R state (f_R) should equal the fraction containing bound substrate (Y). The results in the figure therefore are best explained by the *concerted* model.

32. As in problem 14, this experiment also supports a concerted mechanism. The change in the absorbance at 430 nm reports a conformational change in response to substrate binding at a distant site (on another trimer). (Substrate is prevented from binding to the same trimer that reports the 430 nm absorbance change.) Thus, the binding of succinate to the active sites of a native trimer alters the structure of a different trimer (that carries the reporter nitrotyrosine group).

33. The binding of ATP to the regulatory subunits produces the same absorbance change at 430 nm as did substrate binding in problem 17. ATP therefore is an allosteric activator that drives the catalytic subunits into the active conformation (R state). CTP has a converse or opposite effect, driving the catalytic subunits into an inactive conformation (T state) and decreasing the absorbance at 430 nm.

34. In the R state, ATCase expands and becomes less dense. The decrease in density results in a decrease in the sedimentation value because the sedimentation coefficient s depends on the density ρ, according to the formula: $s = m(1 - \bar{v}\rho)/f$. (See page 78 of the text.)

35. The interaction between trypsin and the inhibitor is so stable that the transition state is rarely formed. Recall that maximal binding energy is released when an enzyme binds to the transition state. If the substrate-enzyme interaction is too stable, there is little driving force for forming a transition state, and the transition state will rarely form.

36. In step 1, the aspartate amino group carries out a nucleophilic attack on the carbonyl carbon of the carbamoyl phosphate to give a tetrahedral transition state. The histidine in the active site can stabilize the negatively charged oxyanion of this transition state. In step 2, phosphate is the leaving group to generate the N-carbamoylaspartate.

37. The reaction is equivalent to a "hydrolysis" (or transfer) of the γ-phosphate of ATP, with the serine-OH group taking the role of water and accepting the phosphate. The enzyme's active site will need a group to accept the proton from the serine oxygen during the attack on the γ-phosphate in step 1. (Histidine plays such a role in the serine proteases (e.g., trypsin and chymotrypsin) and could play a similar role here.) Another valuable functional group at the active site would be one that could stabilize the extra negative charge on the pentacoordinate phosphate intermediate between steps 1 and 2 (before ADP is lost as the leaving group in step 2).

CHAPTER 11

1. Although it can be risky business, chemists have always tried to gain some insight into molecular structure from knowledge of the empirical formula. Since the empirical formula for carbohydrates is $(CH_2O)_n$, it is not surprising that in the latter half of the nineteenth century the name *carbohydrate* was coined.

2. To begin, there are six different $((3)*(2)*(1))$ ways to specify the order of monosaccharide units. Then the first glycosidic bond can join the first two monomers in any of 2^5 or 32 ways, α or β from the C1 oxygen of the first sugar to OH #2, 3, 4, or 6 of the second sugar. Finally, the second glycosidic bond can join the second and third monomers in any of 2^6 or 64 ways, α or β from the C1 oxygen of the second sugar to OH #1 (nonreducing) or #2, 3, 4, or 6 of the third sugar. Putting this all together, one has $(6)*(32)*(64)$ or 12,288 possible trisaccharides.

   For tripeptides, there are only 6 different sequences that use exactly one each of three different amino acids $((3)*(2)*(1) = 6)$.

3. To answer this problem, one must know the structures of the molecules in question and a couple of definitions. By definition, *epimers* are a pair of molecules that differ from each other only in their configuration at a single asymmetric center. *Anomers* are special epimers that differ only in their configuration at a carbonyl carbon; hence, they are usually acetals or hemiacetals. An aldose-ketose pair is obvious. Inspection of Fischer representations of the molecular pairs leads to the conclusion that (a), (c), and (e) are aldose-ketose pairs; (b) and (f) are epimers; and (e) are anomers.

4. Erythrose: tetrose aldose; Ribose: pentose aldose; Glyceraldehyde: triose aldose; Dihydroxyacetone: triose ketose; Erythrulose: tetrose ketose; Ribulose: pentose ketose; Fructose: hexose ketose.

5.

D-Allose   D-Altrose   D-Mannose

D-Gulose   D-Idose   D-Galactose

D-Talose

6. The reason the specific rotation of α-D-glucopyranose changes after it is dissolved in water is that the ring form is in equilibrium with a small amount of the straight-chain form of glucose. The straight-chain form then converts to either α-D-glucopyranose or β-D-glucopyranose. This process, called *mutarotation*, continues until after 1–2 hours a thermodynamically stable mixture of the α and β anomers is obtained. Its specific rotation is 52.7°. The difference in the specific rotations of the two anomers is 93.3° $(112° - 18.7°)$, and the difference between the equilibrium value and that of the β anomer is 34° $(52.7° - 18.7°)$. Since the optical rotation of the equilibrium mixture is closer to that of the β anomer than it is to that of the α anomer, obviously more than half the equilibrium mixture is in the β configuration. The fraction present in the α configuration is $34° ÷ 93.3° = 0.36$. The fraction in the β configuration is $1 - 0.36 = 0.64$.

7. Glucose reacts slowly because the predominant hemiacetal ring form (which is inactive) is in equilibrium with the active straight-chain free aldehyde. The latter can react with terminal amino groups to form a Schiff base, which can then rearrange to the stable amino ketone, sometimes referred to as Hb A_{Ic}, which accounts for approximately 3% to 5% of the hemoglobin in normal adult human red cells. In the diabetic, its concentration may rise 6% to 15% owing to the elevated concentrations of glucose.

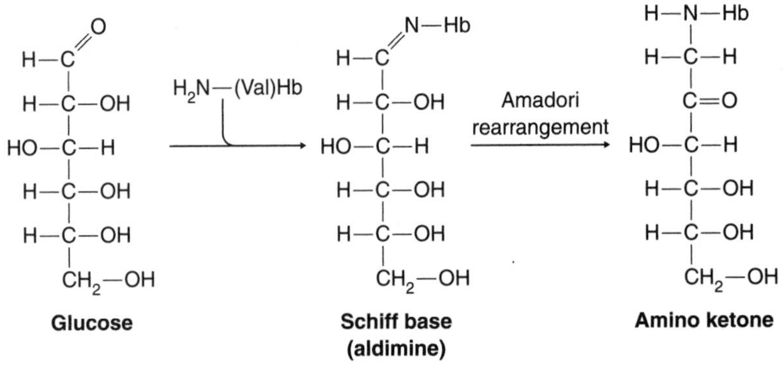

Glucose   **Schiff base (aldimine)**   **Amino ketone**

8. Whereas pyranosides have a series of three adjacent hydroxyls, furanosides have only two. Therefore, oxidation of pyranosides uses *two* equivalents of periodate and yields *one* mole of formic acid, whereas oxidation of furanosides uses only *one* equivalent of periodate and yields *no* formic acid.

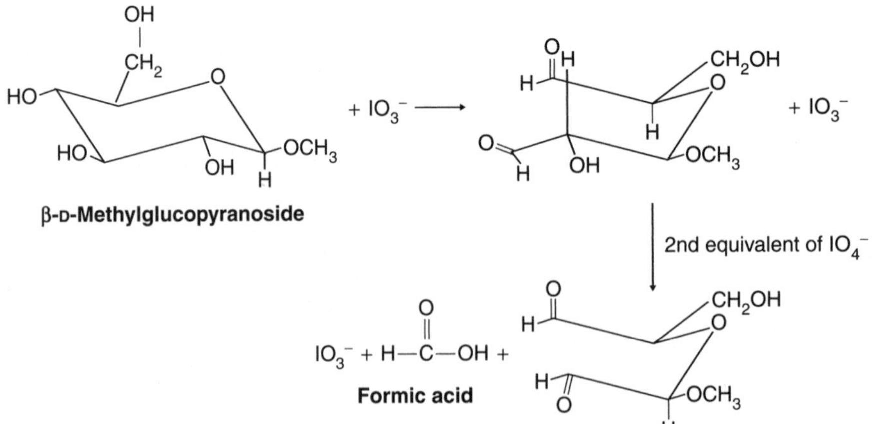

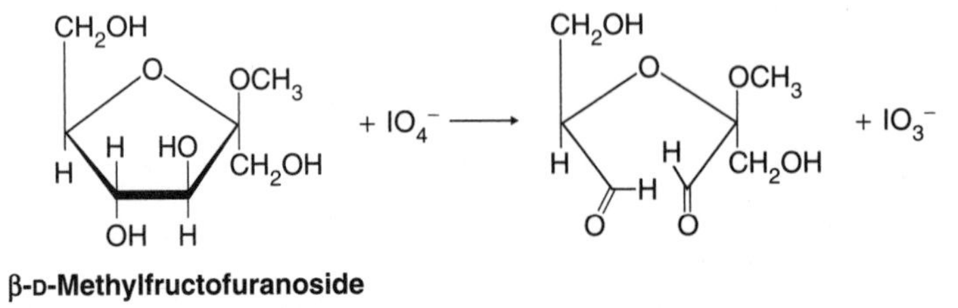

β-D-Methylglucopyranoside

Formic acid

2nd equivalent of IO_4^-

β-D-Methylfructofuranoside

9. The formation of acetals (such as methylglucoside) is acid-catalyzed. In a mechanism similar to that of the esterification of carboxylic acids (shown in most organic chemistry texts), the anomeric hydroxyl group is replaced. The resulting carbocation is susceptible to attack by the nucleophilic oxygen of methanol, leading to the incorporation of this oxygen into the methylglucoside molecule.

D-Glucose
(β-pyranose form)

Electron pair on ring oxygen
can stabilize carbocation at
anomeric position only

10. By inspection, A, B, and D are the pyranosyl forms of D-aldohexoses because the CH$_2$OH is above the plane of the ring. In Haworth projections, OH's above the ring are to the left in Fischer projections, and those below the ring are to the right. Therefore, A is β-D-mannose, B is β-D-galactose, and D is β-D-glucosamine. By similar use of the Haworth projection, C can be identified as β-D-fructose. All these sugars are β because, in Haworth projections, when the CH$_2$OH attached to the C-5 carbon (the carbon that determines whether the sugar is D or L) is above the ring, if the anomeric hydroxyl is also above the ring, the sugar is β.

11. The trisaccharide itself should be a competitive inhibitor of cell adhesion if the trisaccharide unit of the glycoprotein is critical for the interaction.

12. The nonreducing carbon-1 oxygens cannot be methylated, whereas the carbon-1 hydroxyls at the reducing ends can be methylated. Conversely, most of the carbon-6 hydroxyls can be methylated, but not at the branch points. Therefore, the ratio of methylated to nonmethylated C-1 hydroxyls in the final digestion mixture will indicate the relative proportion of reducing ends. Likewise, the ratio of *non*methylated to methylated C-6 hydroxyls in the digestion mixture will indicate the relative proportion of branch points.

13. (a) No. There is no hemiacetal linkage in raffinose, but rather two acetal linkages.
    (b) galactose, glucose, and fructose.
    (c) Galactose and sucrose. (After digestion, the released galactose—in water solution—will establish an equilibrium among the α, β, and open-chain forms. See also the answer to problem 14, below.)

14. The hemiacetal of the α anomer opens in water to give the open-chain aldehyde/alcohol form. The open form then can reclose the ring with either the α or β configuration. In water solution, an equilibrium will be established among the β anomer, the α anomer, and a small amount of the open-chain form, through which the two pyranose ring forms interconvert.

α-D-mannose   open form   β-D-mannose

15. The sweeter β-D-pyranose form of fructose will convert via an open-chain form to the more stable α-D-furanose form in water solution. This process will be accelerated by heat. (The β-D-furanose and α-D-pyranose forms are also accessible from the open-chain molecule.)

β-D-fructopyranose   Open-chain (rotameric) forms   α-D-fructofuranose
(sweeter form)   (more stable form)

16. (a) Regardless of its length or the number of branches, a glycogen molecule will have only *one* reducing end. (All of the glycosidic—acetal—linkages are either 1 → 4 or 1 → 6; therefore, only one C-1 hydroxyl group will not be involved in a glycosidic linkage to either C-4 or C-6.) By contrast, each branch of a glycogen molecule will terminate with a nonreducing end. The number of nonreducing ends, therefore, will be one more than the number of branch points.

(b) For rapid mobilization of glucose monomers, most metabolism should be predicted to take place at the nonreducing ends.

17. No, sucrose is not a reducing sugar. The anomeric carbon atom acts as the reducing agent in both glucose and fructose but, in sucrose, the anomeric carbon atoms of fructose and glucose are joined by a covalent bond and are thus not available to react.

18. Glycogen is polymer of glucose linked by α-1,4-glycosidic bonds with branches formed approximately every 10 glucose units by α-1,6-glycosidic bonds. Starch consists of two polymers of glucose. Amylose is a straight-chain polymer formed by α-1,4-glycosidic bonds. Amylopectin is similar to glycogen, but amylopectin has fewer branches, one branch per 30 or so glucose units.

19. Cellulose is a linear polymer of glucose joined by β-1,4 linkages. Glycogen is a branched polymer with the main chain being formed by α-1,4-glycosidic bonds. The β-1,4 linkages allow the formation of a linear polymer ideal for structural roles. The α-1,4 linkages of glycogen form a helical structure, which allows the storage of many glucose moieties in a small space. Furthermore, different enzymes are required to digest the α-1,4- and β-1,4-glycosidic bonds.

20. Simple glycoproteins are often secreted proteins and thus play a variety of roles. Usually, the protein component constitutes the bulk of the glycoprotein by mass. In contrast, proteoglycans and mucoproteins are predominantly carbohydrates. Proteoglycans have glycosaminoglycans attached, and play structural roles as in cartilage and the extracellular matrix. Mucoproteins often serve as lubricants and have multiple carbohydrates attached through an N-acetylgalactosamine moiety.

21. The attached carbohydrate moiety extends the lifetime of EPO in circulation and thus enables the EPO to function for longer periods of time than would be possible with an equivalent carbohydrate-free protein.

22. The heavily charged glycosaminoglycan binds many water molecules. When cartilage is stressed, such as when a human heel hits the ground while a person is walking, the water is released, thus cushioning the impact. The water binds again when the heel is lifted.

23. The binding site for the carbohydrate marker might be defective on the lectin receptor. The receptor would then not recognize a correctly addressed protein.

24. Glycoproteins are different molecular forms of a glycoprotein that differ in the amount of carbohydrate attached or the location of attachment or both.

25. The glycome is the total collection of carbohydrates synthesized by a particular cell at particular times and under particular environmental conditions.

26. The genome comprises all of the genes present in an organism. The proteome includes all of the possible protein products and modified proteins that a cell could express in response to changing circumstances. The glycome consists of all of the carbohydrates that a cell could synthesize under varying circumstances. Because the genome is static, but any given protein can be variously expressed and modified, the proteome is more complex than the genome. The glycome, which includes not only glycoforms of proteins, but also many possible carbohydrate structures, must be even more complex.

27. The universal observation of lectins in all organisms suggests that carbohydrates are found on the cell surfaces of all organisms for the purpose of recognition by other organisms or by the environment.

28. A glycoprotein is a protein that is decorated with carbohydrates. A lectin is a protein that specifically recognizes carbohydrates. A particular lectin may in some cases also be a glycoprotein.

29. Each of six sites has two choices: to be glycosylated, or not. The number of possible proteins with different extents of glycosylation is therefore $2^6 = 64$. (Diversity within each carbohydrate chain would further increase this number.)

30. The set of enzymes that synthesize sucrose are specific for (only) the α-anomer of D-glucose and the β-anomer of D-fructose. As these particular anomers are drawn from the solution, Le Chatelier's principle dictates that they will be replenished from the supply of β-D-glucose and α-D-fructose, as well as corresponding acyclic compounds, that are also present in the solution, based on equilibrium thermodynamics.

31. If the carbohydrate specificity of the lectin is known, an affinity column could be constructed with the appropriate carbohydrate that would bind the lectin. The protein preparation containing the lectin of interest could be passed over the column. This method was indeed used to purify the glucose-binding lectin, concanavalin A.

32. (a) Aggrecan is heavily decorated with glycosaminoglycans. If glycosaminoglycans are released into the media, aggrecan must be undergoing degradation.
    (b) Another enzyme might cleave glycosaminoglycans from aggrecan without degrading it.
    (c) The control provides a baseline to indicate whether inherent "background" degradation is taking place when no enzymes or factors are added.
    (d) Aggrecan degradation is greatly enhanced.
    (e) Aggrecan degradation is reduced to the background level.
    (f) Not all factors that contribute to cartilage stabilization *in vivo* are present in the *in vitro* system that is used for the assay.

CHAPTER 12

1. $1 \, \mu m^2 = (10^{-6} \, m)^2 = 10^{-12} \, m^2$ $70 \, \text{Å}^2 = 70 \, (10^{-10} \, m)^2 = 70 \times 10^{-20} \, m^2$. Since the bilayer has two sides, $2 \times 10^{-12}/(70 \times 10^{-20}) = 2.86 \times 10^6$ molecules.

2. Phospholipids in an organic solvent would essentially form an inverted or "inside-out" membrane. The hydrophilic groups would come together on the interior of the structure, away from the solvent, and the hydrocarbon chains on the exterior of the structure would interact with the solvent.

3. Using the diffusion coefficient equation, $s = (4Dt)^{1/2}$, one gets $s = (4 \times 10^{-8} \times 10^{-6})^{1/2}$ or $(4 \times 10^{-8} \times 10^{-3})^{1/2}$ or $(4 \times 10^{-8} \times 1)^{1/2}$. Solving for s gives 2×10^{-7} cm, 6.32×10^{-6} cm, and 2×10^{-4} cm, respectively.

4. The gram molecular weight of the protein divided by Avogadro's number $= 10^5$ g/(6.02×10^{23}) $= 1.66 \times 10^{-19}$ g/molecule. 1.66×10^{-19} g/1.35 (density) $= 1.23 \times 10^{-19}$ cm^3/molecule. The volume of a sphere equals $4/3 \, \pi r^3 = 1.23 \times 10^{-19}$ cm^3. Solving for r, one gets 3.08×10^{-7} cm. By substituting this value into the equation given,

$$D = \frac{1.38 \times 10^{-16} \times 310}{6 \times 3.14 \times 1 \times 3.08 \times 10^{-7}}$$

$$= 7.37 \times 10^{-9} \text{ cm}^2/\text{s}$$

By substituting this value for D and the times given in the problem into the equation shown in the answer to problem 2, one obtains the distances traversed: 1.72×10^{-7} cm in 1 μs, 5.42×10^{-6} cm in 1 ms, and 1.72×10^{-4} cm in 1 s.

5. As its name implies, a *carrier* antibiotic must move from side to side when it shuttles ions across a membrane. By contrast, *channel formers* allow ions to pass through their pores much like water through a pipe. Lowering the temperature caused a phase transition from a fluid to a nearly frozen membrane. In the nearly frozen state the *carrier* is immobilized, whereas the pore of the *channel former* remains intact, allowing ions to pass through it.

6. The presence of a *cis* double bond introduces a kink in the fatty acid chain that prevents tight packing and reduces the number of atoms in van der Waals contact. The disruption of packing lowers the melting point compared with that of a saturated fatty acid. Trans fatty acids do not have the kink, and so their melting temperatures are higher, more similar to those of saturated fatty acids. Because *trans* fatty acids have no structural effect and confer no special biological properties, they are rarely observed.

7. Palmitic acid is two carbons shorter than stearic acid. When the chains pack together, there is less opportunity for van der Waals interaction. The melting point of the 16-carbon fatty acid is therefore lower than that of the longer 18-carbon stearic acid.

8. It should be expected that hibernating animals will have body fats with melting temperatures lower than those of their nonhibernating cousins. Indeed, the hibernators achieve this desired effect by selectively feeding on plants that have a high proportion of polyunsaturated fatty acids.

9. The initial decrease in fluorescence with the first addition of sodium dithionite results from the quenching of NBD-PS molecules in the outer leaflet of the bilayer. Sodium dithionite does not traverse the membrane under these experimental conditions; hence, it does not quench the labeled phospholipids in the inner leaflet. An immediate second addition of sodium dithionite has no effect, as the lipid flip-flop rate is rather slow and the NBD-PS molecules in the outer leaflet remain quenched. However, after a 6.5-hour incubation, the remaining NBD-PS has essentially established (by means of flip-flop) a new equilibrium between the inner and outer leaflets, with about 50% on each side. Therefore, about 50% of the remaining fluorescence is quenched when sodium dithionite is added after a 6.5-hour delay. The half-time for lipid flip-flop is therefore slower than several minutes and faster than 6.5 hours.

10. The initial decrease in the amplitude of the paramagnetic resonance spectrum results from the reduction of spin-labeled phosphatidyl cholines in the outer leaflet of the

bilayer. Ascorbate does not traverse the membrane under these experimental conditions, and so it does not reduce the phospholipids in the inner leaflet. The slow decay of the residual spectrum is due to the reduction of phospholipids that have flipped over to the outer leaflet of the bilayer.

11. There are two differences from the structures of common phospholipids. First, the C_{16} fatty acid is attached by an ether linkage. Second, the C-2 carbon atom of glycerol has only a short 2-carbon acetyl group attached by an ester linkage instead of a fatty acid.

12. The helix formation would be more likely in the hydrophobic medium. Water molecules would compete with peptide backbone NH and C=O groups for hydrogen bond formation; this competition would reduce the relative helix propensity in water. Additionally, the isolated NH and C=O groups would be quite unstable if not hydrogen bonded in a hydrophobic medium, and so would be driven to maximize their participation in hydrogen bonds.

13. The protein may contain an α helix that passes through the hydrophobic core of the protein. This helix, though not crossing a membrane, is likely to feature a stretch of hydrophobic amino acids similar to those observed in transmembrane helices.

14. Double bonds *(cis)* in the lipid acyl chains will increase the membrane fluidity. To maintain a similar membrane fluidity at the lower temperature of 25°C, the bacteria would incorporate more of the unsaturated fatty acids in their membrane phospholipids than at 37°C.

15. Each of the 21 v-SNARE proteins could interact with each of 7 t-SNARE partners. The total number of different interacting pairs is found by multiplication.

    $7 * 21 = \underline{\mathbf{147}}$ different v-SNARE−t-SNARE pairs.

16. (a) The main effect is to broaden the phase transition of the lipid bilayer. The relative change in fluidity near T_m is much less dramatic when cholesterol is present. The effect could be biologically important in maintaining the functions of proteins that may be sensitive to membrane fluidity. In particular, such proteins will be less sensitive to small local fluctuations in temperature when cholesterol is present in the membrane.

17. We will presume that the hydropathy plots were constructed using 20-residue windows. Plot c shows several peaks that surpass the criterion level of 20 kcal/mol (84 kJ/mol) for the hydropathy index, indicating possible regions for membrane-spanning α helices. Therefore plot c is likely to predict a membrane protein with about four (possibly five) membrane-spanning α helices. However, there are ambiguities: Membrane-spanning β-strands will escape detection by these hydropathy plots. (Plots a and b, in fact, are somewhat similar to Figure 12.18.) Additionally, a highly nonpolar segment of a protein sequence is not necessarily a transmembrane segment, but may simply be a hydrophobic segment that is buried in the core of the folded protein.

18. Membrane proteins are not soluble in water, and they require lipids for folding into their proper functional states. Lipid/protein complexes are difficult to crystallize. In some cases, the lipids may be replaced by detergents that may solubilize particular membrane proteins (with retention of their biological functions), but some detergents may alter the folded state of the membrane protein. Furthermore, detergent/protein complexes also are difficult to crystallize (though easier than lipid/protein complexes). Key advances in the development of synthetic detergents and of methods for crystallization have led to several dozen crystal structures of important membrane proteins.

CHAPTER 13

1. In simple diffusion, the substance in question can travel down its concentration gradient from one location to another, for example, across a lipid-bilayer membrane. With respect to facilitated diffusion across a lipid bilayer, the substance is not lipophilic and so the bilayer membrane presents a barrier. The substance therefore cannot diffuse directly through the membrane. Rather, a channel or carrier is required to enable or "facilitate" the movement down the concentration gradient.

2. The two forms of energy are (1) chemical bond energy from the hydrolysis of ATP, and (2) potential energy from a concentration gradient of a second substance. In case (2) the energetically favorable movement of one molecule down its concentration gradient is coupled to the movement of another molecule up its concentration gradient.

3. The three types of carriers are symporters, antiporters, and uniporters. Symporters and antiporters involve coupled transport of two substances and therefore can mediate secondary active transport.

4. Since $\Delta G = RT \log c_2/c_1 + ZF\Delta V$, we can substitute and get the following:

$$\Delta G = RT \log \frac{1.5 \times 10^{-3}}{4 \times 10^{-7}} + 2 \times 23 \times 6 \times 10^{-2}$$

$$= 4.86 \text{ kcal/mol (chemical work)} + 2.76 \text{ kcal/mol (electrical work)}$$

5. For chloride, z is -1, and for calcium z is $+2$. Assume that the intracellular space is side 2 and the extracellular space is side 1. At equilibrium $\Delta G = 0$, and $zF\Delta V = -RT \ln(c_2/c_1)$. Rearrange to get $\Delta V = -(RT/zF) \ln(c_2/c_1)$, with c_2 referring to intracellular concentration, and c_1 being the extracellular concentration.

   Using $F = 96{,}500 \text{ J V}^{-1} \text{ mol}^{-1}$, $R = 8.315 \text{ J mol}^{-1} \text{ K}$, and $T = 310 \text{ K}$; substituting gives:

   an equilibrium potential for chloride of $+(RT/F) \ln(4 \text{ mM}/150 \text{ mM}) = -96.8 \text{ mV}$; and

   an equilibrium potential for calcium of $-(RT/2F) \ln(0.0002 \text{ mM}/1.8 \text{ mM}) = +121.6 \text{ mV}$.

6. When a free-energy input of $-\Delta G$ is available, the concentration gradient (c2/c1) that can be supported is $e^{(\Delta G/RT)}$. When ΔG is 10.8 kJ mol^{-1} ($2.6 \text{ kcal mol}^{-1}$), (c2/c1) is **66** at 310 K.

   (*Note*: The sign of ΔG is negative for the Na^+ gradient, positive for the glucose gradient; such that the Na^+ gradient is driving the glucose gradient.)

   Note: Checking the Na^+ gradient: $\Delta G = RT \ln(c2/c1) + zF\Delta V$

   $= RT \ln (143/14) + (1)(96.5 \text{ kJ V}^{-1} \text{ mol}^{-1})(0.05 \text{ V})$

   $= -6.0 \text{ kJ mol}^{-1} - 4.8 \text{ kJ mol}^{-1} = -10.8 \text{ kJ mol}^{-1}$ when T is 310 K.

7. By analogy with the calcium ATPase, with three Na^+ ions binding from inside the cell to the E1 conformation, and two K^+ ions binding from outside the cell to the E2 conformation, here is a plausible mechanism:

   (a) A catalytic cycle could begin with the enzyme in its unphosphorylated state (E1) with three sodium ions bound.

   (b) The E1 conformation binds ATP. A conformational change traps sodium ions inside the enzyme.

   (c) The phosphoryl group is transferred from ATP to an aspartyl residue.

   (d) On ADP release, the enzyme changes its overall conformation, including the membrane domain. This new conformation, E2, releases the sodium ions to the side of the membrane opposite that at which they entered and binds two potassium ions from the side where sodium is released.

(e) The phosphorylaspartate residue is hydrolyzed to release inorganic phosphate.

(f) With the release of phosphate, the interactions stabilizing E2 are lost, and the enzyme reverts back to the E1 conformation. Potassium ions are released to the cytoplasmic side of the membrane. The binding of three sodium ions from the cytoplasmic side of the membrane completes the cycle.

8. Membrane vesicles containing a high concentration of lactose in their inner volume could be formed. The binding of lactose to the inner face of the permease would be followed by the binding of a proton from the inside of the vesicle. The permease would then evert. Because the lactose concentration on the outside is low, lactose and the proton will dissociate from the permease. The downhill flux of lactose will drive the uphill flux of protons in this in vitro system.

9. Ligand-gated channels open in response to the binding of a small molecule that is specifically recognized by the channel. Voltage-gated channels open in response to changes in the membrane potential.

10. An ion channel must transport ions in either direction at the same rate. The net flow of ions is determined only by the composition of the solutions on either side of the membrane.

11. Uniporters act as enzymes do; their transport cycles include large conformational changes, and only a small number of the molecules being transported interact with the protein during each transport cycle. In contrast, an open channel provides a pore in the membrane through which many thousands or even millions of ions per second may pass during the time between single channel opening and closing events. As such, channels mediate transport at much higher rates than do uniporters.

12. By mediating proton transport, FCCP effectively dissipates the proton gradient that would otherwise support uptake of lactose. Protons that are pumped out of the bacteria will return preferentially by means of the low-energy FCCP pathway (the "path of least resistance"), rather than participate in H^+/lactose symport.

13. Cardiac muscle must contract in a highly coordinated manner in order to pump blood effectively. The connexins or gap junctions mediate the orderly cell-to-cell propagation of the action potential through the heart during each beat.

14. The positively charged guanidinium group resembles Na^+ and binds to negatively charged carboxylate groups in the mouth of the channel.

15. SERCA, a P-type ATPase, uses a mechanism by which a covalent phosphorylated intermediate (at an aspartate residue) is formed. At steady state, a subset of the SERCA molecules are trapped in the E_2-P state and, as a result, become radiolabeled. The MDR protein is an ABC transporter and operates through a different mechanism without a covalent phosphorylated intermediate. Hence, a radiolabeled band would not be observed for MDR.

16. The blockage of ion channels inhibits action potentials, leading to loss of nervous function. Like tetrodotoxin, these toxin molecules are useful for isolating and specifically inhibiting particular ion channels.

17. After repolarization, the ball domain of an ion channel binds in the channel pore, blocking the pore and rendering the channel inactive for a short period of time. During this time, the channel cannot be reopened until the ball domain disengages and the channel returns from the "inactivated" state to the "closed" state.

18. Because sodium ions are charged and because sodium channels carry only sodium ions (but not anions), the accumulation of excess positive charge on one side of the membrane dominates the chemical gradients.

19. A mutation that impairs the ability of the sodium channel to inactivate would prolong the duration of the depolarizing sodium current, and thus would lengthen the cardiac action potential.

20. No. It is likely that channels will open or close in response to an external stimulus, but that the unit conductance of the open channel will be influenced very little.

21. From the data given, we get $L_0 = [T_0]/[R_0] = 10^5$ and $c = 1/20 = 5 \times 10^{-2}$. The ratio of closed to open channels = $[T]/[R]$. When no ligands are bound, $[T]/[R] = 10^5$. For one ligand, $[T]/[R] = L_0 \times c = 10^5 \times 5 \times 10^{-2} = 5 \times 10^3$.

    For two ligands, $[T]/[R]$ is $10^5 \times (5 \times 10^{-2})^2 = 2.5 \times 10^2$. For three ligands, $[T]/[R]$ is $10^5 \times (5 \times 10^{-2})^3 = 1.25 \times 10$. For four ligands, it is $10^5 \times (5 \times 10^{-2})^4 = 0.625$.

    From these ratios of closed/open, one can calculate that the fractions of open channels are respectively 10^{-5}, 2×10^{-4}, 3.98×10^{-3}, 7.41×10^{-2}, and 0.615.

22. Acetylcholine is degraded when its ester bond is cleaved by a specific enzyme (named acetylcholinesterase) whose active site resembles that of trypsin and chymotrypsin [see page 255–262]. Tabun, sarin, and parathion contain highly reactive phosphoryl groups that react readily with the active-site serine of the enzyme to form stable phosphorylated derivatives. When acetylcholine is not cleaved and does not turn over, synaptic transmission at cholinergic synapses is blocked, causing respiratory paralysis. See also problem 16 below.

23. (a) The binding of the first acetylcholine increases the open/closed channels by a factor of $1.2 \times 10^{-3}/(5 \times 10^{-6}) = 240$, whereas the binding of the second acetylcholine increases this ratio by a factor of $14/(1.2 \times 10^{-3}) = 11.7 \times 10^3$.

    (b) For the free-energy calculation, refer to Table 8.3 and the accompanying discussion on pages 222–224 in the text. Note that for a tenfold change in the equilibrium constant, there is a standard free-energy ($\Delta G^{o\prime}$) change of 5.69 kJ/mol (1.36 kcal/mol.) Also note that the $\Delta G^{o\prime}$ varies with the $\log_{10}$ of $K^\prime_{eq}$. Therefore, the free-energy change during the binding of the first acetylcholine is the $\log_{10}(240) \times 5.69 = 13.54$ kJ/mol (3.24 Kcal/mol). For the second binding, $\Delta G^{o\prime} = \log_{10}(11.7 \times 10^3) \times 5.69 = 23.15$ kJ/mol (5.54 Kcal/mol).

    (c) No. The MWC model predicts that the binding of each ligand will have the same effect on the closed/open ratio. The acetylcholine receptor channel is not perfectly symmetric. The two α chains are not in identical environments. Also, the presence of desensitized states in addition to the open and closed ones indicates that a more complex model is required.

24. Normally the *open state* of sodium channels lasts for only about 1 ms because it spontaneously converts to an *inactive state*. Its return to a closed but activatable state requires repolarization. Since BTX keeps the sodium channels open after depolarization, it apparently blocks the transition from the open to the inactivated state.

25. (a) Open *channels* enable ions to flow rapidly through membranes in a thermodynamically downhill direction, that is, from higher to lower concentration. Therefore, chloride ions will flow into the cell.

    (b) This flow of chloride ions *increases* the membrane polarization. Since depolarization triggers an action potential, the chloride flux is inhibitory.

    (c) If the GABA_A receptor resembles the acetylcholine receptor, its channel must consist of five subunits.

26. After the addition of ATP and calcium, SERCA will pump Ca^{2+} ions into the vesicle. However, the accumulation of only small amounts of Ca^{2+} ions inside the vesicle will

lead rapidly to the formation of an electrical gradient that cannot be overcome by ATP hydrolysis. The transport of Ca^{2+} will therefore cease. The subsequent addition of calcimycin allows the pumped Ca^{2+} ions to flow back out of the vesicle, thereby dissipating the charge buildup and enabling the pump to operate continuously.

27. For proper nerve activity, the change in membrane permeability caused by the opening of acetylcholine receptor channels must be short-lived. In order to close the receptor channels, it is important to remove the source of the stimulation, the acetylcholine. Once initiated, the nerve impulse moves on and the postsynaptic membrane must return to its resting state in order to be ready to receive and propagate another signal. (Some notable nerve poisons such as DIPF—see problem 16—act by inhibiting the acetylcholinesterase.)

28.

Acetylcholinesterase is a serine esterase whose catalytic mechanism is similar to that of the serine proteases. As with chymotrypsin and trypsin, the active site of acetylcholinesterase has serine as part of a Ser-His-Asp catalytic triad. The mechanism will involve covalent tetrahedral and acyl enzyme intermediates in which the substrate is bonded covalently to the active-site Ser. The reaction starts with nucleophilic attack on the substrate carbonyl group by the Ser OH group to give a tetrahdral intermediate (His acts as a base and accepts H^+). Next, choline will be the leaving group and the acetyl group will be left bonded to the Ser in the acyl-enzyme intermediate. Then in the second half of the reaction, water will act as the nucleophile to attack the acyl enzyme, giving a second tetrahedral intermediate. Finally, the free enzyme will be regenerated when acetate leaves as the second product. The process then can repeat.

29. (a) The ASIC1a channels are most sensitive to the toxin (first set of recordings in part (A) of the problem figure). The currents from these channels are completely inhibited for about 60 s following the application of the toxin; then there is a recovery.

    (b) Yes, the effect of the toxin is reversible. Toward the end of the first set of recordings in part (A) the ASIC1a channels are recovering as the PcTX1 is washed from the system.

    (c) From the graph in part (B) of the figure, the concentration for 50% inhibition is slightly less than 1 nM. A good estimate is approximately 0.7 nM (reading from the logarithmic scale).

30. The channels with the βV266M mutation remain open for longer times. There are several possible explanations for the slower channel closing rate. For example, a tighter binding of acetylcholine (slower release) could keep the channels open longer. Alternatively, acetylcholine could be released at the normal rate, but the mutation could slow the conformational transition from the open state to the closed state. (Other explanations are possible.)

31. With fast channel syndrome, the recordings would show channel events that are very brief, that is, with open channel lifetimes that are shorter than those of the control channels in problem 18. Possible explanations could include the converse of those in problem 18: quicker release of acetylcholine, a more rapid conformational transition from the open state to the closed state, and/or other factors.

32. The rate of indole transport is proportional to the indole concentration. This finding suggests that indole may diffuse freely across the cell membrane, without a need for a specific facilitated transport mechanism. By constrast, the rate of glucose transport is saturable and reaches a plateau (with no further rate increase) at high glucose

concentrations. The finding for glucose is consistent with a specific protein-mediated uptake of glucose, in which glucose would bind to a specific membrane protein and then be transported across the membrane. The glucose transport rate would saturate when all of the protein-binding sites are occupied by glucose molecules. *Effect of ouabain:* ouabain is an inhibitor of the Na^+-K^+ ATPase. The inhibition of glucose transport by ouabain indicates that glucose transport requires energy and further suggests that glucose transport may be linked to the transport of Na^+ or K^+.

CHAPTER 14

1. The common feature of glutamate and phospho-Ser (or phospho-Thr) is the presence of a negative charge (at physiological pH). The negative charge on the glutatmate side chain therefore may sometimes fulfill the role of the negative charge on the phosphate.

2. No. Phospho-Ser and phospho-Thr are significantly smaller than phospho-Tyr. The phosphate of these smaller side chains probably will not reach sufficiently far into the deep binding pocket to make a favorable electrostatic interaction with a counter $(+)$ charge.

3. The GTPase activity terminates the signal. Without such activity, an activated pathway would remain activated too long and would be unresponsive to changes in the initial signal. If the GTPase activity were more efficient, the lifetime of the GTP-bound G_α subunit would be too short to achieve downstream signaling.

4. The two identical receptors must recognize different aspects of the same signal molecule.

5. Some growth factors act by binding *two* receptor molecules and causing the receptor to dimerize. Antibodies with two identical binding sites could similarly cause receptor dimerization and initiate the signaling process.

6. The mutated α-subunit would be defective for signaling because it would be turned "on" at all times, even in the absence of an activated receptor. The inability to turn "off" the signaling pathway would be a serious flaw.

7. A G protein is a component of the signal-transduction pathway. The analog GTP$_\gamma$S is not hydrolyzed by the G_α subunit. Therefore GTP$_\gamma$S causes a prolonged activation.

8. Calcium is slowed because its intracellular concentration is slow and it binds tightly to larger molecules, including proteins. The effective molecular weight of the diffusing complex therefore is large.

9. Fura-2 is a highly negatively charged molecule, with five carboxylate groups. Its charge prevents it from effectively crossing the hydrophobic region of the plasma membrane.

10. Epinephrine initiates a pathway that raises the level of cAMP within the muscle cell. The higher level of cAMP ultimately will mobilize glucose (making more glucose available). Inhibitors of cAMP phosphodiestrerase also will raise the level of cAMP within the cell. Therefore the phosphodiesterase inhibitors will act similarly to epinephrine to *increase* the mobilization of glucose.

11. If the two kinase domains are forced into close proximity of each other, the activation loop of one domain, in its inactive conformation, could be displaced by the activation loop of the neighboring kinase, which then can act as a substrate for phosphorylation.

12. The full network of pathways initiated by insulin involves a large number of proteins and is substantially more elaborate than indicated in Figure 14.25 in the text. Furthermore, many additional proteins are involved in the termination of insulin signaling. A defect in any of the proteins involved in the insulin signaling pathways, or in the subsequent termination of the insulin response, could potentially cause problems. While the number 800 may be a mildly unexpected revelation, it is not surprising that many different gene defects can contribute to type 2 diabetes.

13. The mechanism can be similar to those described in the chapter, except that the tyrosine kinase is not part of the receptor itself, but rather a separate protein. In similar fashion to the description in Section 14.3 of the text, the binding of growth hormone causes its monomeric receptor to dimerize. The dimeric receptor could then activate a separate but nearby tyrosine kinase to which the receptor would bind. The signaling pathway could then continue along similar lines to the pathways that are activated by the insulin receptor or other mammalian EGF receptors.

14. The mutant receptors will block normal EGF signaling. When EGF binds, the truncated receptor will be able to dimerize with a full-length monomer, but cross-phosphorylation cannot take place. The truncated receptor possesses neither the substrate for the neighboring kinase domain nor its own kinase domain to phosphorylate the C-terminal tail of the other monomer. Therefore, the signaling pathway is blocked.

15. The chimeric receptor contains a binding domain for insulin, but not for EGF. Therefore, insulin binding should elicit the response that is normally caused by EGF. It is likely that insulin binding will stimulate dimerization and phosphorylation of the chimeric receptor, and thereby signal the downstream events that normally are triggered by EGF binding. Exposure of these cells to EGF should have no effect.

16. Multiply 100 molecules of active $G_{\alpha S}$ by 1000 reactions per second to compute 100,000 (or 10^5) molecules of cAMP in 1 s.

17. It is reasonable to propose that the nerve growth factor will cause dimerization, autophosphorylation, and activation of its receptor protein tyrosine kinase on binding. The active tyrosine kinase then should phosphorylate and activate a gamma (nonbeta) isoform of phospholipase C. Active PLC then would release both diacylglycerol and inositol 1,4,5-trisphosphate from phosphatidyl inositol 4,5-bisphosphate (PIP_2). Therefore, the concentration of the second messenger inositol 1,4,5-trisphosphate, as well as of diacylglycerol, would be expected to increase.

18. Other potential drug targets within the EGF signaling cascade could include the kinase active sites of the EGF receptor, Raf, MEK, ERK, and others.

19. There are several similarities. Both adenylate cyclase and DNA polymerases use ATP as a substrate. In addition, both enzymes release pyrophosphate while forming a new phosphodiester bond. The key difference is that the adenylate cyclase forms a new *intra*molecular bond, whereas DNA polymerases join molecules by forming new *inter*molecular bonds.

20. Drugs that compete with ATP are likely to act on multiple kinases because every kinase domain contains an ATP-binding site. Hence, these drugs may not be selective for just one particular desired kinase target.

21. (a) From the graphs, approximately 10^{-7} M of X, $3 * 10^{-6}$ M of Y, or 10^{-3} M of Z.
    (b) Hormone X achieves maximal binding in the lowest concentration range and therefore has the highest binding affinity.

(c) For each hormone, the trend for the activation of adenylate cyclase is similar to the trend for hormone/receptor binding. Therefore, it is likely that the hormone/receptor complex plays a direct role in the mechanism of activation of adenylate cyclase.

(d) A requirement for GTP in addition to hormone would suggest that a G_s protein may be required. The trend for $G_{\alpha s}$ activity should then be measured as a function of the concentration of hormones X, Y, and Z. This could be done by monitoring GTP/GDP exchange activity. G_s protein, the receptor, and unlabeled GDP should be preincubated in the absence of GTP and hormone. Then labeled GTP could be added together with varying amounts of a hormone X, Y, or Z. One would then test for the association of labeled GTP with protein when the proteins are subjected to precipitation, electrophoresis, or chromatography.

22. (a) The ligand X may "stick" to a few sites other than the specific receptor. These sites should not be counted.

(b) The experiment allows the background nonspecific binding to be determined. The large excess of nonradioactive ligand will bind to all of the authentic receptor sites. The remaining (residual) background binding of the labeled ligand will be revealed as nonspecific binding (line labeled "nonspecific binding").

(c) The plateau indicates that the ligand binding sites can be saturated. The sites can be saturated because in fact there exist only a discrete number of receptor molecules per cell. (Alternatively, if the cell uptake of ligand were to continue to increase without reaching a plateau, the result would indicate the absence of a specific receptor, and a different uptake mechanism would be operating.)

23. Paying attention to the units, we set up an equation to divide the binding activity by the specific activity and by the number of cells, and finally multiply by Avogadro's number to convert moles to molecules:

$$\frac{\left(\dfrac{10^4 \text{ cpm}}{\text{mg protein}}\right)\left(\dfrac{1 \text{ mg protein}}{10^{10} \text{ cells}}\right)\left(\dfrac{6*10^{23} \text{ molecules}}{10^3 \text{ mmole}}\right)}{10^{12} \text{ cpm}} \cong \frac{600 \text{ molecules}}{\text{cell}}.$$

CHAPTER 15

1. Intermediary metabolism concerns the highly integrated patterns of biochemical reactions that take place inside the cell.

2. Anabolism refers to the set of biochemical reactions that use energy to build new molecules and ultimately new cells. Catabolism is the set of biochemical reactions that extract energy from fuel sources or break down biomolecules.

3. The three primary uses of cellular energy are to perform mechanical work, including cell movement; to synthesize necessary biomolecules; and to transport necessary molecules or ions against a gradient (active transport).

4. 1. f;   2. h;   3. i;   4. a;   5. g;   6. b;   7. c;   8. e;   9. j;   10. d.

5. Charge repulsion destabilizes the triphosphate group. When the triphosphate is split, the products have increased resonance stabilization and hydration. Together these factors account for the high phosphoryl group transfer potential of nucleoside triphosphates.

6. Actually the reasons for the choice of ATP are not known. Adenine does appear to form more readily than the other aromatic base components of nucleotides under simulated prebiotic conditions, so it is conceivable that ATP may have predominated in the prebiotic environment.

7. Having only one nucleotide represent the available energy gives the cell a straightforward way to monitor its energy status.

8. Changing the concentrations of ATP, ADP, Pi or Mg^{2+} could raise or lower the value of ΔG for the hydrolysis reaction (see Problems 21 and 30).

9. The free-energy changes of the individual steps in a pathway are summed to determine the overall free-energy change for the entire pathway. Consequently, a reaction with a positive free-energy value can be powered to take place if coupled to a sufficiently exergonic reaction. Removing the products and providing large amounts of substrate molecules can also help to promote reactions that have positive standard free-energy values.

10. The direction of a reaction when the reactants are initially present in equimolar amounts is dependent on $\Delta G^{o\prime}$. Since, by convention, reactions are written from left to right, if $\Delta G^{o\prime}$ is negative, K'_{eq} is positive and the *direction* is to the right because at equilibrium the product concentrations will exceed those of the reactants. If $\Delta G^{o\prime}$ is positive, the reverse is true. The $\Delta G^{o\prime}$ values for these reactions are

   (a) +12.5 kJ/mol (+3.0 kcal/mol; left),
   (b) −21.3 kJ/mol (−5.1 kcal/mol; right),
   (c) +31.4 kJ/mol (+7.5 kcal/mol; left),
   (d) −16.7 kJ/mol (−4.0 kcal/mol; right).

11. Consider a large rock that has been sitting on the side of a mountain for a million years. It has a large amount of potential energy but no kinetic energy—until you push it! Or consider a mixture of H_2 and O_2; it is perfectly stable until you light it! Notice that the thermodynamics of a reaction tell you little, if anything, about its kinetics.

12. (a) Note that phosphoenolpyruvate is *formed* in this reaction; hence, its contribution to $\Delta G^{o\prime}$ is *plus* 61.9 kJ/mol (14.8 kcal/mol). Therefore, $\Delta G^{o\prime}$ for the entire reaction is +61.9 −30.5 = 31.4 kJ/mol (+7.5 kcal/mol). $K'_{eq} = e^{-31.4/2.48} = 3.2 \times 10^{-6}$.

    (b) Substituting into $K'_{eq} = 3.2 \times 10^{-6} = $ [ADP]/[ATP] × [PEP]/[Pyr], we get $3.2 \times 10^{-6} = 1/10 \times$ [PEP]/[Pyr]. [PEP]/[Pyr] $= 3.2 \times 10^{-5}$, and [Pyr]/[PEP] $= 1/(3.2 \times 10^{-5}) = 3.13 \times 10^4$.

13. $\Delta G^{o\prime} = 20.9 - 13.8 = 7.1 \dfrac{kJ}{mol}$ (1.7 kcal / mol)

$$K'_{eq} = \frac{[G-1-P]}{[G-6-P]} = e^{-7.1/2.48} = 5.62 \times 10^{-2}$$

The reciprocal of this is $\dfrac{[G-6-P]}{[G-1-P]}$ or 17.5.

14. (a) $\Delta G^{\circ\prime}$ = 31.4 kJ/mol − 45.6 kJ/mol = −14.2 kJ/mol (−3.4 kcal/mol).

    (b) The hydrolysis of PP_i drives the reaction toward the formation of acetyl CoA by making $\Delta G^{\circ\prime}$ strongly negative (−14.2 − 19.3) kJ/mol = −33.5 kJ/mol)(−8.0 kcal/mol).

15. (a) By definition, $\log_{10} K = -pK$. Since $\Delta G^{\circ\prime} = -2.3\ RT \log_{10} K$, by substitution one obtains $\Delta G^{\circ\prime} = 2.3\ RT\ pK$.

    (b) $\Delta G^{\circ\prime}$ = 2.3 RT (4.8) = 27.4 kJ/mol (6.5kcal/mol) at 25°C.

16. Arginine phosphate, like creatine phosphate, contains a *phosphoguanidino* group; these compounds are called *phosphagens*. The transfer of the phosphoryl group of arginine phosphate to ADP is catalyzed by arginine phosphokinase in a reaction similar to the one catalyzed by creatine kinase.

    To test the hypothesis that arginine phosphate is acting in a manner similar to that of creatine phosphate, you would monitor the concentrations of ATP and arginine phosphate in invertebrate muscle during contraction. If arginine phosphate is indeed serving as a reservoir of high-potential phosphoryl groups, its concentration will decrease while that of ATP will remain constant (or nearly so) during the early stages of contraction.

17. An ADP unit (or a closely related derivative, in the case of CoA).

18. (a) Creatine can be converted to creatine phosphate by creatine kinase. Creatine phosphate is a short-term reservoir of high-potential phosphoryl groups (for the regeneration of ATP) in vertebrate muscle. The amount of ATP in muscle is sufficient for about one second of contraction, and the amount of creatine phosphate in muscle is sufficient for about four seconds of contraction.

    (b) Creatine supplementation could possibly bring benefits only during very brief periods of vigorous exercise.

19. The actual free energy change under intracellular conditions is equal to the standard free energy change ($\Delta G'$) plus a term due to the actual concentrations of the reactants and products. Therefore:

$$\Delta G = \Delta G^{\circ\prime} + RT \ln \frac{[\text{dihydroxyacetone-P}][\text{glyceraldehyde-3-P}]}{[\text{fructose-1,6-bisphosphate}]}$$

Let us assume a physiological temperature of 37°C, which is 310°K.

$\Delta G^{\circ\prime}$ is given as +23.8 kJ mol^{-1} (+5.7 kcal mol^{-1}).

R is 8.31×10^{-3} kJ mol^{-1} °K^{-1} (remember to distinguish kJ from J!).

At equilibrium, $\Delta G = 0$, whereas under intracellular conditions, $\Delta G = -1.3$ kJ mol^{-1}. We must solve for both cases, but we note that the intracellular ΔG is close to the equilibrium value, so we expect that the final answers should be somewhat similar for the two cases.

$$\Delta G = \Delta G^{\circ\prime} + RT \ln \frac{[\text{dihydroxyacetone-P}][\text{glyceraldehyde-3-P}]}{[\text{fructose-1,6-bisphosphate}]}.$$

Let [dihydroxyacetone-P] = x. Then [glyceraldehyde-3-P] = x. By the conservation of mass, [fructose-1,6-bisphosphate] is (100% minus 2x), or (1.0 − 2x).

RT is $(8.31 \times 10^{-3} \times 310)$ kJ mol^{-1} = 2.58 kJ mol^{-1}.

$$\Delta G = \Delta G^{\circ\prime} + RT \ln \frac{x^2}{(1-2x)}.$$

At equilibrium, $\Delta G = 0$ and

$$\frac{-\Delta G^{\circ\prime}}{RT} = \ln \frac{x^2}{(1-2x)}; \frac{-23.8}{2.58} = \ln \frac{x^2}{(1-2x)}.$$

Thus:

$$\frac{x^2}{(1-2x)} = e^{\left(\frac{-23.8}{2.58}\right)} = e^{-9.22} = 9.9*10^{-5}.$$

Now we make an approximation. We will assume that x is very small, and then check the assumption. With x very small, $(1-2x) \cong 1$, and $x^2 \cong 9.9*10^{-5}$. Thus $x \cong 10^{-2}$, and the assumption is reasonable. (Very little of each product is formed at equilibrium.) This gives a mole fraction of fructose-1,6-bisphosphate of $(1-2x) \cong 0.98$.

Therefore, at equilibrium the ratio of the reactant [fructose-1,6-bisphosphate] to either of the product concentrations is 0.98/x, or 0.98/0.01 = 98 to 1. The starting material (reactant) is highly favored at equilibrium.

Now we repeat the calculation for the intracellular conditions when $\Delta G = -1.3$ kJ mol^{-1}. We note that this number is slightly negative, but actually quite close to the equilibrium value.

$$\Delta G - \Delta G^{\circ\prime} = -25.1 \text{ kJ mol}^{-1} = RT \ln \frac{x^2}{(1-2x)}.$$

$$\frac{x^2}{(1-2x)} = e^{\left(\frac{-25.1}{2.58}\right)} = e^{-9.73} = 5.96*10^{-5}.$$

Approximating, as above, we come to $x^2 \cong 6*10^{-5}$, and $x \cong 7.7*10^{-3}$. This gives the mole fraction of fructose-1,6-bisphosphate = $(1-2x) \cong 0.985$ under intracellular conditions. Therefore, under intracellular conditions, the ratio of the reactant [fructose-1,6-bisphosphate] to either of the product concentrations is $\cong 0.985/0.0077 \cong 127$ to 1.

The reaction is exergonic under intracellular conditions because 127 is greater than the equilibrium ratio of 98. The cell drives the reaction by providing a large amount of fructose-1,6-bisphosphate, and by removing the products dihydroxyacetone-P and glyceraldehyde-3-P. (By contrast, under standard conditions—when all products and reactants are present in equal concentrations—the reaction is endergonic because $\Delta G^{\circ\prime}$ is large and positive, and $\Delta G = \Delta G^{\circ\prime}$ under standard conditions.)

20. Under standard conditions

$$K_{eq}' = \frac{[B]_{eq}}{[A]_{eq}} \times \frac{[ADP]_{eq}[P_i]_{eq}}{[ATP]_{eq}} = 267 \text{ (given)}.$$

At equilibrium, the ratio of [B] to [A] is given by

$$\frac{[B]_{eq}}{[A]_{eq}} = K_{eq}' \frac{[ATP]_{eq}}{[ADP]_{eq}[P_i]_{eq}}$$

The ATP-generating system of cells maintains the $[ATP]/[ADP][P_i]$ ratio at a high level, typically of the order of 500 M^{-1} given. For this ratio,

$$\frac{[B]_{eq}}{[A]_{eq}} = 2.67 \times 10^2 \times 500 = 1.34 \times 10^5$$

This equilibrium ratio is strikingly different from the value of 1.15×10^{-3} for the reaction A → B in the absence of ATP hydrolysis. In other words, coupling the hydrolysis of ATP with the conversion of A into B has changed the equilibrium ratio of B to A by a factor of about 10^8.

21. For ATP hydrolysis, the reaction is ATP + H_2O → ADP + P_i + H^+, and $\Delta G^{o\prime}$ is −30.5 kJ mol^{-1} (Table 14.1). As in problem 10, the actual ΔG under intracellular conditions will vary with the relative concentrations of the reactants and products.

$$\Delta G = \Delta G^{o\prime} + RT \ln \frac{[ADP][P_i]}{[ATP]}.$$

Using $\Delta G^{o\prime}$ of −30.5 kJ mol^{-1} and $RT = 2.58$ kcal mol^{-1} at 37°C (from problem 10), we can construct a table of results using molar concentrations.

Tissue	ATP (M)	ADP (M)	P_i (M)	$\frac{[ADP][P_i]}{[ATP]}$	ln $\left(\frac{[ADP][P_i]}{[ATP]}\right)$	ΔG (kJ mol^{-1})	ΔG (kcal mol^{-1})
Liver	0.0035	0.0018	0.0050	0.00257	−5.963	−45.9	−11.0
Muscle	0.0080	0.0009	0.0080	0.00090	−7.013	−48.6	−11.6
Brain	0.0026	0.0007	0.0027	0.00073	−7.227	−49.1	−11.7

The free energy *released* from ATP hydrolysis is greatest in the *brain*.

22. Possible methods of approach involve counting the total number of carbon-oxygen or carbon-hydrogen bonds per carbon atom present in the molecule. A more reduced molecule will have more carbon-hydrogen bonds and/or fewer carbon-oxygen bonds. This reasoning leads to the following results.

a. Ethanol has more C—H bonds and is more reduced. (Aetaldehyde has a C=O double bond and is more oxidized.)

b. Lactate is more reduced because it has more C—H bonds; furthermore pyruvate has one additional C=O double bond.

c. While the number of carbon-oxygen bonds is the same, succinate is more reduced because it has more C—H bonds than fumarate.

d. Isocitrate has a lower density of carbon-oxygen bonds and a higher density of C-H bonds. Isocitrate is the more reduced molecule.

e. Malate is more reduced because it has more C—H bonds.

f. The 2-phosphoglycerate has more C—H bonds and is more reduced.

23. As in problem 19, the actual free energy release under intracellular conditions is equal to the standard free energy change ($\Delta G^{o\prime}$) plus a term due to the actual concentrations of the reactants and products. Under standard conditions, all products and reactants are present in equal concentrations. The free-energy release will be greater under conditions where relatively large amounts of reactants (e.g., glucose) are available, and products (e.g., pyruvate) are removed by further reactions that keep the product concentrations relatively low.

24. Unless the ingested food is converted into molecules capable of being absorbed by the intestine, no energy can ever be extracted by the body.

25. NADH and FADH$_2$ are electron carriers for catabolism, whereas NADPH is usually the carrier for anabolism.

26. In the thioester group, $\overset{O}{\underset{S}{\|}}$, the resonance structures for electron sharing between the carbonyl C=O bond and the C—S bond are less stable than those that form between the C=O bond and the C—O bond of an oxygen ester. Thus, the thioester is less stabilized by resonance than is an oxygen ester.

27. a.   oxidation–reduction reactions
     b.   ligation reactions
     c.   isomerization reactions
     d.   group-transfer reactions
     e.   hydrolytic reactions
     f.   reactions that remove double bonds by adding functional groups, or reactions that form double bonds by removing functional groups

28. a.   controlling the amounts of particular enzymes.
     b.   controlling the activities of particular enzymes.
     c.   controlling the availability of substrates.

29. Although the reaction is thermodynamically favorable, the reactants are kinetically stable because of the large activation energy. Enzymes lower the activation energy so that reactions take place on time scales required by the cell.

30. The activated form of sulfate in most organisms is 3′-phosphoadenosine 5′-phosphosulfate. See P. W. Robbins and F. Lipmann. *J. Biol. Chem.* 229(1957):837.

31. (a) Decreasing [Mg^{2+}] makes the ΔG for ATP hydrolysis *less* negative. (*Less* energy is released when [Mg^{2+}] is low.) The graph has the shape of a titration curve.
     (b) The trend can be explained by assuming a single binding site for Mg^{2+}. When the site is occupied, the ΔG is more negative (more favorable) than when the site is unoccupied. The midpoint (or inflection point) of the S-shaped curve occurs at the [Mg^{2+}] for which half of the binding sites area occupied (binding constant). The binding constant for Mg^{2+} therefore can be calcualted for $\log_{10}$ (1/[Mg^{2+}]) on the x-axis at the midpoint of the curve.

CHAPTER 16

1. Two molecules of ATP are required to convert glucose into fructose 1,6-bisphosphate. Fructose 1,6-bisphosphate is subsequently converted into two molecules of glyceraldehyde 3-phosphate, each of which yields two molecules of ATP. Thus, the total ATP yield is four, but the net yield is only two molecules of ATP.

2. In both cases, the ultimate electron donor is glyceraldehydes 3-phosphate. In lactic acid fermentation, the electron acceptor is pyruvate, which is converted into lactate. In alcoholic fermentation, acetaldehyde is the electron acceptor, which is converted into ethanol. During intermediate steps, the electrons are passed temporarily to NAD+ (forming NADH) before they are returned to the ultimate acceptor, either pyruvate or acetaldehyde, respectively.

3. Glucose or fructose requires "investment" of 2 ATP to form fructose 1,6-bisphosphate. Fructose 1,6-bisphosphate is then split to form dihydroxyacetone phosphate and

glyceraldehyde 3-phosphate, each of which returns 2 ATP. So the net answer would be 4 ATP from fructose 1,6-bisphosphate, and the other answers are:

(a) Glucose 6-phosphate requires "investment" of only 1 ATP to form the fructose 1,6-bisphosphate and hence yields a net of 3 ATP.

(b) Dihydroxyacetone yields 2 ATP.

(c) Glyceraldehyde 3-phosphate yields 2 ATP.

(d) Fructose yields 2 ATP.

(e) Sucrose yields 1 glucose and 1 fructose, each of which yields 2 ATP, so the net yield from sucrose is 4 ATP.

4. Glucokinase enables the liver to remove glucose from the blood when hexokinase is saturated, ensuring that glucose is captured and stored for later use.

5. Glycolysis is a component of alcoholic fermentation, the pathway that produces alcohol for beer and wine. The belief was that understanding the biochemical basis of alcohol production might lead to more efficient means of producing beer.

6. Niacin deficiency could lead to a shortage of NAD^+. The conversion of glyceraldehyde 3-phosphate into 1,3-bisphosphoglycerate would be impaired. Glycolysis would be less effective.

7. Glucose 6-phosphate must have other uses in addition to its role in glycolysis. Indeed, glucose 6-phosphate can be converted into glycogen (Chapter 21) or can be processed to yield reducing power for biosynthesis (Chapter 20). Therefore, it is appropriate that hexokinase is not the pacemaker of glycolysis. The pacemaker step should not only be irreversible but also should be a step that is required for only one pathway.

8. The energy needs of a muscle cell vary widely, from rest to intense exercise. Consequently, the regulation of phosphofructokinase by energy charge is vital for muscle function. In other tissues, such as the liver, the energy needs are much less variable, the ATP concentration is less likely to fluctuate, and there is therefore no need for the ATP concentration to be a key regulator of phosphofructokinase.

9. The $\Delta G^{\circ\prime}$ for the reverse of glycolysis is $+96$ kJ mol^{-1} ($+23$ kcal mol^{-1}), far too endergonic for a direct reversal of the pathway. Instead, key steps that are energetically unfavorable in the reverse pathway must be bypassed through the use of alternate reaction steps.

10. Three steps are not readily reversible: the conversion of glucose into glucose 6-phosphate by hexokinase; the conversion of fructose 6-phosphate into fructose 1,6-bisphosphate by phosphofructokinase; and the formation of pyruvate from phosphoenolpyruvate by pyruvate kinase.

11. Lactic acid is a strong acid. If it would remain in a muscle cell, the pH of the cell would fall, which could lead to the denaturation of muscle proteins and result in muscle damage.

12. GLUT2 will transport glucose only when the blood concentration of glucose is high, which is precisely the condition in which the β cells of the pancreas should secrete insulin.

13. The conversion would involve steps catalyzed by three enzymes: fructokinase, fructose 1-phosphate aldolase, and triose kinase. These are the steps:

- Fructose + ATP $\rightarrow$ fructose 1-phosphate + ADP (fructokinase).
- Fructose 1-phosphate $\rightarrow$ dihydroxyacetone phosphate + glyceraldehyde (fructose 1-phosphate aldolase).
- Glyceraldehyde + ATP $\rightarrow$ glyceraldehyde 3-phosphate + ADP (triose kinase).

The primary controlling step of glycolysis catalyzed by phosphofructokinase is bypassed by the set of three reactions. Glycolysis will therefore proceed in an unregulated fashion under conditions where energy and ATP synthesis are not needed.

14. Without triose phosphate isomerase, only the glyceraldehyde 3-phosphate and not the dihydroxyacetone phosphate generated by aldolase could be used to generate ATP. Therefore, the yield of ATP from the metabolism of fructose 1,6-bisphosphate would be only two molecules of ATP. But two molecules of ATP would still be required to form fructose 1,6-bisphosphate during the early steps of glycolysis. Therefore, the net yield of ATP from fermentation of glucose would be zero, a yield incompatible with life. The organism would not survive.

15. D-glucopyranose is a cyclic hemiacetal, which is in equilibrium with its open-chain form that contains an active aldehyde group. In contrast, the anomeric carbon atoms of glucose and fructose are joined in an α-glycosidic linkage in sucrose. Hence sucrose is not in equilibrium with an active aldehyde or ketone form.

16. (a) The key is the aldolase reaction. Note that the carbons attached to the phosphate in glyceraldehyde 3-phosphate and dihydroxyacetone phosphate are interconverted by triose phosphate isomerase and *both* become the terminal carbon of the glyceric acids. Hence, the label is in the methyl carbon of pyruvate.

(b) By definition, specific activity is radioactivity/mol (or mmol). In this case the specific activity is halved (to 5 mCi/mmol) because the number of moles of product (pyruvate) is twice that of the labeled substrate (glucose).

17. (a) Glucose + 2 P_i + 2 ADP $\longrightarrow$ 2 lactate + 2 ATP

(b) To obtain the overall $\Delta G^{o\prime}$, -123 kJ/mol (-29.4 kcal/mol), you add the $\Delta G^{o\prime}$ values given in Table 16.3 in the text and the value given for the reduction of pyruvate to lactate. Remember that the values for the three carbon molecules must be doubled, since each hexose yields two trioses.

$$\Delta G' = -123.1 + RT \ln \frac{(5 \times 10^{-5})^2(2 \times 10^{-3})}{(5 \times 10^{-3})(10^{-3})^2(2 \times 10^{-4})}$$

$$= -123.1 + 2.47 \ln 50$$

$$= -114 \text{ kJ mol}^{-1} \ (-27.2 \text{ kcal}/\text{mol}).$$

The concentrations of ATP, ADP, P_i, and lactate are squared because, in reactions such as A $\longrightarrow$ 2 B, the $K_{eq} = [B]^2/[A]$.

18.
$$-31.4 = -2.47 \ln \frac{[\text{Pyr}][\text{ATP}]}{[\text{PEP}][\text{ADP}]}$$

$$12.7 = \ln \frac{[\text{Pyr}]}{[\text{PEP}]} + \ln 10$$

$$\frac{[\text{Pyr}]}{[\text{PEP}]} = e^{(12.7-2.3)}$$

$$\frac{[\text{PEP}]}{[\text{Pyr}]} = e^{-10.4} = 3.06 \times 10^{-5}$$

19. Since $\Delta G^{o\prime}$ for the aldolase reaction is $+23.8$ kJ/mol (Table 16.3 in the text), $K_{eq} = e^{-23.8/2.47} = 6.5 \times 10^{-5}$. If we let the concentration of each of the trioses (DHAP

and G-3P) formed during the reaction be X, then the concentration of F-1,6-BP is $(10^{-3} \text{ M} - X)$ at equilibrium, since we started with millimolar F-1,6-BP. Then,

$$\frac{X^2}{10^{-3} - X} = 6.5 \times 10^{-5}$$

Solving this quadratic equation leads to the answer of 2.24×10^{-4} M for X (DHAP and G-3P). Subtracting this from 10^{-3} gives the F-1,6-BP concentration of 7.76×10^{-4} M.

20. The 3-phosphoglycerate labeled with ^{14}C accepts the phosphate attached to C-1 of 1,3-BPG. Therefore, the resulting 2,3-BPG is ^{14}C-labeled in all three-carbon atoms and is ^{32}P-labeled in the phosphorus atom attached to the C-2 hydroxyl.

21. Hexokinase has a low ATPase activity in the absence of a sugar because it is in a catalytically inactive conformation. The addition of xylose closes the cleft between the two lobes of the enzyme. However, the xylose hydroxymethyl group (at C-5) cannot be phosphorylated. Instead, a water molecule at the site normally occupied by the C-6 hydroxymethyl group of glucose acts as the phosphoryl acceptor from ATP.

22. (a) The fructose 1-phosphate pathway forms glyceraldehyde 3-phosphate. Phosphofructokinase, a key control enzyme, is bypassed. Furthermore, fructose 1-phosphate stimulates pyruvate kinase.
    (b) The rapid, unregulated production of lactate can lead to metabolic acidosis.

23. The reverse of glycolysis is highly endergonic under cellular conditions. In particular, the three reactions catalyzed by hexokinase, phosphofructokinase and pyruvate kinase exhibit large energy barriers for the reverse pathway. The expenditure of six NTP molecules in gluconeogenesis overcomes these barriers and renders gluconeogenesis exergonic.

24. Lactic acid is capable of being further oxidized and thus represents useful energy. The conversion of this acid into glucose saves the carbon atoms for future combustion.

25. The three irreversible steps are the conversion of glucose into glucose 6-phosphate by hexokinase; the conversion of fructose 6-phosphate into fructose 1,6-bisphosphate by phosphofructokinase; and the formation of pyruvate from phosphoenolpyruvate by pyruvate kinase.
    The pyruvate kinase step is bypassed by two reactions in gluconeogenesis: (1) the formation of oxaloacetate from pyruvate and CO_2 by pyruvate carboxylase; and (2) the formation of phosphoenolpyruvate from oxaloacetate and GTP by phosphoenolpyruvate carboxykinase.
    The formation of fructose 1,6-bisphosphate by phosphofructokinase is bypassed by using fructose 1,6-bisphosphatase in gluconeogenesis. This enzyme releases inorganic phosphate from fructose 1,6-bisphosphate to give fructose 6-phosphate, but without synthesis of ATP.
    Finally, the hexokinase step is bypassed by using glucose 6-phosphatase in gluconeogenesis. This enzyme releases inorganic phosphate from glucose 6-phosphate to give glucose—without synthesis of ATP—and only in the liver.

26. Reciprocal regulation of the key allosteric enzymes prevents the simultaneous activity of the two pathways. For instance, phosphofructokinase is stimulated by fructose 2,6-bisphosphate and AMP, whereas conversely these signaling molecules inhibit the cognate enzyme, fructose 1,6-bisphosphatase. If both pathways were operating simultaneously, a futile cycle would result. ATP would be hydrolyzed and would yield only heat.

27. Muscle is likely to produce lactic acid during contraction. Lactic acid is a strong acid and would be detrimental if allowed to accumulate in muscle or blood. Liver takes

care of the problem by removing the lactic acid from the blood and converting it into glucose. The glucose can then either be released into the bloodstream or stored as glycogen for later use.

28. Glucose 6-phosphate produced in the liver could not be converted to glucose for release into the bloodstream. The glucose 6-phosphate would be trapped in the liver. Tissues that rely on glucose as an energy source would not function well, unless a sufficient and relatively constant supply of glucose was provided in the diet.

29. Glucose is an important energy source for both muscle and brain, and is essentially the only source of energy for the brain. Consequently, these tissues should never release glucose. The release of glucose is prevented by the absence of glucose 6-phosphatase.

30. A total of 6 NTP molecules are required. Two ATP and two GTP are required to synthesize two molecules of phosphoenolpyruvate from two pyruvate molecules. Two additional ATP are required to synthesize two 1,3-bisphosphoglycerate from two molecules of 3-phosphoglycerate during gluconeogenesis.

    Two molecules of NADH are required to produce two molecules of glyceraldehyde 3-phosphate from two molecules of 1,3-bisphosphoglycerate.

31. (a) None. Glucose 6-phosphatase will remove the phosphate from glucose-6-phosphate, yielding glucose with no involvement of NTP.
    (b) None. Fructose 1,6-bisphosphatase and glucose 6-phosphatase together will sequentially remove the two phosphate groups from fructose 1,6-bisphosphate, yielding glucose with no involvement of NTP.
    (c) Four NTP molecules. Two GTP are required to synthesize two molecules of phosphoenolpyruvate from two oxaloacetate molecules. In a later step, two ATP molecules are required to synthesize two 1,3-bisphosphoglycerate from two molecules of 3-phosphoglycerate.
    (d) None.

32. The ketoacid pyruvate is produced by removal of the amino group from alanine. The ketoacid oxaloacetate is produced by removal of the amino group from aspartate. Both pyruvate and oxaloacetate are components of the gluconeogenic pathway.

33. (a) Glycolysis increases because ATP can no longer inhibit PFK, phosphofructokinase.
    (b) Glycolysis increases because citrate can no longer inhibit PFK.
    (c) Glycolysis will increase because in the absence of fructose 2,6-bisphosphatase, the level of fructose 2,6-bisphosphate will increase, resulting in activation of PFK.
    (d) Glycolysis will decrease because fructose 1,6-bisphosphate can no longer activate pyruvate kinase.

34. The normal condition is for the level of fructose-2,6-bisphosphate to be high in the fed state, thereby inhibiting fructose-1,6-bisphosphatase. If the fructose-1,6-bisphosphatase is less sensitive to the small regulatory molecule, then fructose-1,6-bisphosphatase will be *less inhibited* when the genetic disorder is present. Consequently, gluconeogenesis will proceed even in the fed state. The net result will be either an oversupply of glucose (hyperglycemia), or nonproductive metabolic cycling through the combined gluconeogenesis and glycolysis pathways to produce heat at the expense of ATP and GTP.

35. Biotin is a cofactor for the synthesis of oxaloacetate from pyruvate by pyruvate carboxylase. Therefore, metabolic conversions, which require pyruvate carboxylase, will be inhibited. These will include only reaction (e) pyruvate $\longrightarrow$ oxaloacetate, and conversion (b) pyruvate $\longrightarrow$ glucose (which must begin with the pyruvate $\longrightarrow$ oxaloacetate reaction). The other listed conversions (a, c, d, f) are independent of pyruvate carboxylase and independent of biotin.

36. The glucose will *not* be labeled. After lactate is oxidized to pyruvate and the resulting pyruvate is carboxylated with labeled CO_2 by pyruvate carboxylase to yield

oxaloacetate, then the same CO_2 will be released during the phosphorylation and decarboxylation of oxaloacetate by phosphoenolpyruvate carboxykinase. (The CO_2 serves to make the phosphorylation reaction energetically feasible, but the CO_2 does not remain in the final product.)

37. Energy generation will be inhibited because arsenate will uncouple oxidation and phosphorylation. The arsenate will establish a small futile cycle that will shuttle between 3-phosphoglycerate and 1-arseno-3-phosphoglycerate. If the conditions also are anaerobic, NADH will accumulate, and NAD^+ will become unavailable for the continuation of sustained glycolysis.

38. The synthesis of lactate is an emergency stop-gap measure that is undertaken because of (a) a local shortage of oxygen in a tissue, and (b) an immediate need for energy. A "quick fix" for the situation is to regenerate NAD^+ from NADH using lactic acid dehydrogenase so that glycolysis can continue. When the emergency passes and oxygen is more plentiful, then the lactate can be reoxidized.

    New synthesis of NAD^+ would be too slow to provide the necessary rapid response. Furthermore, the cell would waste energy in accumulating larger pools of pyridine nucleotides than are needed. Catalytic enzymes, by contrast, are needed in only small molar amounts.

39. The equilibrium constant K_{eq}, $[ATP] \cdot [AMP]/([ADP] \cdot [ADP])$, is directly proportional to both the ATP and AMP concentrations. However, the intracellular concentration of AMP is much smaller than the intracellular concentration of ATP. Therefore, the same absolute changes in the ATP and AMP concentrations (due to adenylate kinase activity) will result in much larger percentage changes for the level of AMP. The [AMP] is therefore a more sensitive signal.

    As an example, let us consider an ATP concentration of 1 mM and an AMP concentration of 0.1 mM. Let us then assume that [ATP] decreases transiently to 0.95 mM, a 5% drop due to metabolic activity. This difference could be compensated by adenylate kinase activity (with a constant pool of total adenylate, i.e. ([ATP] + [ADP] + [AMP]) constant). Then adenylate kinase activity to make up the 0.05 mM of spent ATP would also produce an additional 0.05 mM of AMP, or a 50% increase in the level of AMP, from 0.10 mM to 0.15 mM. This increase in [AMP] would signal a low-energy state for the cell. The small change in [ATP] (e.g., 5%) therefore is magnified into a much larger signal, namely a 50% change in [AMP] in this hypothetical example.

40. The sites of glucose synthesis and glucose breakdown are different. During intense exercise, glycolysis proceeds to lactate in active skeletal muscle, with insufficient oxygen for the complete oxidation, as well as in erythrocytes. Meanwhile, gluconeogenesis proceeds in the liver, using the major raw materials of lactate and alanine produced by the active skeletal muscle and erythrocytes. The glucose that is produced by the liver enters the blood stream and becomes available to the muscles for continued exercise. The advantages to the organism are to buy time and to shift part of the metabolic burden from muscle to liver.

41. Glycolysis yields two net molecules of ATP, whereas gluconeogenesis hydrolyzes four molecules of ATP and two molecules of GTP. The sum of gluconeogenesis plus glycolysis therefore is: 2 ATP + 2 GTP + 4 H_2O $\longrightarrow$ 2 ADP + 2 GDP + 4 P_i. The effects of the additional high phosphoryl-transfer equivalents multiply together to alter the equilibrium constant by a factor of $(10^8)^4 = 10^{32}$.

42. The conversion of glucose-6-phosphate to fructose-6-phosphate is analogous to the conversion of glyceraldehyde-3-phosphate to dihydroxyacetone phosphate. Both of these isomerization reactions interconvert an aldose and a ketose. Key features of the triose phosphate isomerase mechanism include the hydrogen transfer between

carbon 2 and carbon 1 (intramolecular oxidation/reduction), and the enediol inter-mediate (Figure 16.7). Both of these features can be used also for the isomerization of glucose-6-phosphate to fructose-6-phosphate, as shown in the drawing below:

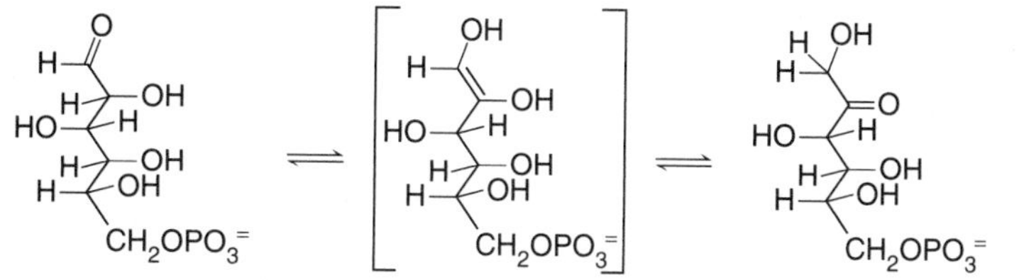

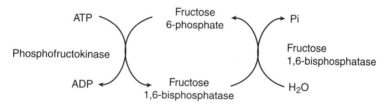

43. Several plausible answers may be possible here. There could likely be alternative non-dietary sources of galactose that pose problems. For example, galactose derivatives may arise from epimerization of the equivalent glucose derivatives. Subsequent meta-bolic breakdown of one such derivative conceivably could produce free galactose in the galactosemic patient and lead to peripheral damage (e.g., in the nervous system).

44. Fructose 2,6-bisphosphate stabilizes the active R state of phosphofructokinase.

45. (a) The graph suggests that ADP rather that ATP is the phosphate donor for *P. furio-sus* phosphofructokinase. Furthermore, AMP and ATP have similar regulatory ef-fects on this enzyme, rather than the opposing effects discussed in the chapter. (See the legend to Figure 16.18.)

    (b) AMP and ATP are both inhibitors. They both convert the hyperbolic binding curve of ADP into a sigmoidal one, probably by allosterically decreasing the affin-ity of the enzyme for ADP.

46. (a) If both enzymes operated simultaneously, the following reactions would take place:

```
         ATP            Fructose            Pi
          \            6-phosphate         /
           \          /        \          /
Phosphofructokinase                 Fructose
           /          \        /    1,6-bisphosphatase
          /            \      /      \
         ADP        Fructose            H₂O
                 1,6-bisphosphatase
```

The net result would be hydrolysis of ATP:

    (b) Not really. For the cycle to generate (only) heat, both enzymes must be func-tional (to the same extent) at the same time in the same cell.

    (c) The species *B. terrestris* and *B. rufocinctus* might show some futile cycling because both enzymes are active to a substantial degree.

    (d) No. These results simply suggest that simultaneous activity of phosphofructok-inase and fructose 1,6-bisphosphatase is unlikely to be employed to generate heat in the species shown.

CHAPTER 17

1. (a) Pyruvate dehydrogenase catalyzes the decarboxylation of pyruvate and the forma-tion of acetyllipoamide.

    (b) Dihydrolipoyl transacetylase catalyzes the formation of acetyl CoA.

    (c) Dihydrolipoyl dehydrogenase catalyzes the reduction of the oxidized lipoic acid.

    (d) The kinase associated with the complex phosphorylates and inactivates the complex.

    (e) The phosphatase associated with the complex dephosphorylates and activates the complex.

2. (a) To achieve the decarboxylation of pyruvate, thiamine pyrophosphate becomes co-valently bound to pyruvate and then releases carbon dioxide and the acetyl group.
   (b) Lipoic acid (as lipoamide) transfers the acetyl group.
   (c) Coenzyme A accepts the acetyl group from lipoic acid to form acetyl CoA.
   (d) FAD accepts the electrons and hydrogen ions when oxidized lipoic acid is reduced.
   (e) NAD^+ accepts electrons from $FADH_2$.

3. The catalytic coenzymes, thiamine pyrophosphate, lipoic acid, and FAD, are modified and then regenerated in each reaction cycle. Thus, they can play a role in the processing of many molecules of pyruvate. The stoichiometric coenzymes, coenzyme A and NAD^+, are components of products of the reaction and therefore can participate in the processing of only one molecule of pyruvate. New stoichiometric coenzyme molecules must be provided in each reaction cycle.

4. Several of the key advantages are these:
   (a) The reaction is facilitated by having the active sites in close proximity to each other.
   (b) All of the enzymes are present in the correct amounts.
   (c) The reactants do not leave the enzyme complex until the final product is formed. Constraining the reactants in this way minimizes side reactions and losses due to diffusion.
   (d) Regulation is more efficient because the regulatory enzymes—the kinase and the phosphatase—are also part of the complex.

5. To answer this problem one must follow carbon atoms around the citric acid cycle as shown by Figure 17.15 of your text. Remember that the randomization of carbon occurs at succinate, a truly symmetrical molecule. Also, this problem (and the answers given) assumes that all pyruvate goes to acetyl CoA. In fact, this is not necessarily true, since pyruvate can also enter the cycle at oxaloacetate.
   (a) After one round of the citric acid cycle, the label emerges in C-2 and C-3 of oxaloacetate.
   (b) The label emerges in CO_2 in the formation of acetyl CoA from pyruvate.
   (c) After one round of the citric acid cycle, the label emerges in C-1 and C-4 of oxaloacetate.
   (d) The fate is the same as in (a).
   (e) C-1 of G-6-P becomes the methyl carbon of pyruvate and hence has the same fate as in (a).

6. (a) Isocitrate lyase and malate synthase are required in addition to the enzymes of the citric acid cycle.
   (b) 2 acetyl CoA + 2 NAD^+ + FAD + 3 $H_2O \rightarrow$
       oxaloacetate + 2 CoA + 2 NADH + $FADH_2$ + $3H^+$
   (c) No, because they lack these two enzymes and hence cannot carry out the glyoxylate cycle.

7. Addition of the $\Delta G^{o'}$ values in Table 17.2 of the text gives the answer -41.0 kJ/mol (-9.8 kcal/mol).

8. As with enzymes, the small-molecule intermediates in the citric acid cycle are not consumed. Rather the cycle intermediates "turn over," that is, each of them is regenerated at a particular point during each turn of the cycle. The cycle as a whole catalyzes the conversion of acetyl-coenzyme A into two molecules of CO_2, with the release of free coenzyme A and the concomitant production of GTP, $FADH_2$, and three molecules of NADH.

9. Thiamine thiazolone pyrophosphate is a transition state analog. The sulfur-containing ring of this analog is uncharged, and so it closely resembles the transition state of the normal coenzyme in thiamine-catalyzed reactions. See J. A. Gutowski and G. E. Lienhard, *J. Biol. Chem.* 251(1976):2863, for a discussion of this analog.

10. Without O_2 as a terminal acceptor for electrons from NADH, the citric acid cycle cannot operate in a sustained manner. Rather, the pyruvate that is produced by glycolysis must be reduced to lactate (in muscle; or ethanol in yeast) so that the NADH produced in glycolysis can be oxidized to NAD^+. Oxygen deficiency is made worse by the presence of carbon dioxide, which, along with acetyl-CoA, is a product of the pyruvate dehydrogenase complex. Therefore, inhibiting pyruvate dehydrogenase will decrease the production of CO_2 and lessen the severity of the shock.

11. First, acetyllipoamide, and then acetyl CoA.

12. In muscle, the acetyl CoA generated by the complex is used for energy generation. Consequently, signals that indicate an energy-rich state (high ratios of ATP/ADP and $NADH/NAD^+$) inhibit the complex, whereas the reverse conditions stimulate the enzyme. Calcium as the signal for muscle contraction (and, hence, energy need) also stimulates the enzyme.

    In liver, the acetyl CoA derived from pyruvate is used for biosynthetic purposes, such as fatty acid synthesis. In this case, insulin, the hormone denoting the fed state, stimulates the complex.

13. (a) Enhanced kinase activity will decrease the activity of the PDH complex because phosphorylation by the kinase inhibits the complex.
    (b) Phosphatase activates the complex by removing a phosphate. If the phosphatase activity is diminished, the activity of the PDH complex also will decrease.

14. She might have been ingesting, in some fashion, the arsenite from the peeling paint or the wallpaper. Also, she might have been breathing arsine gas from the wallpaper, which would be oxidized to arsenite in her body. In any of these circumstances, the ingested arsenite would inhibit enzymes that require lipoic acid—notably, the PDH complex.

15. The TCA cycle depends on a steady supply of NAD^+, which is typically generated from the reaction of NADH with oxygen. If there is no oxygen to accept the electrons, NADH will accumulate. The citric acid cycle will cease to operate because there will be a shortage of NAD^+.

16. (a) The steady-state concentrations of the products (OAA + NADH) are much lower than those of the substrates (Mal + NAD^+). Hence, the reaction is pushed "uphill" by the overwhelming mass action caused by the differences in concentration.

    (b) $$\frac{[OAA][NADH]}{[Mal][NAD^+]} = e^{-29/2.47} = 7.96 \times 10^{-6}$$

    Since $[NADH]/[NAD^+] = 1/8$,

    $$\frac{[OAA]}{[Mal]} = 7.96 \times 10^{-6} \times 8 = 6.37 \times 10^{-5}$$

    The reciprocal of this is 1.57×10^4, the smallest [Mal]/[OAA] ratio permitting net OAA formation.

17. We need a scheme for the *net* synthesis of α-ketoglutarate from pyruvate, that is, we seek reactions that will allow all of the carbons in α-ketoglutarate to come from pyruvate. This will be possible *only* if half of the available pyruvate is converted to

oxaloacetate by the anaplerotic reaction (pyruvate carboxylase), while the other half is converted to acetyl-CoA by pyruvate dehydrogenase. Here is the set of reactions that must be summed:

Pyruvate + CO_2 + ATP + H_2O $\longrightarrow$ oxaloacetate + ADP + P_i + $2H^+$

Pyruvate + CoA + NAD^+ $\longrightarrow$ acetyl-CoA + CO_2 + NADH

Acetyl-CoA + oxaloacetate + H_2O $\longrightarrow$ citrate + CoA + H^+

Citrate $\longrightarrow$ isocitrate

Isocitrate + NAD^+ $\longrightarrow$ α-ketoglutarate + CO_2 + NADH

Sum: 2 Pyruvate + ATP + 2 NAD^+ + 2 H_2O $\longrightarrow$
  α-ketoglutarate + CO_2 + ADP + P_i + 2 NADH + $3H^+$

18. Due to the inhibition, succinate will increase in concentration, followed by α-ketoglutarate and then the other intermediates "upstream" of the site of inhibition. Succinate has two methylene groups that are required for the dehydrogenation, whereas malonate has only one.

19. Pyruvate carboxylase should be active only when the acetyl CoA concentration is high. Acetyl CoA might accumulate if the energy needs of the cell are not being met, for example because of a deficiency of oxaloacetate. Under these conditions the pyruvate carboxylase catalyzes an anapleurotic reaction to provide more oxaloacetate. Alternatively, acetyl CoA might accumulate because the energy needs of the cell have been met. In this circumstance, pyruvate will be converted back into glucose, and the first step in this conversion is the formation of oxaloacetate.

20. The energy released when succinate is reduced to fumarate is not sufficient to power the synthesis of NADH but is sufficient for synthesis of $FADH_2$ from FAD.

21. Citrate is a tertiary alcohol that cannot be oxidized because oxidation requires a hydrogen atom to be removed from the carbon atom bonded to the OH group. No such hydrogen exists in citrate. The isomerization converts the tertiary alcohol into isocitrate, which is a secondary alcohol that can be oxidized.

22. The enzyme nucleoside diphosphokinase is able to transfer a phosphoryl group from GTP (or any nucleoside triphosphate) to ADP, with no energy cost, according to the reversible reaction:

GTP + ADP $\rightleftharpoons$ GDP + ATP

23. The reaction includes the formation of a new carbon-carbon bond and is powered by the hydrolysis of a thioester. Acetyl CoA provides the thioester that is converted initially into citryl CoA. When this thioester is hydrolyzed, citrate is formed in an irreversible reaction.

24. We cannot get the net conversion of fats into glucose because the only means to get the carbons from fats into oxaloacetate, the precursor to glucose, is through the citric acid cycle. However, although two carbon atoms enter the cycle as acetyl CoA, two carbon atoms are lost as CO_2 before the oxaloacetate is formed. Thus, although some carbon atoms from fats may end up as carbon atoms in glucose, we cannot obtain a net synthesis of glucose from fats.

25. As a product of the pyruvate dehydrogenase complex, acetyl CoA from other sources will inhibit the complex. Glucose metabolism to pyruvate also will be slowed because acetyl CoA is being derived from an alternative source.

26. The enolate anion of acetyl CoA attacks the carbonyl carbon atom of glyoxylate to form a C—C bond. This reaction is like the condensation of oxaloacetate with the enolate anion of acetyl CoA. Glyoxylate contains a hydrogen atom in place of the $-CH_2COO^-$ of oxaloacetate; the reactions are otherwise nearly identical.

27. The labeled carbon will be incorporated into citrate at carbon 5 (only):

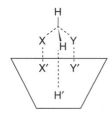

*citrate labeled at carbon 5

In the drawing, carbons 1 and 2 of citrate come from acetyl-CoA. Carbon 6 is lost in the formation of α-ketoglutarate, so none of the label from carbon 5 is lost in that step. Early investigators (until Ogston in 1948; see next problem) thought that carbons 1 and 5 of citrate were indistinguishable and so were surprised when *all* of carbon 5 and *none* of carbon 1 was lost in the decarboxylation of α-ketoglutarate. In fact, citrate is *pro*chiral and so the two ends are distinguishable (see next problem).

28. The enzyme can provide a "3-point landing" at sites X′, Y′, and H′, to bind groups X, Y, and always the lower H (never the upper H on the small molecule). The two hydrogens on the tetrahedral carbon atom in the drawing are therefore distinguishable based on their relative orientations with respect to X and Y. The molecule $CXYH_2$ is "prochiral."

29. (a) A balanced equation for the oxidation of citrate would be

$$C_6H_8O_7 + 4.5\ O_2 \rightarrow 4\ H_2O + 6\ CO_2$$

From the stoichiometry of the balanced equation, 4.5 moles of O_2 would be consumed per mole of citrate, corresponding to 13.5 μmol O_2 per 3 μmol citrate.

(b) The consumption of oxygen is *higher* than a stoichiometric oxidation of citrate would suggest. The result could suggest that the citrate is not being consumed, but rather is acting "catalytically" or is being regenerated in a cycle (as Krebs correctly hypothesized).

30. (a) The presence of arsenite correlates with the disappearance of citrate.

(b) When more citrate is present, a smaller fraction of the total citrate disappears (38% of 90 μmol disappears, whereas 95% of 22 μmol disappears).

(c) A site subsequent to citrate (and more than one step removed) in the citric acid cycle is inhibited by arsenite. (If the immediate step citrate ⟷ isocitrate step were inhibited, then citrate would accumulate, but this is not observed.) At low citrate concentrations, the citrate disappears almost completely because its regeneration is blocked (as some "downstream" step of the cycle is blocked by arsenite). At higher citrate concentrations, some steps between citrate and the block may reversibly approach equilibrium; in this case not all citrate would be depleted.

31. (a) The number of bacterial colony-forming units is much lower in the absence of the gene for isocitrate lyase. (After 15 weeks, the difference is a factor of 100, i.e., fewer than 10^5 cfu without the gene, compared to $>10^7$ cfu when the gene is present.)

    (b) Yes. When the isocitrate lyase gene is restored, then the number of cfu also is restored.

    (c) The experiment in part b confirms the direct influence of the gene for isocitrate lyase. (Because replacing the gene restores the number of CFU, other possible indirect factors or unexpected side effects of removing the gene can be excluded.)

    (d) The glyoxalate cycle will allow the bacteria to subsist on acetate from the breakdown of fatty acids from the lipid-rich environment. Without the glyoxalate cycle, the synthesis of carbohydrates from lipids is not possible. One can speculate that without the glyoxalate cycle, the bacteria will lack carbohydrates or other key metabolic intermediates.

CHAPTER 18

1. In fermentations, organic compounds are both the donors and the acceptors of electrons. In respiration, the electron donor is usually an organic compound, whereas the electron acceptor is an inorganic molecule, such as oxygen.

2. Biochemists use E'_0, the value at pH 7, whereas chemists use E_0, the value in 1 M H^+ (when the pH is 0). The prime denotes that pH 7 is designated as the standard state for biochemistry.

3. The reduction potential of $FADH_2$ is less than that of NADH (see Table 18.1). Consequently, when those electrons are passed along to oxygen, less energy is released. The consequence of the difference is that electron flow from $FADH_2$ to O_2 pumps fewer protons than do the electrons from NADH.

4. $FADH_2 + \frac{1}{2} O_2 \rightarrow FAD + H_2O$. $\Delta E'_0 = 0.82\ V - (-0.22)\ V = +1.04\ V$.

   $G^{o\prime} = -nF(\Delta E'_0) = (-2)(96.48\ kJ\ mol^{-1}\ V^{-1})(1.04\ V) = -200.7\ kJ\ mol^{-1}$
   $(-48.0\ kcal\ mol^{-1})$.

5. For succinate + FAD $\rightarrow$ fumarate + $FADH_2$, $\Delta E'_0$ is ~ 0.00 V $-(0.03)$ V $= -0.03$ V, and $\Delta G^{o\prime}$ is ~$(-2)(96.48\ kJ\ mol^{-1}\ V^{-1}\ (-0.03\ V) = \underline{+5.8\ kJ\ mol^{-1}}$.

   For succinate + NAD^+ $\rightarrow$ fumarate + NADH + H^+, $\Delta E'_0 = -0.32\ V - (0.03)\ V$
   $= -0.35\ V$,
   and $\Delta G^{o\prime}$ is $= (-2)(96.48\ kJ\ mol^{-1}\ V^{-1})(-0.35\ V) = \underline{+67.5\ kJ\ mol^{-1}}$.

   The oxidation of succinate by NAD^+ is not thermodynamically feasible; the energy cost is too high. When E'_0 is assumed to be nearly 0 V for the FAD-$FADH_2$ redox couple, then the value of $\Delta G^{o\prime}$ is only mildly positive (slightly unfavorable yet still feasible).

6. Pyruvate accepts electrons and is thus the oxidant.

   NADH gives up electrons and is the reductant.

7. $\Delta G^{o\prime} = -nF\Delta E'_0$

8. The $\Delta E'_0$ value can be altered by changing the environment around the iron ion.

9. Based on the E'_0 values and the role of ubiquinone, the order is:
   (c) NADH-Q reductase ($E'_0 = -0.32$ V).
   (e) Ubiquinone carries electrons from NADH-Q reductase to Q-cytochrome c oxido-reductase.
   (b) Q-cytochrome c oxidoreductase ($E'_0 = +0.04$ V).
   (a) cytochrome c ($E'_0 = +0.22$ V).
   (d) cytochrome c oxidase(/O_2) ($E'_0 = +0.82$ V).

10. (a) Complex I is NADH-Q oxidoreductase (answer 4).
    (b) Complex II is succinate-Q reductase (answer 3).
    (c) Complex III is Q-cytochrome c oxidoreductase (answer 1).
    (d) Complex IV is cytochrome c oxidase (answer 5).
    (e) Ubiquinone is a mobile electron carrier, coenzyme Q (answer 2).

11. The ten isoprene units render coenzyme Q soluble in the hydrophobic environment of the inner mitochondrial membrane. The two oxygen atoms of coenzyme Q can reversibly bind two electrons and two protons as the molecule transitions between the quinone form and the quinol form.

12. The results with each inhibitor follow the order of components in the electron-transport chain. Components before the block point will be reduced, whereas components after the block point will be oxidized. Therefore:
    - with rotenone, NADH and NADH-Q oxidoreductase will be reduced; and the remainder will be oxidized.
    - with antimycin A, NADH, NADH-Q oxidoreductase and coenzyme Q will be reduced; and the remainder will be oxidized.
    - with cyanide, all will be reduced.

13. Following the logic in problem 12 (above), complex I would be reduced, whereas Complexes II, III, and IV would be oxidized. The citric acid cycle components would become reduced because there is no way to oxidize the NADH that is produced in the cycle.

14. The respirasome is yet another example illustrating the advantages of using supramolecular complexes in biochemistry. Having the three complexes that are proton pumps associated with one another will enhance the efficiency of electron flow from complex to complex, which in turn will cause more-efficient proton pumping.

15. Hydroxyl radical, (OH •) hydrogen peroxide (H_2O_2), superoxide ion (O_2^-), and peroxide (O_2^{2-}) are small molecules that react with a host of macromolecules—including proteins, nucleotides, and lipids. The reactions damage the respective macromolecules and disrupt cell structure and function.

16. The ATP is recycled by processes that generate ATP, most notably oxidative phosphorylation.

17. (a) 12.5 (b) 15 (c) 32 (d) 13.5 (e) 30 (f) 16
    These answers are readily obtained if one remembers the ATP yields from the various parts of glycolysis and remembers that (1) cytoplasmic NADH yields only 1.5 ATP, (2) pyruvate $\rightarrow$ acetyl CoA + 1 NADH yields 2.5 ATPs, (3) each acetyl CoA traversing the citric acid cycle yields 10 ATP (7.5 from 3 NADH, 1.5 from 1 $FADH_2$, and 1 from GTP), and (4) galactose requires 1 ATP for activation, just as glucose does.

18. (a) Blocks electron transport and proton pumping at site 3.
    (b) Blocks electron transport and ATP synthesis by inhibiting the exchange of ATP and ADP across the inner mitochondrial membrane.
    (c) Blocks electron transport and proton pumping at site 1.
    (d) Blocks ATP synthesis without inhibiting electron transport by dissipating the proton gradient.
    (e) Blocks electron transport and proton pumping at site 3.
    (f) Blocks electron transport and proton pumping at site 2.

19. The electron transport chain and the ATP synthase are coupled because both involve a proton gradient across a membrane. The electron transport chain established the proton gradient, while the ATP synthase dissipates the proton gradient. If the ATP synthease is inhibited, the proton gradient will remain large, the barrier to further proton pumping will be high, and consequently electron transport will cease.

20. The subunits are jostled by background thermal energy (Brownian motion). The proton gradient makes clockwise rotation more likely because that direction results in protons flowing down their concentration gradient.

21. If the mitochondria function poorly, the only means of generating ATP is by anaerobic glycolysis, which will lead to an accumulation of lactic acid in blood.

22. If ADP cannot get into mitochondria, the electron-transport chain will cease to function because there will be no acceptor for the energy. A corollary is that NADH will build up in the mitochondrial matrix. Recall that NADH inhibits some citric acid cycle enzymes and that NAD^+ is required by several citric acid cycle enzymes. Therefore, the citric acid cycle will cease operation. Furthermore, glycolysis will stop functioning aerobically but will switch to anaerobic glycolysis so that the NADH can be re-oxidized to NAD^+ by lactate dehydrogenase.

23. Initially, the mitochondrial suspension (if fresh) will be respiring at a slow rate. The rate of oxygen consumption (and electron transport) will increase when glucose is added, increase further when ADP and P_i are added, and increase still further when citrate is added. Then the addition of oligomycin will stop ATP synthesis, electron transport, and the uptake of oxygen. The subsequent additions will have no effect because the system already is inhibited by oligomycin. The graph below summarizes these effects.

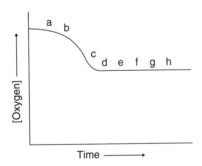

24. (a) Since P:O is the number of high-energy phosphate bonds formed per atom of oxygen and oxygen accepts two electrons, $P/O = {\sim}P/O = ATP/O = ATP/2e^-$. The amount of ATP synthesized is directly proportional to the number of protons pumped (H^+_{pu}) and inversely proportional to the number of protons required for ATP synthesis (H^+_{re}). Hence ATP synthesized = H^+_{pu}/H^+_{re}.
    (b) Since NADH donates two electrons to one atom of oxygen, the P/O = ATP formed = $10H^+_{pu}/4H^+_{re} = 2.5$. For succinate (FADH$_2$) oxidation the P/O = $6H^+_{pu}/4H^+_{re} = 1.5$.

25. Cyanide can be lethal because it binds to the ferric form of cytochrome oxidase and thereby inhibits oxidative phosphorylation. Nitrite converts ferrohemoglobin to ferrihemoglobin, which also binds cyanide. Thus, ferrihemoglobin competes with cytochrome oxidase for cyanide. This competition is therapeutically effective because the amount of ferrihemoglobin that can be formed without impairing oxygen transport is much greater than the amount of cytochrome oxidase.

26. Such a defect (called *Luft's syndrome*) was found in a 38-year-old woman who was incapable of performing prolonged physical work. Her basal metabolic rate was more than twice normal, but her thyroid function was normal. A muscle biopsy showed that her mitochondria were highly variable and atypical in structure. Biochemical studies then revealed that oxidation and phosphorylation were not tightly coupled in these mitochondria. In this patient, much of the energy of fuel molecules was converted into heat rather than ATP. The development of mitochondrial medicine is lucidly reviewed in R. Luft. *Proc. Nat. Acad. Sci.* 91(1994):8731.

27. Recall that during the cleavage of fructose 1,6-bisphosphate, glyceraldehyde 3-phosphate and dihydroxyacetone phosphate are formed. Only glyceraldehyde 3-phosphate can proceed through glycolysis to pyruvate. Although it can be converted to glycerol by some cells, dihydroxyacetone phosphate is a dead-end in glycolysis; to proceed further it must be converted to glyceraldehyde 3-phosphate by triose phosphate isomerase.

28. Inhibitors of electron transport cause the electron carriers between the source of electrons (e.g., NADH, $FADH_2$) and the point of inhibition to become more reduced while the electron carriers between the point of inhibition and O_2 become more oxidized. Thus there is a "crossover point" from reduced carriers to oxidized carriers. Therefore, from the information given we conclude that the inhibitor acts somewhere between QH_2 and cytochrome *c*; it prevents the reduction of cytochrome *c* by QH_2.

29. Uncouplers allow food energy to be wasted. Glucose is oxidized and electrons flow to oxygen, but ATP is not synthesized, and the energy is not available for biosynthesis. The problem is that the energy will appear as heat. Producers of antiperspirants may like this effect, but the difficulty in controlling body temperature will cause the uncouplers to be toxic.

30. If the proton gradient cannot be dissipated by flow through the ATP synthase, the proton gradient will eventually become so large that the energy released by the electron-transport chain will not be sufficient to pump protons against the larger-than-normal gradient.

31. One possible approach would be to prepare vesicles that encapsulate ADP + P_i and contain ATP synthase properly oriented in their membranes. Impose an artificial proton gradient (e.g., using a light-driven proton pump; see Figure 18.25), and test for the synthesis of ATP in the presence and absence of the new chemical. If ATP synthesis is inhibited, then the chemical inhibits ATP synthase. If not, then the chemical could be an inhibitor of electron transport and should be tested for its ability to inhibit oxygen uptake in respiring mitochondria.

32. Presumably, the muscle requires more ATP because it has greater energy needs, especially during exercise. This requirement for ATP means that more sites of oxidative phosphorylation are needed, and these sites can be provided by an increase in the number of cristae.

33. The arginine, with its positive charge, will facilitate proton release from aspartic acid by stabilizing the negatively charged aspartate.

34. Twelve protons per 3 ATP would be 12/3 = 4 protons per ATP.
Fourteen protons per 3 ATP would be 14/3 = 4.7 protons per ATP.

35. In reverse, the ATP synthase would pump protons at the expense of ATP hydrolysis, thus maintaining the proton-motive force. The synthase would function as an ATPase. There is some evidence that damaged mitochondria actually use this tactic to maintain the proton-motive force, at least temporarily.

36. The finding with regard to rotenone suggests that malfunctioning mitochondria may play a role in the development of Parkinson disease. Specifically, it implicates Complex I.

37. Remember that the extra negative charge on ATP relative to ADP accounts for its more rapid translocation out of the mitochondrial matrix. If the charge differences between ATP and ADP were lessened by the binding of the Mg^{2+}, ADP might more readily compete with ATP for transport to the cytoplasm.

38. When all of the available ADP has been converted into ATP, the ATP synthase lacks a necessary substrate and can no longer function. The proton gradient becomes so large that the energy of the electron-transport chain is no longer sufficient to pump against the gradient. Therefore, electron transport activity decreases and, as a consequence, oxygen consumption falls.

39. Because of the H^+/OH^- equilibrium, the effect on the proton gradient is the same in each case.

40. The proton-motive force can be used to drive other transport processes, for example, ATP export from the matrix, or phosphate import into the matrix.

41. Recall that enzymes catalyze reactions in both directions. The direction in which a reaction proceeds is determined thermodynamically by the ΔG difference between substrates and products. An enzyme will speed up the rates of both the forward and the backward reactions. The hydrolysis of ATP is exergonic, and so in the absence of a driving force, isolated and uncoupled subunits of ATP synthase will enhance the hydrolytic reaction.

42. The cytoplasmic kinases thereby obtain preferential access to the ATP that is exported by VDAC.

43. Much of the ATP will be trapped inside mitochondria because ATP/ADP exchange is not facilitated in the absence of the translocase. Electron transport will stop because of low levels of ADP and high levels of ATP in the mitochondria (and due to coupling; see problem 4, above). With electron transport slowed, NADH and $FADH_2$ will be in excess, and dehydrogenases such as pyruvate dehydrogenase and succinate dehydrogenase will be inhibited. The low dehydrogenase activity will cause products such as lactate and alanine (from pyruvate, by reduction or transamination, respectively) as well as succinate to accumulate. Furthermore, with electron transport to oxygen incomplete, intermediate oxidation states of oxygen such as H_2O_2 also will accumulate.

44. (a) Vitamins C and E.
    (b) Exercise induces superoxide dismutase, which converts ROS into hydrogen peroxide and oxygen.
    (c) The answer to this question is not fully established. Two possibilities are (1) the suppression of ROS by vitamins may prevent the expression of more superoxide dismutase, or (2) some ROS may be signal molecules that could be required to stimulate insulin-sensitivity pathways.

CHAPTER 19

1. Photosystem I generates ferredoxin, which reduces $NADP^+$ to NADPH, which provides reducing power for biosynthesis. Photosystem II activates the manganese

complex, an oxidant capable of oxidizing water, generating electrons for photosynthesis, and generating protons to form a proton gradient.

2. The light reactions take place on thylakoid membranes. Increasing the available thylakoid membrane surface area increases the number of ATP- and NADH-generating sites.

3. The idea is to capture more photons. The light-harvesting complexes work with the reaction centers to absorb more light than would be absorbed by the reaction centers alone. The light-harvesting complexes funnel light to the reaction centers.

4. $NADP^+$ is the ultimate electron acceptor, and H_2O is the ultimate electron donor. Light energy powers the electron flow between the donor and the acceptor.

5. A greater pH gradient is required because the charge gradient, a component of the proton-motive force, is less in chloroplasts than in mitochondria. In chloroplasts, the charge gradient is neutralized by the influx of Mg^{2+} into the lumen of the thylakoid membranes.

6. Because chlorophyll is a hydrophobic molecule, it is readily inserted into the hydrophobic interior of the thylakoid membranes.

7. Protons for the proton gradient in chloroplasts arise from several sources, including (a) those released during the oxidation of water, (b) those pumped into the lumen by the cytochrome bf complex, and (c) protons removed from the stroma during the reduction of $NADP^+$ and plastoquinone.

8. The energy of 1 mol of 700-nm photons is Avodagro's number times hc/λ, with h being Planck's constant, c the speed of light, and $\lambda = 700$ nm. Substituting yields:

$(6.022 * 10^{23})(6.626 * 10^{-34}$ J s$)(2.998 * 10^8$ m s$^{-1})/(700 * 10^{-9}$ m$) =$

$(0.171 * 10^6)$ J mol$^{-1} = 171$ kJ mol^{-1}.

The absorption of light by photosystem I results in $\Delta E'_0$ of about -1.0 V.

Recall that $\Delta G^{o\prime} = -nF\Delta E'_0$, where $F = 96.48$ kJ mol^{-1} V^{-1} (or 23.06 kcal mol^{-1} V^{-1}).

Under standard conditions, the energy change for one mol of electrons is therefore about 96.5 kJ (or 23.1 kcal). Thus, the efficiency is about 96.5/171 = <u>56.4%</u>.

9. The electron flow from PS II to PS I is uphill, or endergonic. For this uphill flow, ATP would need to be consumed, thereby detracting from the ATP yield and working against the purpose of photosynthesis.

10. $\Delta E'_0 = -0.32 - (-0.43) = +0.11$ V. $\Delta G^{o\prime}$ (to reduce 1 mol of $NADP^+$) $= -2 \times 96.5 \times 0.11 = -21.2$ kJ/mol^{-1} (-5.1 kcal mol^{-1}).

11. (a) Some process for trapping energy from an external source is necessary for life as we know it. Because radiation is an efficient mechanism for transferring the energy, it is likely that photons would be involved in bringing the energy to the places where life exists. (Alternatively, a local and stable long-term source of energy would be needed. Nevertheless, *time* as well as energy is a critical factor because the evolution of life is a slow process. Therefore, the hypothetical energy source would need to be reliable for a long time.)

 (b) No. Electron donors other than water can be used for photosynthesis, for example, H_2, H_2S, or other small organic molelcules (see Table 19.1).

12. DCMU inhibits electron transfer between Q and plastoquinone in the link between photosystems II and I. O_2 evolution can occur in the presence of DCMU if an artificial electron acceptor such as ferricyanide can accept electrons from Q.

13. Cyclic photophosphorylation could occur. Electrons would go from P700* to ferrodoxin to cytochrome $b6/f$ (generating a proton gradient for ATP synthesis), and finally to

plastocyanin and back to P700. The site of DCMU inhibition (Q to plastoquinone) is outside this cycle.

14. (a) The energy of photons is inversely proportional to the wavelength. Since 600-nm photons have an energy content of 47.6 kcal/einstein, 1000-nm light will have an energy content of 600/1000 × 47.6 kcal/einstein. Since 1 cal = 4.184 joule, 28.7 kcal × 4.184 = 120 kJ/einstein.

   (b) -120 kJ/mol ($\Delta G^{o\prime}$) = $-1 \times 96.5 \times V$. Therefore, $V = -120/-96.5 = 1.24$ volts.

   (c) If 1000-nm photons have a free-energy content of 120 kJ/einstein and ATP has a free-energy content of 50 kJ/mol, then 1000-nm photon has the free-energy content of 120/50, or 2.4 ATP. Therefore, a minimum of 0.42 (1/2.4) photon is needed to drive the synthesis of an ATP.

15. From Section 18.2.3, the electron transfer rate is about 10^{13} s^{-1} for groups in contact and falls by a factor of 10 for every 1.7 Å through a protein environment. Dividing 22 Å by 1.7 Å gives an estimated decrease of about 12.94 orders of magnitude. The estimated rate is therefore $10^{13}/10^{12.94} = 10^{0.06} \simeq 1$ event per second. (This is much too slow for photosynthesis! Charge recombination at the chlorophyll would dominate.)

16. The transfer rate depends upon the sixth power of the distance of separation. In this example, the distance doubles, so the rate should decrease by a factor of $2^6 = 64$. 10 ps × 64 = 640 ps.

17. The cristae.

18. There are a number of similarities. Both processes take place in membranes. In eukaryotes, these membranes reside inside specialized organelles. Both processes depend on high-energy electrons to generate ATP. Photosynthesis and oxidative phosphorylation both use redox reactions to generate a proton gradient, and the enzymes that convert the proton gradient into ATP are very similar in both processes.

   The differences concern the sources of the high-energy electrons. In oxidative phosphorylation, the high-energy electrons originate in fuels and are extracted as reducing power in the form of NADH. In photosynthesis, the high-energy electrons are generated by light and are captured as reducing power in the form of NADPH.

19. One needs to factor in the NADPH from photosynthesis because it is an energy-rich molecule. Recall, from Chapter 18, that NADH is worth 2.5 ATP if oxidized by the electron-transport chain. Therefore, 12 molecules of NADPH represent the energetic equivalent of 30 ATP molecules. The synthesis of glucose requires 18 molecules of ATP (directly) and 12 molecules of NADPH (representing 30 additional molecules of ATP; indirectly). In summary, the equivalent of 48 molecules of ATP is required for the synthesis of glucose.

20. Both photosynthesis and cellular respiration are powered by high-energy electrons that flow toward more-stable states. In cellular respiration, the high-energy electrons are derived from the oxidation of carbon fuels as NADH and FADH$_2$. These electrons release their energy as they reduce oxygen, and they find stability in the water molecules that are formed. In photosynthesis, high-energy electrons are generated when the photopigments absorb light energy, and they find stability in photosystem I and ferredoxin.

21. The Hill reaction uses photosystem I. Electrons from P680 are excited and are replenished by electrons from water (leading to evolution of O$_2$). The excited electrons in P680* pass to pheophytin and then to Q and finally to the artificial acceptor such as ferricyanide.

22. (a) Thioredoxin is the natural regulator in vivo.

   (b) There is no effect on the control mitochondrial enzyme, but increasing the reducing power increases the activity of the modified (chimeric) enzyme.

(c) Thioredoxin enhances the effect of DTT on the modified enzyme by an additional factor of approximately two. Since the DTT alone, especially at the higher concentrations, should provide sufficient reducing power, the additional enhancement with thioredoxin could be due to another effect. For example, thioredoxin could bind to the enzyme and induce a conformational change to a more active state. (In vivo—without DTT—the thioredoxin also would serve a reducing role.)

(d) Yes. The segment that was removed and replaced is responsible for the redox regulation that is observed in chloroplasts but not in mitochondria.

(e) For chloroplasts, the redox potential of the stroma provides a way to link the activities of key enzymes to the level of illumination. Enzymes that do not respond to light directly are thereby able to respond to the levels of reducing agents and have their activities coordinated with the extent of ongoing photosynthesis.

(f) The sulfhydryl groups of Cys are likely to be influenced. The Cys side chains can exist in $-SH$ (reduced) and disulfide ($-S-S-$; oxidized) forms.

(g) Directed mutagenesis experiments to change selected cysteines to alanine or serine could confirm their importance in the regulatory mechanism.

CHAPTER 20

1. The Calvin cycle is the primary means of converting gaseous CO_2 into organic matter—that is, biomolecules that contain carbon-carbon bonds. Essentially, every carbon atom in the human body has passed through rubisco and the Calvin cycle at some time in the past.

2. • The Calvin cycle takes place in the stroma of chloroplasts, whereas the Krebs cycle takes place in the matrix of mitochondria.

   • The Calvin cycle involves anabolic carbon chemistry for photosynthesis, whereas the Krebs cycle involves catabolic carbon chemistry for oxidative phosphorylation.

   • The Calvin cycle fixes CO_2, whereas the Krebs cycle releases CO_2.

   • The Calvin cycle requires high-energy electrons (NADPH), whereas the Krebs cycle generates high-energy electrons in the form of NADH.

   • The Calvin cycle requires ATP, whereas the Krebs cycle generates ATP and GTP.

   • The Calvin cycle embodies a complex stoichiometry, whereas the Krebs cycle utilizes a straightforward stoichiometry.

   • Both cycles regenerate a crucial starting compound, namely ribulose 1,5-bisphosphate for the Calvin cycle, and oxaloacetate for the Krebs cycle.

3. (a) Initially, ribulose 1,5-bisphosphate reacts with CO_2 to form two molecules of 3-phosphoglycerate. Within the first few seconds of the experiments, only 3-phosphoglycerate will be labeled.

   (b) As time passes, the ^{14}C label from CO_2 is distributed around the Calvin cycle, and the other molecules in the Calvin cycle become labeled.

4. Stage 1 is the fixation of CO_2 with ribulose 1,5-bisphosphate and the subsequent formation of 3-phosphoglycerate.

   Stage 2 involves the conversion of some of the 3-phosphoglycerate into hexose.

   Stage 3 is the regeneration of ribulose 1,5-bisphosphate, so that a new cycle can begin.

5. Rubisco catalyzes a crucial reaction that is widely needed, and the enzyme is highly inefficient. Consequently, it is required in large amounts to overcome its slow catalysis.

6. In the absence of CO_2, the carbamate cannot be formed on lysine 201 of rubisco. Therefore, rubisco is prevented from catalyzing the oxygenase reaction when CO_2 is absent.

7. NADPH is available in the chloroplasts because it is generated by the light reactions of photosynthesis. Therefore it is advantageous for the chloroplast isozyme of glyceraldehyde-3-phosphate dehydrogenase to make use of NADPH as a cofactor.

8. The conversion of ribulose 1,5-bisphosphate to 3-phosphoglycerate does not require ATP, so it will continue until the ribulose 1,5-bisphosphate is largely depleted.

9. When the concentration of CO_2 is drastically decreased, the rate of conversion of ribulose 1,5-bisphosphate to 3-phosphoglycerate will greatly decrease, whereas the rate of utilization of 3-phosphoglycerate will not be diminished.

10. Glyoxalate + glutamate $\longrightarrow$ glycine + α-ketoglutarate

11. The crabgrass adapts better to the hot and dry conditions. One could speculate that crabgrass may close the stomata of their leaves during the day and use CO_2 that has been stored as malate in vacuoles the previous night (*Crassulacean* acid metabolism, Section 20.2).

12. The C_4 pathway allows the CO_2 concentration to increase at the site of carbon fixation. High concentrations of CO_2 inhibit the oxygenase reaction of rubisco. This inhibition is important for tropical plants because the oxygenase activity increases more rapidly with temperature than does the carboxylase activity.

13. The net reaction of the C_4 pathway is active transport of CO_2 at the expense of ATP:

CO_2 (is mesophyll cell) + ATP + 2 H_2O $\longrightarrow$

CO_2 (is bundle-sheath cell) + AMP + 2 P_i + 2 H^+.

The energy equivalent of <u>two</u> ATP molecules is consumed in the production of AMP, with release of <u>two</u> molecules of inorganic phosphate to drive the active transport. (The AMP can be converted to ADP by use of a second ATP in the reaction: ATP + AMP $\longrightarrow$ 2 ADP.)

14. Photorespiration is the consumption of oxygen by plants with the production of CO_2, but it does not generate energy. Photorespiration is due to the oxygenase activity of rubisco. It is wasteful because, instead of fixing CO_2 for conversion into hexoses, rubisco is generating CO_2.

15. C_4 plants have the advantage in hotter environments and so may become more prominent at higher latitudes as well as lower latitudes under the influence of global warming. C_3 plants will retreat to cooler regions.

16. The light reactions lead to increased stromal concentrations of NADPH, reduced ferredoxin, and Mg^{2+}. By pumping protons out of the stroma, the light reactions also increase the pH.

17. Transaldolase and transketolase catalyze the transformation of the five-carbon sugar formed by the oxidative phase of the pentose phosphate pathway into the six-carbon fructose 6-phosphate and the three-carbon glyceraldehyde 3-phosphate. The latter sugars are necessary intermediates in glycolysis (and gluconeogenesis).

18. Since the C-1 of glucose is lost during the conversion to pentose, carbon atoms 2 through 6 of glucose become carbon atoms 1 through 5 of the pentose. That is, each pentose carbon is numerically 1 less than its counterpart in glucose.

19. Note that in the oxidative decarboxylation of 6-phosphogluconate, oxidation occurs at the carbon β to the carboxyl group. A similar β-oxidation occurs during the decarboxylation of isocitrate in the citric acid cycle. In both cases a β-keto acid intermediate is formed. Since β-keto acids are relatively unstable, they are easily decarboxylated.

20. (a) To make six pentoses, four glucose 6-phosphates must be converted to fructose 6-phosphate (no ATP required), and one glucose 6-phosphate must be converted to two molecules of glyceraldehyde 3-phosphate (this requires one ATP). These are converted to pentoses by the following reactions (Table 20.3 in the text).

2 Fructose 6-phosphate + 2 glyceraldehyde 3-phosphate → 2 erythrose 4-phosphate + 2 *xylulose 5-phosphate*

2 Fructose 6-phosphate + 2 erythrose 4-phosphate → 2 glyceraldehyde 3-phosphate + 2 sedoheptulose 7-phosphate

2 Glyceraldehyde 3-phosphate + 2 sedoheptulose 7-phosphate → *2 xylulose 5-phosphate + 2 ribose 5-phosphate*

(b) What really happens is that six molecules of glucose 6-phosphate are converted to $6 CO_2$ + 6 ribulose 5-phosphates + 12 NADPH + 12 H^+ (see Table 20.3 in the text). The ribulose phosphates are then converted back to five molecules of glucose 6-phosphate by the action of transketolase and transaldolase. By these reactions three pentoses are converted to two hexoses and one triose. Thus six pentoses can be converted to four hexoses plus two trioses, and the latter can be converted to the fifth hexose.

21. The nonoxidative phase of the pentose phosphate pathway can be used to convert three molecules of ribose 5-phosphate into two molecules of fructose 6-phosphate and one molecule of glyceraldehyde 3-phosphate. These molecules are components of the glycolytic pathway.

22. The conversion of fructose 6-phosphate into fructose 1,6-bisphosphate by phosphofructokinase requires ATP.

23. When much NADPH is required, the oxidative phase of the pentose phosphate pathway will be followed by the nonoxidative phase. The resulting fructose 6-phosphate and glyceraldehyde 3-phosphate are used to generate glucose 6-phosphate through gluconeogenesis, and the cycle is repeated until the equivalent of one glucose molecule is oxidized to CO_2.

24. Fava beans contain pamaquine, a purine glycoside that can lead to the generation of peroxides—reactive oxygen species (ROS) that can damage membranes as well as other biomolecules. Glutathione is used to detoxify the ROS. The regeneration of glutathione, however, depends on an adequate supply of NADPH, which is synthesized by the oxidative phase of the pentose phosphate pathway. People with low levels of glucose 6-phosphate dehydrogenase will be susceptible to pamaquine toxicity.

25. The deficiency appears to be anemia because red blood cells do not have mitochondria and the only means for them to obtain NADPH is through the pentose phosphate pathway. Other cells will seem not susceptible because they have mitochondrial NADH, which can be converted by indirect means into cytoplasmic NADPH.

26. Reactive peroxides are a type of reactive oxygen species. The enzyme glutathione per-oxidase uses reduced glutathione to neutralize peroxides by converting them into al-cohols while generating oxidized glutathione. Reduced glutathione is regenerated by glutathione reductase with the use of NADPH, the product of the oxidative phase of the pentose phosphate pathway.

27. The $\Delta E'_0$ for the reduction of glutathione by NADPH is $+$ 0.09 V. Then $\Delta G^{o\prime} = -nF\Delta E'_0 = -2 \times 96.5 \times 0.09 = -17.5$ kJ/mol^{-1} (-4.15 kcal/mol). Also, $K'_{eq} = e^{-\Delta G^{o\prime}/RT} = e^{17.5/2.47} = 1.126 \times 10^3$. Thus,

$$K_{eq} = \frac{[GSH]^2[NADP^+]}{[GSSG][NADPH]} = 1126$$

After substituting the given concentrations for GSH and GSSG,

$$\frac{[0.01M]^2[NADP^+]}{[0.001M][NADPH]} = 1.126 \times 10^3$$

Therefore,

$$\frac{[NADP^+]}{[NADPH]} = 1.126 \times 10^4$$

and

$$\frac{[NADPH]}{[NADP^+]} = \frac{1}{1.126 \times 10^4} = 8.9 \times 10^{-5}.$$

Remember, in equilibrium constants the molar concentrations of the reactants are raised to a power equal to the number of moles taking part in the reaction. Therefore, in this problem the [GSH] is squared because, for each mole of GSSG, NADP$^+$, and NADPH, two moles of GSH are involved.

28. In similar fashion to the traditional mechanism, the enolate form of dihydroxyace-tone phosphate could be used. A metal ion instead of a protonated Schiff base could stabilize an enolate anion intermediate. The enolate anion could then add to the alde-hyde of glyceraldehyde-3 phosphate:

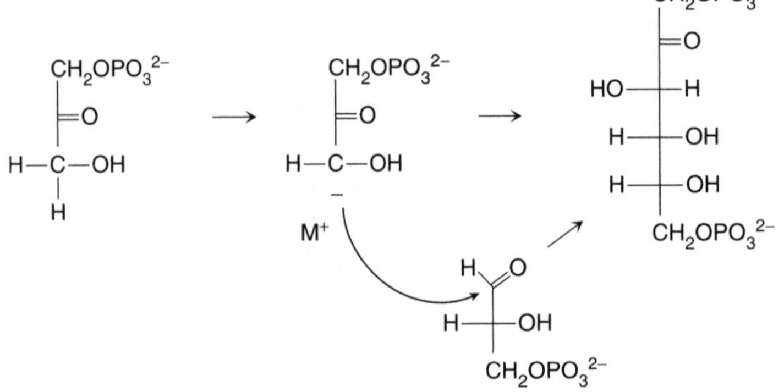

29. The reaction is similar to the hexose phosphate isomerase and triose phosphate isomerase reactions of glycolysis and probably proceeds through an enediol intermediate:

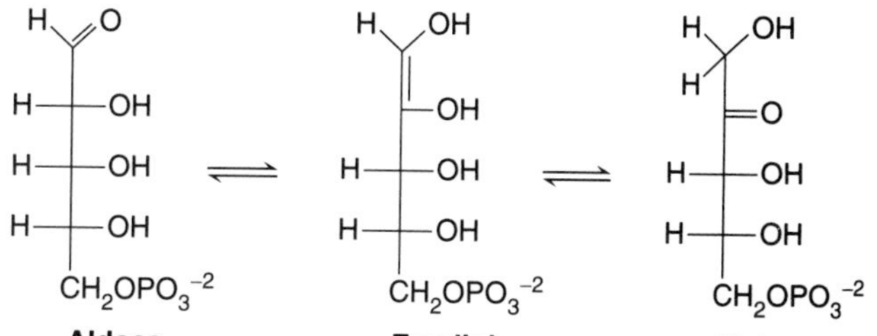

Aldose ⇌ **Enediol** ⇌ **Ketose**

30. Labels at C-1 and C-6 of glucose will behave identically in glycolysis (both emerging at C-3 of pyruvate) and the citric acid cycle. Both labels will transfer to acetyl-CoA (methyl group) and will remain in the citric acid cycle for two rounds. Only with the third turn of the cycle will the C-1 and C-6 labels from glucose finally begin to be released as CO_2 (50% of remaining C-1 and C-6 during the third and each subsequent turn). It is important to note that none of the C-1 or C-6 label will be released in the early stages of glycolysis or the citric acid cycle. By contrast, in the pentose phosphate pathway, *all* of the C-1 label (and *none* of the C-6 label) will be released very quickly as CO_2 at the step where ribulose-5-phosphate is formed. We can put all of these facts together to propose our experiment: incubate a portion each tissue with each labeled glucose sample, and measure the specific activity of CO_2 that is released as a function of time in each experiment. The extent by which release of C-1 label precedes the release of C-6 label will reflect the level of activity of the pentose phosphate pathway. If both labels are released at the same rate by a particular tissue, then the dominant pathway follows glycolysis and the citric acid cycle.

31. From the stoichiometry of the Calvin cycle, two moles of NADPH are needed for every mole of CO_2 that is incorporated into glucose:

$$6\ CO_2 + 18\ ATP + 12\ NADPH + 12\ H_2O \rightarrow$$
$$C_6H_{12}O_6 + 18\ ADP + 18\ P_i + 12\ NADP^+ + 6\ H^+$$

The production of each molecule of NADPH requires illumination from four photons (to activate photosystems I and II, which also produce the necessary ATP). Therefore, the energy of eight photons is needed for every CO_2 that is reduced to the level of hexose. The efficiency is: (477 kJ)/(8 * 199 kJ) = 30%.

32. It is neither a violation nor a miracle. See also problem 19 in Chapter 19. One needs to consider not only the direct need for 18 ATP but also the need for 12 NADPH. Recall, from Chapter 18, that NADH is worth 2.5 ATP if oxidized by the electron-transport chain. Therefore, 12 molecules of NADPH represent the energetic equivalent of 30 ATP molecules. The synthesis of glucose requires 18 molecules of ATP (directly) and 12 molecules of NADPH (representing 30 additional molecules of ATP, indirectly). In summary, the equivalent of 48 molecules of ATP is required for the synthesis of glucose.

33. (a) The C_4 plant is more efficient at higher temperature. Therefore, the curve (on the right) that peaks sharply at about 39°C represents the C_4 plant.

(b) The oxygenase activity of rubisco increases with temperature. Other key enzymes may become inactive at high temperature.

(c) C_4 plants are able to accumulate high concentrations of CO_2 in their bundle-sheath cells.

(d) The C_3 activity depends on passive diffusion of CO_2, whereas the C_4 activity depends on the active transport of CO_2 into the bundle-sheath cells. Once the transport system is saturated (working at maximum rate), then no further increase in photosynthetic activity is possible. By contrast, higher CO_2 concentrations continue to enhance the rate of diffusion and cause increased availability of CO_2 for the C_3 plants.

CHAPTER 21

1. Glycogen is an important fuel reserve for several reasons. The controlled breakdown of glycogen and release of glucose increase the amount of glucose that is available between meals. Hence, glycogen serves to buffer the level of glucose in the blood. Glycogen's role in maintaining blood-glucose levels is especially important because glucose is virtually the only fuel used by the brain, except during prolonged starvation. Moreover, the glucose from glycogen is readily mobilized and is therefore a good source of energy for sudden, strenuous activity. Unlike fatty acids, the released glucose can provide energy in the absence of oxygen and can thus supply energy for anaerobic activity.

2. Glucose is mobilized from the *non*reducing ends of glycogen. The unbranched α-amylose has only one nonreducing end, whereas glycogen has many of them. Therefore, glucose monomers can be released much more quickly from glycogen than from α-amylose.

3. In normal glycogen, branches occur about once in 10 units. Therefore, degradation of this glycogen is expected to give a ratio of glucose 1-phosphate to glucose of about 10:1. An increased ratio (100:1) indicates that the glycogen has a much lower degree of branching, suggesting a deficiency of the branching enzyme.

4. The difference corresponds to the difference in the metabolic role of glycogen in each tissue. Muscle uses glycogen as a fuel for contraction, so when AMP accumulates, the muscle isozyme (the *b* form of phosphorylase) is activated by AMP, signaling a need for more energy and hence more glucose. Liver uses glycogen to maintain blood-glucose levels, such that the liver isozyme (the *a* form of phosphorylase) is inhibited by glucose, signaling that the blood-glucose level is sufficient and breakdown of glycogen should cease.

5. By altering the ratio of substrate and product, a cell can alter the net free-energy change in order to favor either a forward or reverse reaction. Typically, cells maintain the $[P_i]/[\text{glucose 1-phosphate}]$ ratio at greater than 100, thereby substantially favoring the phosphorolysis.

6. The enzymatic defect in von Gierke's disease is the absence of liver glucose 6-phosphatase. The resulting high concentrations of glucose 6-phosphate allosterically activate the inactive glycogen synthase *b*, causing a net increase in liver glycogen.

7. Since glucose 1,6-bisphosphate is an intermediate in the reaction catalyzed by phosphoglucomutase and phosphorylates this enzyme during catalysis, it is reasonable to expect that it can phosphorylate the dephosphoenzyme. Glucose 1,6-bisphosphate is formed from glucose 1-phosphate and ATP by phosphoglucokinase.

8. The different manifestations correspond to the different roles of the liver and muscle. Liver glycogen phosphorylase plays a crucial role in the maintenance of blood-glucose levels. Recall that glucose is the primary fuel for the brain. Muscle glycogen phosphorylase provides glucose only for the muscle and, even then, only when the energy needs of the muscle are high, as during exercise. The fact that there are two different diseases suggests that there are two different isozymic forms of the glycogen phosphorylase—a liver-specific isozyme and a muscle-specific isozyme. An absence of the liver isozyme is much more serious than absence of the muscle isozyme.

9. Water is excluded from the active site to prevent hydrolysis. The entry of water could lead to the formation of glucose rather than glucose 1-phosphate. A site-specific mutagenesis experiment is revealing in this regard. In phosphorylase, Tyr 573 is hydrogen-bonded to the 2'-OH of a glucose residue. The ratio of glucose 1-phosphate to glucose product is 9000:1 for the wild-type enzyme, and 500:1 for the Phe 573 mutant. Model building suggests that a water molecule occupies the site normally filled by the phenolic OH of tyrosine and occasionally attacks the oxocarbonium ion intermediate to form glucose. See D. Palm, H. W. Klein, R. Schinzel, M. Buehner, and E. J. M. Helmreich, *Biochemistry* 29(1990):1099.

10. Glycogenin performs the priming function for glycogen synthesis. Without α-amylase to degrade pre-existing chains, the glycogenin activity would be masked by the more prominent activity of glycogen synthase. The α-amylase treatment halts the activity of glycogen synthase by shortening existing glucose chains below the threshold size required for them to be substrates of glycogen synthase.

11. When two soluble enzymes catalyze consecutive reactions, the product formed by the first enzyme must leave and diffuse to the second enzyme. Catalytic efficiency is substantially increased if both active sites are in close proximity in the same enzyme molecule. A similar advantage is obtained when consecutive enzymes are held close to each other in multienzyme complexes.

12. The mice will be unable to generate phosphorylase *a* from phosphorylase *b*, but phosphorylase *b* will still have a low level of activity and will degrade glycogen, especially during exercise. Although the T state of phosphorylase *b* is favored, accumulation of AMP during exercise will convert some of the phosphorylase *b* to the active R state.

13. Although glucose 1-phosphate is the actual product of the phosphorylase reaction, glucose 6-phosphate is a more versatile molecule with respect to metabolism. Among other fates, glucose-6-phosphate can be processed to yield energy or building blocks. In the liver, glucose 6-phosphate can be converted into glucose and released into the blood. Therefore, the metabolic needs are better reflected by the level of glucose-6-phosphate.

14. Epinephrine binds to its G-protein-coupled receptor. The resulting structural changes activate a G_{α} protein, which in turn activates adenyl cyclase. Adenyl cyclase synthesizes cAMP, which activates protein kinase A. Protein kinase A partially activates phosphoryl kinase, which phosphorylates and activates glycogen phosphorylase. The calcium released during muscle contraction further activates the phosphorylase kinase, leading to further stimulation of glycogen phosphorylase.

15. Glycogen breakdown is inhibited in several ways. First, the signal-transduction pathway stops when the initiating hormone is no longer present. Second, the inherent GTPase activity of the G protein converts the bound GTP into inactive GDP. Third, the level of cyclic AMP becomes depleted as phosphodiesterases convert cyclic AMP into AMP. Fourth, the enzyme PP1 removes a phosphoryl group from glycogen phosphorylase, converting the enzyme into the usually inactive *b* form.

16. The opposing effects of phosphorylation prevent glycogen breakdown and synthesis from operating simultaneously, which would lead to a useless expenditure of energy. See also the answer to Problem 24.

17. The symptoms suggest central nervous system issues. If exercise is exhaustive enough or the athlete has not prepared well enough, or both, liver glycogen in addition to muscle glycogen can be depleted. The brain depends on glucose derived from liver glycogen. The symptoms suggest that the brain is not getting enough fuel.

18. Liver phosphorylase a is inhibited by glucose, which facilitates the R $\longrightarrow$ T transition for the enzyme. This transition releases protein phosphatase 1, which inactivates glycogen breakdown and stimulates glycogen synthesis. Muscle phosphorylase is not sensitive to glucose.

19. The presence of high concentrations of glucose 6-phosphate indicates that glucose is abundant and that it is not being used by glycolysis. Therefore, it makes sense that glycogen synthase should be activated in order that the valuable resource of glucose 6-phosphate be saved by incorporation into glycogen.

20. Free glucose must be phosphorylated at the expense of one molecule of ATP. Glucose 6-phosphate derived from glycogen, by contrast, is formed by phosphorolytic cleavage, thus sparing one molecule of ATP. Therefore, the net yield of ATP when glycogen-derived glucose is processed to pyruvate is three molecules of ATP compared with two molecules of ATP from free glucose.

21. During glycogen breakdown, phosphoglucomutase converts glucose 1-phosphate, liberated from the glycogen, into glucose 6-phosphate, which can be either released as free glucose (liver) or processed in glycolysis (muscle and liver). During glycogen synthesis, phosphoglucomutase converts glucose 6-phosphate into glucose 1-phosphate, which reacts with UTP to form UDP-glucose, the substrate for glycogen synthase.

22. $\text{Glycogen}_n + P_i \longrightarrow \text{glycogen}_{n-1} + \text{glucose 6-phosphate}$

    $\text{Glucose 6-phosphate} \longrightarrow \text{glucose 1-phosphate}$

    $\text{UTP} + \text{glucose 1-phosphate} \longrightarrow \text{UDP-glucose} + 2\ P_i$

    $\text{Glycogen}_{n-1} + \text{UDP-glucose} \longrightarrow \text{glycogen}_n + \text{UDP}$

    _____

    Sum: $\text{UTP} \longrightarrow \text{UDP} + P_i$

23. In principle, having glycogen be the only primer for the further synthesis of glycogen should be a successful strategy. However, if the glycogen granules were not evenly divided between daughter cells, glycogen stores for future generations of cells might be compromised. To avoid this possibility, it is important that glycogenin synthesizes the primer for glycogen synthase.

24. Insulin binds to its receptor and activates the tyrosine kinase activity of the receptor, which in turn triggers a pathway that activates protein kinases. The kinases phosphorylate and inactivate glycogen synthase kinase. Protein phosphatase 1 then removes the phosphate from glycogen synthase and thereby activates the synthase.

25. An enzyme-bound intermediate is likely for amylase, and for the transferase and α-1,6 glucosidase (debranching enzyme). A nucleophile on the enzyme would need to break the α-1,4 bond (transferase) or the α-1,6 bond (debranching enzyme) and form a bond to one part of the carbohydrate chain (at C1). A second nucleophile would then attack and release the enzyme-bound chain. For the case of transferase, the second nucleophile would be a terminal C4-OH group of glycogen to receive the trans-

ferred tri-glucose unit, whereas debranching enzyme would use a water molecule as the second nucleophile to release a free glucose monomer.

26. Recall that galactose enters metabolism by reacting with ATP in the presence of galactokinase to yield galactose 1-phosphate and subsequently glucose 1-phosphate. On the way to glycogen the latter reacts with UTP to give UDP–glucose. Hence:

Galactose + ATP + UTP + H_2O + (glycogen)n $\rightarrow$ (glycogen)$_{n+1}$ + ADP + UDP + 2 P_i + H^+

27. Phosphorylase, transferase, glucosidase, phosphoglucomutase, and glucose 6-phosphatase working together will carry out the job of releasing glucose into the bloodstream when the organism is asleep and fasting.

28. Glucose is an allosteric inhibitor of phosphorylase *a*. Hence, crystals grown in its presence are in the T state. The addition of glucose 1-phosphate, a substrate, shifts the R-to-T equilibrium toward the R state. The conformational differences between these states are sufficiently large that the crystal shatters unless it is stabilized by chemical cross-links.

29. In similar fashion, galactose is converted into UDP-galactose to eventually form glucose 6-phosphate.

30. Similar symptoms could also be produced by a mutation in the gene that encodes the glucose 6-phosphate transporter. Recall that glucose 6-phosphate must be transported into the lumen of the endoplasmic reticulum to be hydrolyzed by phosphatase. Mutations in any of the other three essential proteins of this system can likewise lead to von Gierke disease.

31. (a) The antibodies will detect only glycogenin, and the glycogenin will be bound to glycogen. Without α-amylase treatment, the glycogen will have a high molecular weight and will remain at the top of the gel.
    (b) The glycogen is digested into small pieces that remain bound to the glycogenin. The glycogenin migration distance now will reflect approximately its true molecular weight plus that of a small bound carbohydrate oligomer.
    (c) Proteins such as glycogen phosphorylase, synthase, or debranching enzymes could also be present, but they were not stained with specific antibodies in the Western blot.

32. (a) The pattern reflects glycogenin bound to carbohydrate chains of varying sizes.
    (b) When starved for glucose, the cells use most of their glycogen and the supply is depleted.
    (c) When the cells are given glucose again, the supply of glycogen is replenished, so that lane 3 resembles lane 1.
    (d) The glycogen supply is replenished within one hour and does not further increase in three hours.
    (e) Amylase digests the glycogen in all samples to small fragments that are bound to the glycogenin, whose size is ~66 kD.

CHAPTER 22

1. Glycerol + 2 NAD$^+$ + P_i + ADP $\longrightarrow$ pyruvate + ATP + H_2O + 2 NADH + H^+
   Glycerol kinase and glycerol phosphate dehydrogenase are required. Remember, glycerol enters glycolysis as dihydroxyacetone phosphate; hence, the need for a kinase and a dehydrogenase. For the conversion of dihydroxyacetone phosphate to pyruvate, see Chapter 16.

2. The ready reversibility is due to the high-energy nature of the thioester in the acyl CoA. In effect, a high-energy bond in ATP is traded for a high-energy bond in acyl CoA.

3. To return the AMP to a form that can be phosphorylated by oxidative phosphorylation or substrate-level phosphorylation, another molecule of ATP must be expended in the reaction:

$$ATP + AMP \longrightarrow 2\ ADP.$$

4. The order of events for β-oxidation of fatty acids is:
    (b) Fatty acid in the cytoplasm
    (c) Activation of fatty acid by joining to CoA
    (a) Reaction with carnitine
    (g) Acyl CoA in mitochondrion
    (h) FAD-linked oxidation
    (d) Hydration
    (e) NAD$^+$-linked oxidation, and finally
    (f) Thiolysis.

5. The reactions of the citric acid cycle that take succinate to oxaloacetate, or the reverse, are similar to those of fatty acid metabolism.

6. The eighth CoA moiety comes from the palmitoyl-CoA itself. The next-to-last degradation product, acetoacetyl CoA, yields two molecules of acetyl CoA by means of the thiolysis reaction with input of only one molecule of CoA.

7. Palmitic acid yields 106 molecules of ATP. This number is derived as follows. The activation of palmitic acid to palmitoyl-CoA requires the equivalent of two molecules of ATP, as ATP is split into AMP and two molecules of orthophosphate. The oxidation of palmitoyl CoA then yields seven FADH$_2$, seven NADH, and eight acetyl-CoA molecules. The ATP yield in the subsequent respiratory chain is 10.5 from the seven FADH$_2$ and 17.5 from the seven NADH. The oxidation of eight acetyl CoA molecules by the citric acid cycle yields the equivalent of 80 molecules of ATP. Thus, the complete oxidation of a molecule of palmitate yields $(-2 + 10.5 + 17.5 + 80) = 106$ molecules of ATP.

   Palmitoleic acid has a double bond between carbons C-9 and C-10. When palmitoleic acid is processed in oxidation, one of the oxidation steps (to introduce a double bond before the addition of water) will not take place, because a double bond already exists. Thus, one of the FADH$_2$ molecules will not be generated, and palmitoleic acid will yield 1.5 fewer molecules of ATP than palmitic acid, for a total of 104.5 molecules of ATP.

8. Consider the balance sheet below.

Activation to form heptadecanoyl CoA	-2 ATP
Seven rounds of β-oxidation:	
7 acetyl CoA at 10 ATP/acetyl CoA	$+70$ ATP
7 NADH at 2.5 ATP/NADH	$+17.5$ ATP
7 FADH$_2$ at 1.5 ATP/FADH$_2$	$+10.5$ ATP
Propionyl CoA, which requires an ATP to be converted into succinyl CoA	-1 ATP
Succinyl CoA $\longrightarrow$ succinate	$+1$ ATP (GTP)
Succinate $\longrightarrow$ fumarate + FADH$_2$ (at 1.5 ATP/FADH$_2$)	$+1.5$ ATP
Fumarate $\longrightarrow$ malate	
Malate $\longrightarrow$ oxaloacetate + NADH (at 2.5 ATP/NADH)	$+2.5$ ATP
Total	100 ATP

9. To form stearoyl CoA requires the equivalent of two molecules of ATP. Then
   Stearoyl CoA + 8 FAD + 8 NAD$^+$ + 8 CoA + 8 H$_2$O →
   $\qquad\qquad\qquad$ 9 acetyl CoA + 8 FADH$_2$ + 8 NADH + 8 H$^+$.

9 acetyl CoA at 10 ATP/acetyl CoA	+90 ATP
8 NADH at 2.5 ATP/NADH	+20 ATP
8 FADH$_2$ at 1.5 ATP/FADH$_2$	+12 ATP
Activation to form stearoyl CoA	−2.0
Total	120 ATP

10. Fats are more highly reduced than carbohydrates and yield more energy per gram and per carbon atom.

    First, consider the glucose. Two molecules of ATP are produced when glucose is converted to two molecules of pyruvate during glycolysis. Two molecules of NADH also are produced, but the electrons are transferred to FADH$_2$ for entry into the mitochondria. Each molecule of pyruvate will produce one molecule of NADH and one molecule of acetyl CoA. Each acetyl CoA generates three molecules of NADH, one molecule of FADH$_2$, and one molecule of ATP. Recall that each molecule of FADH$_2$ can generate 1.5 ATP, while each molecule of NADH can generate 2.5 ATP. So, we have a total of 12.5 ATP per pyruvate, or 25 for the two molecules of pyruvate. The production of pyruvate during glycolysis (above) netted two ATP directly plus three more from the two molecules of FADH$_2$. The total ATP yield per molecule of glucose is therefore (5 + 25) = 30 ATP.

    (Recall furthermore that the net result of the citric acid cycle is that one molecule of acetyl CoA yields 10 ATP [from 2 GTP, 2 NADH, and 2 FADH$_2$].)

    Now, consider the hexanoic acid. Caprioic acid is activated to caprioyl CoA at the expense of 2 ATP, and so we begin the accounting with −2 ATP. The first cycle of β oxidation generates 1 FADH$_2$, 1 NADH, and 1 acetyl CoA; leading to (1.5 + 2.5 + 10) = 14 ATP. The second cycle of β oxidation generates 1 FADH$_2$ and 1 NADH but 2 acetyl CoA; leading to (1.5 + 2.5 + 20) = 24 ATP. The total yield is therefore (2 + 14 + 24) = 36 ATP. Thus, the foul-smelling caprioic acid has a net yield of 36 ATP. So on a per-carbon basis, this fat yields 20% more ATP than does glucose, a manifestation of the fact that fats are more reduced than carbohydrates.

11. Stearate + ATP + 13½ H$_2$O + 8 FAD + 8 NAD$^+$ →
    4½ acetoacetate + 14½ H$^+$ + 8 FADH$_2$ + 8 NADH + AMP + 2 P$_i$

    Note that this equation is the sum of the following three equations:

    (1)   Stearate + CoA + ATP + H$_2$O → stearoyl CoA + AMP + 2 P$_i$ + 2H$^+$

    (2)   Stearoyl CoA + 8 FAD + 8 NAD$^+$ + 3½ CoA + 8 H$_2$O →
    $\qquad$ 4½ acetoacetyl CoA + 8 FADH$_2$ + 8 NADH$^+$ + 8 H$^+$

    (3)   4½ acetoacetyl CoA + 4½ H$_2$O → 4½ acetoacetate + 4½ H$^+$ + 4½ CoA

12. Palmitate is activated and then processed by β oxidation according to the following reactions.

    Palmitate + CoA + ATP → palmitoyl CoA + AMP + 2 P$_i$.

    Palmitoyl CoA + 7 FAD + 7 NAD + 7 CoASH + H$_2$O →
    $\qquad\qquad\qquad$ 8 acetyl CoA + 7 FADH$_2$ + 7 NADH + 7H$^+$.

    The eight molecules of acetyl CoA combine to form four molecules of acetoacetate for release into the blood, and so they do not contribute to the energy yield in the liver. However, the FADH$_2$ and NADH generated in the preparation of acetyl CoA can be processed by oxidative phosphorylation to yield ATP.

1.5 ATP/FADH$_2$ × 7 FADH$_2$ = 10.5 ATP.

2.5 ATP/NADH × 7 NADH = 17.5 ATP.

The equivalent of 2 ATP were used to form palmitoyl CoA. Thus, 26 ATP were generated for use by the liver.

13. NADH is produced along with the oxidation of 3-hydroxybutyrate to acetoacetate, and the NADH yields 2.5 ATP molecules. Meanwhile, the acetoacetate is converted into acetoacetyl CoA, which is hydrolyzed to produce two molecules of acetyl CoA, each of which is worth 10 ATP when processed by the citric acid cycle. Therefore, the total ATP yield is (2.5 +10 + 10) = 22.5 ATP.

14. It is costly to produce acetoacetyl CoA because a molecule of succinyl CoA must be used to form the acetoacetyl CoA. The succinyl CoA could otherwise be used to generate one equivalent of ATP (as GTP), and so someone could argue that the yield is reduced by one.

15. For the combustion of fats, not only must they be converted into acetyl CoA, but the acetyl CoA must be processed by the citric acid cycle. In order for acetyl CoA to enter the citric acid cycle, there must be a supply of oxaloacetate. Oxaloacetate can be formed from carbohydrate sources, for example the metabolism of glucose to pyruvate and the subsequent carboxylation of pyruvate to form oxaloacetate.

16. (a) The methyl groups on the phytanic acid will block β oxidation. Because the β oxidation cannot take place, phytanic acid accumulates.

(b) How to solve the problem? One strategy would be to remove the methyl groups. Our livers use a different strategy, α oxidation, in which the OH and carbonyl groups are introduced by oxidizing the carbon. One round of oxidation releases CO_2 and converts phytanic acid into a β-oxidation substrate for the next round.

When the next methyl group is encountered, another round of α oxidation can be implemented.

17. The first oxidation removes two tritium atoms. The hydration adds nonradioactive H and OH. The second oxidation removes another tritium atom from the β-carbon atom. Thiolysis removes an acetyl CoA with only one tritium atom; so the tritium-to-carbon ratio is ½. This ratio will be the same for two of the acetates. The last one, however, does not undergo oxidation, and so all tritium remains. The ratio for this acetate is 3/2. The average ratio for all of the acetates is then (½ + ½ + ³⁄₂)/3 = ⁵⁄₆.

18. In the absence of insulin, lipid mobilization will take place to an extent that it overwhelms the ability of the liver to convert the lipids into ketone bodies. When this happens, triacylglycerols will accumulate.

19. (a) oxidation in mitochondria, synthesis in the cytosol
    (b) CoA in oxidation, acyl carrier protein for synthesis
    (c) FAD and NAD$^+$ in oxidation, NADPH for synthesis

    (d) L isomer of 3-hydroxyacyl CoA in oxidation, D isomer in synthesis

    (e) carboxyl to methyl in oxidation, methyl to carboxyl in synthesis

    (f) The enzymes of fatty acid synthesis, but not those of oxidation, are organized in a multienzyme complex.

20. By analogy with the balanced equation for synthesis of the 16-carbon palmitate (see text), the balanced equation for synthesis of the shorter 14-carbon myristate is:

$$7 \text{ acetyl CoA} + 6 \text{ ATP} + 12 \text{ NADPH} + 5 \text{ H}^+ \longrightarrow$$
$$\text{CH}_3(\text{CH}_2)_{12}\text{COOH} + 7 \text{ CoA} + 6 \text{ ADP} + 6 \text{ P}_i + 12 \text{ NADP}^+ + 5 \text{ H}_2\text{O}.$$

21. By analogy with the synthesis of palmitate and myristate (see problem 20), a balanced equation for synthesis of the still shorter 12-carbon laurate is:

$$6 \text{ acetyl CoA} + 5 \text{ ATP} + 10 \text{ NADPH} + 4 \text{ H}^+ \longrightarrow$$
$$\text{CH}_3(\text{CH}_2)_{10}\text{COOH} + 6 \text{ CoA} + 5 \text{ ADP} + 5 \text{ P}_i + 10 \text{ NADP}^+ + 4 \text{ H}_2\text{O}.$$

Therefore, five molecules of ATP and 10 molecules of NADPH are required to synthesize lauric acid (or sodium laurate).

22. The order of steps for synthesis of a fatty acid is:

    (e) Formation of malonyl ACP

    (b) Condensation

    (d) Reduction of a carbonyl group

    (a) Dehydration

    (c) Release of a complete fatty acid

23. The mutation would inhibit fatty acid synthesis because of a shortage of acetyl CoA. The enzyme cleaves cytoplasmic citrate to yield acetyl CoA for fatty acid synthesis.

24. (a) False. Biotin is required for acetyl CoA carboxylase activity.

    (b) True.

    (c) False. ATP is required to synthesize malonyl CoA for fatty acid synthesis.

    (d) True. (Palmitate is one of the end products.)

    (e) True.

    (f) False. Fatty acid synthase is a dimer.

    (g) True.

    (h) False. Acetyl CoA carboxylase is stimulated by citrate, which also is cleaved to yield the substrate acetyl CoA.

25. Fatty acids with odd numbers of carbon atoms are synthesized starting with propionyl ACP (instead of acetyl ACP). By starting with a 3-carbon precursor, and then adding 2-carbon units, a product with an odd number of carbons can be achieved. The propionyl ACP is formed from propionyl CoA by acetyl transacetylase.

26. All of the labeled carbon atoms will be retained. Because we need eight acetyl CoA molecules and only one carbon atom is labeled in the acetyl group, we will have eight labeled carbon atoms in the 16-carbon palmitic acid.

    One molecule of acetyl CoA will be used directly and will retain all three tritium atoms. The other seven acetyl CoA molecules will be used to make malonyl CoA; each of these will lose one tritium atom on addition of CO_2 and another tritium atom at the dehydration step. Each of the seven malonyl CoA molecules therefore will retain one tritium atom. Thus, the total number of retained tritium will be $(3 + 7) = 10$ atoms. The ratio of tritium to carbon is $(10/8) = 1.25$.

27. The avidin in the raw eggs will inhibit fatty acid synthesis by reducing the amount of free biotin that is available as a required cofactor for the acetyl CoA carboxylase. Cooking the eggs will denature the avidin protein so that it will no longer bind biotin.

28. During the last step of the synthesis, the only acetyl CoA to be used directly, not in the form of malonyl CoA, provides the final two carbon atoms at the end of the fatty acid chain. Because palmitic acid is a C_{16} fatty acid, acetyl CoA will have provided carbons 15 and 16.

29. The involvement of HCO_3^- is temporary. HCO_3^- is attached to acetyl CoA to form malonyl CoA. Subsequently, when malonyl CoA condenses with growing acyl CoA chain to form the keto acyl CoA that is two carbons longer, the HCO_3^- from malonyl CoA is lost as CO_2.

30. Phosphofructokinase is a key regulatory enzyme that controls the flux through the glycolytic pathway. The purpose of glycolysis is to generate ATP and/or building blocks for biosynthesis, depending on the particular tissue. The presence of citrate in the cytoplasm indicates that those needs are met, and there is no need to metabolize additional glucose.

31. During fatty acid biosynthesis, the carbon chain grows two carbons at a time by the condensation of an acyl-ACP with malonyl-ACP, with the malonyl-ACP becoming, in every case, the carboxyl end of the new acyl-ACP. Thus, the chain grows from methyl to carboxyl. Since ^{14}C-labeled malonyl CoA was added a short time before synthesis was stopped, the fatty acids whose synthesis was completed during this short period will be heavily labeled toward the carboxyl end (the last portion synthesized) and less heavily labeled, if at all, on the methyl end.

32. The mutant enzyme would be persistently active because it could not be inhibited by phosphorylation. Fatty acid synthesis would be abnormally active. Such a mutation might lead to obesity.

33. Because mammals lack the enzymes to introduce double bonds at carbon atoms beyond C-9 but can increase the length of the fatty acid chain at the carboxyl end, the easiest way to determine which unsaturated fatty acid is the precursor is to note the number of carbons from the ω end (CH_3 end) to the nearest double bond. Thus, in (a) this number is seven carbons; hence, palmitoleate is the precursor. In (b) it is six carbon atoms; hence, linoleate. In (e) it is nine carbon atoms; hence, oleate; etc.

34. Decarboxylation drives the condensation of malonyl-ACP and acetyl-ACP. In contrast, the condensation of two molecules of acetyl-ACP is energetically unfavorable. In gluconeogenesis, decarboxylation drives the formation of phosphoenolpyruvate from oxaloacetate.

35. Adipose-cell lipase is activated by phosphorylation. Hence, overproduction of the cAMP-activated kinase will lead to accelerated breakdown of triacylglycerols and depletion of fat stores.

36. Carnitine translocase deficiency and glucose 6-phosphate transporter deficiency. For an explanation, see Section 22.5 of the text.

37. In the fifth round of β oxidation, cis-Δ^2-enoyl CoA is formed. Hydration by the classic hydratase yields D-3-hydroxyacyl CoA, the wrong isomer for the next enzyme in β oxidation. This dead end is circumvented by a second hydratase that removes water to give $trans$-Δ^2-enoyl CoA. Addition of water by the classic hydratase then yields L-3-hydroxyacyl CoA, the appropriate isomer. Thus, hydratases of opposite stereospecificities serve to *epimerize* (invert the configuration of) the 3-hydroxyl group of the acyl CoA intermediate. See J. K. Hiltunen, P. M. Palosaari, and W.-H. Kunau. *J. Biol. Chem.* 264(1989):13536.

38. The probability of synthesizing an error-free polypeptide chain decreases as the length of the chain increases. A single mistake can make the entire polypeptide ineffective. In contrast, a defective subunit can be spurned in forming a noncovalent multienzyme complex; the good subunits are not wasted.

39. The person will be unable to oxidize fatty acids to begin their degradation. With acetyl-CoA not available from fatty acid degradation, available glucose (and ketogenic amino

acids) will be used to produce acetyl-CoA for the citric acid cycle. Therefore, glucose will be in short supply. Ketone bodies will not form because acetyl-CoA also will be in short supply (as there is an "energy crisis" with energy from fatty acids not available).

40. Peroxisomes oxidize fatty acids that have more than 18 carbons and reduce their lengths to C18. The shorter chains are better substrates for β oxidation in the mitochondria. Therefore, clofibrate probably aids the degradation of fatty acids generally and thereby will lower the level of triglycerides.

41. Citrate increases the activity of especially the phosphorylated acetyl-CoA carboxylase by allosteric regulation. The level of citrate is high when both acetyl-CoA and ATP are abundant. The abundance of ATP indicates that energy is not needed, and acetyl-CoA can be stored as fatty acids (in triacylglycerols). The presence of palmitoyl-CoA would signify that fatty acid degradation is occurring, and so acetyl-CoA carboxylase, because it is an early step in fatty acid synthesis, should be inhibited.

42. Acetyl-CoA is a product of the thiolase reaction and a substrate for condensing enzyme. The mechanism for condensing enzyme is given in Figure 17.11 of the text. By analogy, a mechanism for thiolase would be similar to the reverse of the condensing reaction, with CoASH acting as a nucleophile:

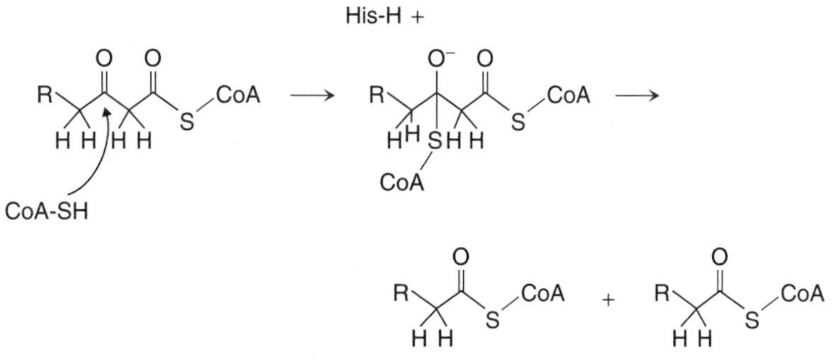

43. The enolate anion of one thioester attacks the carbonyl carbon atom of the other thioester to form a C-C bond.

44. (a) The entry of acetyl-CoA into the citric acid cycle will be inefficient because fat and carbohydrate degradation will not be appropriately balanced. The shortage of pyruvate, oxaloacetate, and cycle intermediates cannot be compensated by fats because mammals are unable to accomplish net synthesis of cycle intermediates from fats. The ability to derive energy from fats therefore will be impaired.
    (b) Acetyl-CoA will be converted to ketone bodies in the blood, and the breath will smell of acetone, from the decarboxylation of acetoacetate.
    (c) Yes. The activated three-carbon units from odd-chain fatty acids can be converted to succinyl-CoA and enter the citric acid cycle to allow some net synthesis of cycle intermediates.

45. Glucose and glycogen can be labeled by exchange without net synthesis of carbohydrate from fats. For example, the labeled stearic will be degraded to acetyl-CoA, which will enter the citric acid cycle and produce some labeled oxaloacetate (by scrambling of the carbons, but not net synthesis). Some of this oxaloacetate can be used for gluconeogenesis, which would lead to some incorporation of ^{14}C into glucose and glycogen.

46. (a) We can use the data in the figure to construct a double-reciprocal plot (see below). From the slope and intercept of each line (see Chapter 8), we can estimate that K_M is about 45 µM and V_{max} about 13 nmol/(mg-min) for the wild-type enzyme. For the mutant enzyme, K_M is about 75 µM and V_{max} about 8 nmol/(mg-min).

The respective values are comparable, and the mutation has little effect on the enzyme activity when the concentration of carnitine is varied.

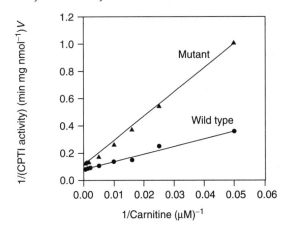

(b) In similar fashion to part (a), we use a double-reciprocal plot to estimate K_M of about 105 μM and V_{max} about 41 nmol/(mg-min) for the wild-type enzyme, and K_M of about 70 μM and V_{max} about 23 nmol/(mg-min) for the mutant enzyme. Once again, the respective values are similar (of the same order of magnitude).

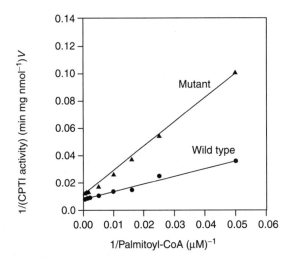

(c) The wild-type enzyme is much more sensitive to inhibition by malonyl CoA.

(d) The mutant enzyme will be more active under these conditions because it retains more than 90% of the activity than it has in the absence of malonyl-CoA. Although the wild-type enzyme is more active without malonyl-CoA, its activity is reduced to about 20% of normal when 10 μM malonyl CoA is present.

(e) Glutamate 3 probably participates in the binding of malonyl-CoA and enables malonyl CoA to be an inhibitor, but glutamate 3 is not necessary for catalysis.

CHAPTER 23

1. When the proteins are denatured, they are unfolded, such that all of the peptide bonds are accessible to proteolytic enzymes. In the folded three-dimensional structure, many of the peptide bonds in the protein would not be accessible to the proteolytic enzymes.

2. First, the ubiquitin-activating enzyme (E1) must link ubiquitin to a sulfhydryl group on the E1 itself. Next, a second ubiquitin-conjugating enzyme (E2) transfers the ubiquitin to a sulfhydryl group of a cysteine residue on E2 itself. Finally, the ubiquitin–protein ligase (E3) uses the ubiquitinated E2 as a substrate and transfers ubiquitin to the target protein.

3. (a) Pepsin is a stomach proteolytic enzyme (7).
   (b) The N-terminal rule determines the half-life of a protein (4).
   (c) Ubiquitin marks a protein for destruction (2)
   (d) The PEST sequence is Pro-Glu-Ser-Thr (10).
   (e) Threonine nucleophiles reside in the 20S catalytic core (5) of the proteasome.
   (f) ATP-dependent protein unfolding is accomplished by a 19S regulatory unit (3).
   (g) The proteasome is a protein degrading machine (9).
   (h) Ubiquitin-activating enzyme requires an adenylate intermediate (1).
   (i) Ubiquitin-conjugating enzyme becomes a substrate for the ligase (6).
   (j) Ubiquitin-ligase recognizes the protein to be degraded (8).

4. (a) The energy from ATP hydrolysis could be used for maintaining the specificity of peptide-bond hydrolysis, unfolding the target protein, threading the target protein through the barrel, or other related functions.
   (b) Small peptides would not need to be unfolded or threaded, and would not need to have any of their bonds protected (e.g., remember that the ubiquitin tags are not hydrolyzed).

5. (a) pyruvate, (b) oxaloacetate, (c) α-ketoglutarate, (d) α-ketoisocaproate, (e) phenylpyruvate, and (f) hydroxyphenylpyruvate

6. (a) Aspartate + α-ketoglutarate + GTP + ATP + 2 H_2O + NADH + H^+ →
   ½ glucose + glutamate + CO_2 + ADP + GDP + NAD^+ + 2P_i

   The glucogenic route for aspartate involves transamination to oxaloacetate, conversion of the latter to phosphoenolpyruvate, which is then converted to glucose (see Chapter 20 of the text). PLP and NADH participate as coenzymes in the conversion of aspartate to glucose.

   (b) Aspartate + CO_2 + NH_4^+ + 3 ATP + NAD^+ + 4 H_2O →
   oxaloacetate + urea + 2 ADP + 4 P_i + AMP + NADH + H^+

   This equation represents the summation of the stoichiometry of urea synthesis, the hydrolysis of PP_i, and the conversion of fumarate to oxaloacetate in the citric acid cycle.

7. Different sites could specialize in the hydrolysis of different categories of peptide bonds (e.g., those that join hydrophobic/hydrophilic, hydrophobic/hydrophobic, or hydrophilic/hydrophilic amino acids, etc.) and thereby optimize the overall kinetics of degrading diverse protein sequences.

8. By analogy with the 20S proteasome (Figure 23.7), the overall architecture could be conserved. Perhaps the six different AAA ATPase subunits from the 19S regulatory complex associate into a hexamer with pseudo six-fold symmetry. If these hexamers could be separated from the 19S complexes, they might be visulaized by electron microscopy or crystallography; suitable crosslinking experiments might then reveal which pairs of AAA ATPase subunits border each other.

9. thiamine pyrophosphate

10. Aminotransferases transfer the α-amino group of a given amino acid to α-ketoglutarate to form glutamate. The glutamate is then oxidatively deaminated to again form α-ketoglutarate, with release of an ammonium ion.

11. Aspartate is deaminated to form oxaloacetate.

    Glutamate is deaminated to form α-ketoglutarate.

    Alanine is deaminated to form pyruvate.

12. Only serine and threonine can be deaminated directly.

13. The carbon skeletons of the amino acids are either fuels for the citric acid cycle, components of the citric acid cycle, or molecules that can be converted into a fuel for the citric acid cycle in only one step.

14. It acts as an electron sink. See C. Walsh (1979), *Enzymatic Reaction Mechanisms*, p. 178 (W. H. Freeman).

15. Carbamoyl phosphate and aspartate are the sources for the two nitrogen atoms in urea.

16. The statements refer to nitrogen metabolism:

    1. Aspartate provides a second source of nitrogen (c) when arginosuccinate is formed.
    2. Urea is a final product (g).
    3. Ornithine accepts the first nitrogen, which is included within a carbamoyl group from carbamoyl phosphate (f).
    4. Carbamoyl phosphate is formed from NH_4^+ (a), together with bicarbonate and phosphate.
    5. Arginine is hydrolyzed to yield urea (b) [and ornithine].
    6. Citrulline reacts with aspartate (d) [to form arginosuccinate].
    7. Cleavage of arginosuccinate yields fumarate (e) [and arginine].

17. The identities and order of appearance in the urea cycle are:

    (c) ornithine.
    (b) citrulline.
    (d) arginosuccinate.
    (a) arginine.

18. $CO_2 + NH_4^+ + 3\,ATP + NAD^+ + 3H_2O + glutamate \rightarrow$
    urea $+ 2\,ADP + 2\,P_i + AMP + PP_i + NADH + H^+ + $ α-ketoglutarate

    The answer in the text is for the conversion of fumarate to oxaloacetate rather than to aspartate. The equation above gives the correct stoichiometry for the synthesis of urea from NH_4^+ and glutamate. It includes a non-energy-requiring transamination reaction. Hence, the number of ~P spent remains at four. Note that aspartate does not appear in this equation, since it is resynthesized. Thus aspartate can be considered a nitrogen-carrying cofactor in the synthesis of urea.

19. The synthesis of fumarate by the urea cycle is important because it links the urea cycle to the citric acid cycle. Fumarate is hydrated to malate, which, in turn, is oxidized to oxaloacetate. Oxaloacetate has several possible fates: (1) transamination to aspartate, (2) conversion into glucose by the gluconeogenic pathway, (3) condensation with acetyl CoA to form citrate, or (4) conversion into pyruvate. You can collect!

20. The compound should inhibit ornithine transcarbamoylase because it appears to be a nonhydrolyzable analogue of an intermediate that should be formed between ornithine and carbamoyl phosphate. (The CH_2 group in compound A will prevent the release of the phosphate.)

21. High concentrations of ammonia could increase the ratio of glutamate/α-ketoglutarate (glutamate dehydrogenase reaction) and *increase* the level of glutamate in the brain. Ammonia also could increase the ratio of asparatate/oxaloacetate; the resulting lower level of oxaloacetate would decrease the availability of all citric acid cycle intermediates.

22. The mass spectrometric analysis strongly suggests that three enzymes—pyruvate dehydrogenase, α-ketoglutarate dehydrogenase, and the branched-chain α-keto dehydrogenase—are deficient. Most likely, the common E3 component of these enzymes is missing or defective. This proposal could be tested by purifying these three enzymes and assaying their capacity to catalyze the regeneration of lipoamide.

23. Benzoate, phenylacetate, and arginine would be given in a protein-restricted diet. Nitrogen would emerge in hippurate, phenylacetylglutamine, and citrulline. See S. W. Brusilow and A. L. Horwich in C. R. Scriver, A. L. Beaudet, W. S. Sly, and D. Valle (eds.) (1989), *The Metabolic Basis of Inherited Disease,* 6th ed., pp. 629–663 (McGraw-Hill), for a detailed discussion of the therapeutic rationale.

    This therapy is designed to facilitate nitrogen excretion in the absence of urea synthesis. Note that much of the nitrogen would be excreted as glycine (hippurate is benzoylglycine), glutamine, and citrulline.

24. The liver is the primary tissue for capturing nitrogen as urea. If the liver is damaged (for instance, by hepatitis or the excessive consumption of alcohol), free ammonia will be released into the bloodstream.

25. An appropriate strategy is to bypass fumarate in the urea cycle while still achieving excretion of nitrogen. The defect thereby can be partly bypassed by providing a surplus of arginine in the diet while restricting the total protein intake. In the liver, arginine is split into urea and ornithine, which then will react with carbamoyl phosphate to form citrulline. This urea-cycle intermediate condenses with aspartate to yield argininosuccinate, which is then excreted (because of the enzyme deficiency). As a result, two nitrogen atoms—one from carbamoyl phosphate and the other from aspartate—will be eliminated from the body per molecule of arginine provided in the diet. In essence, the argininosuccinate substitutes for urea in carrying nitrogen out of the body. The formation of argininosuccinate removes the nitrogen. The dietary arginine is needed in order to form the argininosuccinate for the purpose of removing the nitrogen. The restriction on protein intake serves to relieve the aciduria.

26. Aspartame, a dipeptide ester (aspartylphenylalanine methyl ester), is hydrolyzed to L-aspartate and L-phenylalanine. High levels of phenylalanine are harmful in phenylketonurics.

27. N-acetylglutamate is synthesized from acetyl CoA and glutamate. Once again, acetyl CoA serves as an activated acetyl donor. This reaction is catalyzed by N-acetylglutamate synthase.

28. When one amino acid is missing for protein synthesis, the only source of the essential amino acid will be other proteins. Some proteins therefore would need to be degraded in order to provide the missing amino acid. The nitrogen from the other amino acids in the proteins undergoing degradation would be excreted as urea. Consequently, more nitrogen would be excreted than ingested. Negative nitrogen balance is the result.

29. The carbon skeletons of ketogenic amino acids can be converted into ketone bodies or fatty acids. Only leucine and lysine are purely ketogenic. Glucogenic amino acids are those whose carbon skeletons can be converted into glucose. Four of the amino acids contain both ketogenic and glucogenic carbon atoms. Fourteen amino acids are purely glucogenic.

30. The enzyme is the "branched-chain α-ketoacid dehydrogenase complex." The complex is required for synthesis of the branched-chain amino acids leucine, isoleucine, and valine.

31. The end products of amino acid degradation are:
    - pyruvate (used for glycolysis and gluconeogenesis)
    - acetyl CoA (used in the citric acid cycle and fatty acid synthesis)
    - acetoacetyl CoA (used for ketone-body formation)
    - α-ketoglutarate (used in the citric acid cycle)
    - succinyl CoA (used in the citric acid cycle)
    - fumarate (used in the citric acid cycle)
    - oxaloacetate (used in the citric acid cycle and gluconeogenesis)

32. First serine forms a protonated Schiff base (external aldimine) with pyridoxal-5'-phosphate. Removal of the serine α-hydrogen leads to a quinonoid intermediate, which then can eliminate the β-OH to generate the Schiff base of aminoacrylate. The reaction is completed by the transfer of pyridoxal phosphate back to a Schiff base with a lysine of the enzyme (internal aldimine), with the concomitant release of aminoacrylate. (The aminoacrylate will react with water to give pyruvate and ammonia.)

(*Note:* Another family of serine dehydratases does not use pyridoxal phosphate, but rather an iron-sulfur cluster as the cofactor. These enzymes may use a mechanism similar to the dehydration of citrate catalyzed by aconitase. See *Trends Biochem. Sci.* 18[1993]:297–300.)

33. As in problem 32, an external aldimine (protonated Schiff base) is formed between serine and pyridoxal-5'-phosphate, and the serine α-hydrogen is removed. The α-hydrogen then can be reattached from either of two faces to give the Schiff base of either D-serine or L-serine (see below). The equilibrium constant for the reaction will be one.

34. Protein-protein interactions often relate directly to biological function. Two or more aspects of these interactions may be relevant for degradation: (1) If an interaction domain becomes chemically damaged so that interaction with a partner protein is no longer feasible, then it would be appropriate to turn over (degrade) the protein. (2) If a partner protein is in short supply within the cell, then the interacting protein may not be needed and should be degraded. (Without the partner protein, the interaction domain could be exposed as a possible signal for degradation.)

35. (a) The initial surge originates from a need to supply glucose to the brain. When carbohydrates are depleted, mammals cannot resupply glucose from fatty acids. Glycerol provides a small source of carbohydrate, but mammals also must break down amino acids to meet the short-term demands of the brain for glucose. The ammonia byproduct from amino acid degradation accounts for the initial surge of nitrogen excretion.

    (b) Over time, the liver begins to metabolize acetyl-CoA (from fats) to ketone bodies, and the brain adapts to using ketone bodies. During this period, fats can provide many of the energy needs of the brain, and little nitrogen is excreted.

    (c) When lipid stores have been depleted, the organism once again must metabolize amino acids to provide glucose for the brain, and nitrogen excretion again increases.

36. Isoleucine can give its amino group to α-ketoglutarate in a transamination reaction and then be oxidatively decarboxylated and dehydrogenated to form the corresponding (α,β)-unsaturated acyl-CoA derivative. Further reactions (see the figure below) then are identical to fatty acid oxidation until the carbon skeleton is split into acetyl-S-CoA and propionyl-S-CoA. The three subsequent steps for the conversion of the (odd-chain) propionyl-S-CoA to succinyl-S-CoA have been discussed for the oxidation of odd-chain fatty acids (see Chapter 22).

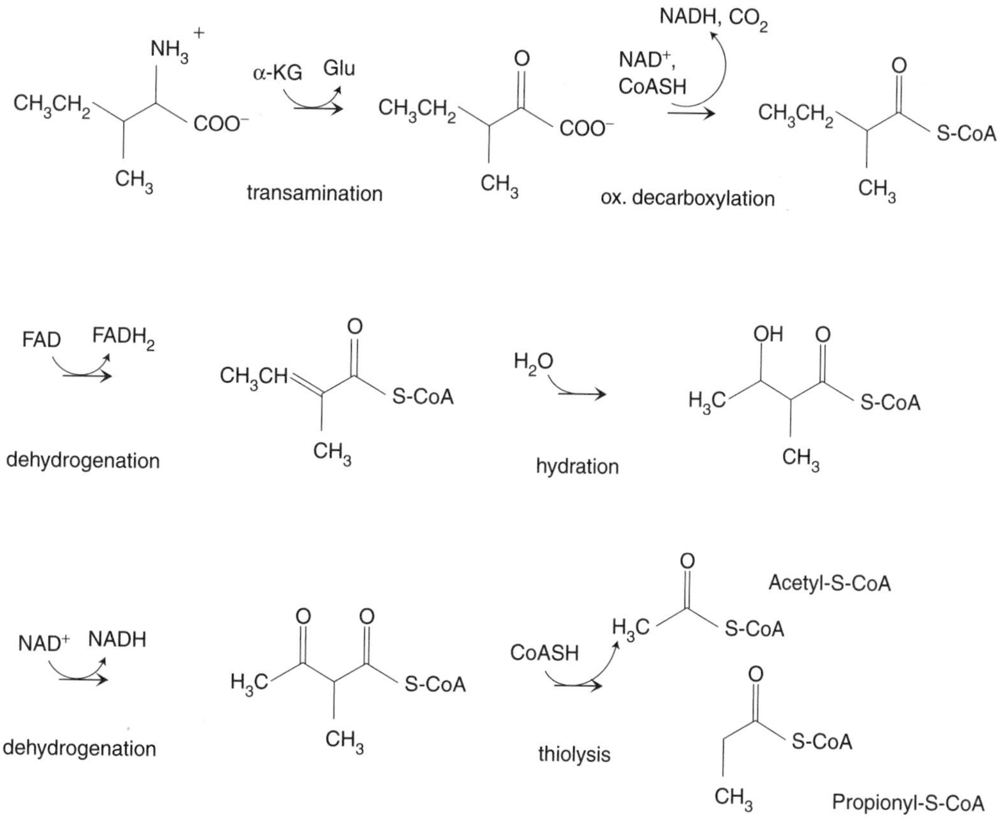

37. Pyridoxal phosphate serves as an acid–base catalyst in the glycogen phosphorylase reaction, under conditions where water is excluded from the active site and ortho-phosphate cleaves the glycogen chain.

38. In the Cori cycle, the carbon atoms are transferred from muscle to liver as lactate. For lactate to be of any use, it must first be reduced to pyruvate. This reduction requires high-energy electrons from NADH. When the carbon atoms are transferred as alanine, the transamination yields pyruvate directly without a need for high-energy electrons.

39. (a) PAN has no effect in the absence of nucleotides.
    (b) ATP is required. Neither ADP nor AMP-PNP is effective in stimulating protein digestion.
    (c) AMP-PNP is a nonhydrolyzable analogue of ATP. The finding that AMP-PNP does not stimulate the digestion suggests that ATP hydrolysis to ADP and P_i is required.
    (d) The peptide digestion does not require PAN or ATP.
    (e) See answer 1(b), above. Small peptides do *not* need to be unfolded or threaded through a superstructure in order to facilitate digestion.
    (f) Although *Thermoplasma* PAN is not as effective with the other proteasomes, it nonetheless results in threefold to fourfold stimulation of digestion.
    (g) In light of the fact that the archaea and eukarya diverged several billion years ago, the fact that *Thermoplasma* PAN can stimulate rabbit muscle suggests homology not only between the proteasomes, but also between PAN and the 19S subunit (most likely the ATPases) of the mammalian 26S proteasome.

CHAPTER 24

1. Nitrogen fixation is the conversion of atmospheric N_2 into ammonium ion, NH_4^+. Only microorganisms possessing the diazotrophic (nitrogen-fixing) capability are able to fix nitrogen.

2. The carbon precursors of the 20 amino acids are the keto acids pyruvate, oxaloacetate, and α-ketoglutarate; along with 3-phosphoglycerate, phosphoenolpyruvate, ribose-5-phosphate, and erythrose-4-phosphate.

3. Human beings can synthesize some of the amino acids, but not others. We do not have particular biochemical pathways to synthesize certain amino acids from simpler precursors. Consequently, these amino acids that humans cannot synthesize are "essential" and must be obtained from the diet.

4. Glucose $+\ 2\,ADP\ +\ 2\,P_i\ +\ 2\,NAD^+\ +\ 2\,glutamate\ \longrightarrow\ 2\,alanine\ +\ 2\,\alpha\text{-ketoglutarate}$ $+\ 2\,ATP\ +\ 2\,NADH\ +\ H^+$

   Glucose is converted to pyruvate via glycolysis, and pyruvate is converted to alanine by transamination.

5. $N_2 \longrightarrow NH_4^+ \longrightarrow$ glutamate $\longrightarrow$ serine $\longrightarrow$ glycine $\longrightarrow \delta$-aminolevulinate $\longrightarrow$ porphobilinogen $\longrightarrow$ heme

6. The statement is false. Nitrogen fixation is thermodynamically favorable. Nitrogenase is required because there is a high activation barrier that renders the process kinetically disfavored.

7. Pyridoxal phosphate is required by all transaminases.

8. Cofactors that are one-carbon carriers include S-adenosylmethionine, tetrahydrofolate, and methylcobalamin.

9. (a) N^5,N^{10}-methylenetetrahydrofolate, (b) N^5-methyltetrahydrofolate

10. γ-Glutamyl phosphate is a likely reaction intermediate.

11. The synthesis of asparagine from aspartate passes through an acyl-adenylate intermediate. Therefore, one of the products of the reaction will be ^{18}O-labeled AMP.

12. The administration of glycine leads to the formation of isovalerylglycine. This water-soluble conjugate, in contrast with isovaleric acid, is excreted very rapidly by the kidneys. See R. M. Cohn, M. Yudkoff, R. Rothman, and S. Segal. *New Engl. J. Med.* 299(1978):996.

13. The nitrogen atom shaded red is derived from glutamine. The carbon atom shaded blue is derived from serine.

14. They carry out nitrogen fixation. The absence of photosystem II provides an environment in which O_2 is not produced. Recall that the nitrogenase is very rapidly inactivated by O_2. See R. Y. Stanier, J. L. Ingraham, M. L. Wheelis, and P. R. Painter, *The Microbial World*, 5th ed. (Prentice-Hall, 1986), pp. 356–359, for a discussion of heterocysts.

15. The cytosol is a reducing environment, whereas the extracellular milieu is an oxidizing environment.

16. The mirror-image enzyme would recognize (only) mirror-image substrates and would produce (only) mirror-image products. The specificity for the achiral molecules α-ketoglutarate and oxaolacetate would not be affected. Therefore, the answers are:

    (a) None.   (b) D-glutamate and oxaloacetate.

17. The synthesis of δ-aminolevulinate requires succinyl CoA, an intermediate in the citric acid cycle, which occurs in the mitochondrial matrix. Thus it makes sense to have the first step in porphyrin biosynthesis occur in the matrix.

18. Alanine and aspartate can be synthesized directly from glutamate by transamination of pyruvate and oxaloacetate, respectively. Glutamate is the common intermediate that serves as the amino group donor in these reactions. (α-Ketoglutarate, the side product, can condense with another molecule of ammonia to regenerate the glutamate for further transamination reactions.)

19. By analogy with ornithine cyclodeaminase, lysine cyclodeaminase should produce a six-membered ring. Indeed, lysine cyclodeaminase converts L-lysine into the six-membered ring analog of proline, also referred to as L-homoproline or L-pipecolate:

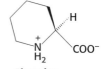

Pipecolate

20. Each final product should inhibit the first *unique* step toward its synthesis, that is, the first step following the branch point of the pathway. Thus, Y should inhibit the conversion of C to D, and Z should inhibit the conversion of C to F. (To avoid wasteful accumulation of C or B if neither Y nor Z is needed, high levels of C should also inhibit the conversion of A to B at the beginning of the pathway.)

21. The effects would multiply, so that the net rate would equal $(100 \times 0.6 \times 0.4 \text{ s}^{-1})$, or 24 s^{-1}.

22. The treatment of the sulfhydryl with 2-bromoethylamine restores a side chain that has a shape and pK_a similar to the original lysine side chain. Hence, some of the

original catalytic activity is restored. Cysteine cannot be expected to occupy the same space or perform the same function as lysine, but the resulting thioether side chain, following the treatment with 2-bromoethylamine, is quite similar to that of lysine.

23. The reaction will begin with the formation of a Schiff base between pyridoxal-5′-phosphate and the amino group of *S*-adenosylmethionine (external aldimine, as opposed to the internal aldimine in which PLP is connected to a lysine on the enzyme). In usual fashion for PLP enzymes, the α-hydrogen will be extracted from *S*-adenosylmethionine (step 2, below). Then the next series of π-electron transfers (step 3) will eliminate *S*-methylthioadenosine and form the three-membered ring. Finally, 1-aminocyclopropane-1-carboxylate (ACC) will be released, and the original internal aldimine between PLP and lysine will be restored, so that the enzyme and cofactor are ready for another round of synthesis. The second product is *S*-methylthioadenosine.

24. As in problem 32 in Chapter 23 of this manual, an external aldimine (protonated Schiff base) is formed between serine and pyridoxal-5′-phosphate, and the serine α-hydrogen is removed. The α-hydrogen then can be reattached from either of two faces to give the Schiff base of either D-serine or L-serine (see below). The equilibrium constant for the reaction will be one.

25. (a) In the first step, histidine attacks the methylene group from the methionine sub-group of SAM (rather than the usual methyl substituent), resulting in the transfer of an aminocarboxypropyl group. Three subsequent conventional SAM-mediated methylations of the primary amine yield diphthine.

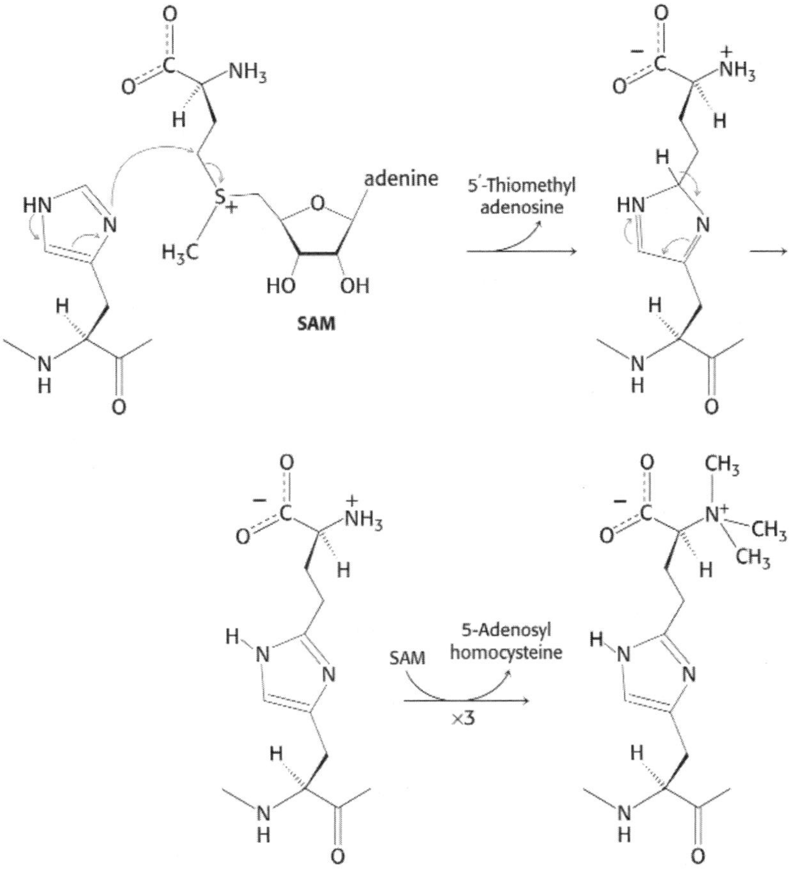

Diphthine

(b) In this chapter, we have observed two examples of an ATP-dependent conversion of a carboxylate into an amide: glutamine synthetase, which uses an acyl-phosphate intermediate, and asparagine synthetase, which uses an acyl-adenylate intermediate. Either mechanism is possible in formation of diphthamide from diphthine.

26. Aspartate and glutamate would be synthesized from the citric acid cycle intermediates oxaloacetate and α-ketoglutarate. Increased synthesis of aspartate and glutamate therefore could begin to deplete citric acid cycle intermediates. The cell would need to respond by breaking down carbohydrates to replenish the supply by net synthesis of new cycle intermediates.

27. Because S-adenosylmethionine (SAM) is the methyl donor for the methylation of DNA (Figure 24.15), a deficiency in SAM could diminish the extent of the methylation of the mutated bacteria's DNA. The lower level of methylation would render the DNA more susceptible to digestion by restriction enzymes.

28. Acetate ⟶ acetyl-CoA, and then via the citric acid cycle
acetyl-CoA ⟶ citrate ⟶ isocitrate ⟶ α-ketoglutarate ⟶ succinyl-CoA.

29. (a) Asn, Gln, and Gly are affected. Asn is much more concentrated in dark-adapted plants, whereas Gln and Gly are present in somewhat elevated levels in light-adapted plants.

(b) The transcription of specific mRNA, translation or enzymatic activity of several enzymes, particularly nitrogen-utilizing enzymes such as asparagine synthetase and glutamine synthetase, could be regulated by light.

(c) From the graph, asparagine would appear to be a likely candidate. (The name of the plant would also suit this interpretation!)

CHAPTER 25

1. In de novo synthesis, the nucleotides are synthesized from simpler precursor compounds, in essence from scratch. In salvage pathways, preformed bases are recovered and attached to riboses.

2. The sources of the respective atoms in the pyrimidine ring are labeled in Figure 25.2: Aspartate provides nitrogen 1 and carbons 4, 5, and 6. Carbon 2 and nitrogen 3 are derived from carbamoyl phosphate.

3. The sources of the respective atoms in the purine ring are labeled in Figure 25.5:

   Nitrogen 1 is from aspartate.
   Carbons 2 and 8 are from N^{10}-formyltetrahydrofolate.
   Nitrogen 3 is from glutamine.
   Carbons 4 and 5 and nitrogen 7 are from glycine.
   Carbon 6 is from CO_2.
   Nitrogen 9 is from glutamine.

4. Some of the uses of nucleotides are:
   - Energy currency: ATP
   - Signal transduction: ATP and GTP.
   - RNA synthesis: ATP, GTP, CTP, and UTP.
   - DNA synthesis: dATP, dCTP, dGTP, and TTP.
   - Components of coenzymes: adenine nucleotides in CoA, FAD, and $NAD(P)^+$;
   - Use in carbohydrate synthesis: UDP-glucose.

5. A nucleoside is a base attached to ribose. A nucleotide is a phosphorylated nucleoside in which the ribose ring bears one or more phosphates.

6. (a) Excessive urate can cause gout (9).
   (b) Lack of adenosine deaminase will lead to immunodeficiency (7).
   (c) Lack of GPRT causes Lesch-Nyhan disease (6).
   (d) Carbamoyl phosphate is coupled during the first step in pyrimidine synthesis (10).
   (e) Inosinate is a precursor to both ATP and GTP (2).
   (f) Ribonucleotide reductase is required for deoxynucleotide synthesis (4).
   (g) Lack of folic acid can cause spina bifida (1).
   (h) Glutamine phosphoribosyl transferase catalyzes the committed step in purine synthesis (11).
   (i) A pyrimidine (8) contains a single aromatic ring.
   (j) A purine (3) contains a bicyclic aromatic ring.
   (k) UTP (5) is a precursor to CTP.

7. Substrate channeling is a process whereby the product of one active site moves to become a substrate at another active site without ever leaving the enzyme. A channel connects the active sites. Substrate channeling greatly enhances enzyme efficiency and minimizes the loss of substrate due to diffusion during transfer from one active site to another.

8. Glucose + 2 ATP + 2 NADP$^+$ + H$_2$O → PRPP + CO$_2$ + ADP + AMP + 2 NADPH + H$^+$

   This equation is the summation of the following conversions: glucose → glucose 6-phosphate → ribose 5-phosphate → PRPP (see Chapter 20 of the text).

9. Glutamine + aspartate + CO$_2$ + 2 ATP + NAD$^+$ → orotate + 2 ADP + 2 P$_i$ + glutamate + NADH + H$^+$

10. (a, c, d) PRPP, (b) carbamoyl phosphate

11. PRPP and formylglycinamide ribonucleotide accumulate because they are reactions 1 and 2 in the first stage of purine biosynthesis (Figure 25.6 in the text). If these glutamine-requiring amidotransferase reactions are inhibited, the precursor molecules will accumulate. However, since the synthesis of formylglycinamide requires that PRPP be converted to phosphoribosylamine, probably only PRPP will accumulate significantly.

12. The serine side chain is reduced from CH$_2$OH to CH$_2$/(CH$_3$) as the carbon is transferred first to N^5, N^{10}-methylene-tetrahydrofolate, and then to carbon 5 of the uracil ring in dUMP to form dTMP, in the overall net reaction:

    dUMP + serine + NADPH + H$^+$ → dTMP + NADP$^+$ + glycine.

13. There is a deficiency of N^{10}-formyltetrahydrofolate (see Figure 25.6 in the text). Sulfanilamide inhibits the synthesis of folate by acting as an analog of *p*-aminobenzoate, one of the precursors of folate.

14. (a) Cell A cannot grow in a HAT medium, because it cannot synthesize dTMP either from thymidine or from dUMP. Cell B cannot grow in this medium, because it cannot synthesize purines by either the de novo pathway or the salvage pathway. Cell C can grow in a HAT medium because it contains active thymidine kinase from cell B (enabling it to phosphorylate thymidine to dTMP) and hypoxanthine-guanine phosphoribosyl transferase from cell A (enabling it to synthesize purines from hypoxanthine by the salvage pathway).

    (b) Transform cell A with a plasmid containing foreign genes of interest and a functional thymidine kinase gene. The only cells that will grow in a *HAT* medium are those that have acquired a thymidylate kinase gene; nearly all these transformed cells will also contain the other genes on the plasmid.

15. The reciprocal substrate relation refers to the fact that AMP synthesis requires GTP, whereas GMP synthesis requires ATP. These requirements tend to balance the synthesis of ATP and GTP.

16. As shown below for cytidylate, ring carbons 4, 5, and 6 in cytosine originate from Asp and will be labeled with ^{13}C. (Carbon 2 comes from carbamoyl phosphate.)

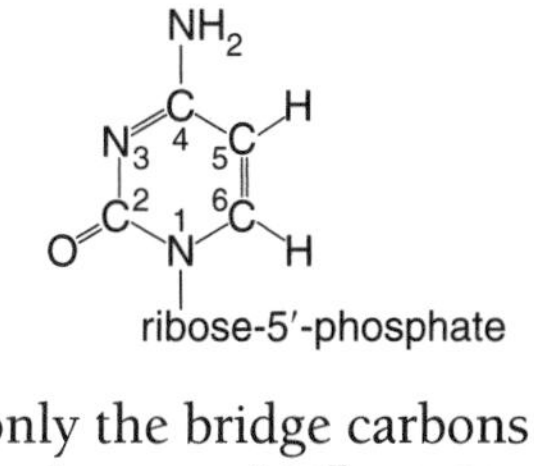

ribose-5'-phosphate

In guanylate (shown below), only the bridge carbons #4 and #5 of guanine will be labeled with ^{13}C. These bridge carbons and N^7 originate from glycine. (Carbons 2 and 8 come from formyl-tetrahydrofolate, while carbon 6 comes from bicarbonate.)

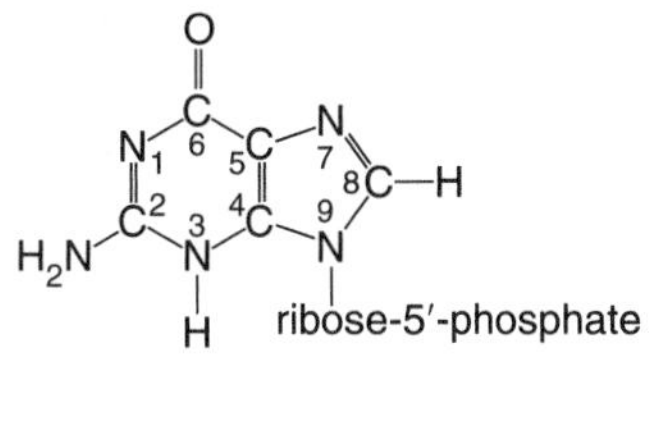

ribose-5'-phosphate

17. The enzyme that uses ammonia assists in the preparation of carbamoyl phosphate for the first step of the urea cycle. H catalyzes two sequential reactions. First, carbamic acid is synthesized from carboxyphosphate and ammonia (Section 23.4 of the text). Then ATP is used to phosphorylate carbamic acid to carbamoyl phosphate, in preparation for reaction with ornithine in the first step of the urea cycle.

   The enzyme that uses glutamine produces carbamoyl phosphate for a different purpose: the first step of pyrimidine biosynthesis (Section 25.1 of the text). After the glutamine is hydrolyzed to glutamate and ammonia, the reactions to synthesize carbamoyl phosphate are the same as those described above.

18. These patients have a high level of urate because of the breakdown of nucleic acids. Allopurinol prevents the formation of kidney stones and blocks other deleterious consequences of hyperuricemia by inhibiting the formation of urate.

19. Though often termed *binding constants,* the values given are really *dissociation constants.* Therefore, to calculate the free energy of binding, one uses the reciprocals of the values given. Thus, for the wild type,

$$\Delta G^{\circ\prime} = -RT \ln \frac{1}{7 \times 10^{-11}} = (-2.47)(23.4)$$

$$= -57.8 \text{ kJ/mol} (-13.8 \text{ kcal/mol})$$

Similar calculations for Asn 27 and Ser 27 will give the results in the answer in the text.

20. IMP is the product of the pathway for de novo purine biosynthesis and the precursor of AMP and GMP. With the de novo pathway for purines not working effectively, it would be helpful to stimulate the salvage pathway, perhaps with a diet that is rich in nucleotides that would then be a source of the preformed purine bases hypoxanthine, adenine, and guanine.

21. N^1 in the purine ring of IMP, AMP, ATP, GMP, and GTP will be labeled, following the scheme below:

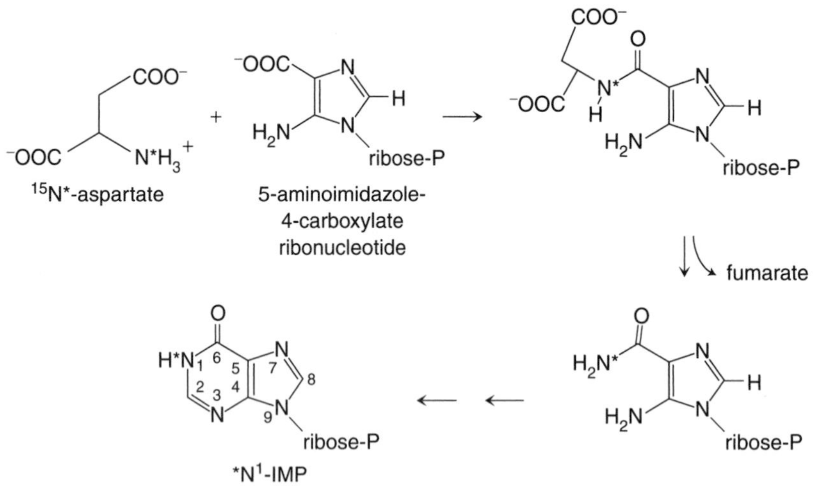

22. Glutamine supplies the nitrogen (via ammonia) for 5-phosphoribosyl-1-amine (figure in margin on page 741); this nitrogen becomes atom 9 in the final purine ring (see standard numbering in Figure 4.4). Glutamine also supplies the incoming (third) nitrogen for formation of formylglycineamidine ribonucleotide (step 3 in Figure 25.6); this nitrogen becomes atom 3 in the final purine ring. Therefore nitrogen atoms 3 and 9 in the purine ring become labeled.

23. Allopurinol, an analog of hypoxanthine, is a suicide inhibitor of xanthine oxidase. Inhibiting this enzyme would help to reduce the oversupply of purine nucleotides.

24. Allopurinol is an analog of hypoxanthine in which the N and C atoms at positions 7 and 8 are interchanged. Xanthine oxidase will hydroxylate C2 of allopurinol in similar manner to its normal reaction with hypoxanthine, but this C-2 hydroxylation of allopurinol gives the new inhibitor:

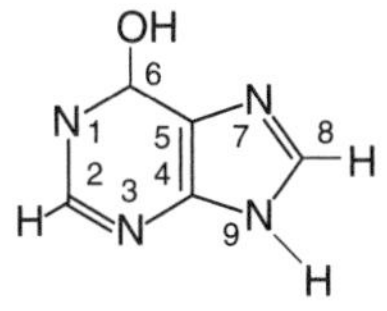

Hypoxanthine 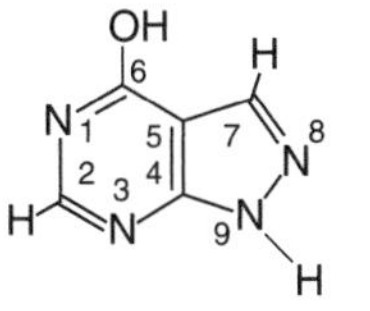   Allopurinol 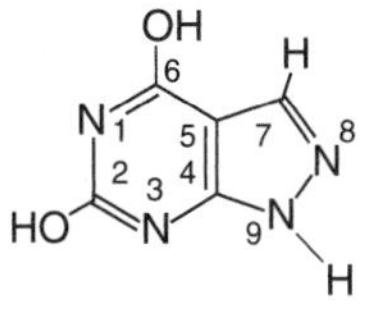   New inhibitor

25. Seven high-energy phosphate bonds are hydrolyzed during synthesis of CTP:

The synthesis of carbamoyl phosphate requires 2 ATP	2 ATP
The formation of PRPP from ribose 5-phosphate yields an AMP*	2 ATP
The conversion of UMP to UTP requires 2 ATP	2 ATP
The conversion of UTP to CTP requires 1 ATP	1 ATP
Total	7 ATP

*Remember that two high-energy phosphate bonds are expended when AMP is produced, since the AMP must be converted first to ADP and then to ATP during the resynthesis.

26. (a) Lack of aspartate will cause carboxyaminoimidazole ribonucleotide to accumulate (blocking step 6 in Figure 25.6).
    (b) Lack of tetrahydrofolate will cause glycinamide ribonucleotide to accumulate (blocking step 2 in Figure 25.6).
    (c) Lack of glycine will cause phosphoribosyl amine to accumulate (blocking step 1 in Figure 25.6).
    (d) Lack of glutamine will inhibit the initial committed step in purine biosynthesis, formation of 5-phosphoribosyl-1-amine, and also will cause formylglycinamide ribonucleotide to accumulate (blocking step 3 in Figure 25.6).

27. For the production of glycinamide ribonucleotide, an acyl phosphate (anhydride) intermediate is formed by reaction of ATP with a carboxylic acid of an amino acid (glycine), whereas in the production of guanylate (GMP) a phosphoryl ester intermediate is formed by reaction of ATP with an enol alcohol on the purine ring of the nucleotide.

Glycinamide ribonucleotide. Step 1.

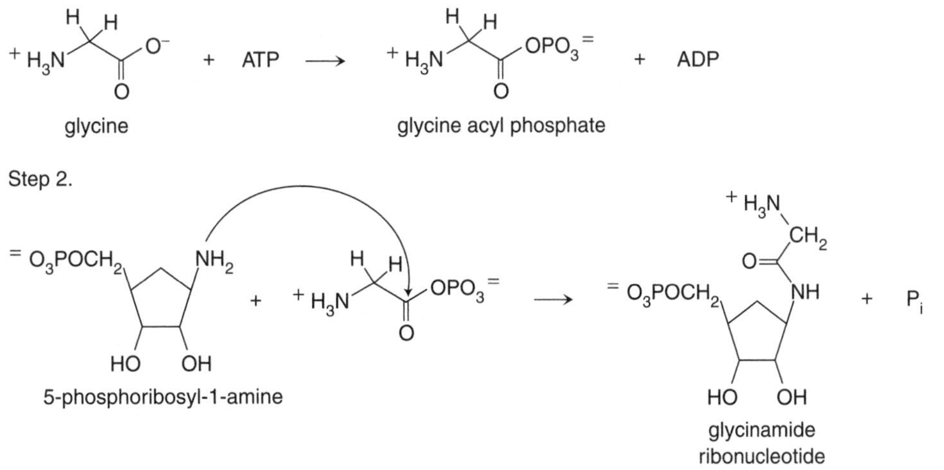

glycine   glycine acyl phosphate

Step 2.

5-phosphoribosyl-1-amine   glycinamide ribonucleotide

Guanylate. Step 1.

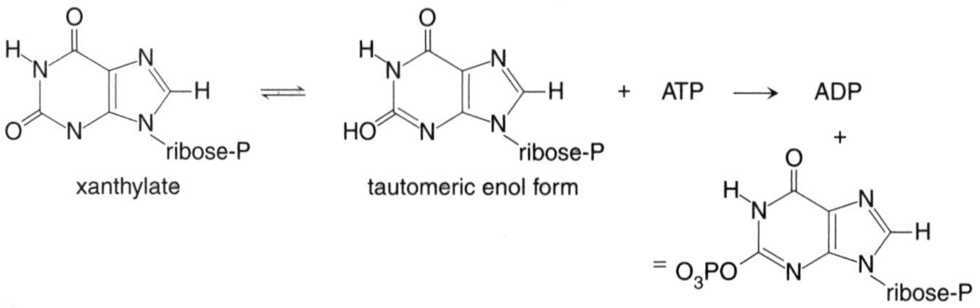

xanthylate ⇌ tautomeric enol form + ATP ⟶ ADP

Step 2.

glutamine + ⟶ glutamate (Glu)

HOPO₃ =

guanylate (GMP)

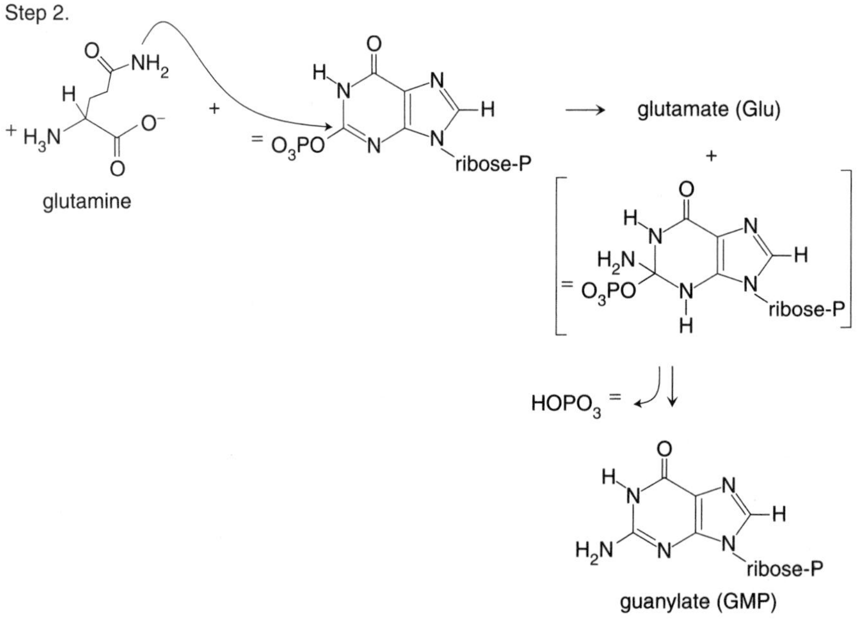

28. The reaction involves a dehydration and ring closure. The amino group that was introduced from aspartate and that will become N-1 in inosinate (see problem 14, above) should be activated (at an enzyme active site) to make a nucleophilic attack on the nearby formyl carbonyl carbon to close the six-membered ring and give a tetrahedral intermediate. The tetrahedral intermediate can then lose water to render the six-membered ring aromatic and generate the inosinate (IMP):

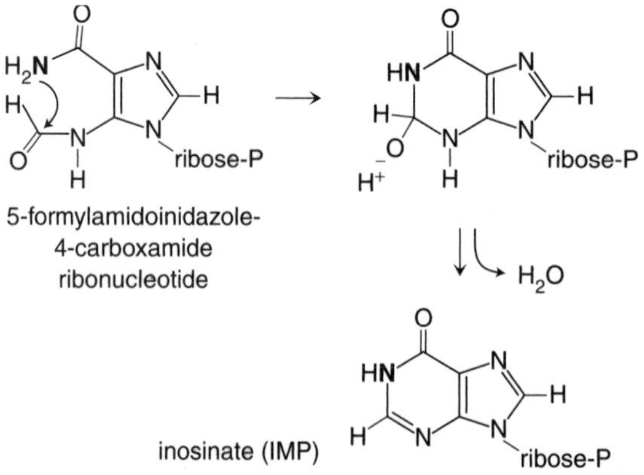

5-formylamidoinidazole-
4-carboxamide
ribonucleotide

inosinate (IMP)

29. PRPP is the activated intermediate in the synthesis of (a) phosphoribosylamine in the de novo pathway of purine formation, (b) purine nucleotides from free bases by the salvage pathway, (c) orotidylate in the formation of pyrimidines, (d) nicotinate ribonucleotide, (e) phosphoribosyl-ATP in the pathway leading to histidine, and (f) phosphoribosylanthranilate in the pathway leading to tryptophan.

30. (a) Cyclic-(5′-3′)-AMP and cyclic GMP influence many intracellular processes.

(b) ATP is the prototype energy-storage molecule with a high phosphoryl-group transfer potential.

(c) ATP is a phosphate donor for generating glucose-6-phosphate and fructose-1,6-bisphosphate.

(d) Flavin adenine dinucleotide (FAD) and nicotinamide adenine dinucleotide (NAD^+) participate in the production of acetyl-CoA ("active acetate") from pyruvate by the pyruvate dehydrogenase complex.

(e) NADH and $FADH_2$ are reduced molecules that have a high electron transfer potential.

(f) Synthetic (2′,3′)-dideoxynucleoside triphosphates serve as chain terminators for DNA sequencing. (The four natural deoxynucleoside triphosphates—dGTP, dCTP, dATP, and dTTP—are the monomers that are precursors for chain elongation.)

(g) The thymine analogue 5-fluorouracil (converted to 5-fluorodeoxyuridylate in vivo) is a potent anticancer drug.

(h) ATP and CTP reciprocally regulate the activity of aspartate transcarbamoylase.

31. In vitamin B_{12} deficiency, methyltetrahydrofolate cannot donate its methyl group to homocysteine to regenerate methionine. Because the synthesis of methyltetrahydrofolate is irreversible (Figure 24.10), the cell's tetrahydrofolate ultimately will be converted into this form. No formyl or methylene tetrahydrofolate will be left for nucleotide synthesis. Pernicious anemia illustrates the intimate connection between amino acid metabolism and nucleotide metabolism. The metabolism of fatty acids that have odd numbers of carbons also will be affected because methylmalonyl-CoA mutase requires vitamin B_{12} for the production of succinyl-CoA. A further connection is that methylmalonyl-CoA mutase also is involved in the degradation of valine and isoleucine.

32. Because folate is required for nucleotide synthesis, cells that are dividing rapidly would be most readily affected. They would include cells of the intestine, which are constantly replaced, and precursors to blood cells. A lack of intestinal cells and blood cells would account for the symptoms often observed.

33. The cytosolic level of ATP in liver falls and that of AMP rises above normal in all three conditions. The excess AMP is degraded to urate. See C. R. Scriver, A. L. Beaudet, W. S. Sly, and D. Vale (eds.), *The Metabolic Basis of Inherited Disease*, 6th ed. (McGraw-Hill, 1989), pp. 984–988, for an illuminating discussion.

34. Succinate can be converted to oxaloacetate by the citric acid cycle. The oxaloacetate can then be transaminated to yield asparate, a key precursor of pyrimidines. The carbons of aspartate then will label positions 4, 5, and 6 in the pyrimidine rings:

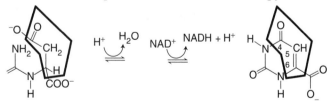

N-carbamoylaspartate

35. Glucose will most likely be converted into two molecules of pyruvate, one of which will be labeled in the 2 position:

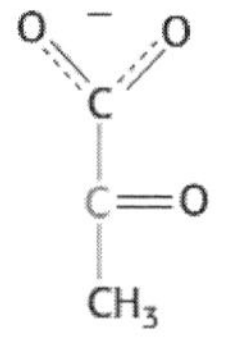

Now consider two common fates of pyruvate—conversion into acetyl CoA and subsequent processing by the citric acid cycle or carboxylation by pyruvate carboxylase to form oxaloacetate. Formation of citrate by condensing the labeled pyruvate with oxaloacetate will yield labeled citrate:

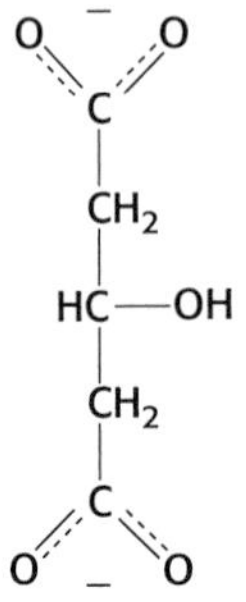

The labeled carbon will be retained through one round of the citric acid cycle but, on the formation of the symmetric succinate, the label will appear in two different positions. Thus, when succinate is metabolized to oxaloacetate, which may be aminated to form aspartate, two carbons will be labeled:

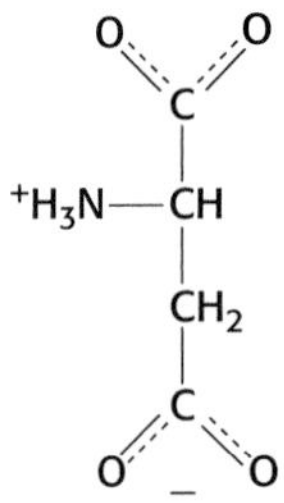

When this aspartate is used to form uracil, the labeled COO^- attached to the α-carbon is lost and the other COO^- becomes incorporated into uracil as carbon 4.

Suppose, instead, that labeled 2-$[_{14}C]$pyruvate is carboxylated to form oxaloacetate and processed to form aspartate. In this case, the α-carbon of aspartate bears the label.

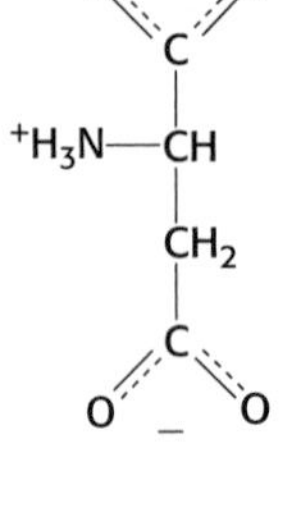

When this aspartate is used to synthesize uracil, carbon 6 bears the label.

36. (i) The formation of 5-aminoimidazole-4-carboxamide ribonucleotide from 5-aminoimidazole-4-(N-succinylcarboxamide) ribonucleotide in the synthesis of IMP (see step 7 in Figure 25.6).
    (ii) The formation of AMP from adenylosuccinate (see Figure 25.7).
    (iii) The formation of arginine from argininosuccinate in the urea cycle (see Figure 23.17 in Chapter 23).

37. Allopurinol is an inhibitor of xanthine oxidase, which is on the pathway for urate synthesis. In the duck, this pathway is the means by which excess nitrogen is excreted. If xanthine oxidase were inhibited in the duck, nitrogen could not be excreted, with severe consequences such as ammonia toxicity and the formation of a dead duck.

38. (a) The ADP from muscle contraction is a ready source of additional ATP for additional contraction. Half of the ADP can be converted immediately to ATP at the expense of the other half (being converted to AMP).

   (b) The reactants and products have the same number of high-energy phosphate bonds. The interconversion of (2 ADP) with (ATP + AMP) therefore is essentially isoenergetic.

   (c) In the reaction 2 ADP $\rightarrow$ ATP + AMP, the removal of one of the products (AMP) will shift the equilibrium to the right and favor the production of additional ATP.

   (d) By first removing and then replacing AMP, the cycle buys time (at the expense of GTP) until aerobic metabolism can "catch up" and reconvert available AMP as well as ADP back into ATP.

CHAPTER 26

1. Glycerol 3-phosphate is the foundation for both triacylglycerol and phospholipid synthesis. Glycerol 3-phosphate is acylated twice to form phosphatidate. In triacylglycerol synthesis, the phosphoryl group is removed from glycerol 3-phosphate to form diacylglycerol, which is then acylated to form triacylglycerol. In phospholipid synthesis, phosphatidate commonly reacts with CTP to form CDP-diacylglycerol, which then reacts with an alcohol to form a phospholipid. Alternatively, diacylglycerol may react with a CDP-alcohol to form a phospholipid.

2. Glycerol 3-phosphate is formed primarily by the reduction of dihydroxyacetone phosphate, an intermediate in the glycolytic and gluconeogenic pathways, and to a lesser extent by the phosphorylation of glycerol.

3. Glycerol + 4 ATP + 3 fatty acids + 4 H_2O $\rightarrow$ triacylglycerol + ADP + 3 AMP + 7 P_i + 4 H^+

   One ATP is used in the formation of glycerol 3-phosphate and three ATPs are used to convert three fatty acids to acyl CoAs. The three PPi formed during fatty acid activation are converted to P_i, hence, the total of seven P_i in the equation above.

4. Glycerol + 3 ATP + 2 fatty acids + 2 H_2O + CTP + serine $\rightarrow$ phosphatidyl serine + CMP + ADP + 2 AMP + 6 P_i + 3 H^+

   The equation above is a summation of equations shown in Section 26.1 of the text.

5. Three. One molecule of ATP is needed to form phosphorylethanolamine and two molecules of ATP are used to regenerate CTP from CMP.

6. Each of the three molecules is synthesized from ceramide, and in each case, the terminal hydroxyl group of ceramide becomes modified. In sphingomyelin, the modification involves attachment of phosphorylcholine. In a cerebroside, a glucose or galactose becomes attached to the hydroxyl group. In a ganglioside, oligosaccharide chains are attached to the hydroxyl group.

7. (i) To begin, synthesize *phosphatidate,* either by adding two fatty acids to glycerol-3-phosphate, or by the ATP-dependent phosphorylation of diacylglycerol (catalyzed by diacylglycerate kinase). Then activate the phosphatidate by means of reaction with CTP to generate CDP-diacylglycerol, which can then react with an alcohol to form a phosphodiester linkage.

   (ii) Alternatively, an alcohol such as ethanolamine can be phosphorylated by ATP and then activated by reaction with CTP to form a CDP-alcohol. The phosphoryl-alcohol unit can then be transferred to diacylglycerol to form the phospholipid.

   (iii) A third possibility is to use the base-exchange reaction to generate phospholipid diversity by exchanging one alcohol moiety for another. In essence, an existing phospholipid is converted into a related phospholipid molecule by the exchange reaction. For example:

   phosphatidylethanolamine + serine $\longrightarrow$ phosphatidylserine + ethanolamine.

8. (a) A high proportion of fatty acids in the blood are bound to albumin. Cerebrospinal fluid has a low content of fatty acids because it has little albumin.

   (b) Glucose is highly hydrophilic and soluble in aqueous media, in contrast to fatty acids, which must be carried by transport proteins such as albumin. Micelles of fatty acids would disrupt membrane structure.

   (c) Fatty acids, not glucose, are the major fuel of resting muscle.

9. Normally, diacylglycerol is acylated to form triacylglycerols. With deficient phosphatidic acid phosphatase activity, there would be a shortage of diacylglycerols that would lead to a severe shortage of triacylglycerols. The amount of adipose tissue would decrease dramatically.

10. The three stages of cholesterol synthesis involve: (i) synthesis of activated isoprene (isopentyl pyrophosphate) units, (ii) condensation of six of the activated isoprene units to form the 30-carbon squalene molecule, and (iii) the cyclization of squalene to form cholesterol.

11. HMG-CoA reductase catalyzes the commited irreversible step in the biosynthesis of cholesterol. Therefore, cholesterol biosynthesis is regulated by the availability and activity the HMG-CoA reductase enzyme. The regulation is implemented at several levels. Transcriptional control is mediated by the sterol regulatory element (SRE) binding protein, a transcription factor. Translation of the reductase mRNA also is controlled. Once formed, the mature reductase may undergo regulated proteolytic degradation. Finally, the activity of the reductase is inhibited by phosphorylation, catalyzed by AMP kinase when ATP levels are low.

12. (a) A watt is equal to 1 joule (J) per second (0.239 calorie per second). Hence, 70 W is equivalent to 0.07 kJ/s or 0.017 kcal/s.

   (b) A watt is a current of 1 ampere (A) across a potential of 1 volt (V). For simplicity, let us assume that all the electron flow is from NADH to O_2 (a potential drop of 1.14 V). Hence, the current is 61.4 A, which corresponds to $3.86 * 10^{20}$ electrons per second (1 A = 1 coulomb/s = $6.28 * 10^{18}$ charges/s).

   (c) About 2.5 molecules of ATP are formed per NADH oxidized (two electrons). Hence, one molecule of ATP is formed per 0.8 electron transferred. A flow of $3.86 * 10^{20}$ electrons per second therefore leads to the generation of $4.83 * 10^{20}$ molecules of ATP per second or 0.80 mmol s^{-1}.

   (d) The molecular weight of ATP is 507. The total body content of ATP of 50 g is equal to 0.099 mol. Hence, ATP turns over about once per 125 seconds when the body is at rest.

13. The hallmark of this devastating genetic disease is elevated cholesterol levels in the blood of even young children. The excess cholesterol is taken up by marcrophages, which eventually leads to the formation of plaques and heart disease. There are many mutations that can cause the disease, all of which result in malfunctioning of the low-density lipoprotein (LDL) receptor.

14. (a) The stoichiometry of complete oxidation of glucose is

$$C_6H_{12}O_6 + 6\,O_2 \rightarrow 6\,CO_2 + 6\,H_2O$$

and that of tripalmitoylglycerol is

$$C_{51}H_{98}O_6 + 72.5\,O_2 \rightarrow 51\,CO_2 + 49\,H_2O$$

Hence, the RQ values are 1.0 (6/6) and 0.703 (51/72.5), respectively.

(b) An RQ value reveals the relative usage of carbohydrate and fats as fuels. The RQ of a marathon runner typically decreases from 0.97 to 0.77 during the race. The lowering of the RQ reflects the shift in fuel from carbohydrate to fat.

15. A reasonable answer is: "Although it is true that cholesterol is essential and is a precursor to steroid hormones, other aspects of the statement are oversimplified. Cholesterol is a component of lipid membranes. While membranes isolate and protect cell components and define cell boundaries, and cells make up tissues, it is nevertheless wrong to say that cholesterol 'makes' cells."

16. Statins are competitive inhibitors of HMG-CoA reductase. They are used as drugs to inhibit cholesterol synthesis in patients with high levels of cholesterol.

17. No. Cholesterol is essential for membrane function and as a precursor for bile salts and steroid hormones. The complete lack of cholesterol would be lethal.

18. One gram of glucose (molecular weight 180.2) is equal to 5.55 mmol, and one gram of tripalmitoylglycerol (molecular weight 807.3) is equal to 1.24 mmol. The reaction stoichiometries (see problem 14) indicate that 6 mol of H_2O are produced per mole of glucose oxidized, and 49 mol of H_2O per mole of tripalmitoylglycerol oxidized. Hence, the H_2O yields per gram of fuel are 33.3 mmol (0.6 g) for glucose and 60.8 mmol (1.09 g) for tripalmitoylglycerol. Thus, complete oxidation of this fat gives 1.82 times as much water as does glucose. Another advantage of triacylglycerols is that they can be stored in essentially anhydrous form, whereas glucose is stored as glycogen, a highly hydrated polymer (see Section 22.1 of the text). A hump consisting mainly of glycogen would be an intolerable burden—far more than the straw that broke the camel's back!

19. The apolipoprotein B-100 component of LDL binds to an LDL receptor, an integral membrane protein. The distribution of LDL receptors within the cell membrane is not random, but rather the receptors cluster into regions of the cell surface known as coated pits. On binding, the LDL/receptor complex is internalized by first folding in and then pinching off a small portion of the cell membrane around the coated pit but in a process known as endocytosis, thereby forming an internal vesicle that contains the LDL-bound cholesteryl esters. The vesicle is separated into two components. One component, with the receptor, is transported back to the cell surface and where it fuses with the cell membrane, allowing continued use ("recycling") of the receptor. The other vesicle component fuses with a lysosome inside the cell. Within the lysozome, the cholesteryl esters are hydrolyzed, and free cholesterol is made available for cellular use. The LDL protein itself is not recycled but rather is hydrolyzed to produce free amino acids.

20. Benign prostatic hypertrophy can be treated by inhibiting the 5 α-reductase. *Finasteride*, the 4-aza steroid analog of dihydrotestosterone, competitively inhibits the reductase but does not act on androgen receptors. Patients taking finasteride have a

markedly lower plasma level of dihydrotestosterone and a nearly normal level of testosterone. The prostate becomes smaller, whereas testosterone-dependent processes such as fertility, libido, and muscle strength appear to be unaffected (see E. Stoner, *Steroid Biochem. Molec. Biol.* 37(1990):375–378). Genetic deficiencies of 5α-reductase are discussed by J. E. Griffin and J. D. Wilson in C. R. Scriver, A. L. Beaudet, W. S. Sly, and D. Valle (eds.), *The Metabolic Basis of Inherited Disease*, 6th ed. (McGraw-Hill, 1989), pp. 1919–1944.

Finasteride

21. Patients who are most sensitive to debrisoquine have a deficiency of a liver P450 enzyme encoded by a member of the *CYP2* subfamily. This characteristic is inherited as an autosomal recessive trait. The capacity to degrade other drugs may be impaired in people who hydroxylate debrisoquine at a slow rate because a single P450 enzyme usually handles a broad range of substrates. See W. B. Pratt and P. Taylor, *Principles of Drug Action: The Basis of Pharmacology,* 3d ed. (Churchill Livingstone, 1990), pp. 496–500; and F. J. Gonzalez and D. W. Nebert. *Trends Genet.* 6(1990):182.

22. Many hydrophobic odorants are deactivated by hydroxylation. O_2 is activated by a cytochrome P450 monooxygenase. NADPH serves as the reductant. One oxygen atom of O_2 goes into the odorant substrate, whereas the other is reduced to water.

23. Propecia effectively lowers the plasma level of dihydrotestosterone (see also problem 7). But dihydrotestosterone is an important embryonic androgen that instigates the development and differeniation of the male phenotype. Pregnant women who had contact with Propecia therefore would risk developmental abnormalities for their unborn male children.

24. The various P450 isozymes probably serve two major categories of function in plants. First, some of them will be important in the detoxification of foreign substances that may arise in the plant's environment. Second, plants may use P450 enzymes for the synthesis of useful molecules such as toxins to fight pests, or pigments to attract organisms that aid in dispersing pollen or seeds.

25. Individual polymorphisms in some of the P450 isozyme genes likely would alter the rates of metabolic degradation (or conversely activation) of particular clinical drugs. Knowledge of the individual differences therefore would help in prescribing appropriately different clinical doses of particular medicines for different individual patients.

26. The small number of cytochrome P450 genes provides an important clue. Indeed, the honey bees may be especially sensitive to environmental toxins, including pesticides. With the minimal P450 system, the honey bees may be less able to detoxify such molecules and could therefore be more susceptible to poisoning.

27. The core structure of a steroid is composed of four fused rings, consisting of three cyclohexane rings and one cyclopentane ring. In vitamin D, the second cyclohexane ring (B) is split by ultraviolet light.

28. Phosphorylation of the serine will place a negative charge adjacent to the key histidine side chain, thereby stabilizing the positively charged form of the histidine and preventing (or markedly slowing) the donation of the proton to the thiolate. A key step of the reaction therefore will be inhibited.

29. In a classic monooxygenase fashion, one oxygen from O_2 will hydroxylate a substrate methyl group, and the other oxygen from O_2 will be reduced to water. The elimination of formyaldehyde will then lead to the products: methylamine and formyaldehyde (see *J. Biol. Chem.* 271[1996]:27321–27329).

30. Cytidine nucleotides play the same role in phosphoglyceride synthesis as a uridine nucleotide plays in the formation of glycogen (Section 21.4). In each case, an activated intermediate (UDP-glucose, or CDP-diacylglycerol, or CDP-alcohol) is formed by reaction of a phosphorylated substrate (glucose 1-phosphate, or phosphatidate, or a phosphorylalcohol) with a nucleoside triphosphate (UTP or CTP). The activated intermediate diphosphate then reacts with a hydroxyl group (the terminus of glycogen; a small alcohol such as, for example, inositol or the side chain of serine; or a diacylglycerol). The parallel chemistry involving UTP or CTP is evident at each step.

31. Proteins having such modifications are targeted to membranes. The isoprenoid side chains confer hydrophobic character and insert into lipid bilayer membranes, where they serve to "anchor" a protein to the membrane.

32. 3-Hydroxy-3-methylglutaryl CoA is also a precursor for synthesis of ketone bodies. If fuel is needed elsewhere in the body, as might be the case during a fast, 3-hydroxy-3-methylglutaryl CoA will be converted into acetoacetate, a ketone. If energy needs are met, the liver will synthesize cholesterol.

33. One way in which phosphatidylcholine can be synthesized is by the addition of three methyl groups to phosphatidylethanolamine. The methyl donor for this process is a modified form of methionine, *S*-adenosylmethionine or "SAM" (see Section 24.2).

34. When citrate is plentiful in the mitochondria, citrate is transported out of the mitochondria. ATP-citrate lyase breaks the citrate into acetyl CoA and oxaloacetate. The acetyl CoA can then be used to synthesize cholesterol.

35. (a) Cholesterol feeding has no effect on the amount of mRNA for HMG-CoA reductase.
    (b) The actin mRNA is a positive control to (1) verify that RNA can be effectively recovered from all samples, and (2) allow normalization of the results, if necessary, to correct for variations in the extent of overall RNA recovery from sample to sample.
    (c) The amount of HMG-CoA reductase protein is greatly reduced when the animals are fed a cholesterol diet.
    (d) Although the amount of specific HMG-CoA reductase mRNA is not affected by cholesterol, the level of HMG-CoA reductase protein decreases to near zero for the cholesterol-fed mice.
    (e) Several mechanisms could explain the presence of the specific mRNA and yet the absence of the specific protein that the mRNA encodes: 1. HMG-CoA reductase could be subject to translational control, so that translation of the message and

synthesis of HMG-CoA reductase by ribosomes are inhibited by cholesterol. 2. Alternatively, the protein could be synthesized but then rapidly degraded in the cholesterol-fed mice.

CHAPTER 27

1. In addition to being a storage site for fat, adipose tissue is now known to be also an active endocrine organ that secretes signal molecules called adipokines.

2. Caloric homeostasis is the condition in which the energy expenditure of an organism is equal to the energy intake.

3. Leptin and insulin are the key hormones responsible for maintaining caloric homeostasis.

4. CCK produces a feeling of having eaten a sufficient amount and stimulates digestion by promoting the secretion of digestive enzymes by the pancreas and of bile salts by the gall bladder. GLP-1 also produces a feeling of satisfied fullness and, in addition, potentiates the glucose-induced secretion of insulin by the β cells of the pancreas.

5. Something is not working properly. Although the answer is not known, the leptin-signaling pathway appears to be inhibited by the group of regulatory proteins known as the "suppressors of cytokine signaling."

6. 1. Leptin (a) and adiponectin (b) are secreted by adipose tissue.
   2. Glucagon (f) stimulates liver gluconeogenesis.
   3. The receptors for GLP-1 (c) and CCK (d) are G-protein-coupled receptors (gate-keepers for GPCR-regulated pathways).
   4. GLP-1 (c) and CCD (d) are satiety signals.
   5. GLP-1 (c) enhances insulin secretion.
   6. Glucagon (f) is secreted by the pancreas during a fast.
   7. Insulin (e) is secreted after a meal.
   8. Insulin (e) stimulates glycogen synthesis.
   9. Insulin (e) is missing in type 1 diabetes.

7. The three major sources of glucose 6-phosphate in the liver are (i) phosphorylation of dietary glucose after it enters the liver; (ii) synthesis of glucose 6-phosphate during gluconeogenesis; and (iii) breakdown of liver glycogen.

8. Type 1 diabetes is due to autoimmune destruction of the insulin-producing cells of the pancreas. Type 1 diabetes is also called insulin-dependent diabetes because affected individuals require insulin to survive. Type 2 diabetes is characterized by insulin resistance. Insulin is produced, but the tissues that should respond to insulin, such as muscle, do not.

9. Leptin stimulates processes that are impaired due to diabetes. For instance, leptin stimulates fatty acid oxidation, inhibits triacylglycerol synthesis, and increases the sensitivity of muscle and the liver to insulin.

10. (a) A watt is equal to 1 joule (J) per second (0.239 calorie per second). Hence, 70 W is equivalent to 0.07 kJ/s or 0.017 kcal/s.
    (b) A watt is a current of 1 ampere (A) across a potential of 1 volt (V). For simplicity, let us assume that all the electron flow is from NADH to O_2 (a potential drop of 1.14 V). Hence, the current is 61.4 A, which corresponds to 3.86×10^{20} electrons per second (1 A = 1 coulomb/s = 6.28×10^{18} charges/s).

(c) About 2.5 molecules of ATP are formed per NADH oxidized (two electrons). Hence, one molecule of ATP is formed per 0.8 electron transferred. A flow of 3.86×10^{20} electrons per second therefore leads to the generation of 4.83×10^{20} molecules of ATP per second or 0.80 mmol s^{-1}.

(d) The molecular weight of ATP is 507. The total body content of ATP of 50 g is equal to 0.099 mol. Hence, ATP turns over about once per 125 seconds when the body is at rest.

11. (a) The stoichiometry of complete oxidation of glucose is

$$C_6H_{12}O_6 + 6\ O_2 \rightarrow 6\ CO_2 + 6\ H_2O$$

and that of tripalmitoylglycerol is

$$C_{51}H_{98}O_6 + 72.5\ O_2 \rightarrow 51\ CO_2 + 49\ H_2O$$

Hence, the RQ values are 1.0 (6/6) and 0.703 (51/72.5), respectively.

(b) An RQ value reveals the relative usage of carbohydrate and fats as fuels. The RQ of a marathon runner typically decreases from 0.97 to 0.77 during the race. The lowering of the RQ reflects the shift in fuel from carbohydrate to fat.

12. One gram of glucose (molecular weight 180.2) is equal to 5.55 mmol, and one gram of tripalmitoylglycerol (molecular weight 807.3) is equal to 1.24 mmol. The reaction stoichiometries (see problem 5) indicate that 6 mol of H_2O are produced per mole of glucose oxidized, and 49 mol of H_2O per mole of tripalmitoylglycerol oxidized. Hence, the H_2O yields per gram of fuel are 33.3 mmol (0.6 g) for glucose and 60.8 mmol (1.09 g) for tripalmitoylglycerol. Thus, complete oxidation of this fat gives 1.82 times as much water as does glucose. Another advantage of triacylglycerols is that they can be stored in essentially anhydrous form, whereas glucose is stored as glycogen, a highly hydrated polymer (see Section 22.1 of the text). A hump consisting mainly of glycogen would be an intolerable burden—far more than the straw that broke the camel's back!

13. The starved–fed cycle is the nightly (and daily) hormonal cycle that humans experience during sleep and on eating. The purpose of the cycle is to maintain adequate amounts of blood glucose, particularly for brain function. Periods of starvation (sleep) are characterized by increased glucagon secretion and decreased insulin secretion. After a meal, the glucagon concentration falls and the insulin concentration rises.

14. Ethanol is oxidized by reaction with NAD$^+$, catalyzed by alcohol dehydrogenase, to yield acetaldehyde, which is subsequently oxidized to acetate. Both reactions generate NADH, which accumulates and inhibits gluconeogenesis and fatty acid oxidation. Ethanol is also metabolized to acetaldehyde by the microsomal ethanol-oxidizing system (MEOS), which uses oxygen and depletes NADPH. Free radical side products and a diminished ability to regenerate glutathione increase the oxidative stress.

15. First, fatty liver develops from the increased amounts of NADH that inhibit fatty acid oxidation and stimulate fatty acid synthesis. Second, alcoholic hepatitis develops because of damage induced by oxidation (free radicals) and excess acetaldehyde, conditions that lead to cell death. Finally, fibrous tissues form, creating scars that impair blood flow and biochemical function. Ammonia cannot be converted into urea, and its toxicity leads to coma and death.

16. A typical macadamia nut has a mass of about 2 g. Because it consists mainly of fats (-37 kJ/g, -9 kcal/g), a nut has a value of about 75 kJ (18 kcal). The ingestion of 10 nuts results in an intake of about 753 kJ (180 kcal). As stated in the answer to problem 4, a power consumption of 1 W corresponds to 1 J s^{-1} (0.239 cal s^{-1}), and so

400-W running requires 0.4 kJ s^{-1} (0.0956 kcal s^{-1}). Hence, one would have to run 1882 s, or about 31 minutes, to spend the calories provided by 10 nuts.

17. A high blood-glucose level would trigger the secretion of insulin, which would stimulate the synthesis of glycogen and triacylglycerols. A high insulin level would impede the mobilization of fuel reserves during the marathon.

18. A lack of adipose tissue leads to an accumulation of fats in the muscle, with the generation of insulin resistance. The experiment shows that adipokines secreted by the adipose tissue, here leptin, facilitate in some fashion the action of insulin in muscle.

19. Such a mutation would increase the phosphorylation of the insulin receptor and insulin-receptor substrates in the muscle and would improve insulin sensitivity. Indeed, PTP1B is an attractive therapeutic target for type 2 diabetes.

20. Insulin-dependent diabetes is characterized by high levels of blood glucose due to poor entry of glucose into cells. The impaired carbohydrate utilization leads to uncontrolled breakdown of lipids to acetyl-CoA. However, much of the acetyl-CoA cannot enter the citric acid cycle because of a shortage of oxaloacetate; furthermore, the acetyl-CoA cannot be converted to pyruvate or glucose. Acetyl-CoA therefore will be converted back to triacylglycerides, some of which will accumulate in the bloodstream.

21. Glycolysis is inhibited in the liver so that available glucose can be saved for use by the brain. Meanwhile, the liver supplies its energy needs by oxidizing fatty acids.

22. Electron transfer pathways depend on reactions in both compartments. For example, NADH is produced in both the cytoplasm and the mitochondria. NADH equivalents from glycolysis must be transported into the mitochondria by the glycerol–phosphate shuttle or malate–aspartate shuttle. Furthermore, ATP that is produced in the mitochondria must be transported specifically to the cytoplasm to support the energy needs of many reactions.

23. (a) Insulin inhibits lipolysis. An abundance of glucose and fatty acids in adipose tissue will lead to synthesis and storage of triacylglycerols.

    (b) Insulin promotes uptake of branched amino acids and has a general stimulating effect on protein synthesis and inhibitory effect on protein degradation. Nevertheless, the individual will continue to be protein deficient due to the poor diet.

    (c) Nonspecific damage to cell membranes (including transport systems) could cause fluid to leak into extracellular spaces.

24. During strenuous exercise, muscle converts glucose into pyruvate through glycolysis. Some of the pyruvate is processed by cellular respiration. However, some of it is converted into lactate and released into the blood. The liver takes up the lactate and converts it into glucose through gluconeogenesis. Muscle may furthermore process the carbon skeletons of branched-chain amino acids aerobically. The nitrogens of these amino acids are transferred to pyruvate to form alanine, which is released into the blood and taken up by the liver. After transfer of the amino group from alanine to α-ketoglutarate, the resulting pyruvate can be converted into glucose and released from the liver back into the bloodstream, for transport to the muscle. Finally, muscle glycogen may be mobilized, and the resulting glucose released into the bloodstream for uptake and use by the muscles.

25. The conversion of pyruvate to lactate allows muscle to function anaerobically. NAD^{+} is regenerated for glycolysis when pyruvate is reduced to lactate, and so energy can continue to be extracted from glucose during strenuous exercise. The lactate is transported to the liver, where it is converted back into glucose.

26. When time is not an issue, fatty acids are the major fuel for muscle. Under the emergency conditions of strenuous work, glucose becomes the major fuel for muscle.

27. The practice is called carbo-loading. Depleting the glycogen stores initially will cause the muscles to synthesize large amounts of glycogen. Later when dietary carbohydrates are restored, the condition will lead to a rich supply of glycogen, or a "super-compensation" of the glycogen stores.

28. The oxygen will serve as the ultimate acceptor of electrons from NADH, as important recovery reactions take place. During the recovery, lactate that was produced during exercise will be converted back into pyruvate (primarily in the liver), with the concomitant production of NADH from NAD^+ by lactate dehydrogenase. Electrons from NADH will pass through the electron-transport chain to oxygen, producing NAD^+ in addition to ATP. The NAD^+ will be used to oxidize more lactate and will be available as an electron acceptor for future glycolysis, when needed. Some of the ATP will be used to regenerate phosphocreatine and some for gluconeogenesis, to replenish the expended supplies of glucose and glycogen, in the liver and muscle.

29. Excess oxygen is needed because thermodynamic machines, including mammalian bodies, are less than 100% efficient. Some of the energy is lost as heat, and additional energy is expended because gluconeogenesis (to replenish muscle glycogen) is not the thermodynamic or chemical reverse of glycolysis. Rather, the resynthesis of glucose from lactate requires more ATP than is produced by anaerobic glycolysis. The amount of excess oxygen consumed typically is about 15% of the total oxygen consumed during exercise (see *J. Appl. Physiol.*, 62[1987]:485–490).

30. Many brain functions depend on a balance between excitatory and inhibitory neurotransmission. It is likely that the diverse effects of ethanol result from alterations in this balance. Although ethanol interacts with several receptor and channel systems, the detailed mechanisms are not yet understood.

31. One possible approach would be to attempt to fix samples for microscopy under aerobic and anaerobic conditions. Perhaps differences in fiber morphology or crossbridge formation could be observed (particularly for type I fibers) in the presence and absence of oxygen. Alternatively, the subclasses of myosin differ in type I and type II fibers and can be distinguished using specific antibodies as labels that can be viewed bound to the respective fiber types in the electron microscope (see *J. Cell. Biol.*, 90[1981]:128–144).

32. (a)
$$\frac{(836,000 \text{ kJ})(503 \text{ g mol}^{-1})}{50 \text{ kJ mol}^{-1}} = 8.4 \times 10^6 \text{ g.}$$

(b) Multiply the answer from part (a) by the cost per gram to obtain a total cost of $1.2 billion!

33. Converting the units, $(55 \text{ pounds})(453.6 \text{ g pound}^{-1}) = -25,000 \text{ g}$ total weight gain.

Additionally, $(40 \text{ years})(365 \text{ days year}^{-1}) = 14,600$ days.

$25,000 \text{ g}/14,600 \text{ days} = 1.7 \text{ g day}^{-1}$. One gram of fat is about 9 kcal,

so 1.7 g day^{-1} is about **15 kcal day^{-1}**.

With mass of 175 lb (79.4 kg) and height of 66 in (167.6 cm),

the BMI is $(79.4 \text{ kg})/(1.676 \text{ m})^2 = $ **28.2**.

The individual would be considered overweight but not obese.

34. Exercise greatly enhances the ATP needs of muscle cells. To more efficiently meet these needs, more mitochondria are synthesized.

35. The inability of muscle mitochondria to process all of the fatty acids produced when excess nutrients are present leads to excessive levels of diacylglycerol and ceramide in the muscle cytoplasm. These second-messenger molecules then activate enzymes that impair insulin signaling.

36. Both diseases are caused by a lack of thiamine (vitamin B_1). Thiamine, which is sometimes called aneurin, is required most notably for the proper functioning of pyruvate dehydrogenase.

37. (a) Red blood cells always produce lactate, and fast-twitch muscle fibers (see Problem 31) also produce a large amount of lactate.

   (b) When the lactate threshold is crossed, there is a major switch from aerobic exercise to primarily anaerobic conditions in the muscle. Most of the energy then is produced by anaerobic glycolysis.

   (c) The lactate threshold is essentially the point at which the athlete switches from aerobic exercise, which can be done for extended periods, to anaerobic exercise, essentially sprinting, which can be done for only short periods. The idea is to race at the extreme of his or her aerobic capacity until the finish line is in sight and then to switch to a final burst of anaerobic activity in order to gain a short but unsustainable advantage at the end of the race.

   (d) Training increases the amount of blood vessels and the number of muscle mitochondria. Together, they increase the ability to process glucose aerobically. Consequently, a greater effort can be expended under aerobic conditions before the switch to anaerobic energy production.

CHAPTER 28

1. DNA polymerase I uses deoxyribonucleoside triphosphates; pyrophosphate is the leaving group. DNA ligase uses DNA-adenylate (AMP joined to the 5′-phosphate) as a reaction partner; AMP is the leaving group. Topoisomerase I uses a DNA-tyrosyl intermediate (5′-phosphate linked to the phenolic OH); the tyrosine residue of the enzyme is the leaving group.

2. Positive supercoiling resists the unwinding of DNA. The melting temperature of DNA increases in going from negatively supercoiled to relaxed to positively supercoiled DNA. Positive supercoiling is probably an adaptation to high temperature, where the unwinding (melting, denaturing) of DNA is markedly increased.

3. The nucleotides used for DNA synthesis have the triphosphate attached to the 5′-hydroxyl group with free 3′-hydroxyl groups. Such nucleotides can be utilized only for 5′-to-3′ DNA synthesis.

4. DNA replication requires RNA primers. Without appropriate ribonucleotides, the primers cannot be synthesized, and DNA replication will cease.

5. This close contact prevents the incorporation of ribonucleotides. Only 2′-deoxyribonucleotides can be incorporated into the growing chain by DNA polymerase.

6. (a) 1000 nucleotides/s divided by 10.4 nucleotides/turn for B-DNA gives 96.2 revolutions per second.

   (b) 0.34 μm/s. (1000 nucleotides/s corresponds to 3400 Å/s because the axial distance between nucleotides in B-DNA is 3.4 Å.)

7. The unwinding of DNA to expose single-stranded regions at the replication fork causes overwinding (positive supercoils) ahead of the fork. The action of topoisomerase II overcomes this effect by introducing negative supercoils to compensate. Without topoisomerase II, the DNA would become too tightly wound ahead of the fork.

8. The linking number Lk is defined as $Tw + Wr = 48 + 3 = 51$. With a constant linking number Lk of 51, if $Tw = 50$, then $Wr = 1$.

9. Telomerase is required to synthesize the ends of new linear chromosomes during cell division. Because cancer cells are dividing rapidly, it is likely that the telomerase gene must be activated for a cell to become cancerous.

10. No. The strands of DNA run in opposite directions. Helicase can be effective in either direction. If helicase binds to one strand, it may move in the 5'-to-3' direction, for example; but the net effect would be the same if helicase would bind to the other strand and move in the 3'-to-5' direction.

11. The activity will be similar to the replacement of an RNA primer with DNA by DNA polymerase I. One makes use of the combined $5' \to 3'$ exonuclease and $5' \to 3'$ polymerase activities of DNA polymerase I. From the point of the internal nick (of only one strand) by the endonuclease, polymerase I will extend the free 3'-OH using radioactive dNTPs while at the same time digesting from the internal 5'-phosphate to make room for the newly synthesized DNA. The result is a "nick translation" event in which an unlabeled portion of one DNA strand is replaced with a radioactive stretch of DNA. (Over the section of new synthesis, only one strand becomes labeled. The strand used as the "template" remains unlabeled.)

12. If replication were unidirectional, tracks with a low-grain density at one end and a high-grain density at the other end would be seen. On the other hand, if replication were bidirectional, the middle of a track would have a low density, as shown below. For *E. coli*, the grain tracks are denser on both ends than in the middle, indicating that replication is bidirectional.

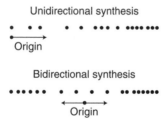

13. (a) Pro (CCC), Ser (UCC), Leu (CUC), and Phe (UUC). Alternatively, the last base of each of these codons could be U.
    (b) These $C \to U$ mutations were produced by nitrous acid.

14. Potentially deleterious side reactions are avoided. The enzyme itself might be damaged by light if it could be activated by light in the absence of bound DNA harboring a pyrimidine dimer. The DNA-induced absorption band is reminiscent of the glucose-induced activation of the phosphotransferase activity of hexokinase.

15. The free DNA ends that appear in the absence of telomeres are repaired by DNA fusions that may join one chromosome to another.

16. The free energy of ATP hydrolysis under standard conditions is -30.5 kJ mol^{-1} (-7.3 kcal mol^{-1}). At 10 kJ mol^{-1} (base pair)$^{-1}$, in principle, the energy of ATP hydrolysis could be used to break a maximum of three base pairs.

17. The oxidation of guanine could lead to DNA repair: DNA strand cleavage could allow looping out of the triplet repeat regions and triplet expansion.

18. The release of DNA topoisomerase II after the enzyme has acted on its DNA substrate requires ATP hydrolysis. Negative supercoiling of the DNA requires only the binding of topoisomerase II and ATP, not hydrolysis of ATP.

19. (a) Different supercoiled forms (topological isomers) migrate through the gel at different rates. The highly supercoiled DNA has a compact shape and migrates rapidly. Relaxed DNA has a more extended shape (a larger radius of gyration) and moves more slowly through the gel matrix.

    (b) The bands in lane B represent different supercoiled isomers of DNA. Neighboring bands differ from each other by ±1 superhelical turn.

    (c) With longer exposure to topoisomerase 1 (a "relaxing" enzyme), the population of DNA molecules shifts toward a distribution that is near thermal equilibrium.

20. (a) The control plate indicates the extent of spontaneous reversion in the absence of an external mutagen.

    (b) The known mutagen is a positive control to show that the procedures are correctly implemented and the test is working.

    (c) The experimental compound by itself (plate C) gives results that are only marginally above background (plate A). The experimental compound itself therefore should be classified as either non-mutagenic or only very slightly mutagenic. However, a metabolic product derived from the experimental compound is mutagenic (plate D).

    (d) One or more enzymes from the liver probably are responsible for the metabolic conversion of the experimental compound into a mutagenic compound.

CHAPTER 29

1. The sequence of the coding (+, sense) strand is

   5′-ATGGGGAACAG CAAGAGTGGGGCCCTGTCCAAGGAG-3′

   and the sequence of the template (−, antisense) strand is

   3′-TACCCCTTGTCGTTCTCACCCCGGGAC AGGTTCCTC-5′

   Note that the *coding* strand has the same sequence as mRNA (except U → T), whereas the *template* strand is complementary to the coding strand.

2. RNA turns over in the cell, whereas DNA is the "nearly" permanent record of inherited information. The consequences of most errors in RNA synthesis will be short-lived, perhaps a protein synthesized with a mistake in its sequence for a short time during the life of the cell, but then the RNA in question may be degraded and the error will disappear. By contrast, an uncorrected error in DNA replication will be passed to the next generation.

3. RNA synthesis involves copying only short segments of chromosomes, whereas DNA replication must involve copying entire chromosomes at the time of cell division. The length of consecutive nucleotide sequence that must be copied is many times greater for DNA replication than for RNA transcription.

4. The results suggest that the similar active sites of DNA and RNA polymerases have been formed by means of convergent evolution within otherwise unrelated protein sequences.

5. Heparin, a glycosaminoglycan, is highly anionic. Its negative charges, like the phosphodiester bridges of DNA templates, bind to lysine and arginine residues of β′.

6. This mutant sigma would competitively inhibit the binding of holoenzyme and prevent the specific initiation of RNA chains at promoter sites.

7. The core enzyme without sigma binds more tightly to the DNA template than does the holoenzyme. The retention of sigma after chain initiation would make the mutant RNA polymerase less processive. Hence, RNA synthesis would be much slower than normal.

8. A 100-kd protein contains about 910 residues, which are encoded by 2730 nucleotides. At a maximal transcription rate of 50 nucleotides per second, the protein would be synthesized in 54.6 seconds.

9. The faster rate is due to one-dimensional diffusion instead of three-dimensional diffusion. The RNA polymerase holoenzyme will bind initially with low affinity to any random site on the duplex DNA. The polymerase then reaches a promoter site rapidly by sliding along the DNA, rather than needing to diffuse through three-dimensional space.

10. The characteristic "-35" and "-10" prokaryotic promoter sequences are underlined and labeled below. Transcription is likely to start at the underlined *A* base:

    5′-GCCG<u>TTGACA</u>CCGTTCGGCGATCGATCCGC<u>TATAAT</u>GTGTGG*A*TCCGCTT-3′
            −35   −10

    (See also Figure 29.4)

11. Initiation at strong promoters occurs every two seconds. In this interval, 100 nucleotides are transcribed. Hence, centers of transcription bubbles are 34 nm (340 Å) apart (100 × 3.4 Å = 340 Å).

12. (a) The lowest band on the gel will be that of (i), whereas the highest will be that of (v). Band (ii) will be at the same position as (i) because the RNA is not complementary to the nontemplate strand, whereas band (iii) will be higher because a complex is formed between RNA and the template strand. Band (iv) will be higher than the others because strand 1 is complexed to 2, and strand 2 to 3. Band (v) is the highest because core polymerase associates with the three strands.

    (b) None, because rifampicin acts before the formation of the open complex.

    (c) RNA polymerase is processive. Once the template is bound, heparin cannot enter the DNA-binding site.

    (d) When GTP is absent, synthesis stops when the first cytosine residue downstream of the bubble is encountered in the template strand. In contrast, with all four nucleoside triphosphates present, synthesis will continue to the end of the template.

13. RNA polymerase must backtrack before it can cleave an incorrect sequence in the RNA. The mechanism leads to dinucleotide products.

14. Segments of double helix that are shorter than tetranucleotides (having fewer than about 4 base pairs) are unstable at physiological temperatures due to their small extent of base stacking and small number of interstrand hydrogen bonds. Until the critical length of RNA is synthesized for a stable RNA/DNA double helix, the short initial di- and trinucleotides are susceptible to release.

15. (a) The lack of a 3′-OH group will cause cordycepin to be a chain terminator for RNA synthesis. Cordycepin will be incorporated at the 3′-end of a chain, and further elongation will be blocked.

    (b) The substrate specificity of poly(A) polymerase is higher than that of RNA polymerase because the RNA polymerase uses four nucleotides (ATP, UTP, GTP, CTP), whereas poly(A) polymerase uses only ATP. The result suggests that poly(A) poly-merase has a higher apparent affinity for 3'-deoxy-ATP than does RNA polymerase.

(c) Yes. It must receive a 5'-triphosphate in order to be a substrate for poly(A) poly-merase or RNA polymerase.

16. Two alternative outcomes exist at each of eight sites. Therefore, a choice between two possible outcomes must be made eight times. The result is $2^8 = 256$ different possi-ble products.

17. The mechanism could involve small adjustments in the precise distances between the -35 and -10 sequences, and/or between the -10 and start sequences for a particu-lar gene. The relative three-dimensional placement of these respective sequences can be influenced by insertion or deletion of single nucleotide bases, as well as by the tor-sional strain related to supercoiling and unwinding. The fact that negative supercoil-ing decreases the rate of transcription of topoisomerase II itself provides a good safety feature, namely by preventing the topoisomerase II from overstimulating the expres-sion of its own gene.

18. If every T is changed to U, the polarity and base sequence of the DNA strands shown are identical with the mRNA synthesized from the complementary strand. Thus, TCTATTTTCCACCCTTAG becomes UCUAUUUUCCACCCUUAG and encodes Ser-Ile-Phe-His-Pro-stop.

19. A mutation that disrupted the normal AAUAA recognition sequence for the endonu-clease could account for this finding. In fact, a change from U to C in this sequence caused this defect in a thalassemic patient. Cleavage occurred at the AAUAAA 900 nucleotides downstream from this mutant AACAAA site. See S. H. Orkin, T.-C. Cheng, S. E. Antonarakis, and H. H. Kazazian, Jr. *EMBO J.* 4(1985):453.

20. One possibility is that the 3' of the poly(U) donor strand cleaves the phosphodiester bond on the 5' side of the insertion site. The newly formed 3' terminus of the accep-tor strand then cleaves the poly(U) strand on the 5' side of the nucleotide that initi-ated the attack. In other words, a U could be added by two transesterification reactions. This postulated mechanism is akin to the one in RNA splicing (see Figure 28.29 in the text). See T. R. Cech. *Cell* 64(1991):667.

21. The possibilities of alternative splicing and of RNA editing allow the final protein products to be more complex and more highly varied than the genes that encode them. Additionally, posttranslational modification of proteins (e.g., glycosylation, phosphorylation) will further enhance the complexity.

22. One could make use of the A–T base pairing potential of the poly-A sequence that is characteristic of eukaryotic mRNAs. One would construct an affinity chromatogra-phy column in which an oligo-dT nucleotide is covalently linked to a resin. Eukaroytic mRNAs would bind to this column, whereas other RNAs would not. After washing the other RNAs away from the column, one would elute the eukaroytic mRNAs by weakening the A–T hydrogen bonds, for example, by changing the tem-perature, or by washing the column with a solution containing an excess of soluble oligo-dA (to displace the mRNA).

23. (a) The different genes are expressed to differing extents. Only those genes for which mRNA is actively being transcribed will give positive hybridization signals.

(b) Gene expression patterns differ in the different tissues. Some of the mRNAs are transcribed in some tissues but not others.

(c) The genes that are expressed in all three tissues could be essential for funda-mental metabolic processes that are common to most if not all cells.

(d) Including the initiation inhibitor allows counting and comparison of the number of ongoing mRNA chains being synthesized among the different gene types and from tissue to tissue at the given moment in time when the cells are broken. (Without such an inhibitor, the results could be skewed by differing initiation rates from gene to gene, or by the possibility of some artificial initiation events that could be induced when the cells are broken to isolate the nuclei.)

24. The long strand that goes entirely across the picture from left to right is the DNA. The strands of increasing length are molecules of mRNA that are beginning to be transcribed. Transcription begins just ahead of the site on the DNA where shortest strands of mRNA are seen. Transcription ends just after the site on the DNA where the longest strands of mRNA are attached. (As RNA polymerase passes this site, the primary transcript mRNA is released.) On the page, the direction of RNA synthesis is from left to right. Many different enzymes are simultaneously making many different RNA molecules on a single gene.

CHAPTER 30

1. The Oxford English Dictionary defines translation as the action or process of turning one language into another. Protein synthesis converts nucleic acid sequence information into amino acid sequence information.

2. An error frequency of one incorrect amino acid every 10^4 incorporations allows for the rapid and accurate synthesis of proteins as large as 1000 amino acids. Higher error rates would result in too many defective proteins. Lower error rates would likely slow the rate of protein synthesis without a significant gain in accuracy.

3. All transfer RNA molecules have these features:
   (i)   Each is a single chain of RNA.
   (ii) They contain unusual bases.
   (iii) Approximately half of the bases are base-paired to form double helices. The strands that comprise the double-helical segments are antiparallel to each other.
   (iv) The 5′ end of the tRNA is phosphorylated and is usually pG.
   (v)   The 3′ end of the tRNA terminates with the sequence CCA, and the hydroxyl group of the A residue is the acceptor site for attachment of the incoming cognate amino acid.
   (vi) The anticodon is located in a loop near the center of the tRNA sequence.

4. The first step involves the formation of an aminoacyl adenylate, an intermediate that is formed by reaction between an amino acid and ATP, with release of pyrophosphate. The aminoacyl adenylate then reacts with tRNA to form the aminoacyl-tRNA. Both steps are catalyzed by an aminoacyl-tRNA synthetase that is specific for a particular amino acid and its cognate tRNA molecules.

5. Unique features are required so that the aminoacyl-tRNA synthetases can distinguish among the different tRNA molecules and attach the correct amino acid to the proper tRNA. Common features are required because all of the aminoacyl-tRNAs then must interact with the same protein-synthesizing machinery in the ribosomes.

6. An activated amino acid is one that is covalently coupled to its appropriate tRNA molecule.

7. The enzyme-bound Ile-AMP intermediate is necessary for the $^{32}PP_i$ exchange into ATP. Since isoleucine is a requirement, labeled ATP will be formed only in (c).

8. When the single ATP molecule is cleaved to AMP and pyrophosphate, the pyrophosphate is subsequently hydrolyzed to give two molecules of orthophosphate. The effective cost is two high-energy phosphate bonds. A second ATP molecule is then required to convert the AMP into ADP, the substrate for oxidative phosphorylation.

9. Amino acids larger than the correct amino acid cannot fit into the active site of the tRNA. Smaller but incorrect amino acids that may become attached to the tRNA fit into the editing site and are cleaved from the tRNA. For example, serine may become attached to threonyl-tRNA, but the serine will fit into the editing site and be cleaved from the threonyl-tRNA. When the correct threonine is attached, it will not fit the editing site and will not be cleaved.

10. Recognition sites on both faces of the tRNA molecules may be required to uniquely and correctly identify all members of the set of 20 different tRNA molecules.

11. The first two bases in a codon form Watson–Crick base pairs that are checked for fidelity by bases of the 16S rRNA. The third base is not inspected for accuracy, and so some variation is tolerated. Non-canonical base pairs are sometimes permitted in the third position.

12. Four bands: light, heavy, a hybrid of light 30S and heavy 50S, and a hybrid of heavy 30S and light 50S. Recall that "ribosomes dissociate into 30S and 50S subunits after the polypeptide product is released." As protein synthesis continues, 70S ribosomes are re-formed from the various heavy and light subunits.

13. About 799 high-energy phosphate bonds are consumed—400 to activate the 200 amino acids, 1 for initiation, and 398 to form 199 pepitide bonds.

14. Type I: b, c, and f; type 2: a, d, and e.

15. The reading frame is a set of contiguous, nonoverlapping three-nucleotide codons that begins with a start codon and ends with a stop codon. The position of the start codon serves to define a choice among three possible reading frames. Insertion or deletion of a single nucleotide base will cause a shift in the reading frame. Insertion or deletion of three consecutive bases will insert or remove one codon without changing the reading frame for the rest of the sequence.

16. A mutation caused by the insertion of an extra base can be suppressed by a tRNA that contains a fourth base in its anticodon. For example, UUUC rather than UUU is read as the codon for phenylalanine by a tRNA that contains 3′-AAAG-5′ as its anticodon.

17. One approach is to synthesize a tRNA that is acylated with a reactive amino acid analog. For example, bromoacetylphenylalanyl-tRNA is an affinity-labeling reagent for the P site of *E. coli* ribosomes. See H. Oen, M. Pellegrini, D. Eilat, and C. R. Cantor. *Proc. Nat. Acad. Sci.* 70(1973):2799.

18. The sequence GAGGU is complementary to a sequence of five bases at the 3′ end of 16S rRNA and is located several bases on the 5′ side of an AUG codon. Hence this region is a start signal for protein synthesis. The replacement of G by A would be expected to weaken the interaction of this mRNA with the 16S rRNA and thereby diminish its effectiveness as an initiation signal. In fact, this mutation results in a tenfold decrease in the rate of synthesis of the protein specified by this mRNA. For a discussion of this informative mutant, see J. J. Dunn, E. Buzash-Pollert, and F. W. Studier. *Proc. Nat. Acad. Sci.* 75(1978):2741.

19. The triplets codons 5′-UUU-<u>UGC</u>-CAU-GUU-<u>UGU</u>-GCU-3′ would be translated in sequence as: Phe-<u>Ala</u>-His-Val-<u>Ala</u>-Ala. The codons UGC and UGU encode cysteine

but, because the cysteine has been modified to alanine on the aminoacyl-tRNA, alanine is incorporated in place of cysteine at the underlined positions in the sequence.

20. Proteins are synthesized from the amino to the carboxyl end on ribosomes, and in the reverse direction in the solid-phase method. The activated intermediate in ribosomal synthesis is an aminoacyl-tRNA; in the solid-phase method, it is the adduct of the amino acid and dicyclohexylcarbodiimide.

21. The error rates of DNA, RNA, and protein synthesis are of the order of 10^{-10}, 10^{-5}, and 10^{-4} per nucleotide (or amino acid) incorporated. The fidelity of all three processes depends on the precision of base-pairing to the DNA or mRNA template. No error correction occurs in RNA synthesis. In contrast, the fidelity of DNA synthesis is markedly increased by the $3' \rightarrow 5'$ proofreading nuclease activity and by postreplicative repair. In protein synthesis, the mischarging of some tRNAs is corrected by the hydrolytic action of the aminoacyl-tRNA synthetase. Proofreading also takes place when aminoacyl-tRNA occupies the A site on the ribosome; the GTPase activity of EF-Tu sets the pace of this final stage of editing.

22. GTP is not hydrolyzed until aminoacyl-tRNA is delivered to the A site of the ribosome. An earlier hydrolysis of GTP would be wasteful because EF-Tu–GDP has little affinity for aminoacyl-tRNA.

23. The translation of an mRNA molecule can be blocked by antisense RNA, an RNA molecule with the complementary sequence. The antisense–sense RNA duplex is degraded by nucleases. Antisense RNA added to the external medium is spontaneously taken up by many cells. A precise quantity can be delivered by microinjection. Alternatively, a plasmid encoding the antisense RNA can be introduced into target cells. For an interesting discussion of antisense RNA and DNA as research tools and drug candidates, see H. M. Weintraub. *Sci. Amer.* 262(January 1990):40.

24. (a) Intact protein isolated after only one minute will have been started with unlabeled amino acids. (Only during the last minute of synthesis will label have been incorporated into the protein.) Therefore, the carboxyl-terminal peptide A_5, the last segment to be synthesized, will be most heavily labeled.

    (b) Due to the continuation of previously initiated chains, the order, from most labeled to least, will reflect the reverse order of synthesis:

    $$A_5 > A_4 > A_3 > A_2 > A_1$$

    (c) Synthesis begins at the amino terminal and proceeds to the carboxyl terminal.

25. Aminoacyl-tRNA synthetases are the only components that actually match a nucleotide sequence (the three-base RNA anticodon) with a particular amino acid to define the genetic code. All the other interactions of genetic code components involve simply "Watson-Crick" pairing between complementary bases.

26. The rate of protein synthesis would be slower because the cycling of EF-Tu between its GTP-bound and GDP-bound forms would be slowed.

27. Protein factors are required for initiation, elongation, and termination of protein synthesis. Initiation factors IF1 and IF3 prevent premature binding of the 30S and 50S ribosomal subunits, whereas IF2 delivers Met-tRNA$_{fMet}$ to the ribosome. Together these factors modulate the initiation of protein synthesis. Additional protein factors are required for elongation (EF-G and EF-Tu), for termination (release factors, RFs), and for ribosome dissociation (ribosome release factors, RRFs).

28. Protein translocation across the endoplasmic reticulum membrane usually requires a signal sequence, signal-recognition particle (SRP), SRP receptor, and the translocon.

29. The formation of peptide bonds, which in turn is powered by the hydrolysis of the aminoacyl-tRNAs, also drives the co-translational movement of an emerging protein across the endoplasmic reticulum.

30. The Shine–Dalgarno sequence of the mRNA base-pairs with a part of the 16S rRNA of the 30S subunit. This base pairing positions the 30S ribosomal subunit so that the initiator AUG in the mRNA is recognized as distinct from other AUG codons in the sequence.

31.

	Prokaryote	Eukaryote
Ribosome size	60S	80S
mRNA	polycistronic	Not polycistronic
Initiation	Shine–Dalgarno is required	First AUG is used
Protein factors	Required	Many more required
Relation to transcription	Translation can start before transcription is completed	Transcription and translation are spatially separated
First amino acid	fMet	Met

32. The signal-recognition particle (SRP) binds to the signal sequence and inhibits further translation. The SRP then ushers the arrested ribosome to the endoplasmic reticulum (ER) membrane, where it interacts with the SRP receptor (SR). The SRP–SR complex then binds the translocon and simultaneously hydrolyzes GTP. On hydrolysis of GTP, the SRP and SR dissociate from each other and from the ribosome. Protein synthesis resumes and the nascent protein is channeled through the translocon.

33. Without polysomes, the alternative would be to have a single ribosome translating a single mRNA molecule. The use of polysomes is more efficient because polysomes enable more protein synthesis per mRNA molecule in a given period of time. Protein production is thereby enhanced.

34. (a) Initiation requires GTP (1) for energy, the AUG codon (2), fMet (3) as the first amino acid, factor IF2 (5), and a Shine-Dalgarno sequence (6).
    (b) Elongation requires GTP (1) for energy, factor EF-Tu (7), and peptidyl transferase (8).
    (c) Termination requires GTP (1), factor RRF (4), a termination codon such as UGA (9), and transformylase (10).

35. Transfer RNAs need to be rather large to enable specific recognition by particular protein partners. The set of different tRNAs must be specifically recognized by their appropriate cognate aminoacyl-tRNA synthetases. Each tRNA also must interact with the ribosome and, in particular, with the peptidyl transferase.

36. The nitrogen atom of the deprotonated α-amino group of aminoacyl-tRNA makes a nucleophilic attack on the ester bond of peptidyl-tRNA to form the new peptide bond. As a result, the growing peptide chain is transferred to the tRNA that bears the new amino acid. The tRNA that formerly held the peptide is released:

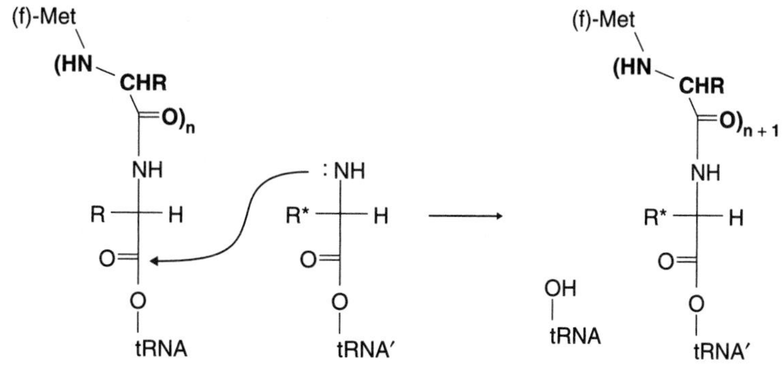

(f)-Met
(HN — CHR =O)$_n$ ... (f)-Met (HN — CHR =O)$_{n+1}$

37. The ornithinyl-tRNA is unstable because the nitrogen of the side chain will serve as a nucleophile to hydrolyze the ester bond to tRNA. The hydrolysis is facile because the transition state involves a six-membered ring. Lysyl-tRNA, by contrast, is more stable because a similar internal reaction would require a less favorable transition state with a seven-membered ring.

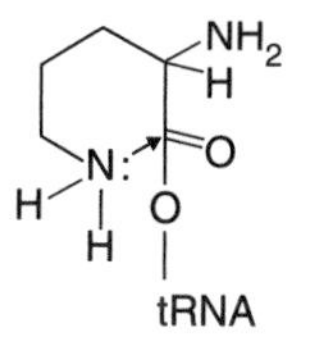

Self-hydrolysis of putative ornithinyl-tRNA

38. EF-Ts catalyzes the exchange of GTP for GDP bound to EF-Tu. In G protein cascades, an activated seven-helix receptor catalyzes GTP–GDP exchange in a G protein. For example, photoexcited rhodopsin triggers GTP–GDP exchange in transducin (see Section 32.3 of the text for further discussion).

39. Many G proteins are sensitive to ADP ribosylation by cholera toxin or pertussis toxin (see Chapter 14). In each case, an ADP-ribose unit is transferred from NAD^+, but the acceptor residue varies. For the modification of EF2 by diphtheria toxin, the acceptor is diphthamide (a derivative of histidine; see Figure 30.31 in the text), whereas the acceptors for the cholera toxin or pertussis toxin modifications of G proteins are arginine or cysteine.

40. Initially Glu-tRNA^{Gln} is formed by misacylation. Then the activated glutamate is subsequently amidated to form Gln-tRNA^{Gln}. In regard to $H.$ $pylori$, a specific enzyme, Glu-tRNA^{Gln} amidotransferase, catalyzes the reaction:

Gln + Glu-tRNA^{Gln} + ATP $\rightarrow$ Gln-tRNA^{Gln} + Glu + ADP + Pi

Glu-tRNA^{Glu} is not a substrate for the enzyme; so the transferase must also recognize aspects of the structure of tRNA^{Gln}.

41. The folded protein structure determines its function. The primary structure determines the three-dimensional structure of the protein. Thus, the final phase of information transfer from DNA to RNA to protein synthesis involves the folding of the protein into its functional state.

42. (a) In Graph A, eIF4H exhibits two effects: (1) The higher slope observed at early reaction times shows that the rate of helix unwinding increases. (2) The extent of helix unwinding in the plateau region at late reaction times also increases.

(b) To establish that eIF4H by itself does not have inherent helicase activity.

(c) Half-maximal activity was achieved with about 0.11 μM eIF4H, that is, about half of the concentration of eIF4. Depending on the relative kinetics of association and dissociation, this result may suggest a stoichiometric 1:1 binding of the helper to the initiation factor.

(d) The upward displacement of the straight line indicates that eIF4H enhances the rate of unwinding of all helices. The smaller slope when eIF4H is present indicates that the helper effect is greater for the more stable helices.

(e) Several answers are possible. Graph A shows that the helper enhances both the rate and extent of helix unwinding. Both of these effects would result if the helper would slow the dissociation of eIF4 from the RNA helix. Such a mechanism would increase the processivity and also would be consistent with the energetics shown in Graph C.

43. (a) The three peaks represent, from left to right, the 40S ribosomal subunit, the 60S ribosomal subunit, and the intact 80S ribosome.

(b) Not only are ribosomal subunits and the 80S ribosome present, but polysomes of various lengths also are apparent. The individual peaks in the polysome region near the bottom of the tube represent polysomes of discrete lengths. Under the influence of RNase, the polysomes are digested into monosomes (part A).

(c) Hypoxic conditions significantly inhibited the number of polysomes while increasing the number of free ribosomal subunits. This outcome could be due to lack of mRNA caused by an inhibition of transcription, or to inhibition of the initiation of protein synthesis on mRNA.

CHAPTER 31

1. (a) Without the *lac* repressor gene, the repressor protein will not be produced. Without a repressor, the *lac z, y,* and *a* proteins will be produced constitutively (independently of the presence or absence of lactose, albeit at low levels when glucose is present, due to catabolite repression).

(b) Provided that the *lac* promoter remains intact, the effect would be the same as in (a): no repression, and constitutive production of the *lac z, y,* and *a* proteins.

(c) Without CAP to stimulate transcription, the levels of the *lac z, y,* and *a* proteins that are produced will remain low, even in the presence of lactose.

2. A liter contains 1000 cm^3, so a cell volume of about 10^{-12} cm^3 is 10^{-15} liter. One molecule divided by Avogadro's number in 10^{-15} liter corresponds to a concentration of about $1.7 * 10^{-9}$ M.

$$\frac{(1 \text{ molecule})}{(6.02 * 10^{23} \text{ molecules mol}^{-1})(10^{-15} \text{ liter})} = 1.7 * 10^{-9} \underline{M}$$

Since the repressor concentration is much higher than the dissociation constant for the repressor/operator complex (10^{-13} M), the single molecule will be bound to (operator) DNA.

3. The probability of having a particular chosen nucleotide sequence at a given site is ($\frac{1}{4}$) for a single base, ($\frac{1}{4}$)^2 for a two-base sequence, and ($\frac{1}{4}$)^n for a sequence of n bases. The number of statistically expected occurrences of a particular sequence of length n in a genome of L base pairs is L times ($\frac{1}{4}$)^n. The table below summarizes the results for the *E. coli* genome, which contains about $4.8 * 10^6$ base pairs.

Length of sequence, n	$(1/4)^n$	Predicted number of sites in E. coli genome
8	$1.5 * 10^{-5}$	73
10	$9.5 * 10^{-7}$	4.6
12	$6.0 * 10^{-8}$	0.3

4. Whereas the *lac* repressor is released from DNA by binding to a small molecule, the *pur* repressor is induced to associate with DNA by the binding of a small corepressor molecule, either guanine or hypoxanthine. An additional difference is the number of respective binding sites in the E. coli genome, which contains only one binding site for the *lac* repressor but about 20 sites for the *pur* repressor (see Figure 31.14 in the text).

5. The anti-inducer could be a competitive inhibitor of the inducer. As such, the anti-inducer would bind to the repressor at a similar or overlapping site to that of the inducer, but would not cause the conformational change necessary to release the repressor from the operator DNA. Higher concentrations of inducer would then be needed to displace the competitively bound anti-inducer from its site on the repressor.

6. Because symmetry is a recurring theme for protein–DNA interactions, the DNA sequence may have functional importance. One possibility is that the DNA sequence could be a binding site for a dimeric regulatory protein. Alternatively, inverted repeat sequences sometimes serve as hot spots for genetic rearrangements because they may form hairpin secondary structures that block DNA polymerases or are processed by structure-specific endonucleases.

7. With λ repressor unable to bind to the mutant O_R2, the cooperative binding of the λ repressor to O_R2 and O_R1, which supports the lysogenic pathway, would be disrupted. Therefore, bacteriophage λ would be more likely to enter the lytic phase.

8. Let us compare the sequences:

   λ repressor gene   −10 region **GATTTA**   −35 region **TAGATA**.

   Cro gene   −10 region **TAATGG**   −35 region **TTGACT**.

   The different bases are underlined in the Cro sequences. Note that there are four differences in the −10 region and three differences in the −35 region.

9. Lambda repressor and Cro protein repress each other, such that Cro blocks λ repressor production and λ repressor blocks production of Cro. Therefore, increased Cro concentrations reduce the expression of the λ repressor gene. In similar fashion, increased λ repressor concentrations reduce the expression of the Cro gene. (See Figure 31.17.)

   The influence of λ repressor on production of λ repressor itself depends on the repressor concentration. At low concentration, the λ repressor binds to sites O_R1 and O_R2, and stimulates transcription of its own mRNA, thus enhancing λ repressor expression. However, at higher concentration, the λ repressor binds also to site O_R3, thus blocking access to its own promoter and inhibiting its own expression.

10. The translation will not be efficient. Normally, bacterial mRNAs have a leader sequence in which a Shine–Delgarno sequence precedes the ATG start codon and promotes ribosome binding and the initiation of translation. The absence of such a leader sequence is expected to lead to inefficient translation.

11. Add each compound to a culture of *V. fischeri* at low density and look for the development of luminescence. Compounds that show effective autoinducer activity will cause the cells to express luciferase and to become luminescent even at low cell density.

12. Counting from the sequence in Figure 31.22A, one finds that ACC is used seven times and ACA once. The codons ACU and ACG for threonine are not used in the given leader sequence.

13. The figure in the problem statement illustrates overall retention of configuration. Each elementary step is likely to involve inversion of configuration, which suggests that the reaction mechanism consists of an even number of steps.

    A possible mechanism is nucleophilic attack by the carboxylate group of Glu 537 on the C-1 carbon atom of the galactose moiety within lactose, thereby releasing glucose and forming a covalent intermediate with the galactose linked to the enzyme by means of an ester bond. In a second step, water could attack the carbonyl carbon of the ester, in order to displace the carboxylate of Glu 537 and release galactose as the second product.

14. It appears that about half of the DNA is protected when the λ repressor concentration is about 3.7 nM. Thus, at 298°K,

    $\Delta G°$ is approximately RT ln $(3.7 * 10^{-9})$ = −48.1 kJ/mol (−11.5 kcal/mol).

CHAPTER 32

1. If one ignores the histidine residues, one counts negative charges for each aspartic and glutamic acid (D and E in the sequences) and positive charges for each lysine and arginine (K and R in the sequences).

   H2A contributes +13 for K, +13 for R, –2 for D, and –7 for E, for a net charge of +17.

   H2B contributes +20 for K, +8 for R, –3 for D, and –7 for E, for a net charge of +18.

   H3 contributes +13 for K, +18 for R, –4 for D, and –7 for E, for a net charge of +20.

   H4 contributes +11 for K, +14 for R, –3 for D, and –4 for E, for a net charge of +18.

   The histone octamer contains two each of H2A, H2B, H3, and H4. The total charge on the histone octamer is therefore estimated to be 2 × (17 + 18 + 20 + 18) = +146. The total charge on 150 base pairs of DNA is −300. Therefore, the histone octamer neutralizes approximately one-half of the charge on the DNA.

2. The isolated mixture of DNA fragments could be tested for fragments that would hybridize to a single-stranded probe corresponding to a portion of the known sequence of interest. To prepare for the analysis, the known probe DNA could be attached to a filter. The DNA fragments isolated from the immunoprecipitation could be amplified by the polymerase chain reaction, if necessary, labeled with ^{32}P using 5′-polynucleotide kinase and α-^{32}P-ATP, and heated to separate the strands of the DNA double helix. Labeled fragments would then be incubated with the filter-attached probe (under "stringent" hybridization conditions). After washing the filter, the extent of binding could be determined by counting the specific radioactivity that the filter acquired during the hybridization reaction. For a *lac* repressor immunoprecipation experiment, only one unique DNA fragment is expected to be protected by the binding of the repressor. The *pur* repressor, by contrast, should protect about 20 or more different sites on the *E. coli* chromosome (see Figure 31.14 in the text).

3. The faster rate is due to one-dimensional diffusion instead of three-dimensional diffusion. The RNA polymerase holoenzyme will bind initially with low affinity to any random site on the duplex DNA. The polymerase then reaches a promoter site rapidly by sliding along the DNA, rather than needing to diffuse through three-dimensional space.

4. Transcriptionally inactive regions of DNA have a high content of 5-methylcytosine. Incorporating 5-azacytidine into DNA will prevent methylation. The lack of methylation will lead to the activation of some normally inactive genes.

5. Because 5-methylcytosine often is a signal for gene inactivity, the protein domain might play a role in gene inactivation. The domain could perhaps block transcription by binding to regulatory regions of double-stranded DNA that contain 5-methylcytosine. The protein domain would bind in the major groove. The 5-methyl group will be on the "outside" of a GC base pair (see diagram below) and will protrude into the major groove of double-stranded DNA. (To view an example, examine the C5 positions on the cytosines in structure 1D64 in the Protein Data Bank.)

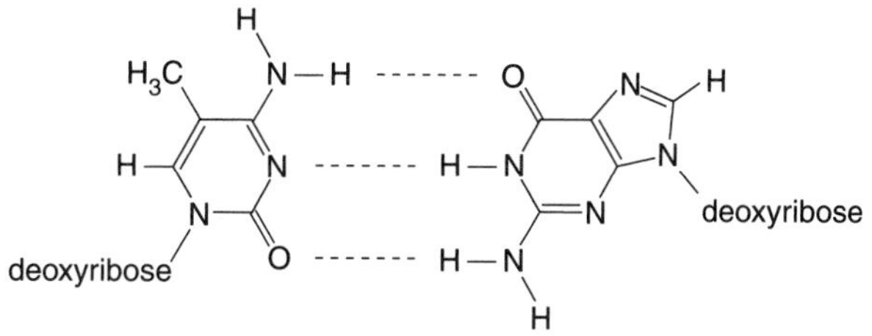

base pair between 5-methyl-C and G

6. Because estrogen will not bind to the hybrid receptor, gene expression is expected to give no response to the presence of estrogen. However, genes for which expression normally responds to estrogen will respond instead to the presence of progesterone.

7. Acetylation converts the lysine ε-ammonium group to an amide, thereby decreasing the charge from +1 to 0. Methylation produces methyl ammonium and does not change the charge.

8. Let us examine the eleven histidine and six cysteine residues (underlined) in the given sequence for the transcription factor:

HTCDYAGCGKTYTKSSHLKAHLRTHTGEKPYHCDWDGCGWKF

ARSDELTRHYRKHTGHRPFQCQKCDRAFSRSDHLALHMKRHF.

There are three pairs of cysteines (in which four or fewer other residues separate the members of each pair). Additionally, there are multiple pairs of candidate histidine residues which could coordinate a zinc ion, in combination with one of the cysteine pairs. The pattern therefore suggests that the region contains three zinc-finger domains.

9. Chromatin structure shields a large number of the potential binding sites for transcription factors in eukaryotic cells. For example, only about ten of 4000 potential GAL4 binding sites in yeast are actually occupied by GAL4 when the yeast are growing on galactose. The fraction 10/4000 is 0.25%. Within the 12 Mb yeast genome, 1.0% of 12 million base pairs is 120,000 base pairs, such that 0.25% of 12 million base pairs is 30,000 base pairs.

10. An iron response element (IRE) at the 5′ end of an mRNA will introduce a stem-loop structure that will bind a specific protein and block translation in the absence of iron. The same IRE at the downstream 3′ end of the mRNA would not be expected to block translation, but it could affect the stability of the mRNA.

11. Sequences complementary to the given micro RNA sequence could be regulated. A prediction strategy would involve searching a database of known human mRNA sequences to identify sequences that are fully or nearly complementary to the sequence of the miRNA. These sequences would be candidates for regulation by this miRNA.

12. The lysine amino group can make a nucleophilic attack on the carbonyl carbon of the thioester of acetyl-CoA to give a tetrahedral intermediate. The tetrahedral intermediate then could eliminate CoASH as a leaving group to yield acetyl-lysine.

13. A large percentage of the cytosine residues in mouse DNA are methylated, whereas very few C's in *Drosophila* or *E. coli* DNA are methylated. Therefore, the *Drosophila* and *E. coli* DNA are cut by HpaII into pieces of average size about 256 base pairs, whereas the mouse DNA is cut into pieces of average size about 50,000 base pairs.

CHAPTER 33

1. The specificity would be switched, and the "attractant" would become a "repellant." The AWB neurons induce avoidance behavior when their receptors encounter the corresponding ligands. This general behavior would be expected to remain true in the transgenic nematode.

2. At least two compounds must be present: C_5-COOH (receptor 5) and HOOC-C_7-COOH (receptor 9, as well as receptors 3, 12, and 13). Other compounds that activate receptor 5 cannot be present because of the pattern of non-activated receptors. Each of the following additional compounds *could* be present, but is not necessarily present: Br-C_3-COOH, Br-C_4-COOH, HOOC-C_4-COOH, HOOC-C_5-COOH (all activating receptor 13 only), and HOOC-C_6-COOH (receptors 3, 12, and 13).

3. Sour and salty taste responses result from the direct action of hydrogen ions or sodium ions on channels; these responses therefore have the potential for very rapid time resolution. Taste responses (bitter and sweet) that are likely to require 7TM receptors and second messengers will exhibit slower time resolution.

4. Dividing $(0.15 \text{ m})/(350 \text{ m s}^{-1}) = 428.6 \text{ }\mu\text{s}$.

   The human hearing system is capable of sensing time differences of close to a microsecond, and so the difference in arrival times at the two ears is substantial. G proteins typically respond in milliseconds, so a system based on G protein receptors would be unlikely to be able to reliably distinguish between signals arriving at the two ears.

5. If a plant tastes bitter, animals will avoid eating it. Even if the plant is nontoxic, the bitterness may provide a selective advantage to the plant.

6. Test mice in which either the T1R1 gene or the T1R3 gene (or genes for both T1R1 and T1R3) has (have) been disrupted. Monitor the taste responses of each strain of mice to glutamate and aspartate, and also to a wide variety of other amino acids. The pattern of the taste responses will reveal which encoded subunit is responsible for the specificity.

7. The women have four functional color receptors: blue, red, green, and a red–green hybrid. The additional color receptor allows the individuals to distinguish some colors that appear essentially identical to most other people.

8. For a set of 380 different receptors:

- If each receptor binds one odorant, 380 odorants can be detected.

- If each receptor binds two odorants, the number of possible pairwise binding combinations is (380 379)/(2!), so 72,010 odorants can be detected.

- If each receptor binds three odorants, the number of possible combinations for the three independent binding events is (380 379 378)/(3!), so 9,073,260 odorants can be detected.

9. The absorption of light catalyzes the isomerization of the rhodopsin-bound 11-cis-retinal into all-trans-retinal.

10. The compounds are enantiomers. Each will interact differently with protein receptors to elicit a smell. Subtle structural differences will be reflected in the receptor binding sites and will influence the relative receptor binding affinities and, hence, the elicited odor.

11. Ion channels are important for many of our sensory functions.
    An example for vision is the cGMP-gated channel.
    An example for taste is the amiloride-sensitive sodium channel.
    An example for hearing is the tip-link channel.

12. For all senses, ATP hydrolysis is required to generate and maintain ion gradients and membrane potential. Olfaction: ATP is required for the synthesis of cAMP. Gustation: ATP is required for the synthesis of cyclic nucleotides, and GTP is required for the action of gustducin in the detection of bitter and sweet tastes. Vision: GTP is required for the synthesis of cGMP and for the action of transducin. Hearing and touch: ATP hydrolysis is required to generate and maintain ion gradients and membrane potential and may be required for other roles as well.

13.

CHAPTER 34

1. The innate immune system responds rapidly to common features present in many pathogens. The genes for the innate immune system's key molecules are expressed without substantial modification. In contrast, the adaptive immune system responds to specific features present only in a given pathogen. Its genes undergo significant rearrangement and mutation to enable specific recognition of a vast number of potential binding surfaces, including new surfaces that may be presented as the pathogens adapt and change over time. The adaptive immune system may become involved in a "tug of war" that indeed can develop into a survival struggle with respect to particular pathogens.

2. The main mechanisms used by B cells to generate antibody diversity involve:

   - VJ and V(D)J recombination,
   - variable joining of segments by the action of terminal deoxyribonucleotidyl transferase, and
   - somatic mutation.

3. *Affinity* refers to the strength of a single interaction. *Avidity* refers to the cumulative strength of multiple independent binding interactions. Because the IgM immunoglobulin class features 10 binding sites, *avidity* may play a significant role in the interactions between IgM molecules and their antigens.

4. The intracellular TIR signaling domain common to each of the TLRs is responsible for docking other proteins and reporting that a targeted pathogen-associated molecular pattern ("PAMP"), such as LPS, has been detected. If a mutation within the TIR domain interfered with the intracellular docking and signal transduction, then TLR-4 would not respond to LPS. (It is possible that the actual binding of LPS could be normal, but the subsequent signaling events could be impaired; indeed, the location of the mutation—within the TIR domain—suggests that this is the case.)

5. Because TLR-3 recognizes double-stranded RNA, viruses that contain dsRNA genomes would be expected to stimulate a TLR-3-mediated immune response.

6. (a) $\Delta G^{\circ\prime} = -RT \ln K'_{dis} = -2.47 \times \ln (3 \times 10^{-7}) = 37$ kJ/mol (8.9 kcal/mol). Note that this *positive* value is the $\Delta G^{\circ\prime}$ for *dissociation* of the F_{ab}-hapten complex. Since binding is the reverse of dissociation, the $\Delta G^{\circ\prime}$ for binding $= -37$ kJ mol^{-1} (-8.9 kcal/mol).

   (b) $K_a = 1/$dissociation constant $= 1/3 \times 10^{-7}$ M $= 3.3 \times 10^6$ M. *Note:* see part (a)—that the $\Delta G^{\circ\prime}$ of binding $= -RT \ln K_a$.

   (c) An equilibrium constant is equal to the ratio of the rate constant of the forward (off) reaction to the rate constant of the reverse (on) reaction. Therefore, $k_{off}/k_{on} = 120\,s^{-1}/k_{on} = 3 \times 10^{-7}$ M. Solving this equation gives $k_{on} = 4 \times 10^8$ M^{-1}s^{-1}. This value is close to the diffusion-controlled limit for combination of a small molecule with a protein. Hence, the extent of structural change is likely to be small because extensive conformational transitions take time.

7. The fluorescence enhancement and shift to the blue indicate that water is largely excluded from the combining site when the hapten is bound. Hydrophobic interactions contribute significantly to the formation of most antigen-antibody complexes.

8. (a) An antibody-combining site is formed by CDRs from both the H and L chains. The V_H and V_L domains are essential. A small portion of F_{ab} fragments can be further digested to produce F_V, a fragment that contains just these two domains. C_H1 and C_L contribute to the stability of F_{ab} but not to antigen binding.

   (b) A synthetic F_V analog 248 residues long was prepared by expressing a synthetic gene consisting of a V_H gene joined to a V_L gene through a linker. See J. S. Huston, et al., *Proc. Nat. Acad. Sci.* 85(1988):5879.

9. (a) Multivalent antigens lead to the dimerization or oligomerization of transmembrane immunoglobulins, an essential step in their activation. This mode of activation is reminiscent of that of receptor tyrosine kinases.

   (b) An antibody specific for a transmembrane immunoglobulin will activate a B cell by cross-linking these receptors. This experiment can be carried out using, for example, a goat antibody to cross-link receptors on a mouse B cell.

10. B cells do not express T-cell receptors. Hybridization of T-cell cDNAs with B-cell mRNAs removes cDNAs that are expressed in both cells. Hence, the mixture of cDNAs

following this hybridization are enriched in those encoding T-cell receptors. This procedure, called *subtractive hybridization,* is generally useful in isolating low-abundance cDNAs. Hybridization should be carried out using mRNAs from a closely related cell that does not express the gene of interest. See S. M. Hedrick, M. M. Davis, D. I. Cohen, E. A. Nielsen, and M. M. Davis, *Nature* 308(1984):149, for an interesting account of how this method was used to obtain genes for T-cell receptors.

11. TLR-4 is the receptor for LPS, a toxin found specifically in the walls of gram-negative bacteria. Mutations that inhibit the function of TLR4 will impair an affected person's defenses against this class of bacteria.

12. For an organ transplant, if the HLA alleles are not matched, then the recipient's T cell receptors will identify the MHC proteins of the transplanted tissue as non-self proteins and will unleash defenses against them. Transplant rejection is likely.

13. The model could be tested by an unfolding/refolding experiment. An antibody would be reversibly unfolded using high temperature or a chemical denaturant such as guanidine hydrochloride (see Chapter 2, Section 2.6). Then a slow refolding would be attempted by gradually lowering the temperature or removing the denaturant in the presence of different putative small-molecule haptens. The model would suggest that different folding should be produced by refolding the presence of different antigens.

14. The gene rearrangements (see Section 33.4 of the text) that generate antibody diversity inevitably may introduce some premature termination codons. If expressed, the resulting truncated proteins or peptides could adversely affect the immune response. Nonsense-mediated RNA decay would provide a way to prevent or down-regulate the expression of the mistakenly terminated genes (see also *J. Immunology* 167[2001]:6901).

15. Under fortunate circumstances, the mutant bacteria may stimulate an immune response without causing the disease. In such a case, members of the set of mutant bacteria may be valuable starting points for the design of a live attenuated vaccine which could provide protection from the original pathogenic strain.

16. The most likely peptide for HLA-A2 presentation will have 8–10 residues, with L in the second position and V in the last position. Such a peptide is underlined within the sequence below:

MSRLASKNLIRSDHAGG <u>LLQATYSAV</u> SSIKNTMSFGAWSNAALNDSRDA.

17. The transition-state analog matches the geometry of the tetrahedral transition state that is formed during the hydrolysis of amides or esters by serine proteases (or cysteine proteases). For catalytic activity, therefore, one would expect an antibody to have an appropriate nucleophile such as serine (or cysteine) in its binding site. (A neighboring general base such as histidine would be useful to enhance the nucleophilicity of the Ser or Cys.)

18. A distinct intramolecular domain (e.g., a domain homologous to SH2; see Section 15.4 of the text) may bind to the critical phosphotyrosine residue and maintain the enzyme in an inactive conformation. If a phosphatase such as CD45 would remove the phosphate from the phosphotyrosine, then the second domain may not bind to the tyrosine and the inhibition could be relieved. (See also *J. Biol. Chem.* 276[2001]:23173, and references therein.)

19. (a) K_d, the antibody concentration needed to achieve 50% binding, is about 10^{-7} M for antibody A.
    (b) From the graph, K_d is about 10^{-9} M for antibody B.
    (c) Antibody B results from repeated immunizations, binds more tightly to the antigen and is therefore improved over antibody A. We recall that antibody B has been

produced by a process known as "affinity maturation." Somatic mutation is a likely mechanism for this process because a single codon change has led to the selection of antibodies that more precisely fit the antigen (see Section 33.4 of the text).

CHAPTER 35

1. (a) ATP is the energy source for skeletal muscle and eukaryotic cilia, and proton-motive force is the energy source for bacterial flagella.
    (b) There are two essential components in each system: myosin–actin, dynein–microtubule, and motA–motB, respectively.

2. $\dfrac{6400}{80} = 80$ "body lengths" per second

    For a 10-foot automobile, 80 body lengths per second would correspond to 800 ft s^{-1}, or:

    $$\frac{(800 \text{ ft s}^{-1}) \times (3600 \text{ s hr}^{-1})}{(5280 \text{ ft mile}^{-1})} = 545 \text{ miles per hour!}$$

3. The force generated, in grams, is

    $(0.22 \text{ lb}) \times (4 \times 10^{-12}) \times (454 \text{ g lb}^{-1}) = 4 \times 10^{-10} \text{ g.}$

    The protein mass, in grams, is

    $$\frac{(100,000 \text{ g mole}^{-1})}{(6.02 \times 10^{23} \text{ molecules mole}^{-1})} = 1.66 \times 10^{-19} \text{ g molecule}^{-1}.$$

    The ratio, force lifted per myosin molecule, obtained by dividing the above two numbers, is: 2.4×10^9 molecular "bodyweights" lifted. Heavy lifting, indeed!

4. Both actin filaments and microtubles are built from subunits. In both cases the subunits bind and hydrolyze nucleoside triphosphates. Actin filaments utilize ATP and are constructed using a single type of subunit. Microtubules utilize GTP and are constructed from two different types of subunits.

5. The light chains in myosin stiffen the lever arm. The light chains in kinesin bind the cargo that is to be transported.

6. The rapid decrease in the level of ATP following death has two consequences. First, the cytosolic level of calcium rises rapidly because the Ca^{2+}-ATPase pumps in the plasma membrane and sarcoplasmic reticulum membrane no longer operate. High Ca^{2+}, through troponin and tropomyosin, enables myosin to interact with actin. Second, a large proportion of S1 heads will be associated with actin. Recall that ATP is required to dissociate the actomyosin complex. In the absence of ATP, skeletal muscle is locked in the contracted (rigor) state.

7. The critical concentration for polymerization is 20-fold lower for actin–ATP than for actin–ADP (see Section 34.2 of the text). For monomer concentrations in between the critical concentrations for actin–ATP and actin–ADP, therefore, the actin–ATP will polymerize and later will gradually depolymerize, as the bound ATP becomes hydrolyzed to ADP.

8. Assuming a step size of 3.4 Å (corresponding to a conformation similar to that in double-stranded DNA), a one-base step is approximately $3.4 \text{ Å} = 3.4 \times 10^{-4}$ µm. If a stoichiometry of one molecule of ATP per step is assumed, the hydrolysis of 50 ATP molecules per second would correspond to a movement along 50 bases over a distance of $(3.4 \times 50) = 170$ Å. This distance corresponds to a velocity of 170 Å s^{-1}, or 0.017 µm s^{-1}. Kinesin moves at a velocity of 6400 Å per second, or 0.64 µm s^{-1}.

9. A protonmotive force across the plasma membrane is necessary to drive the flagellar motor. Under conditions of starvation, this protonmotive force is depleted. In acidic solution, the pH difference across the membrane is sufficient to power the motor.

10. Under the influence of a chemo-attractant, the mean distance between tumbles would become longer whenever the (randomly determined) direction of movement of the bacterium is up the concentration gradient (toward the chemo-attractant).

11. (a) A particle of 2 µm diameter has a radius of 1 µm. Substituting into the equation for the force (F), one obtains $F = 6\pi$ (0.01 g cm^{-1} s^{-1})(0.0001 cm)(0.6 × 10^{-4} cm s^{-1}) = 1.13×10^{-9} g cm s^{-2} = 1.13×10^{-9} dyne.

(b) If one erg is one dyne cm and the bead moves 0.6 µm (6 × 10^{-5} cm), the work performed per second is $(1.13 \times 10^{-9}$ dyne$)(6 \times 10^{-5}$ cm$) = 6.8 \times 10^{-14}$ erg.

(c) To calculate the energy content of ATP in ergs, one needs to know that 1 kJ is 10^{10} ergs. Then for ATP hydrolysis within the cell $\Delta G = -50$ kJ mol^{-1} = -50×10^{10} ergs per mole. Dividing ergs per mole by Avogadro's number (6.02×10^{23}), one gets 8.3×10^{-13} ergs per molecule. In one second, 80 molecules of ATP are hydrolyzed, corresponding to 6.6×10^{-11} ergs, much more energy than the actual work performed in moving the 2-µm-diameter bead. Thus, the hydrolysis of ATP by a single kinesin motor provides more than enough free energy to power the transport of micrometer-size cargoes at micrometer-per-second velocities.

12. A step size of only 6 nm would be inconsistent with the actual distance of 8 nm between equivalent binding sites on tubulin subunits.

13. One or more additional tether domains might allow the KIF1A protein to remain bound to a microtubule during times when the motor domain needed to detach. An alternation between tether and motor attachement could enable the protein to be processive. (For a discussion see *Proc. Natl. Acad. Sci., USA*, 97[2000]:640.)

14. (i) Filaments built from subunits can be arbitrarily long.
(ii) The assembly process can be dynamic and easily reversible.
(iii) Only a small amount of genetic material is required to encode the necessary information for building the filaments from subunits.

15. Because protons must still flow from outside to inside the cell, it is conceivable that neighboring half-channel sites may cooperate with each other. Each proton might pass into the outer half-channel of one MotA−MotB complex, bind to the MS ring, rotate clockwise, and then pass into the inner half-channel of the neighboring MotA−MotB complex.

16. The effect is mediated through Ca^{2+}-calmodulin, which stimulates myosin light-chain kinase (MLCK). Phosphoryl groups introduced by MLCK are removed from myosin by a Ca^{2+}-independent phosphatase.

17. (a) From the graph, the maximum velocity is about 13 molecules of ATP hydrolyzed per myosin molecule per second, that is, $k_{cat} \simeq 13$ s^{-1}. K_M is the ATP concentration required to yield half maximal activity (6.5 s^{-1}), that is, $K_M \simeq 15$ micromolar ATP.

(b) There are about 7 major steps between 120 nm and 380 nm (y-axis). (380 – 120) mm/7 = 37 nm per step.

(c) It is plausible that the two heads of myosin may "walk" along the actin filament as two feet alternately cycle past each other, and alternately exchange leading and trailing positions, when a person walks. The respective cycles of binding, ATP hydrolysis, ADP release, and movement would alternate between the two myosin heads. Processivity could be achieved by one head always remaining attached when the other was released or moving, and vice versa.

CHAPTER 36

1. The physiological effects of penicillin (a) and aspirin (e) were discovered before the respective drug targets were identified. For the other examples listed, the drugs were designed against specific targets, and the physiological effects were characterized or confirmed later. In some cases, the drugs proved to have unexpected effects or side effects.

   a. before   b. after   c. after
   d. after   e. before   f. after.

2. Lipinsky's rules for a drug candidate state that the molecular weight should be greater than 500, the number of hydrogen-bond donors greater than 5, the number of hydrogen-bond acceptors greater than 10, and $\log(P)$ greater than 5. These rules are all satisfied for (a) atenolol, and (b) sildenafil. For indinavir (compound c), the molecular weight is greater than 600, so the rules are not (all) satisfied.

3. If computer programs could estimate $\log(P)$ values on the basis of chemical structure, then the required laboratory time for drug development could be shortened. It would no longer be necessary to determine the relative solubilities of pharmaceutical candidates by allowing each compound to equilibrate between water and an organic phase.

4. Perhaps the N-acetylcysteine would conjugate to some of the N-acetyl-p-benzoquinone imine that is produced by metabolism of acetaminophen, thereby preventing the depletion of the liver's supply of glutathione. (The −SH on N-acetylcysteine could compete with the −SH on glutathione and thereby protect the majority of the liver's glutathione.)

5. In phase 1 clinical trials, approximately 10 to 100 healthy volunteers are typically enrolled in a study designed to assess product safety. A typical phase 2 trial requires a much larger number of subjects. Moreover, during phase 2, the persons who receive the drug may benefit from the drug that is being administered. In a phase 2 trial, efficacy, dosage and safety can be assessed.

6. As a general carrier for hydrophobic molecules, albumin is likely to sequester some of the coumadin. The extent of such sequestering will affect a patient's recommended daily dose of coumadin and the resulting clotting time. Binding of other drugs to albumin then could cause extra coumadin to be released; this release of coumadin would reduce the clotting time and could perhaps even lead to uncontrolled bleeding.

7. A drug that inhibits a P450 enzyme may dramatically affect the disposition of other drugs that are metabolized by the same enzyme. If the inhibited metabolism is not considered when dosing, other medications may reach very high levels in the bloodstream, such that the levels of some other drugs may become toxic.

8. Unlike competitive inhibition, noncompetitive inhibition cannot be overcome with additional substrate. This feature can provide a particular advantage because a drug

that acts by a noncompetitive mechanism will not be affected by variations in the available level of the physiological substrate.

9. An inhibitor of MDR could prevent the efflux of a chemotherapeutic drug from tumor cells. Hence, this type of an inhibitor could be useful as part of a combination drug strategy. The purpose of the MDR inhibitor would be to avert development of a resistance to the cancer chemotherapy.

10. Based upon the fact that trypanosomes derive all of their energy from glycolysis, any of the enzymes of the glycolysis pathway could be a potential target for treating sleeping sickness. Agents that inhibit one or more of the glycolytic enzymes would therefore act to deprive trypanosomes of energy. A difficulty is that glycolysis in the host cells would also be inhibited.

11. Imatinib is an inhibitor of the Bcr-Abl kinase, a mutant kinase present only in tumor cells that have undergone a translocation between chromosomes 9 and 22 (see Figure 14.33). Before initiating treatment with imatinib, one could sequence the DNA of the tumor cells in order to determine (a) whether the characteristic translocation has taken place, and (b) whether the sequence of *bcr-abl* carries mutations that could render the kinase resistant to imatinib. If the translocation has not taken place or if the gene carries known resistance mutations, then imatinib would likely not be an effective treatment for the patient carrying the particular tumor.

12. Sildenafil increases cGMP levels by inhibiting the phosphodiesterase-mediated breakdown of cGMP to GMP. Intracellular cGMP levels can also be increased by activating its synthesis. This activation can be achieved with the use of NO donors (such as sodium nitroprusside and nitroglycerin) or compounds that activate guanylate cyclase activity. Either stimulating the synthesis or inhibiting the breakdown will increase the level of cGMP.

13. A reasonable mechanism would be an oxidative deamination following an overall mechanism similar to that in Figure 35.9 in the text, with release of ammonia as an extra product:

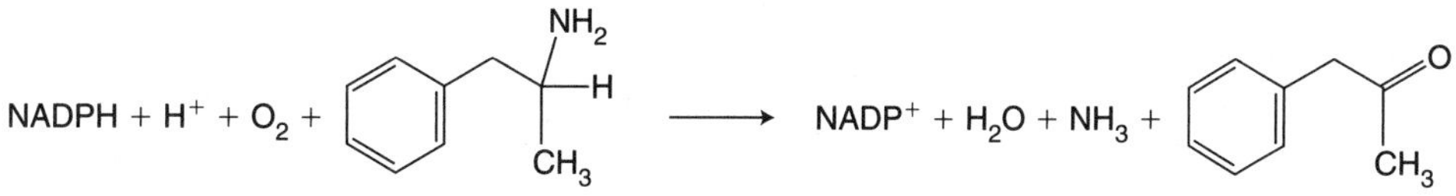

14. For a competitive inhibitor, $v_o/V_{max} = [S]/([S] + K_M(1 + [I]/K_I))$. With $[S] = K_M$,

$v_o/V_{max} = [S]/([S] + [S](1 + [I]/K_I)) = 1/(2 + [I]/K_I) = K_I/(2K_I + [I])$.

Therefore, $V_{max}/v_o = (2K_I + [I])/(K_I) = 2 + [I]/K_I$, and a plot of $1/v_o$ versus $[I]$ should yield a straight line with a slope of $1/K_I$. Such a graph is shown below:

Compound A (nM)	Protease activity	
	v	1/v
0	11.2	0.09
0.2	9.9	0.10
0.4	7.4	0.14
0.6	5.6	0.18
0.8	4.8	0.21
1	4	0.25
2	2.2	0.45
10	0.9	1.11*
100	0.2	5.00*

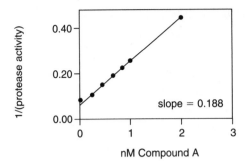

The slope of the line in the graph above is 0.19 nM^{-1}, such that K_I is approximately 5.3 nM.

For IC$_{50}$, one notes from the table that viral RNA production is reduced to 50% (380/760) when 2.0 nM compound A is present. The graph below confirms this value of IC$_{50}$ $\simeq$ 2.0 nM. To prepare the graph, the % inhibition is calculated as:

(100% − (viral RNA production)/760).

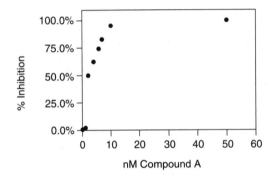

Compound A (nM)	Viral RNA	% Inhibition
0	760	0.0
1	740	2.6
2	380	50.0
3	280	63.2
4	180	76.3
5	100	86.8
10	30	96.1
50	20	97.4

Finally, based upon the oral dosage data, compound A should be effective when taken orally, because 400 nM is much greater than the estimated values of both K_I and IC$_{50}$.